Microwave and Radar Engineering

Microwave and Radar Engineering

Gottapu Sasi Bhushana Rao

Department of Electronics and Communications Engineering
Andhra University Engineering College
Visakhapatnam, Andhra Pradesh

Pearson

Assistant Editor—Acquisitions: R. Vijay Pritha
Editor—Production: C. Purushothaman

ISBN 978-81-317-9944-4

First Impression, 2014

Fifth Impression, 2022

Sixth Impression

Published by Pearson India Education Services Pvt. Ltd, CIN: U72200TN2005PTC057128.

Head Office: 15th Floor, Tower-B, World Trade Tower, Plot No. 1, Block-C, Sector 16, Noida 201 301, Uttar Pradesh, India.
Registered Office: 7th Floor, SDB2, ODC 7, 8 & 9, Survey No. 1, Block-C, Sector 16, Sholinganallur, Chennai – 600 119, Tamilnadu, India.
Phone: 044-66540100
Website: in.pearson.com; Email: companysecretary.india@pearson.com

Printed in India by: Gopsons Papers Pvt. Ltd. Noida

To my wife, son, daughter and son-in-law:

Smt. K. V. Ratnam

Gottapu Santosh Kumar

Gottapu Prashanti

Marpina Naveen Kumar

To my mother and brothers:

Smt. Gottapu Parvatamma

Gottapu Chinnam Naidu

Gottapu Satyannarayana

Gottapu Krishna Murthy

In fond memory of my father and grandfather:

Gottapu Appala Naidu

Gottapu Chinnam Naidu

About the Author

Prof. Gottapu Sasi Bhushana Rao is Professor and Head of the Department of Electronics and Communication Engineering, Andhra University College of Engineering, Visakhapatnam, India. A seasoned academician with 30 years of R&D, industrial and teaching experience, he has published 300 research papers in various journals of international and national repute and participated in scholastic conferences. He is a senior member of IEEE, fellow of IETE, member of IEEE Communication Society, Indian Geophysical Union (IGU) and International Global Navigation Satellite System (IGNSS), Australia. He is also the chartered engineer for IETE.

Prof. Rao was also the Indian member in the International Civil Aviation organization (ICAO), Canada Working Group for developing the SARPS. His current research areas include cellular and mobile communication, GPS and microwave engineering. A recipient of the Best Researcher Award and Dr Survepalli Radhakrishnan Award for Best Academician, he is the Chief Editor of *Journal of Communication, Navigation and Signal Processing*, Andhra University College of Engineering. He is also an editorial member of *International Journal of Positioning (POS),* Scientific Research Publishing (USA), Journals of Scientific and Academic Publishing (USA) and International Journal of Electrical and Electronic Engineering and Telecommunications (IJEETC). He has authored three books *Global Navigation Satellite Systems (GNSSs)*, McGraw Hill; *Mobile and Cellular Communications*, Pearson Education; and *Electromagnetic Field Theory and Transmission Lines*, John-Wiley; and guided ten students to their PhD degrees.

Brief Contents

Contents

Preface

Microwave and Radar Engineering is an important fundamental course with great academic relevance. Progress in this exciting field made possible the advent of many technologies like wireless communication, antennas and wave propagation etc. Interference and electrical noise problems that affect industry can also be better understood and solutions provided using Microwave Engineering.

This book covers, in a comprehensive manner, the theory to gain better insight into the underlying principles and technological spin-offs. Each chapter begins with a treatment of the fundamentals and progresses gradually to discuss advanced concepts. Though Microwave Engineering involves mathematical analysis for better comprehension, the readers do not necessarily need a mathematical background as the relevant details are covered in the first chapter and wherever required.

OBJECTIVE

Microwave and Radar Engineering is a subject of great academic and industrial importance, in which active research has produced a large number of applications. Many books that are available on this subject involve complex mathematics and are in a language that is difficult to comprehend. Hence, a need was felt to bring out a comprehensive text book in easy language, which would cover all the topics of EM field theory, transmission lines, waveguides, waveguide components, *S*-parameters, microwave tubes, solid state microwave devices, MMICs, microwave measurements and radar engineering with relevant mathematical treatment. This book is an outcome of my effort to fill this lacuna.

ABOUT THE BOOK

This book covers the relevant topics of EM field theory, transmission lines, waveguides, waveguide components, *S*-parameters of various microwave networks, microwave tubes, microwave solid state devices, MMICs, microwave measurements, MTI, MTD, Pulse Doppler Radar and tracking radar in a comprehensive manner. It features worked examples in every chapter that explain how to use the theory presented in the text to solve different kinds of problems. It helps students to get a good grasp of waveguide theory, microwave devices and networks and radar engineering while making them realize the utility and indispensability of knowledge in this subject.

The text is easy-to-read, logical, and has a step-by-step approach that enables the reader to follow new and complex ideas clearly. Ample number of illustrations helps students to visualize the phenomena. Spread over 17 Chapters, the topics covered in the sections of each chapter are structured in a modular fashion.

This book is designed as a textbook for UG and PG students, and research scholars of Electronics and Communication Engineering. At the same time, it is suitable for researchers, professionals in the fields of microwave engineering, radar and wireless communications. It also serves as a good reference for practising professionals and scientists.

SALIENT FEATURES OF THE BOOK

- Deals at length with all significant topics of waveguide theory, waveguide components, microwave devices, networks, *S*-parameters of all microwave networks and radar theory.

- The book starts with the basics and builds progressively on the concepts to cover transmission lines, waveguides, waveguide components, S-matrices, microwave tubes, solid state microwave electronics, MMICs, microwave measurements and Radar.
- The analysis, presentation and physical interpretation of theory is purposefully lucid to ensure that students are able to grasp the subject with ease.
- All the concepts are well-explained with help of appropriate illustrations and examples.
- Questions of previous GATE and IES examinations are solved and presented chapter-wise to enable students preparing for these tests.
- Key equations and formulae have been highlighted.
- Derivations / Topics are presented in a modular fashion to facilitate quick revision.
- In addition, each chapter includes summary, multiple-choice questions and review questions to enable quick recapitulation of the discussed topics.

CHAPTER ORGANIZATION

This book may be divided into three major parts. The first three chapters provide an overview of vector algebra, Maxwell's equations and transmission lines. Chapters 4 to 11 describe the waveguides and Microwave Engineering concepts in detail. Chapters 12 to 16 deal with the fundamentals of Radar Engineering. The topics covered in each chapter are as follows:

Chapter 1 reviews the basics of vector algebra which include unit vector, scalar mathematics, vector components, addition and multiplication of vectors, dot and cross products, scalar and vector triple product, etc. It also discusses various coordinate systems and transformations from one coordinate system into another. It also covers vector calculus, which includes line, surface and volume integrals and vector operators such as gradient, divergence, curl and Laplacian operators. Divergence and Stoke's theorem are also discussed.

Chapter 2 explains Faraday's law, introduces displacement current density and then brings up the most important topic of electromagnetic field theory—Maxwell's four equations. These equations, in various forms, for free space, dielectrics and conductors are examined. It then proceeds to several other important topics before focusing on polarization and its different kinds, the Poyinting vector and Poyinting theorem.

Chapter 3 probes the transmission of EM energy through a transmission line. It analyzes how a wave exists in transmission lines and shows how its wave equation, phase and group velocity is determined. It also presents the behaviour of a wave in a lossless and low-loss case. The characteristic impedance in an open and short circuited line, reflection coefficient and voltage standing wave ratio are discussed. Transmission line design with help of the Smith Chart and its other applications are also given due importance in the discussion. Further, this chapter includes topics about impedance transformations and stub matching, which are useful in practical applications.

Chapter 4 introduces the Microwave Engineering. It touches upon the history of microwave technology before elucidating the microwave spectrum along with its IEEE frequency band designations. It also gives an insight into the advantages and applications of microwaves.

Chapter 5 is devoted to Waveguides. A detailed discussion about wave propagation in rectangular and circular wave guides, which include various modes (TE, TM), field equations, filter characteristics, power contained in the wave etc is presented. The chapter ends with a discussion on microstrip lines and cavity resonators.

Chapter 6 deals with various coupling mechanisms, waveguide components and applications. The important waveguide components discussed in this chapter are attenuators, phase shifters, ferrite components, wave guide bends and joints, microwave Tee junctions, bethe-hole and two-hole directional couplers.

Chapter 7 expounds on application of the scattering matrix for various waveguide components. The properties of scattering matrix and scattering matrix calculations for 1-port, 2-port, 3-port and 4-port junctions are presented are presented in this chapter.

Chapter 8 sheds light on the limitations of conventional vacuum tubes at microwave frequencies and velocity modulation. It explains the principle of operation of vacuum tubes before presenting detailed description of performance (i.e., power output and efficiency) of all important linear beam tubes (Klystron, Reflex Klystron, TWT, BWO) and crossed field tubes (Magnetron and CFA).

Chapter 9 unravels the negative resistance phenomenon, which is a common characteristic of all active two-terminal devices. Further, this chapter gives a detailed descriptions of all microwave solid-state devices including two-terminal devices such as transferred electron devices (Gunn diodes), avalanche transit time devices (IMPATT, TRAPATT, BARITT diodes) and tunnel diodes, and three-terminal devices such as bipolar junction transistors and field effect transistors (MESFETs and HEMTs). The chapter ends with the brief discussion of the PIN diode, Schottky diode, Varactor diode, parametric amplifier, step-recovery diode and crystal diodes.

Chapter 10 is about MMICs. This chapter begins with the classification of MICs, general applications and technology of MMICs. The advantages and disadvantages of MMICs, comparison with hybrid MICs, materials used for MMICs, NMOS and CMOS fabrication techniques and formation of thin film are described in this chapter.

Chapter 11 discusses the basic microwave bench used to measure frequency, power, attenuation, VSWR, impedance and cavity Q parameters at microwave frequencies. It explores the various techniques used for measuring these parameters.

Chapter 12 serves as an introduction to Radar. It begins with the history of radar and deals at length with the frequencies and applications of radar. This chapter also describes the classification of radars, radar range equation, radar block diagram, pulse radar characteristics, and so on.

Chapter 13 elucidates the principle of operation of CW radar, FM CW radar and pulse radars.

Chapter 14 explains Doppler frequency and presents Pulse, MTI and Pulse Doppler radars. This chapter also delineates pulse characteristics, Doppler processing in MTI, pulse Doppler radar and MTDs. The chapter analyzes MTI radar performances and limitations, and compares Moving Target Indicator with MTD.

Chapter 15 deals with the fundamentals of search and tracking radar system. It discusses various scanning and tracking techniques such as range and angle tracking techniques. It elaborates on the various angle tracking methods (sequential, conical and monopulse), track while scan radar, phased array radar and radar displays.

Chapter 16 the concluding chapter, discusses the linear receiver characteristics of a matched filter that improves the SNR, various methods for the detection of desired signals and the rejection of undesired noise, noise figure, phased array radars and its characteristics

Chapter 17 presents a total of nine microwave laboratory experiments. Each experiment first spells out its aim and lists the equipment required. It then describes the microwave bench setup, initial settings of the equipment, experimental procedure and ends with observations and results.

Comments and suggestions to enhance the contents of this are welcome.

Acknowledgements

I express my gratitude to MNVS Santosh Kumar, Idam Silpa, G. Sateesh Kumar, R. Madhu, B. Anurag, Malladi Madhavi and B. Sowjanya for their enthusiasm in reviewing the manuscript and providing necessary suggestions to improve the text.

I thank A. Lakshmi Narayana and K. V. Ramakrishna for giving their best to add the pictorial representations to the text, which greatly enhanced the value of the book.

I am obliged to P. Devi Sowjanya, P. Vijayalakshmi, K. Sree Lakshmi, R. Santosh, K. Rajasekhar Reddy and K. Theja Durga for correcting and editing the text, which helped greatly in bringing out this book in record time.

I am indebted to my research scholars Rajkumar Goswami, M. Rajan Babu, Solomon Gotham, S. Neeraja, T. Madhavi and Ch. Usha for verifying the calculations and minimizing the errors.

Finally, I would like to thank my family members wholeheartedly, without whose cooperation and constant encouragement this milestone could have not been achieved.

List of Symbols

Symbol	Quantity	Unit
$\vec{A}, \underline{A}, \bar{A}, \mathbf{A}$	Vector A	
A_x, A_y, A_z	x, y, z scalar components of $\vec{A}$	
$\vec{A}_x, \vec{A}_y, \vec{A}_z$	x, y, z vector components of $\vec{A}$	
$\vec{A}\cdot\hat{a}_x$	Projection of $\vec{A}$ in the x direction	
$\vec{A}\cdot\hat{a}_y$	Projection of $\vec{A}$ in the y direction	
$\vec{A}\cdot\hat{a}_z$	Projection of $\vec{A}$ in the z direction	
$\hat{a}_x, a_x, \hat{i}, \hat{x}$	Unit vector along x direction	
$\hat{a}_y, a_y, \hat{j}, \hat{y}$	Unit vector along y direction	
$\hat{a}_z, a_z, \hat{k}, \hat{z}$	Unit vector along z direction	
$\hat{a}_\rho, \hat{a}_\phi$ and $\hat{a}_z$	Unit vectors along ρ, ϕ, and z directions, respectively	
$\hat{a}_r, \hat{a}_\theta$ and $\hat{a}_\phi$	Unit vectors along r, θ, and ϕ directions, respectively	
∇	Gradient	
$\nabla\cdot$	Divergence	
$\nabla\times$	Curl	
dl	Differential length (for line integral)	meter
ds	Differential area (for surface integral)	meter^2 (m^2)
dv	Differential volume (for volume integral)	meter^3 (m^3)
$d\vec{S}$	Elementary vector area	meter^2 (m^2)

Symbol	Quantity	Unit
Φ_E	Electric Flux	Volt-meter or Coulomb (V-m or C)
$\vec{D}$	Electric Flux Density	Coulomb/meter^2 (C/m^2)
emf	Electromotive force	Volt
Φ_B	Magnetic Flux	Weber (Wb)
$\vec{B}$	Magnetic Flux Density	Weber/meter^2 (Wb/m^2)
$\vec{F}$	Force Vector	Newton (N)
$\vec{E}$	Electric field intensity vector	Newton/ Coulomb(N/C) or V/m
E_x, E_y and E_z	$\vec{E}$ components along x, y, and z directions, respectively	Newton/ Coulomb(N/C) or V/m
Q	Charge	Coulomb (C)
ε_0	Permittivity of free space	Farad/meter (F/m)
ε_r	Relative Permittivity	No unit
ε	Absolute Permittivity	Farad/meter (F/m)
μ_0	Permeability of free space	Henry/meter (H/m)
μ	Absolute Permeability	Henry/meter(H/m)
μ_r	Relative Permeability	No unit
ρ_L	Line charge density	Coulomb/m (C/m)
ρ_s	Surface charge density	Coulomb/m^2 (C/m^2)
ρ_v	Volume charge density	Coulomb/m^3 (C/m^3)
V	Electrical Potential	Volt(V)
∇V	Potential Gradient	Volt(V)
I	Current	Ampere(A)
$\vec{J}$	Current density	Ampere/meter^2(A/m^2)
v	Velocity	meter/second (m/s)

Symbol	Quantity	Unit
σ	Conductivity	mho/m
J	Current density	Ampere/meter^2 (A/m^2)
$\vec{H}$	Magnetic field intensity vector	Newton / Weber or A/m
γ	Propagation constant	dB/m
β	Phase constant or wave number	rad/m
α	Attenuation constant	dB/m
ω	Angular velocity or radian frequency	radian/second (rad/s)
Γ	Reflection coefficient	No unit
υ_p	Phase velocity	meter/second (m/s)
υ_g	Group Velocity	meter/second (m/s)
f	Frequency	Hertz (Hz)
f_0	Cutoff frequency	Hertz (Hz)
λ	Wavelength	meter (m)
λ_0	Free space wavelength	meter (m)
λ_c	Cutoff wavelength	meter (m)
λ_g	Guided wavelength	meter (m)
η	Intrinsic impedance	Ohm (Ω)
Z_0	Characteristic impedance	Ohm (Ω)
$\oint$	A path integral	
g_m	Transconductance	
ω_e	Angular frequency of velocity fluctuatiations in axial wave	
V_0	Anode to cathode voltage	
P_{tr}	Transmitted power	
f_r	Resonance frequency	

Symbol	Quantity	Unit
β_1	Beam coupling coefficient	
θ_g	Average gap transit angle	
λ_n	Wavelength in direction of the normal to the reflecting wave	
λ_p	Wavelength parallel to the direction of the plane	
e_r	Reference electron	
e_e	Early electron	
E	Electric field strength	
H·	Magnetic field strength	
η_0	Intrinsic impedance of free space	
δ	Skin depth	
$J_n(nX)$	Bessel's function	
n	Mode number	
c	3×10^8 m/s is the velocity of light in the free space	
r_m	Mean radius of spoke	
V_H	Hartree voltage	
P_i	Power flowing into the device	Watts
P_0	Power flowing out of the device	Watts
IL	Insertion loss	
v_d	Drift velocity	
n_1	Electron density in lower valley	
n_2	Electron density in upper valley	
ε_s	Semiconductor dielectric permittivity	
μ_n	Electron mobility	

Symbol	Quantity	Unit
L_b, L_e	Base and emitter bond-wire inductances, respectively	
P_t	Peak transmit power is the average power when the radar is transmitting a signal.	Watts
$S_{\min}$	Minimum received power that the radar receiver can sense and is referred to as *Minimum Detectable Signal*	Watts
τ_p	Pulse width	
B	Bandwidth	Hertz
B_0	DC magnetic field	Weber/m^2
T_r	Pulse repetition interval	
f_r	Pulse repetition frequency	
v_{bn}	Blind speed	
f_{LO}	Local oscillator frequency when the track error is zero	Hz
f_{c}	The COHO frequency, which is the nominal IF (Hz)	
f_T	Transmit frequency	
A_e	Effective area of the antenna	
Ω_A	The beam solid angle	
K	Boltzmann's constant	
T_0	Absolute temperature of the receiver input	°Kelvin = 290°K
Δf	Frequency offset or excursion in the turn around region of modulated signal	Hz

Microwave and Radar Engineering

1 Vector Analysis

1.1 INTRODUCTION

All physical quantities in electromagnetism are classified as scalars or vectors. Physical laws can be simplified using vectors. Therefore, knowledge of vector analysis and coordinate systems is essential for a better understanding of the Electromagnetics and Microwave engineering. To explain the spatial variations of the quantities, an appropriate coordinate system is required. Coordinate systems are very useful in solving problems in electromagnetics such as the Maxwell's equation. The three most important coordinate systems used in electromagnetic are as follows:

- Cartesian or rectangular (x, y, z)
- Circular or cylindrical (ρ, ϕ, z)
- Spherical (r, θ, ϕ).

This is an introductory chapter and provides reviews of mathematical concepts required for understanding the Electromagnetics.

1.2 SCALAR AND VECTOR

In microwave engineering, the quantities used are either a **scalar** or a **vector**. A field is a quantity that can be specified everywhere in space as a function of position(x, y, z) and time (t). The quantity that is specified may be a scalar or a vector. In this section the basic ideas of vector analysis are developed in a comprehensive manner.

1.2.1 Scalar Field

A scalar is a quantity that has *magnitude* but *no direction*. A *scalar field* is a function that has a different value at every point in space and time. A scalar quantity does not have any direction. For example, when we say 'the temperature of the room is 30°C, we do not specify the direction.

Time, distance, mass, temperature, electric potential, volume, speed are some examples of scalar quantities. Scalar quantity can represented simply by using a letter for example, A, B.

At a given point in space, scalar quantities are the same whether expressed in Cartesian, cylindrical, or spherical coordinates. Therefore, scalar quantities are *invariant* to a transformation of coordinates.

1.2.2 Vector Field

A *Vector* is a quantity that has both magnitude and direction in space. It provides information about how large a measurement is and the direction of the measurement. Examples of vectors are: velocity (**v**), electric field (**E**) intensity and so on.

A vector is usually indicated by a boldfaced letters, such as **A**, *or* an arrow over a letter $(\vec{A})$ *or* under bar notation $(\underline{A})$ *or* a bar over a letter $(\bar{A})$ and ordinary type for scalars (A).

A *vector* is represented by an arrow whose length indicates the magnitude and whose direction indicates the direction of the quantity. Figure 1.1 shows a vector $(\vec{A})$ and its negative $(-\vec{A})$. As can be seen from the figure, the negative of vector is a vector of the same magnitude, but with the opposite direction.

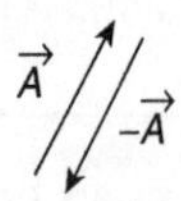

Figure 1.1 *Vector* $(\bar{A})$ *and its negative*

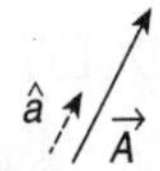

Figure 1.2 *Unit vector*

The magnitude of vector $\vec{A}$ is a scalar and is represented by $|\vec{A}|$ or A. A vector can also be written in terms of unit vector $(\hat{a})$ as, $\vec{A} = |\vec{A}|\hat{a} = A\hat{a}$.

The function that assigns a vector to every point in a space is called as vector field and these fields are used to model force fields like gravity, fluid flow, electric and magnetic fields.

1.2.3 Unit Vector

A unit vector can be created by dividing a simple vector with its own magnitude.

A unit vector along $\vec{A}$ is defined as a vector whose magnitude is unity and whose direction is along $\vec{A}$. It is denoted by $\hat{a}_A$ or $\hat{a}$, that is the symbol (^) will also be used over a particular symbol to indicate the unit vector (Figure 1.2). Suppose we are given a vector $\vec{A}$ and are asked to determine a unit vector ($\hat{a}_A$ or $\hat{a}$) which is parallel to $\vec{A}$. Mathematically, it can be written as Eq.(1.1).

$$\hat{a}_A = \hat{a} = \frac{\vec{A}}{|\vec{A}|} \tag{1.1}$$

Thus

$$\vec{A} = |\vec{A}|\hat{a}_A \tag{1.2}$$

From the above Eq. 1.2 we can say that any vector $\vec{A}$ can be represented as product of its magnitude and unit vector i.e., $|\vec{A}|$ and $\hat{a}$ respectively.

$\hat{a}_x, \hat{a}_y, \hat{a}_z$ are referred to as the base vectors.

For example, unit vectors along *x*, *y* and *z* directions are indicated by a_x, a_y and a_z or $\hat{a}_x$, $\hat{a}_y$ and $\hat{a}_z$ or $\hat{i}, \hat{j}$ and $\hat{k}$ or $\hat{x}, \hat{y}$ and $\hat{z}$ respectively.

EXAMPLE PROBLEM 1.1

For the given vector $\vec{A} = (x+y)\hat{a}_x + 3\hat{a}_y + z^2\hat{a}_z$, determine the unit vector in the direction of $\vec{A}$ at the Cartesian coordinate point (1,1,1).

Solution

From the given vector, we can write the component vectors of $\vec{A}$ as

$$A_x = x + y,\ A_y = 3,\ A_z = z^2$$

At the point (1, 1, 1) the unit vector is $\hat{a} = \dfrac{\vec{A}}{|A|} = \dfrac{(x+y)\hat{a}_x + 3\hat{a}_y + z^2\hat{a}_z}{\sqrt{(x+y)^2 + 9 + z^4}}$

$$\hat{a} = \frac{1}{\sqrt{14}}(2\hat{x} + 3\hat{y} + \hat{z})$$

■

1.3 VECTOR ALGEBRA

Vector algebra refers to simple operations on vectors such as addition, subtraction and multiplication of vectors. Various scalars and vectors properties and their operations are explained in this section.

1.3.1 Vector Addition and Vector Subtraction

Vector addition for more than two vectors can be performed by polygon method. To illustrate vector addition refer to Figure 1.3. Consider two vectors $\vec{A}$ and $\vec{B}$ in the plane or space. There are two ways to construct the vector sum $\vec{A} + \vec{B}$.

(i) By attaching the origin of the vector $\vec{B}$ to the tip of vector $\vec{A}$ (see Figure 1.3b). In this the vector sum is the vector extending from the origin of vector $\vec{A}$ to the tip of vector $\vec{B}$.

(ii) By showing the vectors $\vec{A}$ and $\vec{B}$ with a common origin and completing the parallelogram resulting from drawing lines parallel to vectors $\vec{A}$ and $\vec{B}$ at the tips of vectors $\vec{B}$ and $\vec{A}$, respectively, as shown in Figure 1.3c. The vector sum, also known as the resultant, is the diagonal of the parallelogram that starts at the common origin of vectors $\vec{A}$ and $\vec{B}$.

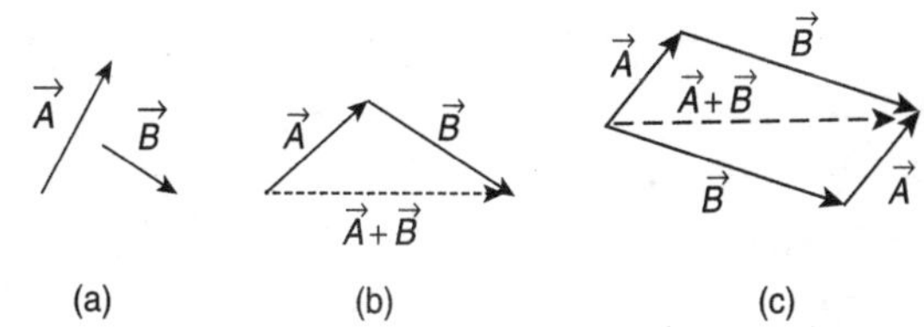

Figure 1.3 *(a) Vector representation; (b) Vector addition; (c) Vector sum is the diagonal of parallelogram*

Vector subtraction

Vector subtraction may be accomplished by multiplying the subtracted vector by –1 and using the technique for adding. Therefore, vector subtraction (Figure 1.4) is a special case of vector addition because:

Subtraction is accomplished using the definition: $\vec{A} - \vec{B} = \vec{A} + (-\vec{B})$.

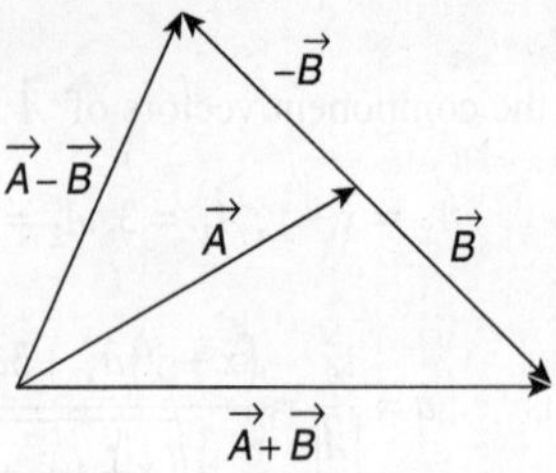

Figure 1.4 *Vector subtraction* ($\vec{A}-\vec{B}$)

1.3.2 Position and Distance Vectors

A vector $\vec{A}$ in space can be represented as a vector sum of three vectors ($\vec{A}_x$, $\vec{A}_y$ and $\vec{A}_z$). If these three vectors are parallel to the coordinate axes, they are known as the three vector components. Figure 1.5a illustrates vector $\vec{A}$ and its three vector components.

Position vector

Let P be a point in space (Figure 1.5(b)). The position vector of point P is the directed distance from the origin O to the point P and is written as

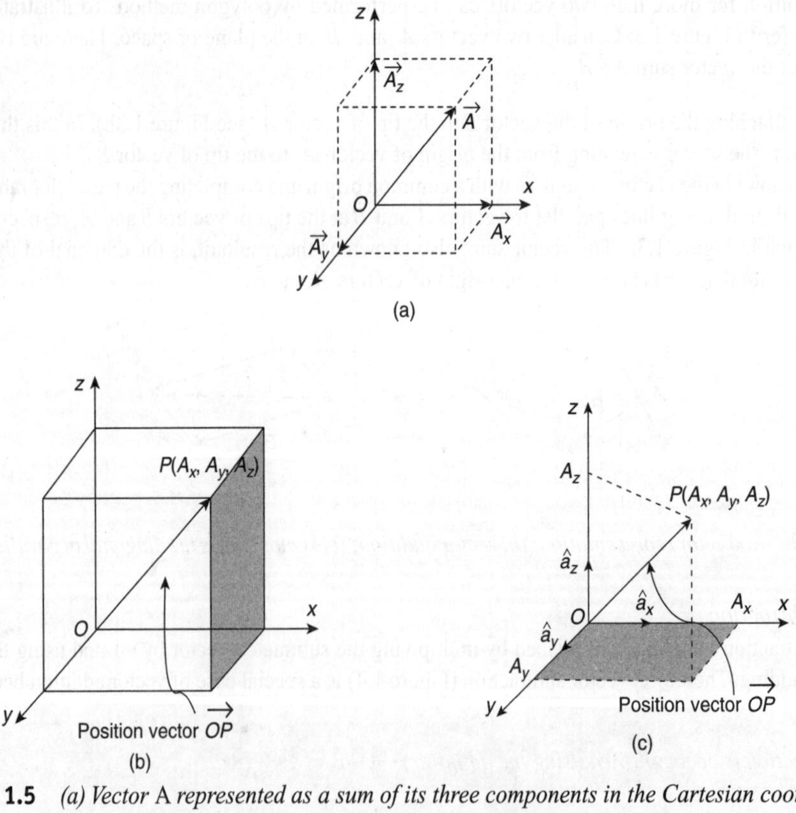

Figure 1.5 *(a) Vector* A *represented as a sum of its three components in the Cartesian coordinate system; (b) Position vector; (c) Scalar and vector components of a position vector*

$$\text{OP} = \vec{A} = \vec{A}_x + \vec{A}_y + \vec{A}_z = A_x\hat{a}_x + A_y\hat{a}_y + A_z\hat{a}_z$$

where $\vec{A}_x, \vec{A}_y$ and $\vec{A}_z$ are the three vector components and A_x, A_y, and A_z are the three scalars or the values of the *vector* $\vec{A}$ in the direction of x-, y- and z-axes.

Coordinate directions are represented by unit vectors $\hat{a}_x, \hat{a}_y$ and $\hat{a}_z$, each of which has a unit length and points in the direction along one of the coordinate axes (Figure 1.5c). The magnitude of a vector can be expressed in terms of components as

$$A = \left|\vec{A}\right| = \sqrt{A_x^2 + A_y^2 + A_z^2} \tag{1.3}$$

1.3.3 Vector Multiplication

Another important operation that is carried out on vectors is the multiplication. One must be thorough with the fundamentals of these operations for better understanding of the field behavior of EMFs. The following are the important vector multiplication operations.

- Scalar multiplication
- Dot product/Scalar product $\left(\vec{A}\cdot\vec{B}\right)$
- Cross product or vector product $\left(\vec{A}\times\vec{B}\right)$
- Scalar triple product $\left(\vec{A}\cdot\left(\vec{B}\times\vec{C}\right)\right)$
- Vector triple product $\left(\vec{A}\times\left(\vec{B}\times\vec{C}\right)\right)$

Scalar multiplication: A vector $\vec{A}$ can be *multiplied by a scalar c*, resulting in a vector $c\vec{A}$ which is oriented in the same direction as $\vec{A}$, and whose magnitude is c times that of $\vec{A}$. The magnitude of the vector changes, but not the direction.

Figure 1.6 illustrates the case in which $c = 3$. If the scalar c is negative, then the orientation of the vector $c \cdot A$ will be opposite to that of $\vec{A}$.

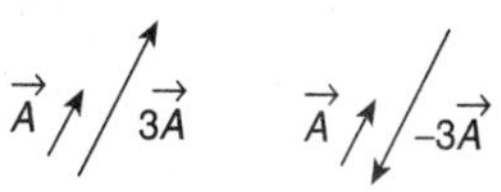

Figure 1.6 *Multiplication of a vector by a scalar*

An example of a vector multiplied by a scalar is Newton's second law, which relates two vectors force (N) and acceleration (m/s^2) by a scalar mass (kg).

$$\vec{F} = m\vec{a}$$

Dot product (scalar product), cross product (vector product) and scalar triple and vector triple products are discussed in the coming sections as they need thorough understanding of coordinate systems.

1.4 COORDINATE SYSTEMS

A coordinate system is a way of uniquely specifying the location of any position in space with respect to a reference origin. The mathematical representation of a vector typically requires it to be referred to a specific coordinate system. Each coordinate system has a distinct set of principle axes, represented by three surfaces. For Cartesian, the set of axes is represented as x, y, and z, for cylindrical, ρ, ϕ, and z, for spherical, r, θ, and ϕ. The choice of using a particular coordinate system depends on the geometry of the application. The geometry of many electromagnetic problems often dictates which coordinate system is more appropriate in facilitating a solution.

1.4.1 Cartesian Coordinate System

Any vector in space can be uniquely expressed in terms of x-, y- and z- coordinates using a rectangular coordinate system. Using a Cartesian coordinate system, the unit vectors $\hat{a}_x, \hat{a}_y, \hat{a}_z$ or $(\hat{x}, \hat{y}, \hat{z})$ are introduced corresponding to the x, y and z directions, respectively. The unit vectors are shown in Figure 1.7.

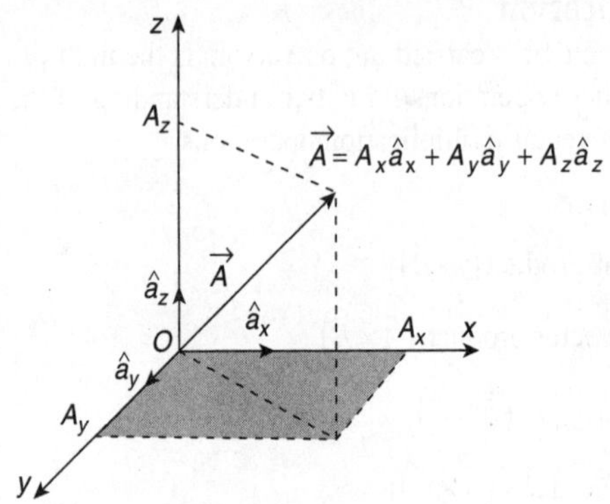

Figure 1.7 *Vectors represented in a rectangular coordinate system*

Hence,

$$\vec{A} = \mathbf{A}_x + \mathbf{A}_y + \mathbf{A}_z = A_x\hat{a}_x + A_y\hat{a}_y + A_z\hat{a}_z \tag{1.4}$$

where

$\vec{A}_X = A_X\hat{\mathbf{a}}_x$ is the x-component of the vector **A** and $A_X = |\vec{A}_X|$

$\vec{A}_Y = A_Y\hat{\mathbf{a}}_y$ is the y-component of the vector **A** and $A_Y = |\vec{A}_Y|$

$\vec{A}_Z = A_Z\hat{\mathbf{a}}_z$ is the z-component of the vector **A** and $A_Z = |\vec{A}_Z|$

$$\hat{a}_x = \text{unit vector in } x\text{-direction} = \frac{\vec{A}_x}{|\vec{A}_x|}$$

$$\hat{a}_y = \text{unit vector in } y\text{-direction} = \frac{\vec{A}_y}{|\vec{A}_y|}$$

$$\hat{a}_z = \text{unit vector in } z\text{-direction} = \frac{\vec{A}_z}{|\vec{A}_z|}$$

where $\mathbf{A}_X$, $\mathbf{A}_Y$ and $\mathbf{A}_Z$ are **projections** of $\vec{A}$ on x-, y- and z-axes, respectively.

Thus, the unit vector of $\vec{A}$ is given by $\hat{a}_A = \dfrac{A_X\hat{a}_x + A_Y\hat{a}_y + A_Z\hat{a}_z}{\sqrt{A_X^2 + A_Y^2 + A_Z^2}}$ (1.5)

Vector Addition and Subtraction

If $\vec{A} = A_x\hat{a}_x + A_y\hat{a}_y + A_z\hat{a}_z$ and $\vec{B} = B_X\hat{a}_x + B_Y\hat{a}_y + B_z\hat{a}_z$, then addition and subtraction of given two vectors are defined as

$$\vec{C} = \vec{A} + \vec{B} = (A_X + B_X)\hat{a}_x + (A_Y + B_Y)\hat{a}_y + (A_Z + B_Z)\hat{a}_z$$

$$\vec{D} = \vec{A} - \vec{B} = (A_X - B_X)\hat{a}_x + (A_Y - B_Y)\hat{a}_y + (A_Z - B_Z)\hat{a}_z$$

Vector Multiplication

In this section, scalar and vector products are continued; different multiplication operations are specified and scalar multiplication is discussed.

Scalar Product or Dot Product

The dot product of two vectors $\vec{A}$ and $\vec{B}$ gives a scalar quantity and is defined as the product of magnitudes of the two vectors and the cosine of the angle between them.

$$\vec{A}\cdot\vec{B} = AB\cos\theta_{AB} \tag{1.6}$$

where θ_{AB} is the smaller angle between them as shown in Figure 1.8.

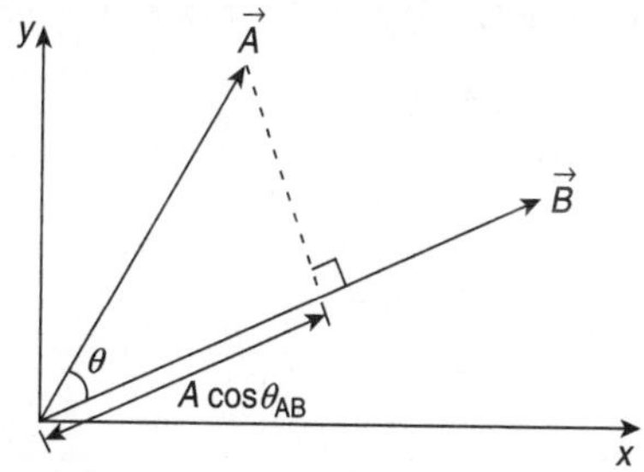

Figure 1.8 *The dot product between two vectors*

The result of $\vec{A}\cdot\vec{B}$ is a scalar, hence dot product is also known as Scalar Product

The properties of vectors are as follows:

(i) Commutative property: $\vec{A}\cdot\vec{B} = \vec{B}\cdot\vec{A}$

(ii) When two vectors are perpendicular, the angle between them is $\theta = 90°$ and $\cos 90° = 0$

$$\vec{A}\cdot\vec{B} = AB\cos 90° = 0$$

(iii) If $\hat{a}_x, \hat{a}_y$ and $\hat{a}_z$ are mutually perpendicular then

$$\hat{a}_x \cdot \hat{a}_y = \hat{a}_y \cdot \hat{a}_z = \hat{a}_z \cdot \hat{a}_x = 0$$

$$\hat{a}_x \cdot \hat{a}_x = \hat{a}_y \cdot \hat{a}_y = \hat{a}_z \cdot \hat{a}_z = 1$$

(iv) When two vectors are parallel the angle between them is either 0° or 180°.

$$\vec{A} \cdot \vec{B} = AB\cos 0° = AB$$

$$\vec{A} \cdot \vec{B} = AB\cos 180° = -AB$$

(v) The dot product of a vector with itself is the square of its magnitude

$$\vec{A} \cdot \vec{A} = AA\cos 0° = A^2$$

$$\vec{A}^2 = A^2$$

(vi) The scalar product is equal to the sum of products of their corresponding components.

If $\vec{A} = A_x\hat{a}_x + A_y\hat{a}_y + A_z\hat{a}_z$ and $\vec{B} = B_x\hat{a}_x + B_y\hat{a}_y + B_z\hat{a}_z$

$$\vec{A} \cdot \vec{B} = \left(A_x\hat{a}_x + A_y\hat{a}_y + A_z\hat{a}_z\right)\left(B_x\hat{a}_x + B_y\hat{a}_y + B_z\hat{a}_z\right)$$

from the property (iii), the above equation reduces to

$$\vec{A} \cdot \vec{B} = A_xB_x + A_yB_y + A_zB_z$$

Uses of Dot Product

Dot product is used for the following purposes:

- To determine the angle between the two vectors.
- To find the projection (or component) of a vector in a given direction.

Projection of a vector on another vector

The scalar component of $\vec{A}$ along $\vec{B}$ is called the scalar projection of $\vec{A}$ on $\vec{B}$ and is given by (from Figure 1.8 and Eq. (1.6))

$$A\cos\theta_{AB} = \frac{\vec{A}.\vec{B}}{|\vec{B}|} \tag{1.7}$$

($\because$ dot product of $\vec{A}.\vec{B} = |\vec{A}||\vec{B}|\cos\theta_{AB}$)

where as the vector projection of $\vec{A}$ on $\vec{B}$ is given by

$$A\cos\theta_{AB}(\hat{a}_B) = \frac{\vec{A}\cdot\vec{B}}{|\vec{B}|}\hat{a}_B \tag{1.8}$$

The above two Eqs (1.7) and (1.8) can be written as

Scalar component of $\vec{A}$ along $\vec{B}$ = dot product of $\vec{A}$ and $\hat{a}_B = \vec{A}\cdot\hat{a}_B$

Vector component of $\vec{A}$ along $\vec{B} = \left(\vec{A}\cdot\hat{a}_B\right)\hat{a}_B$

Vector product or cross product

The *cross product* of two vectors $\vec{A}$ and $\vec{B}$ results in a vector quantity and is defined as the product of magnitudes of the two vectors, the sine of the angle between them and unit normal vector, $\hat{a}_n$.

$$\vec{A}\times\vec{B} = |\vec{A}||\vec{B}| \sin\theta_{AB}\,\hat{a}_n = AB \sin\theta_{AB}\,\hat{a}_n \tag{1.9}$$

where θ_{AB} is the smaller angle between the vectors $\vec{A}$ and $\vec{B}$ in the plane of $\vec{A}$ and $\vec{B}$, $\hat{a}_n$ is unit vector perpendicular to the plane containing $\vec{A}$ and $\vec{B}$ and the direction of $\hat{a}_n$ is taken as the direction of the right thumb (using right-hand rule). $\vec{A}\times\vec{B}$ is perpendicular to both $\vec{A}$ and $\vec{B}$ (Figure 1.9).

$AB\sin\theta_{AB}$ = area of the parallelogram formed by the $\vec{A}$ and $\vec{B}$

A cross product of a vector results in another vector.

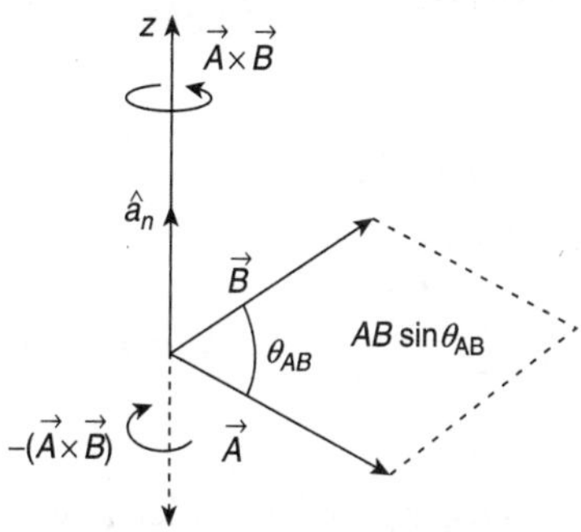

Figure 1.9 *Vector cross product*

Properties of vector cross product are as follows:

(i) Anti-commutative: $\vec{A}\times\vec{B} = -\vec{B}\times\vec{A}$

(ii) Distributive: $\vec{A}\times\left(\vec{B}+\vec{C}\right) = \left(\vec{A}\times\vec{B}\right)+\left(\vec{A}\times\vec{C}\right)$

(iii) Not associate: $\vec{A}\times\left(\vec{B}\times\vec{C}\right) \neq \left(\vec{A}\times\vec{B}\right)\times\vec{C}$

(iv) Vector product of two parallel vectors is zero.

$$\vec{A}\times\vec{B} = |\vec{A}||\vec{B}|\sin\theta_{AB}\hat{a}_n = |\vec{A}||\vec{B}|\sin 0°\hat{a}_n = 0$$

(v) $\vec{A}\times\vec{A} = |\vec{A}||\vec{A}|\sin 0°\hat{a}_n = 0$

(vi) $\hat{a}_x\times\hat{a}_x = \hat{a}_y\times\hat{a}_y = \hat{a}_z\times\hat{a}_z = 0$

(vii) $\hat{a}_x\times\hat{a}_y = 1.1\sin 90°\;\hat{a}_z = \hat{a}_z$

Similarly,

$$\hat{a}_x\times\hat{a}_y = \hat{a}_z,\ \hat{a}_y\times\hat{a}_z = \hat{a}_x,\ \hat{a}_z\times\hat{a}_x = \hat{a}_y,\ \hat{a}_y\times\hat{a}_x = -\hat{a}_z,\ \hat{a}_z\times\hat{a}_y = -\hat{a}_x,\ \hat{a}_x\times\hat{a}_z = -\hat{a}_y$$
$$\text{and } \hat{a}_x\times\hat{a}_x = \hat{a}_y\times\hat{a}_y = \hat{a}_z\times\hat{a}_z = 0 \tag{1.10}$$

If $\vec{A} = A_x\hat{a}_x + A_y\hat{a}_y + A_z\hat{a}_z$ and $\vec{B} = B_x\hat{a}_x + B_y\hat{a}_y + B_z\hat{a}_z$

Then
$$\vec{A}\times\vec{B} = \begin{vmatrix} \hat{a}_x & \hat{a}_y & \hat{a}_z \\ A_x & A_y & A_z \\ B_x & B_y & B_z \end{vmatrix} \quad (1.11)$$

EXAMPLE PROBLEM 1.2

Show that the vector fields $\vec{A} = \rho\,\text{cosec}\,\phi\,\hat{a}_\rho + \rho\cot\phi\,\hat{a}_\phi + \rho\,\hat{a}_z$ and $\vec{B} = \rho\,\text{cosec}\,\phi\hat{a}_\rho - \rho\cot\phi\hat{a}_\phi - \rho\hat{a}_z$ are everywhere perpendicular to each other.

Solution

Given that

$$\vec{A} = \rho\,\text{cosec}\,\phi\,\hat{a}_\rho + \rho\cot\phi\,\hat{a}_\phi + \rho\,\hat{a}_z$$
$$\vec{B} = \rho\,\text{cosec}\,\phi\hat{a}_\rho - \rho\cot\phi\hat{a}_\phi - \rho\hat{a}_z$$

We know that,

$$\cos\theta = \frac{\vec{A}\cdot\vec{B}}{|A|\cdot|B|}$$

$$\Rightarrow \vec{A}\cdot\vec{B} = \rho^2\left(\text{cosec}^2\phi - \cot^2\phi\right) - \rho^2 = 0. \text{ Therefore, } \cos\theta = 0 \text{ or } \theta = 90°$$

■

EXAMPLE PROBLEM 1.3

Given vector, $\vec{A} = 3\hat{a}_y + 4\hat{a}_z$ and $\vec{B} = 4\hat{a}_x - 10\hat{a}_y + 5\hat{a}_z$, determine

(a) The vector component of **A** in the direction of **B**.
(b) A unit vector perpendicular to both **A** and **B**.

Solution

(a) To find the vector component of **A** in the direction of **B**, we must dot the vector **A** with the unit vector in the direction of **B**.

$$\hat{a}_B = \frac{\vec{B}}{|\vec{B}|} = \frac{4\hat{a}_x - 10\hat{a}_y + 5\hat{a}_z}{\sqrt{4^2 + 10^2 + 5^2}} = \frac{1}{\sqrt{141}}\left(4\hat{a}_x - 10\hat{a}_y + 5\hat{a}_z\right)$$

The dot product of **A** and $\hat{a}_B$ (i.e. scalar component of **A** along **B**) is

$$\vec{A}\cdot\hat{a}_B = \left(3\hat{a}_y + 4\hat{a}_z\right)\cdot\frac{1}{\sqrt{141}}\left(4\hat{a}_x - 10\hat{a}_y + 5\hat{a}_z\right) = \frac{1}{\sqrt{141}}[(3)(-10)+(4)(5)] = -\frac{10}{\sqrt{141}}$$

The vector component of **A** along **B** is $= \left(\vec{A}\cdot\hat{a}_B\right)\hat{a}_B = -\frac{10}{141}\left(4\hat{a}_x - 10\hat{a}_y + 5\hat{a}_z\right)$

(b) To find a unit vector normal to both **A** and **B**, we use the cross product. The result of the cross product is a vector that is normal to both **A** and **B**.

$$\vec{A}\times\vec{B}=\begin{vmatrix}\vec{a}_x & \vec{a}_y & \vec{a}_z\\ 0 & 3 & 4\\ 4 & -10 & 5\end{vmatrix}=\left(55\vec{a}_x+16\vec{a}_y-12\vec{a}_z\right)$$

We then divide this vector by its magnitude to find the unit vector.

$$\hat{a}=\frac{\vec{A}\times\vec{B}}{\left|\vec{A}\times\vec{B}\right|}=\frac{55\hat{a}_x+16\hat{a}_y-12\hat{a}_z}{\sqrt{55^2+16^2+12^2}}=\frac{1}{\sqrt{3425}}\left(55\hat{a}_x+16\hat{a}_y-12\hat{a}_z\right)$$

The negative of this unit vector is also normal to both **A** and **B.** ■

EXAMPLE PROBLEM 1.4

A Vector field is given by $\vec{A}=y\hat{a}_x+(x-1)^2\hat{a}_y+zx\hat{a}_z$. Determine a unit vector in the direction of $\vec{A}$ at point P (2, 0, 3).

Solution

Given that $\vec{A}=y\hat{a}_x+(x-1)^2\hat{a}_y+zx\hat{a}_z$

Vector $\vec{A}$ at P (2, 0, 3) is $\hat{a}_y+6\hat{a}_z$

The unit vector in the direction of $\vec{A}$, at point P is $\dfrac{\hat{a}_y+6\hat{a}_z}{\sqrt{1+36}}=0.16\hat{a}_y+0.98\hat{a}_z$

■

1.4.2 Cylindrical Coordinate System

The cylindrical coordinate system is convenient to use when dealing with problems having cylindrical symmetry. Cylindrical coordinates just adds a z-coordinate to the polar coordinates (ρ,ϕ). Consider Figure 1.10, an arbitrary point P in space is defined by the intersection of the three perpendicular surfaces of (i) circular cylinder of radius r, (ii) a plane at constant ϕ and (iii) a plane at constant z. *The coordinates* of P are represented as (ρ,ϕ,z). Where ρ is the radius of the cylinder; and is a radial displacement from the z-axis; ϕ is the azimuthal angle in the xy-plane or angular displacement from x-axis; z is the vertical displacement (as in the Cartesian system).

The ranges of the variables are: $0\le\rho<\infty$, $0\le\phi<2\pi$, $-\infty<z<\infty$.

A vector at the point P in cylindrical coordinates can be written as $\left(A_\rho, A_\phi, A_z\right)$ or can be expressed in terms of three mutually perpendicular unit vectors $\hat{a}_\rho$, $\hat{a}_\phi$ and $\hat{a}_z$ as shown in Figure 1.11 as

$$\vec{A}=A_\rho\hat{a}_\rho+A_\phi\hat{a}_\phi+A_z\hat{a}_z \tag{1.14}$$

The magnitude of $\vec{A}$ is $\left|\vec{A}\right|=\sqrt{A_\rho^2+A_\phi^2+A_z^2}$ where $\hat{a}_\rho$ = unit vector normal to the cylindrical surface and pointing in the direction of increasing ρ ($\hat{a}_\rho$ is parallel to the xy-plane); $\hat{a}_\phi$ = unit vector (tangential to the cylindrical surface), perpendicular to both of these and in the direction of increasing ϕ ($\hat{a}_\phi$ is also parallel to the xy-plane); $\hat{a}_z$ = unit vector pointing in the direction of increasing z (it is same as in Cartesian coordinates).

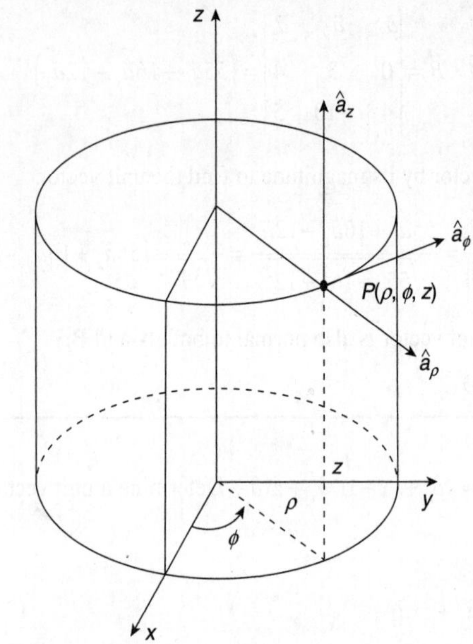

Figure 1.10 *Geometry of a cylindrical system*

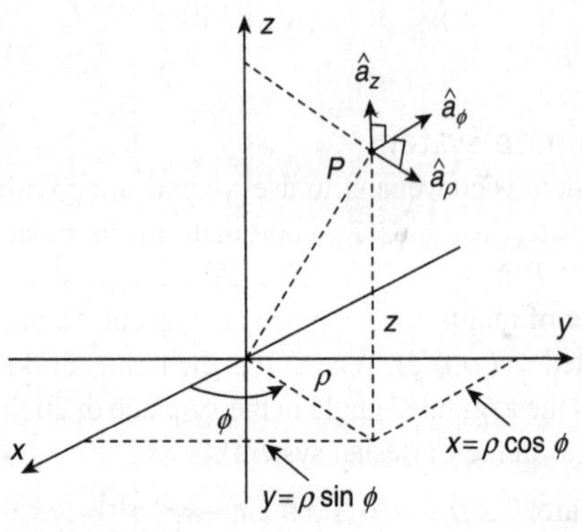

Figure 1.11 *Representation of unit vectors in cylindrical coordinates*

The unit vectors satisfy the relations given below.

$$\hat{a}_\rho \times \hat{a}_\phi = \hat{a}_z, \hat{a}_\phi \times \hat{a}_z = \hat{a}_\rho, \hat{a}_z \times \hat{a}_\rho = \hat{a}_\phi \text{ and}$$
$$\hat{a}_\rho \cdot \hat{a}_\phi = \hat{a}_\phi \cdot \hat{a}_z = \hat{a}_\rho \cdot \hat{a}_z = 0$$
$$\hat{a}_\rho \cdot \hat{a}_\rho = \hat{a}_\phi \cdot \hat{a}_\phi = \hat{a}_z \cdot \hat{a}_z = 1$$

Vector differential area and differential volume in cylindrical coordinates

A section of differential size cylindrical volume is shown in Figure1.12. It is formed when one moves from a point at coordinate (ρ, ϕ, z) by an incremental distance $d\rho$, $\rho\, d\phi$, and dz in each of the three coordinate directions.

(i) The differential volume in cylindrical coordinates as shown Figure 1.13 is given by $dv = \rho d\rho d\phi dz$

(ii) Consider an infinitesimal area element on the surface of a cylinder of radius ρ (Figure 1.14). The area magnitude is $dA = \rho d\phi dz$

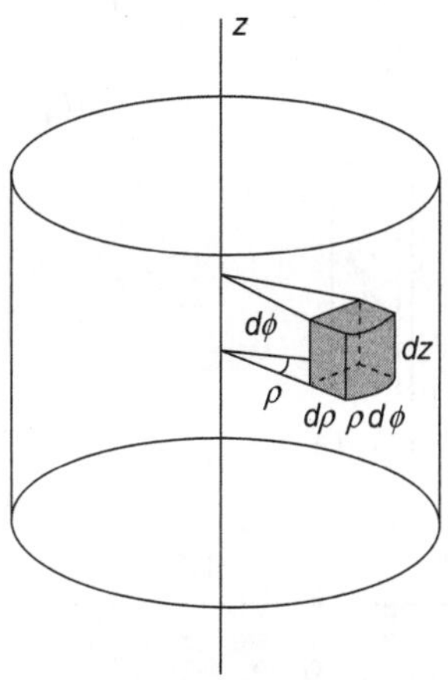

Figure 1.12 *Differential volume in cylindrical coordinates*

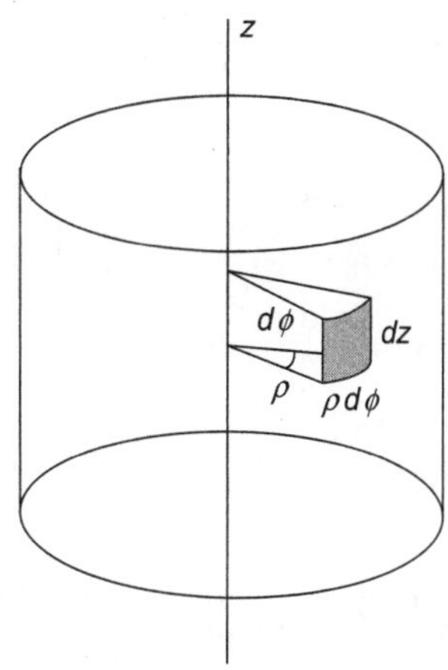

Figure 1.13 *Area element for cylinder of radius ρ*

Relation between cartesian and cylindrical coordinates

The relationship between rectangular and cylindrical coordinates is summarised as follows:

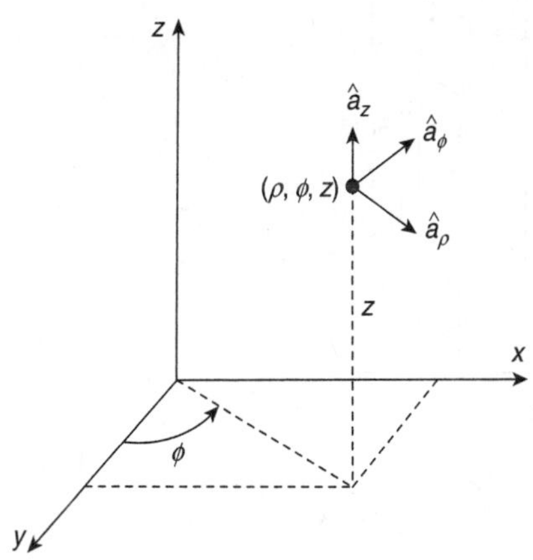

$x = \rho \cos\phi$ $\quad \rho = \sqrt{x^2 + y^2}$

$y = \rho \sin\phi$ $\quad \phi = \tan^{-1}\left(\frac{y}{x}\right)$

$z = z$ $\quad z = z$

Figure 1.14 *Cylindrical coordinate system and its unit vectors*

Figure 1.14 shows the unit vectors $\hat{a}_\rho = d\rho$, $\hat{a}_\phi = d\phi$ and $\hat{a}_z = dz$. These unit vectors are mutually orthogonal every where. Unlike rectangular coordinates, the unit vectors $\hat{a}_\rho$ and $\hat{a}_\phi$ change direction depending on the particular point in space. Hence care should be taken while differentiating in cylindrical coordinates.

The rectangular to cylindrical and cylindrical to rectangular transformation in matrix form is given below.

$$\begin{bmatrix} A_\rho \\ A_\phi \\ A_z \end{bmatrix} = \begin{bmatrix} \cos\phi & \sin\phi & 0 \\ -\sin\phi & \cos\phi & 0 \\ 0 & 0 & 1 \end{bmatrix} \begin{bmatrix} A_x \\ A_y \\ A_z \end{bmatrix} \tag{1.15}$$

and

$$\begin{bmatrix} A_x \\ A_y \\ A_z \end{bmatrix} = \begin{bmatrix} \frac{x}{\sqrt{x^2+y^2}} & \frac{y}{\sqrt{x^2+y^2}} & 0 \\ \frac{y}{\sqrt{x^2+y^2}} & \frac{x}{\sqrt{x^2+y^2}} & 0 \\ 0 & 0 & 1 \end{bmatrix} \begin{bmatrix} A_\rho \\ A_\phi \\ A_z \end{bmatrix} \tag{1.16}$$

EXAMPLE PROBLEM 1.5

Convert the following Cartesian coordinates point *P* (3, 4, 5) to Cylindrical coordinates and sketch the location of the point *P*.

Solution

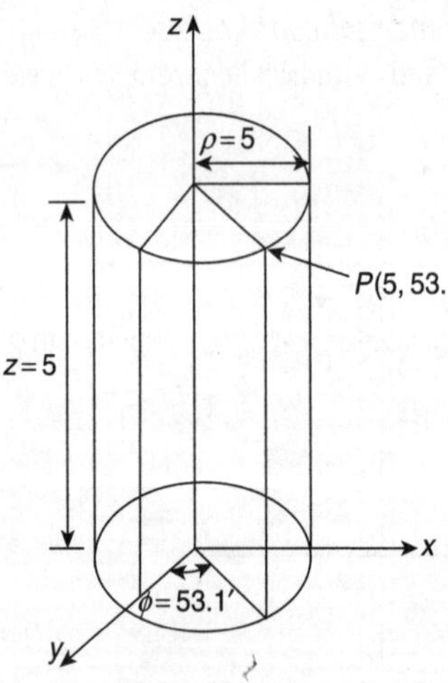

Figure 1.15 *Representation of point P in cylindrical coordinate system*

We know that $\rho = \sqrt{X^2+Y^2} = \sqrt{3^2+4^2} = 5, \phi = \tan^{-1}\left({}^{y}\!/\!_{x}\right) = \tan^{-1}\left({}^{4}\!/\!_{3}\right) = 53.1', z = 5.$

The point in cylindrical coordinate system is expressed as $P\,(3, 4, 5) \Rightarrow P\,(5{,}53.1{,}5)$.

■

EXAMPLE PROBLEM 1.6

Convert the following points expressed in cylindrical coordinates to rectangular coordinate system.

(i) $P(2, 5\pi/6, 3)$ (ii) $Q(4, 4\pi/3, -1)$

Solution

We know that(i)

$$x = \rho\cos\phi,\ y = \rho\sin\phi,\ z = z$$

(i) $x = \rho\cos\phi = 2\cos\dfrac{5\pi}{6} = -1.732,\ y = \rho\sin\phi = 2\sin\dfrac{5\pi}{6} = 1$

$$P\left(4, \frac{5\pi}{6}, 3\right) \Rightarrow P(-1.732, 1, 3)$$

(ii) $P\left(4, \dfrac{4\pi}{3}, -1\right) \Rightarrow P(-2, -3.464, -1)$

■

EXAMPLE PROBLEM 1.7

Convert the vector $\bar{F} = 4\hat{a}_x - 2\hat{a}_y - 4\hat{a}_z$ located at A(2, 3, 5) into cylindrical coordinates.

Solution

Using the matrix form of the rectangular to cylindrical transformation equations:

Using Eq.15, we can write $\begin{bmatrix} A_\rho \\ A_\phi \\ A_z \end{bmatrix} = \begin{bmatrix} \cos\phi & \sin\phi & 0 \\ -\sin\phi & \cos\phi & 0 \\ 0 & 0 & 1 \end{bmatrix} \begin{bmatrix} 4 \\ -2 \\ -4 \end{bmatrix}$

$$A_\rho = 4\cos\phi - 2\sin\ \phi,\ A_\phi = -4\sin\phi - 2\cos\phi,\ A_z = -4$$

At the point $P(2, 3, 5)$ $\phi = \tan^{-1}\left(\dfrac{y}{x}\right) = \tan^{-1}\left(\dfrac{3}{2}\right) = 56.3^\circ$

$$A_\rho = (4\cos 56.3^\circ - 2\sin 56.3^\circ),\ A_\phi = -4\sin 56.3^\circ - 2\cos 56.3^\circ,\ A_z = -4$$

$$\vec{F} = (4\cos 56.3^\circ - 2\sin 56.3^\circ)\hat{a}_\rho + (-4\sin 56.3^\circ - 2\cos 56.3^\circ)\hat{a}_\phi - 4\hat{a}_z$$

$$\vec{F} = 0.556a_\rho - 4.44a_\phi - 4a_z$$

■

1.4.3 Spherical Coordinate System

A spherical coordinate system is useful when there is a point of symmetry that is taken as the origin. In a spherical coordinate system any point in space is represented as the intersection of three surfaces as shown in the Figure 1.16.

- A sphere of radius r from the origin (r = constant)
- A cone centered on the z-axis (θ = constant)
- A vertical plane (ϕ = constant)

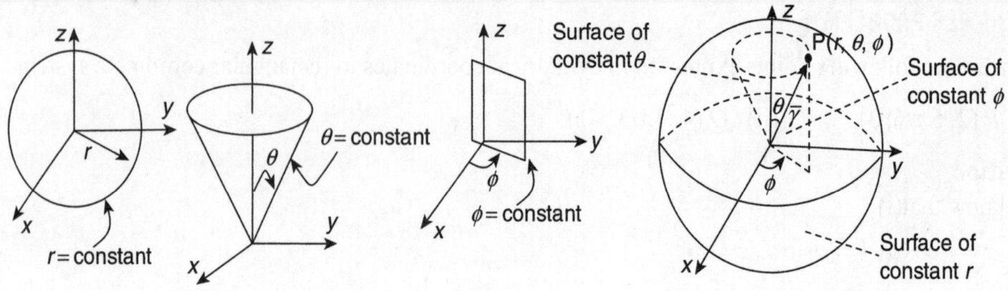

Figure 1.16 *Planes in spherical coordinate system*

Specification of the coordinate system: Figure1.17 shows a general point, P, in space is located on a spherical surface centered at (0, 0, 0) of the Cartesian coordinate system.

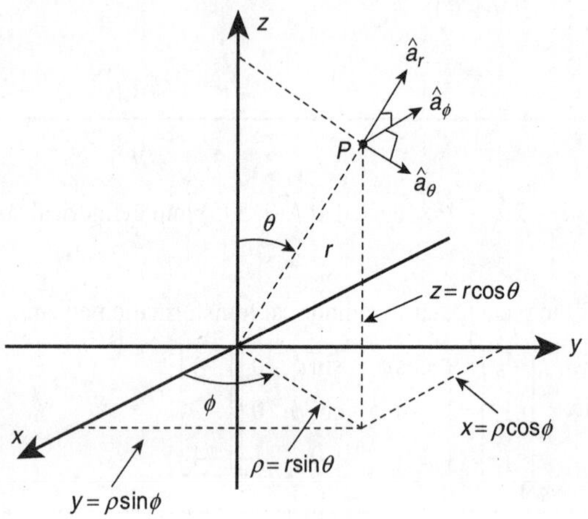

Figure 1.17 *Representation of a point in a spherical coordinate system*

Spherical coordinates (r, θ, ϕ):

The location of a point P is specified by three *spherical coordinates* (r, θ, ϕ) as shown in Figure 1.18. They are defined as follows:

$$r = \text{distance from origin to observation point } (P) = \sqrt{x^2 + y^2 + z^2}$$

$$\theta = \text{angle made by } r \text{ with respect to positive } z\text{-axis} = \cos^{-1}\left(\frac{z}{\sqrt{x^2 + y^2 + z^2}}\right)$$

$$\phi = \text{angle made by positive } x\text{-axis by projection of '}r\text{' on to } xy\text{- plane} = \tan^{-1} y/x$$

The limits on r, θ, ϕ are: $0 < r < \infty$, $0 < \theta < \pi$, $0 < \phi < 2\pi$.

Spherical to Cartesian coordinates transformation equations are:

$$x = r\sin\theta\cos\phi \quad y = r\sin\theta\sin\phi \quad z = r\cos\theta$$

Three unit vectors of the spherical coordinate system are shown in Figure 1.17. The unit vector $\hat{a}_r$ lies along the radially outward direction to the spherical surface. It lies on the cone $\theta =$ constant and the plane $\phi =$ constant. The unit vector $\hat{a}_\theta$ is normal to the conical surface and lies in $\phi =$ constant plane and is tangential to the spherical surface. The unit vector $\hat{a}_\theta$ is the same as in the cylindrical coordinate system. It is normal to $\phi =$ constant plane and is tangential to both the cone and the sphere. The unit vectors are mutually perpendicular and forms a right handed set.

A vector $\vec{A}$ and its magnitude in spherical coordinate system may be expressed as

$$\vec{A} = A_r\hat{a}_r + A_\theta\hat{a}_\theta + A_\phi\hat{a}_\phi \qquad |\vec{A}| = \sqrt{A_r^2 + A_\theta^2 + A_\phi^2} \tag{1.17}$$

The unit vectors $\hat{a}_r$, $\hat{a}_\theta$, $\hat{a}_\phi$ are mutually orthogonal. Thus

$$\hat{a}_r \cdot \hat{a}_r = \hat{a}_\theta \cdot \hat{a}_\theta = \hat{a}_\phi \cdot \hat{a}_\phi = 1$$

$$\hat{a}_r \cdot \hat{a}_\theta = \hat{a}_\theta \cdot \hat{a}_\phi = \hat{a}_\phi \cdot \hat{a}_r = 0$$

1.4.4 Conversion of Vector between the Coordinate Systems

Very often, we need to transform a vector from one coordinate system to another while solving the EMF theory problems.

For example, suppose vector $\vec{A}$ is given in Cartesian system as $= \vec{A} = A_x\hat{a}_x + A_y\hat{a}_y + A_z\hat{a}_z$ and now it is required to transform it into cylindrical coordinate system, that is $\vec{A} = A_\rho\hat{a}_\rho + A_\phi\hat{a}_\phi + A_z\hat{a}_z$.

So, the values of A_ρ, A_ϕ and A_z are required to be computed using the following steps.

(i) To find 'A_ρ', we take the dot product between $\vec{A}$ (Cartesian) and unit vector $\hat{a}_\rho$ (which is of the desired direction).

$$A_\rho = \vec{A} \cdot \hat{a}_\rho = A_x\hat{a}_x \cdot \hat{a}_\rho + A_y\hat{a}_y \cdot \hat{a}_\rho + A_z\hat{a}_z \cdot \hat{a}_\rho \tag{1.18}$$

(ii) To find 'A_ϕ' we take the dot product between $\vec{A}$ (Cartesian) and unit vector $\hat{a}_\phi$.

$$A_\phi = \vec{A} \cdot \hat{a}_\phi = A_x\hat{a}_x \cdot \hat{a}_\phi + A_y\hat{a}_y \cdot \hat{a}_\phi + A_z\hat{a}_z \cdot \hat{a}_\phi \tag{1.19}$$

(iii) To find 'A_z' we take the dot product between $\vec{A}$ (Cartesian) and unit vector $\hat{a}_z$.

$$A_z = \vec{A} \cdot \hat{a}_z = A_x\hat{a}_x \cdot \hat{a}_z + A_y\hat{a}_y \cdot \hat{a}_z + A_z\hat{a}_z \cdot \hat{a}_z \tag{1.20}$$

Therefore, the above three equations shows that there is dot product between unit vector of different coordinates systems. These are tabulated as below (Table 1.1).

Table 1.1 Relationship between Cartesian, cylindrical and spherical coordinate unit vectors

	Rectangular			Cylindrical			Spherical		
•	$\hat{a}_x$	$\hat{a}_y$	$\hat{a}_z$	$\hat{a}_\rho$	$\hat{a}_\phi$	$\hat{a}_z$	$\hat{a}_r$	$\hat{a}_\theta$	$\hat{a}_\phi$
$\hat{a}_x$	1	0	0	$\cos\phi$	$-\sin\phi$	0	$\sin\theta\cos\phi$	$\cos\theta\cos\phi$	$-\sin\phi$
$\hat{a}_y$	0	1	0	$\sin\phi$	$\cos\phi$	0	$\sin\theta\sin\phi$	$\cos\theta\sin\phi$	$\cos\phi$
$\hat{a}_z$	0	0	1	0	0	1	$\cos\theta$	$-\sin\theta$	0
$\hat{a}_\rho$	$\cos\phi$	$\sin\phi$	0	1	0	0	$\sin\theta$	$\cos\theta$	0
$\hat{a}_\phi$	$-\sin\phi$	$\cos\phi$	0	0	1	0	0	0	1
$\hat{a}_z$	0	0	1	0	0	1	$\cos\theta$	$-\sin\theta$	0
$\hat{a}_r$	$\sin\theta\cos\phi$	$\sin\phi\sin\theta$	$\cos\theta$	$\sin\theta$	0	$\cos\theta$	1	0	0
$\hat{a}_\theta$	$\cos\theta\cos\phi$	$\cos\theta\sin\phi$	$-\sin\theta$	$\cos\theta$	0	$-\sin\theta$	0	1	0
$\hat{a}_\phi$	$-\sin\phi$	$\cos\phi$	0	0	1	0	0	0	1

So, by using Table 1.1, Eqs (1.18), (1.19) and (1.20) become,

$$A_\rho = \vec{A}\cdot\hat{a}_\rho = A_x\hat{a}_x\cdot\hat{a}_\rho + A_y\hat{a}_y\cdot\hat{a}_\rho + A_z\hat{a}_z\cdot\hat{a}_\rho = A_x\cos\phi + A_y\sin\phi$$

$$A_\phi = \vec{A}\cdot\hat{a}_\phi = A_x\hat{a}_x\cdot\hat{a}_\phi + A_y\hat{a}_y\cdot\hat{a}_\phi + A_z\hat{a}_z\cdot\hat{a}_\phi = -A_x\sin\phi + A_y\cos\phi$$

$$A_z = \vec{A}\cdot\hat{a}_z = A_x\hat{a}_x\cdot\hat{a}_z + A_y\hat{a}_y\cdot\hat{a}_z + A_z\hat{a}_z\cdot\hat{a}_z = A_z$$

1.5 VECTOR CALCULUS

1.5.1 Differential Length, Area, Volume

Differential length

Line integrals are defined by using differential lengths or *differential displacements* in any arbitrary directions along a given path. In rectangular, cylindrical and spherical coordinates, differential lengths are given by

$$\left.\begin{aligned} d\vec{L} &= dx\,\hat{a}_x + dy\,\hat{a}_y + dz\,\hat{a}_z \text{(Rectangular)} \\ d\vec{L} &= d\rho\,\hat{a}_\rho + \rho\,d\phi\,\hat{a}_\phi + dz\,\hat{a}_z \text{(cylindrical)} \\ d\vec{L} &= dr\,\hat{a}_r + r\,d\theta\,\hat{a}_\theta + r\sin\theta\,d\phi\,\hat{a}_\phi \text{(spherical)} \end{aligned}\right\} \tag{1.21}$$

The above mentioned differential lengths are acceptable for integration in any given general direction but must be parameterized in terms of one variable only.

Differential area

The differential surface element is **dA** or **dS** and is defined as

$$\left.\begin{aligned} dA_x &= dydz\ \hat{a}_x \\ dA_y &= dzdx\ \hat{a}_y \\ dA_z &= dxdy\ \hat{a}_z \end{aligned}\right\}\text{(Cartesian coordinates)} \tag{1.22}$$

$$\left.\begin{aligned} dA_\rho &= \rho d\phi dz\ \hat{a}_\rho \\ dA_\phi &= d\rho dz\ \hat{a}_\phi \\ dA_z &= \rho d\rho d\phi\ \hat{a}_z \end{aligned}\right\}\text{(cylindrical coordinates)} \tag{1.23}$$

$$\left.\begin{aligned} dA_r &= r^2 \sin\theta d\theta d\phi\ \hat{a}_r \\ dA_\theta &= r\sin\theta dr d\phi\ \hat{a}_\theta \\ dA_\Phi &= r dr d\theta\ \hat{a}_\phi \end{aligned}\right\}\text{(spherical coordinates)} \tag{1.24}$$

$$d\vec{S} = d\vec{A} = \begin{cases} dydz\ \hat{a}_x + dx\,dz\ \hat{a}_y + dxdy\ \hat{a}_z \\ \rho d\phi dz\ \hat{a}_\rho + d\rho\ dz\ \hat{a}_\phi + \rho d\rho\ d\phi\ \hat{a}_z \\ r^2 \sin\theta d\theta\ d\phi \hat{a}_r + r\sin\theta\ dr\ \hat{a}_\phi + r\,dr\,d\theta\ \hat{a}_\theta \end{cases}$$

(differential elements of vector area) (1.25)

$$dv = \begin{cases} dx\,dy\,dz \\ \rho d\rho d\phi dz \\ r^2 \sin\theta dr\ d\theta\ d\phi \end{cases} \quad \text{(differential elements of volume)} \tag{1.26}$$

1.5.2 Line, Surface and Volume Integrals

Line integral

Line integral is defined as any integral that is to be evaluated along a line. A line indicates a path along a curve in space. The line integral is an expression of the form

$$\int_a^b \vec{A}\cdot d\vec{L} \text{ and } \int_a^b \vec{F}\cdot d\vec{L}$$

Where, A is a vector function, d$\vec{L}$ is the infinitesimal displacement vector and the integral is to be carried out along a prescribed path i.e. from the point a to the point b.

A vector field $\boldsymbol{A}$ is conservative if the line integral of $\vec{A}\cdot d\vec{L}$ round any closed curve is zero:

$$\oint \vec{A}\cdot d\vec{L} = 0$$

The circle on the integral sign indicates that the path of integration is closed.

Surface integral

Consider a vector field $\vec{A}$ continuous in a region of space containing a smooth surface S (Figure 1.18). The surface integral of $\vec{A}$ can be defined as

$$\psi = \int_S \vec{A}\cdot d\vec{S}$$

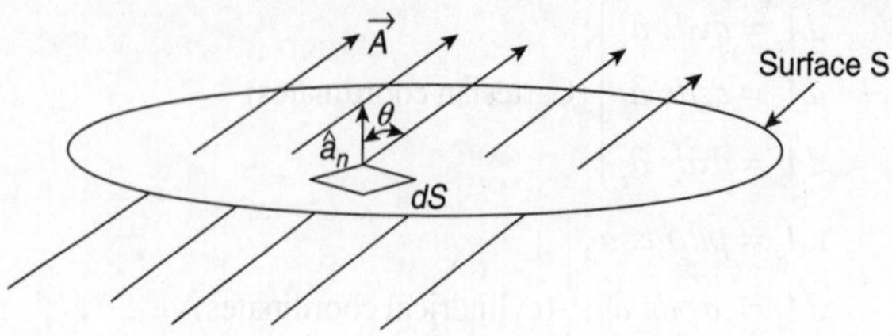

Figure 1.18 *Representation of surface integral*

Consider a small incremental surface area on the surface S denoted by dS. Let $\hat{a}_n$ be a unit normal to the surface dS. The magnitude of flux crossing the unit surface normally is given by

$$\left|\vec{A}\right|\cos\theta\, dS = \vec{A}\cdot\hat{a}_n\, dS = \vec{A}\cdot dS\hat{a}_n = \vec{A}\cdot d\vec{S}$$

$$\because \vec{A}\cdot\hat{a}_n = |\vec{A}||\hat{a}_n|\cos\theta = |\vec{A}|\cos\theta$$

where $d\vec{S}$ denotes the vector area having magnitude equal to dS and whose direction is in the direction of the unit normal.

$$d\vec{S} = dS\hat{a}_n$$

The total flux crossing the surface is obtained by integrating $\vec{A}\cdot d\bar{S}$

$$\Phi_E = \int_S \vec{A}\cdot d\vec{S}$$

For a closed surface defining a volume, the surface integral becomes closed surface integral and is denoted by

$$\Phi_E = \oint_s \vec{A}\cdot d\vec{S} \tag{1.27}$$

Volume integral

A volume integral is of the form $\int_v v dv$, where v is a scalar function and dv is an infinitesimal volume element (Figure 1.19). For example, dv can be computed in Cartesian coordinates as

$$dv = (dx)(dy)(dz)$$

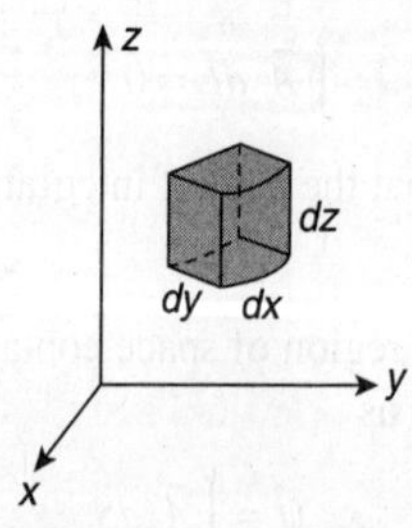

Figure 1.19 *Representation of volume integral*

1.5.3 Del Operator ∇

An inverted uppercase delta symbol represents a **vector differential operator called '*nabla*' or '*del*', "∇"**. It indicates that the derivatives of the quantity should be derived on which the operator is acting. This del operator is followed by symbols indicating the exact form of those derivatives as given below:

- "∇·" signifying **divergence**,
- "∇ ×" indicating **curl**, and
- "∇" signifying **gradient**

∇ is an instruction to take partial derivatives in three directions and is defined as,

$$\nabla = \hat{i}\frac{\partial}{\partial x} + \hat{j}\frac{\partial}{\partial y} + \hat{k}\frac{\partial}{\partial z} \tag{1.28}$$

where $\hat{i}$, $\hat{j}$ and $\hat{k}$ are the unit vectors in x, y and z axes and they can be also represented as $\hat{x}$, $\hat{y}$ and $\hat{z}$.

Del (∇) represents a differential operator that can operate on scalar or vector fields and results a scalar or vector.

1.5.4 Gradient of a Scalar Field

The space rate of change of scalar function is defined as the gradient of that scalar function.

The gradient of the scalar field ψ is defined as

$$\text{grad}\,(\psi) = \nabla\psi = \left(\hat{i}\frac{\partial}{\partial x} + \hat{j}\frac{\partial}{\partial y} + \hat{k}\frac{\partial}{\partial z}\right)\psi \text{ (Cartesian)} \tag{1.29}$$

and can be expanded as

$$\nabla\psi = \hat{i}\frac{\partial\psi}{\partial x} + \hat{j}\frac{\partial\psi}{\partial y} + \hat{k}\frac{\partial\psi}{\partial z} = \frac{\partial\psi}{\partial x}\hat{i} + \frac{\partial\psi}{\partial y}\hat{j} + \frac{\partial\psi}{\partial z}\hat{k}$$

Thus the slope of the scalar field in x-direction, of the gradient ψ is indicated by the x-component, similarly the slope in the y-direction is indicated by y-component, and the slope in z-direction indicated by z-component. The total steepness of the slope at gradient location can be derived by taking the sum of the squares of x, y and z components.

In cylindrical and spherical coordinates, the gradient or 'del' operator is

$$\text{grad}(\psi) = \nabla\psi = \left(\hat{a}_\rho\frac{\partial}{\partial\rho} + \hat{a}_\phi\frac{1}{\rho}\frac{\partial}{\partial\phi} + \hat{a}_z\frac{\partial}{\partial z}\right)\psi \text{ (Cylindrical)} \tag{1.30}$$

and

$$\text{grad}(\psi) = \nabla\psi = \left(\hat{a}_r\frac{\partial}{\partial\rho} + \hat{a}_\theta\frac{1}{r}\frac{\partial}{\partial\theta} + \hat{a}_\varphi\frac{1}{r\sin\theta}\frac{\partial}{\partial\varphi}\right)\psi \text{ (Spherical)} \tag{1.31}$$

Gradient, functions on a scalar field which results in a vector. This shows the rate of spatial change of the field at a point and in the direction of extreme increase from that point.

EXAMPLE PROBLEM 1.8

Find the gradient of $\Psi = x^2$.

Solution

We know that

$$\text{grad}\,(\psi) = \nabla\psi = \left(\hat{i}\frac{\partial}{\partial x} + \hat{j}\frac{\partial}{\partial y} + \hat{k}\frac{\partial}{\partial z}\right)\psi = \left(\hat{i}\frac{\partial}{\partial x} + \hat{j}\frac{\partial}{\partial y} + \hat{k}\frac{\partial}{\partial z}\right)x^2 = 2x\hat{i}$$

Therefore, from the result, we can say that the grad (ψ) has the property that the rate of change of Ψ with respect to distance in a particular direction ($\hat{i}$) is the projection of grad (ψ) onto that direction.

1.5.5 Divergence of a Vector Field

In Physics and engineering, divergence is an important concept as it is used to explain the behaviour of vector fields. Divergence arises when there is a flow of something like liquid or gas flowing in a tube or electric or magnetic flux lines flowing in a vector space.

Divergence of a vector field $\vec{A}$ is a measure of how much a vector field converges to or diverges from a given point.

Divergence of vector field becomes positive when the vector diverges out from the given point and it becomes negative when the vector field converges at that point.

For example, the vector function in Figure 1.20a has a large (positive divergence). If the arrows pointed in, it would be a large negative divergence as shown in the Figure 1.20b.

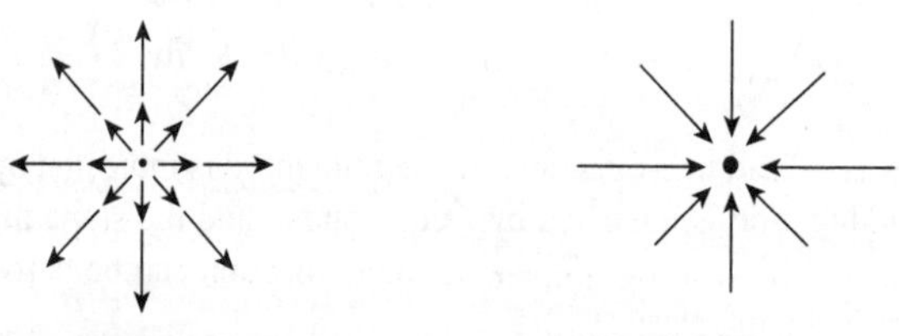

Figure 1.20 *(a) Positive divergence; (b) Negative divergence*

Like the *dot* product the divergence produces a scalar result. In Cartesian coordinates, the divergence of a vector **A** (pronounced 'del dot **A**') is defined as

$$\nabla\cdot\vec{A} = \left(\hat{i}\frac{\partial}{\partial x} + \hat{j}\frac{\partial}{\partial y} + \hat{k}\frac{\partial}{\partial z}\right)\cdot\left(\hat{i}A_x + \hat{j}A_y + \hat{k}A_z\right) \tag{1.32}$$

and since $\quad \hat{i}\cdot\hat{i} = \hat{j}\cdot\hat{j} = \hat{k}\cdot\hat{k} = 1$

$$\nabla\cdot\vec{A}=\left(\frac{\partial A_x}{\partial x}+\frac{\partial A_y}{\partial y}+\frac{\partial A_z}{\partial z}\right) \tag{1.33}$$

where, the symbol 'del' $=\nabla=\left(\hat{i}\frac{\partial}{\partial x}+\hat{j}\frac{\partial}{\partial y}+\hat{k}\frac{\partial}{\partial z}\right)$

vector $\vec{A}=\left(A_x\hat{i}+A_y\hat{j}+A_z\hat{k}\right)$ and is any vector in terms of the unit vectors $\hat{i},\hat{j}$ and $\hat{k}$ corresponding to the x-, y-, and z- directions. A_x, A_y, and A_z are called the Cartesian components of the vector $\vec{A}$.

Thus, the divergence of the vector field $\vec{A}$ is the sum of the change in its x-component, y-component, z-component along their respective axis i.e., x-axis, y-axis and z-axis respectively.

Note that the divergence of a vector field is a scalar quantity; it has magnitude but no direction.

1.5.6 Divergence Theorem

The *divergence theorem* is also known as *Gauss' theorem*. Gauss theorem and Stokes theorem can be used to convert between the differential and integral forms of Maxwell's equations.

Gauss' theorem converts a surface integral to a volume integral

For any vector field $\vec{A}$:

$$\oint_S \vec{A}\cdot d\vec{S}=\int_v \nabla\cdot\vec{A}\,dv \tag{1.34}$$

where S is the closed surface bounding the volume v, and the surface area element $d\vec{S}$ is directed out of the volume v.

For example, consider Maxwell's first equation ($\nabla\cdot\vec{D}=\rho_v$). By applying the divergence theorem, it relates the flux of a vector field through a closed surface S to the divergence of the vector field in the enclosed volume ($\oint_S \vec{D}\cdot d\vec{S}=\int_V \rho_v\,dv$).

1.5.7 Curl of a Vector Field

The curl of a vector field can be defined as *measure of the field's tendency to circulate about a point.*

The cross product of the gradient operator and a vector field ($\nabla\times\vec{A}$) is called the 'curl' of the vector field $\vec{A}$ or 'del cross'$(\nabla\times)$. For the given vector $\vec{A}=A_x\mathbf{i}+A_y\mathbf{j}+A_z\mathbf{k}$. The curl of a vector field is itself a vector field

$$\text{Curl }\mathbf{A}=\nabla\times\vec{A}=\begin{vmatrix} i & j & k \\ \frac{\partial}{\partial x} & \frac{\partial}{\partial y} & \frac{\partial}{\partial z} \\ A_x & A_y & A_z \end{vmatrix}$$

By expanding the equation

$$\nabla\times\vec{A}=\hat{i}\left(\frac{\partial A_z}{\partial y}-\frac{\partial A_y}{\partial z}\right)-\left(\frac{\partial A_z}{\partial x}-\frac{\partial Ax}{\partial z}\right)\hat{j}+\left(\frac{\partial A_y}{\partial x}-\frac{\partial Ax}{\partial y}\right)\hat{k}$$

By separating the vector operator (del) and the vector field A components from the vector cross product, the above equation can be written as

$$\nabla\times\vec{A}=\left(\hat{i}\frac{\partial}{\partial x}+\hat{j}\frac{\partial}{\partial y}+\hat{k}\frac{\partial}{\partial z}\right)\times\left(\hat{i}A_x+\hat{j}A_y+\hat{k}A_z\right) \tag{1.35}$$

The property of the field to rotate in one of the coordinate planes is denoted by each component of the curl of A. It means that the field has considerable circulation about a point in the y-z plane only when the curl of the field at that point has a large x-component. Using the right-hand rule we can sense the overall direction of the curl that symbolizes the axis about which the rotation is maximum.

The curl of vector A in cylindrical and spherical coordinates is:

$$\nabla\times\vec{A}=\left(\frac{1}{\rho}\frac{\partial A_z}{\partial\phi}-\frac{\partial A_\phi}{\partial z}\right)\hat{\rho}+\left(\frac{\partial A_\rho}{\partial z}-\frac{\partial A_z}{\partial\rho}\right)\hat{\phi}+\frac{1}{\rho}\left(\frac{\partial(\rho A_\phi)}{\partial\rho}-\frac{\partial A_\rho}{\partial\phi}\right)\hat{z}\quad\text{(cylindrical)}$$

$$\nabla\times\vec{A}=\frac{1}{r\sin\theta}\left(\frac{\partial(A_\phi\sin\theta)}{\partial\theta}-\frac{\partial A_\theta}{\partial\phi}\right)\hat{r}+\frac{1}{r}\left(\frac{1}{\sin\theta}\frac{\partial A_r}{\partial\phi}-\frac{\partial(rA_\phi)}{\partial r}\right)\hat{\theta}+\frac{1}{r}\left(\frac{\partial(rA_\theta)}{\partial r}-\frac{\partial A_r}{\partial\theta}\right)\hat{\phi}\quad\text{(spherical)}$$

EXAMPLE PROBLEM 1.9

If $\vec{A}=(1+z^2)\hat{i}+xy\hat{j}+x^2y\hat{k}$ then calculate curl A

Solution

By definition curl $\vec{A}=\nabla\times\vec{A}=\begin{vmatrix}\hat{i} & \hat{j} & \hat{k}\\ \frac{\partial}{\partial x} & \frac{\partial}{\partial y} & \frac{\partial}{\partial z}\\ 1+z^2 & xy & x^2y\end{vmatrix}$

$$=\hat{i}\left\{\frac{\partial}{\partial y}(x^2y)-\frac{\partial}{\partial z}(xy)\right\}-\hat{j}\left\{\frac{\partial}{\partial x}(x^2y)-\frac{\partial}{\partial z}(1+z^2)\right\}+\hat{k}\left\{\frac{\partial}{\partial x}(xy)-\frac{\partial}{\partial y}(1+z^2)\right\}$$

$$=x^2\hat{i}-(2xy-2z)\hat{j}+y\hat{k}$$

■

EXAMPLE PROBLEM 1.10

If $\vec{A}=xyz\hat{i}+\cos(xyz)\hat{j}+xy^2z^3\hat{k}$ then calculate curl $\vec{A}$

Solution

$$\nabla\times\vec{A}=\begin{vmatrix}\hat{i} & \hat{j} & \hat{k}\\ \dfrac{\partial}{\partial x} & \dfrac{\partial}{\partial y} & \dfrac{\partial}{\partial z}\\ xyz & \cos(xyz) & xy^2z^3\end{vmatrix}$$

$$= (2xyz^3+\sin(xyz)xy)\hat{i}-(y^2z^3-xy)\hat{j}+(-yz\sin(xyz)-xz)\hat{k}$$

■

EXAMPLE PROBLEM 1.11

If $\vec{A}=xy\,\hat{a}_x+x^2\hat{a}_y$ then find $\oint_C \vec{A}\cdot d\vec{L}$ over the path shown in Figure 1.21. **(GATE 2010)**

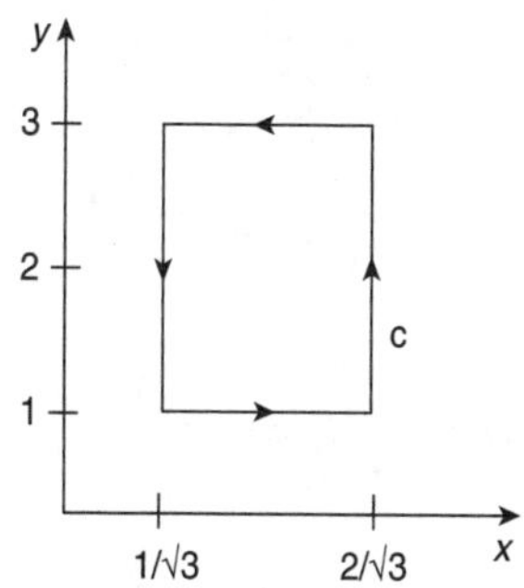

Figure 1.21 *Line integral of a vector field A around closed path C.*

Solution

Given that $\vec{A}=xy\,\hat{a}_x+x^2\hat{a}_y$.

$$\oint_C \vec{A}\cdot d\vec{L}=\oint(xy\,\hat{a}_x+x^2\hat{a}_y)\cdot(dx\,\hat{a}_x+dy\,\hat{a}_y)=\oint(xydx+x^2dy)=\oint xydx+\oint x^2dy$$

$$=\int_{1/\sqrt{3}}^{2/\sqrt{3}} x\cdot 1dx+\int_{2/\sqrt{3}}^{1/\sqrt{3}} x\cdot 3dx+\int_3^1\left(\frac{1}{\sqrt{3}}\right)^2 dy+\int_1^3\left(\frac{2}{\sqrt{3}}\right)^2 dy=\frac{1}{2}-\frac{3}{2}-\frac{2}{3}+\frac{8}{3}=-1+2=1$$

■

1.5.8 Stokes's Theorem

The line integral of a vector field **A** around a closed path C is equal to the surface integral of the curl of **A** on any surface whose only edge is C. It can be expressed as

$$\oint_C \vec{A}\cdot d\vec{L}=\iint_S \nabla\times\vec{A}\cdot d\vec{S}$$

Here, *S* indicates the surface attached to the closed path or open surface integral

Stokes theorem relates a surface integral to a closed line integral.

Consider any vector **A** on an open surface. Consider that the vector has some finite amount of curl at each point on the surface. Then the summing of these individual curls at each point will result in the net flow of that vector at the boundary of the surface (Figure 1.22a). Generally, the direction of a surface integral is outward normal. But, for an open surface there are two outward normals. To avoid confusion, a simple right-hand rule can be remembered, which tells that if the line integral is in the direction of the fingers then the direction of the thumb gives the direction of the surface integral (Figure 1.22b).

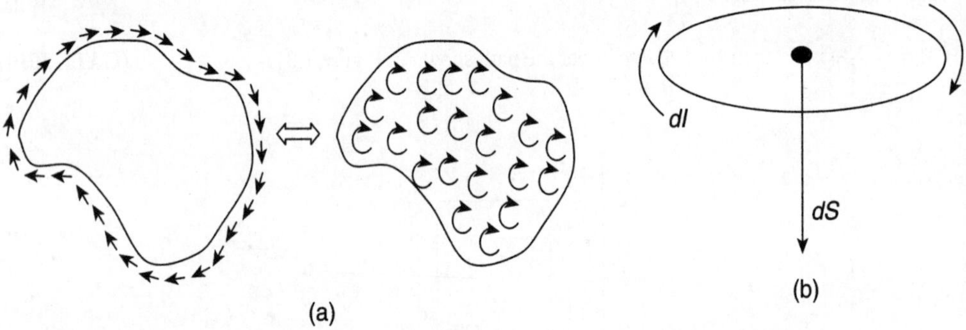

Figure 1.22 *(a) Geometrical interpretation of Stokes theorem; (b) Direction of surface and line integrals*

Laplacian operator

We have seen that gradient of *a* scalar field is a vector field and also we can compute the divergence of any vector field. So we can certainly compute divergence of the gradient of a scalar field.

Another useful relation involves the ***divergence of the gradient of a scalar field****; this is called the Laplacian of the field:*

For a scalar ψ,

$$\nabla \cdot \nabla \psi = \nabla^2 \psi = \frac{\partial^2 \psi}{\partial x^2} + \frac{\partial^2 \psi}{\partial y^2} + \frac{\partial^2 \psi}{\partial z^2} \quad \text{(Cartesian)}$$

The usefulness of these relations can be illustrated by applying them to the electric field as described by Maxwell's Equations.

1.5.9 Laplacian of a Scalar

When the Laplacian is applied to a scalar field, the result obtained will also be a scalar, because Laplacian is a scalar operator.

$$\nabla^2 = \frac{\partial^2}{\partial x^2} + \frac{\partial^2}{\partial y^2} + \frac{\partial^2}{\partial z^2}$$

Laplacian operator is defined as the divergence of the gradient of the scalar field. It is a second order differential equation. Laplacian of a scalar field V is written as $\nabla^2 V$.

Laplacian of a scalar is used in:

(i) Electrostatics to represent the charge associated to a given potential.
(ii) Defining the Helmholtz equation of propagation of EM wave.
(iii) Laplace's and Poisson's equation.

Laplacian of a scalar field V in Cartesian coordinate system is given as

$$\nabla^2 V = \frac{\partial^2 V}{\partial x^2} + \frac{\partial^2 V}{\partial y^2} + \frac{\partial^2 V}{\partial z^2}$$

Laplacian of a scalar field V in cylindrical coordinate system is given as

$$\nabla^2 V = \frac{1}{\rho}\frac{\partial}{\partial \rho}\left(\rho \frac{\partial V}{\partial \rho}\right) + \frac{1}{\rho^2}\frac{\partial^2 V}{\partial \phi^2} + \frac{\partial^2 V}{\partial z^2}$$

Laplacian of a scalar field V in spherical coordinate system is given as

$$\nabla^2 V = \frac{1}{r^2}\frac{\partial}{\partial r}\left(r^2 \frac{\partial V}{\partial r}\right) + \frac{1}{r^2 \sin\theta}\frac{\partial}{\partial \theta}\left(\sin\theta \frac{\partial V}{\partial \theta}\right) + \frac{1}{r^2 \sin^2\theta}\frac{\partial^2 V}{\partial \phi^2}$$

Note: If Laplacian, $\nabla^2 V = 0$, then ∇V is a vector field that is both divergence less and curl free.

EXAMPLE PROBLEM 1.12

Determine the Laplacian of a scalar function $V = xy^2 + x^2yz + xyz^2$.

Solution

Given that $V = xy^2 + x^2yz + xyz^2$

Laplacian of V is given by

$$\nabla^2 V = \frac{\partial^2 V}{\partial x^2} + \frac{\partial^2 V}{\partial y^2} + \frac{\partial^2 V}{\partial z^2}$$

$$= \frac{\partial^2 (xy^2 + x^2yz + xyz^2)}{\partial x^2} + \frac{\partial^2 (xy^2 + x^2yz + xyz^2)}{\partial y^2} + \frac{\partial^2 (xy^2 + x^2yz + xyz^2)}{\partial z^2} = 2yz + 2x + 2xy$$

■

SUMMARY

1. The quantities that we deal in EMF theory are either scalars or vectors.

2. A vector $\vec{A}$ is described by its components along the three coordinate directions that is

$$\vec{A} = \vec{A}_x + \vec{A}_y + \vec{A}_z = A_x\hat{a}_x + A_y\hat{a}_y + A_z\hat{a}_z$$

3. To define a unit vector in the direction of A, we simply divide the vector by its magnitude.

4. The x-component of a vector can be written in terms of unit vector as $A_x = |\vec{A}|\,\hat{a}_x$.

5. Vector, $\vec{A}$ can be multiplied in the following three ways:

- with a scalar, $c \cdot \vec{A} = (c\vec{A})$
- dot product with another vector $\vec{B}$, $\vec{A} \cdot \vec{B} = |A||B|\cos\theta$
- multiply another vector via cross product $\vec{A} \times \vec{B} = |A||B|\sin\theta\,\hat{a}_n$.

Correspondingly, there are three ways the ∇ operator can act: ∇A (the gradient), $\nabla \cdot \vec{A}$ (the divergence) and $\nabla \times \vec{A}$ (the curl)

6. Dot product is commonly used to determine the component of a vector in a particular direction. The dot product of a vector with a unit vector gives the component of the vector in the direction of the unit vector. The vector component of $\vec{A}$ along $\vec{B} = (\vec{A} \cdot \hat{a}_B)\hat{a}_B$

7. The magnitude of the cross product of a vector with a unit vector yields the component of vector perpendicular to the direction of unit vector. The vector component of $\vec{A}$ perpendicular to $\vec{B}$ = $|\vec{A} \times \hat{a}_B|$

8. A coordinate system is a way of uniquely specifying the location of any position in space along the principle axes with respect to a reference origin.

9. Surface is defined by fixing one space variable, similarly fixing two space variables defines a line and fixing three space variables defines a point.

10. Each coordinate system has a distinct set of principle axes, represented by the three surfaces.

- For Cartesian x, y, and z.
- For cylindrical ρ, ϕ and z.
- For spherical r, θ, and ϕ.

11. Coordinates, Differential lengths, Differential volumes of the three coordinate systems

	Cartesian	**Cylindrical**	**Spherical**
Coordinates	x, y, z	ρ, ϕ, and z	r, θ, and ϕ.
Differential lengths	dx, dy, dz	$d\rho$, $\rho d\phi$ and dz	dr, $rd\theta$, and $r\sin\theta\,d\phi$.
Differential volume dv	$dxdydz$	$\rho d\rho\,d\phi\,dz$	$r^2\sin\theta\,dr\,d\theta\,d\phi$

For example, the three differential elements in case of Cartesian coordinates are:

differential displacement $d\vec{L} = dx\,\hat{a}_x + dy\,\hat{a}_y + dz\,\hat{a}_z$

differential normal area $d\vec{S} = dydz\;\hat{a}_x, d\vec{S} = dxdz\;\hat{a}_y, d\vec{S} = dxdy\;\hat{a}_z$

differential volume $dv = dxdydz$

12. Basic *Vector Operators in Rectangular Coordinates*

Name	Symbol	Operates on	Yields	Formula
Scalar(dot) product	$\bar{A}\cdot\bar{B}$	2 Vectors	Scalar Function	$\bar{A}\cdot\bar{B} = A_xB_x + A_yB_y + A_zB_z$
Vector(cross) product	$\bar{A}\times\bar{B}$	2 Vectors	1 Vector	$\bar{A}\times\bar{B} = \begin{vmatrix} \hat{a}_x & \hat{a}_y & \hat{a}_z \\ A_x & A_y & A_z \\ B_x & B_y & B_z \end{vmatrix}$
Divergence	$\nabla\cdot\bar{A}$	1 Vector	Scalar Function	$\nabla\cdot\vec{A} = \left(\frac{\partial A_x}{\partial x} + \frac{\partial A_y}{\partial y} + \frac{\partial A_z}{\partial z}\right)$
Curl	$\times\vec{A}$	1 Vector	1 Vector	$\nabla\times\vec{A} = \begin{vmatrix} \hat{i} & \hat{j} & \hat{k} \\ \frac{\partial}{\partial x} & \frac{\partial}{\partial y} & \frac{\partial}{\partial z} \\ A_x & A_y & A_z \end{vmatrix}$
Gradient	$\nabla\psi$	Scalar Function	1 Vector	$\nabla\psi = \frac{\partial\psi}{\partial x}\hat{i} + \frac{\partial\psi}{\partial y}\hat{j} + \frac{\partial\psi}{\partial z}\hat{k}$

where $\vec{A}$ and $\vec{B}$ are vectors $\vec{A} = A_x\hat{a}_x + A_y\hat{a}_y + A_z\hat{a}_z$ and $\vec{B} = B_x\hat{a}_x + B_y\hat{a}_y + B_z\hat{a}_z$ and $\psi\,(x, y, z)$ is a scalar function.

13. Line integral is defined as any integral that is to be evaluated along a line. A line indicates a path along a curve in space is given by $\int \vec{A}\cdot d\vec{L}$

14. Divergence measures the tendency of a vector field to flow away from a point and the curl indicates the circulation of a vector field around a point, the *gradient applies to scalar fields.*

15. Divergence theorem converts a surface integral to a volume integral

$$\oint_S \vec{A}\cdot d\vec{S} = \int_V \nabla\cdot\vec{A}\,dv$$

16. Stoke's theorem relates a line integral over a closed path to a surface integral

$$\oint_c \vec{A}\cdot d\vec{l} = \iint_S \nabla\times\vec{A}\cdot d\vec{S}$$

17. Laplacian is a scalar operator. Hence, when applied to a scalar field, the result is also a scalar field.

$$\nabla^2 = \frac{\partial^2}{\partial x^2} + \frac{\partial^2}{\partial y^2} + \frac{\partial^2}{\partial z^2}$$

OBJECTIVE-TYPE QUESTIONS

1. The distance between point A (4, –6, 3) and B (2, 3, –1) is

(a) $\sqrt{133}$ (b) 10 (c) $\sqrt{29}$ (d) $\sqrt{101}$

2. Given the vector $\vec{M} = 2u_x + 3u_y + 5u_z$ and the vector is $\vec{N} = -4u_x + 4u_y + 3u_z$. The magnitude of $3u_x + 4\vec{M} - \vec{N}$ is

(a) 25.4 (b) 28.8 (c) 25.5 (d) 15.0

3. Given the vector $\vec{M} = 8u_x + 4u_y - 8u_z$ and $\vec{N} = 8u_x + 6u_y - 2u_z$. The unit vector in the direction of $-\vec{M} + 2\vec{N}$ is

(a) $\left(\frac{2}{3}, \frac{2}{3}, \frac{1}{3}\right)$ (b) $\left(\frac{2}{3}, \frac{1}{3}, \frac{2}{3}\right)$ (c) $\left(\frac{1}{3}, \frac{2}{3}, \frac{2}{3}\right)$ (d) $\left(\frac{2}{3}, \frac{1}{3}, \frac{1}{3}\right)$

4. The vector field $\vec{E}$ is given by $\vec{E} = 6zy^2 \cos 2xu_x + 4xy \sin 2xu_y + y^2 \sin 2xu_z$.
The region in which $\vec{E} = 0$ is

(a) $y = 0$ (b) $x = 0$ (c) $z = 0$ (d) $x = \frac{n\pi}{2}$

5. The divergence of vector $\vec{A} = yzu_x + 4xyu_y + yu_z$ at point P(1, –2, 3) is

(a) 2 (b) –2 (c) 0 (d) 4

6. The value of $\nabla^2 V$ at point p (3, 60°, –2) if $V = \rho^2 z\,(\cos\phi + \sin\phi)$

(a) –8.2 (b) 12.3 (c) –12.3 (d) 0

7. Consider the two points A (1, –6, 4) and B (7, –2, 0). The unit vector extending from the origin to the midpoint of line AB is ________.

8. The surface $\rho = 3$, $\rho = 5$, $\phi = 100°$, $\phi = 130°$, $z = 3$ and $z = 4.5$ define a closed volume. The length of the longest straight line that lies entirely within the volume is ________.

9. A vector is $\vec{A} = yu_x + (x + z)u_y$ at point P (–2, 6, 3), $\vec{A}$ in cylindrical coordinates is

10. The gradient of the field $f = \rho^2 z\cos 2\phi$ at point (1, 45°, 2) is

11. The curl $\nabla \times \vec{A}$ of a vector field $\vec{A} = x^2 yu_x + y^2 zu_y - 2xzu_z$ is

12. If a vector field $\vec{V}$ is related to another vector field $\vec{A}$ through $\vec{V} = \nabla X\vec{A}$, which of the following is true? **(GATE 2009)**

(a) $\oint_C \vec{V} \cdot d\vec{L} = \iint_S \vec{A} \cdot d\vec{S}$ (b) $\oint \mathbf{A} \cdot d\vec{L} = \iint_S \nabla X\vec{A} \cdot d\vec{S}$

(c) $\oint_c \nabla X\vec{V} \cdot d\vec{L} = \iint_S \nabla X\vec{A} \cdot d\vec{S}$ (d) $\oint \nabla X\vec{A} d\vec{L} = \iint_S \vec{V} \cdot d\vec{S}$

13. If $\vec{r} = xu_x + yu_y + zu_z$ then $(\vec{r}\cdot\nabla)r^2$ is equal to _______.

14. The angle between the normal to the surface $x^2y + z = 3$ and $x\ln z - y^2 = -4$ at the point of intersection (–1, 2, 1) is _______.

15. The curl of vector $\vec{A} = e^{xy}u_x + \sin xyu_y + \cos^2 xzu_z$ is _______.

16. A vector field is **A** is said to be irrotational if _____.

17. $\nabla\times\nabla\times\vec{P}$, where **P** is a vector, is equal to **(GATE 2006)**

(a) $\vec{P}\times\nabla\times\vec{P} - \nabla^2\vec{P}$ (b) $\nabla^2\vec{P} + \nabla(\nabla\cdot\vec{P})$

(c) $\nabla^2\vec{P} + \nabla\times\vec{P}$ (d) $\nabla(\nabla\cdot\vec{P}) - \nabla^2\vec{P}$

18. $\iint(\nabla\times\vec{P})\cdot d\vec{S}$, where **P** is a vector, is equal to **(GATE 2006)**

(a) $\oint_c \vec{P}\cdot d\vec{L}$ (b) $\oint_c \nabla\times\nabla\times\vec{P}\cdot d\vec{L}$

(c) $\oint_c \nabla\times\vec{P}\cdot d\vec{L}$ (d) $\iiint \nabla\cdot\vec{P}dv$

19. The direction of vector A is radially outward from the origin, with $|\vec{A}| = kr^n$, where $r^2 = x^2 + y^2 + z^2$ and k is a constant. The value of n, for which $\nabla\cdot\vec{A} = 0$ is **(GATE 2012)**

(a) –2 (b) 2 (c) 1 (d) 0

ANSWERS TO OBJECTIVE-TYPE QUESTIONS

(1) d (2) b (3) a (4) a

(5) d (6) a (7) $\frac{2}{3}U_x - \frac{1}{3}U_z$ (8) 3.21

(9) $-0.949u_\rho - 6.008u_\phi$ (10) $-4u_\phi$ (11) 0 (12) b

(13) $3r^2$ (14) 73.4° (15) $z\sin 2xyu_y + (y\cos xy - xe^{xy})u_z$ (16) $\nabla\times\vec{A} = 0$

(17) d (18) a (19) a

REVIEW QUESTIONS

1. Find the dot product of the vectors $\vec{A}$ and $\vec{B}$ if $\vec{A} = 2a_x - 3a_y + 4a_z$, $\vec{B} = -a_x + 2a_y + 2a_z$

2. How are the unit vectors defined in three coordinate systems?

3. Check validity of the divergence theorem considering the field $\vec{D} = 2xya_x + 2xa_y$ and the rectangular parallelepiped formed by the planes $x = 0$, $x = 1$, $y = 0$, $y = 2$ & $z = 0$, $z = 3$.

4. Given two points A (3, 2, 3) and B (1, 3, 6)

Find: (i) $\bar{A} + \bar{B}$; (ii) $\bar{A}\cdot\bar{B}$; (iii) θ_{AB}; (iv) $\bar{A}\times\bar{B}$

5. Given the two coplanar vectors $\bar{A} = 4\hat{a}_x + 5\hat{a}_y + \hat{a}_z$, $\bar{B} = 4\hat{a}_x - 2\hat{a}_y + 3\hat{a}_z$, obtain the unit vector normal to the plane containg the vectors $\bar{A}, \bar{B}$.

6. Consider $\mathbf{E} = 3a_x + 4a_z$ and $\mathbf{F} = 4a_x - 10a_y + 5a_z$

   (a) Find the component of **E** along **F**.

   (b) Determine a unit vector perpendicular to both **E** and **F**.

7. If the two vectors are $\bar{A} = 4\hat{a}_y + 10\hat{a}_z$ and $\bar{B} = 2\hat{a}_x + 3\hat{a}_y$, find the projection of $\bar{A}$ and $\bar{B}$.

8. A vector field $D = [5r^2/4]I_r$ is given in spherical coordinates. Evaluate both sides of divergence theorem for the volume enclosed between $r = 1$ and $r = 2$.

9. Explain three coordinate systems.

10. Show that the surface area of the sphere of radius R is $4\pi R^2$ by integration.

11. State divergence theorem and what is physical significance of divergence of **D**?

12. Express the value of differential volume in rectangular and cylindrical coordinate systems.

13. Discuss the conversion from the cylindrical coordinates of a point to its Cartesian coordinates.

14. Find a unit vector extending from (0, 0, 0) to (1, 2, 3).

15. Convert vector $\vec{A} = za_x - 2xa_y + ya_z$ from Cartesian to cylindrical coordinates.

16. Convert the vector $\vec{A} = 2a_x + 3a_z$ from Cartesian to spherical coordinate system.

17. Find gradient of $x^3 + y^3 + z^3 - 3xyz$.

18. Find curl of vector $\vec{A} = xa_x + ya_y + za_z$

19. Find curl of vector $\vec{A} = (x^2 - y^2 + x)a_x - (2xy + y)a_y$

20. Show that the vector fields $\mathbf{A} = \rho \sec\phi \mathbf{a}_\rho + \rho \tan\phi \mathbf{a}_\varphi + \rho\, \mathbf{a}_z$ and $\mathbf{B} = \rho \sec\phi \mathbf{a}_\rho - \rho \tan\phi \mathbf{a}_\varphi - \rho\, \mathbf{a}_z$ are everywhere perpendicular to each other.

21. Evaluate $\int_C xdx + (2x - y)dy + (2y + z)dz$, where C is boundary of triangle with vertices (1, 0, 0), (0, 3, 0) and (0, 0, 2).

Review of Maxwell's Equations and EM Wave Characteristics

2

2.1 INTRODUCTION

Electrostatics and Magnetostatics are two different theories developed from two different laws *Coulomb's law* and *Ampere's law* that are derived seperately. *Electrostatics* deals with the stationary charges that produce the electric field ($\vec{E}$ and $\vec{D}$). Magnetostatics deals with steady currents (charges in constant motion) that produces magnetic field ($\vec{B}$ and $\vec{H}$). In this chapter, we will consider time-varying electric and magnetic fields. We will observe two new phenomena: (i) time-varying $\vec{B}$ produces $\vec{E}$, and (ii) time-varying $\vec{E}$ produces $\vec{B}$.

One of the amazing things about electromagnetic fields is that the entire theory is adequately described in the organized, component set of equations called *Maxwell's* equations. The four laws, namely Gauss' law for $\vec{E}$, Gauss' law for $\vec{B}$, Faraday's law, and modified Ampere's law, are collectively known as *Maxwell's four equations*. These equations specify the relationship between the variations of the electric field $\vec{E}$ and the magnetic field $\vec{H}$ in time and space within a medium.

Faraday's law and Ampere's law are responsible for electromagnetic radiation. When the E field is changed in space, it produces a time-varying magnetic field. A time-varying magnetic field, when varied as a function of (space), results in a time-varying electric field, which is also a function of space. These equations together gives wave equation.

Maxwell's equation is important in the prediction of the presence of electromagnetic waves that travel at the speed of light. This is due to two reasons. First, changing an electric field produces a magnetic field and vice versa, and second, coupling between the two fields.

Integral form of Maxwell's equations is employed at media interfaces where the medium properties change suddenly. Such conditions are called *boundary conditions*. It is used to show the relationship between fields in the two media.

Electromagnetic wave (EM) propagation can be best understood using Maxwell's equations. In this chapter, the *basic EM wave equation* will also be studied, followed by a special case of EM wave known as *uniform plane wave*. The propagation of uniform plane wave in various kinds of media, its absorption, attenuation, and so on followed by different kinds of polarization will also be studied.

We know that EM waves transport energy from one point to another. Poynting theorem, which quantifies the rate of such energy transportation, is also discussed in this chapter.

2.2 FARADAY'S LAW OF INDUCTION

A magnetic field ($\vec{B}$) is produced when a steady current flows through a wire, as shown in Figure 2.1. The magnitude of $\vec{B}$ varies proportionally with current I and is inversely proportional to the distance from the conductor.

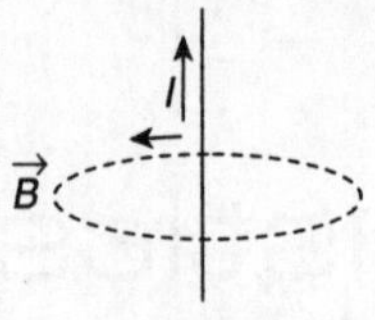

Figure 2.1 *Magnetic field around a straight conductor carrying a steady current I*

Faraday observed the following in his experiment shown in Figure 2.2. He used an Ammeter and a small bar magnet in the experiment. A loop of wire is connected to a sensitive ammeter. We will start the discussion of Faraday's law with a description of the experiment.

(i) Even though the magnet is in the loop, no current will be induced if the magnet is stationary. Hence there is no deflection in the ammeter.

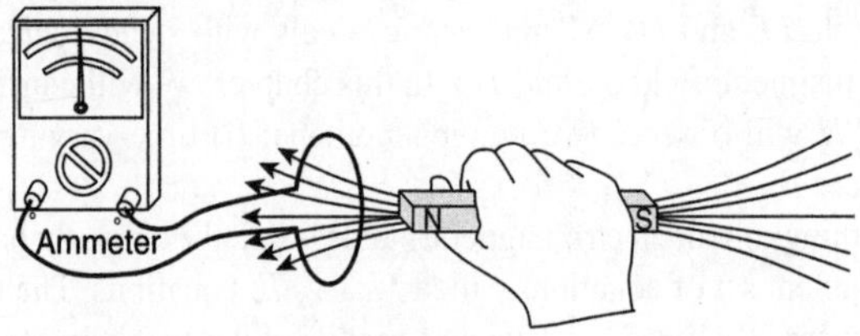

Figure 2.2 (a) *Bar magnet is stationary*

(ii) The ammeter deflects when the magnet is moved towards the loop, and the direction was randomly selected to be right.

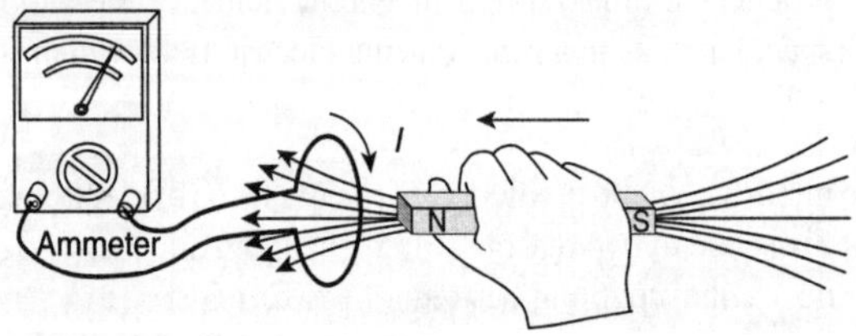

Figure 2.2 (b) *Bar magnet is moved toward the loop*

(iii) The deflection in the ammeter will be in opposite direction when the magnet is moved away from the loop.

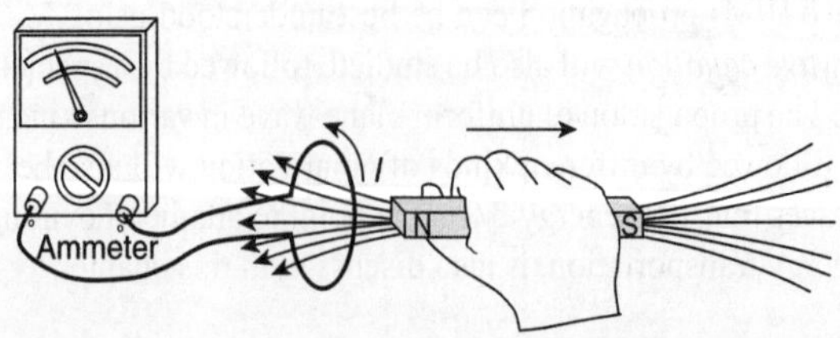

Figure 2.2 (c) *Bar magnet is moved away from the loop*

From the figures above, Faraday's experiment shows that a time-varying magnetic field produces (or induces) a current (I) in a closed loop. The coil behaves as if it were connected to an emf source.

Faraday's Law – Statements

Faraday was able to formulate a law that accounted for all his experiments—*The emf generated around a loop of wire in a magnetic field is proportional to the rate of change of time-varying magnetic field ($\vec{B}(t)$) through the loop.*

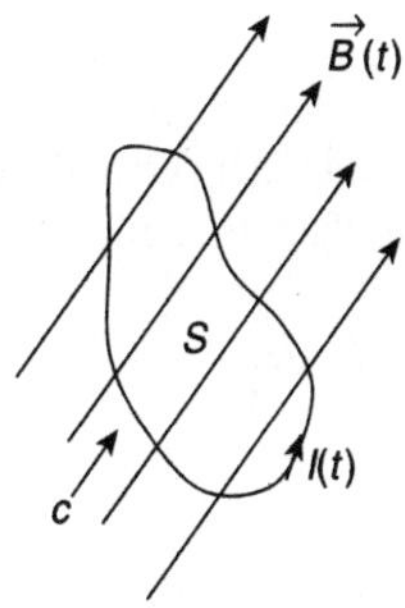

Figure 2.3 *Time-varying magnetic field*

So, if C is the loop and S is some surface attached to the loop, as shown in Figure 2.3, then Faraday's law can be written as

$$\underbrace{\int_C \vec{E}\cdot d\vec{L}}_{\text{emf in loop integral of electric field around } L} = \underbrace{-\int_S \frac{\partial}{\partial t}\overbrace{\vec{B}\cdot d\vec{S}}^{\text{flux through } S}}_{\text{rate of change of magnetic flux through surface defined by loop } L} \tag{2.1}$$

where $\oint \vec{E}\cdot d\vec{L} = V$ is the electromotive force (emf). It is the net push causing charges to move, and $\int_S \vec{B}\cdot d\vec{S} = \Phi_B$ is the magnetic flux through the surface S.

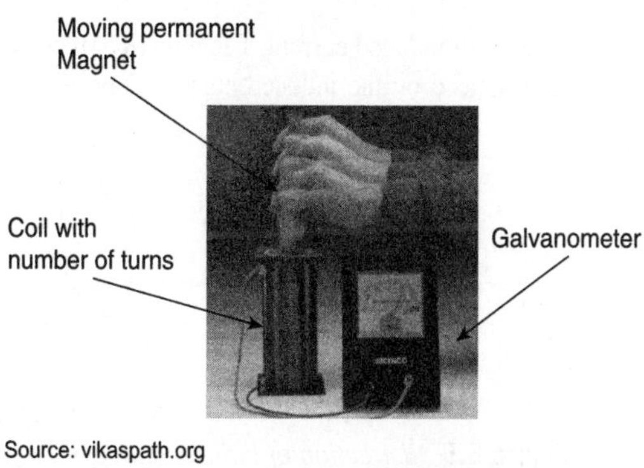

Figure 2.4 *Changing flux due to moving permanent magnet*

Substituting definitions of *emf* and Φ_B in the above equation, mathematically, Faraday's law states that

$$emf = -\frac{\partial \Phi_B}{\partial t} \text{ Volts (Faraday's law for a single loop)} \tag{2.2}$$

Faraday's law for a coil having N turns (Figure 2.4): $emf = -N\frac{d\Phi_B}{dt}$ (2.3)

Point Form of Faraday's Law

By applying Stokes' theorem to Eq. (2.1), we can derive the point form of Faraday's law. Specifically, applying Stokes' theorem to the left-hand side of Eq. (2.1) gives

$$\int_C \vec{E}\cdot d\bar{L} = \int_S (\nabla\times\vec{E})\cdot d\bar{S} \tag{2.4}$$

Substituting the above in Eq. (2.1), we get

$$\int_S (\nabla\times\vec{E} + \frac{\partial \vec{B}}{\partial t})\cdot d\bar{S} = 0 \tag{2.5}$$

Since this result is valid for any surface *S* and any loop *C*, the integrand vanishes, leaving the point form of Faraday's law

$$\nabla\times\vec{E} = -\frac{\partial \vec{B}}{\partial t} \tag{2.6}$$

Faraday's law shows that the time-varying magnetic field ($\partial\vec{B}/\partial t$) is a source of the electric field ($\vec{E}$).

Lenz's Law

Lenz's law gives the reason for introducing the minus sign in Faraday's law of induction $\left(emf = -\frac{\partial \Phi_B}{\partial t}\right)$.

The minus sign indicates the direction of induced current. The direction of current flow in the circuit is such that the induced magnetic field produced by the induced current opposes the original magnetic field, as shown in Figure 2.5.

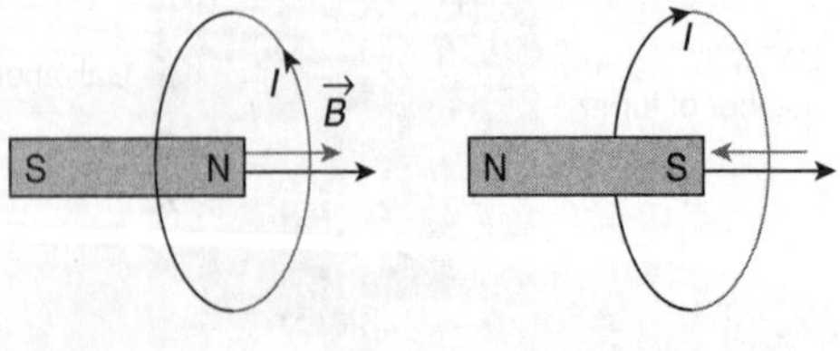

Figure 2.5 *Direction of induced current (I)*

Lenz's law states that the magnetic field produced by an induced current (Let us call this $\vec{B}$) opposes the change in the magnetic field that produced the induced current.

2.3 TRANSFORMER EMF

A transformer is a device that transforms voltages, currents, and impedances from one value to another. It is a multiport ac device. Time-varying signals are transformed through inductive coupling. We will consider the transformer circuit shown next (Figure 2.6). It consists of 2 coils (known as *primary* and *secondary* coils) of wire wound around a soft iron core. The wires are tightly wound on a magnetic core—the total induced magnetic field will flow inside the core only.

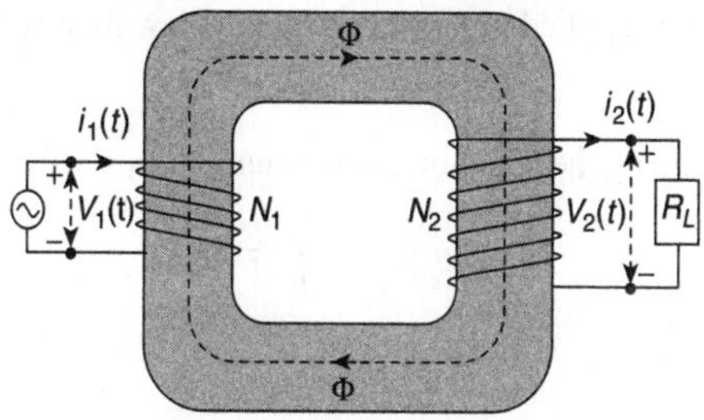

Figure 2.6 *Transformer*

On the primary side, when an alternating voltage (V_1) is applied, the time-varying current (I_1) in the primary circuit generates a time-varying magnetic field $\left(\frac{d\Phi_B}{dt}\right)$ around this coil. As we know, a time-varying magnetic field through a coil of wire produces a voltage between the ends of the coil. This phenomenon was discovered by Faraday and is mathematically stated in Faraday's law as

$$V_{emf} = -N\frac{d\Phi_B}{dt} \text{ (volts)}$$

where N is the number of turns of the coil.
On the secondary side, a time-varying magnetic field will induce voltage.
Using Faraday's law, primary and secondary voltages can be written as:

$$V_1 = N_1\frac{d\Phi_B}{dt},\ V_2 = N_2\frac{d\Phi_B}{dt} \Rightarrow \frac{d\Phi_B}{dt} = \frac{V_1}{N_1} = \frac{V_2}{N_2}$$

Interestingly, we see here that the "output" voltage V_2 can be different in amplitude from the "input" voltage V_1, which is given by

$$V_2 = \frac{N_2}{N_1}V_1$$

2.4 INCONSISTENCY OF AMPERE'S LAW

For time-varying fields, if we try to check the validity of Ampere's circuital law for a static field: Ampere's circuital law for a static field is given below in point form (differential form):

$$\nabla \times \overline{H} = \overline{J} \tag{2.7}$$

Application of the divergence on both sides yields $\nabla \cdot \left(\nabla \times \vec{H}\right) = \nabla \cdot \mathbf{J}$.

According to vector identities, for any vector $\vec{A}, \nabla \cdot \left(\nabla \times \vec{A}\right) = 0$

$$\nabla \cdot \left(\nabla \times \vec{H}\right) = 0$$

However, this expression that is applied to Ampere's circuital law for a static field results in

$$\nabla \cdot \mathbf{J} = 0 \tag{2.8}$$

However, Eq. (2.8) disagrees with the equation of continuity, which says that

$$\nabla \cdot \overline{J} = -\frac{\partial \rho_V}{\partial t} \neq 0$$

$\nabla \cdot \mathbf{J} = 0$ can be true only if $\dfrac{\partial \rho_v}{\partial t} = 0$

That is, the form of Ampere's law in Eq. (2.7) is only valid if fields are not time varying. For the time-varying case, the law needs some modifications. To overcome this contradiction, Maxwell suggested that the definition of total current density of Ampere's circuital law is incomplete and advised an additional current called a *displacement current*, whose current density will be $\vec{J}_d$.

Hence, the magnetostatic curl equation should be modified so that it is matched to the continuity equation. It is done by adding a term to the continuity as given below

$$\nabla \times \overline{H} = \overline{J} + \overline{J}_d \tag{2.9}$$

where $\vec{J}$ is the conduction current density and is given by $\vec{J} = \sigma_E \vec{E}$

In the next section, we derive an equation for the displacement current density ($\vec{J}_d$) that was added by Maxwell, and we prove the modified Ampere's law.

2.5 DISPLACEMENT CURRENT DENSITY AND PROOF OF MODIFIED AMPERE'S LAW

Modified Ampere's law by Maxwell is

$$\nabla \times \overline{H} = \overline{J} + \overline{J}_d$$

By applying the divergence on both sides of the modified Ampere's law, we have

$$\nabla \cdot (\nabla \times \overline{H}) = \nabla \cdot (\overline{\text{J}} + \overline{J}_d)$$

$$\nabla \cdot (\nabla \times \overline{H}) = \nabla \cdot \overline{J} + \nabla \cdot \overline{J}_d$$

$$0 = \nabla \cdot \overline{J} + \nabla \cdot \overline{J}_d$$

$$\nabla\cdot\bar{J}_d=-\nabla\cdot\bar{J}$$

In order for this equation to agree with the continuity equation,

$$\nabla\cdot\bar{J}=-\frac{\partial\rho_V}{\partial t},\nabla\cdot\bar{J}_d=-\nabla\cdot\bar{J}=\frac{\partial\rho_V}{\partial t}$$

We know that

$$\nabla\cdot\bar{D}=\rho_V \text{ (Gauss' law)}$$

$$\frac{\partial}{\partial t}(\nabla\cdot\bar{D})=\frac{\partial(\rho_V)}{\partial t}$$

$$\nabla\cdot\frac{\partial\bar{D}}{\partial t}=\nabla\cdot\bar{J}_d \quad\Rightarrow\quad \bar{J}_d=\frac{\partial\bar{D}}{\partial t}$$

The relationship between the sources, charge, and current is depicted in the continuity equation that represents the principle of conservation of charge.

Since $\vec{J}_d$ arises due to the variation of electric displacement (electric flux density) $\bar{D}$ with time, it is termed as displacement current density. The modified Ampere's Circuital Law (Maxwell's equation), therefore, for the time-varying field assumes the following differential form:

$$\nabla\times\vec{H}=\vec{J}+\frac{\partial\vec{D}}{\partial t} \tag{2.10}$$

Applying Stokes' theorem to the above equation, the modified Ampere's law (Figure 2.7) can be given in integral form as

$$\oint_L\bar{H}\cdot d\bar{L}=\int_S\vec{J}\cdot d\bar{S}\ +\ \int_S\frac{\partial\bar{D}}{\partial t}d\bar{S}$$

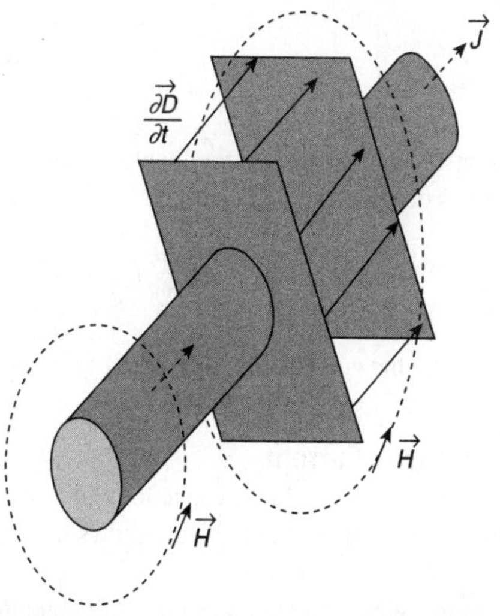

Figure 2.7 *Illustration of Maxwell's modified Ampere's law*

The first term on the right-hand side of the above equation represents the conduction current (the net transfer of electric charge), and the second term is the time rate of change of electric flux leaving the surfaces.

$$\oint_L \overline{H} \cdot d\overline{L} = I + \int_S \frac{\partial \overline{D}}{\partial t} \cdot d\overline{S} \tag{2.11}$$

The important conclusion that can be drawn now is that, since displacement current is related to the electric field, it is not possible in case of time-varying fields to deal separately with electric and magnetic fields; instead, the two fields are interlinked, thereby giving rise to electromagnetic fields. It should be noted that in a good conductor $\vec{J}_d$ is negligible compared with $\vec{J}$ at a frequency lower than light frequencies (10^{15} Hz).

Note: The *charge volume density* is defined as the amount of charge stored in a unit volume:

$$\rho = \underset{\Delta V \to 0}{Lt} \frac{\Delta q}{\Delta V} \text{ [C/m}^3\text{]}$$

The *charge surface density* is defined by analogy:

$$\sigma = \underset{\Delta S \to 0}{Lt} \frac{\Delta q}{\Delta S} \text{ [C/m}^2\text{]}$$

The *charge linear density* is defined as the charge stored along a line or a curve of unit length:

$$\tau = \underset{\Delta l \to 0}{Lt} \frac{\Delta q}{\Delta l} \text{ [C/m]}$$

Electric current is created by a moving charge. This is also defined as the *passing charge* per time interval Δt:

$$I = \underset{\Delta t \to 0}{Lt} \frac{\Delta q}{\Delta t} \text{ [A = C/sec]}$$

It is useful to define current densities, which are vector quantities, as it is necessary to define the current flow direction. *Current density* is defined as

$$J = \underset{\Delta S \to 0}{Lt} \frac{\Delta I}{\Delta S} \hat{i}_0 \text{ [A/m}^2\text{]}$$

where $\hat{i}_0$ is a unit vector describing the current flow direction.

2.5.1 Example of Displacement Current

When an alternating voltage source is applied to the capacitor plates, then the current passing through it is known as the displacement current. The example below shows the requirement for the displacement current.

Consider an example of a simple electrical circuit (Figure 2.8) containing a sinusoidal voltage source (V_s) and a capacitor.

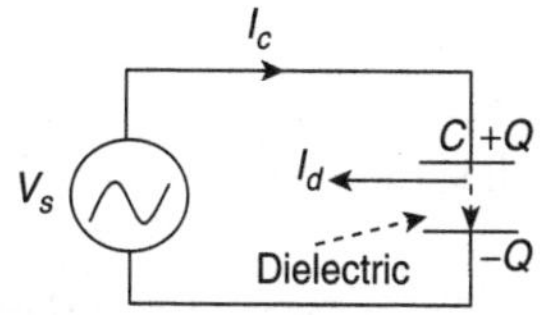

Figure 2.8 *Simple electric circuit*

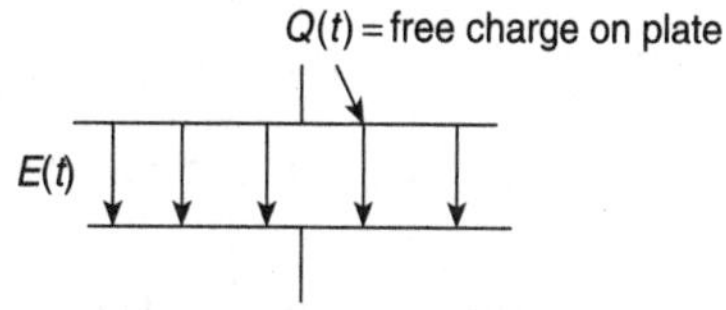

Figure 2.9 *Capacitor*

If the voltage source is DC, then the current flowing through the circuit, $I = 0$. However, if the voltage source is AC, we will observe and measure some nonzero time varying current I. This current is neither the conduction current nor the convection current, but it is actually the displacement electric current, which completes the circuit even when there is no direct connectivity between the plates of the capacitor, as shown in Figure 2.9.

We interpret the current in the circuit (Figure 2.8) as consisting of two types of current:

- $I_c(t) = \dfrac{dQ(t)}{dt}$ = conduction current in the wires
- $I_d(t) = C\dfrac{dV}{dt}$ = displacement current in the capacitor

2.6 MAXWELL'S EQUATIONS IN DIFFERENT FORMS

There are three static sources and two dynamic sources for generating the electric and magnetic fields.

Static sources $\left(\dfrac{\partial}{\partial t} = 0\right)$

The static or non-time-varying sources are:

(i) Static positive or negative charge that produces an electric field $\vec{E}$;

(ii) For example, the phasor can be written for the sinusoidally varying electric field as follows:

$$E(r, t) = \text{Re}\left\{\vec{E}(\vec{r})e^{j\omega t}\right\};$$

which denotes the real part of what is in the brackets. For example, if the phasor field has a phase of 900, the time dependence would be $\vec{E}(\vec{r})$.

Figure 2.10 *Permanent magnet*

(iii) A permanent magnet (Figure 2.10) that produces the magnetic field $(\vec{H})$.

A static electric field can exist even if there is no magnetic field $(\vec{H})$ (Example: a capacitor with a static charge Q). Likewise, a conductor with a constant current I_{DC} has a magnetic field $(\vec{H})$ without an $(\vec{E})$ field.

In a conducting medium, both static electric and magnetic fields can exist together and form an electromagnetostatic field. The magnetic field is a consequence; it does not affect the calculation of the electric field.

Time-varying field sources $\left(\frac{\partial}{\partial t} \neq 0\right)$

If the source that produces the fields changes with time (time varying), then it is known as a *time-varying source.*

Time-varying electric field ($\vec{E}$) produces → magnetic field ($\vec{H}$);

Time-varying magnetic field ($\vec{H}$) produces → electric field ($\vec{E}$).

However, when fields are time variable, then neither $\vec{H}$ can exist without a $\vec{E}$ field nor $\vec{E}$ can exist without a corresponding $\vec{H}$ field.

Hence, the fundamental relations for electrostatic and magnetostatic models should be modified to show the mutual dependence of both the fields in the time-varying case.

These equations can be grouped and are called *Maxwell's equations*. They can be expressed in different forms such as point form or differential form, phasor and integral forms. The next three sections describe Maxwell's equations separately.

2.6.1 Maxwell's Equations in Differential Form for Time-varying Fields

The differential form of Maxwell's equations is point relations, that is, they gives a relationship between fields and sources at any point in space and hence maxwells equation is called point relation. The differential form involves derivatives of the fields with regard to space and time. The general form of the time-varying Maxwell's equations can be written in differential form as follows:

Differntial Form	*Name*	
$\nabla \cdot \vec{D} = \rho_v$	Gauss Law for electric fields	(2.12)
$\nabla \cdot \vec{B} = 0$	Gauss Lawfor magnetic fields	(2.13)
$\nabla \times \vec{E} = -\frac{\partial \vec{B}}{\partial t}$	Faraday's law	(2.14)
$\nabla \times \vec{H} = \vec{J} + \frac{\partial \vec{D}}{\partial t}$	Modified Ampere's Law	(2.15)

All these field quantities $-\vec{E}, \vec{B}, \vec{H}, \vec{D}, \vec{J}$ and ρ_v are assumed to be time varying and each is a function of space coordinates and time, that is, $\vec{E} = \vec{E}(x, y, z; t)$. Since these field quantities are real functions of position and time, they can also be represented as $\vec{E}(\vec{r}, t), \vec{B}(\vec{r}, t), \vec{H}(\vec{r}, t), \vec{D}(\vec{r}, t), \vec{J}(\vec{r}, t)$, and $\rho(\vec{r}, t)$.

The definitions and units of the quantities are as follows:

$\vec{E}$ = Electric Field Intensity (volt/m), $\vec{H}$ = Magnetic Field Intensity (Ampere/m)

$\vec{J}$ = Electric Current Density (Ampere/m^2), $\vec{D}$ = Electric Flux Density (Coulomb/m^2)

$\vec{B}$ = Magnetic Flux Density (Wb/m^2), ρ_v = Volume Charge Density (in C/m^3)

These equations prove the principle of conservation of charge. It is given by the continuity equation for current in point form and integral forms as follows:

$$\nabla \cdot \vec{J} = -\frac{\partial \rho_V}{\partial t}$$

$$\oint_S \vec{J}\cdot d\vec{S} = -\frac{\partial}{\partial t}\oint_v \rho_v dv$$

2.6.2 Integral Form of Maxwell's Equations for Time-varying Fields

The integral form of Maxwell's equations explains the relationship between the field vectors, charge, and current densities over an extended region of space. Their applications are limited and are generally used to work out electromagnetic problems with absolute symmetry. The integral form can be derived from the differential form through the use of Stokes' and Divergence theorems.

$\nabla\cdot\vec{D} = \rho_v$	$\oint_S D\cdot d\vec{S} = \int_v \rho_v dv$	Gauss's Law
$\nabla\cdot\vec{B} = 0$	$\Rightarrow \oint_S B\cdot d\vec{S} = 0$	Nonexistence of isolated magnetic charge
$\nabla\times\vec{E} = -\frac{\partial\vec{B}}{\partial t}$	$\oint_c E\cdot d\vec{L} = -\int_S \frac{\partial}{\partial t}B\cdot d\vec{S}$	Faraday's Law
$\nabla\times\vec{H} = \frac{\partial\vec{D}}{\partial t} + \vec{J}$	$\oint_L H\cdot d\vec{L} = \int_S\left(\frac{\partial\vec{D}}{\partial t} + J\right)\cdot d\vec{S}$	Ampere's circuital law

Note: In case of Gauss' law for electric and magnetic fields, *S* is a closed surface which encloses the volume *V* and in case of Faraday's and Ampere's laws, *L* is a closed path that bounds the surface *S*.

Note:

$$\oint_S () = \iint () \text{ and } \int_v () = \iiint ()$$

Any surface, *S* is oriented by normal vector $\hat{n}$. The surface is closed for Gauss' law and open for Ampere's and Faraday's laws. In Ampere's and Faraday's laws the surface is bounded by closed curve *C*. The components *dl* indicates the differential length and is tangential to closed curve *C*, *ds* is the differential surface element and *dv* is the differential volume.

2.6.2.1 Maxwell's equations in word statements

Maxwell's first law is known as *Gauss' law for electric fields*. Consider a closed surface *S* that bounds a volume *V*, as shown in Figure 2.11. There is an electric charge distributed in volume *V* with a charge density of ρ_v. This electrical charge sets up a displacement flux with *D* as its density.

Figure 2.11 *(a) Divergence of the E field is proportional to the contained charge density, and its direction is related to the sign of the net charge, as shown;*
(b) Electric flux density D originating from a volume v bounded by a surfaces

$$\oint_S D \cdot d\bar{S} = \oint_v \rho_v dv$$

The net electric displacement flux originating from a closed surface S is equal to the total charge ρ_v inside the volume v. This is the first law.

The divergence of the electric flux density $(\vec{D})$ field equals the volume charge density.

Maxwell's second law is known as *Gauss' law for magnetic fields*. Since the magnetic flux lines are always closed and magnetic charges do not exist, unlike positive or negative electric charges, net magnetic charge in a volume v should be zero. As illustrated in Figure 2.12, the magnetic flux emanating from the closed surface s is, therefore, equal to zero.

$$\oint_S B \cdot d\bar{S} = 0$$

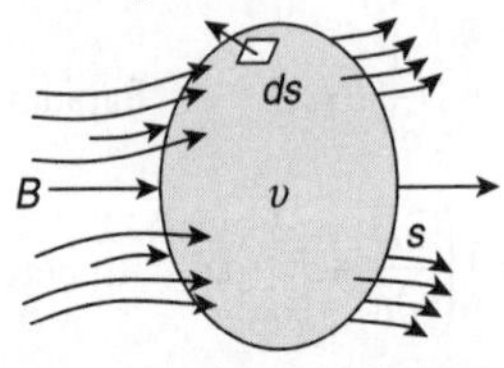

Figure 2.12 (a) *Magnetic flux density B emanating from a volume v bounded by a surface s.*

Figure 2.12 (b) gives an illustration for divergence of the B field being zero.

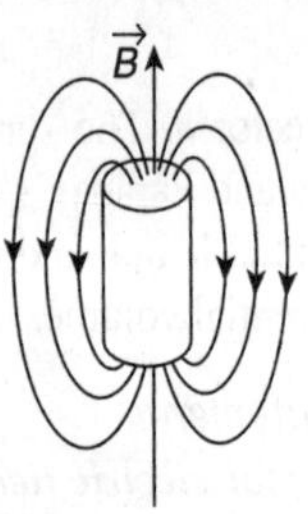

Figure 2.12 (b) *Bar magnet illustrating that since the divergence of the B field is zero, the field lines, when sufficiently extended, always close on themselves.*

The second law states that the total magnetic flux passing through any closed surface is zero (as there are no magnetic charges or monopoles).

Maxwell's third law is known as *Faraday's law of induction*. Faraday discovered that if a conducting wire loop is placed in the magnetic flux, electromotive force is induced (Figure 2.13). The induced emf depends on the time rate of change of magnetic flux that leaves the surface, bound by the loop.

$$\oint_L \vec{E} \cdot d\bar{L} = -\int_S \frac{\partial}{\partial t} \vec{B} \cdot d\bar{S}$$

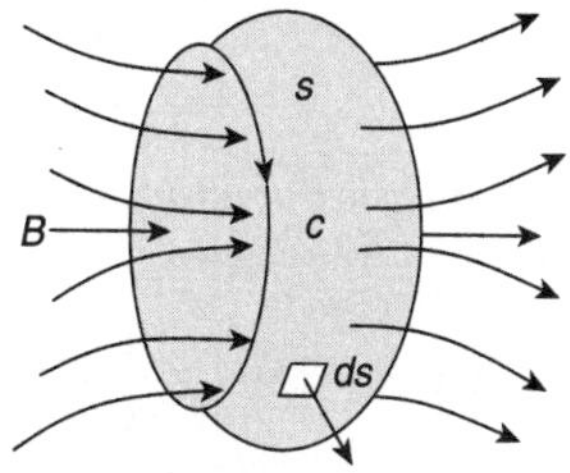

Figure 2.13 *Magnetic flux density B passing through an area s bounded by a curve c.*

The third law states that the electromotive force (voltage) around a closed path is equal to the inflow of magnetic current through any surface bounded by the path.

The curl of the *E* field is equal to negative time rate of change of the magnetic flux density.

Maxwell's fourth law is known as *modified Ampere's law.* Consider Figure 2.14 which shows a closed path *c* that bounds a surface *s*. The tangential component of the magnetic field intensity when integrated along *c* i.e. the line integral of vector magnetic field intensity along *c*, gives the flow of the magnetic field intensity, called the *magneto motive force* (mmf) which is equal to the net current enclosed by the closed path *c*.

$$\oint_L \vec{H}\cdot d\bar{L} = \int_S \left(\frac{\partial \vec{D}}{\partial t} + \vec{J}\right)\cdot d\bar{S}$$

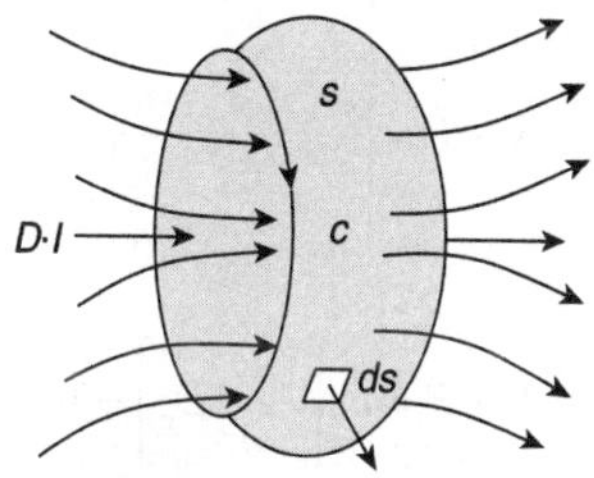

Figure 2.14 *Electric flux density D and current density J passing through an area S bounded by a curve c.*

The magneto motive force surrounding a closed path is equal to the summation of electric displacement and conduction currents through any surface enclosed by the path. This is the fourth law.

The curl of the *H* field equals the combined conduction and displacement current densities.

EXAMPLE PROBLEM 2.1

Find the electric flux density on the surface of a hollow sphere of radius 3 m and having a volume charge density of $\rho = 20/r$ C/m^3. Find the total charge enclosed within the sphere.

Solution

The total charge enclosed

$$Q=\int_V \rho_v dv=\int_{\phi=0}^{2\pi}\int_{\theta=0}^{\pi}\int_{r=0}^{3}\rho_v r^2 \sin\theta\, dr\, d\theta\, d\phi=\int_0^{2\pi} d\phi\int_0^{\pi}\sin\theta\, d\theta\int_0^3 \frac{20}{r}r^2 dr$$

$$=80\pi\int_0^3 r\,dr = 80\pi\times\frac{9}{2}=360\pi$$

The total electric flux from the surface of the sphere can be obtained from Gauss' law:

$$D=\frac{Q}{4\pi r^2}=\frac{360\pi}{4\pi 3^2}=10\,\frac{\text{C}}{\text{m}^2}$$

■

EXAMPLE PROBLEM 2.2

Given $E=10\sin(\omega t-\beta z)a_y,\ \frac{V}{m}$, in free space, determine $\vec{D},\vec{B},\vec{H}$.

Solution

$$E=10\sin(\omega t-\beta z)a_y,\ \frac{V}{m},\quad D=\epsilon_0 E,\ \epsilon_0=8.854\times10^{-12}\ \frac{\text{F}}{\text{m}}$$

$$D=10\,\epsilon_0\sin\left(\omega t-\beta z\right)a_y,\ \frac{C}{m^2}$$

Maxwell's second equation is $\nabla\times\vec{E}=-\frac{\partial\vec{B}}{\partial t}$

That is,

$$\nabla\times\vec{E}=\begin{vmatrix} a_x & a_y & a_z\\ \frac{\partial}{\partial x} & \frac{\partial}{\partial y} & \frac{\partial}{\partial z}\\ 0 & E_y & 0\end{vmatrix}$$

or,

$$\nabla\times\vec{E}=a_x\left[-\frac{\partial}{\partial z}E_y\right]+0+a_z\left[\frac{\partial}{\partial x}E_y\right]$$

As

$$\nabla\times\vec{E}=a_x\left[-\frac{\partial}{\partial z}E_y\right]+0+a_z\left[\frac{\partial}{\partial x}E_y\right],\qquad \frac{\partial E_y}{\partial x}=0$$

Now, $\nabla\times E$ becomes

$$\nabla\times\vec{E}=-\frac{\partial E_y}{\partial z}a_x$$

$$=10\beta\cos(\omega t-\beta z)a_x=-\frac{\partial\vec{B}}{\partial t}$$

$$\vec{B}=-\int 10\beta\cos\,(\omega t-\beta z)\ dt\ a_x=\frac{10\beta}{\omega}\sin(\omega t-\beta z)a_x,\text{wb/m}^2$$

$$\vec{H}=\frac{B}{\mu_0}=\frac{10\beta}{\mu_0\omega}\sin\,(\omega t-\beta z)a_x,\text{A/m}$$

■

2.6.3 Maxwell's Equations for Harmonically Varying Fields

Electromagnetic waves are used to transmit information (audio or data) from one place to another. The information is usually transmitted by imposing the amplitude, frequency, or phase modulation on a sinusoidal carrier.

Maxwell's equations are a set of eight first-order partial differential equations with four independent variables (three space coordinates and time), whose solution is complicated sometimes. The field is dependant on one or more independent variables. This dependency needs to be removed by using Fourier or Laplace transforms equations. The resulting equations in the transform domain should be solved and inverse transformation is applied to get the required field quantities.

The main advantage of a transform technique is to change the dependence of the equations on variable from a differential one to an algebraic one. Thus, the differential system in Maxwell's equation can be changed to an algebraic system in the transform domain with the help of four-fold Fourier transform. Fourier transform allows us to transform the electric field from the time domain (where the appropriate field vector is $\vec{E}(\vec{r}, t)$) to the frequency domain, where the appropriate field vector is $\vec{E}(\vec{r}, \omega)$, and vice versa.

$$\vec{E}(\vec{r}, \omega) = \int_{-\infty}^{\infty} \vec{E}(\vec{r}, t)e^{-j\omega t}\, dt,$$

$$\vec{E}(\vec{r}, t) = \frac{1}{2\pi}\int_{-\infty}^{\infty} \vec{E}(\vec{r}, \omega)e^{j\omega t}\, d\omega,$$

Field quantities calculated with the complex exponential will be complex, and are called *phasors*.

Let us assume that E and H are functions of $e^{j\omega t}$ and $\frac{d}{dt} = j\omega$ to convert from the time domain to the frequency or phasor domain

$$\begin{aligned}
\nabla \times \vec{E} &= -\frac{\partial \vec{B}}{\partial t} & \nabla \times \vec{E} &= -j\omega\vec{B} \\
\nabla \times \vec{H} &= \frac{\partial \vec{D}}{\partial t} + \vec{J} \quad \Rightarrow & \nabla \times \vec{H} &= \vec{J} + j\omega\vec{D} \\
\nabla \cdot \vec{D} &= \rho_v & \nabla \cdot \vec{D} &= \rho_v \\
\nabla \cdot \vec{B} &= 0 & \nabla \cdot \vec{B} &= 0
\end{aligned}$$

Similarly, the continuity equation becomes

$$\nabla \cdot \vec{J} = -j\omega\rho_v.$$

Sinusoidal analysis can be applied to most waveforms using Fourier and Laplace transform techniques. It is also called time-harmonic analysis. But, time-harmonic or sinusoidally varying fields are mostly used, with time-dependent $\cos \omega t$.

It is comfortable to use the complex exponential $e^{j\omega t}$ instead of using real sinusoidal functions directly. By using phasors, the actual or physically meaningful time dependence can be shown. For example, the phasor can be written for the sinusoidally varying electric field as follows:

$$E(r,\ t) = \text{Re}\left\{\vec{E}\ (\vec{r})e^{j\omega t}\right\}$$

where Re {} denotes the real part of the equation. For example, if the phasor field $\vec{E}(\vec{r})$ has a phase of 90°, the time dependence would be $\sin \omega t$.

Similarly, for all field quantities (considering $\vec{r}$ in Cartesian coordinates), we can write

$$\bar{E}(x, y, z, t) = \text{Re}[\bar{E}(x, y, z)e^{j\omega t}]$$

$$\bar{H}(x, y, z, t) = \text{Re}[\bar{H}(x, y, z)e^{j\omega t}]$$

$$\bar{D}(x, y, z, t) = \text{Re}[\bar{D}(x, y, z)e^{j\omega t}]$$

$$\bar{B}(x, y, z, t) = \text{Re}[\bar{B}(x, y, z)e^{j\omega t}]$$

$$\bar{J}(x, y, z, t) = \text{Re}[\bar{J}(x, y, z)e^{j\omega t}]$$

$$\rho(x, y, z, t) = \text{Re}[\rho(x, y, z)e^{j\omega t}]$$

The phasor form or time-harmonic form of Maxwell's equations is obtained by using $e^{j\omega t}$. Hence time dependence in Maxwell's equations makes us to eliminate the time variable entirely.

$$\nabla \times \vec{E} = -\frac{\partial \vec{B}}{\partial t} \qquad \nabla \times \text{Re}(\vec{E}\, e^{j\omega t}) = -\frac{\partial}{\partial t}\text{Re}(\vec{B}\, e^{j\omega t})$$

$$\nabla \times \vec{H} = \frac{\partial \vec{D}}{\partial t} + \vec{J} \;\Rightarrow\; \nabla \times \text{Re}(\vec{H}\, e^{j\omega t}) = \frac{\partial}{\partial t}\text{Re}(\vec{D}\, e^{j\omega t}) + \text{Re}(\vec{J}\, e^{j\omega t})$$

$$\nabla \cdot \vec{D} = \rho \qquad \nabla \cdot \text{Re}(\vec{D}\, e^{j\omega t}) = \text{Re}(\rho\, e^{j\omega t})$$

$$\nabla \cdot \vec{B} = 0 \qquad \nabla \cdot \text{Re}(\vec{B}\, e^{j\omega t}) = 0$$

Finally, we can conclude that for time-varying fields in either sinusoidal or phasor form and the continuity equation as $\nabla \cdot \vec{J} = -j\omega\rho_v$,

$$\nabla \times \vec{E} = -j\omega\vec{B} \qquad \nabla \cdot \vec{D} = \rho$$

$$\nabla \times \vec{H} = \vec{J} + j\omega\vec{D} \qquad \nabla \cdot \vec{B} = 0$$

2.7 MAXWELL'S EQUATIONS IN DIFFERENT MEDIA

Maxwell's equations in different media are given in the following sections:

2.7.1 Maxwell's Equations for Free Space

The significant features of free space are:

- Homogeneous everywhere,
- Contains no electrical charge,
- Holds no current,
- Infinite extent in all dimensions.

The parameters for free space are $\rho_v = 0$ and $J = 0$.

Maxwell's equations in free space are obtained by considering the absence of dielectric or magnetic media (i.e. when $\mu_r = 1, \varepsilon_r = 1$).

Differential form	*Integral form*
$\nabla\cdot\vec{D}=0$	$\oint_s \vec{D}\cdot ds=0$
$\nabla\cdot\vec{B}=0$	$\oint_s \vec{B}\cdot ds=0$
$\nabla\times\vec{E}=-\dfrac{\partial\vec{B}}{\partial t}$	$\oint_c \vec{E}\cdot dl=-\dfrac{\partial}{\partial t}\int_s \vec{B}\cdot ds$
$\nabla\times\vec{H}=\dfrac{\partial\vec{D}}{\partial t}$	$\oint_c \vec{H}\cdot dl=\dfrac{\partial}{\partial t}\int_s \vec{D}\cdot ds$

2.7.2 Maxwell's Equations for Good Conductors

In a conducting medium with conductivity σ and charge density ρ_v,

We can write, conduction current $(\vec{J})$ >> displacement current $\left(\vec{J}_D=\dfrac{\partial\vec{D}}{\partial t}\right)$ (because σ is infinite, and $\vec{J}=\sigma\vec{E}$).

Maxwell's equations are as given below:

$$\nabla\cdot\vec{D}=\rho_v,\ \nabla\cdot\vec{B}=0,\ \nabla\times\vec{E}=-\frac{\partial\vec{B}}{\partial t},\ \nabla\times\vec{H}=\vec{J}$$

2.7.3 Maxwell's Equations in Lossless or Non-conducting Medium

In a lossless medium, current density $\vec{J}$ and charge density ρ are zero, and Maxwell's equations are simplified as below.

$$\nabla\cdot\vec{D}=0,\ \nabla\cdot\vec{B}=0,\ \nabla\times\vec{E}=-\frac{\partial\vec{B}}{\partial t},\ \nabla\times\vec{H}=\frac{\partial\vec{D}}{\partial t}$$

2.7.4 Maxwell's Equations in Charge-free/Current-free Medium

A perfect dielectric is that for which $\sigma = 0$, and thus there is no charge and current. An insulator is a good or near-perfect dielectric.

$$\nabla\cdot\overline{E}=0,\ \nabla\times\overline{E}=-j\omega\mu\overline{H},\ \nabla\cdot\overline{H}=0,\ \nabla\times\overline{H}=j\omega\varepsilon_o\overline{E}$$

2.7.5 Maxwell's Equations for Static Field

Static field laws in differential and integral form are summarized as follows:

Differential	*Integral form*	
$\nabla\cdot\vec{D}=\rho_v$	$\oint\vec{D}\cdot d\vec{S}=\int\rho_v dv$	Charge is a source or sink for $\vec{E}$
$\nabla\cdot\vec{B}=0$	$\oint\vec{B}\cdot d\vec{S}=0$	Continuity of magnetic filedlines, no magnetic monopoles
$\nabla\times\vec{E}=0$	$\oint\vec{E}\cdot d\vec{L}=0$	Irrotational property of static $\vec{E}$ field, potential depends on point not on path
$\nabla\times\vec{H}=\vec{J}$	$\oint\vec{H}\cdot d\vec{L}=\int J\cdot d\vec{S}$	Ampere's circuital law
$\nabla\cdot\vec{J}=0$	$\oint\vec{J}\cdot d\vec{S}=0$	Equation of continuity for steady currents

2.8 BOUNDARY CONDITIONS AT A SURFACE

A *media interface* is a surface across which the media properties $(\varepsilon, \mu, \sigma)$ change abruptly. The discontinuity in the material properties leads to discontinuities in certain electric and magnetic field components across this boundary. At the media interface, there are specific interfacial conditions that the fields should obey. Interestingly, when we apply the integral form of Maxwell's equations at the media interface, we get the relationship between the fields in the two media. These are generally referred to as the *boundary conditions*.

The value of the fields at the boundary surface is known as *boundary conditions.*

Note: Since $\vec{D} = \varepsilon\vec{E}$ and $\vec{B} = \mu\vec{H}$, the differential forms of Maxwell's equations are suitable for a continuous medium only, and they cannot be used for a media interface.

Boundary conditions

The four basic vector fields $\vec{E}$, $\vec{H}$, $\vec{D}$, and $\vec{B}$ consist of a tangential component denoted with subscript "t", and a normal component denoted with subscript "$\hat{n}$". In addition, the numerical in the subscript indicate medium. For example, $\vec{E}_{t1}$ is tangential component of electric field strength in medium 1, and $\vec{D}_{n2}$ is normal component of electric flux density in medium 2 (Figure 2.15 (a)). $\hat{n}$ is a unit vector that is normal to the boundary pointing from medium 2 into medium 1. The quantities ρ_s and $\mathbf{J}_s$ are any external surface charge and surface current densities on the boundary surface, respectively. The four boundary conditions are:

1. The tangential component of the electric field across a media interface (Figure 2.15 (b)) is always continuous.

$$E_{t1} = E_{t2}$$

2. The normal component of the electric flux density is continuous at the boundary (Figure 2.16), in the absence of the surface charge. But, it is discontinuous in the presence of the surface charge, by an amount that is equal to the surface charge density.

$$D_{n1} - D_{n2} = \rho_s$$

3. The normal component of $\vec{B}$ is continuous at a boundary.

$$B_{n_1} = B_{n_2}$$

4. The tangential component of the magnetic field is continuous at the boundary (Figure 2.17) if there is no surface current at the boundary. In the presence of the surface current, the tangential component of the magnetic field is discontinuous by an amount that is equal to the surface current density.

$$H_{t2} - H_{t1} = J_s$$

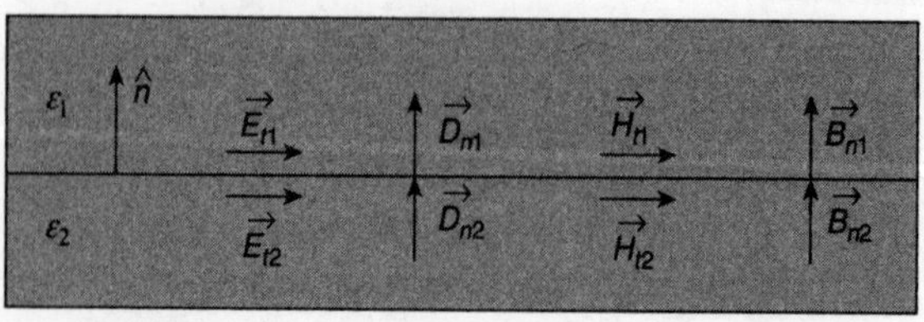

Figure 2.15 (a) *Tangential and normal components of electric and magnetic fields*

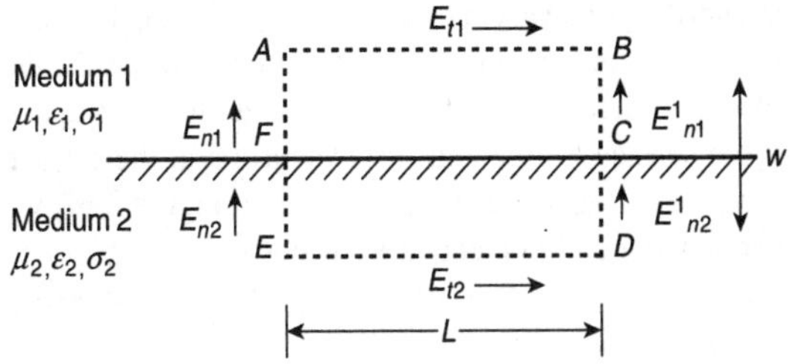

Figure 2.15 (b) *Interface formed by two media*

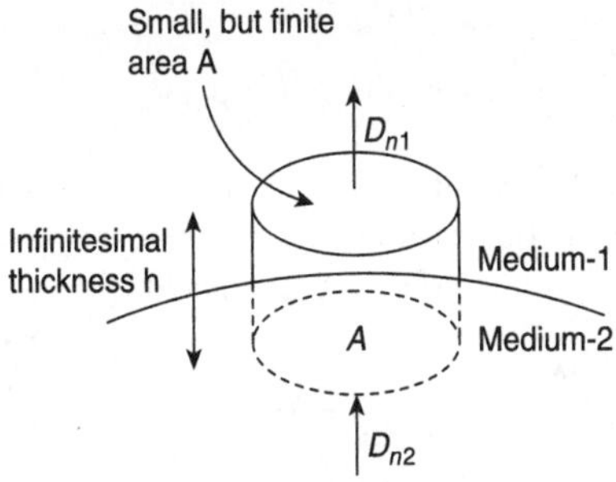

Figure 2.16 *Boundary conditions corresponding to Gauss' law*

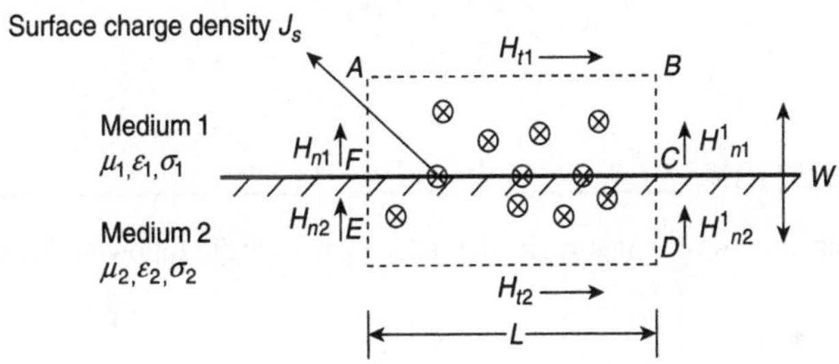

Figure 2.17 *Boundary between two dielectrics*

2.8.1 Boundary Conditions at the Interface between Dielectric and Dielectric

A dielectric is a medium that has zero conductivity $(\sigma = 0)$.

Therefore, current density $\vec{J}(\vec{J} = \sigma\vec{E})$ becomes zero, as $\sigma = 0$, and the surface charge density $\rho_s = 0$.

Substituting $\rho_s = 0$ and $J_s = 0$ in the above equations, we get

$$E_{t_1} = E_{t_2},\ D_{n_1} = D_{n_2}$$

$$H_{t_2} = H_{t_1},\ B_{n_1} = B_{n_2}$$

2.8.2 Boundary Conditions at Interface of Dielectric and Perfect Conductor

A perfect conductor is a medium that has infinite conductivity $(\sigma = \infty)$. A perfect conductor is an imagination, as there is no media which has infinite conductivity. In practice, properties of good conductors such as silver and copper can be nearly approximated by those of a perfect conductor.

The magnetic field H and the time-varying electric field are related for time-varying fields. The time-varying magnetic field does not exist inside a perfect conductor as the electric field is zero inside it. Thus it can be concluded that there are no time-varying electric and magnetic fields inside a perfect conductor, and there may be surface charge and surface current at the surface of a perfect conductor. If we consider medium 2 as the perfect conductor, as shown in Figure 2.17, we have E_2, D_2, H_2, and B_2 identically zero, and the boundary condition equations become

$$E_{t_1} = 0$$

$$D_{n_1} = \rho_s$$

$$B_{n_1} = 0$$

$$-\hat{n}\times\vec{H}_1 = \vec{H}_1\times\hat{n} = \vec{J}_s$$

Hence, we can conclude that, at a perfect conducting surface, the electric field is always normal to the surface ($\vec{E}_{t_1} = 0$), and the magnetic field is always tangential to the surface ($\vec{B}_{n_1} = \mu\vec{H}_{n_1} = 0$). The boundary conditions of dielectric-conductor interface are given below:

Dielectric-conductor interface

$$E_{t_1} = 0$$

$$D_{n_1} = \rho_s$$

$$H_{t_1} = J_s$$

$$B_{n_1} = 0$$

2.9 WAVE EQUATION

To derive the fundamental wave equation in any media, consider the following Maxwell's equations in differential form:

$$\nabla\times\vec{H} = \frac{\partial\vec{D}}{\partial t} + \vec{J} \tag{2.16}$$

$$\nabla\times\vec{E} = -\frac{\partial\vec{B}}{\partial t} \tag{2.17}$$

$$\nabla\cdot\vec{D} = \rho_v \tag{2.18}$$

$$\nabla\cdot\vec{B} = 0 \tag{2.19}$$

Consider a homogenous medium where ε, μ, σ are constants throughout the medium.

Now, taking into account the curl of Eq. (2.16) gives

$$\begin{aligned}\nabla\times\nabla\times\vec{H} &= \frac{\partial(\nabla\times\vec{D})}{\partial t} + \nabla\times\vec{J}\\ &= \varepsilon\frac{\partial(\nabla\times\vec{E})}{\partial t} + \sigma(\nabla\times\vec{E}) \qquad \left(\because \vec{D} = \varepsilon\vec{E} \text{ and } \vec{J} = \sigma\vec{E}\right)\end{aligned}$$

By substituting Eq. (2.17) in the above equation, we get

$$\nabla\times\nabla\times\vec{H} = -\mu\varepsilon\frac{\partial^2\vec{H}}{\partial t^2} - \mu\sigma\frac{\partial\vec{H}}{\partial t} \qquad (\because\ \vec{B} = \mu\vec{H})$$

Using vector identity $\nabla\times\nabla\times\vec{A} = \nabla(\nabla\cdot\vec{A}) - \nabla^2\vec{A}$ gives

$$\nabla^2\vec{H} - \nabla(\nabla\cdot\vec{H}) = \mu\varepsilon\frac{\partial^2\vec{H}}{\partial t^2} + \mu\sigma\frac{\partial\vec{H}}{\partial t} \tag{2.20a}$$

$$\nabla\cdot\vec{H} = 0, \text{ as } \nabla\cdot\vec{B} = 0 \text{ and } \vec{B} = \mu\vec{H}$$

$$\nabla^2\vec{H} = \mu\varepsilon\frac{\partial^2\vec{H}}{\partial t^2} + \mu\sigma\frac{\partial\vec{H}}{\partial t} \tag{2.20b}$$

Following the same procedure for electric field E, that is, taking the curl of equation (2.17) into account, gives

$$\nabla\times\nabla\times\vec{E} = -\frac{\partial(\nabla\times\vec{B})}{\partial t} = -\mu\frac{\partial(\nabla\times H)}{\partial t} \qquad (\because\ B = \mu H)$$

Using Eq. (2.16), $\nabla\times\nabla\times\vec{E} = -\mu\frac{\partial}{\partial t}\left(\frac{\partial\vec{D}}{\partial t} + \vec{J}\right)$

Using vector identity $\nabla\times\nabla\times\vec{A} = \nabla(\nabla\cdot\vec{A}) - \nabla^2\vec{A}$ and the relations $D = \varepsilon E$ and $J = \sigma E$, we get

$$\nabla^2\vec{E} = \mu\varepsilon\frac{\partial^2\vec{E}}{\partial t^2} + \mu\sigma\frac{\partial\vec{E}}{\partial t} \tag{2.21a}$$

$$\nabla\cdot\vec{E} = 0 \text{ for a chargeless medium}$$

$$\nabla^2\vec{E} - \nabla(\nabla\cdot\vec{E}) = \mu\varepsilon\frac{\partial^2\vec{E}}{\partial t^2} + \mu\sigma\frac{\partial\vec{E}}{\partial t} \tag{2.21b}$$

These two second-order differential equations (2.20) and (2.21) are known as *wave equations* for a homogeneous media.

The wave equations for particular cases, that is, for a perfect conducting medium and a perfect dielectric medium, are given in the following sections.

2.9.1 Wave Equation for a Conducting Media

It is already shown in the previous chapters that no net charge exists within a conductor. Hence, the charge density in a uniform conducting media is zero.

Hence, $\nabla\cdot\vec{D} = \rho_v = 0 \Rightarrow \nabla\cdot\varepsilon\vec{E} = 0 \ \Rightarrow \nabla\cdot\vec{E} = 0.$

Substituting the above result in Eq. 2.21(a), gives

$$\nabla^2\vec{E} = \mu\varepsilon\frac{\partial^2\vec{E}}{\partial t^2} + \mu\sigma\frac{\partial\vec{E}}{\partial t} \tag{2.22}$$

Substituting Eq. 2.19 in Eq. 2.20 (a) gives

$$\nabla^2 \vec{H} = \mu\varepsilon \frac{\partial^2 \vec{H}}{\partial t^2} + \mu\sigma \frac{\partial \vec{H}}{\partial t} \tag{2.23}$$

2.9.2 Wave Equation for Perfect Dielectric Media

A perfect dielectric can be characterized by zero conductivity, that is, $\sigma = 0$. Hence, the wave equation for a perfect dielectric is reduced to

$$\nabla^2 \vec{H} = \mu\varepsilon \frac{\partial^2 \vec{H}}{\partial t^2} \tag{2.24}$$

$$\nabla^2 \vec{E} = \mu\varepsilon \frac{\partial^2 \vec{E}}{\partial t^2} \tag{2.25}$$

2.9.3 Wave Equation for Free Space

Free space is a perfect dielectric with no sources, charges, or currents. Hence, both its conductivity and charge density are zero. Combining the results of the earlier two sections yields the following wave equations:

$$\nabla^2 \vec{E} = \mu_0\varepsilon_0 \frac{\partial^2 \vec{E}}{\partial t^2} \tag{2.26}$$

$$\nabla^2 \vec{H} = \mu_0\varepsilon_0 \frac{\partial^2 \vec{H}}{\partial t^2} \tag{2.27}$$

where μ_0 and ε_0 are permeability and permittivity of free space, respectively.

Propagation of EM waves in free space is a very important case practically, and these wave equations will be considered in further sections.

2.10 UNIFORM PLANE WAVES

There can be many solutions to the wave equations depending on the nature of propagation. One such solution is a uniform plane wave, which is an important case of general EM waves.

2.10.1 Uniform Plane Wave Definition

A uniform plane wave is an EM wave travelling in one direction and independent of the other two directions. Here, we consider a uniform plane wave travelling in z direction and independent of x, y directions. Since there is no variation of fields E and H with respect to x and y co-ordinates, $\frac{\partial \vec{E}}{\partial x}$, $\frac{\partial \vec{E}}{\partial y}$, $\frac{\partial \vec{H}}{\partial x}$ and $\frac{\partial \vec{H}}{\partial y}$ all are equal to zero.

Hence, for uniform plane wave, Eqs. (2.26) and (2.27) become

$$\frac{\partial^2 \vec{E}}{\partial z^2} = \mu\varepsilon \frac{\partial^2 \vec{E}}{\partial t^2} \tag{2.29}$$

$$\frac{\partial^2 \vec{H}}{\partial z^2} = \mu\varepsilon \frac{\partial^2 \vec{H}}{\partial t^2} \tag{2.30}$$

It is obvious that in the absence of any volume charges ($\nabla \cdot \vec{D} = 0$), a uniform plane wave does not have field components in the direction of propagation; that is, both E_z and H_z are zero.

$$\Rightarrow \vec{E}_z = 0 \text{ and } \vec{H}_z = 0 \tag{2.31}$$

Respective *x*, *y*, and *z* components of Eq. (2.28) are equal. Hence,

$$\frac{\partial^2 E_z}{\partial z^2} = \mu\varepsilon \frac{\partial^2 E_z}{\partial t^2} \tag{2.32}$$

From Eqs. (2.31) and (2.32),

$$\frac{\partial^2 E_z}{\partial t^2} = 0 \tag{2.33}$$

It follows that E_z should be zero; that is, E_z can neither be a constant nor vary linearly with time in order for it to constitute a wave. A similar explanation using Eqs. (2.19) and (2.30) states that H_z should be zero.

Equation (2.32) is a second order-differential equation that will yield a general solution of form:

$$E(z, t) = F(z - ct) + G(z + ct) \tag{2.34}$$

where *F* is a wave travelling in the positive (+ve) *z* direction, *G* is a wave travelling in the negative (–ve) *z* direction, and *c* is $\dfrac{1}{(\mu\varepsilon)^{1/2}}$. Such a phenomenon can be represented by Figure 2.18.

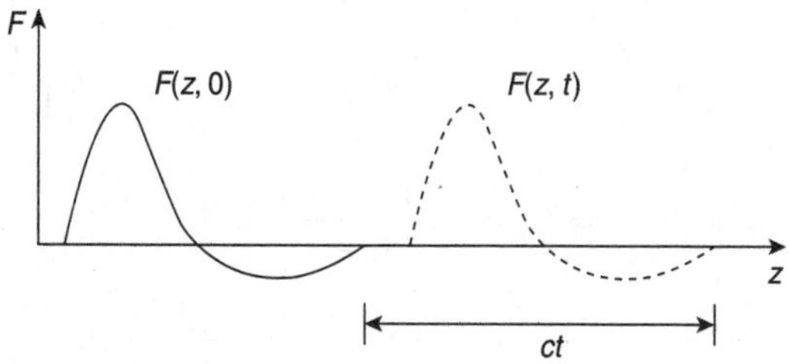

Figure 2.18 *Plane wave travelling in positive z direction*

2.10.2 Relationship Between *E* and *H* in a Uniform Plane Wave

Different properties of the plane wave can be found by analyzing the Maxwell's equations thoroughly.

1. The ratio between the amplitudes of *E* and *H* in a uniform plane wave is known as *intrinsic impedance* η and is given by

$$\frac{E}{H} = \sqrt{\frac{\mu}{\varepsilon}}$$

   Its value for free space is 120π.

Derivation of Ratio Between Amplitudes of *E* and *H* in A Uniform Plane Wave

In order to find the relationship between electric and magnetic fields of a plane wave, let us consider Maxwell's equations Eqs. (2.16) and (2.17) with $\sigma = 0$.

$$\left(\frac{\partial E_z}{\partial y}-\frac{\partial E_y}{\partial z}\right)\hat{x}+\left(\frac{\partial E_x}{\partial z}-\frac{\partial E_z}{\partial x}\right)\hat{y}+\left(\frac{\partial E_y}{\partial x}-\frac{\partial E_x}{\partial y}\right)\hat{z}=-\mu\left(\frac{\partial H_x}{\partial t}\hat{x}+\frac{\partial H_y}{\partial t}\hat{y}+\frac{\partial H_z}{\partial t}\hat{z}\right) \quad (2.35)$$

$$\left(\frac{\partial H_z}{\partial y}-\frac{\partial H_y}{\partial z}\right)\hat{x}+\left(\frac{\partial H_x}{\partial z}-\frac{\partial H_z}{\partial x}\right)\hat{y}+\left(\frac{\partial H_y}{\partial x}-\frac{\partial H_x}{\partial y}\right)\hat{z}=\varepsilon\left(\frac{\partial E_x}{\partial t}\hat{x}+\frac{\partial E_y}{\partial t}\hat{y}+\frac{\partial E_z}{\partial t}\hat{z}\right) \quad (2.36)$$

Using Eq. (2.31) and equating corresponding terms gives

$$\frac{\partial E_y}{\partial z}=\mu\frac{\partial H_x}{\partial t} \quad (2.37)$$

$$\frac{\partial E_x}{\partial z}=-\mu\frac{\partial H_y}{\partial t} \quad (2.38)$$

$$\frac{\partial H_y}{\partial z}=-\varepsilon\frac{\partial E_x}{\partial t} \quad (2.39)$$

$$\frac{\partial H_x}{\partial z}=\varepsilon\frac{\partial E_y}{\partial t} \quad (2.40)$$

From Eq. (2.34), $E_x = F(z-ct)$, as it is a first-order equation; hence.

$$\frac{\partial E_x}{\partial t}=\frac{\partial F(z-ct)}{\partial(z-ct)}\frac{\partial(z-ct)}{\partial t}=\frac{\partial F}{\partial(z-ct)}(-c)$$

Using this in Eq. (2.39), we get $\dfrac{\partial H_y}{\partial z}=\dfrac{\partial F(z-ct)}{\partial(z-ct)}(c\varepsilon)$

$$\Rightarrow H_y = c\varepsilon\int\frac{\partial F(z-ct)}{\partial(z-ct)}\,dz$$

$$= c\varepsilon\int\frac{\partial F(z-ct)}{\partial z}\times\frac{1}{\dfrac{\partial(z-ct)}{\partial z}}\,dz$$

$$= c\varepsilon\, F(z-ct)$$

$$= c\varepsilon E_x$$

Hence, $$\frac{H_Y}{E_x}=\frac{1}{\sqrt{\mu\varepsilon}}\,\varepsilon=\sqrt{\frac{\varepsilon}{\mu}} \quad (2.41)$$

Similarly, it can be shown that $$\frac{E_y}{H_x}=-\sqrt{\frac{\mu}{\varepsilon}} \quad (2.42)$$

The total field values are given by the root mean square value of their respective x and y components:

$$E=\sqrt{E_y^2+E_x^2}$$

Using the above relationship gives

$$E = \sqrt{\frac{\mu}{\varepsilon}(H_y^2 + H_x^2)}$$

$$\frac{E}{H} = \sqrt{\frac{\mu}{\varepsilon}} \tag{2.43}$$

2. Dot product of electric and magnetic fields is 0, showing that E and B are perpendicular to each other:

$$E \cdot H = E_x H_X + E_Y H_Y + E_Z H_Z$$
$$= -H_Y \times \sqrt{\frac{\mu}{\varepsilon}} \times H_x + H_x \times \sqrt{\frac{\mu}{\varepsilon}} \times H_Y + 0 \times 0 = 0$$

3. Cross product of the fields yields the following,

$$E \times H = \hat{z}(E_x H_y - E_y H_x) \quad (\because E_z = H_z = 0)$$

From Eq. 2.41 and Eq. 2.42

$$E \times H = \hat{z}(\eta\, H_y^2 - (-\eta H_x^2)) = \hat{z}\eta H^2$$

Hence cross product of fields gives the direction in which wave travels.
The following points can be summarized from the relationship between E and H in a plane wave:

- The ratio between the amplitudes of electric and magnetic fields is known as *characteristic impedance* (η). Its value for free space is 120π.
- The electric field is normal to the magnetic field.
- The cross product of fields gives the direction in which the wave travels.

2.10.3 Sinusoidal Variations

Sinusoidal waves are the most familiar waves and are used in many applications such as communications. It is known (from Fourier analysis) that any function can be represented in terms of *sines* and *cosines* with appropriate wavelengths. In addition, many applications involving EM waves use sinusoidal variations. Hence, the fields can be expressed as

$$\tilde{E}_z = |E_z| \cos(\omega t) \tag{2.44}$$
$$\tilde{H}_z = |H_z| \cos(\omega t) \tag{2.45}$$

As mentioned in Chapter 1, it is convenient to introduce the complex exponential $e^{j\omega t}$ rather than to use real sinusoidal functions directly as differentiation or integration w.r.t time. It can be replaced by multiplication or division of the phasor form with the factor $j\omega$. Equation (2.44) can also be expressed in phasor notation as

$$\tilde{E}_z = \text{Re}\left(|E_z| e^{j\omega t}\right).$$

Maxwell's equations can be stated in phasor form as follows:
Consider

$$\nabla \times \tilde{H} = \frac{\partial \tilde{D}}{\partial t} + \tilde{J}$$

$$\nabla \times \mathrm{Re}(\vec{H}e^{j\omega t}) = \frac{\partial}{\partial t}\mathrm{Re}(\vec{D}e^{j\omega t}) + \mathrm{Re}(\vec{J}e^{j\omega t})$$

$$\nabla \times \vec{H} = j\omega\vec{D} + \vec{J}$$

Similarly, other equations can be expressed as

$$\nabla \times \vec{E} = -j\omega\vec{B}$$

$$\nabla \cdot \vec{D} = \rho_v$$

$$\nabla \cdot \vec{B} = 0$$

For sinusoidal variations, the wave equation for a lossless medium becomes

$$\nabla^2\vec{E} = -\omega^2\mu\varepsilon\vec{E}$$

and for a conducting medium,

$$\nabla^2\vec{E} + (\omega^2\mu\varepsilon - j\omega\mu\sigma)\vec{E} = 0 \quad \left(\because \nabla^2\vec{E} = \mu\varepsilon\frac{\partial^2\vec{E}}{\partial t^2} + \mu\sigma\frac{\partial\vec{E}}{\partial t}\right)$$

2.11 WAVE PROPAGATION

The most general case of wave propagation is in a lossy dielectric that is in a medium where the EM wave loses power as it propagates. Maxwell's equations can be solved in phasor notation to derive some more new parameters such as attenuation constant and phase constant.
Later, we consider the propagation of EM waves in various kinds of media as special cases.
Using the relation $J = \sigma\vec{E}$ and $\mathbf{D} = \varepsilon\vec{E}$, Maxwell's equation can be written as

$$\nabla \times \vec{H} = \sigma\bar{E} + j\omega\varepsilon\bar{E} \tag{2.46}$$

and

$$\nabla \times \vec{E} = -j\mu\omega\bar{H} \tag{2.47}$$

Differentiating Eq. (2.46) with regard to time and changing the order of time and space differentiation yields

$$\nabla \times \left(\frac{\partial\bar{H}}{\partial t}\right) = j\omega(\sigma + j\omega\varepsilon)\bar{E}$$

Taking into account the curl of Eq. (2.17) and using the above equation yields

$$\nabla \times \nabla \times \bar{E} = -\mu\nabla \times \left(\frac{\partial\bar{H}}{\partial t}\right) = -j\mu\omega(\sigma + j\omega\varepsilon)\bar{E}$$

For charge-free region, $\nabla \times \nabla \times \bar{E} = -\nabla^2\bar{E} + \nabla(\nabla \cdot \bar{E}) = -\nabla^2\bar{E}$

$$\Rightarrow \nabla^2\vec{E} = j\mu\omega(\sigma + j\omega\varepsilon)\vec{E}$$

Hence,

$$\nabla^2\vec{E} = \gamma^2\vec{E} \tag{2.48}$$

where γ is called a *propagation constant* and is given by

$$\gamma^2 = j\mu\omega(\sigma + j\omega\varepsilon) \tag{2.49}$$

Similarly, it can be shown that

$$\nabla^2 \overline{H} = \gamma^2 \overline{H}$$

The propagation constant is complex with real value (α) and complex value (β)

Hence, $\gamma = \alpha + j\beta,$

where

$$\alpha = \omega \sqrt{\frac{\mu\varepsilon}{2}\left(\sqrt{\left(1+\frac{\sigma^2}{\omega^2\varepsilon^2}\right)}-1\right)} \text{ and}$$

$$\beta = \omega \sqrt{\frac{\mu\varepsilon}{2}\left(\sqrt{\left(1+\frac{\sigma^2}{\omega^2\varepsilon^2}\right)}+1\right)}$$

For, a uniform plane wave travelling in the *z* direction, Eq. (2.48) can be written as

$$\frac{\partial^2 \vec{E}}{\partial z^2} = \gamma^2 \vec{E} \tag{2.50}$$

Equation (2.50) can be satisfied by

$$\vec{E}(z) = \vec{E}_0 e^{-\gamma z}$$

Including time for time-varying cases gives

$$\vec{E}(z,t) = \text{Re}(\vec{E}_0 e^{-\gamma z + j\omega t}) = \text{Re}(\vec{E}_0 e^{-\alpha z - j\beta z + j\omega t})$$

$$= e^{-\alpha z}\,\text{Re}(\vec{E}_0 e^{j(\omega t - \beta z)})$$

Here, we see that as *z* increases, the *E* value decreases, because of the term $-\alpha z$ in the exponent. Hence, α is known as *attenuation constant*, and it is a measure of the spatial rate of decay of the wave in the medium (Figure 2.19). It is measured either in nepers per meter (Np/m) or in decibels per meter (dB/m).

In Eq. (2.50), the factor βz causes the phase changes. The quantity β is a measure of the phase shift per length and is called the *phase constant* or *wave number*. In terms of the phase constant, wavelength and wave velocity are

$$\lambda = \frac{2\pi}{\beta} \text{ and } v = \frac{\omega}{\beta}$$

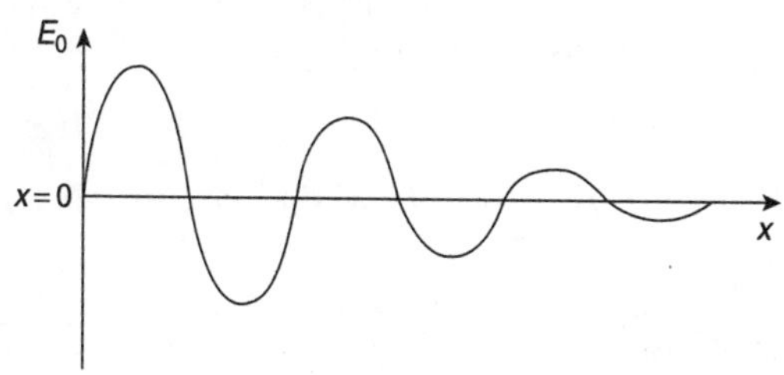

Figure 2.19 *Spatial rate of decay of the wave in the medium*

EXAMPLE PROBLEM 2.3

The electric field in free space is given by $\mathbf{E} = 100\cos(10^8 t + \beta x)\hat{y}$ V/m. Calculate β and the time it takes to travel a distance of $\lambda/4$.

Solution

Given that $E = 100\cos(10^8 t + \beta x)y$ V/m
Hence, $\omega = 10^8$ rad/s
In free space, velocity of the EM wave is

$$v = 3\times10^8 \text{ m/s}$$

$$\beta = \frac{\omega}{v} = \frac{10^8}{3\times10^8} = \frac{1}{3} \text{ rad/m}$$

To travel a distance of $\lambda/4$, it will take 1/4 of its time period.

$$t = \frac{T}{4} = \frac{2\pi}{4\omega} = \frac{\pi}{2}\times10^{-8}\,\text{s} = 15.71 \text{ ns}$$

■

EXAMPLE PROBLEM 2.4

Find the skin depth at 10 GHz for silver with a conductivity of 6.1x10^7 s/m and a relative permittivity of 1.0.

Solution

Skin depth is given by $\delta = \sqrt{\dfrac{2}{\omega\mu\sigma}}$

$$= \sqrt{\frac{2}{2\pi\times10^{10}\times4\pi\times10^{-7}\times6.1\times10^{7}}} = \frac{0.4048}{2\pi\times10^{5}}\text{ m} = 0.6443\ \mu\text{m}$$

■

EXAMPLE PROBLEM 2.5

Find the attenuation constant, phase constant and propagation constant for a large copper conductor ($\sigma = 5.8\times10^7$ s/m, $\varepsilon_r = \mu_r = 1$). Support a uniform plane wave at 50 Hz.

Solution

Given data $\sigma = 5.8\times10^7$ s/m, $\varepsilon_r = \mu_r = 1$ and $f = 50$ Hz

Here, $\dfrac{\sigma}{\omega\varepsilon} = \dfrac{5.8\times10^7}{2\pi\times50\times8.854\times10^{-12}} = 2.08\times10^{16} >> 1$

Hence the medium is a good conductor.

Attenuation constant

$$\alpha = \omega\sqrt{\frac{\mu\varepsilon}{2}\left(\sqrt{\left(1+\frac{\sigma^2}{\omega^2\varepsilon^2}\right)}-1\right)} \cong \sqrt{\frac{\omega\mu\sigma}{2}}$$

$$= \sqrt{\frac{2\pi\times 50\times 4\pi\times 10^{-7}\times 5.8\times 10^{7}}{2}} = 106.94 \text{ Np/m}$$

Phase constant

$$\beta \cong \sqrt{\frac{\omega\mu\sigma}{2}} = 106.94 \text{ rad/m}$$

Propagation constant

We know that $\gamma = \alpha + j\beta$

Hence propagation constant is $106.94 + j106.94$. ■

2.11.1 Derivation of Values of Attenuation Constant and Phase Shift Constant

For a detailed analysis of electromagnetic waves, α and β should be known. To find these values, consider Eq. (2.49):

$$\gamma^2 = j\mu\omega(\sigma + j\omega\varepsilon) = j\omega\mu\sigma - \omega^2\mu\varepsilon$$

$$(\alpha + j\beta)^2 = \alpha^2 - \beta^2 + j2\alpha\beta$$

Real and imaginary parts of the above two equations can be equated, which results in

$$\alpha^2 - \beta^2 = -\omega^2\mu\varepsilon$$

$$2\alpha\beta = \omega\mu\sigma \tag{2.51}$$

Based on the knowledge of algebraic formulae, the following equation can be given:

$$\alpha^2 + \beta^2 = \sqrt{(\alpha^2 - \beta^2)^2 + 4\alpha^2\beta^2} = \sqrt{(\omega^2\mu\varepsilon)^2 + (\omega\mu\sigma)^2}$$

$$= \omega^2\mu\varepsilon\sqrt{1+\left(\frac{\sigma}{\omega\varepsilon}\right)^2} \tag{2.52}$$

Adding Eqs. (2.51) and (2.52) yields

$$2\alpha^2 = \omega^2\mu\varepsilon\left(\sqrt{1+\left(\frac{\sigma}{\omega\varepsilon}\right)^2}-1\right)$$

$$\alpha = \omega\sqrt{\frac{\mu\varepsilon}{2}\left(\sqrt{\left(1+\frac{\sigma^2}{\omega^2\varepsilon^2}\right)}-1\right)} \tag{2.53}$$

Similarly, subtracting Eqs. (2.51) and (2.52) yields

$$\beta = \omega\sqrt{\frac{\mu\varepsilon}{2}\left(\sqrt{\left(1+\frac{\sigma^2}{\omega^2\varepsilon^2}\right)}+1\right)} \tag{2.54}$$

2.11.2 Intrinsic Impedance

The general form of intrinsic impedance (or characteristic impedance) can be easily found out by solving Maxwell's equations in phasor form, and it is given as

$$\frac{\vec{E}}{\vec{H}} = \eta = \sqrt{\frac{j\omega\mu}{\sigma + j\omega\varepsilon}} \tag{2.55}$$

It can be seen that the intrinsic impedance is complex and can be written as $\eta = |\eta|e^{\theta_\eta}$

$$|\eta| = \frac{\sqrt{\frac{\mu}{\varepsilon}}}{\left[1+\left(\frac{\sigma}{\omega\varepsilon}\right)^2\right]^{\frac{1}{4}}} \text{ and } \theta_\eta = \frac{1}{2}\tan^{-1}\left(\frac{\sigma}{\omega\varepsilon}\right), \text{ where } 0 \le \theta_\eta \le 45^\circ$$

Here, we consider propagation in a lossless medium, conduction medium, and free space.

From Eq. (2.55), it can be seen that E and H are out of phase by θ_η, at any instant of time due to the complex intrinsic impedance of the medium. Thus, at any time, E leads H (or H lags E) by θ_η.

2.12 POLARIZATION

As a transducer, an antenna converts radio frequency (RF) electric current into electromagnetic waves and then radiates into space. For the selection and installation of antennas, antenna polarization is an important factor. The polarization of the electric field vector of the radiated wave is nothing but the polarization of the antenna. The direction and position of the electric field with respect to the earth's surface or ground tells us about the wave polarization.

The polarization of an EM wave is, by definition, the direction of the electric field.

The following types of polarizations occurs when the wave is approaching:

- Linear
- Circular
- Elliptical

2.12.1 Linear Polarization

A linearly polarized wave is a transverse EM wave whose electric field vector lies along a straight line at all times.

In a linear polarization, the path of the electric field vector moves back and forth along a line (Figure 2.20).

Most wireless communication systems use either linear (vertical, horizontal) or circular polarization. Linear polarization can be of two types:

- Horizontal polarization
- Vertical polarization

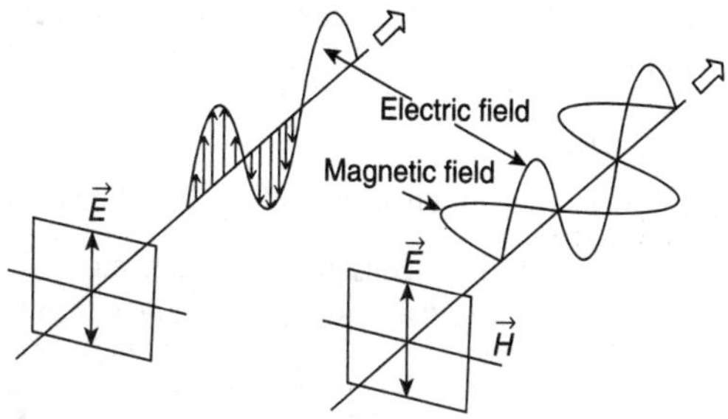

Figure 2.20 *A linearly (vertically) polarized wave*

Horizontal polarization

A linearly polarized wave is said to be horizontally polarized if the electric field is aligned in parallel with the horizontal plane.

Vertical polarization

A linearly polarized wave is said to be vertically polarized if the electric field is aligned in parallel with the vertical plane.

The orientation of the electric field vector in case of vertical and horizontal polarization is illustrated in Figure 2.21:

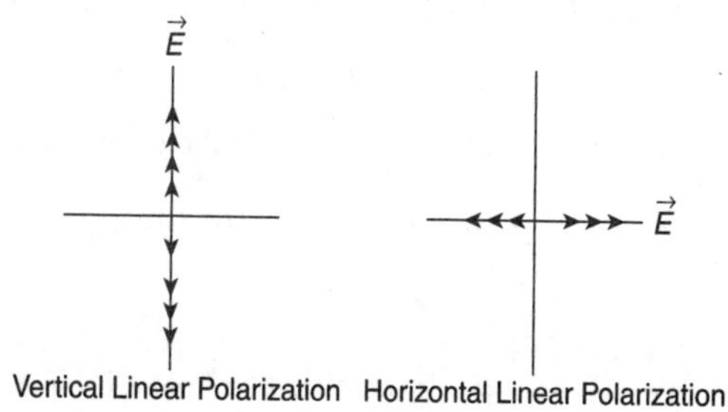

Figure 2.21 *Orientation of electric field vector in vertical and horizontal polarization*

2.12.2 Circular Polarization

In a circularly polarized wave, the electric field vector remains constant in length but rotates around in a circular path. Depending on the direction of rotation, it can be Left-hand polarized wave or Right-hand polarized wave.

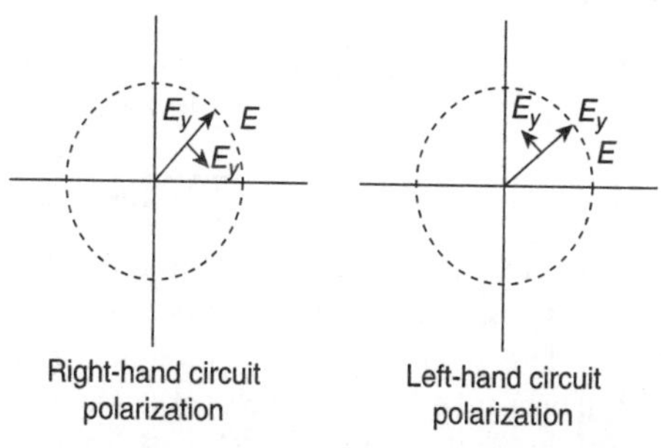

Figure 2.22 *Circular polarization*

If the wave rotates clockwise it is right-hand polarized wave. If it rotates counter clockwise it is left-hand polarized wave, as shown in Figure 2.21.

2.12.3 Elliptical Polarization

A wave is said to be elliptically polarized if either or both of the following conditions are satisfied:

- If E_x and E_y are not in phase, but have a constant phase difference other than 90°.
- If the ratio of amplitudes of E_x and E_y is constant but not one, then the wave is said to be elliptically polarized.

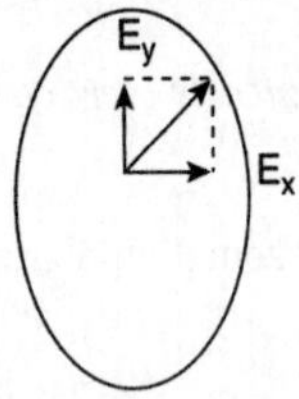

Figure 2.23 *Elliptical polarization*

Here, the resultant electric field vector traces out an ellipse, as shown in Figure 2.23. To clearly understand this, let us consider a wave travelling in z direction, where the x and y components have different amplitudes and a phase difference of 90°. The field satisfying such conditions is

$$E_0 = A\cos\omega t + B\sin\omega t$$

where the field in the x direction is $A\cos\omega t$, and that in the y direction is $B\sin\omega t$.

$$\vec{E}_x = A\cos\omega t \text{ and } \vec{E}_y = B\sin\omega t$$

where $A \neq B$

$$\frac{\vec{E}_x^2}{A^2} + \frac{\vec{E}_y^2}{B^2} = 1 \quad \left(\because \cos^2\omega t + \sin^2\omega t = 1\right)$$

Thus, the wave is elliptically polarized. Elliptical polarization is a general case; whereas if the amplitudes of both the field components is equal and the difference is 90°, then the polarization is said to be circular polarization. Such a field can be represented by

$$\vec{E}_0 = A\cos\omega t + A\sin\omega t$$

$$\vec{E}_x = A\cos\omega t \text{ and } \vec{E}_y = A\sin\omega t$$

$$\vec{E}_x^2 + \vec{E}_y^2 = A^2$$

2.13 POYNTING VECTOR AND POYNTING THEOREM

Electromagnetic fields are used to transmit the signal (or information) over long distances. Electromagnetic waves carry energy when they move from one point to another. The energy per unit time is referred to as *power*. Therefore, when an EM wave travels, power is transported from the source to the destination point. The total outward power from a closed surface can be estimated using Poynting theorem.

Poynting theorem states that the total power (W) leaving the closed surface is equal to the closed surface integral of Poynting vector ($\vec{P}$).

Total power leaving the closed surface = $W = \oint_S \vec{P} \cdot d\vec{S}$

where $\vec{P}$ is the power density vector associated with the electromagnetic field and is defined as

$$\vec{P} = \vec{E} \times \vec{H}$$

where, $\boldsymbol{E} \times \boldsymbol{H}$ represents power flow per unit area (watts/m^2)

Proof of $W = \oint_S (\vec{E} \times \vec{H}) \cdot d\vec{S}$

To derive the relationship between energy transfer (power) and field intensities, Maxwell's equations (Faraday's and Ampere's laws) can be used:

$$\nabla \times \vec{E} = -\frac{\partial \vec{B}}{\partial t} = -\mu \frac{\partial \vec{H}}{\partial t} \text{ (Faraday's law)} \tag{2.56}$$

$$\nabla \times \vec{H} = \frac{\partial \vec{D}}{\partial t} + \vec{J} = \varepsilon \frac{\partial \vec{E}}{\partial t} + \vec{J} \text{ (Ampere's laws)} \tag{2.57}$$

We have the vector identity

$$\nabla \cdot (\vec{A} \times \vec{B}) = \vec{B} \cdot (\nabla \times \vec{B}) - \vec{A} \cdot (\nabla \times \vec{B})$$

From the above identity, we can write the cross product of the electric and magnetic fields as

$$\nabla \cdot (\vec{E} \times \vec{H}) = \vec{H} \cdot (\nabla \times \vec{E}) - \vec{E} \cdot (\nabla \times \vec{H}) \tag{2.58}$$

Substituting Eqs. (2.56) and (2.57) in (2.58), we get

$$\nabla \cdot (\vec{E} \times \vec{H}) = \vec{H} \cdot \left(-\mu \frac{\partial \vec{H}}{\partial t}\right) - \vec{E} \cdot \left(\varepsilon \frac{\partial \vec{E}}{\partial t} + \vec{J}\right) \tag{2.59}$$

In addition, we have the following vector identity:

$$\frac{\partial (\vec{A} \cdot \vec{B})}{\partial t} = \vec{A} \cdot \frac{\partial \vec{B}}{\partial t} + \vec{B} \cdot \frac{\partial \vec{A}}{\partial t}$$

$$\vec{A} \cdot \frac{\partial \vec{A}}{\partial t} = \frac{1}{2} \frac{\partial \vec{A}^2}{\partial t}, \quad \text{(if } \vec{B} = \vec{A}\text{)}$$

By applying the above identity for E and H fields, we can write,

$$\vec{E} \cdot \frac{\partial \vec{E}}{\partial t} = \frac{1}{2} \frac{\partial \vec{E}^2}{\partial t}, \quad \vec{H} \cdot \frac{\partial \vec{H}}{\partial t} = \frac{1}{2} \frac{\partial \vec{H}^2}{\partial t}$$

Now, Eq. (2.59) becomes

$$\nabla \cdot (\vec{E} \times \vec{H}) = -\frac{\mu}{2} \frac{\partial \vec{H}^2}{\partial t} - \frac{\varepsilon}{2} \frac{\partial \vec{E}^2}{\partial t} - \vec{E} \cdot \vec{J} \tag{2.60}$$

The above equation gives the relationship between E and H at a point, but it should be valid for every point in space. So, by taking into account the volume integral for equation (2.60), we get

$$\int_v \nabla\cdot(\vec{E}\times\vec{H})\,dv = \int_v\left(-\frac{\mu}{2}\frac{\partial\vec{H}^2}{\partial t}-\frac{\varepsilon}{2}\frac{\partial\vec{E}^2}{\partial t}-\vec{E}\cdot\vec{J}\right)dv \tag{2.61}$$

By divergence theorem, we have

$$\oint_S(\vec{E}\times\vec{H})\cdot d\bar{S} = \int_v \nabla\cdot(\vec{E}\times\vec{H})\,dv \tag{2.62}$$

Substituting Eq. (2.62) in (2.61) and by interchanging the integral and $\frac{\partial}{\partial t}$, we have

$$\oint_S(\vec{E}\times\vec{H})\cdot d\bar{S} = -\frac{\partial}{\partial t}\int_v\left(\frac{1}{2}\varepsilon\vec{E}^2+\frac{1}{2}\mu\vec{H}^2\right)dv-\int_v\vec{E}\cdot J\,dv \tag{2.63}$$

The first term in the RHS of the above equation represents the rate at which the stored energy in the magnetic and electric fields is changing. The minus sign indicates a decrease in energy.

The second term on the right-hand side represents the ohmic power loss in the volume v.

The two terms on the right-hand side represent the total decrease in the EM energy per unit time.

Therefore, by the law of conservation of energy, the decrease in the EM energy per unit time (power loss) is equal to the rate at which the energy is leaving (power coming out of) the volume.

$$\text{Total outward power} = W = \oint_S(\vec{E}\times\vec{H})\cdot d\bar{S} = \oint_S\vec{P}\cdot d\bar{S}$$

where Poynting vector,

$$\vec{P} = \vec{E}\times\vec{H} \tag{2.64}$$

2.13.1 Average Power through a Surface

The average power density or average Poynting vector is another parameter of more practical importance that can be computed from the instantaneous Poynting vector or instantaneous power. The instantaneous power does not give the actual power at a location. Instead, it gives the power oscillations around a point and the actual power at that point. However, the average Poynting vector gives the actual power at a location. For simultaneously time-varying fields, the Poynting vector can be expressed in terms of phasors as complex Poynting vector $P = E\times H^*$, which gives the instantaneous power density. The average power or average Poynting vector is given as

$$\vec{P}_{\text{avg}} = \frac{1}{2}\,\text{Re}\left\{\vec{E}\times\vec{H}^*\right\}$$

$\text{Re}\left\{E\times H^*\right\}$ denotes the real part of the exiting complex power.

From the above equation, we can also get the total average power through a surface, which is given as

$$P_{\text{tot\_avg}} = \int\vec{P}_{\text{avg}}\cdot d\vec{S}$$

Derivation for Average Power through a surface

It is very useful to have average power density or average Poynting vector at a point for the time-varying fields. The E and H fields can be given as

$$\vec{E} = \vec{E}_{\text{real}} + j\vec{E}_{\text{img}} \text{ and } \vec{H} = \vec{H}_{\text{real}} + j\vec{H}_{\text{img}}$$

Now, considering time-varying fields,

$$\vec{E} = \text{Re}(\vec{E}e^{j\omega t}) = \frac{1}{2}(\vec{E}e^{j\omega t} + (\vec{E}e^{j\omega t})^*) \tag{2.65}$$

and

$$\vec{H} = \text{Re}(\vec{H}e^{j\omega t}) = \frac{1}{2}(\vec{H}e^{j\omega t} + (\vec{H}e^{j\omega t})^*) \tag{2.66}$$

From Poynting theorem, Poynting vector is given as

$$\vec{P} = \vec{E} \times \vec{H} = \frac{1}{2}(\vec{E}e^{j\omega t} + (\vec{E}e^{j\omega t})^*) \times \frac{1}{2}(\vec{H}e^{j\omega t} + (\vec{H}e^{j\omega t})^*) = \frac{1}{2}\text{Re}(\vec{E} \times \vec{H}^*) + \frac{1}{2}\text{Re}(\vec{E} \times \vec{H}e^{j2\omega t}).$$

In the above equation, the first term is a time-varying function, and the second is a function of space. Hence, the second term is averaged to zero. So, the average power density or average Poynting vector is given as

$$\vec{P}_{\text{avg}} = \frac{1}{2}\text{Re}(\vec{E} \times \vec{H}^*).$$

Total time average power through a surface ds is

$$P_{\text{tot\_avg}} = \int \vec{P}_{\text{avg}} \cdot d\vec{S} = \frac{1}{2}\int \text{Re}(\vec{E} \times \vec{H}^*) \cdot d\vec{S} \tag{2.67}$$

2.13.2 Application of Poynting Theorem

- Rate of energy flow in a uniform plane wave travelling in free space can be easily obtained using Eq. (2.64).
- Power flow in some standard conductors such as coaxial cable can be found. Let V be the voltage between inner and outer conductors with negligible resistance; let I be the current flowing in inner and outer conductors. Now, if we apply Ampere's law, we get

$$\oint H \cdot d\vec{S} = I = 2\pi rH$$

because H is constant along any circular path. The magnetic field strength and electric field strength in such conductors will be of the form

$$H = \frac{I}{2\pi r} \text{ and } E = \frac{V}{r\log\left(\frac{b}{a}\right)}.$$

The Poynting vector is directed parallel to the axis of the cable; hence, (2.64) becomes

$$P = E \times H$$

The total power can be obtained by integrating the Poynting vector over the cross-sectional area. Then,

$$\text{power} = \int_s E \times H \cdot d\overline{S}$$

$$= \int_a^b \frac{V}{r\log\left(\frac{b}{a}\right)}\left(\frac{I}{2\pi r}\right) 2\pi r\, dr$$

$$= \text{VI} \tag{2.68}$$

2.14 POWER LOSS IN A PLANE CONDUCTOR

In this section, power flow per unit area through the surface of a conductor will be derived. Consider a large plate having a thickness greater than skin depth. Let its surface contain the tangential component of the magnetic field $H_{\tan}$. Then, the tangential component of the electric field will be $E_{\tan} = \eta H_{\tan}$, where η is the intrinsic impedance of metal.
The average power flow per unit area normal to the surface will be

$$P = \frac{1}{2} \text{Re}\,(E_{\tan} \times H_{\tan}^{*}) \tag{2.69}$$

However, for a good conductor, $E_{\tan}$ leads $H_{\tan}$ by 45°; hence,

$$P = \frac{1}{2}|E_{\tan}||H_{\tan}|\cos 45°$$

$$= \frac{1}{2\sqrt{2}}|\eta||H_{\tan}|^2 \tag{2.70}$$

$$= \frac{1}{2\sqrt{2}}\frac{|E_{\tan}|^2}{|\eta|} \text{ Watt/sq m} \tag{2.71}$$

For a thick conductor, the surface impedance Z_s is equal to η of the conductor. In addition, the linear current density is equal to H_{tan}. Hence, Eq. (2.69) becomes

$$P = \frac{1}{2\sqrt{2}}|z_s||J_s|^2 \text{ Watt/sq m} \tag{2.72}$$

$$= R_s\, J_{s(\text{eff})}^2 \text{ Watt/ sq m} \tag{2.73}$$

SUMMARY

1. Electromagnetic induction was discovered by both Faraday and Henry.

2. Faraday's law of induction states that in a conducting circuit when the magnetic field linking the circuit is altered, a voltage which is proportional to the time rate of change magnetic flux linking the circuit is induced in the circuit. For a circuit of N turns, the induced voltage $V = -N\frac{\partial \phi_B}{\partial t}$, where N is the number of turns and ϕ_B is the flux through each turn.

3. Lenz' law states that the magnetic field produced by an induced current will be such as to oppose the change in the magnetic field $\vec{B}$ which produced the induced current. Hence, the direction in which an induced current flows in a circuit is given by Lenz' law.

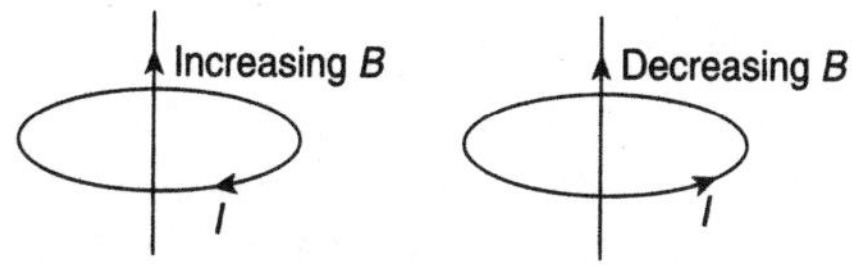

4. In a transformer, the emf induced in a stationary loop is caused by a time-varying magnetic field.

5. In the static case, the electric field (specified by $\vec{E}$ and $\vec{D}$) and the magnetic field (specified by $\vec{B}$ and $\vec{H}$) are described by separate and independent sets of equations.

6. In a conducting medium, both electrostatic and magnetostatic fields can exist, and are coupled through the Ohm's law ($\vec{J} = \sigma \vec{E}$). Such a field is called an electromagnetostatic field.

7. The relationship between the sources, charge, and current is the continuity equation representing the principle of conservation of charge and is given by

$$\nabla \cdot \vec{J} = -\frac{\partial \rho_V}{\partial t}$$

It is important, because it should be satisfied by all real currents and charges.

8. Maxwell's equations for static fields ($\partial/\partial t = 0$) are as follows: $\nabla \cdot \vec{D} = \rho$; $\nabla \cdot \vec{B} = 0$; $\nabla \times \vec{E} = 0$; and $\nabla \times \vec{H} = J$, where $\vec{D} = \varepsilon \vec{E}$ and $\vec{B} = \mu \vec{H}$ (ρ and J are the charge and current densities, respectively).

9. Maxwell's equations for time-varying fields are as follows:

$$\nabla \cdot \vec{D} = \rho;\ \nabla \cdot \vec{B} = 0;\ \nabla \times E = -\frac{\partial \vec{B}}{\partial t};\ \text{and}\ \nabla \times \vec{H} = J + \frac{\partial \vec{D}}{\partial t}$$

(a) The first two equations are derived from the Gaussian theorem, one for the electrical field and the other for the magnetic field.

(b) Faraday's law $\left(\nabla \times \vec{E} = -\frac{\partial \vec{B}}{\partial t}\right)$ shows that the time-varying magnetic field $\left(\frac{\partial \vec{B}}{\partial t}\right)$ is a source of the electric field ($\vec{E}$).

(c) Ampere's law $\left(\nabla \times \vec{H} = \mathbf{J} + \frac{\partial \vec{D}}{\partial t}\right)$ shows that both electric-current (J) and time-varying E field $\left(\frac{\partial \vec{D}}{\partial t}\right)$ are sources for the magnetic field ($\vec{H}$).

10. The expression $\left(\frac{\partial \vec{D}}{\partial t}\right)$ has the same dimension of current density. Therefore, it is termed *displacement current density.*

11. A moving electromagnetic field is called the EM wave which once created by charges and currents, continues to exist with no connection whatsoever to the charges and currents which generated it.

12. Quantities which vary sinusoidally in time are called *time harmonic*. To relate the instantaneous value of the vector fields to their complex spatial equivalents Time-harmonic representation is required. This is done by using the exponential function, $e^{j\omega t}$ as a basic function.

13. General boundary conditions at the interface:
    (a) $\vec{E}$ field tangential component is continuous across the boundary.
    (b) $\vec{H}$ field tangential component is discontinuous across the boundary at a point where a surface current exists. The amount of discontinuity and the surface current density are equal.
    (c) $\vec{D}$ field normal component is discontinuous across the boundary at a point where a surface current exists. The amount of discontinuity and the surface current density are equal.
    (d) $\vec{B}$ field normal component is continuous across the boundary.

14. Wave equations are second-order differential equations that are obtained by solving Maxwell's equations.

15. A uniform plane wave propagates in only one direction, and it has no field components in that direction.

16. The electric and magnetic fields are perpendicular to each other in a uniform plane wave.

17. The direction of the plane wave can be obtained from the cross product of the two fields.

18. The ratio of amplitudes of electric and magnetic fields in an EM wave is intrinsic impedance.

19. As an EM wave progresses in a conductor, the wave attenuates very rapidly and the phase shift per unit length increases.

20. In a conductor, the depth at which the wave becomes $\left(\frac{1}{e}\right)$ of its initial value is known as *skin depth* or *depth of penetration*.

21. The electric field leads the magnetic field by 45° at all frequencies in a good conductor.

22. The ratio of the magnitude of the conduction current density J to that of the displacement current density J_D in a lossy medium is known as *loss tangent*.

23. The polarization of a plane wave refers to its electric field orientation as a function of time at some fixed point in space.

24. Poynting theorem states that the net power flowing out of a given volume V is equal to the time rate of decrease in energy within V minus the conduction losses.

OBJECTIVE-TYPE QUESTIONS

1. Displacement current density is
    (a) Time rate of change of electric flux density
    (b) Time rate of change of magnetic flux density
    (c) Time rate of change of potential
    (d) Time rate of change of magnetic potential

2. Identify the modified form of ampere's law is

(a) $\nabla \times \bar{H} = \bar{J}_C + \bar{J}_D$
(b) $\nabla \times \bar{H} = \bar{I}_C + \bar{J}_D$
(c) $\nabla \times \bar{E} = \bar{J}_C + \bar{J}_D$
(d) $\nabla \times \bar{E} = \bar{I}_C + \bar{I}_D$

3. The electrostatic field $\nabla \times \bar{E} =$

(a) zero
(b) $-\dfrac{\partial \bar{B}}{\partial t}$
(c) both a and
(d) none

4. Let $\bar{E} = 20\cos(\omega t - 50x)\hat{a}_y \; v/m$ For this field calculate $\bar{J}_D$

(a) $-20\,\varepsilon_0 \cos(\omega t - 50x)\hat{a}_y \; A/m^2$
(b) $-20\,\omega\,\varepsilon_0 \sin(\omega t - 50x)\hat{a}_y \; A/m^2$
(c) $-20\omega\,\varepsilon_0 \sin(\omega t - 50x)\hat{a}_x \; A/m^2$
(d) $-20\,\hat{a}_x \; A/m^2$

5. Which of the following is not true for a dielectric to dielectric interface

(a) $B_{n1} = B_{n2}$
(b) $E_{t1} = E_{t2}$
(c) $D_{n1} = D_{n2}$
(d) $H_{t1} - H_{t2} = Js$

6. Dielectric constant and dielectric strength are
(b) Same
(c) different
(d) one is a number and the other indicates when the breakdown of the dielectric occurs when a p.d. is applied
(e) one is dimensionless and the other is KV/mm

7. A time varying field is applied to a circuit with a capacitor.
(a) The current in the circuit consists of conduction current and displacement current.
(b) The displacement current passes through the capacitor.
(c) The conduction current density is J and displacement current density $\dfrac{\partial D}{\partial t}$
(d) All the above statements are correct.

8. The properties of a medium are
(a) permittivity, permeability, insulation
(b) permittivity, permeability, conductivity
(c) permeability, resistivity, inductivity
(d) permeability, flux, magnetism

9. For static electric and magnetic fields in an inhomogeneous source-free medium, which of the following represents the correct form of two Maxwell's equations?

(a) $\nabla \cdot \vec{E} = 0 \quad \nabla \times \vec{B} = 0$
(b) $\nabla \cdot \vec{E} = 0 \quad \nabla \cdot \vec{B} = 0$
(c) $\nabla \times \vec{E} = 0 \quad \nabla \times \vec{B} = 0$
(d) $\nabla \times \vec{E} = 0 \quad \nabla \cdot \vec{B} = 0$

10. The Maxwell equation is $\nabla \times \vec{H} = \vec{J} + \dfrac{\partial \vec{D}}{\partial t}$ based on
(a) Ampere's law
(b) Gauss' law
(c) Faraday's law
(d) Coulomb's law

11. In uniform plane $\vec{E}$ and $\vec{H}$ wave are
    (a) Orthogonal
    (b) having same direction
    (c) Neither orthogonal nor having same direction
    (d) not possible to defined

12. Value of the intrinsic impedance in free space
    (a) 377 Ω (b) 277 Ω
    (c) 177 Ω (d) 477 Ω

13. In a good conductor
    (a) *E* leads *H* by 45° (b) *E* leads *H* by 90°
    (c) *H* leads *E* by 45° (d) *H* leads *E* by 90°

14. In uniform plane $\vec{E} \times \vec{H}$ wave gives
    (a) Direction in which wave travels (b) direction in which wave $\bar{E}$ field exists
    (c) Direction in which wave $\bar{H}$ field exists (d) none

15. In a circularly polarized uniform wave, travelling in *x* direction, the phase difference between E_z and E_y is
    (a) 30° (b) 45°
    (c) 90° (d) 180°

16. If the frequency of incident wave increases by a factor 9 then the depth to which an EM wave propagates in a conducting material
    (a) increases by a factor of 2 (b) decreases by a factor of 2
    (b) increases by a factor of 3 (d) decreases by a factor of 3

17. The Brewster angle for an perpendicularly polarized EM wave incident obliquely on a dielectric, is
    (a) $\tan^{-1}\sqrt{\frac{\varepsilon_2}{\varepsilon_1}}$ (b) $\sin^{-1}\sqrt{\frac{\varepsilon_2}{\varepsilon_1}}$
    (c) $\cos^{-1}\sqrt{\frac{\varepsilon_2}{\varepsilon_1}}$ (d) does not exist

18. The Brewster angle for an parallely polarized EM wave incident obliquely on a dielectric, is
    (a) $\tan^{-1}\sqrt{\frac{\varepsilon_2}{\varepsilon_1}}$ (b) $\sin^{-1}\sqrt{\frac{\varepsilon_2}{\varepsilon_1}}$
    (c) $\cos^{-1}\sqrt{\frac{\varepsilon_2}{\varepsilon_1}}$ (d) does not exist

19. $\frac{\sin\theta_1}{\sin\theta_2} = \sqrt{\frac{\varepsilon_2}{\varepsilon_1}}$ is known as
    (a) Snell's law (b) law of reflection
    (c) Law of refraction (d) Both A and C

20. Following is the condition for total internal reflection

(a) the angle of incidence is very high
(b) medium 1 is denser than medium 2
(c) $\sin\theta_i > \sqrt{\frac{\varepsilon_2}{\varepsilon_1}}$
(d) All of the above

21. Following has its units as Watt/m^2

(a) Poynting vector
(b) Reflection coefficient
(c) Transmission coefficient
(d) None

22. Following is true for an EM wave incident on a dielectric normally

(a) Wave is totally reflected back
(b) Results in a standing wave distribution
(c) Part of the wave is reflected and part of it is transmitted
(d) None of the above

ANSWERS TO OBJECTIVE-TYPE QUESTIONS

1. (a)	2. (a)	3. (a)	4. (b)	5. (d)	6. (a)	7. (d)	8. (b)	9. (b)	10. (a)	11. (a)
12. (a)	13. (a)	14. (a)	15. (b)	16. (d)	17. (d)	18. (a)	19. (d)	20. (d)	21. (a)	22. (c)

REVIEW QUESTIONS

1. State Faraday's law of induction.

2. State Lenz' law.

3. Give the equation of transformer emf.

4. What are displacement current and displacement current density? What is the emf produced by a moving loop in a time-varying field?

5. What is a time-harmonic field? Give time-harmonic Maxwell's equations in point form. Assume time factor: $e^{-j\omega t}$.

6. Write Maxwell's equation in point and integral form for good conductors.

7. What is the significance of displacement current density?

8. With a necessary explanation, derive the Maxwell's equation in differential and integral forms.

9. (a) Write short notes on Faraday's law of electromagnetic induction.
(b) The magnetic field intensity in free space is given as $\vec{H} = H_0 \sin\theta\, \hat{a}_y\, t$ A/m, where $\theta = \omega t - \beta z$ and is a constant quantity. Determine the displacement current density.

10. Write down Maxwell's equations in their simplest form in terms of $\vec{E}$, $\vec{D}$, $\vec{B}$, and $\vec{H}$. State the physical origin of each equation. By using the relationships between $\vec{E}$, $\vec{D}$ and $\vec{B}$, $\vec{H}$write Maxwell's equations in terms of $\vec{E}$ and $\vec{B}$.

11. Mention Maxwell's equations for free space where there exists no dielectrics or magnetic materials and no free charge or conduction currents.

12. Write down the integral forms of Maxwell's equations. For each equation, write the name of the law that it describes, and explain in words (perhaps with the aid of illustrations) your interpretation and understanding of each law.

13. Derive the differential forms from the integral forms using the Divergence theorem and Stokes' theorem.

14. Show that the differential form of Ampere's circuital law is not consistent with the continuity (of charge) equation.

15. What modification did Maxwell make to Ampere's law? Identify the term known as the *displacement current density*.

16. Show that the modified version of Maxwell's law satisfies the continuity equation.

17. What is meant by a uniform plane wave?

18. Derive the equation for attenuation constant and phase shift constant in a lossy medium. (or) Define the intrinsic impedance of free space and phase constant β.

19. What is skin depth; derive an expression for it.

20. Show that in a uniform plane wave E and H are perpendicular and the ratio of their magnitudes is constant.

21. Derive wave equations for sinusoidal time variations.

22. What is meant by the *polarization of the wave*? When is the wave linearly polarized and when is it circularly polarized?

Review of Transmission Lines

3

3.1 INTRODUCTION

A transmission line consists of two or more conductors embedded in a system of dielectric media that are used to convey the energy or information from one point to other, specifically from source end to the load end at frequencies starting from 0 Hz to 300 MHz. Cables used for TV, computer and telephone, printed circuit boards, antenna feed lines, micro strip and power lines are few applications of common transmission-line.

In a basic circuit, if the length of the circuit/element is negligibly small, then the time delay effects are negligible. Such elements are called *lumped elements*. If the length of the circuits is large enough, then they may be called *distributed elements;* this means that the behavior of the elements is distributed throughout the line and can be evaluated on a per-unit-distance basis. Transmission lines come under distributed phenomena. Field theory is required for an exact analysis of this two-conductor transmission line.

In this chapter, we consider the TEM waves that is present in the region which are bounded by two or more conductors. The wave phenomenon along the line, that is, how the wave reflects and standing waves exist in transmission lines, is discussed here. The use of the Smith chart is also discussed here.

3.1.1 Definition of Transmission Lines

Transmission lines may be defined as the metallic conductors which are used to guide energy from one point to another, that is, from source to load.

Using Maxwell's equations energy can be transmitted in the form of unguided waves (plane wave) in space and guided waves on a transmission line.

Transmission line having cross geometry that is invariable in the direction of propagation along the line is called *uniform transmission line.*

3.1.2 Types of Transmission Lines

The various types of transmission lines are

(i) Coaxial line
(ii) Two-wire parallel transmission line
(iii) Parallel-plate transmission line
(iv) Strip lines
(v) Microstrip lines

Two concentric cylindrical conductors when parted by a dielectric material like air or polyethylene forms a coaxial cable. Figure 3.1 (a) shows the geometry of the coaxial line with a center conductor of radius "*a*" and an outer conductor of radius "*c*" and shows the electric and magnetic field patterns of a wave. Coaxial cables are used as antenna feed lines and as input cables to high-frequency precision measurement equipment such as oscilloscopes, spectrum analyzers, and network analyzers.

Figure 3.1 (b) shows the cross section as well as the field patterns of a parallel-plate transmission line of length '*a*' whose conducting plates are separated at a distance "*d*".

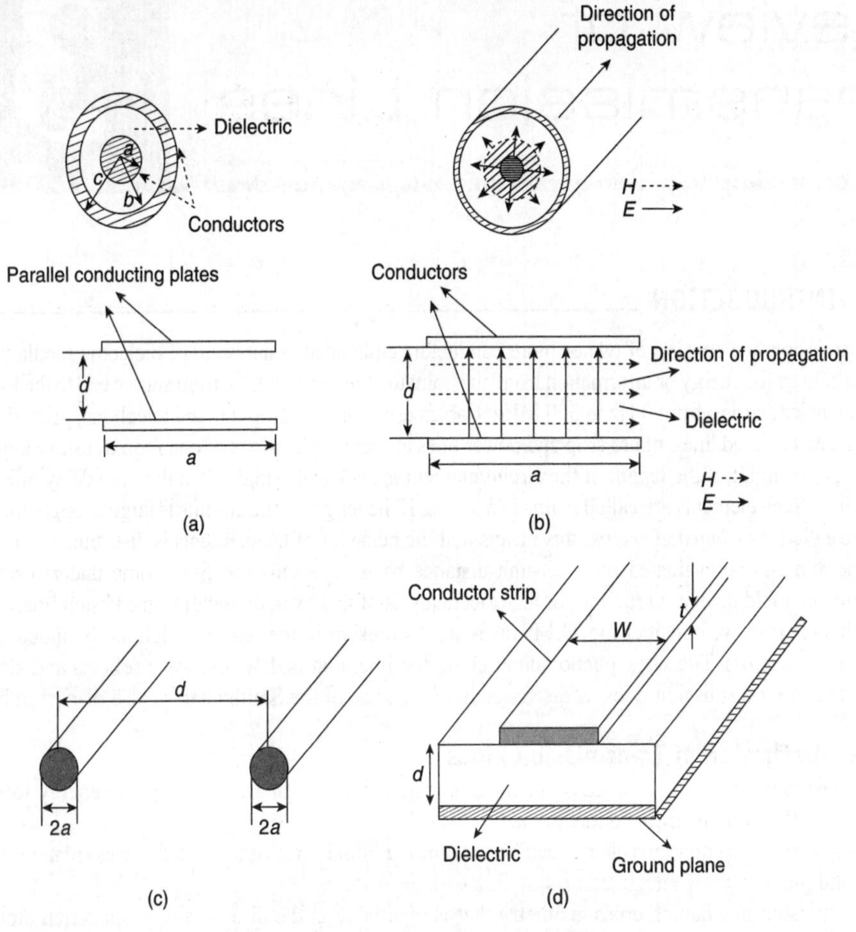

Figure 3.1 *(a) Coaxial line and its field patterns; (b) Parallel plate line; (c) Two-wire parallel line; (d) Microstrip line and its field patterns*

Two-wire line is another type of transmission-line which is widely used. It is shown in Figure 3.1 (c). The two wire parallel lines are separated by a distance of "*d*", in which each wire has a radius of "*a*". In this type, the signal current flows down one wire and returns to the source on the other. The two-wire line is an open transmission-line structure. Hence, it is more prone to electromagnetic interference which can be minimized by using wires that are shielded and twisted pair wires. Power and telephone lines are the most common example of two-wire lines.

A microstrip line consists of a single ground plane and an open strip conductor with a thickness of "*t*" and a width of "*w*" separated by a dielectric substrate, as shown in Figure 3.1 (d). These types of lines are used in printed circuit boards to connect components.

3.2 LUMPED VERSUS DISTRIBUTED ELEMENT CIRCUITS

Lumped

At low frequencies, the length (l) of the circuit is small compared with the wavelength(λ), that is, ($l << \lambda$) because of which there is propagation delay $t_d \approx 0$; so, the V and I waves will affect the entire circuit at the same time. The equivqlent circuit of a two conductor transmission line at low frequencies is show n in Figure 3.2 (a). With increase in frequency, the lumped elements become lossy which leads to increase in parasitic reactance and radiation loss. For example, due to the parasitic inductances of the connecting wires a capacitor turns into a resonant circuit.

For analysing lumped circuits we can apply KVL and KCL.

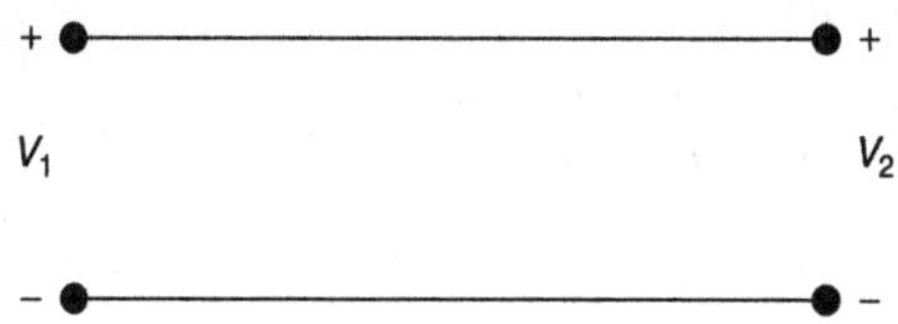

Figure 3.2 (a) *Equivalent circuit of a two conductor transmission line at low frequencies*

Distributed

Distributed components contains sections of transmission lines or waveguides (Figure 3.2 (b)). The phase differences between the parts of a component are important and their size is equivalent to a wavelength. Due to this, the propagation delay (t_d) cannot be neglected, and the V and I waves do not affect the entire circuit at the same time.

In case of disturbed elements Maxwell's equations can be applied.

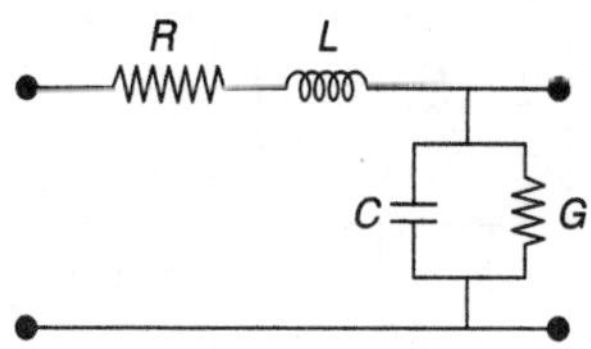

Figure. 3.2 (b) *Equivalent circuit of a two conductor transmission line at high frequencies*

3.2.1 Transmission-Line Parameters

Consider a two-wire transmission line as shown in Figure 3.3:

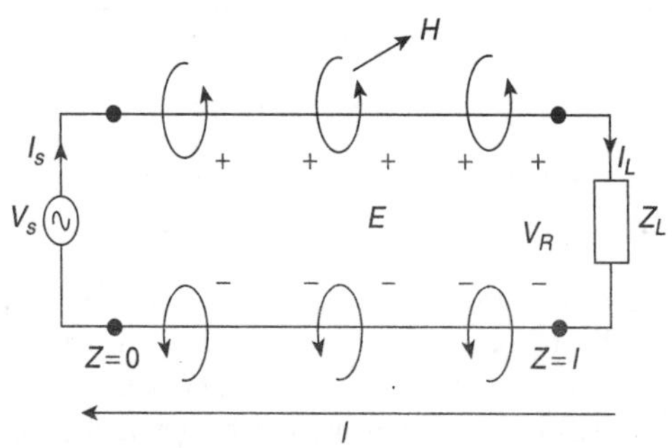

Figure 3.3 *Transmission line with a load*

When a current is passed through the line due to applied voltage across the two conducting wires, then a voltage drop exists within a conductor. This indicates that the line has some series resistance R. In addition, a magnetic field exists around the conductor, due to the flow of current, which indicates that the line has the series inductance L.

Due to the applied voltage across the two conducting wires, the charges are built on them, which indicates that the line has the shunt capacitance C. The capacitance can never be an ideal one, and it has some leakage conductance G. These four parameters (resistance, inductance, capacitance, and conductance) are the primary constants, which are distributed throughout the transmission line as shown in the Figure 3.2(b). If they are distributed uniformly, then the line is said to be a *uniform transmission line.*

Therefore, any transmission line may be described in terms of the following distributed circuit parameters, which are also called *line parameters*:

(i) The inductance per unit length L (H/m)
(ii) The resistance per unit length R (Ω / m)
(iii) The capacitance per unit length C (F/m)
(iv) The conductance per unit length G (1/ Ω m)

Table 3.1. Distributed line parameters for different transmission lines

Parameters	Coaxial line	Two wire line	Parallel plate line	Unit
R	$\frac{1}{2}\sqrt{\frac{f\mu_c}{\pi\sigma_c}}\left(\frac{1}{a}+\frac{1}{b}\right)$	$\frac{1}{a}\sqrt{\frac{f\mu_c}{\sigma_c\,\pi}}$	$\frac{2}{a}\sqrt{\frac{\pi f\mu_c}{\sigma_c}}$	Ω/m
L	$\frac{\mu}{2\pi}\ln(b/a)$	$\frac{\mu}{\pi}\cosh^{-1}\left(\frac{d}{2a}\right)$	$\frac{\mu d}{a}$	$\mathrm{H/m}$
C	$\frac{2\pi\varepsilon_d}{\ln(b/a)}$	$\frac{\pi\varepsilon_d}{\cosh^{-1}\left(\frac{d}{2a}\right)}$	$\frac{\varepsilon_d a}{d}$	$\mathrm{F/m}$
G	$\frac{2\pi\sigma_d}{\ln(b/a)}$	$\frac{\pi\sigma_d}{\cosh^{-1}\left(\frac{d}{2a}\right)}$	$\frac{\sigma_d a}{d}$	$\frac{1}{\Omega\mathrm{m}}$

For each line, the conductors are specified by σ_c, μ_c and ε_d and the dielectric media separating the conductors are characterized by σ_d, μ_d and ε_d. The formulae for calculating R, L, C, and G for the coaxial line, two-wire parallel line, and parallelplate line, are tabulated in Table 3.1. From this table, it can be seen that R, L, C, and G depend on the geometry of the transmission line, the characteristics of the conductor and the dielectric, and the frequency of operation.

3.3 TRANSMISSION LINE EQUATIONS

The wave equations for voltage and current at any point on uniform line of transmission lines are second order differential equations. They are used to explain the voltage and current components on the line.

$$\frac{d^2V}{dz^2} = \gamma^2 V$$

$$\frac{d^2I}{dz^2} = \gamma^2 I$$

where propagation constant, $\gamma = \alpha + j\beta$

The general solution of the wave equations can be expressed in a form of exponential functions.

$$V = V^+ e^{-\gamma z} + V^- e^{\gamma z}$$

$$I = I^+ e^{-\gamma z} + I^- e^{\gamma z}$$

where $V^+ e^{-\alpha z}$ represents the amplitude of the wave travelling in positive Z direction at location Z on the line, βz is the phase of the wave as a function of Z, and β represents the phase change per unit length of the transmission line for a travelling wave. The constants V^+, V^-, I^+, and I^- are the forward voltage, reverse voltage, forward current and reverse current respectively.

The above equations are said to be transmission-line equations.

Proof of Transmission Line-Equations

The behavior of a line when a voltage source is connected is described by a set of differential equations known as *transmission line equations*. Let us examine an incremental portion of length Δz of a two-conductor transmission line. The infinitesimal section of the line compromises of resistance ($R\Delta z$), inductance ($L\Delta z$), conductance ($G\Delta z$) and capacitance ($C\Delta z$). Consider sinusoidal voltage with a frequency ω is applied between AA' and a current I flows into terminal A (Figure 3.4).

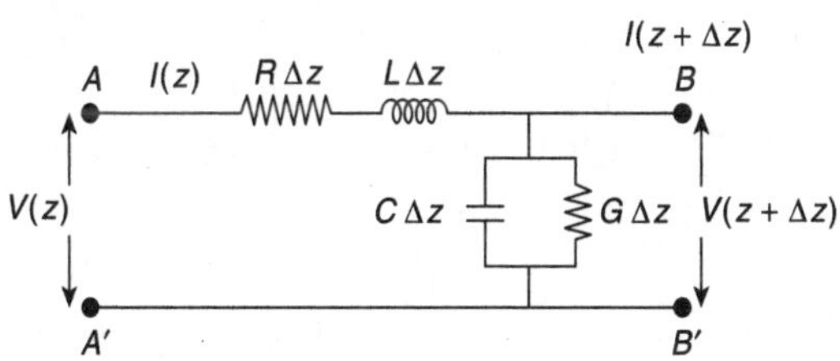

Figure 3.4 *Voltage and current on infinitesimal section of transmission line*

Now, the voltage at BB' will not be equal to that at AA' as there is a voltage drop in the series elements, R and L. Similarly, the output current at the point B will not be same as that at A as some part of the input current will be passed through the shunt elements, C and G. Let $I(z + \Delta z)$ and $V(z + \Delta z)$ represent the current and the voltage at BB' respectively.

By applying Kirchhoff's voltage law to the outer loop of the circuit in Figure 3.4, we obtain

$$V(z) = R\Delta z\, I(z) + L\Delta z \frac{\partial I(z)}{\partial t} + V(z + \Delta z)$$

$$V(z) - R\Delta z\, I(z) - L\Delta z \frac{\partial I(z)}{\partial t} - V(z+\Delta z) = 0$$

Dividing the above equation with Δz, we get

$$-\left[\frac{V(z+\Delta z) - V(z)}{\Delta z}\right] = R\, I(z) + L\frac{\partial I(z)}{\partial t}$$

Taking the limit of the above equation as $\Delta z \to 0$ leads to

$$-\underset{\Delta z \to 0}{Lt} \frac{\Delta V(z)}{\Delta z} = R\, I(z) + L\frac{\partial I(z)}{\partial t}$$

Here, $\Delta V(z)$ represents the voltage difference between input and output.
We applied the limit in order to make sure that Δz is small

$$-\frac{\partial V(z)}{\partial z} = R\, I(z) + L\frac{\partial I(z)}{\partial t} \tag{3.1}$$

Similarly, applying Kirchhoff's current law to the main node of the circuit in Figure 3.4 gives

$$I(z) = I(z+\Delta z) + G\Delta z\, V(z+\Delta z) + C\Delta z \frac{\partial V(z+\Delta z)}{\partial t}$$

Dividing the above equation with Δz, we get

$$-\frac{I(z+\Delta z) - I(z)}{\Delta z} = G\, V(z+\Delta z) + C\frac{\partial V(z+\Delta z)}{\partial t}$$

Taking the limit of the above equation as $\Delta z \to 0$ leads to

$$-\underset{\Delta z \to 0}{Lt} \frac{\Delta I(z)}{\Delta z} = G\, V(z+\Delta z) + C\frac{\partial V(z+\Delta z)}{\partial t}$$

$$-\frac{\partial I(z)}{\partial z} = G\, V(z) + C\frac{\partial V(z)}{\partial t} \tag{3.2}$$

If a differential length dz is considered at this point, then the series impedance of this section will be $(R + j\,\omega\, L)\, dz$ and the shunt impedance will be $(G + j\,\omega\, C)\, dz$. Let the voltage and current at the transmitting end of the differential length be V and I, respectively, and those at the receiving end be $V + dV$ and $I + dI$, respectively; then,

$$V - (V + dV) = I(R + j\omega L)dz$$

$$I - (I + dI) = V(G + j\omega C)dz$$

Therefore,

$$\frac{dV}{dz} = -(R + j\omega L)I \tag{3.3}$$

$$\frac{dI}{dz} = -(G + j\omega C)V \tag{3.4}$$

We find that Eqs. (3.3) and (3.4) are the voltage and current, respectively, on a transmission line governed by two first-order differential equations. These two equations are called *telegrapher's equations*. Differentiating Eq. (3.3) with regard to z, we get

$$\frac{d^2V}{dz^2} = -(R + j\omega L)\frac{dI}{dz} \tag{3.5}$$

Now, substituting Eq. (3.4) in (3.5), we get

$$\frac{d^2V}{dz^2} = (R + j\omega L)(G + j\omega C)V \tag{3.6}$$

Similarly, differentiating Eq. (3.4) with regard to z, and substituting for dV/dz from Eq. (3.3), we get

$$\frac{d^2I}{dz^2} = (R + j\omega L)(G + j\omega C)I \tag{3.7}$$

Now, let us define the propagation constant γ of the transmission line as

$$\gamma^2 = (R + j\omega L)(G + j\omega C) \tag{3.8}$$

Substituting Eq. (3.8) in Eqs. (3.6) and (3.7), we get

$$\frac{d^2V}{dz^2} = \gamma^2 V \tag{3.9}$$

$$\frac{d^2I}{dz^2} = \gamma^2 I \tag{3.10}$$

It is interesting to note that both voltage and current are governed by the same differential equation. These two equations are wave equations of the transmission line. Since, for a given operating frequency γ is constant, Eqs. (3.9) and (3.10) are homogenous equations with constant coefficients, and their solutions can be written as

$$V = V^+e^{-\gamma z} + V^-e^{\gamma z} \tag{3.11}$$

$$I = I^+e^{-\gamma z} + I^-e^{\gamma z} \tag{3.12}$$

where the constants V^+, V^-, I^+, and I^- are the forward voltage, reverse voltage, forward current and reverse current respectively.

3.4 PRIMARY AND SECONDARY CONSTANTS

For a transmission line, R, L, C, and G are the primary constants. The secondary constants are as follows:

(i) Characteristic impedance (Z_0)
(ii) Propagation constant (γ)

3.5 CHARACTERISTIC IMPEDENCE (Z_0)

The ratio of the voltage and the current at any point "z" on an infinitely long line is called the *characteristic impedance*, which is independent of "z"; that is, the direction in which the wave is travelling. The *characteristic impedance* is also defined as the ratio of the forward voltage wave to the forward current wave at any point on the line.

$$Z_0 = \frac{V^+}{I^+}$$

The *characteristic impedance* is also defined as the negative of the ratio of reverse voltage wave to the reverse current wave at any point on the line.

$$Z_0 = -\frac{V^-}{I^-}$$

The derivation of the Z_0 is given in the above section.
Characteristic impedance of two basic types of transmission lines is given in Table 3.2

Table 3.2 Characteristic impedances of various transmission lines

Type of the line	Characteristic impedance
Coaxial (filled with dielectric medium of relative permittivity ε_r)	$\frac{138}{\sqrt{\varepsilon_r}}\log\frac{b}{a}$
Coaxial (filled with air) $\varepsilon_r = 1$	$138\log\frac{b}{a}$
Two wire (in the medium of relative permittivity ε_r)	$\frac{276}{\sqrt{\varepsilon_r}}\log\frac{d}{a}$
Two wire (in air) $\varepsilon_r = 1$	$276\log\frac{d}{a}$

Propagation Constant (γ)

The natural logarithm of the ratio of the sending end voltage or current to the receiving end voltage or current of the line is known as the *propagation constant.*

Propagation constant details the variation of current and voltage in the line as a function of distance. It also gives the way in which the wave is propagated.

$$\gamma = \ln\left(\frac{V_S}{V_L}\right) = \ln\left(\frac{I_S}{I_L}\right)$$

In general, the propagation constant is complex, and is given by

$$\gamma = \sqrt{ZY}$$

where Z and Y are series impedance and shunt admittance, respectively

$$\gamma = \sqrt{(R + j\omega L)(G + j\omega C)}$$
$$= \alpha + j\beta \tag{3.13}$$

where α is the attenuation constant in Nepers/m and the imaginary part β is called the *phase constant* in rad/m.

$$\alpha = \text{Re}\sqrt{ZY} = \sqrt{\frac{1}{2}\left[(RG - \omega^2 LC) + \sqrt{(R^2 + \omega^2 L^2)(G^2 + \omega^2 C^2)}\right]} \tag{3.14}$$

$$\beta = \text{Im g}\sqrt{ZY} = \sqrt{\frac{1}{2}\left[(\omega^2 LC - RG) + \sqrt{(R^2 + \omega^2 L^2)(G^2 + \omega^2 C^2)}\right]} \tag{3.15}$$

Derivation of Attenuation Constant and Phase Constant

From Eq. (3.13), we know

$$\alpha + j\beta = \gamma = \sqrt{(R + j\omega L)(G + j\omega C)}$$
$$\Rightarrow (\alpha + j\beta)^2 = \alpha^2 - \beta^2 + 2j\alpha\beta = (R + j\omega L)(G + j\omega C)$$
$$= RG + j\omega(LG + RC) - \omega^2 LC$$

By equating real parts of the equations, we get

$$\alpha^2 - \beta^2 = RG - \omega^2 LC \tag{3.16}$$

In addition, from Eq. (3.13),

$$|\gamma| = \sqrt{(\alpha^2 + \beta^2)} = \sqrt{\sqrt{(R^2 + \omega^2 L^2)(G^2 + \omega^2 C^2)}}$$
$$\Rightarrow (\alpha^2 + \beta^2) = \sqrt{(R^2 + \omega^2 L^2)(G^2 + \omega^2 C^2)} \tag{3.17}$$

By adding Eqs. (3.16) and (3.17), we get

$$2\alpha^2 = \left[(RG - \omega^2 LC) + \sqrt{(R^2 + \omega^2 L^2)(G^2 + \omega^2 C^2)}\right]$$

$$\therefore \alpha = \text{Re}\sqrt{ZY} = \sqrt{\frac{1}{2}\left[(RG - \omega^2 LC) + \sqrt{(R^2 + \omega^2 L^2)(G^2 + \omega^2 C^2)}\right]}$$

By subtracting Eqs. (3.16) and (3.17), we get

$$2\beta^2 = \left[-(RG - \omega^2 LC) + \sqrt{(R^2 + \omega^2 L^2)(G^2 + \omega^2 C^2)}\right]$$

$$\therefore \beta = \text{Im g}\sqrt{ZY} = \sqrt{\left[\frac{1}{2}(\omega^2 LC - RG) + \sqrt{(R^2 + \omega^2 L^2)(G^2 + \omega^2 C^2)}\right]}$$

Voltage and Current at any Given Point on the Transmission Line in Terms of Phase and Attenuation Constants

The first term in Eqs. (3.11) and (3.12) indicates the incident wave that progresses from the transmitting end to the receiving end in $+z$ direction. In contrast, the second term indicates the reflected wave that progresses from the receiving end to the transmitting end in $-z$ direction.

The instantaneous values of the voltage and current are calculated by multiplying $e^{j\omega t}$ with Eqs. (3.11) and (3.12):

$$V = V^+ e^{j\omega t - \gamma z} + V^- e^{j\omega t + \gamma z} \quad (3.18)$$

$$I = I^+ e^{j\omega t - \gamma z} + I^- e^{j\omega t + \gamma z} \quad (3.19)$$

Substituting $\gamma = \alpha + j\beta$ in Eqs. (3.18) and (3.19), we get

$$V = V^+ e^{-\alpha z} e^{j\omega t - j\beta z} + V^- e^{\alpha z} e^{j\omega t + j\beta z} \quad (3.20)$$

$$I = I^+ e^{-\alpha z} e^{j\omega t - j\beta z} + I^- e^{\alpha z} e^{j\omega t + j\beta z} \quad (3.21)$$

Physical Significance of Propagation Constant (α and β)

The complex propagation constant γ describes the wave propagation characteristics of the medium. It is a complex quantity with both real and imaginary parts, that is, $\gamma = \alpha + j\beta$. The term α represents the attenuation of the wave on the transmission line, and it is called the *attenuation constant* of the line. The unit of α is Neper per meter. The term β represents the phase change per unit length of the transmission line for a travelling wave, and it is called the *phase constant* of the line. The unit of the β is radians per meter.

The attenuation constant of a line is defined as the natural logarithm of the magnitude of the ratio of the sending end current to the receiving end current. The phase constant is defined as the phase difference between the sending end current to the receiving end current. Sometimes, it is useful to have the expressions for attenuation and phase constants in terms of the primary constants of the line.

3.5.1 Voltage and Current at any Given Point on The Transmission Line in Terms of Characteristic Impedance

The differential Eqs. (3.3) and (3.4) at every point on the transmission line should be satisfied by the voltage and current equations given by Eqs (3.11) and (3.1) to minimize the number of arbitrary constants.

Derivation of Characteristic Impedance (Z_0)

Substituting Eqs. (3.11) and (3.12) in Eq. (3.3), we get

$$\frac{d}{dz}\left[V^+ e^{-\gamma z} + V^- e^{\gamma z}\right] = -(R + j\omega L)\left[I^+ e^{-\gamma z} + I^- e^{\gamma z}\right]$$

$$-\gamma V^+ e^{-\gamma z} + \gamma V^- e^{\gamma z} = -(R + j\omega L)\left[I^+ e^{-\gamma z} + I^- e^{\gamma z}\right] \quad (3.22)$$

The coefficients of $e^{-\gamma z}$ and $e^{\gamma z}$ on the two sides of the equality sign should be separately equated, and we get

$$\text{Coefficients of } e^{-\gamma z}: -\gamma V^+ = -(R + j\omega L) I^+ \quad (3.23)$$

$$\text{Coefficients of } e^{\gamma z}: \gamma V^- = -(R + j\omega L) I^- \quad (3.24)$$

We know that from Eq. (3.8), $\gamma = \sqrt{(R + j\omega L)(G + j\omega C)}$, from Eqs. (3.22) and (3.23), we get

$$\frac{V^+}{I^+} = \frac{R + j\omega L}{\gamma} = \sqrt{\frac{R + j\omega L}{G + j\omega C}} \quad (3.25)$$

$$\frac{V^-}{I^-} = -\frac{R + j\omega L}{\gamma} = -\sqrt{\frac{R + j\omega L}{G + j\omega C}} \tag{3.26}$$

It is, therefore, called the *characteristic equation* of the transmission line and is usually denoted by Z_0.

$$Z_0 = \sqrt{\frac{R + j\omega L}{G + j\omega C}} \tag{3.27}$$

Substituting the values of I^+ and I^- from Eqs. 3.25 and 3.26 in Eqs. (3.11) and (3.12), we get

$$V = V^+ e^{-\gamma z} + V^- e^{\gamma z} \tag{3.28}$$

$$I = \frac{V^+}{Z_0} e^{-\gamma z} - \frac{V^-}{Z_0} e^{\gamma z} \tag{3.29}$$

3.5.2 Voltage and Current at any Given Point on the Transmission Line in Terms of Reflection Coefficent

Reflection coefficient gives the relative amplitudes of the two waves at any point on the line.

$$\Gamma_L = \frac{Z_L - Z_0}{Z_L + Z_0} \tag{3.30}$$

Γ_L denotes the reflection coefficient at the load end and $\Gamma_L = \frac{V^-}{V^+}$ (measured from load end).

Now, the voltage and current equations at any location on the transmission line can be written as

$$V = V^+ e^{-\gamma z}[1 + \Gamma_L e^{-2\gamma z}] \tag{3.31}$$

$$I = \frac{V^+}{Z_0} e^{-\gamma z}[1 - \Gamma_L e^{-2\gamma z}]$$

Phase Velocity and Group Velocity

Energy is propagated along a transmission line in the form of a TEM wave. Along with the primary and secondary constants, transmission lines have phase velocity and group velocity. Here, we discuss the phase velocity and group velocity.

Phase Velocity

The velocity by which a plane of a constant phase is propagated in a transmission line is called the *phase velocity*, or the rate at which the wave changes its phase in terms of guide wavelength.

$$v_p = \frac{\lambda_g}{\text{unit time}} = \lambda_g \cdot f$$

$$v_p = \frac{2\pi f \cdot \lambda_g}{2\pi} = \frac{2\pi f}{\frac{2\pi}{\lambda_g}}$$

i.e.,

$$v_p = \frac{\omega}{\beta}$$

where,

$$\omega = 2\pi f,\ \beta = \frac{2\pi}{\lambda_g}.$$

In general, it can be greater, equal to, or less than the velocity of light in free space. The phase velocity is the virtual velocity and does not determine the velocity of transmission of energy or signal pulse.

Transit time or time delay is the time elapsed for the wave to travel from one end to another.

$$t_d = \frac{1}{v_p} \tag{3.32}$$

Group Velocity

In lossless systems, the group velocity is the velocity at which a pulse travels or at which energy travels, and it is less than the velocity of light in free space. If β is not proportional to ω and if the wave components travel with different velocities, the envelope of the wave travels with a velocity, known as the *group velocity* v_g.

$$v_g = \frac{d\omega}{d\beta} \tag{3.33}$$

Energy and information propagate with the group velocity.

3.5.3 Input Impedence of a Uniform Transmission Line

Input impedance is defined as the ratio of the voltage and current at the sending end of the line.

$$Z_{\text{in}} = \frac{V}{I}$$

and the impedance at any point on the line is

$$Z_{\text{in}} = Z_0 \frac{[1+\Gamma_L e^{-2\gamma z}]}{[1-\Gamma_L e^{-2\gamma z}]} \tag{3.34}$$

Substituting Eq. (3.30) into Eq. (3.34) and taking $e^{-\gamma z}$ common from numerator and

$$Z_{\text{in}} = Z_0 \frac{[e^{\gamma z} + \frac{Z_L - Z_0}{Z_L + Z_0} e^{-\gamma z}]}{[e^{\gamma z} - \frac{Z_L - Z_0}{Z_L + Z_0} e^{-\gamma z}]}$$

$$Z_{\text{in}} = Z_0 \frac{[(Z_L + Z_0)e^{\gamma z} + (Z_L - Z_0)e^{-\gamma z}]}{[(Z_L + Z_0)e^{\gamma z} - (Z_L - Z_0)e^{-\gamma z}]}$$

Rearranging terms of Z_L and Z_0, we get

$$Z_{\text{in}} = Z_0 \frac{[Z_L(e^{\gamma z} + e^{-\gamma z}) + Z_0(e^{\gamma z} - e^{-\gamma z})]}{[Z_L(e^{\gamma z} - e^{-\gamma z}) + Z_0(e^{\gamma z} + e^{-\gamma z})]}$$

Since $(e^x + e^{-x})/2 = \cosh x$ and $(e^x - e^{-x})/2 = \sinh x$, the above equation can be written as

$$Z_{\text{in}} = Z_0 \frac{[Z_L \cosh \gamma z + Z_0 \sinh \gamma z]}{[Z_L \sinh \gamma z + Z_0 \cosh \gamma z]}$$

The input impedance can finally be written as

$$Z_{\text{in}} = Z_0 \frac{[Z_L + Z_0 \tanh \gamma z]}{[Z_0 + Z_L \tanh \gamma z]} \tag{3.35}$$

Note 1: If a transmission line is terminated by characteristic impedance, then the impedance at any point on the transmission line looking toward the load is similar to the characteristic impedance.

$$Z_L = Z_0 \Rightarrow Z_{\text{in}} = Z_0$$

Note 2: When the transmission line is not terminated by characteristic impedance, then the impedance looking toward the load changes from point to point.

EXAMPLE PROBLEM 3.1

A 100 m long lossless transmission line has a total inductance and capacitance of 100 µH and 10 nF, respectively. Determine the velocity of propagation, phase constant and characteristic impedance of the transmission line at the operating frequency of 100 KHz.

Solution

The inductance and capacitance per unit length of the 100 m transmission line are

$$L = \frac{100}{100} = 1\ \mu\text{H/m}$$

$$C = \frac{10}{100} = 0.1\ \text{n F/m}$$

The velocity of propagation is $v_P = 1/\sqrt{LC}$

$$= \frac{1}{\sqrt{10^{-6} \times 0.1 \times 10^{-9}}} = 10^8\ \text{m/s}$$

The Phase constant is

$$\beta = \frac{\omega}{v_p}$$

$$\beta = \frac{2\pi f}{v_P} = \frac{2 \times 3.14 \times 100 \times 10^3}{10^8} = 6.28 \times 10^{-3}\ \text{rad/m}$$

The characteristic impedance is

$$Z_0 = \sqrt{\frac{L}{C}} = \sqrt{\frac{10^{-6}}{10^{-10}}} = \sqrt{10^4} = 100\ \Omega$$

■

3.6 INPUT IMPEDANCE RELATIONS

Let us consider a transmission line whose length l, propagation constant γ and characteristic impedance Z_0, is connected to a load with impedance Z_L, as shown in Figure 3.5. The line has an input impedance Z_{in} and a voltage source V_s, is connected to it as shown.

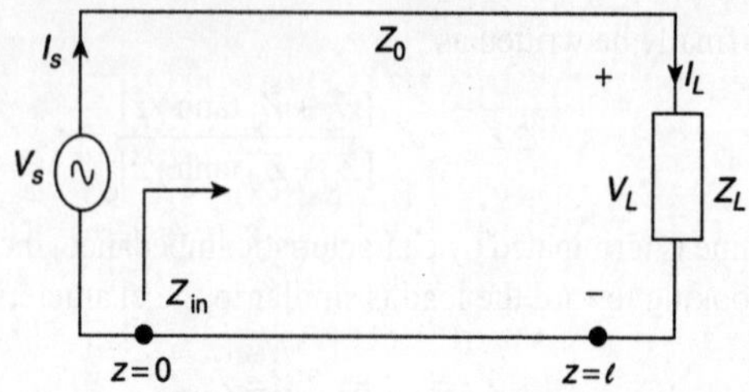

Figure 3.5 *Input impedance of a transmission line with a load*

The following are the various special cases of transmission lines that are obtained by varying the load of the line.

(i) Open circuited line ($Z_L = \infty$)
(ii) Short circuited line ($Z_L = 0$)
(iii) Matched line ($Z_L = Z_0$)

3.6.1 Short Circuited Line (Z_L= 0)

The short circuit is formed when the load impedance is characterized by a zero value, as shown in Figure 3.6.
In such a case, the input impedance of the transmission line is given by

$$Z_{\text{in}|S.C} = Z_0 \tanh(\gamma l)$$

If the transmission line happens to be lossless, then input impedance becomes

$$Z_{\text{in}|S.C} = jZ_0 \tan(\beta l)$$

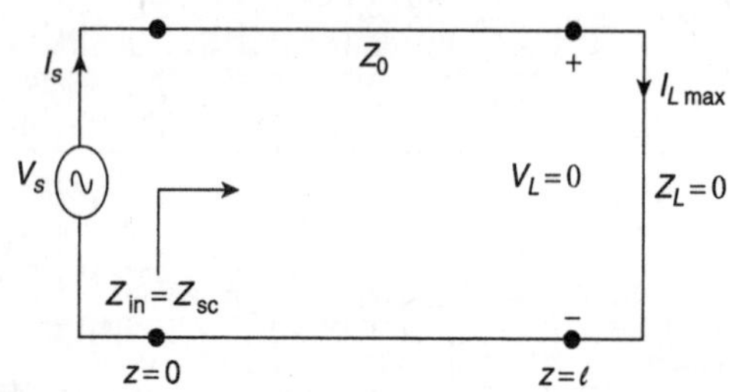

Figure 3.6 *Short circuited transmission line*

The variation of input impedance of the short-circuited line with the physical length is shown by solid lines in Figure 3.8.

Derivation of Input Impedance for a Short Circuited Transmission Line

The load impedance of a short-circuited line is given by

$$Z_L = 0 \tag{3.36}$$

The input impedance of a transmission line is given by

$$Z_{\text{in}} = Z_0 \left[\frac{Z_L + Z_0 \tanh(\gamma l)}{Z_0 + Z_L \tanh(\gamma l)} \right] \tag{3.37}$$

where γ is the propagation constant and Z_0 the characteristic impedance of the transmission line. For the lossy short-circuited line ($z = l$) the input impedance is obtained from Eqs. (3.36) and (3.37) as

$$Z_{\text{in}|S.C} = Z_0\left[\frac{Z_0 \tanh(\gamma l)}{Z_0}\right] = Z_0 \tanh(\gamma l) \tag{3.38}$$

For the lossless line, $\alpha = 0$, when the propagation constant $\gamma = j\beta$, then the input impedance of the short-circuited line is

$$Z_{\text{in}|S.C} = jZ_0 \tan(\beta l) \tag{3.39}$$

3.6.2 Open Circuited Line ($Z_L = \infty$)

The open-circuited line is formed when the load impedance has infinite value, as shown in Figure 3.7.

The input impedance of such an open-circuited transmission line is

$$Z_{\text{in}|O.C} = Z_0 \coth(\gamma l) \tag{3.40}$$

If the transmission line is lossless, then input impedance in this case becomes

$$Z_{\text{in}|O.C} = -jZ_0 \cot(\beta l) \tag{3.41}$$

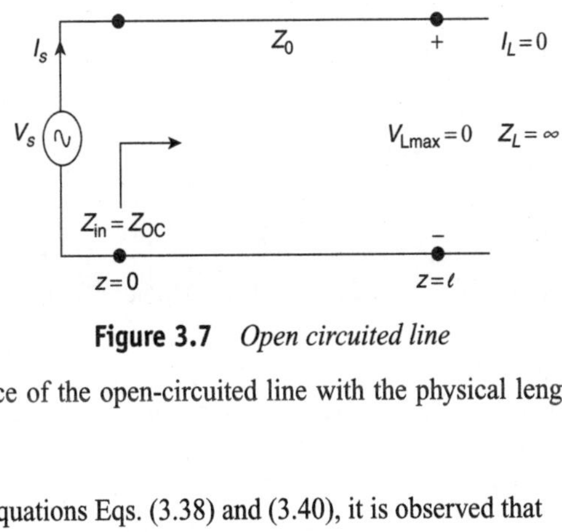

Figure 3.7 *Open circuited line*

The variation of impedance of the open-circuited line with the physical length is as shown by dotted lines in Figure 3.8.

Note:

1. For the above two equations Eqs. (3.38) and (3.40), it is observed that

$$Z_o = \sqrt{Z_{\text{in}|S.C}\, Z_{\text{in}|O.C}} \tag{3.42}$$

Derivation of Input Impedance for an Open Circuited Transmission Line

The load impedance of the open-circuited transmission line is $Z_L = \infty$, for the lossy open-circuited line, the input impedance is given by Eq. (3.37). It can be written as

$$Z_{\text{in}|o.C} = Z_0\left[\frac{1+\left(\frac{Z_0}{Z_L}\right)\tanh(\gamma l)}{\left(\frac{Z_0}{Z_L}\right)+\tanh(\gamma l)}\right] = Z_0 \coth(\gamma l)$$

For the lossless line, $\alpha = 0$, then the propagation constant $\gamma = j\beta$, then the input impedance of the open-circuited line is $Z_{\text{in}|\text{o}\cdot\text{C}} = -jZ_0 \cot(\beta l)$

Figure 3.8 shows the variations of Z_{OC} and Z_{SC} as a function of physical length or electrical length.

- Consider the variation of Z_{SC} only
  At even multiples of $\lambda/4$, the line is offering zero reactance; that is, it is behaving similar to a series resonant circuit. However, at odd multiples of $\lambda/4$, the line is offering infinite reactance; that is, it is behaving similar to a parallel resonant circuit.
- Consider the variation of Z_{OC} only
  At even multiples of $\lambda/4$, the line is offering infinite reactance; that is, it is behaving similar to a parallel resonant circuit. However, at odd multiples of $\lambda/4$, the line is offering zero reactance; that is, it is acting similar to a series resonant circuit.

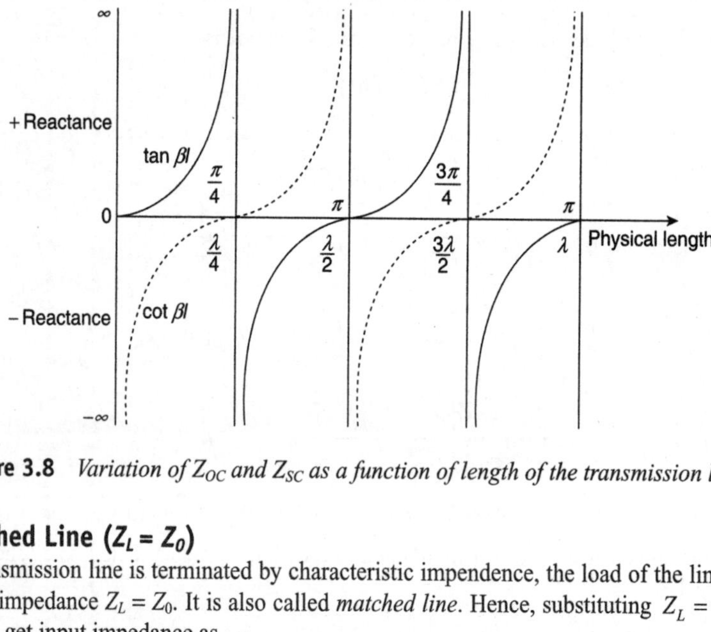

Figure 3.8 *Variation of Z_{OC} and Z_{SC} as a function of length of the transmission line*

3.6.3 Matched Line ($Z_L = Z_0$)

When the transmission line is terminated by characteristic impendence, the load of the line is equal to characteristic impedance $Z_L = Z_0$. It is also called *matched line*. Hence, substituting $Z_L = Z_0$ in Eq. in Eq. (3.37), we get input impedance as

$$Z_{\text{in}} = Z_0 \left[\frac{Z_0 + Z_0 \tanh(\gamma l)}{Z_0 + Z_0 \tanh(\gamma l)} \right]$$

$$Z_{\text{in}} = Z_0 \tag{3.43}$$

Therefore, for a matched line, input impedance is equal to characteristic impedance.

3.7 REFLECTION COEFFICIENT OF A TRANSMISSION LINE

In a uniform transmission line, when the incident wave encounters a second medium such as mismatched load or a discontinuity, then under these conditions, the reflection will exist as shown in Figure 3.9, and the ratio of reflected wave strength to incident wave strength is known as the *reflection coefficient* and is denoted by Γ.

$$\Gamma(Z) = \frac{\text{Reflected wave voltage}}{\text{Incident wave voltage}}$$

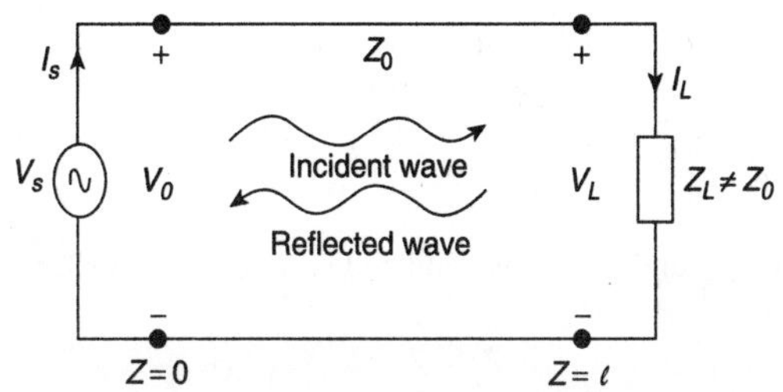

Figure 3.9 *Incident and reflected waves on a transmission line*

3.7.1 Voltage Reflection Coefficient

The voltage reflection coefficient is the ratio of the magnitude of reflected voltage to the incident voltage of the wave at any point on the transmission line. The voltage reflection coefficient at the load end $Z=\ell$ is defined as

$$\Gamma_L = \frac{V^- e^{\gamma l}}{V^+ e^{-\gamma l}} = \frac{V^-}{V^+} e^{2\gamma l} = \frac{V^-}{V^+} e^{2\alpha l} e^{2j\beta l} = |\Gamma_L| \angle \phi_r \tag{3.44}$$

where V^+ indicates the incident wave; whereas the second term V^- indicates the reflected wave, and ϕ_r is the angle of the reflection coefficient. The voltage reflection coefficient in terms of load impedance and characteristic impedance is given by

$$\Gamma_L = \frac{Z_L - Z_0}{Z_L + Z_0} \tag{3.45}$$

Derivation of reflection coefficient in terms of Z_L and Z_0

The general solution for the voltage and current equations for the transmission line where the impedance mismatch occurs is

$$V = V^+ e^{-\gamma z} + V^- e^{\gamma z} \tag{3.46}$$

$$I = \frac{V^+}{Z_0} e^{-\gamma z} - \frac{V^-}{Z_0} e^{\gamma z} \tag{3.47}$$

where V^+ and V^- are complex constants, Z_0 is the characteristic impedance, and γ is the propagation constant of the transmission line. To find V^+ and V^-, the terminal conditions should be given. First, we give the condition at input, that is, $V_0 = V(z)$ at $z = 0$ and $I_0 = I(z)$ at $z = 0$, substituting these equations in Eqs. (3.46) and (3.47) yield

$$V^+ = \frac{1}{2}(V_0 + Z_0 I_0) \tag{3.48}$$

$$V^- = \frac{1}{2}(V_0 - Z_0 I_0) \tag{3.49}$$

If we give the condition at load end, that is, $V_L = V(z)$ at $z = \ell$ and $I_L = I(z)$ at $z = \ell$, substituting these equations in Eqs. (3.46) and (3.47), we get

$$V^+ = \frac{1}{2}(V_L + Z_0 I_L)e^{\gamma l} \tag{3.50}$$

$$V^- = \frac{1}{2}(V_L - Z_0 I_L)e^{-\gamma l} \tag{3. 51}$$

Substituting the values of V^+ and V^- at load end and $V_L = Z_L I_L$ in the Eq. (3.44), gives

$$\Gamma_L = \frac{Z_L - Z_0}{Z_L + Z_0}$$

The input impedance of the line can be expressed in terms of the reflection coefficient and characteristic impedance as

$$Z_{\text{in}} = \frac{V(z)}{I(z)} = Z_0 \frac{[1 + \Gamma_L(z)]}{[1 - \Gamma_L(z)]}$$

3.7.2 Derivation of Input Impedance in Terms of Reflection Coefficient

By definition, the reflection coefficient is given by Eq. (3.44):

$$\Gamma_L = \frac{V^- e^{\gamma l}}{V^+ e^{-\gamma l}} = \frac{V^-}{V^+} e^{2\gamma l} = \frac{V^-}{V^+} e^{2\alpha l} e^{2j\beta l} = |\Gamma_L| \angle \phi_r \tag{3.52}$$

For a lossless transmission line,

$$\Gamma_L = \frac{V^-}{V^+} e^{2j\beta l} \quad \because \alpha = 0 \text{ for a lossless transmission line} \tag{3.53}$$

From Eqs. (3.46) and (3.47), the input impedance of the transmission line at any point is

$$Z_{in}(z) = \frac{V(z)}{I(z)} = \frac{V^+ e^{-\gamma z} + V^- e^{\gamma z}}{\dfrac{V^+}{Z_0} e^{-\gamma z} - \dfrac{V^-}{Z_0} e^{\gamma z}} = Z_0 \left(\frac{1 + \dfrac{V^-}{V^+} e^{2\gamma z}}{1 - \dfrac{V^-}{V^+} e^{2\gamma z}} \right) = Z_0 \frac{[1 + \Gamma_L(z)]}{[1 - \Gamma_L(z)]} \tag{3.54}$$

3.8 STANDING WAVE RATIO

In a transmission line, two waves with same frequency traveling in the opposite directions results in a standing wave. A standing wave pattern of voltage and current is generated on a transmission line when these two waves of same frequency meet; that is, the incident wave and reflected wave will combine to form a standing wave.

3.8.1 Voltage Standing Wave Ratio

The ratio of the maximum voltage to the minimum voltage is called the *voltage standing wave ratio.* $V_{\max}$ is the voltage maximum observed on the transmission line when the incident and reflected voltage wave adds in phase. Therefore, the maximum voltage along the line is given as

$$V_{max} = |V^+| + |V^-| \tag{3.55}$$

$$= |V^+|(1 + |\Gamma_L|) \tag{3.56}$$

V_{min} is the minimum voltage observed on the transmission line when the incident and reflected voltage wave adds in the opposite phase. Therefore, the minimum voltage along the line is given as

$$V_{min} = |V^+| - |V^-| \tag{3.57}$$

$$= |V^+|(1 - |\Gamma_L|) \tag{3.58}$$

Then, the voltage standing wave ratio

$$VSWR = \frac{V_{max}}{V_{min}} = \frac{1 + |\Gamma_L|}{1 - |\Gamma_L|} \tag{3.59}$$

$$|\Gamma_L| = \frac{VSWR - 1}{VSWR + 1} \tag{3.60}$$

Note:

- For a matched transmission line the VSWR is

$$Z_L = Z_0 \Rightarrow \Gamma_L = 0$$

$$VSWR = \frac{1+0}{1-0} = 1$$

In this condition, the standing wave pattern is simply a line representing constant amplitude.

- For an open-load transmission line

$$Z_L = \infty \Rightarrow \Gamma_L = 1$$

$$VSWR = \frac{1+1}{1-1} = \infty$$

The standing wave pattern poses perfect nulls. This is the case for complete standing waves

- For a short-circuited line $Z_L = 0$

$$Z_L = 0 \Rightarrow \Gamma_L = -1$$

$$VSWR = \frac{1+1}{1-1} = \infty$$

From the above three results, it is obvious that the standing wave ratio lies between 1 and ∞, and the reflection coefficient lies between 0 and 1.

$$1 \le VSWR \le \infty \text{ and } 0 \le \Gamma_L \le 1$$

3.8.2 Current Standing Wave Ratio

Similarly, for the current standing wave, I_{max} is the current maximum observed on the transmission line when the incident and reflected current wave adds in phase. Therefore, the maximum current along the line is given as

$$I_{max} = |I^+| + |I^-| \tag{3.61}$$

$$= \frac{|V^+|}{Z_0}(1+|\Gamma_L|) \tag{3.62}$$

I_{min} is the current minimum observed on the transmission line when the incident and reflected current wave adds in the opposite phase. Therefore, the minimum current along the line is given as

$$I_{min} = |I^+| - |I^-| \tag{3.63}$$

$$= \frac{|V^+|}{Z_0}(1-|\Gamma_L|) \tag{3.64}$$

Then, the current standing wave ratio

$$CSWR = \frac{I_{max}}{I_{min}} = \frac{1+|\Gamma_L|}{1-|\Gamma_L|}$$

EXAMPLE PROBLEM 3.2

A transmission line of characteristic impedance 100 Ω is connected to a load of 200 Ω. Calculate the reflection coefficient and standing wave ratio.

Solution

Given that characteristic impedance $Z_0 = 100\,\Omega$, load impedance $Z_L = 200\,\Omega$

Reflection coefficient

$$\Gamma_L = \frac{Z_L - Z_0}{Z_L + Z_0} = \frac{200-100}{200+100} = \frac{1}{3} = 0.33$$

Standing wave ratio is defined as

$$VSWR = \frac{1+|\Gamma_L|}{1-|\Gamma_L|} = \frac{1+0.33}{1-0.33} = 1.98$$

■

3.9 SMITH CHART

The Smith chart is basically used for measuring the reflection coefficient and impedances at various points on a (lossless) transmission line system. In general Smith chart is a polar plot of the complex reflection coefficient Γ(*Z*) [ratio of the reflected wave voltage to the forward voltage] superimposed with the corresponding *Z*(*z*) [ratio of the overall voltage to the overall current].

A plot of Γ for different normalized resistance and reactance values is known as Smith chart, assuming circuit is to be passive. This chart is the superimpose of the locus of constant resistance values, which are circles centered on the real axis, and the locus of constant reactance values, which are circles centered on the imaginary axis.

The Smith chart consists of two families of circles:

(i) Constant- r circles
(ii) Constant- x circles

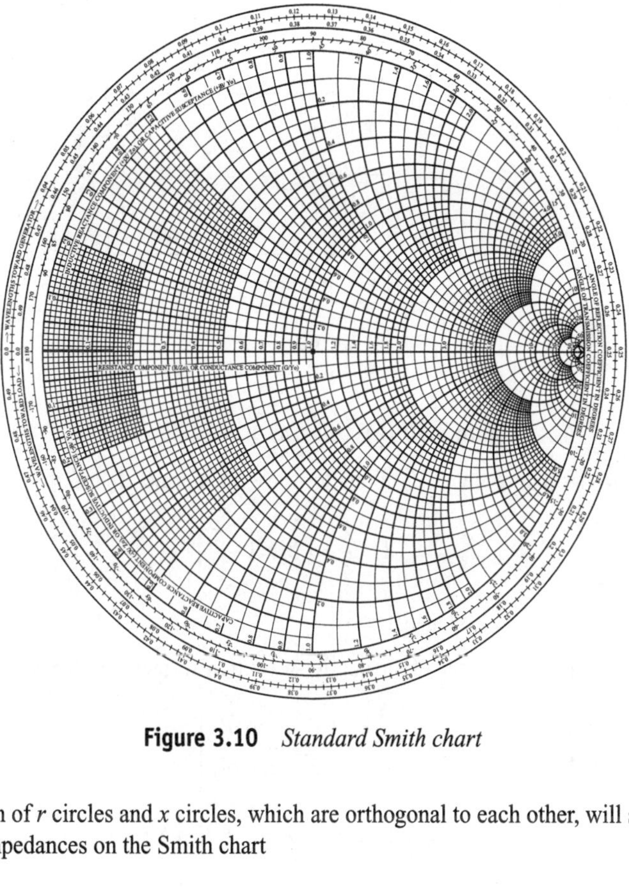

Figure 3.10 *Standard Smith chart*

The intersection of r circles and x circles, which are orthogonal to each other, will specify the different normalized impedances on the Smith chart

3.10 IMPEDANCE MATCHING

Reflection will exist when the load impedance is not matched with source impedance in a transmission line. Reflections will cause standing waves of voltage and current along the transmission line. So, the maximum power cannot be delivered from source to load in a transmission line when this mismatch occurs.

It is not feasible to select all impedances such that they satisfy the overall matched conditions. Hence in order to eliminate reflections, matching networks are required.

Various techniques are used in order to achieve impedance matching. They are

(i) Quarter wave transformer
(ii) Single Stub
(iii) Double Stub

3.10.1 Quarter-wave Transformer

Reflections exist when the impedance mismatch occurs, that is, $Z_L \neq Z_0$ on the transmission line. No reflection $(\Gamma = 0)$ exists when $Z_L = Z_0$. A simple method for matching load impedance to the transmission line involves the use of a quarter-wave transformer, which is a piece of transmission line having a $\lambda/4$ length and a characteristic impedance of Z_Q. The quarter-wave transformer is shown in Figure 3.10 (a).

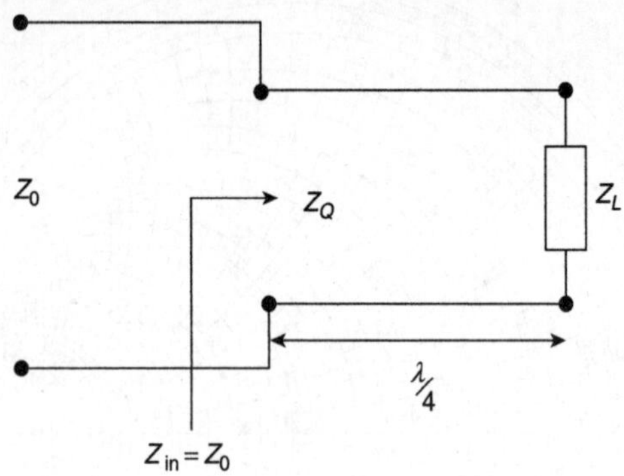

Figure 3.10 (a) *Quarter-wave transformer*

Derivation: Characteristic impedance Z_0 of quarter wave transformer Z_Q inserted

$$l = \frac{\lambda}{4} \Rightarrow \beta l = \left(\frac{2\pi}{\lambda}\right)\left(\frac{\lambda}{4}\right) = \frac{\pi}{2}$$

$$\Rightarrow \tan(\beta l) = \infty$$

From the figure, the input impedance of the quarter-wave transformer with the load Z_L is

$$Z_{\text{in}} = Z_Q \left[\frac{Z_L + Z_Q j\tan(\beta l)}{Z_Q + Z_L j\tan(\beta l)}\right]$$

$$\Rightarrow Z_{\text{in}} = \frac{Z_Q^2}{Z_L} \Rightarrow Z_Q^2 = Z_{\text{in}} Z_L$$

$$z_Q = \sqrt{z_{\text{in}} z_L}$$

EXAMPLE PROBLEM 3.3

What is the characteristic impedance (Z_Q) of a quarter-wave transformer with a load of 100 Ω to a 25 Ω feed line.

Solution

To create a match, we require that Z_{in} to be same as the characteristic impedance of the line $Z_{\text{in}} = 25\ \Omega$

$$Z_Q = \sqrt{Z_{\text{in}} Z_L}$$

$$Z_Q = \sqrt{25 \times 100}$$
$$Z_Q = 50\ \Omega$$

So, for a 100 Ω load to be matched to a 25 feed line, the quarter-wave transform should have a characteristic impedance of 50 Ω. ■

SUMMARY

1. Transmission lines are used to transfer energy from one point to another.
2. Transmission lines may take many forms such as coaxial cables, two-wire parallel lines, microstrip lines, and parallel-plate lines.
3. Transmission lines are used as impedance-matching devices.
4. A transmission line is described by its distributed parameters R, L, C, and G, called line parameters.
5. The secondary constants of the line are propagation constant and characteristic impedance.
6. The characteristic impedance of a transmission line is
$$Z_0 = \sqrt{\frac{R + j\omega L}{G + j\omega C}}$$
7. The characteristic impedance does not depend on the length of the line.
8. The propagation constant is given by $\gamma = \alpha + j\beta = \sqrt{(R + j\omega L)(G + j\omega C)}$
9. The velocity of the propagation of the wave in the transmission line is
$$v_P = \frac{\omega}{\beta}$$
10. The transmission line is terminated by characteristic impedance, and then, the input impedance is the same as the characteristic impedance.
11. The input impedance of the transmission line is $Z_{\text{in}}(z) = Z_0 \left(\frac{Z_L + Z_0 \tanh \gamma z}{Z_0 + Z_L \tanh \gamma z} \right)$
12. Reflection coefficient is defined as the ratio of the reflected wave phasor to the incident wave phasor.
13. Impedance mismatch exists when the load impedance is not equal to the source impedance; then, all the power that is transmitted from the source will not reach the load end.
14. Reflections exist when there is impedance mismatch.
15. The ratio of the maximum to minimum magnitudes of voltage or current on a transmission line is called standing wave ratio.

16. The range of the reflection coefficient and SWR is

$$0 \le \Gamma \le 1,\ 1 \le VSWR \le \infty$$

17. The relationship between the VSWR and reflection coefficient is

$$SWR = \frac{1+|\Gamma|}{1-|\Gamma|},\quad |\Gamma_L| = \frac{VSWR-1}{VSWR+1}$$

18. For a lossless line, $Z_{\text{in}|o.c|} = -jZ_0 \cot(\beta l), Z_{\text{in}|S.C} = jZ_0 \tan(\beta l)$ and $Z_0 = \sqrt{Z_{\text{in}|S.C}\ Z_{\text{in}|O.C}}$.

19. The Smith chart contains both resistance (r) circles and reactance (x) circles.

20. Smith chart is used to find input impedance and input admittance of the transmission line.

21. The Smith chart is used in the analysis of distributed elements (transmission lines) as well as lumped elements.

22. In the Smith chart, the distance toward the load and generator is always anti-clockwise and clockwise directions, respectively.

OBJECTIVE-TYPE QUESTIONS

1. In a coaxial transmission line, the electric and magnetic fields are
   (a) Confined to the inner conductor
   (b) Confined to the outer conductor
   (c) Confined to a dielectric medium
   (d) Not confined to a dielectric medium

2. In the transmission line, which one is the correct
   (a) $R = G$
   (b) $R \neq \frac{1}{G}$
   (c) $R = \frac{1}{G}$
   (d) $L = C$

3. The line parameters R, L, C and G are
   (a) Lumped
   (b) Distributed
   (c) Discrete
   (d) None of these

4. The ratio of positively travelling voltage wave to current wave at any point of the line is
   (a) Characteristic impedance
   (b) Load impedance
   (c) Source impedance
   (d) Shunt admittance

5. TEM mode means
   (a) neither electric nor magnetic field components in the longitudinal direction
   (b) electric field components along direction of propagation
   (c) magnetic field components along direction of propagation
   (d) both electric and magnetic field components in the longitudinal direction

6. A two conductor transmission line supports
   (a) TE wave
   (b) TM wave
   (c) TEM wave
   (d) Both TE and TM waves

7. Two wire transmission lines consists of two parallel lines separated at a
   (a) Uniform distance
   (b) Non uniform distance
   (c) Zero distance
   (d) Infinite distance

8. For a lossless transmission line
   (a) Characteristic impedance does not depends on the frequency
   (b) Attenuation coefficient is zero
   (c) $R = 0 = G$
   (d) All the above are right.

9. For a lossy transmission line, the characteristic impedance does not depend on
   (a) Operating frequency
   (b) Inductance
   (c) Capacitance
   (d) Length of the line

10. Which one of the statements is false?
    (a) G and C are shunt elements.
    (b) R and L are series elements
    (c) Both R and G are depend on the conductivity of the conductors
    (d) The parameters are lumped.

11. A 10 km long transmission line has an inductance of 56 mH. Its distributed inductance is
    (a) 5.6 μH
    (b) 56 μH
    (c) 0.56 μH
    (d) 0.056 μH

12. A 10 km long transmission line has a capacitance of 56 mF. Its distributed capacitance is
    (a) 5.6 μF
    (b) 56 μF
    (c) 0.56 μF
    (d) 0.056 μF

13. The expression for phase velocity in lossless line is
    (a) $v_p = \sqrt{LC}$
    (b) $v_p = 1/\sqrt{LC}$
    (c) $v_p = \omega/\sqrt{LC}$
    (d) $v_p = 0$

14. The distributed parameters of a distortion less line are $L = 0.25\ \mu\text{H}$ and $C = 100\ \text{pF}$, then the characteristic impedance in ohms is
    (a) 100
    (b) 75
    (c) 50
    (d) 25

15. The distributed parameters of a loss less line are $L = 0.25\ \mu\text{H}$ and $C = 100\ \text{pF}$, then the phase constant in rad/sec is
    (a) 18.85
    (b) 20.6
    (c) 12.44
    (d) 10.3

16. The propagation constant of the transmission line is
    (a) $\sqrt{ZY}$
    (b) $\sqrt{Z/Y}$
    (c) $\sqrt{Z}$
    (d) None of these

17. The unit for attenuation constant
(a) H/m (b) F/m
(c) Np/m (d) m/s

18. The unit for phase constant is
(a) rad/s (b) rad
(c) H/m (d) m/s

19. The condition for distortion less line is
(a) $\frac{R}{L} = \frac{G}{C}$ (b) $\frac{R}{L} = G$
(c) $RG = LC$ (d) $R = 0 = G$

20. Time delay depends on
(a) Operating frequency (b) Both L and C
(c) Both R and G (d) None of these

21. The range of reflection coefficient is
(a) 0 to 1 (b) 0.5 to + 1
(c) $-\infty$ to ∞ (d) 0

22. The range of standing wave ratio is
(a) 0 to 1 (b) 1 to ∞
(c) $-\infty$ to ∞ (d) 0

23. If the VSWR is 1, then the reflection coefficient is
(a) 0 (b) 1
(c) ∞ (d) -1

24. For a quarter wave transformer βl is equals to
(a) $\pi/2$ (b) π (c) 2π (d) $\pi/4$

25. Stub is used to neutralize
(a) Resistance of the load (b) impedance of the load
(c) susceptance of the load (d) Admittance of the load

26. The smith chart can be used as
(a) Impedance chart only (b) Both impedance and admittance charts
(c) Admittance chart only (d) none of these

27. λ/4 line acts as
(a) impedance inverter (b) Current inverter
(c) same impedance transformation (d) Voltage inverter.

28. On the smith chart, the distance between normalized input impedance and normalized input admittance is
(a) λ/2 (b) λ/4
(c) λ (d) 2λ

29. The input impedance of short circuited loss less line of length less than a quarter wavelength is
(a) purely resistive (b) purely inductive
(c) purely capacitive (d) complex

30. For a transmission line, the open circuit and short circuit impedances are 20 and 5 ohms respectively, then the characteristic impedance of the line is
(a) 100 Ω (b) 50 Ω
(c) 25 Ω (d) 10 Ω

31. All r-circles pass through the point
(a) $A = 1, B = 0.$ (b) $A = 1, B = 1.$
(c) $A = 0, B = 1.$ (d) $A = 0, B = 0.$

32. For a open circuited load, the VSWR for a line is
(a) 0 (b) 1
(c) ∞ (d) – ∞

33. In a double stub tuner, the spacing between the two stubs is
(a) λ/2 (b) λ/4
(c) λ (d) 2λ

34. The real axis and imaginary axis of smith chart represents
(a) Pure resistive, pure imaginary (b) pure imaginary only
(b) pure imaginary, Pure resistive (d) pure resistive.

35. The center of the smith chart represents
(a) Matched load (b) Short circuit
(b) open circuit (d) none of these

36. The condition that the maximum power transforms in the line is
(a) short circuited (c) matched with the load
(b) open circuited (d) not matched with the load.

37. Reflection occurs because of the following cases
(a) when the load end is open circuited
(b) when the load end is short-circuited
(c) when the line is not terminated in its characteristic impedance
(d) All of the above.

38. VSWR in terms of reflection coefficient is
(a) $\dfrac{1+|\Gamma|}{1-|\Gamma|}$ (b) $\dfrac{1-|\Gamma|}{1+|\Gamma|}$
(c) $\dfrac{|\Gamma|-1}{1-|\Gamma|}$ (d) $\dfrac{|\Gamma|-1}{|\Gamma|+1}$

39. The impedance relation of a quarter wave transform
(a) $Z_Q = \sqrt{Z_{\text{in}} Z_L}$ (b) $Z_Q = \sqrt{Z_{\text{in}}}$
(c) $Z_{\text{in}} = \sqrt{Z_Q Z_L}$ (d) $Z_L = \sqrt{Z_{\text{in}} Z_Q}$

40. For a loss less transmission line, the characteristic impedance is
(a) $\sqrt{LC}$ (b) $\sqrt{L/C}$
(c) $\sqrt{C/L}$ (d) L/C

41. What is the input impedance for open circuit load
(a) $-jZ_o \cot(\beta l)$ (b) $jZ_o \cot(\beta l)$
(c) $-jZ_o \tan(\beta l)$ (d) $jZ_o \tan(\beta l)$

42. The input impedance of a $\lambda/8$ long short circuited section of a loss less transmission line is
(a) ∞ (b) 0
(c) inductive (d) capacitive

43. Characteristic impedance of a transmission line is 50 Ω. Input impedance of the open circuited line is $Z_{oc} = 100 + j150\ \Omega$. When the transmission line is short-circuited the value of the input impedance will be **(GATE 2005)**
(a) 50 Ω (b) $100 + j150\ \Omega$
(c) $7.69 + j11.54\ \Omega$ (d) $7.69 - j11.54\ \Omega$

44. Voltage standing wave pattern in a lossless transmission line with characteristic impedance 50 Ω and a resistive load is shown in figure. **(GATE 2005)**

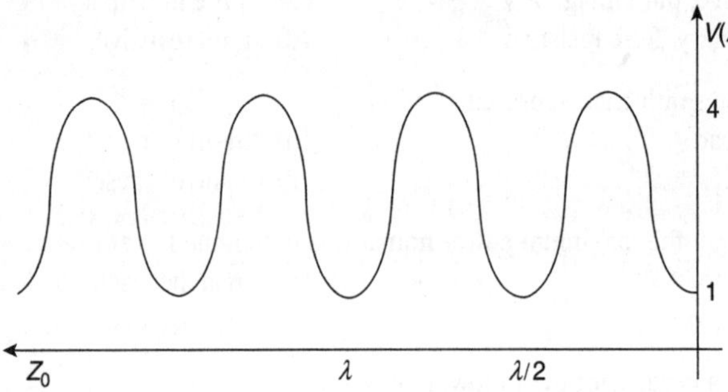

(A) The value of the load resistance is
(a) 50 Ω (b) 200 Ω
(c) 12.5 Ω (d) 0 Ω

(B) The reflection coefficient is given by
(a) – 0.6 (b) –1
(c) 0.6 (d) 0

45. In an impedance smith chart, a clockwise movement along a constant resistance circle gives rise to. **(GATE 2002)**
(a) A decrease in the value of reactance
(b) An increase in value of reactance
(c) No change in reactance value
(d) No change in impedance value

46. A transmission line is distortion-less if **(GATE 2001)**

(a) $RL = 1/GC$ (b) $RL = GC$

(c) $LG = RC$ (d) $RG = LC$

47. Assuming perfect conductors of a transmission line, pure TEM propagation is NOT possible in **(GATE 1999)**

(a) coaxial cable (b) air-filled cylindrical wave guide

(c) parallel twin-wire line in air (d) semi-infinite parallel plate wave guide

48. All transmission line sections shown in Figure below have characteristic impedance $R_0 + j_0$ The input impedance Z_{in} equals **(GATE 1998)**

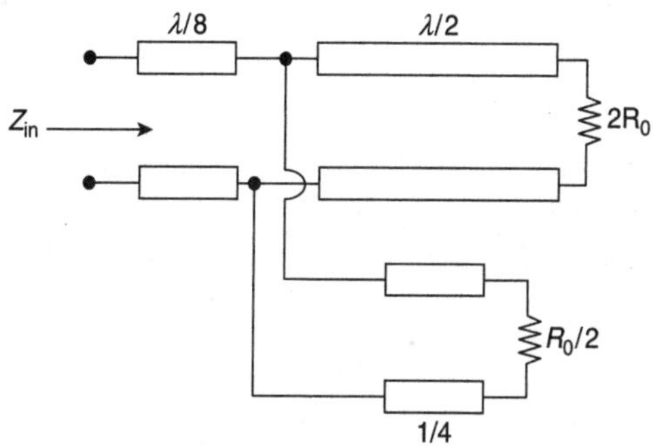

(a) $\frac{2}{3}R_0$ (b) R_0

(c) $\frac{3}{2}R_0$ (d) $2\ R_0$

ANSWERS TO OBJECTIVE-TYPE QUESTIONS

1. (c)	2. (b)	3. (b)	4. (a)	5. (a)	6. (d)	7. (a)	8. (d)	9. (d)	10. (d)	11. (a)
12. (c)	13. (b)	14. (c)	15. (a)	16. (a)	17. (c)	18. (a)	19. (a)	20. (b)	21. (a)	22. (b)
23. (a)	24. (a)	25. (c)	26. (b)	27. (a)	28. (a)	29. (b)	30. (d)	31. (a)	32. (c)	33. (b)
34. (a)	35. (b)	36. (c)	37. (d)	38. (a)	39. (a)	40. (b)	41. (a)	42. (c)	43. (d)	44. A. (c)
44.B. (a)	45. (b)	46. (c)	47. (d)	48. (b)						

REVIEW QUESTIONS

1. Consider a transmission line whose characteristic impedance is $(40 - j2)$ ohm at 8 MHz. The propagation constant is given as $(0.01 + j0.18)$ per meter. Find the primary constants R, L, C, and G.

2. A lossy coaxial cable, operated at a frequency of 500 MHz, has primary constants- R of 2.25 ohm, L of 1 μH/m, C of 100 pF/m and G of zero mho/m. Determine the propagation constant and attenuation constant.

3. Determine the sending-end impedance of a transmission line which is terminated with an impedance of $Z_l = 225 + j24$. The length of transmission line, characteristic impedance and frequency are 80 m, 220 ohm and 3.5 MHz respectively.

4. A 100 ohm line of 1 Km is terminated by 200 ohm load and is fed by a generator of 10 V and an internal impedance of 50 ohm. Find the load voltage and load current at 3×10^5 rad/s.

5. A 100 ohm lossless line connects a signal of 100 KHz to a load of 140 ohm. The load power is 100 mW. Calculate (a) Voltage reflection coefficient; (b) VSWR (voltage standing wave ratio)

6. A distortion-less line of 60 ohm and an attenuation constant of 20 mNp/m and velocity is given as 60% of light velocity. Find the primary constants at 100 MHz.

7. A 75 ohm line is terminated by a load of $(120 + j80)$ ohm. Find the maximum and minimum impedances over the transmission line.

8. Find the input impedance for a lossless line with a characteristic impedance of 75 ohm and a termination impedance of $45 + j60$ ohm with the following

   $f = 50$ MHz, length = 3 m
   $f = 15$ MHz, length = 5 m

Introduction to Microwave Engineering

4

4.1 INTRODUCTION

In communication engineering, a signal occupies a finite bandwidth. For transmitting information from one place to another, a carrier frequency has to be modulated by the information to be carried, and, thus, sidebands will be produced. The resulting modulated signal occupies a certain amount of bandwidth in the RF spectrum. In addition, carrier frequencies have to be selected in such a way that they do not interfere with adjacent channel frequencies. Nowadays, the number of RF channels has been dramatically increased, and the spectrum has become extremely congested. The data transmission rate and frequency band are directly related. With increase in frequencies the channel capacity increases which in turn increases the transmission rate of data. Therefore, to solve the problem of spectrum congestion (eroding), we have to move to a portion of the electromagnetic spectrum that is situated between radio waves and infrared radiation. This particular spectrum is known as a *microwave band*. The *microwave frequencies* refer to the frequency range starting from 300 MHz up to 300 GHz.

The important characteristics of microwave frequencies are *high frequency and short wavelength.* These two features offer wide bandwidth and high antenna gain with a narrow beam.

In lumped circuit analysis such as dc or low-frequency circuits, the voltage and current magnitudes along the line are assumed to be constant. But this assumption fails at high frequencies because of the propagation delay of the V and I waves. Therefore, magnitudes of the voltage and current at high frequencies depend on the position along the transmission line (as resistance varies along the line). This chapter discuss the microwave frequency band in the electromagnetic spectrum and its IEEE frequency band designations as well as the advantages and applications of the microwave frequencies.

4.2 HISTORY OF MICROWAVE TECHNOLOGY

Scientists of the nineteenth century laid the foundation of communication using RF waves and wireless technology, which has affected the modern environment. In 1845, Michel Faraday observed the effect of the magnetic field on light propagation. In 1864, James C. Maxwell unified all previous experimental and theoretical known results of electromagnetic waves into four equations and predicted the existence of these waves. In 1893, Henrich Hertz experimentally confirmed Maxwell's prediction. G. Marconi transmitted information on an experimental basis at microwave frequencies. In 1930, George G. Southworth actually carried out Marconi's experiments on a commercial basis. During the Second World War (1945), based on the previous experiments, radar was invented for military applications. At that time, many experiments were conducted to investigate the operation of devices in Ultra High Frequencies and microwave bands with larger powers. The best suited device was conventional vacuum tube; however it had several difficulties at these frequencies, such as interelectrode capacitance (IEC) between elements within the vacuum tube, lead reactance and longer electron transit time. For example, the IEC causes a short between two electrodes at high frequencies. Because of these limitations vacuum tubes operation is limited to a maximum of 1GHz frequency only.

In 1920, German scientists K. Kurz and H. Barkhausen developed a special vacuum tube called *Barkhausen-Kurz Oscillator* (BKO), which solved the problem of transit time effect in vacuum tube. High frequency oscillations were generated by the special vacuum tube, but the output power was limited. In 1921, A. W. Hull used a magnetic field to control the flow of electrons and developed a popular magnetron device. In 1939, Randoll and Boots improved the performance of the magnetron. For several years, there was a deadlock between power and frequency which became a major problem until W. W. Hansen and D. Heil proposed velocity modulation. In this mechanism, electron transit time is used as an advantage and developed *velocity modulation*. In 1937, the Varian brothers investigated velocity modulation proposed by Hansen and developed a klystron vacuum tube, which could be used as a power amplifier as well as an oscillator. With the advent of these vacuum tubes, radars were also developed for military use. Semiconductor devices were also produced at microwave frequencies. By the 1960s, microwave communication had almost replaced 40% of the telephone communication between major cities of the world.

In the 1990s, microwaves became common consumer market products with the development of microwave ovens, personal communication systems, network television, cell phones, and so on. They also found applications in other areas such as medicine, surveying land, industrial quality control, radio astronomy, global positioning system, power transmission, and space shuttle.

4.3 MICROWAVE SPECTRUM AND BANDS

The electromagnetic spectrum mainly contains three bands, namely *Radio and Microwaves, Infrared,* and *Light waves*. Figure 4.1 illustrates the electromagnetic radiation spectrum.

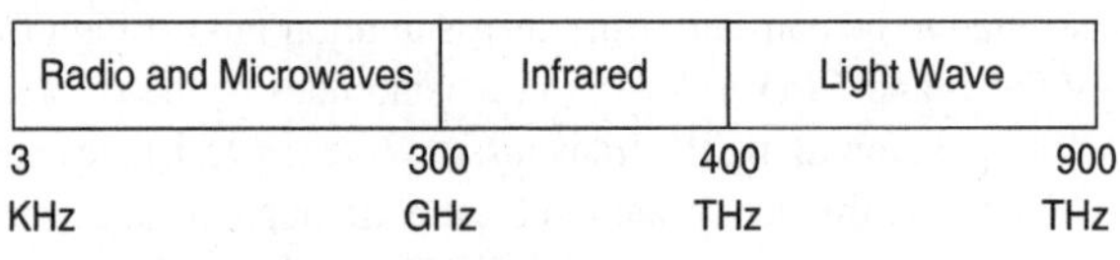

Figure 4.1 *Electromagnetic Spectrum*

The relationship between frequency and wavelength is given as

$$\text{speed of light } (c) = \text{wavelength } (\lambda) \times \text{frequency } (f)$$

To convert from frequency (f) to wavelength (λ) and vice versa, apply $f = c/\lambda$, or $\lambda = c/f$; (where c = speed of light).

The radio and microwaves band is subdivided into Radio Frequency (RF) and Microwave Frequency bands. The most fundamental characteristic that distinguishes RF from microwaves is directly related to the frequency of the electronic signals being processed.

- ***RF frequency spectrum:*** The term *RF* is used to refer frequencies in the range from approximately 300 KHz to 300 MHz. The RF frequency spectrum covers the Medium Frequency (MF), High Frequency (HF), and Very High Frequency (VHF) bands.
- ***Microwave frequency spectrum:*** The term *microwave* is used to refer operating frequencies in the range from 300 MHz to 300 GHz. The microwave frequency spectrum covers the UHF, SHF, and EHF bands.

4.3.1 Microwave Frequency Band Designations

The radio and microwave frequency bands can be further divided into various frequency bands. The frequency range, mode of propagation, and their applications are listed in Table 4.1.

Table 4.1 RF and microwave frequency bands

Band	Description	Frequency (f) range	Propagation mode	Application
RF bands				
VLF	Very low frequency	3–30 KHz	Ground	Long-range radio navigation
LF	Low frequency	30–300 KHz	Ground	Radio beacons and navigational locators
MF	Medium frequency	300 KHz–3MHz	Sky	AM radio
HF	High frequency	3–30 MHz	Sky	Citizens band (CB), ship/aircraft communication
VHF	Very high frequency	30–300 MHz	Sky and line-of-sight	VHF TV, FM radio
Microwave frequency bands				
UHF	Ultra-high frequency	300 MHz–3GHz	Line of sight	UHF TV, cellular phones, paging, and satellites
SHF	Super high frequency	3–30 GHz	Line of sight	Satellite communication
EHF	Extremely high frequency (millimeter waves)	30–300 GHz	Line of sight	Long-range radio navigation

4.3.2 IEEE Frequency Band Designations

The Institute of Electrical and Electronics Engineers (IEEE) recommended microwave frequency band designations, which are given in Table 4.2. In this representation, a letter is used to designate the various bands.

Table 4.2 IEEE frequency band designations

	Letter band designator	Frequency range (GHz)
Microwave Region	L	1 to 2
	S	2 to 4
	C	4 to 8
	X	8 to 12
	K_u	12 to 18
	K	18 to 26
	K_a	27 to 40
Millimeter wave Region	V	40 to 75
	W	75 to 110
	Millimeter waves	30 to 300
	Submillimeter waves	300 to 3000

4.4 ADVANTAGES OF MICROWAVES

Microwaves are useful for communication and radar applications because of their *high frequency* and *short wavelength*. These two features have the following advantages:

- **High bandwidth capability**

  Higher bandwidths are realized at higher frequencies, which provide more information-carrying capacity; that is, thousands of telephone channels and billions of data bits can be sent. A channel with 1% bandwidth provides more frequency range at microwaves than at HF.

  Example: The video signal of TV transmission requires a bandwidth of 6 MHz. If 1% of bandwidth is allocated at a 600-MHz carrier, then it corresponds to 6 MHz, which is the bandwidth of a TV channel, and we can accommodate only 1 TV channel. Similarly, if 1% of bandwidth is allocated at a 60-GHz carrier, it corresponds to 600 MHz, and we can accommodate 100 TV channels.

- **High antenna gain**

  For a given antenna size, more antenna gain is possible at higher frequencies because of a shorter wavelength (λ).

$$G = \frac{4\pi A_e}{\lambda^2}$$

G = gain, A_e = antenna aperture

  We know that the beamwidth is a function of λ (i.e. beamwidth α λ/antenna diameter (D)). For a given antenna diameter, shorter λ provides narrow beamwidth. Therefore, shorter λ allows microwave energy to be concentrated in a small area (Example applications are microwave oven, industrial heating).

- **Line-of-Sight (LOS) propagation**
  Unlike HF signals, microwaves are not affected by the ionosphere because they travel by line of sight. Therefore, satellite and terrestrial communication links that use microwaves provides high capacities.
- **Fading effect and reliability**
  Variations in the received signal strength due to atmospheric changes and/or ground reflections in the signal propagation path are called fading. Fading effect is observed to be severe at low frequencies than at high frequencies. Due to LOS propagation, microwave communication is reliable because of less fading effects.
- **Transparency property of microwaves**
  A study of atmospheric layers, ionosphere, sun, other planet characteristics, and remote sensing is possible with microwaves. At microwave frequencies, the electromagnetic properties of many materials change with frequency. This is due to molecular, atomic, and nuclear resonances of conducting materials and substances when they are exposed to microwave fields.
- **Low-Power requirements**
  The transmitter /receiver power requirements at microwave frequencies are less compared with low frequencies due to narrow beam widths.

4.5 APPLICATIONS OF MICROWAVES

Microwaves are preferred in communication and radar engineering applications because of the following two main advantages:

- High frequency, which provides a high bandwidth
- Shorter wavelength

The main applications of microwaves are:

- **Long-distance communication**
  TV programs are transmitted by communication satellites using LOS microwave propagation.
- **Terrestrial communication**
  Telephone and data signals are transmitted by microwave relay stations.
- **Radars**
  Radar systems use microwaves as the radar cross-section (σ) of target, which is many times greater than λ, resulting in greater reflection power and larger probability of detection.
- **Defence applications**
  Missiles, war planes, and ships are controlled and guided by microwaves.
- **Air traffic controlling and Navigation**
  Ground-based systems (VOR, ILS etc.,) and satellite-based navigation systems (GPS, Galileo, GLONASS, and GAGAN) use microwaves for enroute navigation and landing purposes in air traffic controlling.
- **Microwave heating**
  Microwave heating is used in industrial processes for drying and curing products. It is also used to treat cancer patients.
- **Microwave oven**
  In microwave oven, microwave radiation (at 2.45 GHz) passes through food. Water, sugar and fats in the food absorb energy and causes dielectric heating.

- **Wireless Data Networks**
  Microwaves are used in all Wireless LANs (e.g. Bluetooth (IEEE802.11), WiMax (IEEE 802.16), and also in broadband Internet links.
- **Remote sensing**
  Satellites are used to monitor on or above the earth's surface by using microwave signals to detect the weather conditions, ozone, soil moisture, forest, and exploration of natural resources.
- **Astronomy**
  We can obtain information about other planets, stars, meteors, and phenomena in galaxies by receiving microwave signals from outer space by using gigantic dish antennas.
- **Medical applications**
  Microwaves are used in medical fields such as heart stimulation, haemorrhage control, sterilization etc.

SUMMARY

1. A microwave is an EM wave whose frequency ranges from 300 MHz to 300 GHz.
2. The microwave frequency spectrum covers the UHF, SHF, and EHF bands.
3. The important properties of microwaves are: *High frequency*, which provides high bandwidth and shorter wavelength; LOS transmission/reception; high antenna gain; and directivity.
4. Microwaves that travel by LOS are not affected by the ionosphere. Therefore, high capacity satellite and terrestrial communication links are achievable.
5. Microwave frequencies give the solutions to the problems that arise at higher frequencies during generation, transmission and circuit design.
6. Microwaves have many advantages compared with lower frequencies such as high bandwidth capability, high antenna gain, more directivity, and less power for transmission.
7. Other than communication and radar, microwave technology also found applications in other areas such as medicine, surveying land, industrial quality control, radio astronomy, GPS, mobile communication, and space shuttle.

OBJECTIVE-TYPE QUESTIONS

1. Microwave frequencies ranges from
   (a) 30 MHz–3 GHz (b) 300 MHz–300 GHz
   (c) 3 MHz–3 GHz (d) 300 MHz–3 GHz
2. The relation between frequency and wavelength of microwave is
   (a) $f = \frac{c}{\lambda}$ (b) $f = \frac{\lambda}{c}$
   (c) $f = c\lambda$ (d) $f = \frac{c}{\lambda^2}$

3. One of the important characteristic of microwave frequencies is
   (a) Higher bandwidth (b) lower bandwidth
   (c) low antenna gain (d) high power requirements

4. In microwave ovens the microwave frequency used is
   (a) 2.3 GHz (b) 2.45 GHz
   (c) 2.6 GHz (d) 2.5 GHz

5. In domestic houses, microwaves are used in
   (a) hair driers (b) mixes
   (c) refrigerators (d) microwave ovens

6. The microwave frequency spectrum covers
   (a) UHF (b) SHF
   (c) EHF (d) above all

ANSWERS TO OBJECTIVE-TYPE QUESTIONS

1. (b) 2. (a) 3. (a) 4. (b) 5. (d) 6. (d)

REVIEW QUESTIONS

1. What are microwaves? Explain the reason for using microwave frequencies.
2. What are the different frequency band designations recommended by IEEE?
3. Discuss the advantages of microwave frequencies compared with low-frequency waves.
4. List out the various applications of microwaves.
5. If a microwave having a frequency of 10 GHz is moving out with speed of light, calculate the corresponding wavelength.
6. Give two applications of microwaves where the shorter wavelength can be used.

Waveguides

5

5.1 INTRODUCTION

In Chapter 4, we explained the two conductor structures, usually referred to as *transmission lines*, which can support a transverse electromagnetic (TEM) mode of propagation. The electric and magnetic fields of TEM waves have only transverse components; that is, in the direction of propagation, $\mathbf{E_z} = \mathbf{H_z} = 0$. Coaxial cables, parallel plates, and two-wire lines are examples of practical transmission lines. This chapter investigates a special form of transmission line known as a *waveguide*. In general, the term *waveguide* is applied to structures consisting of a single conductor. Unlike the two-conductor structures, the single-conductor structures cannot support a TEM mode. A waveguide is just a hollow metallic tube that may be rectangular or circular in shape and is used to guide the microwaves. Waveguides are constructed of brass, copper, or aluminum. The inner surface of the waveguide is usually coated with either gold or silver.

Waveguides are primarily used at microwave and optical frequency ranges, whereas transmission lines are used at lower frequencies. In practice, the required operating frequency band and the amount of power to be transferred are the main basis for the choosing the structure. The cause for the inefficiency of Transmission lines is the skin effect and dielectric losses. In waveguides, the electromagnetic (EM) waves are propagated in bounded medium. So, no power is lost due to radiation. Since the guides are generally air filled, the dielectric loss is negligible. However, some power is lost as heat in the walls of the guides owing to the skin effect. This loss in the walls is negligible.

The EM waves can be propagated in several modes within a waveguide, namely, in the Transverse Electric (TE) and Transverse Magnetic (TM) modes. The TE wave has the electric field only in the plane that is transverse to the direction of propagation (i.e. the longitudinal components, $E_z = 0$ and $H_z \neq 0$).The TM wave has only the magnetic field in the transverse plane (i.e. in the direction of propagation, $H_z = 0$, and $E_z \neq 0$).These modes are dependent on the solutions of Maxwell's equation for the given waveguide. Every mode has a particular cut-off frequency, and this cut-off frequency is dependent on physical dimensions of the waveguides. Below the cut-off frequency, the waveguide does not transmit signals. The dominant mode is the mode having the lowest cut-off frequency. In further sections, different types of modes and their corresponding cut-off frequencies are explained for various types of waveguides.

5.2 TYPES OF WAVEGUIDES

The shape of the waveguide decides the functionality of the given waveguide. The cross-section of the waveguide can be of any shape. However, since irregular shapes are difficult to analyze, they are rarely used. The three most commonly used shapes are as follows (Figure 5.1):

1. **Rectangular waveguides** Both TE and TM modes can be supported by these waveguides. The electric field is transverse to the direction of propagation in TE modes. The magnetic field is transverse to the direction of propagation in TM modes.
2. **Circular waveguides** They tend to twist the waves as they travel through them and are used with rotating antennas in radars.
3. **Elliptical waveguides** An elliptical shape is often preferred in flexible waveguides. These waveguides will be required whenever the waveguide section is capable of movement, such as bending, stretching, or twisting.

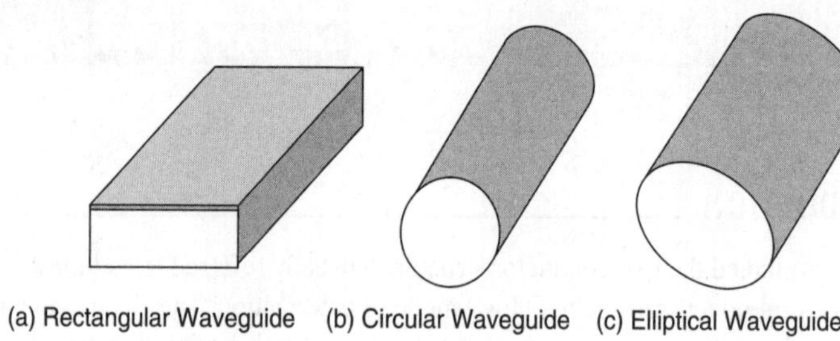

Figure 5.1 *Waveguides of various shapes*

5.3 RECTANGULAR WAVEGUIDES

A waveguide which is a hollow metallic tube of a rectangular cross-section is known as the rectangular waveguide. The EM fields can be confined. Therefore, the EM waves can be guided by the walls of the guide through reflections. Rectangular waveguides are usually made in standard sizes with breadth "a" (along x direction) approximately twice the height "b"(along y direction) (Figure 5.2 (a)). The "a" dimension cannot be less than one-half wavelength. This can be seen, as the guide is made up of two-quarter wavelength stubs separated by a small distance. Any frequency that makes the "a" dimension less than one-half wavelength allows no propagation of energy down the waveguide.

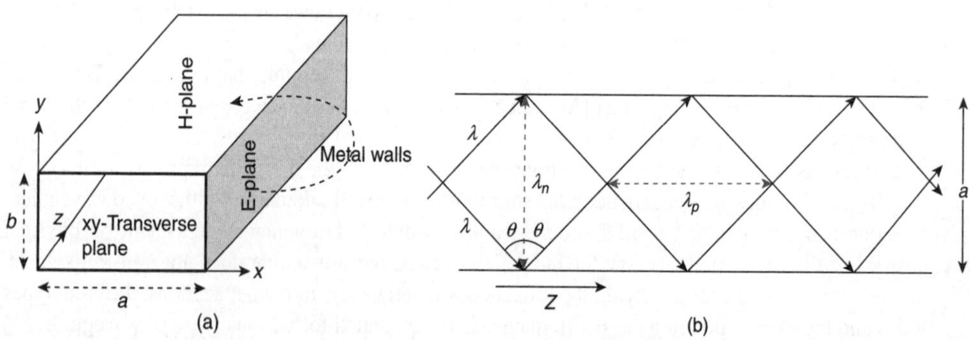

Figure 5.2 *(a) Rectangular waveguide; (b) EM field propagation in waveguide*

While the plane wave travels through the waveguide, it gets reflected from wall to wall in a zigzag manner (Figure 5.2 (b)). During this process, a component of either the electric or magnetic field travels in the direction of propagation of the resultant wave, and the other field (along breadth) will be normal to the direction of propagation of the wave.

λ is the wavelength of the incident wave (actual wave). λ_n is the wavelength in the direction that is normal to the reflecting plane, and λ_p is the wavelength parallel to the direction of the plane. Mathematically, λ_n and λ_p are given by

$$\lambda_n = \frac{\lambda}{\cos\theta} \text{ and } \lambda_p = \frac{\lambda}{\sin\theta}.$$

A plane wave in a waveguide can be resolved into two components as shown in Figure 5.2 (b). One is normal to the direction of propagation. This component is the standing wave and is perpendicular to the direction of propagation. The other component is the travelling wave and is parallel to the direction of propagation. Since only one field is parallel to the direction of propagation, the TEM mode does not exist.

Note: λ is designated as λ_o when the wave is outside the waveguide
λ is designated as λ_g when the wave is inside the waveguide

The boundary conditions at the surface of the conductor are as follows:

- The tangential component of the electric field is continuous. Therefore, in order for the electric field to exist at the surface of the conductor, it should be normal to the conductor.
- The normal component of the magnetic field is continuous. Therefore, the magnetic field should not be normal to the surface of the waveguide.

The above conditions can be satisfied by either a *TE* or a transverse magnetic wave *TM*. However, the TEM wave violates these conditions and, hence, cannot exist in a waveguide.

So, *TE* and *TM* modes are the basic modes in rectangular waveguides. These modes are designated as TE_{mn} or TM_{mn}. Here, *m* and *n* are integers and are defined as

m = number of one-half wavelength variations of fields along *x* direction (breadth of waveguide or "*a*" dimension)

n = number of one-half wavelength variations of fields along *y* direction (height of waveguide or "*b*" dimension).

Example of E and H field patterns in TE_{mn} and TM_{mn} modes

For example, the number of one-half wavelength variations of electric fields along *x* and *y* directions are shown below for TE_{10}, TE_{11}, and TE_{21} configurations separately (Figure 5.3). In case of TE_{10} configuration, the number of ½ wave variations of electric fields along *x* and *y* directions are 1 and 0, respectively. Figure 5.4 (a) shows the end view and typical 3D view of transverse electric and longitudinal magnetic field patterns of the dominant mode (TE_{10}) of a rectangular waveguide, and Figure 5.4 (b) shows the typical 3D view of transverse magnetic and longitudinal electric field patterns of the dominant mode (TM_{11}) of a rectangular waveguide.

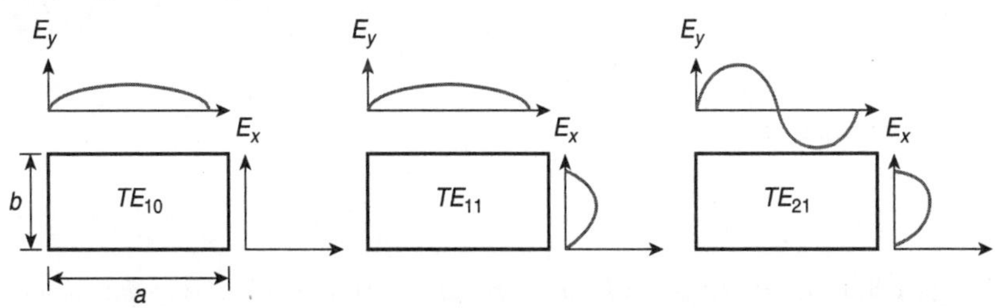

Figure 5.3 *TE_{mn} mode field configurations for various m, n values.*

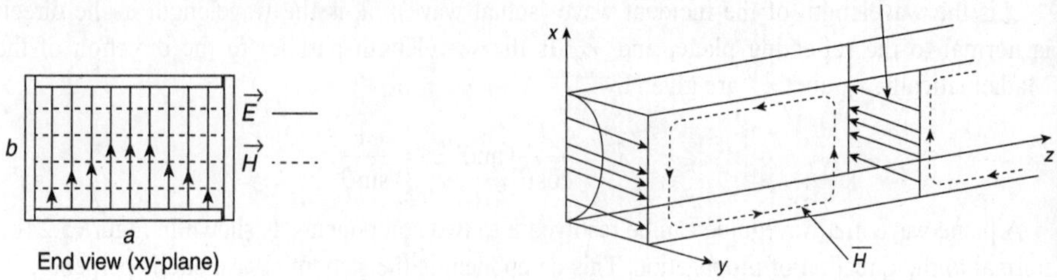

Figure 5.4 (a) *Electric and magnetic field patterns in TE_{10} mode configuration*

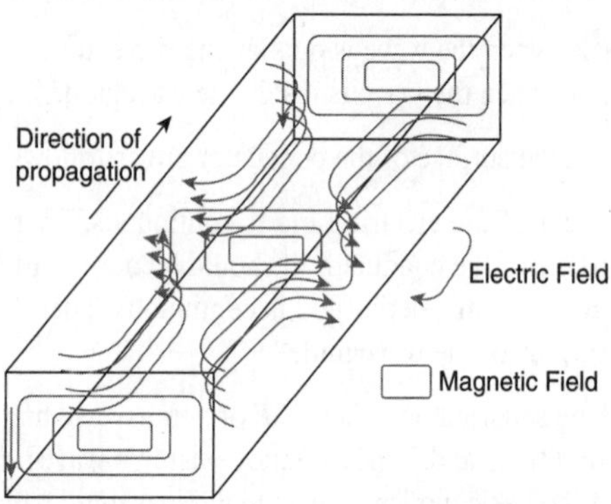

Figure 5.4 (b) *Electric and magnetic field patterns in TM_{11} mode configuration*

These modes will be discussed in detail in the next few sections.

> The "*a*" dimension determines the frequency range of the waveguide, and the "*b*" dimension determines the power-handling capability. The physical size of the waveguides depends on the travelling frequency.

5.3.1 Field Equations in a Rectangular Waveguide

Consider a rectangular waveguide in the rectangular co-ordinate system with its breadth, "*a*", along the *x* axis; height, "*b*", along the *y* axis; and length along the *z*-axis along which the wave propagates. The waveguide is filled with dielectric air, as shown in Figure 5.5 (a).

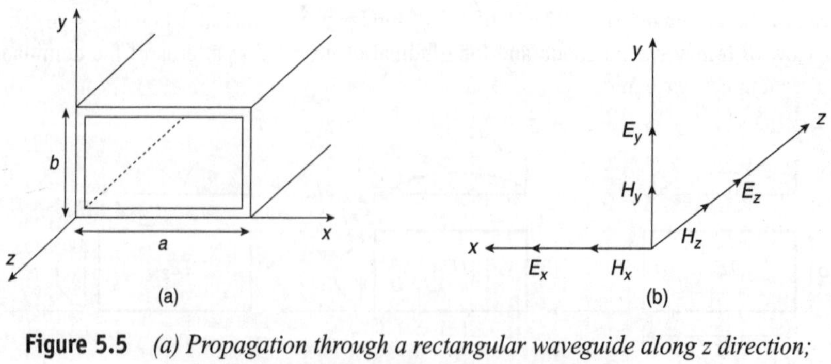

Figure 5.5 *(a) Propagation through a rectangular waveguide along z direction; (b) Field components along x, y, and z directions*

The x, y, and z components of the electric and magnetic fields are E_x, H_x, E_y, H_y, and E_z, H_z. The components along the x and y axes are transverse field components, and the components along the z axis are in the direction of propagation (Figure 5.5 (b)). In this section, we present two derivations. They are as follows:

(i) Wave equations in TE and TM modes
(ii) General expressions for transverse components (E_x, E_y, and H_x, H_y) in terms of the longitudinal components (E_z, H_z)

These two derivations are used to find the solution for the z-directed electric field (E_z) and /or magnetic field (H_z) and field component expressions of TM and/or TE modes of rectangular waveguide.

Derivation of Wave Equations for TE and TM Modes

The electric and magnetic wave equations in the frequency domain along the z direction are given by

$$\left.\begin{aligned}\nabla^2 E_z &= \gamma^2 E_z\\ \nabla^2 H_z &= \gamma^2 H_z\end{aligned}\right\}\text{(Wave equations)}$$

where γ= propagation constant $= \sqrt{j\omega\mu(\sigma + j\omega\varepsilon)} = \alpha + j\beta$

α and β are attenuation and phase shift constants, respectively

Since attenuation is zero along the z direction, α (attenuation constant) is zero

Substitute this in the above equation $\gamma = \sqrt{j\omega\mu(\sigma + j\omega\varepsilon)} = j\beta$

Squaring on both sides, we get $\gamma^2 = j\omega\mu(\sigma + j\omega\varepsilon) = -\beta^2$

Since there is no current in waveguide, $\sigma = 0$

$$\gamma^2 = -\omega^2\mu\varepsilon = -\beta^2$$

Substitute this γ^2 in the wave equations given above; then, we get

For TM wave ($H_z = 0$) $$\nabla^2 E_z = -\omega^2\mu\varepsilon E_z \tag{5.1}$$

For TE wave ($E_z = 0$) $$\nabla^2 H_z = -\omega^2\mu\varepsilon H_z \tag{5.2}$$

Expanding the Laplacian, $\nabla^2 E_z$ in terms of the rectangular coordinate system

$$\frac{\partial^2 E_z}{\partial x^2} + \frac{\partial^2 E_z}{\partial y^2} + \frac{\partial^2 E_z}{\partial z^2} = -\omega^2\mu\varepsilon E_z \tag{5.3}$$

Since the wave propagates in the "z" direction, the spatial variations in the z axis are known so that $\frac{\partial}{\partial z} = -\gamma$ (by differentiating the field equation $E = |E|e^{-\gamma z}$ with regard to z).

Then, we have the operator $$\frac{\partial^2}{\partial z^2} = \gamma^2 \tag{5.4}$$

Substituting Eq. (5.4) in Eq. (5.3), we get

$$\frac{\partial^2 E_z}{\partial x^2}+\frac{\partial^2 E_z}{\partial y^2}+\gamma^2 E_z=-\omega^2\mu\varepsilon E_z \tag{5.5}$$

$$\frac{\partial^2 E_z}{\partial x^2}+\frac{\partial^2 E_z}{\partial y^2}+\left(\gamma^2+\omega^2\mu\varepsilon\right)E_z=0 \tag{5.6}$$

Let us consider "h" as a constant such that

$$h^2=\left(\gamma^2+\omega^2\mu\varepsilon\right)$$

The parameter, h is also denoted with k_c.

Note: For a lossless waveguide, γ has to be imaginary; that is, $\gamma = j\beta$, and, hence, the above equation becomes $h^2=(-\beta^2+\omega^2\mu\varepsilon)$. Here, $\omega^2\mu\varepsilon$ is the square of the phase constant in an ideal medium, and is often denoted as k; β is the phase constant of the wave in the waveguide along the direction of propagation, and it is given as $\beta=\omega\sqrt{\mu\varepsilon}\left(\sqrt{1-(f_c/f)^2}\right)$ which is explained in the next few sections (Eq. 5.81), and h represents the phase constant of the wave in the transverse plane.

The parameter h can be determined from the guide dimensions, as will be explained in the next few sections (Eq.5.70), and k can be determined from the frequency of operation. Therefore, the parameter β can be determined using the relation, $h^2=(-\beta^2+\omega^2\mu\varepsilon)$.

Then, Eq. (5.6) is rewritten as

For TM waves
$$\frac{\partial^2 E_z}{\partial x^2}+\frac{\partial^2 E_z}{\partial y^2}+h^2 E_z=0 \tag{5.7}$$

Similarly, for TE waves
$$\frac{\partial^2 H_z}{\partial x^2}+\frac{\partial^2 H_z}{\partial y^2}+h^2 H_z=0 \tag{5.8}$$

The transverse field components can be expressed in terms of longitudinal components (**E$_z$** and **H$_z$**) along the waveguide (Figure 5.5b). The x and y components (i.e. transverse components) of the electric and magnetic fields (E_x, E_y, and H_x, H_y) in phasor form for the rectangular waveguide shown above are given by

$$E_x=-\frac{\gamma}{h^2}\frac{\partial E_z}{\partial x}-\frac{j\omega\mu}{h^2}\frac{\partial H_z}{\partial y}$$

$$E_y=-\frac{\gamma}{h^2}\frac{\partial E_z}{\partial y}+\frac{j\omega\mu}{h^2}\frac{\partial H_z}{\partial x}$$

$$H_x=\frac{j\omega\varepsilon}{h^2}\frac{\partial E_z}{\partial y}-\frac{\gamma}{h^2}\frac{\partial H_z}{\partial x}$$

$$H_y=-\frac{j\omega\varepsilon}{h^2}\frac{\partial E_z}{\partial x}-\frac{\gamma}{h^2}\frac{\partial H_z}{\partial y}$$

(where γ = propagation constant, and $h^2 = \gamma^2 + \omega^2\mu\varepsilon$ is called the *characteristic equation*)
The values of h are known as the *Eigen values* or characteristic values.

Derivation of Expressions for Field Components (E_x, H_x and E_y, H_y)

Using Maxwell's equation, it is possible to find the various components along x and y directions (E_x, H_x and E_y, H_y).

From Ampere's equation, we get

$$\nabla \times \vec{H} = j\omega\varepsilon\vec{E} \quad (\because \sigma = 0, \mathbf{J} = 0) \tag{5.9}$$

Expanding Eq. (5.9),

$$\nabla \times \vec{H} = \begin{vmatrix} \hat{i} & \hat{j} & \hat{k} \\ \dfrac{\partial}{\partial x} & \dfrac{\partial}{\partial y} & \dfrac{\partial}{\partial z} \\ H_x & H_y & H_z \end{vmatrix} = j\omega\varepsilon\left[\hat{i}E_x + \hat{j}E_y + \hat{k}E_z\right] \tag{5.10}$$

Replacing $\dfrac{\partial}{\partial z} = -\gamma$ (an operator), we get

$$\begin{vmatrix} \hat{i} & \hat{j} & \hat{k} \\ \dfrac{\partial}{\partial x} & \dfrac{\partial}{\partial y} & -\gamma \\ H_x & H_y & H_z \end{vmatrix} = j\omega\varepsilon\left[\hat{i}E_x + \hat{j}E_y + \hat{k}E_z\right] \tag{5.11}$$

Equating coefficients of $\hat{i}, \hat{j},$ and $\hat{k}$ (after expansion), we get

$$\frac{\partial H_z}{\partial y} + \gamma H_y = j\omega\varepsilon E_x \tag{5.12}$$

$$\frac{\partial H_z}{\partial x} + \gamma H_x = -j\omega\varepsilon E_y \tag{5.13}$$

$$\frac{\partial H_y}{\partial x} - \frac{\partial H_x}{\partial y} = j\omega\varepsilon E_z \tag{5.14}$$

Similarly, from the Faraday's law equation,

$$\nabla \times \vec{E} = -j\omega\mu\vec{H}\left(\because \nabla \times \vec{E} = -\frac{\partial \vec{B}}{\partial t}, \vec{B} = \mu\vec{H}\right) \tag{5.15}$$

$$\nabla \times \vec{E} = \begin{vmatrix} \hat{i} & \hat{j} & \hat{k} \\ \dfrac{\partial}{\partial x} & \dfrac{\partial}{\partial y} & \dfrac{\partial}{\partial z} \\ E_x & E_y & E_z \end{vmatrix} = -j\omega\mu\left[\hat{i}H_x + \hat{j}H_y + \hat{k}H_z\right] \tag{5.16}$$

Replacing $\dfrac{\partial}{\partial z} = -\gamma$ (an operator), we get

$$\begin{vmatrix} \hat{i} & \hat{j} & \hat{k} \\ \dfrac{\partial}{\partial x} & \dfrac{\partial}{\partial y} & -\gamma \\ E_x & E_y & E_z \end{vmatrix} = -j\omega\mu\left[\hat{i}H_x + \hat{j}H_y + \hat{k}H_z\right] \tag{5.17}$$

Equating coefficients of $\hat{i}, \hat{j}$, and $\hat{k}$ (after expansion), we get

$$\frac{\partial E_z}{\partial y} + \gamma E_y = -j\omega\mu H_x \tag{5.18}$$

$$\frac{\partial E_z}{\partial x} + \gamma E_x = j\omega\mu H_y \tag{5.19}$$

$$\frac{\partial E_y}{\partial x} - \frac{\partial E_x}{\partial y} = -j\omega\mu H_z \tag{5.20}$$

Eqs.(5.12–5.14) and Eqs.(5.18–5.20) are used to produce simple algebraic equations for the transverse components of **E** and **H** (x and y). Using these equations, we can find expressions for the four transverse components [E_x, H_x, E_y, and H_y] in terms of the z-directed components (E_z and H_z). For instance, solving Eq. (5.19) for H_y, we find

$$H_y = \frac{1}{j\omega\mu}\frac{\partial E_z}{\partial x} + \frac{\gamma}{j\omega\mu}E_x \tag{5.21}$$

Substituting Eq. (5.21) in Eq. (5.12), we get

$$\frac{\partial H_z}{\partial y} + \gamma\left(\frac{1}{j\omega\mu}\frac{\partial E_z}{\partial x} + \frac{\gamma}{j\omega\mu}E_x\right) = j\omega\varepsilon E_x$$

$$\frac{\partial H_z}{\partial y} + \frac{\gamma}{j\omega\mu}\frac{\partial E_z}{\partial x} + \frac{\gamma^2}{j\omega\mu}E_x = j\omega\varepsilon E_x$$

$$E_x\left(j\omega\varepsilon - \frac{\gamma^2}{j\omega\mu}\right) = \frac{\gamma}{j\omega\mu}\frac{\partial E_z}{\partial x} + \frac{\partial H_z}{\partial y}$$

Multiplying by $j\,\omega\,\mu$, we get

$$E_x(-\omega^2\mu\varepsilon-\gamma^2)=\gamma\frac{\partial E_z}{\partial x}+j\omega\mu\frac{\partial H_z}{\partial y}$$

$$\gamma\frac{\partial E_z}{\partial x}+j\omega\mu\frac{\partial H_z}{\partial y}=E_x(-(\gamma^2+\omega^2\mu\varepsilon))$$

where $$h^2=\left(\gamma^2+\omega^2\mu\varepsilon\right).$$

Dividing the above equation by $-h^2$, we get

$$E_x=-\frac{\gamma}{h^2}\frac{\partial E_z}{\partial x}-\frac{j\omega\mu}{h^2}\frac{\partial H_z}{\partial y} \tag{5.22}$$

Similarly, $$E_y=-\frac{\gamma}{h^2}\frac{\partial E_z}{\partial y}+\frac{j\omega\mu}{h^2}\frac{\partial H_z}{\partial x} \tag{5.23}$$

$$H_x=\frac{j\omega\varepsilon}{h^2}\frac{\partial E_z}{\partial y}-\frac{\gamma}{h^2}\frac{\partial H_z}{\partial x} \tag{5.24}$$

$$H_y=-\frac{j\omega\varepsilon}{h^2}\frac{\partial E_z}{\partial x}-\frac{\gamma}{h^2}\frac{\partial H_z}{\partial y} \tag{5.25}$$

These equations give a general relationship for field components within a waveguide. Equations (5.22) to (5.25) are solutions to Maxwell's equations. Note that the propagation coefficient γ is not known. So, there are seven scalar unknowns (the six field components and γ). It is evident that the axial components E_z and H_z can determine all transverse components of E and H. Due to this fact the mode designations *TEM*, *TE*, and *TM* are allowed. From equations (5.22) to (5.25), we find out that there are different types of field patterns or configurations. A *mode* is nothing but a distinct field pattern. There are four different mode categories. They are,

1. TEM Mode or Principal Mode ($E_z = 0$ and $H_z = 0$)

In this mode, both the **E** and **H** fields are transverse to the direction of wave propagation, and this is known as the *transverse electromagnetic* (TEM) mode. By substituting $E_z = 0$ and $H_z = 0$ in equations (5.22 to 5.25), all field components are reduced to zero, and, hence, we can conclude that there is no field component along the direction of propagation, as shown in Figure 5.6. Thus, from the result, a rectangular waveguide cannot support the TEM mode.

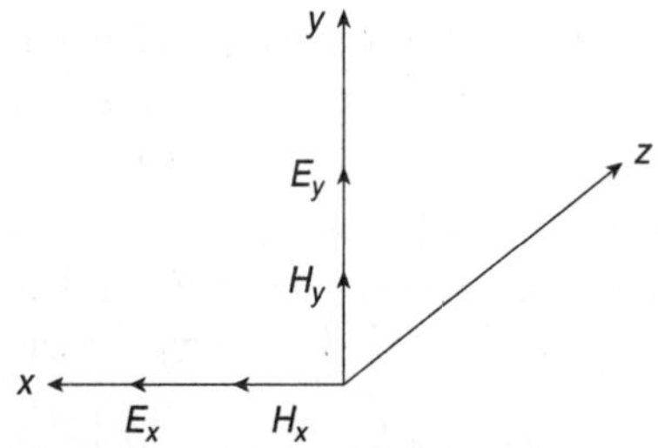

Figure 5.6 *TEM mode*

2. TE modes ($E_z = 0$ and $H_z \neq 0$)

These modes have $E_z = 0$, at all points within the waveguide. This means that there is no electric field vector component along the direction of propagation, and the magnetic field vector is along the direction of propagation, as shown in Figure 5.7.

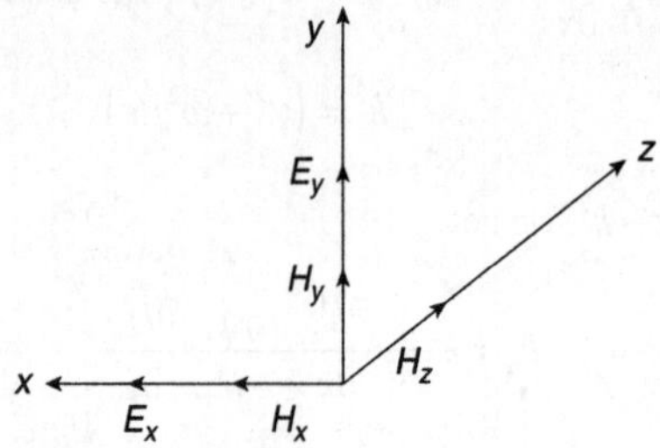

Figure 5.7 *TE mode:* $\mathbf{E} = (E_x, E_y, \text{ and } 0)$ *and* $\mathbf{H} = (H_x, H_y, \text{ and } H_z)$

3. TM modes ($E_z \neq 0$ and $H_z = 0$)

These modes have $H_z = 0$ at all points within the waveguide. This means that there is no magnetic field vector component along the direction of propagation, and the electric field vector is parallel to the length axis, as shown in Figure5.8.

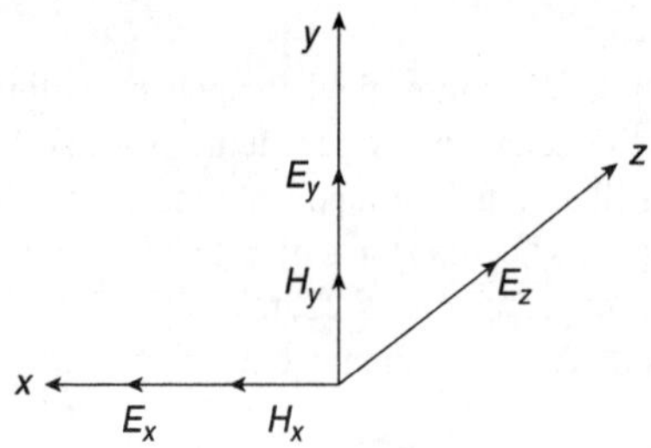

Figure 5.8 *TM mode:* $\mathbf{E} = (E_x, E_y, \text{ and } E_z)$ *and* $\mathbf{H} = (H_x, H_y, \text{ and } 0)$

4. HE modes($E_z \neq 0$ and $H_z \neq 0$)

In this case, neither ***E*** nor ***H*** field is transverse to the direction of wave propagation, and they are known as *hybrid* modes.

5.3.1.1 Field components of TM and TE waves for rectangular waveguides

To find the expressions for field components of rectangular waveguides, the following steps are to be carried out:

1. Use the wave equations of TM and /or TE modes.
2. Solve the wave equations to get the general solution of E_z and/or H_z.
3. Apply appropriate boundary conditions to get the complete solutions for a particular mode (TM and/or TE).
4. Substitute E_z and /or H_z in the field component expressions (Eqs. 5.22–5.25) to find the transverse field components of TM and/or TE modes.

5.3.1.1.1 Field components of TM waves

Transverse magnetic (TM) modes in a rectangular waveguide are characterized by $H_z = 0$ and $E_z \neq 0$. The transverse field components of a TM wave in a rectangular waveguide are given by

$$E_x = -\frac{\gamma}{h^2}C\left(\frac{m\pi}{a}\right)\cos\left(\frac{m\pi}{a}\right)x\sin\left(\frac{n\pi}{b}\right)ye^{j\omega t-\gamma z}$$

$$E_y = -\frac{\gamma}{h^2}C\left(\frac{n\pi}{b}\right)\sin\left(\frac{m\pi}{a}\right)x\cos\left(\frac{n\pi}{b}\right)ye^{j\omega t-\gamma z}$$

$$H_x = \frac{j\omega\varepsilon}{h^2}C\left(\frac{n\pi}{b}\right)\sin\left(\frac{m\pi}{a}\right)x\cos\left(\frac{n\pi}{b}\right)ye^{j\omega t-\gamma z}$$

$$H_y = -\frac{j\omega\varepsilon}{h^2}C\left(\frac{m\pi}{a}\right)\cos\left(\frac{m\pi}{a}\right)x\sin\left(\frac{n\pi}{b}\right)ye^{j\omega t-\gamma z}$$

Derivation of Transverse Field Components of a TM Wave in Rectangular Waveguides

The Helmholtz-wave equation of a TM wave for propagation of the electric field in a lossless medium can be written as

$$\nabla^2 E_z = -\omega^2\mu\varepsilon E_z$$

Expanding this equation for z-propagating fields, we get

$$\frac{\partial^2 E_z}{\partial x^2} + \frac{\partial^2 E_z}{\partial y^2} + h^2 E_z = 0 \quad (5.26)$$

This is a partial differential equation that can be solved to get different field components E_x, H_x, E_y, and H_y. To solve this Eq. (5.26), we employ the method of separation of variables by assuming a solution:

$$E_z = \mathrm{XY} \quad (5.27)$$

Here, E_z can be expressed as the product of the function X and the function Y, where X is a pure function of "x" only

Y is a pure function of "y" only

Since X and Y are independent variables,

$$\frac{\partial^2 E_z}{\partial x^2} = \frac{\partial^2 (XY)}{\partial x^2} = Y\frac{d^2 X}{dx^2} \quad (5.28)$$

$$\frac{\partial^2 E_z}{\partial y^2} = \frac{\partial^2 (XY)}{\partial y^2} = X\frac{d^2 Y}{dy^2} \quad (5.29)$$

Substituting Eqs. (5.27–5.29) in Eq. (5.26), we get

$$Y\frac{d^2 X}{dx^2} + X\frac{d^2 Y}{dy^2} + h^2 XY = 0 \quad (5.30)$$

Dividing Eq. (5.30) throughout by XY, we get

$$\frac{1}{X}\frac{d^2X}{dx^2}+\frac{1}{Y}\frac{d^2Y}{dy^2}+h^2=0 \tag{5.31}$$

$\frac{1}{X}\frac{d^2X}{dx^2}$ is a pure function of x only

$\frac{1}{Y}\frac{d^2Y}{dy^2}$ is a pure function of y only

The sum of these terms is a constant. Hence, each term should be equal to a constant separately, as X and Y are independent variables. We use the separation of variables method to solve the differentialequation, Eq. (5.31).

Let
$$\frac{1}{X}\frac{d^2X}{dx^2}=-B^2 \tag{5.32}$$

and
$$\frac{1}{Y}\frac{d^2Y}{dy^2}=-A^2 \tag{5.33}$$

where A and B are constants
Substituting Eqs. (5.32) and (5.33) in Eq. (5.31), we get $-B^2-A^2+h^2=0$

$$h^2=B^2+A^2 \tag{5.34}$$

Equations (5.32) and (5.33) are ordinary second-order differential equations; the differential equation has a general solution, and the solutions of X, Y are given by

$$X=C_1\cos Bx+C_2\sin Bx \tag{5.35}$$
$$Y=C_3\cos Ay+C_4\sin Ay \tag{5.36}$$

where C_1, C_2, C_3, and C_4 are constants, and they can be evaluated by applying boundary conditions. The complete solution is given by $E_z=XY$.

Substituting solutions of X and Y, we get the total solution of the Helmholtz equation in rectangular coordinates, which is

$$E_z=[C_1\cos Bx+C_2\sin Bx][C_3\cos Ay+C_4\sin Ay] \tag{5.37}$$

To investigate the electromagnetic field in the waveguide, the wave equations and the Maxwell's equations are solved according to boundary conditions. We know that tangential components of the electric field and normal components of magnetic field components vanish across conductor interfaces.

$$E_{\text{tangential}}=0 \text{ and } H_{\text{normal}}=0$$

As the entire surface of the rectangular waveguide acts as a ground for the electric field, $E_z=0$ all along the boundary walls of the waveguide. There are four boundary conditions as there are four walls, as shown in Figure 5.9.

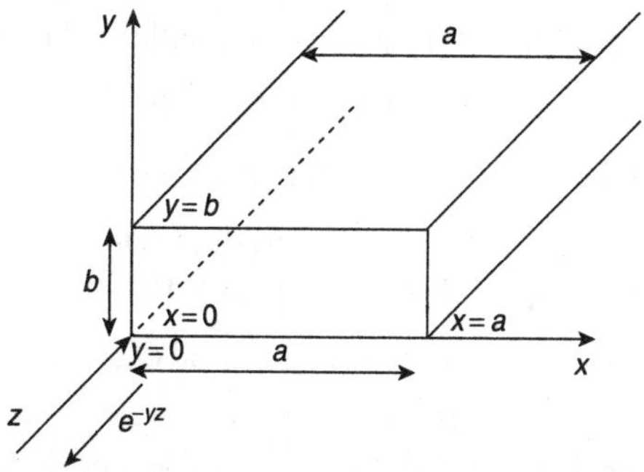

Figure 5.9 *Representing the boundary conditions*

First boundary condition: - [bottom wall]

From the Figure 5.9, along the bottom wall of the waveguide, x and z components are tangential to the bottom wall; so,

$E_z = 0$ along the bottom wall, that is, $E_z = 0$ at $y = 0$ ($\forall\, x \to 0$ to a)

(Note: $\forall$ stands for "for all", and $x \to 0$ to a means that x varies between 0 and a)

Second boundary condition: - [left-side wall]

$$E_z = 0 \text{ at } x = 0 (\forall y \to 0 \text{ to } b)$$

Third boundary condition: - [top wall]

$$E_z = 0 \text{ at } y = \text{b} (\forall x \to 0 \text{ to } a)$$

Fourth boundary condition: - [right-side wall]

$$E_z = 0 \text{ at } x = a (\forall y \to 0 \text{ to } b)$$

(i) By substituting the first boundary condition in Eq. (5.37), we have

$$E_z = 0 \text{ at } y = 0 (\forall x \to 0 \text{ to } a)$$
$$0 = [C_1 \cos Bx + C_2 \sin Bx]\,[C_3 \cos 0 + C_4 \sin 0]$$
$$0 = [C_1 \cos Bx + C_2 \sin Bx]\; C_3 \text{ (since } \cos 0 = 1, \sin 0 = 0)$$

This is true for all $x \to 0$ to a

$$C_1 \cos Bx + C_2 \sin Bx \neq 0$$

Therefore, $C_3 = 0$

Now substituting $C_3 = 0$ in Eq. (5.37), the solution is reduced to

$$E_z = [C_1 \cos Bx + C_2 \sin Bx][C_4 \sin Ay] \tag{5.38}$$

(ii) By substituting the second boundary condition in Eq. (5.38), we have

$$E_z = 0 \text{ at } x = 0 (\forall y \to 0 \text{ to } b)$$
$$[C_1 \cos 0 + C_2 \sin 0][C_4 \sin Ay] = 0$$
$$C_1\, C_4 \sin Ay = 0 \text{ (since } \cos 0 = 1, \sin 0 = 0)$$

This is true for all $y \to 0$ to b

$$\sin Ay \neq 0 \text{ and } C_4 \neq 0$$

Therefore, $C_1 = 0$

Now substituting $C_1 = 0$ in Eq. (5.37), the solution is further reduced to

$$E_z = C_2C_4 \sin Bx \sin Ay \quad (5.39)$$

(iii) By substituting the third boundary condition in Eq. (5.39), we have

$$E_z = 0 \text{ at } y = b \ (\forall x \rightarrow 0 \text{ to } a)$$
$$0 = C_2 C_4 \sin Bx \sin Ab$$

This is true for all $x \rightarrow 0$ to a, as $\sin Bx \neq 0$, $C_4 \neq 0$,and $C_2 \neq 0$; therefore,

$$\sin Ab = 0$$

$Ab = n\pi$, where n is a constant, $n = 0, 1, 2, \ldots$

Therefore,
$$A = \frac{n\pi}{b} \quad (5.40)$$

(iv) Substituting the fourth boundary condition in Eq. (5.39), we have

$$E_z = 0 \text{ at } x = a \ (\forall y \rightarrow 0 \text{ to } b)$$
$$0 = C_2 C_4 \sin Ba \sin Ay$$

This is true for all $y \rightarrow 0$ to b, as $\sin Ay \neq 0$, $C_4 \neq 0$, and $C_2 \neq 0$; therefore,

$$\sin Ba = 0$$

$Ba = m\pi$, where n is a constant, $m = 0, 1, 2, \ldots$

Therefore,
$$B = \frac{m\pi}{a} \quad (5.41)$$

Now, the complete solution is given by substituting A and B values in Eq. (5.39)

$$E_z = C_2 C_4 \sin\left(\left(\frac{m\pi}{a}\right)x\right) \sin\left(\left(\frac{n\pi}{b}\right)y\right) e^{-\gamma z} e^{j\omega t}$$

where $e^{-\gamma z}$ = propagation along the z–direction

$e^{j\omega t}$ = sinusoidal variation with regard to time

The general solution for the z-directed electric field for TM mode propagation is, therefore, given by

$$E_z = C \sin\left(\left(\frac{m\pi}{a}\right)x\right) \sin\left(\left(\frac{n\pi}{b}\right)y\right) e^{j\omega t - \gamma z} \quad (5.42)$$

where C is the product of the C_2 and C_4 constants.

We can find the transverse field components by substituting Eq. (5.42), and $H_Z = 0$ in Eqs. (5.22–5.25) yields the TM field equations in rectangular waveguides as

$$E_x = -\frac{\gamma}{h^2} C\left(\frac{m\pi}{a}\right)\cos\left(\frac{m\pi}{a}\right)x \sin\left(\frac{n\pi}{b}\right)y e^{j\omega t - \gamma z} \quad (5.43)$$

$$E_y = -\frac{\gamma}{h^2} C\left(\frac{n\pi}{b}\right)\sin\left(\frac{m\pi}{a}\right)x \cos\left(\frac{n\pi}{b}\right)y e^{j\omega t - \gamma z} \quad (5.44)$$

$$H_x = \frac{j\omega\varepsilon}{h^2} C\left(\frac{n\pi}{b}\right)\sin\left(\frac{m\pi}{a}\right)x\cos\left(\frac{n\pi}{b}\right)ye^{j\omega t-\gamma z} \tag{5.45}$$

$$H_y = -\frac{j\omega\varepsilon}{h^2} C\left(\frac{m\pi}{a}\right)\cos\left(\frac{m\pi}{a}\right)x\sin\left(\frac{n\pi}{b}\right)ye^{(j\omega t-\gamma z)} \tag{5.46}$$

5.3.1.1.2 Modes of TM waves in rectangular waveguides

The *m* and *n* represent the mode of propagation and indicate the number of variations (half waves) of the field in the *x* and *y* directions. Note that for the TM mode, if *n* or *m* is zero, all fields are zero. The order of the next modes changes depending on the dimensions of the guide. Various modes in TM waves can be obtained, depending on the values of *m* and *n*. Generally, the modes can be representing as TM_{mn}, where *m* and *n* are defined earlier. Various TM_{mn} modes are given below.

TM_{00} mode: $m = 0$ and $n = 0$

Here, $m = 0$ and $n = 0$ are substituted in E_x, E_y, H_x and H_y (Eqs. (5.43) to (5.46)). All these results disappear; that is, the number of half-wave variations along wide and narrow dimensions(a and b, respectively) are zero. Therefore, all the field components vanish inside the waveguide, and the TM_{00} mode cannot exist.

TM_{01} mode: $m = 0$ and $n = 1$

For $m = 0$ and $n = 1$, there is only one half-wave variation of the magnetic field along the narrow dimension, and there is no magnetic field variation along the wide dimension. Again, all the field components vanish inside the waveguide, and, therefore, the TM_{01} mode cannot exist.

TM_{10} mode: $m = 1$ and $n = 0$

For $m = 1$ and $n = 0$, there is only one half-wave variation of the magnetic field along the wide dimension, and there is no magnetic field variation along the narrow dimension. Therefore, again, all the field components vanish inside the waveguide, and the TM_{10} mode cannot exist.

TM_{11} mode: $m = 1$ and $n = 1$

For $m = 1$ and $n = 1$, there is one half-wave variation of the magnetic field along the wide and narrow dimensions. Now, we have all the four components E_x, E_y, H_x, and H_y inside the wave guide. Therefore, the TM_{11} mode exists and for all higher values of m and n, the components exist; that is, all higher modes exist.

5.3.1.1.3 Field components for TE waves

Transverse electric (TE) modes in a rectangular waveguide are characterized by $E_z = 0$ and $H_z \neq 0$. The transverse field components of a TE wave in a rectangular waveguide are as follows:

$$E_x = \frac{j\omega\mu}{h^2} C\left(\frac{n\pi}{b}\right)\cos\left(\frac{m\pi}{a}\right)x\sin\left(\frac{n\pi}{b}\right)ye^{(j\omega t-\gamma z)}$$

$$E_y = -\frac{j\omega\mu}{h^2} C\left(\frac{m\pi}{a}\right)\sin\left(\frac{m\pi}{a}\right)x\cos\left(\frac{n\pi}{b}\right)ye^{(j\omega t-\gamma z)}$$

$$H_x = \frac{\gamma}{h^2} C\left(\frac{m\pi}{a}\right)\sin\left(\frac{m\pi}{a}\right)x\cos\left(\frac{n\pi}{b}\right)ye^{(j\omega t-\gamma z)}$$

$$H_y = \frac{\gamma}{h^2} C\left(\frac{n\pi}{a}\right)\cos\left(\frac{m\pi}{a}\right)x\sin\left(\frac{n\pi}{b}\right)ye^{j\omega t-\gamma z}$$

Derivation of Transverse Field Components of a TE Wave in Rectangular Waveguides

The Helmholtz-wave equation of a TE wave for propagation of the electric field in a lossless medium can be written as

$$\nabla^2 H_z = -\omega^2 \mu\varepsilon H_z$$

Expanding this equation for the z-propagating fields, we have

$$\frac{\partial^2 H_z}{\partial x^2} + \frac{\partial^2 H_z}{\partial y^2} + h^2 H_z = 0 \tag{5.47}$$

This is a partial differential equation that can be solved to get the different field components E_x, H_x, E_y, and H_y. To solve this equation, we employ the method of separation of variable, by assuming a solution of H_z as

$$H_z = XY \tag{5.48}$$

Here, H_z can be expressed as the product of the function X and the function Y.
where X is a pure function of "x" only
Y is a pure function of "y" only
Since X and Y are independent variables,

$$\frac{\partial^2 H_z}{\partial x^2} = \frac{\partial^2 (XY)}{\partial x^2} = Y\frac{d^2 X}{dx^2} \tag{5.49}$$

$$\frac{\partial^2 H_z}{\partial y^2} = \frac{\partial^2 (XY)}{\partial y^2} = X\frac{d^2 Y}{dy^2} \tag{5.50}$$

Substituting Eqs. (5.48–5.50) in Eq. (5.47), we get

$$Y\frac{d^2 X}{dx^2} + X\frac{d^2 Y}{dy^2} + h^2 XY = 0 \tag{5.51}$$

The Eq. (5.51) is divided by XY throughout, therefore we get

$$\frac{1}{X}\frac{d^2 X}{dx^2} + \frac{1}{Y}\frac{d^2 Y}{dy^2} + h^2 = 0 \tag{5.52}$$

$\frac{1}{X}\frac{d^2 X}{dx^2}$ is a pure function of x only, and $\frac{1}{Y}\frac{d^2 Y}{dy^2}$ is a pure function of y only.

The sum of these terms is constant. Since X and Y are independent variables, each term must be equal to a constant separately. Separation of variables method is used to solve the differential equation, Eq. (5.52).

Let
$$\frac{1}{X}\frac{d^2 X}{dx^2} = -B^2 \tag{5.53}$$

and
$$\frac{1}{Y}\frac{d^2 Y}{dy^2} = -A^2 \tag{5.54}$$

where A and B are constants.

Substituting Eqs. (5.53) and (5.54) in Eq. (5.52), we get $-B^2 - A^2 + h^2 = 0$

$$h^2 = A^2 + B^2 \tag{5.55}$$

Equations 5.53 and 5.54 are ordinary second-order differential equations.The differential equation has a general solution, and the solutions of X, Y are given by

$$X = C_1 \cos Bx + C_2 \sin Bx \tag{5.56}$$

$$Y = C_3 \cos Ay + C_4 \sin Ay \tag{5.57}$$

where C_1, C_2, C_3, and C_4 are constants by applying boundary conditions these can be evaluated
The complete solution is given by $H_z = XY$.

Substituting the solutions of X and Y, the solution of Helmholtz equation in rectangular coordinates can be obtained as

$$H_z = [C_1 \cos Bx + C_2 \sin Bx][C_3 \cos Ay + C_4 \sin Ay] \tag{5.58}$$

Boundary conditions

$H_z = 0$ along the boundary walls of the waveguide, since the entire surface of the rectangular waveguide acts as ground for electric field. Since there are four walls, as shown in Figure 5.9 there exists four boundary conditions.

Here since a TE wave is considered

$E_z = 0$ but we have components along the x and y direction
$E_x = 0$ along the bottom and top walls of the waveguide
$E_y = 0$ along the left and right walls of the waveguide

First boundary condition (bottom wall)

$$E_x = 0 \text{ at } y = 0 \ (\forall x \to 0 \text{ to } a)$$

Second boundary condition (top wall)

$$E_x = 0 \text{ at } y = b (\forall x \to 0 \text{ to } a)$$

Third boundary condition (left-side wall)

$$E_y = 0 \text{ at } x = 0 \ (\forall y \to 0 \text{ to } b)$$

Fourth boundary condition (right-side wall)

$$E_y = 0 \text{ at } x = a \ (\forall y \to 0 \text{ to } b)$$

(i) From the first boundary conditions, we have

$$E_x = 0 \text{ at } y = 0 \ (\forall x \to 0 \text{ to } a)$$

Let us write E_x in terms of H_z
From Eq. (5.22), we have

$$E_x = -\frac{\gamma}{h^2}\frac{\partial E_z}{\partial x} - \frac{j\omega\mu}{h^2}\frac{\partial H_z}{\partial y}$$

Since $E_z = 0$, the first term in the above equation is zero; therefore,

$$E_x = -\frac{j\omega\mu}{h^2}\frac{\partial}{\partial y}[(C_1 \cos Bx + C_2 \sin Bx)(C_3 \cos Ay + C_4 \sin Ay)]$$

$$E_x = -\frac{j\omega\mu}{h^2}(C_1 \cos Bx + C_2 \sin Bx)(-AC_3 \sin Ay + A\,C_4 \cos Ay)$$

The above equation can be rewritten as below after substituting 1st boundary condition in it

$$0 = -\frac{j\omega\mu}{h^2}(C_1 \cos Bx + C_2 \sin Bx)(0 + A\,C_4)$$

Since $(C_1 \cos Bx + C_2 \sin Bx) \neq 0, A \neq 0\ C_4 = 0$.
Substituting the value of C_4 in Eq. (5.58) to reduce the solution,

$$H_z = (C_1 \cos Bx + C_2 \sin Bx)(C_3 \cos Ay) \tag{5.59}$$

(ii) From the third boundary condition, we have

$$E_y = 0 \text{ at } x = 0\ \forall\, y \to 0 \text{ to } b$$

From Eq. (5.23), we have

$$E_y = -\frac{\gamma}{h^2}\frac{\partial E_z}{\partial y} + \frac{j\omega\mu}{h^2}\frac{\partial H_z}{\partial x}$$

Since $E_z = 0$ and substituting the value of H_z from Eq. (5.59), we get

$$E_y = \frac{j\omega\mu}{h^2}\frac{\partial}{\partial x}[(C_1 \cos Bx + C_2 \sin Bx)\ C_3 \cos Ay]$$

$$E_y = \frac{j\omega\mu}{h^2}[(-BC_1 \sin\ Bx + BC_2 \cos Bx)C_3 \cos Ay]$$

We get the below equation by substituting 3rd boundary condition in the above equation

$$0 = \frac{j\omega\mu}{h^2}(0 + BC_2)C_3\ \cos Ay$$

Since $\cos Ay \neq 0, B \neq 0$, and $C_3 \neq 0$, therefore $C_2 = 0$.

Substituting the value of C_2 in Eq. (5.59), reduce the solution to

$$H_z = C_1 C_3 \cos Bx \cos Ay \tag{5.60}$$

(iii) From the second boundary condition, we have

$$E_x = 0 \text{ at } y = b\ (\forall\, x \to 0 \text{ to } a)$$

From Eq. (5.22), we have

$$E_x = -\frac{\gamma}{h^2}\frac{\partial E_z}{\partial x} - \frac{j\omega\mu}{h^2}\frac{\partial H_z}{\partial y}$$

Since $E_z = 0$ and substituting the value of H_z from Eq. (5.60), we get

$$E_x = -\frac{j\omega\mu}{h^2}\frac{\partial}{\partial y}[C_1 C_3\ \cos Bx\ \cos Ay]\ \text{(since, } E_z = 0)$$

$$E_x = \frac{j\omega\mu}{h^2}[C_1C_3A \cos Bx \sin Ay]$$

Now, substituting the second boundary condition in the above equation, we get

$$0 = \frac{j\omega\mu}{h^2}[C_1C_3A \cos Bx \sin Ab]$$

Since $\cos Bx \neq 0$, $C_1 \neq 0$, and $C_3 \neq 0$, therefore

$$\sin Ab = 0$$

$Ab = n\pi$, where n is a constant, $n = 0, 1, 2, \ldots$

Therefore, $$A = \frac{n\pi}{b} \quad (5.61)$$

(iv) From the fourth boundary condition, we have

$$E_y = 0 \text{ at } x = a (\forall\, y \to 0 \text{ to } b)$$

From Eq. (5.23), we have

$$E_y = -\frac{\gamma}{h^2}\frac{\partial E_z}{\partial y} + \frac{j\omega\mu}{h^2}\frac{\partial H_z}{\partial x}$$

Since $E_z = 0$ and substituting the value of H_z from Eq. (5.60), we get

$$E_y = \frac{j\omega\mu}{h^2}\frac{\partial}{\partial x}[C_1C_3 \cos Bx \cos Ay]$$

$$E_y = -\frac{j\omega\mu}{h^2}C_1C_3\, B \sin Bx \cos Ay$$

Now, substituting the fourth boundary condition in the above equation, we get

$$0 = -\frac{j\omega\mu}{h^2}C_1C_3\, B \sin Ba \cos Ay,$$

as $\cos Ay \neq 0$, $C_1 \neq 0$,and $C_3 \neq 0$; therefore,

$$\sin Ba = 0$$

$Ba = m\pi$, where n is a constant, $m = 0, 1, 2, \ldots$

$$\text{Therefore, } B = \frac{m\pi}{a} \quad (5.62)$$

Now, the complete solution is given by substituting A and B values in Eq. (5.60):

$$H_z = C_1C_3 \cos\left(\frac{m\pi}{a}x\right)\cos\left(\frac{n\pi}{b}y\right)e^{-\gamma z}e^{j\omega t}$$

where $e^{-\gamma z}$ = propagation along the z direction

$e^{j\omega t}$ = sinusoidal variation with regard to "t"

Let $C = C_1\, C_3$, some other constant.

The general solution for the z-directed electric field for the TM mode propagation is, therefore,

$$H_z = C\cos\left(\frac{m\pi}{a}x\right)\cos\left(\frac{n\pi}{b}y\right)e^{j\omega t-\gamma z} \tag{5.63}$$

We can find the transverse field components by substituting Eq. (5.63), and $E_Z = 0$ in Eqs.(5.22) to (5.25) yields the TE field equations in rectangular waveguides as below:

$$E_x = \frac{j\omega\mu}{h^2}C\left(\frac{n\pi}{b}\right)\cos\left(\frac{m\pi}{a}\right)x\sin\left(\frac{n\pi}{b}\right)ye^{j\omega t-\gamma z} \tag{5.64}$$

$$E_y = -\frac{j\omega\mu}{h^2}C\left(\frac{m\pi}{a}\right)\sin\left(\frac{m\pi}{a}\right)x\cos\left(\frac{n\pi}{b}\right)ye^{(j\omega t-\gamma z)} \tag{5.65}$$

$$H_x = \frac{\gamma}{h^2}C\left(\frac{m\pi}{a}\right)\sin\left(\frac{m\pi}{a}\right)x\cos\left(\frac{n\pi}{b}\right)ye^{(j\omega t-\gamma z)} \tag{5.66}$$

$$H_y = \frac{\gamma}{h^2}C\left(\frac{n\pi}{b}\right)\cos\left(\frac{m\pi}{a}\right)x\sin\left(\frac{n\pi}{b}\right)ye^{(j\omega t-\gamma z)} \tag{5.67}$$

5.3.1.1.4 Modes of TE waves in rectangular waveguides

In the transverse electric (TE) mode, the electric field is perpendicular to the direction of wave propagation. A TE wave has $E_z = 0$ and $H_z \neq 0$. The m and n represent the mode of propagation and indicate the number of ½ variations of the electric field in the x and y directions. Depending on the values of m and n, we have various modes in TE waves. In general, we represent the modes as TE_{mn}. Modes that propagate through a given waveguide are the *propagating modes,* and those which do not propagate are the *evanescent modes.* For example, (i) to transmit energy, we use propagating modes and (ii) while making an attenuator out of a waveguide section, we use evanescent modes. Various TE_{mn} modes are given below based on m and n values:

TE_{00} mode: $m = 0$ and $n = 0$
Here, $m = 0$ and $n = 0$ are substituted in E_x, E_y, H_x, and H_y (Eqs. (5.64) to (5.67)). This results in all of them vanishing; that is, the number of half-wave variations on wide and narrow dimensions are zero. Therefore, all the field components vanish inside the waveguide, and the TE_{00} mode cannot exist.

TE_{01} mode: $m = 0$ and $n = 1$
For $m = 0$ and $n = 1$, there is only one half-wave variation of the electric field along the narrow dimension, and there is no electric field variation along the wide dimension. Therefore, the field components $E_y, = 0$, $H_x, = 0$, $E_x \neq 0$, and $H_y \neq 0$; then, the TE_{01} mode can exist.

TE_{10} mode: $m = 1$ and $n = 0$
For $m = 1$ and $n = 0$, there is only one half-wave variation of the electric field along the wide dimension, and there is no electric field variation along the narrow dimension.Therefore, the field components $E_x, = 0$, $H_y, = 0$, $E_y \neq 0$, and $H_x \neq 0$; then, the TE_{10} mode can exist.

TE_{11} mode: $m = 1$ and $n = 1$
For $m = 1$ and $n = 1$, there is only one half-wave variation of the electric field along the narrow and wide dimensions. Now, we have all the four components E_x, E_y, H_x, and H_y inside the waveguide. Therefore the TE_{11} mode exists and for all higher values of m and n, the components exist; that is, all higher modes exist.

5.3.1.1.5 Impossibility of TEM waves

1. If the TEM wave exists in a waveguide, then the components of the electric field and magnetic field lie in the transverse *x-y* plane. No component of either the electric field or magnetic field lies in the *z* direction.
2. The magnetic field lines should be in closed loops in order to satisfy the equation

   $$\nabla \cdot \vec{H} = 0$$

   From this equation, we can deduce that the magnetic field lines will be in closed loops lying in the transverse *x-y* plane.
3. We consider the Maxwell's equation for Ampere's law:

   $$\nabla \times \vec{H} = \vec{J} + \frac{\partial \vec{D}}{\partial t}$$

   In the given equation, the first component denotes the conduction current density. In waveguides, this current is absent due to the absence of an axial conductor. The second component denotes the displacement current. The displacement current occurs due to the time-varying electric field. However, in the case of TEM waves, the electric field along the *z* direction is zero; so, even the displacement current is zero. With both components being zero, the *H* field becomes zero. This is a violation of Ampere's law. So, the TEM mode cannot exist in hollow waveguides.

5.3.2 Cut-off Frequencies of Rectangular Waveguides

The attenuation occurs below the cut-off frequency which is the operating frequency and propagation takes place above the cut-off frequency. The lowest frequency at which a mode will propagate is the cut-off frequency of an electromagnetic waveguide in it (Figure 5.10).

> The physical dimensions of a waveguide determine the cut-off frequency for each mode.

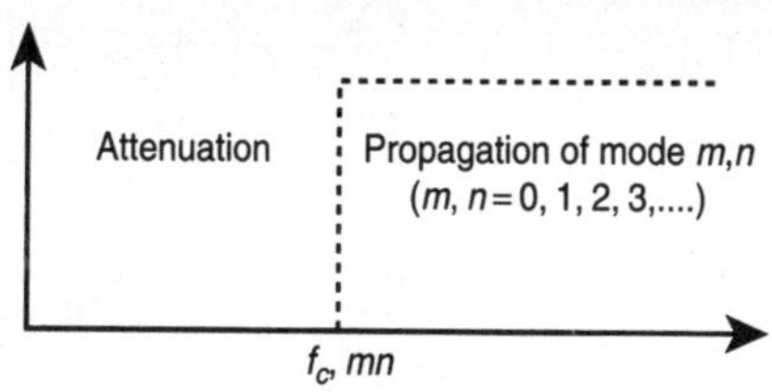

Figure 5.10 *Cut-off frequency of a waveguide*

$$f_c = \frac{c}{2}\left[\left(\frac{m}{a}\right)^2 + \left(\frac{n}{b}\right)^2\right]^{\frac{1}{2}}$$

The cut-off wavelength, λ_c is

$$\lambda_c = \frac{c}{f_c} = \frac{c}{\frac{c}{2}\left[\left(\frac{m}{a}\right)^2 + \left(\frac{n}{b}\right)^2\right]^{\frac{1}{2}}}$$

$$\lambda_c = \frac{2ab}{\sqrt{m^2b^2 + n^2a^2}} \tag{5.68}$$

All wavelengths greater than λ_c are attenuated, and those less than λ_c are transmitted.

Derivation of Cut-off Frequency and Cut-off Wavelength

The cut-off frequency is found in the characteristic Helmholtz equation for electromagnetic waves. From the relation $h^2 = \gamma^2 + \omega^2 \mu\varepsilon$ and $h^2 = A^2 + B^2$, we can write

$$h^2 = \gamma^2 + \omega^2 \mu\varepsilon = A^2 + B^2 \tag{5.69}$$

Now, substituting the values of A and B from Eqs. (5.61) and (5.62) in Eq. (5.69), we get

$$h^2 = \gamma^2 + \omega^2 \mu\varepsilon = A^2 + B^2 = \left(\frac{n\pi}{b}\right)^2 + \left(\frac{m\pi}{a}\right)^2 \tag{5.70}$$

$$\gamma^2 = \left(\frac{m\pi}{a}\right)^2 + \left(\frac{n\pi}{b}\right)^2 - \omega^2 \mu\varepsilon$$

$$\gamma = \sqrt{\left(\frac{m\pi}{a}\right)^2 + \left(\frac{n\pi}{b}\right)^2 - \omega^2 \mu\varepsilon} = \alpha + j\beta \tag{5.71}$$

γ is the propagation constant of the wave in waveguides along the direction of propagation.

At lower frequencies, $\omega^2 \mu\varepsilon < \left(\frac{m\pi}{a}\right)^2 + \left(\frac{n\pi}{b}\right)^2$

i.e. γ then becomes real and positive and equal to the attenuation constant 'α' i.e. there is no phase change and the wave is completely attenuated. Thus the wave cannot propagate

However, at higher frequencies, $\omega^2 \mu\varepsilon > \left(\frac{m\pi}{a}\right)^2 + \left(\frac{n\pi}{b}\right)^2$

That is, when $\omega^2 \mu\varepsilon > \left(\frac{m\pi}{a}\right)^2 + \left(\frac{n\pi}{b}\right)^2 \Rightarrow \gamma = j\beta$ and $\alpha = 0$

There will be phase change β, γ becomes imaginary and hence the wave propagates. At the transition, the propagation just starts and γ becomes zero.

The frequency at which γ just becomes zero is defined as the cut-off frequency (or threshold frequency), 'f_c'.

$$\text{At } \gamma = 0, f = f_c \text{ or } \omega = 2\pi f = 2\pi f_c = \omega_c$$

Therefore, $$0 = \left(\frac{m\pi}{a}\right)^2 + \left(\frac{n\pi}{b}\right)^2 - \omega^2 \mu\varepsilon$$

$$\omega_c = \frac{1}{\sqrt{\mu\varepsilon}}\left[\left(\frac{m\pi}{a}\right)^2 + \left(\frac{n\pi}{b}\right)^2\right]^{\frac{1}{2}} \tag{5.72}$$

In a rectangular wavelength, the cut-off frequencies and cut-off wavelength is given as,

$$f_c = \frac{1}{2\pi\sqrt{\mu\varepsilon}}\left[\left(\frac{m\pi}{a}\right)^2 + \left(\frac{n\pi}{b}\right)^2\right]^{\frac{1}{2}} \tag{5.73}$$

$$f_c = \frac{c}{2\pi}\left[\left(\frac{m\pi}{a}\right)^2 + \left(\frac{n\pi}{b}\right)^2\right]^{\frac{1}{2}} \qquad \left(\therefore \text{Since } c = \frac{1}{\sqrt{\mu\varepsilon}}\right)$$

$$f_c = \frac{c}{2}\left[\left(\frac{m}{a}\right)^2 + \left(\frac{n}{b}\right)^2\right]^{\frac{1}{2}} \tag{5.74}$$

This shows that the cut-off frequency depends only on the physical dimensions (a and b) of the waveguide and the properties of the medium (μ and ε).

The cut-off wavelength, λ_c is

$$\lambda_c = \frac{c}{f_c} = \frac{c}{\frac{c}{2}\left[\left(\frac{m}{a}\right)^2 + \left(\frac{n}{b}\right)^2\right]^{\frac{1}{2}}} \tag{5.75}$$

$$\lambda_c = \frac{2ab}{\sqrt{m^2b^2 + n^2a^2}} \tag{5.76}$$

5.3.3 Filter Characteristics

The cut-off frequency defines the high-pass filter characteristic of the waveguide. Thus, for low frequencies, the waveguide does not allow propagation; whereas for high frequencies, the waveguide allows propagation. Therefore, the waveguide behaves similar to a high-pass filter. The cut-off frequency depends on the shape and size of the cross-section of the waveguide.

The filter characteristics of waveguides can be derived from the phase constant expressed in terms of cut-off frequency (f_c).

The phase constant is given by $\beta = \omega\sqrt{\mu\varepsilon}\left(\sqrt{1-(f_c/f)^2}\right)$

Let us assume that for a given waveguide, $f > f_c$; then, the phase constant is real, and this means that the wave of this frequency propagates along the waveguide. If $f < f_c$, the expression under square root in the above equation is negative. Then, the phase constant becomes imaginary, and the propagation factor becomes real. This means that waves of a frequency lower than f_c cannot propagate along the waveguide.

$$\lambda_o > \lambda_c \text{ or } f < f_c \Rightarrow \text{No propagation}$$

$$\lambda_o < \lambda_c \text{ or } f > f_c \Rightarrow \text{propagation is possible}$$

Derivation of Phase Constant in Terms of f_c

The propagation constant (γ) is related to the frequency, ω, of the wave through the relation $h^2 = \gamma^2 + \omega^2\mu\varepsilon$ and $h^2 = A^2 + B^2$

Therefore,
$$h^2 = \gamma^2 + \omega^2\mu\varepsilon = A^2 + B^2 \tag{5.77}$$

Now, substituting the values of A and B in the above equation, we get

$$h^2 = \gamma^2 + \omega^2 \mu\varepsilon = A^2 + B^2 = \left(\frac{m\pi}{a}\right)^2 + \left(\frac{n\pi}{b}\right)^2$$

$$\gamma^2 = \left(\frac{m\pi}{a}\right)^2 + \left(\frac{n\pi}{b}\right)^2 - \omega^2 \mu\varepsilon \tag{5.78}$$

At $\gamma = 0, f = f_c$ or $\omega = 2\pi f = 2\pi f_c = \omega_c$

Therefore, $$0 = \left(\frac{m\pi}{a}\right)^2 + \left(\frac{n\pi}{b}\right)^2 - \omega_c^{\,2} \mu\varepsilon$$

$$\omega_c^{\,2} \mu\varepsilon = \left(\frac{m\pi}{a}\right)^2 + \left(\frac{n\pi}{b}\right)^2 \tag{5.79}$$

By substituting Eq. (5.79) in Eq. (5.78), we get

$$\gamma^2 = \omega_c^{\,2} \mu\varepsilon - \omega^2 \mu\varepsilon \tag{5.80}$$

For a wave propagation along the lossless waveguide, $\gamma = j\beta$ (as $\alpha = 0$)

$$(j\beta)^2 = \omega_c^{\,2} \mu\varepsilon - \omega^2 \mu\varepsilon$$

$$(\beta)^2 = \omega^2 \mu\varepsilon - \omega_c^{\,2} \mu\varepsilon$$

$$(\beta)^2 = \omega^2 \mu\varepsilon \left(1 - \frac{\omega_c^{\,2} \mu\varepsilon}{\omega^2 \mu\varepsilon}\right)$$

$$\beta = \sqrt{\omega^2 \mu\varepsilon \left(1 - \frac{\omega_c^{\,2}}{\omega^2}\right)}$$

The expression for phase constant is $\beta = \omega\sqrt{\mu\varepsilon}\left(\sqrt{1-(f_c/f)^2}\right)$ (5.81)

5.3.4 Wave Impedance in a Rectangular Waveguide

The *wave impedance* of a waveguide is defined as the ratio of the strength of electric field in one direction and the magnetic field along the other transverse direction at a certain point in the waveguide.

$$Z = \frac{E_x}{H_y} = \frac{-E_y}{H_x} \tag{5.82}$$

5.3.4.1 *Wave impedance for a TM wave in a rectangular waveguide*

Wave impedance for a TM wave in a rectangular wave can be derived from Eq. 5.82

$$Z = Z_{TM} = \frac{E_x}{H_y} = \frac{\dfrac{-\gamma}{h^2}\dfrac{\partial E_z}{\partial x} - \dfrac{j\omega\mu}{h^2}\dfrac{\partial H_z}{\partial y}}{\dfrac{-\gamma}{h^2}\dfrac{\partial H_z}{\partial y} - \dfrac{j\omega\varepsilon}{h^2}\dfrac{\partial E_z}{\partial x}} \tag{5.83}$$

For a TM wave, $H_z = 0$ and $\gamma = j\beta$

$$Z_{TM} = \frac{\dfrac{-\gamma}{h^2}\dfrac{\partial E_z}{\partial x}}{-\dfrac{j\omega\varepsilon}{h^2}\dfrac{\partial E_z}{\partial x}} = \frac{\gamma}{j\omega\varepsilon} = \frac{j\beta}{j\omega\varepsilon}$$

$$Z_{TM} = \frac{\beta}{\omega\varepsilon} \tag{5.84}$$

We know that $\beta = \sqrt{\omega^2\mu\varepsilon - \omega_c^2\mu\varepsilon}$

$$Z_{TM} = \frac{\sqrt{\omega^2\mu\varepsilon - \omega_c^2\mu\varepsilon}}{\omega\varepsilon} = \frac{\sqrt{\mu\varepsilon}.\sqrt{\omega^2 - \omega_c^2}}{\varepsilon.\omega}$$

$$= \sqrt{\frac{\mu}{\varepsilon}}\frac{\sqrt{\omega^2 - \omega_c^2}}{\omega} = \sqrt{\frac{\mu}{\varepsilon}}\sqrt{1-\left(\frac{\omega_c}{\omega}\right)^2}$$

$$\therefore Z_{TM} = \sqrt{\frac{\mu}{\varepsilon}}\sqrt{1-\left(\frac{f_c}{f}\right)^2} \text{ (or)} \sqrt{\frac{\mu}{\varepsilon}}\sqrt{1-\left(\frac{\lambda_0}{\lambda_c}\right)^2} \tag{5.85}$$

$$\sqrt{\frac{\mu}{\varepsilon}} = \sqrt{\frac{\mu_0\mu_r}{\varepsilon_0\varepsilon_r}} = \sqrt{\frac{4\pi\times10^{-7}}{1/36\pi\times10^{-9}}} = \sqrt{4\pi\times36\pi\times10^2} = 120\pi = 377\Omega = \eta$$

where η is the intrinsic impedance of free space

$$\therefore Z_{TM} = \eta\sqrt{1-(\lambda_0/\lambda_c)^2} \tag{5.86}$$

Since λ_0 is always less than λ_c for wave propagation, $Z_{TM} < \eta$.

This shows that the wave impedance for a TM wave is always less than the free space impedance.

5.3.4.2 Wave impedance of TE waves in rectangular waveguides

Wave impedance for a TE wave in a rectangular wave can be derived from Eq. 5.82

$$Z = Z_{TE} = \frac{E_x}{H_y} = \frac{\dfrac{-\gamma}{h^2}\dfrac{\partial E_z}{\partial x} - \dfrac{j\omega\mu}{h^2}\dfrac{\partial H_z}{\partial y}}{\dfrac{-\gamma}{h^2}\dfrac{\partial H_z}{\partial y} - \dfrac{j\omega\varepsilon}{h^2}\dfrac{\partial E_z}{\partial x}} \tag{5.87}$$

For a TE wave, $E_z = 0$ and $\gamma = j\beta$

$$Z_{TE} = \frac{\dfrac{-j\omega\mu}{h^2}\dfrac{\partial H_z}{\partial y}}{-\dfrac{\gamma}{h^2}\dfrac{\partial H_z}{\partial y}} = \frac{j\omega\mu}{\gamma} = \frac{j\omega\mu}{j\beta} = \frac{\omega\mu}{\beta} \tag{5.88}$$

$$Z_{\text{TE}} = \frac{\omega\mu}{\sqrt{\mu\varepsilon}\sqrt{\omega^2 - \omega_c^2}} = \frac{\eta}{\sqrt{1-(\omega_c/\omega)^2}} = \frac{\eta}{\sqrt{1-(f_c/f)^2}} \tag{5.89}$$

$$Z_{\text{TE}} = \frac{\eta}{\sqrt{1-(\lambda_0/\lambda_c)^2}} \tag{5.90}$$

Therefore, $Z_{\text{TE}} > \eta$ as $\lambda_0 < \lambda_c$ for wave propagation. This shows that the wave impedance for a TE wave is always greater than the free space impedance.

$$Z_{\text{TE}} = \eta_{\text{TE}} = \frac{\eta}{\sqrt{1-(f_c/f)^2}} \tag{5.91}$$

$$Z_{\text{TM}} = \eta_{\text{TM}} = \eta\sqrt{1-(f_c/f)^2} \tag{5.92}$$

$$\eta_{\text{TE}}\eta_{\text{TM}} = \eta^2 \tag{5.93}$$

5.3.5 Dominant Mode and Degenerate Modes

A dominant mode in a waveguide is the mode having the lowest cut-off frequency (or highest cut-off wavelength). From previous discussion, it is clear that the walls of the waveguides can be considered nearly perfect conductors. Hence, according to the boundary conditions,

(i) the electric field should be normal (perpendicular), to the waveguide walls.
(ii) the magnetic fields should be tangential to the waveguide walls.

A zero subscript can be in the TE mode but not in the TM mode, according to the above mentioned boundary conditions. For example, the modes that can exist in a rectangular waveguide are TE_{10}, TE_{01}, and TE_{20}; while in TM, only the TM_{11}, TM_{12}, and TM_{21} modes can exist. Besides, as per the cut-off frequency relationship, the propagation of modes is determined by the physical size of the waveguide depending on the values of *m* and *n*.

The minimum cut-off frequency for a rectangular waveguide is obtained for a dimension $a = 2b$ and for $m = 1$ and $n = 0$. Thus, the dominant mode for a rectangular waveguide is the TE_{10} mode and is the most commonly used. This mode also provides low attenuation propagation in the *z* axis. The field equations for the TE_{10} mode are given by

$$H_z = C\cos\left(\frac{\pi}{a}\right)x e^{j\omega t-\gamma z}$$

$$E_y = -\frac{j\omega\mu}{h^2}C\left(\frac{\pi}{a}\right)\sin\left(\frac{\pi}{a}\right)x\, e^{(j\omega t-\gamma z)}$$

$$H_x = \frac{\gamma}{h^2}C\left(\frac{\pi}{a}\right)\sin\left(\frac{\pi}{a}\right)x\, e^{(j\omega t-\gamma z)}$$

$$E_z = H_y = E_x = 0$$

It can be seen that the field H_z is a cosine function of *x* with its zero at the center. It repeats itself with a period of λ_g in the *z* direction. In the *x* direction, E_y has only a sine term and, hence, shows sinusoidal variations with its maximum at the center ($x = a/2$) of the waveguide. In addition, the magnetic field component H_x has the same sinusoidal variation, but the direction will be opposite to that of E_y. Therefore, H_z has a phase difference of 90° from E_y, H_x.

E-field patterns of TE_{10} mode: E-field patterns of TE_{10} mode,as shown in Figure 5.11 (a)

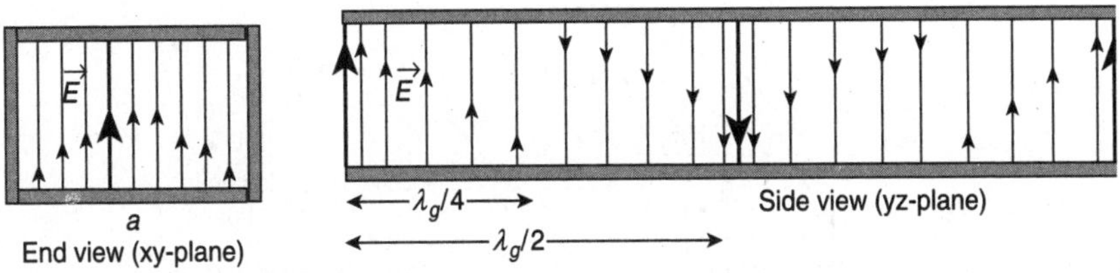

Figure 5.11 (a) *TE_{10} E-field patterns*

H-field patterns of TE_{10} mode: Typical 3D view H-field patterns of TE_{10} mode as shown in Figure 5.11 (b). The magnetic flux lines appear as continuous loops.

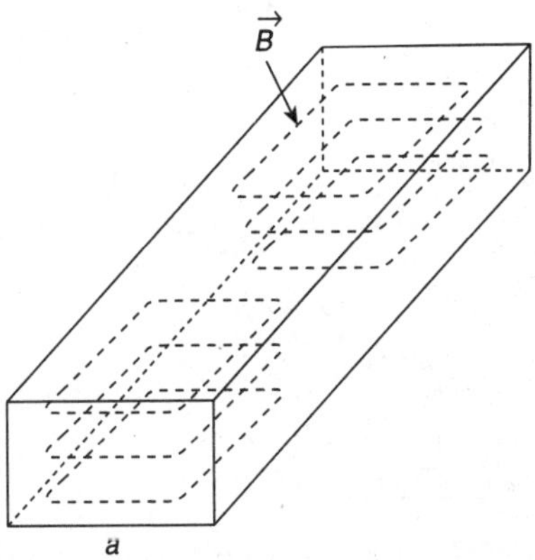

Figure 5.11 (b) *TE_{10} H-field patterns*

Combined E- and H-field patterns in TE_{10}: Figure 5.11 (c) shows the combined *E*-field and *H*-field patterns in TE_{10} mode.

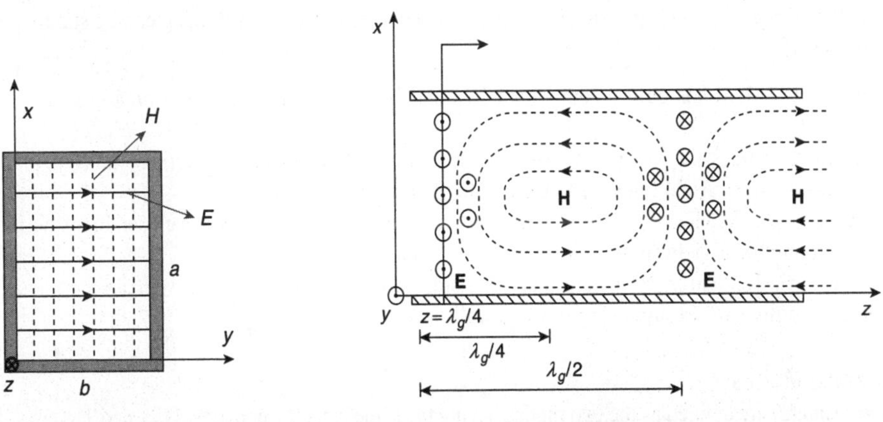

Figure 5.11 (c) *TE_{10} mode end view and top view of E- and H-field patterns*

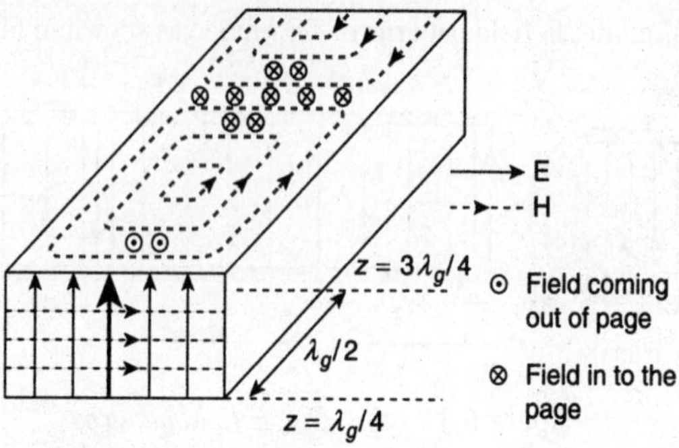

Figure 5.12 (a) *Top and front view of TE_{10} mode E- and H-field patterns in a rectangular waveguide at a given instant of time*

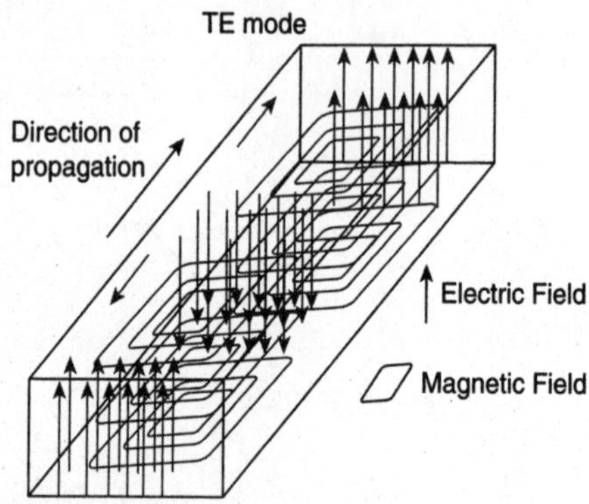

Figure 5.12 (b) *3D view of TE_{10} mode E- and H-field patterns in a rectangular waveguide at a given instant of time*

The field patterns shown above are obtained at an instant of time by substituting $m = 1$, $n = 0$, in Eqs. (5.64 – 5.67) and Eq. (5.77). Though the TE wave mode starts at $z = 0$ here in Figure 5.12 (a) and (b), the initial point of z is taken at $z = \lambda_g/4$, as it is convenient to show the maximum electric field lines. At $z = 0$, the the electric field component (E_y) (Eq. 5.65) becomes zero. The following can be summarized from the above figures:

1. The electric field has only y component (E_y) and is constant. It varies sinusoidally along the z direction.
2. Electric field lines are shown as vectors and are always normal to the conducting walls (x-z plane); whereas magnetic field lines are tangential to the conducting walls.
3. The magnetic field has H_x and H_z components, and the field lines are always closed loops horizontally around the E field (in the z direction or x-z plane).
4. The electric field strength is zero along the side walls of the waveguide, whereas the magnetic field strength is maximum along the side walls.

Degenerate Modes

Degenerate modes are defined as the two modes having the same cut-off frequency. TE_{mn} and TM_{mn} modes are called *degenerate modes* for a rectangular waveguide, when both m and n values corresponding to these modes are equal. It is necessary that higher-order degenerate modes are not supported by the guide in the operating band of frequencies to avoid undesirable components appearing at the output along with losses.

5.3.6 Sketches of TE and TM Mode Fields in the Cross-Section

As discussed, in the *x-y* cross section of the guide, the number of half-cycle variations is indicated by the integers *m* and *n*. Hence, the field configuration as given in Figure 5.13 gives the configuration of a different mode, for a fixed time.

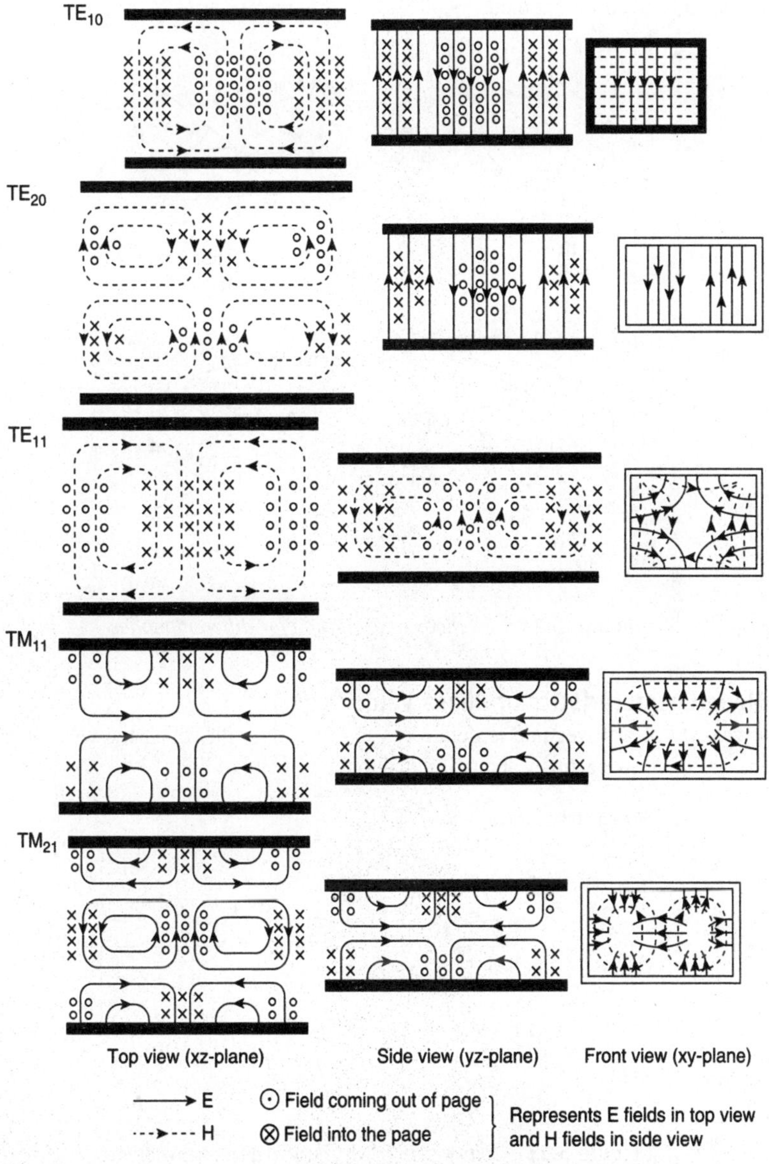

Figure 5.13 *Field configurations of different modes in rectangular waveguides*

5.3.7 Excitation of Modes in Rectangular Waveguides

Excitation of modes in waveguides is done by introducing an excited probe or antenna that is oriented in the direction of the electric field. The probe is often placed near a point where the maximum electric field is present in the mode pattern. The placing of the probe in a waveguide is of utmost importance. For example, in the TE mode, the probe is placed below the waveguide as shown in Figure 5.14. The distance of the probe from the surface of the wall is $\lambda/2$. If this length is not observed, then there is a possibility of improper excitation. Different methods of excitation are given in Figure 5.14. Each method corresponds to a different mode.

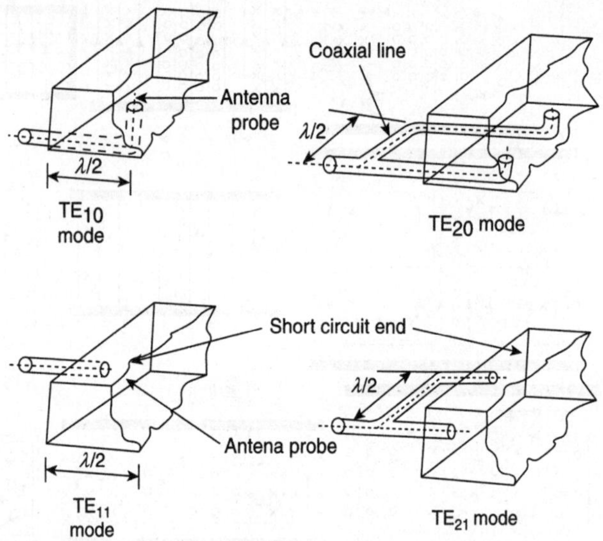

Figure 5.14 *Methods of excitation for different modes*

5.3.8 Mode Characteristics of Phase Velocity (v_p)

The *phase velocity* of a waveguide is defined as the velocity with which the electromagnetic wave changes phase. In free space, this velocity is equal to the velocity of light.

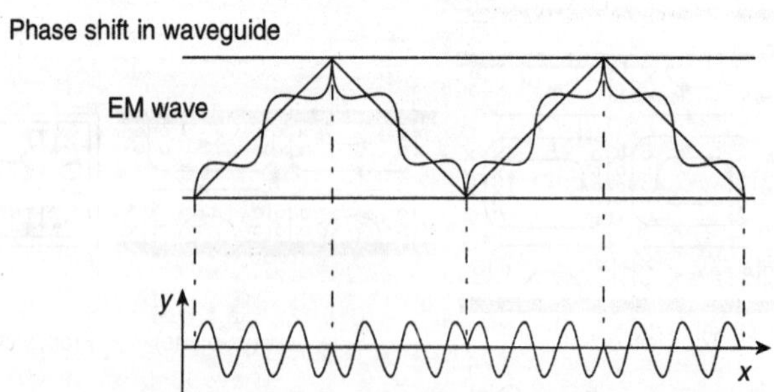

Figure 5.15 *Phase shift of an EM wave in a waveguide*

In waveguides, the conducting walls act as reflectors for incoming EM waves. For every reflection, there is a 180° phase shift. The rate of phase change in the EM wave is greater than the velocity of light,as there is a lead of 180° for every reflection. The change in the phase of an EM wave in a waveguide is shown in Figure. 5.15.

The *phase velocity* is given by

$$v_p = \frac{\omega}{\beta} \tag{5.94}$$

Here, $\omega = 2\pi f, \beta = \frac{2\pi}{\lambda_g}$, where λ_g = the wavelength of the wave inside the waveguide

From Eq. (5.81), $\beta = \omega\sqrt{\mu\varepsilon}\left(\sqrt{1-(f_c/f)^2}\right)$

The expression for v_p is given by

$$v_p = \frac{\omega}{\beta} = \frac{\omega}{\omega\sqrt{\mu\varepsilon}\left(\sqrt{1-(f_c/f)^2}\right)} = \frac{1}{\sqrt{\mu\varepsilon}\left(\sqrt{1-(f_c/f)^2}\right)} \tag{5.95}$$

That is,

$$v_p = \frac{c}{\sqrt{1-(f_c/f)^2}} \tag{5.96}$$

It is also known that f (any frequency) = c/λ_0, where λ_0 is free space wavelength and f_c (cut-off frequency) = c/λ_c, where λ_c is cut-off wavelength

$$\frac{f_c}{f} = \frac{c}{\lambda_c}\frac{\lambda_0}{c} = \frac{\lambda_0}{\lambda_c}$$

$$\therefore v_p = \frac{c}{\sqrt{1-(\lambda_0/\lambda_c)^2}} \tag{5.97}$$

The wavelength along the incident surface $\lambda_p = \frac{\lambda}{\sin\theta}$

Then, the phase velocity of the wave along the incident surface becomes

$$v_p = f\lambda_p = \frac{f\lambda}{\sin\theta} = \frac{c}{\sin\theta} \tag{5.98}$$

5.3.9 Mode Characteristics of Group Velocity (v_g)

Group velocity is defined as the velocity of the wave in the direction parallel to the conducting surface. The time taken by an EM wave to travel through the waveguide is more than the time taken by an EM wave to cover the length of the waveguide. This is due to the fact that the path travelled by the EM wave in the waveguide is much longer than the length of the waveguide. This virtual velocity of the EM wave is known as *group velocity* and is shown in Figure 5.16.

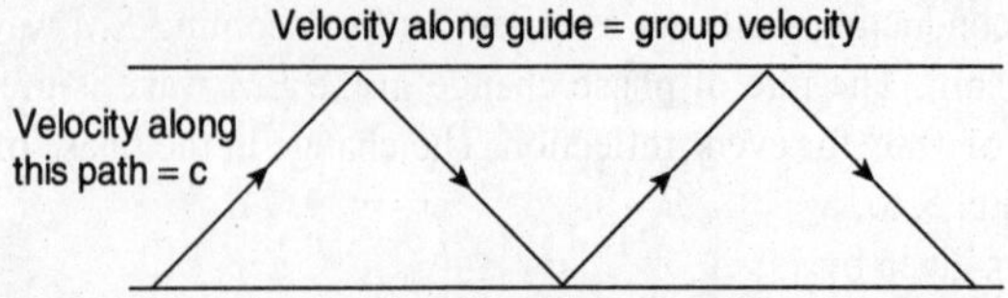

Figure 5.16 *Virtual velocity of EM wave in waveguide*

It is defined as the rate at which the wave propagates through the waveguide and is given by

$$v_g = \frac{d\omega}{d\beta} \tag{5.99}$$

We know $\beta = \sqrt{\mu\varepsilon(\omega^2 - \omega_c^2)}$

Now β is differentiated with respect to 'ω', to obtain the below relation

$$\frac{d\beta}{d\omega} = \frac{1}{2\sqrt{(\omega^2 - \omega_c^2)\mu\varepsilon}} 2\omega\mu\varepsilon$$

$$\frac{d\beta}{d\omega} = \frac{\sqrt{\mu\varepsilon}}{\sqrt{1-(\omega_c/\omega)^2}} = \frac{\sqrt{\mu\varepsilon}}{\sqrt{1-(f_c/f)^2}}$$

$$v_g = \frac{d\omega}{d\beta} = \frac{\sqrt{1-(f_c/f)^2}}{\sqrt{\mu\varepsilon}} \tag{5.100}$$

$$\therefore v_g = c\sqrt{1-\left(\frac{\lambda_0}{\lambda_c}\right)^2} \tag{5.101}$$

Considering the product of v_p and v_g:

i.e., $$v_p \cdot v_g = \frac{c}{\sqrt{1-(\lambda_0/\lambda_c)^2}} c\sqrt{1-(\lambda_0/\lambda_c)^2}$$

$$\therefore \; v_p v_g = c^2 \tag{5.102}$$

5.3.10 Wavelength and Impedance Relations

The relationship between λ_g, λ_0 and λ_c is given by

The phase velocity is $v_p = \lambda_g f = \frac{\lambda_g}{\lambda_0} c$

Also

$$v_p = \frac{c}{\sqrt{1-\left(\frac{\lambda_0}{\lambda_c}\right)^2}}$$

$$\Rightarrow \frac{c}{\sqrt{1-\left(\frac{\lambda_0}{\lambda_c}\right)^2}} = \frac{\lambda_g}{\lambda_0}c$$

$$\therefore \lambda_g = \frac{\lambda_0}{\sqrt{1-\left(\frac{\lambda_0}{\lambda_c}\right)^2}} \tag{5.103}$$

$$\frac{1}{\lambda_0^2} = \frac{1}{\lambda_g^2} + \frac{1}{\lambda_c^2} \tag{5.104}$$

5.3.11 Power Transmission in a Rectangular Waveguide

The power transmitted through the waveguide can be obtained by integrating the Poynting vector over the waveguide cross-section. Assume that there will be no reflections from the receiving end or that the waveguide is infinitely long when compared to its wavelength, if the guide is terminated.

Complex Poynting vector is given as $P = \text{Re}(E \times H^*)$

The power transmitted, P_{tr}, through a waveguide is given by

$$P_{tr} = \oint P_{avg} \cdot ds = \oint \frac{P}{2} \cdot ds = \oint \frac{1}{2} \text{Re}(E \times H^*) \cdot ds \tag{5.105}$$

The average power flow through a rectangular waveguide for a lossless dielectric is

$$P_{tr} = \frac{1}{2Z} \oint |E|^2 \, ds = \frac{Z}{2} \oint |H|^2 \, ds \tag{5.106}$$

where $$Z = \frac{E_x}{H_y} = \frac{-E_y}{H_x}$$

$$|E|^2 = |E_x|^2 + |E_y|^2 \tag{5.107}$$

$$|H|^2 = |H_x|^2 + |H_y|^2 \tag{5.108}$$

For TM_{mn} mode, $$Z_{TM} = \eta\sqrt{1-(\lambda_0/\lambda_c)^2}$$

The relation given below represents the average power transmitted through a rectangular waveguide in TM_{mn} mode of dimensions *a* and *b*.

$$P_{tr} = \frac{1}{2\eta\sqrt{1-(\lambda_0/\lambda_c)^2}} \int_{y=0}^{b}\int_{x=0}^{a} |E_x|^2 + |E_y|^2 \, dxdy \tag{5.109}$$

For the TE_{mn} mode, $$Z_{TE} = \frac{\eta}{\sqrt{1-(\lambda_0/\lambda_c)^2}} \tag{5.110}$$

$$\therefore P_{tr} = \frac{\sqrt{1-(\lambda_0/\lambda_c)^2}}{2\eta} \int_{y=0}^{b}\int_{x=0}^{a} |E_x|^2 + |E_y|^2 \, dxdy \tag{5.111}$$

5.3.12 Transmission Losses in a Rectangular Waveguide

The transmission losses in a waveguide are primarily due to losses in a dielectric medium filling the space between the conductors in which the field propagates, and others are due to losses in conductors. Consider α_d and α_c as the attenuation constants due to the losses in the dielectric and conductors; then, the attenuation constant of the waveguide is

$$\alpha = \alpha_d + \alpha_c$$

The attenuation constant due to losses in the dielectric is given as

$$\alpha_d = \frac{Z_{TE}\sigma}{2} \tag{5.112}$$

where σ is the conductivity of the medium

The attenuation constant due to losses in conductors in the TE_{10} mode is given as

$$\alpha_c = \frac{R_s\left[1+\frac{2b}{a}\left(\frac{f_c}{f}\right)^2\right]}{b\eta\sqrt{1-\left(\frac{f_c}{f}\right)^2}} \tag{5.113}$$

where R_s = surface resistance

The attenuation constant of the waveguide is given by

$$\alpha = (\alpha_d + \alpha_c) = \left\{\frac{Z_{TE}\sigma}{2} + \frac{R_s\left[1+\frac{2b}{a}\left(\frac{f_c}{f}\right)^2\right]}{b\eta\sqrt{1-\left(\frac{f_c}{f}\right)^2}}\right\} \tag{5.114}$$

EXAMPLE PROBLEM 5.1

A rectangular waveguide has a cross-sectional area of 2.29 × 1.45 cm^2, and the operating frequency is 10 GHz. Calculate the following (a) Free space wavelength; (b) Cut-off wavelength; (c) Cut-off frequency; (d) Angle of incidence; (e) Guided wavelength; (f) Phase velocity; (g) Phase shift constant; (h) Wave impedance of the guide.

Solution

Given f = 10 GHz, a = 2.29 cm, and b = 1.45 cm,

(a) Free space wavelength, $\lambda_0 = \frac{c}{f} = \frac{3\times10^{10}}{10\times10^9} = 3\,\text{cm}$

(b) For TE_{10}, cut-off wavelength $\lambda_c = 2a = 2\times2.29 = 4.58$ cm

(c) Cut-off frequency, $f_c = \frac{c}{\lambda_c} = \frac{3\times10^8}{4.58\times10^{-2}} = 6.55\,\text{GHz}$

(d) Angle of incidence, $\theta = \sin^{-1}\left(\frac{\lambda_0}{2a}\right) = \sin^{-1}\left(\frac{3}{4.58}\right) = 40.92°$

(e) Guided wavelength, $\lambda_g = \frac{\lambda_0}{\sqrt{1-((f_c/f))^2}} = \frac{3}{\sqrt{1-\left(6.55\times10^9/10\times10^9\right)^2}} = 3.97\,\text{cm}$

(f) Phase velocity in the guide,

$$v_p = \frac{c}{\sqrt{1-(f_c/f)^2}} = \frac{3\times10^8}{\sqrt{1-\left(6.55\times10^9/10\times10^9\right)^2}} = 3.97\times10^8 \text{ m/sec}.$$

(g) Phase shift constant, $\beta_g = \frac{2\pi}{\lambda_g} = \frac{2\pi}{3.97} = 1.58$

(h) Wave impedance of the waveguide, for TE mode, $Z_g = Z_0\left(\frac{\lambda_g}{\lambda_0}\right) = 377\times\left(\frac{3.97}{3}\right) = 498.89\ \Omega$

for TM mode $Z_g = Z_0\left(\frac{\lambda_0}{\lambda_g}\right) = 377\times\left(\frac{3}{3.97}\right) = 284.88\ \Omega$ ■

EXAMPLE PROBLEM 5.2

An air-filled rectangular waveguide has dimensions a = 6 cm and b = 4 cm. The signal frequency is 4 GHz. Compute the cut-off frequency for the TE_{10}, TE_{01}, TE_{11}, and TM_{11} modes.

Solution

The cut-off frequency for TE_{mn} and TM_{mn} modes is

$$f_c = \frac{c}{2}\left[\left(\frac{m}{a}\right)^2 + \left(\frac{n}{b}\right)^2\right]^{1/2}$$

For the TE_{10} mode, the cut-off frequency is

$$f_{c_{10}} = \frac{3\times10^8}{2}\sqrt{\left(\frac{1}{0.06}\right)^2 + \left(\frac{0}{0.04}\right)^2} = 2.5 \text{ GHz}.$$

For the TE_{01} mode, the cut-off frequency is

$$f_{c_{01}} = \frac{3\times10^8}{2}\sqrt{\left(\frac{0}{0.06}\right)^2 + \left(\frac{1}{0.04}\right)^2} = 3.75 \text{ GHz}.$$

For the TE_{11} and TM_{11}modes, the cut-off frequency is

$$f_{c_{11}} = \frac{3\times10^8}{2}\sqrt{\left(\frac{1}{0.06}\right)^2 + \left(\frac{1}{0.04}\right)^2} = 4.506 \text{ GHz}.$$

Since the signal frequency of 4 GHz is below f_c for TE_{11} and TM_{11}, it will be attenuated. ■

EXAMPLE PROBLEM 5.3

Calculate the cut-off frequency of the following modes in a square waveguide 4 cm × 4 cm TE_{10}, TM_{11}, and TE_{22}.

Solution

A square waveguide is a special case of a rectangular waveguide, where $a = b$.
For TE_{10}, $\lambda_c = 2a = 8$ cm

$$f_c = \frac{3\times10^8}{8\times10^{-2}} = 3.75 \text{ GHz}$$

For TE_{11},

$$\lambda_c = \frac{1}{\sqrt{\left(\frac{m}{2a}\right)^2 + \left(\frac{n}{2b}\right)^2}} = \frac{2a}{\sqrt{2}} = \sqrt{2}a = 4\sqrt{2} \text{ cm}$$

$$f_c = \frac{3\times10^8}{4\sqrt{2}\times10^{-2}} = 5.303 \text{ GHz}$$

For TE_{22},

$$\lambda_c = \frac{a}{\sqrt{2}} = \frac{4}{\sqrt{2}} \text{cm}; f_c = \frac{\sqrt{2}\times3\times10^8}{4\times10^{-2}} = 10.607 \text{ GHz}$$

■

EXAMPLE PROBLEM 5.4

The guided wavelength for a frequency of 20,000 MHz is 6 cm, when the dominant mode is propagated in an air-filled rectangular waveguide. Find the breadth of the guide?

Solution

For a rectangular waveguide, the dominant mode is the TE_{10} mode. TE_{10} mode can propagate at a lower frequency.

Given $f = 20{,}000$ MHz = 20 GHz; $\lambda_g = 6$ cm
We know for the TE_{10} mode, $\lambda_c = 2a$

$$\lambda_0 = \frac{c}{f} = \frac{3\times10^{10}}{20\times10^9} = 1.5 \text{ cm}$$

$$\lambda_g = \frac{\lambda_0}{\sqrt{1-\left(\frac{\lambda_0}{\lambda_c}\right)^2}}$$

Substituting the values of λ_g and λ_0 in the above equation, we get

$$6 = \frac{1.5}{\sqrt{1-\left(1.5/\lambda_c\right)^2}} \quad \text{or} \quad 6\left\{\sqrt{1-\left(\frac{1.5}{\lambda_c}\right)^2}\right\} = 1.5$$

Squaring both sides, we get

$$36\left\{1-\left(\frac{1.5}{\lambda_c}\right)^2\right\}=(1.5)^2$$

$$1-\frac{2.25}{\lambda_c^2}=\frac{2.25}{36}$$

$$\frac{2.25}{\lambda_c^2}=1-\frac{2.25}{36}=\frac{33.75}{36}$$

$$\lambda_c^2=\frac{2.25\times 36}{33.75}=2.4\Rightarrow\lambda_c=1.549$$

$\lambda_c > \lambda_0$ the wave propagates and $\lambda_c = 2a$ for the TE_{10} mode

$$a=\frac{\lambda_c}{2}=\frac{1.549}{2}=0.7745\text{ cm}$$

$$b=\frac{\lambda_c}{4}\quad\because a=2b$$

$$b=\frac{1.549}{4}=0.38725\text{ cm}$$

■

EXAMPLE PROBLEM 5.5

Calculate the phase and group velocities, and the wave impedance of the TE_{10} mode in a rectangular waveguide filled by air with the internal dimensions a = 22 mm, b = 10 mm. The frequency is f = 10 GHz.

Solution

The cut-off frequencies of the TE_{10} mode

$$f_c=\frac{c}{2}\left[\left(\frac{m}{a}\right)^2+\left(\frac{n}{b}\right)^2\right]^{1/2}$$

$$f_c=\frac{(3\times 10^8)}{2}\left[\left(\frac{1}{0.22}\right)^2+\left(\frac{0}{0.1}\right)^2\right]^{1/2}=6.82\text{ GHz}$$

We will first calculate the term $\sqrt{1-(f_c/f)^2}=\sqrt{1-(6.82/10)^2}=0.73$

Phase velocity is given by

$$v_p=\frac{c}{\sqrt{1-(f_c/f)^2}}=\frac{3\times 10^8}{0.73}=4.1\times 10^8\text{ m/sec.}$$

Group velocity is given by

$$v_g=c\sqrt{1-\left(\frac{f_c}{f}\right)^2}=3\times 10^8\times 0.73=2.19\times 10^8\text{ m/sec}$$

Wave impedance of the TE mode is given by

$$Z_{TE}=\frac{\eta}{\sqrt{1-(f_c/f)^2}}=\frac{377}{0.73}=516.2\Omega$$

■

EXAMPLE PROBLEM 5.6

For an angle of incidence of 44° calculate the group and phase velocities.

Solution

Given $\theta = 44°$

(a) Group velocity $v_g = c\sin\theta = (3 \times 10^8)\,(\sin 44°)$
$= (3 \times 10^8)\,(0.6946) = 2.1 \times 10^8$ m/s

(b) Phase velocity $v_p = \dfrac{c}{\sin\theta} = (3 \times 10^8 \text{ m/s})/\sin 44°$
$= (3 \times 10^8 \text{ m/s})/(0.6946) = 4.31 \times 10^8$ m/s

■

5.4 CIRCULAR WAVEGUIDE

A circular waveguide is a hollow metallic tube with a circular cross-section. Figure 5.17(a) shows a circular waveguide. A circular waveguide is basically a tubular, circular conductor. This waveguide is used only for some special applications. For example, it is used in a rotating joint, which transmits an electromagnetic wave to the feeder of a rotating radar antenna. In circular waveguides, the plane of polarization is not stable due to geometry. The frequency band of the single mode operation of a circular waveguide is narrower than the same band of a rectangular waveguide.

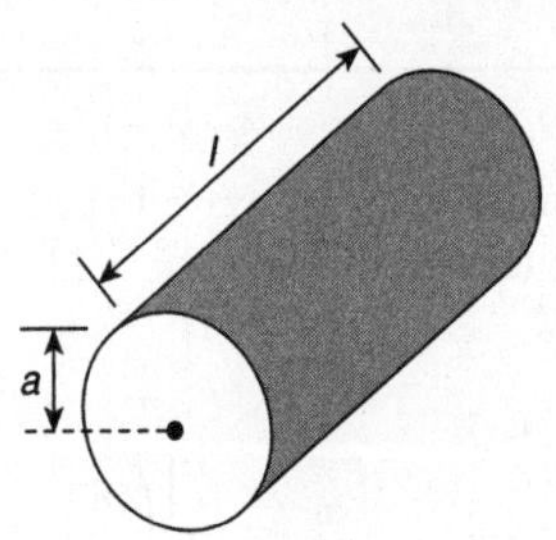

Figure 5.17 (a) *Circular waveguide*

The field in a circular waveguide can be separated into TE and TM modes, which are treated separately. The TEM mode cannot exist in the waveguide. We have to use the cylindrical coordinate system. Figure 5.17(b) shows a circular waveguide of inner radius $\rho = a$ and length l.

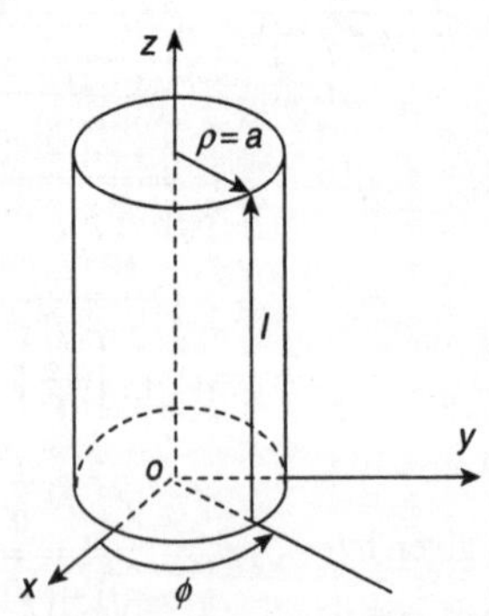

Figure 5.17 (b) *Cylindrical coordinate system in a circular waveguide*

Here, ρ varies from 0 to a, ϕ varies from 0 to 2π and l varies along the z-axis. For TE and TM waves in circular waveguide which are travelling in the z direction, the general wave equations are given by

$$\nabla^2 E_z = \gamma^2 E_z$$

$$\nabla^2 H_z = \gamma^2 H_z \tag{5.115}$$

The various field components of circular waveguides E_ρ, E_ϕ, H_ρ, and H_ϕ can be obtained by using the cylindrical coordinates and the Maxwell's curl equations.

$$E_p = -\frac{\gamma}{h^2}\frac{\partial E_z}{\partial \rho} - \frac{j\omega\mu}{h^2}\frac{1}{\rho}\frac{\partial H_z}{\partial \phi} \tag{5.116}$$

$$E_\phi = -\frac{\gamma}{h^2}\frac{1}{\rho}\frac{\partial E_z}{\partial \phi} + \frac{j\omega\mu}{h^2}\frac{\partial H_z}{\partial \rho} \tag{5.117}$$

$$H_\rho = \frac{j\omega\varepsilon}{h^2}\frac{1}{\rho}\frac{\partial E_z}{\partial \phi} - \frac{\gamma}{h^2}\frac{\partial H_z}{\partial \rho} \tag{5.118}$$

$$H_\phi = -\frac{j\omega\varepsilon}{h^2}\frac{\partial E_z}{\partial \rho} - \frac{\gamma}{h^2}\frac{\partial H_z}{\partial \phi} \tag{5.119}$$

5.4.1 Field Components of TE Waves

To find the expressions for field components of circular waveguides, the following steps are to be carried out:

1. Use the wave equations of TM and/or TE modes.
2. Solve the wave equations in cylindrical coordinates to get the general solution; this is a standard form of Bessel's function. E_z and /or H_z solution is calculated using Bessel's function.
3. Apply appropriate boundary conditions to get the complete solutions for a particular mode (TM and /or TE).
4. Substitute E_z and /or H_z in the general field component expressions to find the transverse field components of TM and /or TE modes.

The field components of a TE wave in a circular waveguide are given by

$$E_\rho = C_{0\rho} J_n\left(\frac{P'_{\mathrm{nm}}}{a}\rho\right)\sin n\ \phi e^{-\gamma z}$$

$$E_\phi = C_{0\phi} J'_n\left(\frac{P'_{\mathrm{nm}}}{a}\rho\right)\cos n\ \phi e^{-\gamma z}$$

$$\boldsymbol{E_z = 0}$$

$$H_\rho = \frac{C_{0\phi}}{Z} J'_n\left(\frac{P'_{\mathrm{nm}}}{a}\rho\right)\cos n\ \phi e^{-\gamma z}$$

$$H_\phi = \frac{C_{0\rho}}{Z} J_n\left(\frac{P'_{\mathrm{nm}}}{a}\rho\right)\sin n\ \phi e^{-\gamma z}$$

$$H_z = C_0 J_n\left(\frac{P'_{\mathrm{nm}}}{a}\rho\right)\cos n\ \phi e^{-\gamma z}$$

Derivation of transverse field components of a TE wave in circular waveguides

For a TE wave to propagate, $E_Z = 0$ and $H_Z \neq 0$

From Maxwell's equation,

$$\nabla^2 H_z = -\omega^2 \mu\varepsilon H_z \tag{5.120}$$

In cylindrical coordinates, $\nabla^2 H_z$ is expanded as

$$\frac{\partial^2 H_z}{\partial \rho^2} + \frac{1}{\rho}\frac{\partial H_z}{\partial \rho} + \frac{1}{\rho^2}\frac{\partial^2 H_z}{\partial \phi^2} + \frac{\partial^2 H_z}{\partial z^2} = -\omega^2 \mu\varepsilon H_z \tag{5.121}$$

It is known that $\frac{\partial^2}{\partial z^2} = \gamma^2$ (an operator)

$$\frac{\partial^2 H_z}{\partial \rho^2} + \frac{1}{\rho}\frac{\partial H_z}{\partial \rho} + \frac{1}{\rho^2}\frac{\partial^2 H_z}{\partial \phi^2} + (\gamma^2 + \omega^2 \mu\varepsilon) H_z = 0 \tag{5.122}$$

We know that

$$\gamma^2 + \omega^2 \mu\varepsilon = h^2$$

Substituting this in Eq. (5.122)

$$\frac{\partial^2 H_z}{\partial \rho^2} + \frac{1}{\rho}\frac{\partial H_z}{\partial \rho} + \frac{1}{\rho^2}\frac{\partial^2 H_z}{\partial \phi^2} + h^2 H_z = 0 \tag{5.123}$$

The solution for this partial differential equation is obtained by variable separation method which is

$$H_z = PQ \tag{5.124}$$

where P is a function of ρ only, and Q is a function of ϕ only.
Substituting H_z in Eq. (5.123), it gets reduced to

$$\frac{\partial^2 (PQ)}{\partial \rho^2} + \frac{1}{\rho}\frac{\partial (PQ)}{\partial \rho} + \frac{1}{\rho^2}\frac{\partial^2 (PQ)}{\partial \phi^2} + h^2 PQ = 0 \tag{5.125}$$

or

$$Q\frac{d^2 P}{d\rho^2} + \frac{Q}{\rho}\frac{dP}{d\rho} + \frac{P}{\rho^2}\frac{d^2 Q}{d\phi^2} + h^2 PQ = 0 \tag{5.126}$$

Multiplying throughout the equation by $\frac{\rho^2}{PQ}$,

$$\frac{\rho^2}{P}\frac{d^2 P}{d\rho^2} + \frac{\rho}{P}\frac{dP}{d\rho} + \frac{1}{Q}\frac{d^2 Q}{d\phi^2} + h^2 \rho^2 = 0 \tag{5.127}$$

Let $\frac{1}{Q}\frac{d^2 Q}{d\phi^2} = -n^2$, where n is a constant. Equation 5.127 becomes

$$\frac{\rho^2}{P}\frac{d^2 P}{d\rho^2} + \frac{\rho}{P}\frac{dP}{d\rho} + (h^2 \rho^2 - n^2) = 0 \tag{5.128}$$

Multiplying Eq. (5.128) by P, we get

$$\rho^2 \frac{d^2P}{d\rho^2} + \rho \frac{dP}{d\rho} + (h^2\rho^2 - n^2)P = 0 \tag{5.129}$$

This is similar to Bessel's equation of the form

$$x^2 \frac{d^2y}{dx^2} + x\frac{dy}{dx} + (x^2 - n^2)y = 0$$

The solution of Bessel's function is $y = C_n J_n(x)$, where $J_n(x)$ represents the nth order. Bessel's function is of the first kind, and C_n is a constant. The above equation is modified to represent the Bessel's function

$$(\rho h)^2 \frac{d^2P}{d(\rho h)^2} + \rho h \frac{dP}{d(\rho h)} + (\rho^2 h^2 - n^2)P = 0 \tag{5.130}$$

Therefore,

$$P = C_n J_n(\rho h) \tag{5.131}$$

Also,

$$\frac{1}{Q}\frac{d^2Q}{d\phi^2} = -n^2 \tag{5.132}$$

The general solution of this equation is given by

$$Q = A_n \cos n\phi + B_n \sin n\phi \tag{5.133}$$

$$Q = \sqrt{A_n^2 + B_n^2} \cos\left(n\phi - \tan^{-1}\frac{B_n}{A_n}\right) \tag{5.134}$$

Therefore, the complete solution becomes as per $H_z = PQ$; hence,

$$H_z = C_n J_n(\rho h).\sqrt{A_n^2 + B_n^2} \cos\left(n\phi - \tan^{-1}\frac{B_n}{A_n}\right)$$

Since $\quad n\phi >> \tan^{-1}\frac{A_n}{B_n}$ and $C'_n = \sqrt{A_n^2 + B_n^2}$

$$H_z = C_n C'_n \, J_n(\rho h)\cos(n\phi) \tag{5.135}$$

$$H_z = C_0 J_n(\rho h)\cos n\phi \quad (\text{where } C_0 = C_n C'_n) \tag{5.136}$$

For a sinusoidal variation along z

$$H_z = C_0 J_n(\rho h)\cos n\phi e^{-\gamma z} \tag{5.137}$$

Boundary Conditions

From the boundary conditions, along the surface of the circular waveguide, at $\rho = a, E_\phi = 0$ for all values of ϕ varying between 0 and 2π, the field component that is given from

Maxwell's curl equations is $E_\phi = \dfrac{j\omega\mu}{h^2}\dfrac{\partial H_z}{\partial \rho}$

Therefore, for the conditions, $\rho = a, E_\phi = 0$

$$\Rightarrow \left.\frac{\partial H_z}{\partial \rho}\right|_{\rho=a} = 0 \tag{5.138}$$

So, we have $J_n(ah) = 0$

Here, the *prime* denotes differentiation with respect to *ah*. The *m*th root of this equation is denoted by P'_{nm}, whose Eigen values are given by

$$P'_{\text{nm}} = ah \tag{5.139}$$

(Note: This equation is only for TE wave).

From the above equation, we have $h = \frac{P'_{\text{nm}}}{a}$ So, Eq. (5.137) is given as

$$H_z = C_0 J_n\left(\frac{P'_{\text{nm}}}{a}\rho\right)\cos n\phi e^{-\gamma z} \tag{5.140}$$

This equation represents all possible solutions of H_z for TE_{mn} waves in circular waveguides. Since J_n is an oscillatory function, the $J_n\left(\frac{P'_{\text{nm}}}{a}\rho\right)$ is also an oscillatory function. P'_{nm} values for TE_{mn} modes in circular waveguides are given in Table 5.1.

Table 5.1 P'_{nm} values for TE_{mn} modes in circular waveguides

n \ *m*	1	2	3
0	3.832	7.016	10.173
1	1.841	5.331	8.536
2	3.054	6.706	9.969
4	4.201	8.015	11.346

Substituting $E_z = 0$ in Eqs. 5.116–5.119, E_r, E_f, E_z, H_r, H_f, and H_z can be obtained

$$E_\rho = \frac{-j\omega\mu}{h^2}\frac{1}{\rho}\frac{\partial H_z}{\partial \phi} \tag{5.141}$$

$$E_\phi = \frac{j\omega\mu}{h^2}\frac{\partial H_z}{\partial \rho} \tag{5.142}$$

$$E_z = 0 \tag{5.143}$$

$$H_\rho = \frac{-\gamma}{h^2}\frac{\partial H_z}{\partial \rho} \tag{5.144}$$

$$H_\phi = \frac{-\gamma}{h^2}\frac{1}{\rho}\frac{\partial H_z}{\partial \phi} \tag{5.145}$$

$$H_z = C_0 J_n(\rho h)\cos n\phi e^{-\gamma z} \tag{5.146}$$

where $$h^2 = \gamma^2 + \omega^2 \mu\varepsilon$$

Substituting H_z in the above equations with $h = \frac{P'_{\text{nm}}}{a}$, the complete field equations for TE_{mn} modes in circular waveguides are obtained as

$$E_\rho = C_{0\rho} J_n\left(\frac{P'_{\text{nm}}}{a}\rho\right)\sin n\phi e^{-\gamma z} \tag{5.147}$$

$$E_\phi = C_{0\phi} J'_n\left(\frac{P'_{\text{nm}}}{a}\rho\right)\cos n\phi e^{-\gamma z} \tag{5.148}$$

$$E_z = 0 \tag{5.149}$$

$$H_\rho = \frac{C_{0\phi}}{Z} J'_n\left(\frac{P'_{\text{nm}}}{a}\rho\right)\cos n\phi e^{-\gamma z} \tag{5.150}$$

$$H_\phi = \frac{C_{0\rho}}{Z} J_n\left(\frac{P'_{\text{nm}}}{a}\rho\right)\sin n\phi e^{-\gamma z} \tag{5.151}$$

$$H_z = C_0 \cdot J_n\left(\frac{P'_{\text{nm}}}{a}\rho\right)\cos n\phi e^{-\gamma z} \tag{5.152}$$

where $Z = \frac{E_\rho}{H_\phi} = \frac{-E_\phi}{H_\rho}$ the wave impedance in the guide, where $n = 0, 1, 2, 3, \ldots$ and $m = 1, 2, 3, 4, \ldots$

5.4.2 Field Components of TM Waves

For a TM wave to propagate in a circular waveguide, $H_z = 0$ and $E_z \neq 0$. The field components of a TM wave in a circular waveguide are given by

$$E_\rho = C_{0\rho} J'_n\left(\frac{P_{\text{nm}}}{a}\rho\right)\cos n\phi e^{-\gamma z}$$

$$E_\phi = C_{0\phi} J_n\left(\frac{P_{\text{nm}}}{a}\rho\right)\sin n\phi e^{-\gamma z}$$

$$E_z = C_{0z} J_n\left(\frac{P_{\text{nm}}}{a}\rho\right)\cos n\phi e^{-\gamma z}$$

$$H_\rho = \frac{C_{0\phi}}{Z} J_n\left(\frac{P_{\text{nm}}}{a}\rho\right)\sin n\phi e^{-\gamma z}$$

$$H_\phi = \frac{C_{0\rho}}{Z} J'_n\left(\frac{P_{\text{nm}}}{a}\rho\right)\cos n\phi e^{-\gamma z}$$

$$H_z = 0$$

Derivation of Transverse Field Components of a TM Wave in Circular Waveguides

The Helmholtz-wave equation is given by

$$\nabla^2 E_z = -\omega^2 \mu\varepsilon E_z$$

The solution of this equation (similar to TE waves) is given by

$$E_z = C_0 J_n(\rho h)\cos n\phi e^{-\gamma z} \tag{5.153}$$

Applying boundary conditions, that is, $E_z = 0$ at $\rho = a$, gives

$$J_n(ah) = 0 \tag{5.154}$$

There are an infinite number of roots for which $J_n(ah) = 0$ which are called Eigen values and are denoted by P_{nm}, where $P_{nm} = ah$, as $J_n(ah)$ are oscillatory functions. For TM waves this relation holds good and should be confused with Eq. (5.139) of TE waves. The roots for some values of n and m are shown in Table 5.2.

Table 5.2 P_{nm} values for TM_{mn} modes in circular waveguides

n \ m	1	2	3
0	2.405	5.520	8.645
1	3.832	7.106	10.173
2	5.135	8.417	11.620
4	6.380	9.761	13.015

Using Maxwell's curl equations, various field components are obtained by substituting

$$h = \frac{P_{mn}}{a} \text{ and } E_z = C_0 J_n\left(\rho \frac{P_{nm}}{a}\right)\cos n\phi e^{-\gamma z} \tag{5.155}$$

The various field components are given by

$$E_\rho = C_{0\rho} J_n\left(\frac{P_{nm}}{a}\rho\right)\cos n\phi e^{-\gamma z} \tag{5.156}$$

$$E_\phi = C_{0\phi} J_n'\left(\frac{P_{nm}}{a}\rho\right)\sin n\phi e^{-\gamma z} \tag{5.157}$$

$$E_z = C_{0z} J_n\left(\frac{P_{nm}}{a}\rho\right)\cos n\phi e^{-\gamma z} \tag{5.158}$$

$$H_\rho = \frac{C_{0\phi}}{Z} J_n'\left(\frac{P_{nm}}{a}\rho\right)\sin n\phi e^{-\gamma z} \tag{5.159}$$

$$H_\phi = \frac{C_{0\rho}}{Z} J_n\left(\frac{P_{nm}}{a}\rho\right)\cos n\phi e^{-\gamma z} \tag{5.160}$$

$$H_z = 0 \tag{5.161}$$

5.4.3 Characteristic Equation and Cut-off Frequency of a Circular Waveguide

The cut-off wavelength in a mode occurs when the mode propagation constant γ becomes zero.

That is, $\gamma = \alpha + j\beta = 0$

We know that $\beta = \sqrt{\mu\varepsilon\left(\omega^2 - \omega_c^2\right)}$

$$\beta = \sqrt{\mu\varepsilon\omega^2 - \mu\varepsilon(2\pi f)^2}\text{, where } f = \frac{c}{\lambda_c},\ c = \frac{1}{\sqrt{\mu\varepsilon}}$$

$$\beta = \sqrt{\omega^2\mu\varepsilon - \left(\frac{2\pi}{\lambda_c}\right)^2}$$

$$= \sqrt{\omega^2\mu\varepsilon - h^2}$$

where $h = \frac{P'_{\text{nm}}}{a}$ for TE waves, and $h = \frac{P_{\text{nm}}}{a}$ for TM waves.

Therefore, the cut-off wavelength for TE wave is given by

$$\lambda_c = \frac{2\pi}{h} = \frac{2\pi}{\left(\frac{P'_{\text{nm}}}{a}\right)} = \frac{2\pi a}{P'_{\text{nm}}} \tag{5.162}$$

λ_c will be maximum if P'_{nm} is minimum.

The TE wave's cut-off frequency can be written as

$$f_c = \frac{c}{\lambda_c}$$

$$f_c = \frac{c}{\lambda_c} = \frac{c}{\left(\frac{2\pi a}{P'_{\text{nm}}}\right)} \tag{5.163}$$

where "c" is the velocity of light.

Similarly, for a TM wave,

$$\lambda_c = \frac{2\pi}{h}$$

where

$$h = \frac{P_{\text{nm}}}{a}$$

$$\therefore \lambda_c = \frac{2\pi a}{P_{\text{nm}}} \tag{5.164}$$

For TM wave the cut-off frequency can be written

$$f_c = \frac{c}{\lambda_c} = \frac{c}{\left(\frac{2\pi a}{P_{\text{nm}}}\right)} \tag{5.165}$$

5.4.4 Dominant Mode and Degenerate Modes

From the above tables (Tables 5.1 and 5.2), when P'_{nm} is least the lowest-order cut-off frequency is obtained, for $n = 1$ and $m = 1$ corresponding to the TE_{11} mode. Hence, TE_{11} is the dominant mode in the circular waveguide. The fields are called as *hybrid mode fields*, if the fields exist in both $E_Z \neq 0$ and $H_Z \neq 0$ and are neither transverse electric nor transverse magnetic. From these tables, it is also shown that $P'_{0m} = P_{1m}$, and, hence degenerate modes in a uniform circular waveguide are all the TE_{0m} and TM_{1m} modes.

5.4.5 Sketches of TE and TM Mode Fields in the Cross-Section

As mentioned earlier, the number of half-cycle variations in the *x-y* cross-section of the guide is indicated by the integers *m* and *n*. Thus, for a fixed time, the field configuration of Figure 5.18 gives the configuration of different modes.

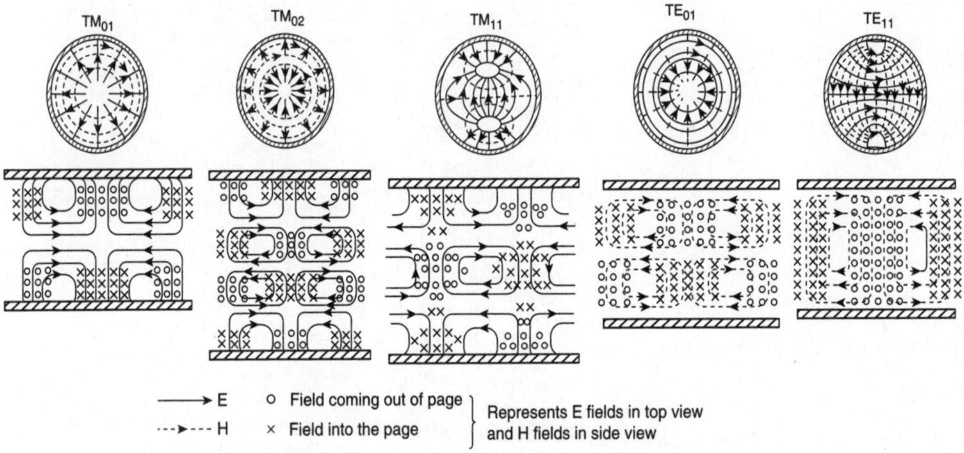

Figure 5.18 *Field configurations of different modes in circular waveguides with solid lines indicating* **E** *lines and dashed lines indicating* **H** *lines*

5.4.6 Excitation of Modes in Circular Waveguides

The excitation of modes in the circular waveguide is similar to that of the rectangular waveguide. Here, an excited probe or antenna is oriented in the direction of the electric field. The probe is placed near the maxima of the electric field that is present in the mode pattern. The excitation method for different modes is shown in Figure 5.19.

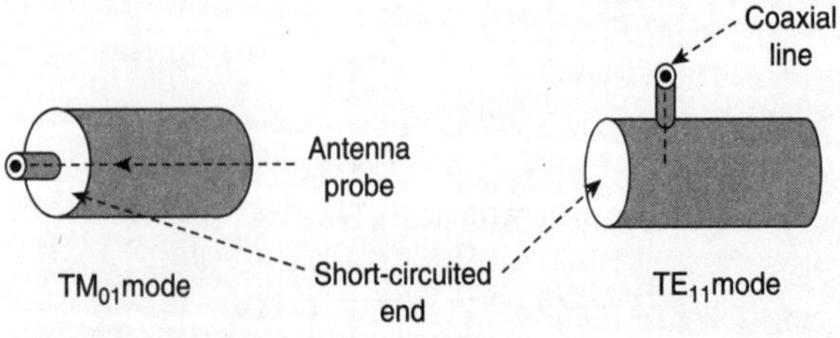

Figure 5.19 *Field excitation methods for different modes*

5.4.7 Advantages and Disadvantages

1. Rectangular waveguides are preferred over circular waveguides due to the following reasons:
   (a) For a similar operating frequency, a rectangular waveguide is not suitable in some applications as it is smaller in size than a circular waveguide.
   (b) The polarization is not maintained by the wave through the circular waveguide; i.e, the circular symmetry of the waveguide may reflect the possibility of the wave without maintaining its polarization throughout the length of the guide.
   (c) The frequency difference between the lowest frequency in the dominant mode and the next mode of a rectangular waveguide is bigger than that in a circular waveguide.
2. The advantages of circular waveguides over rectangular waveguides are as follows:
   (a) First is their shape, the use of circular terminations and connectors is allowed, which are easier to manufacture and attach.
   (b) Due to the TE_{01} mode in the circular waveguide, which has the lower attenuation per unit length of the waveguide these are suitable for long-distance communications.
   (c) The rotary joints are also made by the circular waveguides, which are needed when a section of the waveguide should be able to rotate, such as for the feeds of revolving antennas.

EXAMPLE PROBLEM 5.7

Find (a) cut-off wavelength (b) cut-off frequency (c) wavelength in the guide for the dominant mode of operation in an air filled circular waveguide of inner diameter 6 cms.

Solution

(a) Cut-off wavelength: TE_{11} is the dominant mode in the circular waveguide. Given that $D = 6$ cm, so

$$r = \frac{D}{2} = \frac{6}{2} = 3 \text{ cm}$$

$$\lambda_c = \frac{2\pi r}{P'_{nm}}$$

Now, λ_c is the maximum from Table 5.1.We know that $P'_{nm} = 1.841$

$$\lambda_c = \frac{2\pi r}{1.841} = \frac{2\times\pi\times 3}{1.841} = 10.2336 \text{ cm}$$

(b) Cut-off frequency:

$$\lambda_c = \frac{c}{f_c} \Rightarrow f_c = \frac{c}{\lambda_c} = \frac{3\times 10^{10}}{10.2336} = 2.931 \text{ GHz}$$

Frequencies higher than f_c will be propagated since cut-off frequency is 2.931 GHz. Assume a signal of frequency of 5 GHz is being propagated. Therefore

$$\lambda_0 = \frac{3\times 10^{10}}{5\times 10^9} = 6 \text{ cm}$$

(c) Wavelength in the guide

$$\lambda_g = \frac{\lambda_0}{\sqrt{1-\left(\lambda_0/\lambda_c\right)^2}} = \frac{6}{\sqrt{1-\left(6/10.2336\right)^2}} = 7.41\,\text{cm}$$

■

EXAMPLE PROBLEM 5.8

An air filled circular waveguide is to have dimensions such that $f_c = 0.6f$ for TE_{11} mode and is to be operated at a frequency of 4 GHz. Determine (a) the diameter of the waveguide and (b) guide wavelength.

Solution

The TE_{11} is the dominant mode in the circular waveguide. Let r and D be the radius and the diameter of the waveguide, respectively.

(a) $\lambda_c = \dfrac{2\pi r}{1.841}$

It is given that $f_c = f$; and $f = 4$ GHz. $f_c = 0.6 \times 4 \times 10^9 = 2.4$ GHz

$$\therefore \lambda_c = \frac{c}{f_c} = \frac{3\times10^{10}}{2.4\times10^9} = 12.5 \text{ cm}$$

$$\therefore 12.5 = \frac{2\pi r}{1.841}$$

$$2r = D$$

$$12.5 = \frac{\pi D}{1.841}$$

$$\therefore \quad D = \frac{12.5\times1.841}{\pi} = 7.3288 \text{ cm}$$

(b) The guided wavelength is given by, $\lambda_g = \dfrac{\lambda_0}{\sqrt{1-(\lambda_0/\lambda_c)^2}}$

$$\lambda_0 = \frac{c}{f} = \frac{3\times10^{10}}{4\times10^9} = 7.5 \text{ cm and } \lambda_c = 12.5 \text{ cm}$$

$$\lambda_g = \frac{\lambda_0}{\sqrt{1-(\lambda_0/\lambda_c)^2}} = \frac{7.5}{\sqrt{1-(7.5/12.5)^2}} = 9.375 \text{ cm}$$

■

EXAMPLE PROBLEM 5.9

A TE_{11} wave is propagating through a circular waveguide. If the guide is air-filled and the diameter of the guide is 8 cm. Find (a) cut-off frequency, (b) wavelength λ_g in the guide for frequency of 4 GHz, and (c) the wave impedance in the guide.

Solution

(a) The cut-off frequency of a circular waveguide is given by

$$f_c = \frac{c}{\lambda_c} = \frac{c}{\left(\dfrac{2\pi a}{P'_{nm}}\right)}$$

Where 'a' is the radius of the guide; the diameter of the guide is given as 10 cm; therefore the radius $r = 4$ cm $= 0.04$ m, the value of P_{nm} for a TE_{11} waveguide is 1.841 (from Table 5.1).

$$f_c = \frac{c}{\left(\dfrac{2\pi a}{P'_{nm}}\right)} = \frac{3\times 10^8}{\left(2\pi \times \dfrac{0.04}{1.841}\right)} = 2.199 \text{ GHz}$$

(b) Wavelength is given by $\lambda_g = \dfrac{\lambda_0}{\sqrt{1-(f_c/f)^2}} = \dfrac{3\times 10^8/4\times 10^9}{\sqrt{1-(2.199\times 10^9/4\times 10^9)^2}} = 8.97\,\text{cm}$

(c) Wave impedance of the TE mode is given by $Z_{TE} = \dfrac{\eta}{\sqrt{1-\left(\dfrac{f_c}{f}\right)^2}} = \dfrac{120\pi}{0.84} = 448.57\ \Omega$

■

5.5 CAVITY RESONATORS

A metallic enclosure that bounds electromagnetic energy is a *resonant cavity*. A more attractive approach at microwave frequencies and above is to construct devices that use the constructive and destructive interferences of multiply reflected waves to cause resonances. These reflections occur in enclosures called *cavity resonators*. So, cavity resonators are used in microwave circuits for high frequencies similar to standard resonant LC circuits at lower frequencies. Resonant circuits are used for a variety of applications, including oscillator circuits, filters, and tuned amplifiers.

A cavity resonator is a waveguide that is shorted at both ends, which can hold an electromagnetic oscillation.

(i) When the waveguide's one end is terminated in the shorting plane, there will reflections, resulting in standing waves.
(ii) When the waveguide's other end is also terminated with the plane at $n_g\lambda/2$ distance, the signal bounces back and forth between two shorting plates along the Z axis, and as a result resonance occurs.
(iii) Hollow space is called a *cavity resonator* (i.e., a cavity resonator is a waveguide that is shorted at both ends).
(iii) Modes of operations in a cavity are designated in terms of fields existing in X, Y, and Z directions; m, n, and p are the three subscripts used.
(v) The general mode of propagation in the cavity resonator is TE_{mnp} or TM_{mnp}.

A wave is generated by a resonator (or select the specific frequency of a signal), whereas a repetitive signal is created by an electronic circuit called oscillator. An oscillatory system operates at microwave frequencies; it is the analogy of an oscillator circuit.

5.5.1 Types of Cavity Resonators

In microwave applications, the commonly used cavity resonators are

- Circular Cavity Resonator
- Rectangular Cavity Resonator

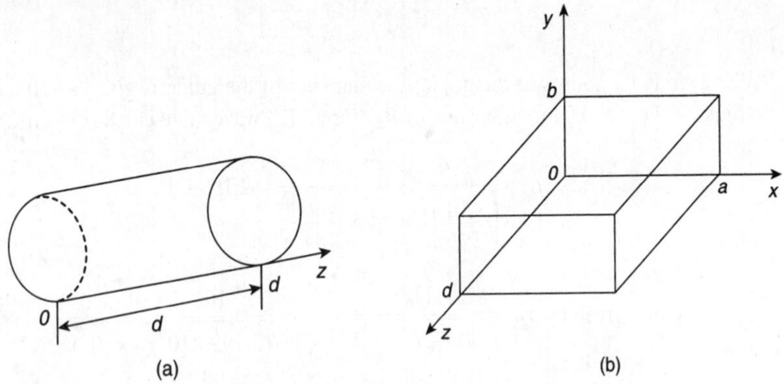

Figure 5.20 *(a) Circular cavity resonator; (b) Rectangular cavity resonator*

The rectangular and circular cavity resonators are shown in Figure 5.20 (a) and (b).

5.5.2 Rectangular Cavity Resonators

Consider a rectangular cavity of breadth a, height b, and length d. Maxwell's equations in the rectangular cavity should be satisfied. Since the walls of the cavity are conducting, the electric field should be zero along these walls.

The boundary condition of the zero tangential should be satisfied by the wave equation; E at the four walls should be satisfied. It is necessary that the harmonic functions in z should be chosen in such a way that the condition $E = 0$ at the remaining two end walls should be satisfied.

For the TE_{mnp} mode,

$$H_z = C\cos\left(\frac{m\pi}{a}\right)x\cos\left(\frac{n\pi}{b}\right)y\sin\left(\frac{p\pi}{d}\right)ze^{j\omega t - j\beta z} \tag{5.166}$$

where $m = 0, 1, 2, 3, \ldots$ represents the number of half waves in x direction
$n = 0, 1, 2, 3, \ldots$ represents the number of half waves in y direction
$p = 1, 2, 3, 4, \ldots$ represents the number of half waves in z direction

For the TM_{mnp} mode,

$$E_z = C\sin\left(\frac{m\pi}{a}\right)x\sin\left(\frac{n\pi}{b}\right)y\cos\left(\frac{p\pi}{d}\right)ze^{j\omega t - j\beta z} \tag{5.167}$$

where $m = 1, 2, 3, 4, \ldots$
$n = 1, 2, 3, 4, \ldots$
$p = 0, 1, 2, 3, \ldots$

$$h^2 = \gamma^2 + \omega^2\mu\varepsilon = A^2 + B^2 = \left(\frac{m\pi}{a}\right)^2 + \left(\frac{n\pi}{b}\right)^2 \tag{5.168}$$

$$\gamma^2 = \left(\frac{m\pi}{a}\right)^2 + \left(\frac{n\pi}{b}\right)^2 - \omega^2 \mu\varepsilon$$

$$\omega^2 \mu\varepsilon = \left(\frac{m\pi}{a}\right)^2 + \left(\frac{n\pi}{b}\right)^2 - \gamma^2$$

$$\omega^2 \mu\varepsilon = \left(\frac{m\pi}{a}\right)^2 + \left(\frac{n\pi}{b}\right)^2 + \beta^2 \qquad \because (\gamma = j\beta) \tag{5.169}$$

For resonance to occur in cavity resonators,

$$\beta = \frac{p\pi}{d} \tag{5.170}$$

By substituting Eq. (5.170) in Eq. (5.169), we get

$$\omega^2 \mu\varepsilon = \left(\frac{m\pi}{a}\right)^2 + \left(\frac{n\pi}{b}\right)^2 + \left(\frac{p\pi}{d}\right)^2$$

We know that $\omega = 2\pi f$, and $c = \dfrac{1}{\sqrt{\mu\varepsilon}}$

Therefore, the resonant frequency is expressed by

$$f_0 = \frac{c}{2}\sqrt{\left(\frac{m}{a}\right)^2 + \left(\frac{n}{b}\right)^2 + \left(\frac{p}{d}\right)^2} \tag{5.171}$$

For $a > b < d$, the dominant mode is the TE_{101} mode.

5.5.3 Circular Cavity Resonators

Consider a circular cavity of radius a and length d. Maxwell's equations in the circular cavity should be satisfied in a similar manner as in the case of a rectangular cavity. Since the walls of the cavity are conducting, the electric field should be zero along these walls. The wave equation should satisfy the boundary condition of the zero tangential; E at four walls should be satisfied. In order to satisfy the condition where $E = 0$ at the two end wall, the harmonic function is selected in z. The H_z is given for the TE_{mnp} mode as

$$H_z = CJ_n\left(\rho \frac{P'_{nm}}{a}\right)\cos(n\phi)\sin\left(\frac{p\pi}{d}\right)ze^{j\omega t - j\beta z} \tag{5.172}$$

where $m = 0, 1, 2, 3, \ldots$ represents the number of half waves in x direction
$N = 1, 2, 3, 4, \ldots$ represents the number of half waves in y direction
$p = 1, 2, 3, 4, \ldots$ represents the number of half waves in z direction
J_n = Bessel's function of the first kind

For the TM_{mnp} mode as

$$E_z = CJ'_n\left(\rho \frac{P_{nm}}{a}\right)\cos(n\phi)\sin\left(\frac{p\pi}{d}\right)ze^{j\omega t - j\beta z} \tag{5.173}$$

where $m = 0, 1, 2, 3, \ldots$ represents the number of half waves in x direction
$n = 1, 2, 3, 4, \ldots$ represents the number of half waves in y direction
$p = 0, 1, 2, 3, \ldots$ represents the number of half waves in z direction
J_n' = Bessel's function of the first kind

The following equation is the characteristic equation

$$h^2 = \gamma^2 + \omega^2 \mu\varepsilon$$

$$h^2 = \omega^2 \mu\varepsilon - \beta^2 \text{ (as } \gamma = j\beta\text{)}$$

where
$$\beta = \sqrt{\omega^2 \mu\varepsilon - \left(\frac{2\pi}{\lambda_c}\right)^2} = \sqrt{\omega^2 \mu\varepsilon - h^2} \tag{5.174}$$

where $h = \dfrac{P'_{\text{nm}}}{a}$ for TE waves, and $h = \dfrac{P_{\text{nm}}}{a}$ for TM waves

The given condition should be satisfied to sustain the resonance in the cavity resonance

$$\beta = \frac{p\pi}{d}$$

Substituting the above equation in Eq. (5.174),

$$\left(\frac{p\pi}{d}\right)^2 = \omega^2 \mu\varepsilon - h^2$$

The equation for the TE mode is given by

$$\omega^2 \mu\varepsilon = \left(\frac{p\pi}{d}\right)^2 + \left(\frac{P'_{\text{nm}}}{a}\right)^2 \tag{5.175}$$

By substituting $\omega = 2\pi f$ and $c = \dfrac{1}{\sqrt{\mu\varepsilon}}$, the resonant frequency for the TE_{mnp} mode is

$$f_0 = \frac{c}{2\pi}\sqrt{\left(\frac{P'_{\text{nm}}}{a}\right)^2 + \left(\frac{p\pi}{d}\right)^2} \tag{5.176}$$

and the TM mode is given by

$$\omega^2 \mu\varepsilon = \left(\frac{p\pi}{d}\right)^2 + \left(\frac{P_{\text{nm}}}{a}\right)^2 \tag{5.177}$$

By substituting $\omega = 2\pi f$ and $c = \dfrac{1}{\sqrt{\mu\varepsilon}}$, the resonant frequency for the TM_{mnp} mode is

$$f_0 = \frac{c}{2\pi}\sqrt{\left(\frac{P_{\text{nm}}}{a}\right)^2 + \left(\frac{p\pi}{d}\right)^2} \tag{5.178}$$

For $2a > d$, the dominant mode is TM_{110} and for $d > 2a$, the dominant mode is the TE_{111} mode.

The dominant mode of the cylindrical waveguide is the TE_{111} mode; so, the resonators are mostly designed to work with this mode.

5.5.4 Applications of Cavity Resonators

In various applications, resonators act as active components. The following are the applications of resonators.

1. Resonators can be used in tuned circuits
2. Used in klystron amplifiers, UHF tubes, or oscillators
3. Used in duplexers of radars
4. Used in cavity wave meters in measuring frequency

5.5.5 Quality Factor (*Q*) and Coupling Coefficient

The measure of the frequency selectivity of a resonant or anti resonant circuit is called quality factor Q, and it is represented as

$$Q = 2\pi \frac{\text{maximum energy stored}}{\text{energy dissipated per cycle}} = \frac{\omega W}{P_L} \tag{5.179}$$

where W = maximum energy stored

P_L = average power loss.

The electric and magnetic energies are equal and in time quadrature, at resonant frequency. The electric energy is zero when the magnetic energy is maximum, and vice versa. By integrating the energy density over the volume of the resonator, the total energy stored in the resonator can be obtained:

$$W_c = \int_v \frac{\varepsilon}{2}|E|^2\, dv = W_m = \int_v \frac{\mu}{2}|H|^2\, dv = W \tag{5.180}$$

where $|E|$ and $|H|$ are the peak values of the field intensities.

By integrating the power density over the inner surface of the resonator, the average power loss in the resonator can be obtained. Thus

$$P_L = \frac{R_s}{2}\int_s |H_t|^2\, da \tag{5.181}$$

Here, the peak value of the tangential magnetic intensity is H_t and the surface resistance of the resonator is R_s.

The below equation can be obtained by substituting Eq. (5.180) and (5.181) in to Eq. (5.179)

$$Q = \frac{\omega\mu\int_v |H|^2\, dv}{R_s\int_s |H_t|^2\, da} \tag{5.182}$$

Since the peak value of the magnetic intensity is related to its tangential and normal components by

$$|H|^2 = |H_t|^2 + |H_n|^2 \tag{5.183}$$

where $|H_t|^2$ at the resonator walls is about two times the value of $|H|^2$ over the volume of resonator and H_n is the peak value of the normal magnetic intensity.

So the Q of a cavity resonator as shown in Eq. (5.182) can be written as

$$Q = \frac{\omega\mu\,(\text{volume})}{2R_s\,(\text{surface areas})} \tag{5.184}$$

Either a series or a parallel resonant circuit can represent an unloaded resonator. The resonant frequency and the unloaded Q_0 of a cavity resonator are

$$f_0 = \frac{c}{2}\sqrt{\left(\frac{m}{a}\right)^2 + \left(\frac{n}{b}\right)^2 + \left(\frac{p}{d}\right)^2} \tag{5.185}$$

$$Q_0 = \frac{\omega_0 L}{R} \tag{5.186}$$

5.5.5.1 *Coupling coefficient*

If the cavity is coupled by an ideal N:1 transformer and a series inductance L_s to a generator with an internal impedance Z_g, then the coupling circuit and its equivalent as shown in Figure 5.21(*a*) and (*b*):

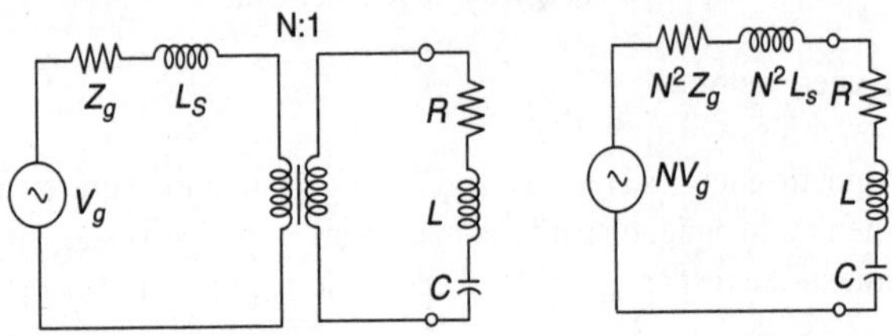

Figure 5.21 *Cavity coupled to a generator. (a) Coupling circuit; (b) Equivalent circuit*

For a system the loaded Q_l is given by

$$Q_l = \frac{\omega_0 L}{R + N^2 Z_g} \quad \text{for } \left|N^2 L_s\right| < \left|R + N^2 Z_g\right| \tag{5.187}$$

The coupling coefficient of the system is represented as

$$K \equiv \frac{N^2 Z_g}{R} \tag{5.188}$$

And the loaded Q_l would become

$$Q_l = \frac{\omega_0 L}{R(1+K)} = \frac{Q_0}{1+K} \tag{5.189}$$

Rearranging of Eq. (5.189) yields

$$\frac{1}{Q_l} = \frac{1}{Q_0} + \frac{1}{Q_{ext}} \tag{5.190}$$

where $Q_{ext} = Q_0/K = \omega_0 L/(KR)$ is the external Q.

There are three types of coupling coefficients:

1. **Critical coupling**: If the resonator is matched to the generator, then $K = 1$. The loaded Q_l is given by

$$Q_l = \frac{1}{2}Q_{ext} = \frac{1}{2}Q_0$$

2. **Over coupling**: At resonance, if $K > 1$, the cavity terminals are at a voltage maximum in the input line. The standing-wave ratio ρ is the normalized impedance at the voltage maximum. That is $K = \rho$. The loaded Q_l is given by

$$Q_l = \frac{Q_0}{1+\rho}$$

3. **Under coupling**: If $K<1$, the cavity terminals are at a voltage minimum, and the input terminal impedance is equal to the reciprocal of the standing-wave ratio. That is,

$$K = \frac{1}{\rho}$$

The loaded Q_t is given by
$$Q_l = \frac{\rho}{\rho+1}Q_0 .$$

EXAMPLE PROBLEM 5.10

An air-filled rectangular cavity resonator has $a = d = 2$ cm, and $b = 1$ cm and is operated in the TE_{101} mode Calculate (i) resonant frequency; (ii) if the cavity is filled with a dielectric of relative permittivity 2.5, what is the resonant frequency?

Solution

Given $a = d = 2$ cm, and $b = 1$ cm:

(i) The resonant frequency of a rectangular cavity resonator is given by

$$f_{0mnp} = \frac{c}{2}\sqrt{\left(\frac{m}{a}\right)^2 + \left(\frac{n}{b}\right)^2 + \left(\frac{p}{d}\right)^2}$$

$$f_{0(101)} = \frac{3\times10^8}{2}\sqrt{\left(\frac{1}{2}\right)^2 + \left(\frac{1}{2}\right)^2} = 10.61 \text{ GHz}$$

(ii) For a dielectric-filled cavity, $f_{101} = \dfrac{10.61}{\sqrt{(2.5)}} = 6.71 \text{ GHz}$

■

EXAMPLE PROBLEM 5.11

Design a resonator tuned to the resonant frequency 15 GHz for the mode TE_{101}. The resonator is formed by the segment of the rectangular waveguide with the dimensions $a = 30$ mm, $b = 15$ mm. The resonator is filled with air.

Solution

Given the dimensions $a = 30$ mm, $b = 15$ mm, and f = 15 MHz:

The dominant mode of the rectangular waveguide has the simplest field distribution; so, we design the resonator for this mode. Thus, for the mode TE_{101}, we have the resonant frequency (5.171)

$$f_{0(101)} = \frac{c}{2}\sqrt{\frac{1}{a^2} + \frac{1}{d^2}},$$

where c is the speed of light in a vacuum. From the above formula, we can determine the necessary length d:

$$d = \frac{1}{\sqrt{\left(\frac{2f_{0(101)}}{c}\right)^2 - \frac{1}{a^2}}} = 10.6 \text{ mm}$$

The resonator should be 10.6 mm long. ■

EXAMPLE PROBLEM 5.12

A rectangular cavity resonator is 10 × 8 × 6 cm. Find the resonating frequency for the TE_{111} mode.

Solution

Given $a = 10$ cm = 0.10 m

$b = 8$ cm = 0.08 m

$d = 6$ cm = 0.06 m, and $m = 1, n = 1, p = 1$

The resonant frequency for a rectangular cavity is given by

$$f_{0mnp} = \frac{1}{2\pi\sqrt{\mu_0\varepsilon_0\varepsilon_r}}\sqrt{\left(\frac{m\pi}{a}\right)^2 + \left(\frac{n\pi}{b}\right)^2 + \left(\frac{p\pi}{d}\right)^2}$$

For $\varepsilon_r = 1$, the equation becomes

$$f_{0mnp} = \frac{c}{2}\sqrt{\left(\frac{m}{a}\right)^2 + \left(\frac{n}{b}\right)^2 + \left(\frac{p}{d}\right)^2}$$

$$f_{0mnp} = \frac{3\times10^8}{2}\sqrt{\left(\frac{1}{0.10}\right)^2 + \left(\frac{1}{0.08}\right)^2 + \left(\frac{1}{0.06}\right)^2} = 3.4 \text{ GHz}$$

■

EXAMPLE PROBLEM 5.13

Find the first five resonances of an air-filled rectangular cavity with dimensions of $a = 5$ cm, $b = 4$ cm, and $c = 10$ cm $(d > a > b)$.

Solution

Given the dimensions of $a = 5$ cm, $b = 4$ cm, and $c = 10$ cm $(d > a > b)$:
The expression for the frequency of a rectangular cavity resonator is given by

$$f_{c_{mn}} = \frac{1}{2\sqrt{\mu\varepsilon}}\sqrt{\left(\frac{m}{a}\right)^2 + \left(\frac{n}{b}\right)^2 + \left(\frac{p}{d}\right)^2}$$

The given dimensions of a rectangular cavity resonator are $a = 5$ cm, $b = 4$ cm, and $c = 10$ cm.

TE_{mnp} modes $m = 0, 1, 2,\ldots$ $n = 0, 1, 2,\ldots$ $P = 1, 2, 3,\ldots$
TM_{mnp} modes $m = 1, 2, 3,\ldots$ $n = 1, 2, 3,\ldots$ $P = 0, 1, 2,\ldots$

The first five resonances of a rectangular cavity resonator are

$f_{r_{101}} = 3.335$ GHz $\quad TE_{101}$

$f_{r_{011}} = 4.040$ GHz $\quad TE_{011}$

$f_{r_{102}} = 4.243$ GHz $\quad TE_{102}$

$f_{r_{110}} = 4.800$ GHz $\quad TE_{110}$

$f_{r_{111}} = 5.031$ GHz $\quad TE_{111}, TM_{111}$

■

EXAMPLE PROBLEM 5.14

For a cavity of dimensions 3 cm × 2 cm × 7 cm filled with air and made of copper, find the resonant frequency.

Solution

Given the cavity of dimensions 3 cm × 2 cm × 7 cm:
The resonant frequency of a rectangular cavity resonator is given by

$$f_{0mnp} = \frac{c}{2}\sqrt{\left(\frac{m}{a}\right)^2 + \left(\frac{n}{b}\right)^2 + \left(\frac{p}{d}\right)^2}$$

$$f_{r110} = \frac{3\times10^8}{2}\sqrt{\left(\frac{1}{3}\right)^2 + \left(\frac{1}{2}\right)^2 + \left(\frac{0}{7}\right)^2} = 9\,\text{GHz}$$

■

5.6 MICROSTRIP LINES

A microstrip line is basically a conductor on top of a dielectric layer with a single ground plane (see Figure 5.22 (a)). The top conductor is a narrow rectangular metallic foil, and the ground plane is a continuous metallic plane. The conductor and the ground plane loosely form a two-wire transmission line. The circuit is completed, because the current flowing through the top conductor returns to the source through the ground plane. Since the dielectrics on the top and bottom of the conductor are different, a microstrip line is an inhomogeneous structure. The microstrip line is also called an *open strip line*, as the dielectric on top of the conductor is air.

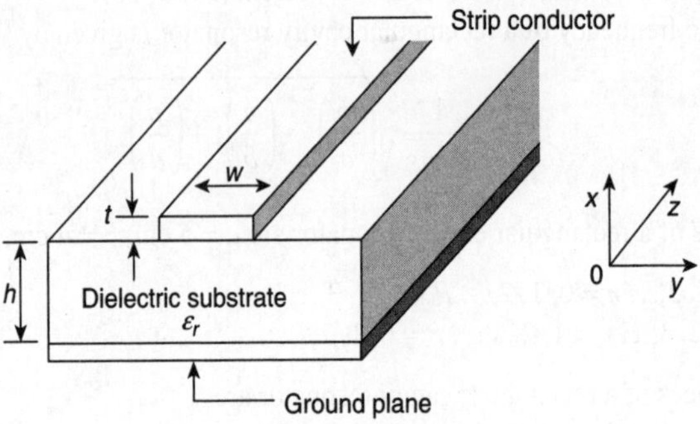

Figure 5.22 (a) *Microstrip line*

Since a microstrip line is inhomogeneous,the TEM mode cannot be supported but a quasi-TEM mode is possible. Fields in microstrip line are shown in Figure 5.22 (b). The quasi-TEM mode has different phase velocities in two dielectric layers, as the phase velocity depends only on material properties. For all practical purposes, the difference between the two velocities is not very large. Due to this, the quasi-TEM mode acts similar to the TEM mode.

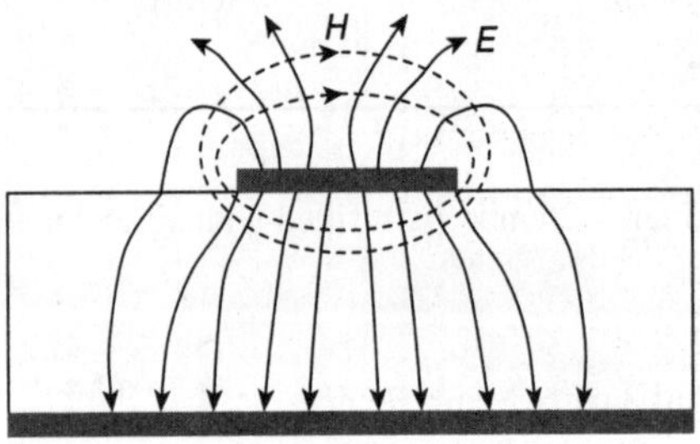

Figure 5.22 (b) *Propagation of electric and magnetic fields in microstrip lines*

Advantages over coaxial lines and wave guides:

1. By using the planar nature of the microstripline, semiconductor chips are conveniently attached to the microstrip element.
2. Fabrication costs are lower.
3. Better interconnection features
4. Easier fabrication
5. Small size and weight
6. Increased reliability and low cost

Disadvantages:

1. They have greater radiation losses due to the open conductor above the dielectric substrate.
2. They are more effective to nearby conductors due to the interference created by the open-ended conductor.
3. A discontinuity in the electric and magnetic fields is generated, due to nearness of the air dielectric interface with the microstrip conductor.

5.6.1 Characteristic Impedance (Z_0) of Microstrip Lines

The field-equation method is one among the several methods employed for calculating an accurate value of the characteristic impedance. However, it is extremely complicated and requires large digital computers for calculation. A method called *comparative* or an *indirect method* can be used to derive the characteristic-impedance equation of a microstrip line by making some changes in a well-known equation.

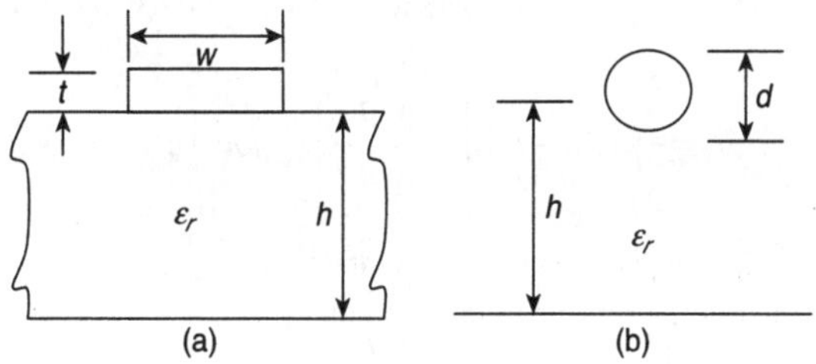

Figure 5.23 *Cross-section of (a) Microstrip line and (b) Wire over ground line*

The characteristic impedance of a wire-over-ground transmission line, as shown in Figure 5.23(b), is given by

$$Z_0 = \frac{60}{\sqrt{\varepsilon_r}} \ln \frac{4h}{d} \quad \text{for } h >> d \tag{5.191}$$

where ε_r = dielectric constant of the ambient medium
h = the height from the center of the wire to the ground plane
d = diameter of the wire

The characteristic impedance of the microstrip line can be calculated, if the diameter d of the wire and the equivalent values of the relative dielectric constant ε_r of the ambient medium.

5.6.2 Effective Dielectric Constant (ε_{re})

The effective relative dielectric constant for a microstrip line can be related to the relative dielectric constant of the board material. The empirical equation for the effective relative dielectric constant of a microstrip line is given by

$$\varepsilon_{re} = 0.475\varepsilon_r + 0.67 \tag{5.192}$$

where ε_r = the relative dielectric constant of the board material
ε_{re} = the effective relative dielectric constant for a microstrip line

5.6.3 Transformation of a Rectangular Conductor into an Equivalent Circular Conductor

The rectangular conductor of a microstrip line can be transformed into an equivalent circular conductor. The empirical equation for this transformation is given by

$$d = 0.67w\left(0.8 + \frac{t}{w}\right) \tag{5.193}$$

where d = diameter of the wire over the ground
w = width of the microstrip line
t = thickness of the microstrip line

The limitation of the ratio of thickness to width is between 0.1 and 0.8.

Characteristic Impedance Equation

The dielectric constant and the equivalent diameter can be obtained by substituting Eq. (5.192) and Eq. (5.193) in Eq. (5.191)

$$Z_0 = \frac{87}{\sqrt{\varepsilon_r + 1.41}} \ln\left[\frac{5.98h}{0.8w + t}\right] \quad for\ (h < 0.8w) \tag{5.194}$$

where ε_r = relative dielectric constant of the board material
h = height from the microstrip line to the ground
w = width of the microstrip line
t = thickness of the microstrip line

characteristic impedance for a narrow microstrip line is given by the equation Eq. (5.194). The velocity of propagation is

$$v = \frac{c}{\sqrt{\varepsilon_{re}}} = \frac{3 \times 10^8}{\sqrt{\varepsilon_{re}}} \text{ m/s} \tag{5.195}$$

The characteristic impedance for a wide microstrip line can be represented as

$$Z_0 = \frac{h}{w}\sqrt{\frac{\mu}{\varepsilon}} = \frac{377}{\sqrt{\varepsilon_r}} \frac{h}{w} \text{ for} (w >> h) \tag{5.196}$$

5.6.4 Losses in Microstrip Lines

The three basic types of losses in microstrip lines are

(a) Dielectric losses
(b) Ohmic losses
(c) Radiation losses

(a) Dielectric Losses

The conductivity of a dielectric cannot be neglected, and, therefore, the electric and magnetic fields in the dielectric are no longer in the time phase. In this case, the dielectric attenuation constant is given by

$$\alpha_d = \frac{\sigma}{2}\sqrt{\frac{\mu}{\varepsilon}} \text{ Np/cm} \tag{5.197}$$

where, σ is the conductivity of the dielectric substrate board in ℧/cm.

(b) Ohmic Losses

Ohmic losses are due to the non-perfect conductors. In a microstrip line the current density is concentrated at a skin depth in the conductor which is thick inside the surface and is exposed to the electric field.

The conducting attenuation constant of a wide microstrip line is given by

$$\alpha_c \approx \frac{8.686R_s}{Z_0 w} \ \text{dB/cm} \quad \text{for} \frac{w}{h} > 1 \tag{5.198}$$

where R_s is the surface skin resistance in Ω.

Z_o is characteristic impedance
w is width of the microstrip

(c) Radiation Losses

The radiation loss depends on the substrate's thickness and dielectric constant, as well as on its geometry. The ratio of radiated power to total dissipated power for an open-circuited microstrip line is

$$\frac{P_{rad}}{P_t} = 240\pi^2 \left(\frac{h}{\lambda_0}\right)^2 \frac{F(\varepsilon_{re})}{Z_0} \tag{5.199}$$

where ε_{re} = the effective dielectric constant, and
$\lambda_0 = c/f$ is the free-space wavelength

The radiation factor decreases by increasing the substrate dielectric constant. For lower dielectric-constant substrates, radiation is significant at higher impedance levels. Radiation becomes important until the impedance levels are very low, for higher dielectric-constant substrates. So, alternatively, Eq. (5.202) can be expressed as

$$\frac{P_{rad}}{P_t} = \frac{R_r}{Z_0} \tag{5.200}$$

where R_r is the radiation resistance of an open-circuited microstrip and is given by

$$R_r = 240\pi^2 \left(\frac{h}{\lambda_0}\right)^2 F(\varepsilon_{re}) \tag{5.201}$$

5.6.5 Quality Factor (*Q*) of Microstrip Lines

High-quality resonant circuits are required by many microwave-integrated circuits. The quality factor Q of a microstrip line is very high. However it is limited with a low dielectric constant and by the radiation losses of the substrates.

The quality factor due to the conductor attenuation constant of a wide microstrip line is

$$Q_c = 0.63h\sqrt{\sigma f_{GHZ}} \tag{5.202}$$

where σ is the conductivity of the dielectric substrate board in ℧/*cm*.

Similarly, the quality factor Q_d is related to the dielectric attenuation constant:

$$Q_d = \frac{27.3}{\alpha_d} \tag{5.203}$$

where α_d is the attenuation constant per wavelength, in dB/λ_g, and is given as

$$\alpha_d = 27.3\left(\frac{q\varepsilon_r}{\varepsilon_{re}}\right)\frac{\tan\theta}{\lambda_g} dB/\lambda_g \tag{5.204}$$

where $\lambda_g = \dfrac{\lambda_0}{\sqrt{\varepsilon_{re}}}$ and λ_0 is the wavelength in free space

Substituting Eq. (5.204) in Eq. (5.203) yields

$$Q_d = \frac{\lambda_0}{\sqrt{\varepsilon_{re}}\tan\theta} \approx \frac{1}{\tan\theta} \tag{5.205}$$

where λ_0 is the free-space wavelength in cm. Note that the Q_d for the dielectric attenuation constant of a microstrip line is approximately the reciprocal of the dielectric loss tangent θ and is relatively constant with frequency.

EXAMPLE PROBLEM 5.15

The parameters of a certain microstrip line are: $\varepsilon_r = 4.31$; $h = 6$ mils; $t = 2.4$ mils; $w = 8$ mils. Obtain the characteristic impedance Z_0 of the line.

Solution

Given parameters $\varepsilon_r = 4.31$, $h = 6$ mils; $t = 2.4$ mils; $w = 8$ mils.

The characteristic impedance

$$Z_0 = \frac{87}{\sqrt{\varepsilon_r + 1.41}}\ln\left[\frac{5.98h}{0.8w+t}\right]$$

$$= \frac{87}{\sqrt{4.31+1.41}}\ln\left[\frac{5.98\times 6}{0.8\times 8+2.4}\right] = 51.18\ \Omega$$

■

EXAMPLE PROBLEM 5.16

A microstrip line is made of a copper conductor its dimensions are as follows: that is 0.362 mm in width on a G-10 fiber glass-epoxy board which is 0.30 mm in height. The relative dielectric constant ε_r of the board material measured at 30 GHz is 5.2. The microstrip line of 0.028 mm thickness is used for 15 GHz. Assume the conductivity of copper is 5.96×10^7 ℧/cm. Determine the parameters given below:

(a) Characteristic impedance Z_0 of the microstrip line
(b) Surface resistivity R_s of the copper conductor
(c) Conductor attenuation constant α_c
(d) Quality factors Q_c

Solution

Given $\varepsilon_r = 5.2$; $w = 0.362$ mm; $h = 0.30$ mm; $t = 0.028$ mm

(a) Characteristic impedance Z_0 of the microstrip line

$$Z_0 = \frac{87}{\sqrt{\varepsilon_r + 1.41}} \ln\left[\frac{5.98h}{0.8w + t}\right] = \frac{87}{\sqrt{5.2 + 1.41}} \ln\left[\frac{5.98 \times 0.30}{0.8 \times 0.362 + 0.028}\right] = 58.591\ \Omega$$

(b) Surface resistivity R_s of the copper conductor

$$R_s = \sqrt{\frac{\pi f \mu}{\sigma}} = 2\pi\sqrt{\frac{f_{GHZ}}{\sigma}}\ \Omega / \text{square}$$

where σ is the conductivity of the dielectric substrate of the copper board in ℧/cm; the typical value is 5.96×10^7 ℧/cm.

$$R_s = 2\pi\sqrt{\frac{15 \times 10^9}{5.96\ \ 10^7}} = 99.679\ \Omega\text{-cm}$$

(c) Conductor attenuation constant α_c

$$\alpha_c = \frac{8.686 R_s}{Z_0 w} = \frac{8.686 \times 99.679}{58.591 \times 0.362} = 40.821\ \text{dB/cm}$$

(d) $Q_c = \dfrac{27.3}{\alpha_c} = \dfrac{27.3}{40.821} = 0.667$ ■

EXAMPLE PROBLEM 5.17

A strip-line transmission has a distance of 0.3175 cm between the ground planes. If the diameter of the equivalent circular conductor is 0.0539 cm, determine the characteristic impedance and velocity of propagation if the dielectric constant is 2.32 for the strip-line material.

Solution

Here, $b = 0.3175$ cm

$d = 0.0539$ cm

$\epsilon_r = 2.32$

Characteristic impedance, $z_0 = \dfrac{60}{\sqrt{\epsilon_r}} \ln\left(\dfrac{4b}{\pi d}\right)$

$$= \frac{60}{\sqrt{2.32}} \ln\left[\frac{4 \times 0.3175}{\pi(0.0539)}\right] = 79.38\ \Omega$$

and velocity of propagation $v = \dfrac{c}{\sqrt{\epsilon_r}} = \dfrac{3 \times 10^8}{\sqrt{2.32}} = 1.97 \times 108$ m/s

■

SUMMARY

1. A waveguide is a hollow conducting tube of a rectangular or circular cross-section through which the energy is transmitted, in the form of EM waves.

2. The waveguide acts similar to a high-pass filter with regard to the frequency.

3. The EM wave inside a waveguide can have an infinite number of field configurations called *modes*. The rectangular waveguide (RWG) operating modes are TE or TM (cannot support a TEM mode,i.e., $E_z = H_z = 0$).
   - TE modes have $E_z = 0$, and $H_z \neq 0$.
   - TM modes have $H_z = 0$, and $E_z \neq 0$.

4. The cut-off frequency is the operating frequency below which attenuation occurs and above which propagation takes place. TE and TM waves can propagate only if their frequency is above the cut-off frequency.

5. The dominant mode is the mode with the lowest cut-off frequency, and it is the only mode that can propagate in the frequency range from its cut-off frequency to the next higher cut-off frequency.

6. The non-propagating mode is known as an *evanescent mode*, in which the operating frequency is less than the cut-off frequency.

7. Whenever two or more modes have the same cut-off frequency, they are called *degenerate modes*.

8. The cut-off frequency of the TEM mode is zero.

9. The wave equations for TE and TM waves

   $\nabla^2 E_z = -\omega^2 \mu\varepsilon E_z$ for TM wave ($H_z = 0$)

   $\nabla^2 H_z = -\omega^2 \mu\varepsilon H_z$ for TE wave ($E_z = 0$)

10. The wave impedance of a waveguide is defined as the ratio of the strength of the electric field in one direction to the magnetic field along the other transverse direction at a certain point in the waveguide.

$$Z = \frac{E_x}{H_y} = \frac{-E_y}{H_x}$$

Wave impedance for a TM wave in a rectangular waveguide is $Z_{TM} = \eta\sqrt{1-\left(\frac{\lambda_0}{\lambda_c}\right)^2}$

Wave impedance for a TE wave in a rectangular waveguide is $Z_{TE} = \eta / \sqrt{1-\left(\frac{\lambda_0}{\lambda_c}\right)^2}$
where η = intrinsic impedance of free space

11. The wavelength along the z axis of a waveguide is guided wavelength, $\lambda_g = 2\pi/\beta$ or the relationship between the λ_0, λ_g, and λ_c is $\lambda_g = \dfrac{\lambda_0}{\sqrt{1-\left(\lambda_0/\lambda_c\right)^2}}$

12. The phase velocity (v_p)of a waveguide is defined as the velocity of a point having a constant phase on the carrier.

$$v_p = \frac{\omega}{\beta} = \frac{c}{\sqrt{1-\left(\frac{\lambda_0}{\lambda_c}\right)^2}}$$

and the velocity of a point having a constant phase on the message or information signalis the

group velocity $v_g = \frac{d\omega}{d\beta} = c\sqrt{1-\left(\frac{\lambda_0}{\lambda_c}\right)^2}$

If $f < f_c \Rightarrow$ both v_p and v_g are imaginary, meaning no propagation

13. Group velocity, phase velocity, and the free-space velocity are related by $v_p v_g = c^2$

14. Group and phase velocities are same in TE and TM waves, respectively.

15. The relationship between the TE, TM, and free-space impedance is $\eta_{TE}\eta_{TM} = \eta^2$.

16. TE_{10} is the dominant mode in the rectangular waveguide, because it has the lowest cut-off frequency ($\lambda_c = 2a$).

Cut-off frequency (f_c) and cut-off wavelength, λ_c of a rectangular waveguide are

$$f_c = \frac{2}{\sqrt{\left(\frac{m}{a}\right)^2+\left(\frac{n}{b}\right)^2}} \text{ and } \lambda_c = \frac{c}{f_c} = \frac{c}{\frac{c}{2}\left[\left(\frac{m}{a}\right)^2+\left(\frac{n}{b}\right)^2\right]^{\frac{1}{2}}} = \frac{2ab}{\sqrt{m^2b^2+n^2a^2}}$$

17. TE_{11} is the dominant mode in the circular waveguide. Circular waveguides are used as attenuators and phase shifters.The advantages of circular waveguides are higher power-handling capacity and lower attenuation for a given cut-off wavelength.

18. Cavity resonators are metallic enclosures that confine the electromagnetic energy within them.

19. A circular resonant cavity is a circular waveguide with both its ends closed.

20. The resonant frequency for a rectangular cavity is given by

$$f_{0mnp} = \frac{1}{2\pi\sqrt{\mu_0\varepsilon_0\varepsilon_r}}\sqrt{\left(\frac{m\pi}{a}\right)^2+\left(\frac{n\pi}{b}\right)^2+\left(\frac{p\pi}{d}\right)^2}$$

21. The quality factor (Q) is a measure of selectivity of the resonant circuit.

$$Q = 2\pi\,\frac{\text{maximum energy stored}}{\text{energy dissipated per cycle}}$$

22. A microstrip line consists of a conductor strip and a ground plane. The electromagnetic wave propagates in the quasi-TEM mode.

- The major advantage of a microstrip over a stripline is that all active components can be mounted on top of the board. The disadvantages are that when high isolation is required such as in a filter or a switch, some external shielding may have to be considered.
- The characteristic impedance of a microstrip line is

$$Z_0 = \frac{87}{\sqrt{\varepsilon_r + 1.41}} \ln\left[\frac{5.98h}{0.8w + t}\right] \quad \text{for} \left(h < 0.8w\right)$$

- There are three types of losses in a microstrip line:
  (i) Dielectric loss
  (ii) Ohmic loss
  (iii) Radiation loss

OBJECTIVE-TYPE QUESTIONS

1. The waves in a waveguide
(a) travel along the border walls of the waveguide
(b) are reflected from the side walls but do not travel along them
(c) travel through the dielectric without touching the walls
(d) travel along all the four walls

2. Waveguides can carry
(a) TE mode
(b) TM mode
(c) mixed mode
(d) all

3. The cut-off frequency of a waveguide depends on
(a) dimensions of the waveguide
(b) wave mode
(c) the dielectric property of the medium in the waveguide
(d) all

4. In RWG, the mode subscripts m and n indicate
(a) no. of half-wave patterns
(b) no. of full-wave patterns
(c) no. of the zeros of the field
(d) none

5. Wave impedance of waveguides in the TE mode can be
(a) $\dfrac{\eta}{\sqrt{1-\left(\dfrac{\lambda}{\lambda_0}\right)^2}}$

(b) $\eta\sqrt{1-\left(\frac{\lambda}{\lambda_0}\right)^2}$

(c) both

(d) none

6. The dominant TE mode in rectangular waveguides is

(a) TE_{01}

(b) TE_{11}

(c) TE_{20}

(d) TE_{10}

7. Cut-off wave length of rectangular waveguides is

(a) $\dfrac{2}{\sqrt{\left(\frac{m}{a}\right)^2+\left(\frac{n}{b}\right)^2}}$

(b) $\dfrac{2}{\sqrt{\left(\frac{m}{b}\right)^2+\left(\frac{n}{a}\right)^2}}$

(c) both

(d) none

8. In RWG, for dominant mode, the cut-off wave length is

(a) 2a

(b) 2b

(c) a

(d) none

9. An air-filled rectangular waveguide has dimensions of 6×4 cm. Its cut-off frequency for the TE_{10} mode is

(a) 2.5 GHz

(b) 25 GHz

(c) 25 MHz

(d) 5 GHz

10. In hollow rectangular waveguides,

(a) the phase velocity is greater than the group velocity

(b) the phase velocity is greater than the velocity of light in free space

(c) both

(d) none

11. The dominant mode in the circular waveguide is

(a) TE_{10}

(b) TE_{11}

(c) TE_{01}
(d) TE_{12}

12. Nonexistent modes in circular waveguides are
(a) TE_{10}
(b) TE_{00}
(c) both
(d) none

13. Degenerate modes in circular waveguides are
(a) TE_{01} and TM_{11}
(b) TE_{22} and TM_{22}
(c) both
(d) none

14. In cylindrical waveguides, Z_{TE} is
(a) $\dfrac{\beta}{\omega\mu}$
(b) $\dfrac{\omega\mu}{\beta}$
(c) $\dfrac{\omega\beta}{\mu}$
(d) $\omega\mu\beta$

15. Theoretically, the number of modes that can exist in cylindrical waveguides are
(a) zero
(b) one
(c) two
(d) infinite

16. Guide wave length of cylindrical waveguides is
(a) $\dfrac{\lambda}{\sqrt{1-\left(\dfrac{\lambda}{\lambda_0}\right)^2}}$
(b) $\dfrac{\lambda}{\sqrt{1-\left(\dfrac{f_0}{f}\right)^2}}$
(c) both
(d) none

ANSWERS TO OBJECTIVE-TYPE QUESTIONS

1. (a) 2. (d) 3. (d) 4. (a) 5. (a) 6. (d) 7. (a) 8. (a) 9. (a) 10. (c) 11. (b)
12. (b) 13. (a) 14. (b) 15. (d) 16. (c)

REVIEW QUESTIONS

1. Why is TEM mode not possible for rectangular waveguides?
2. Explain the wave impedance of a rectangular waveguide and derive the expression for the wave impedance of TE, TM, and TEM modes.
3. Derive the expression for cut-off frequency, phase constant, and phase velocity of a wave in a circular waveguide.
4. Write the advantages and disadvantages of rectangular waveguide over circular waveguide.
5. Write about the filter characteristics.
6. List out the differences between the TE mode and TM mode.
7. What is meant by a cavity resonator? Derive the expression for the resonant frequency of the rectangular cavity resonator.
8. Derive the expression for the resonant frequency of the circular cavity resonator.
9. Derive the expression for the characteristic impedance of microstrip lines.
10. What are the various losses in a microstrip line? Explain.
11. A rectangular waveguide has the following dimensions: a = 5.1cm, b = 2.4 cm. a) Calculate the cut-off frequency of the dominant mode. b) Calculate the lowest frequency and determine the mode closest to the dominant mode.
12. A wave of frequency 6 GHz is propagated in a parallel plane waveguide separated by 3 cm. Calculate a) the cut-off wavelength for the dominant mode; b) the wavelength in the waveguide; c) the group and phase velocities; d) the characteristic wave impedance.
13. A wave of frequency 10GHz is propagated in a circular waveguide of inner diameter 4cm. Calculate (a) the cut-off wavelength, (b) the guided wavelength and (c) characteristic wave impedance.
14. A circular waveguide with a radius of 4 cm is used to propagate an electromagnetic wave in the TM_{01} mode. Determine the wave impedance, phase velocity, and group velocity of the waveguide for the wavelength of 8 cm.
15. A rectangular waveguide with a width of 4 cm and a height of 2 cm is used to propagate an electromagnetic wave in the TE_{10} mode. Determine the wave impedance, phase velocity, and group velocity of the waveguide for the wavelength of 6 cm.
16. An air-filled rectangular waveguide has a cross section of 8 × 4 cm. Find the cut-off frequencies for the following modesTE_{10},TE_{20},TE_{11} and the rates of the guide velocity v_p to the velocity in free space for each of these modes. f_c= (3/2)f.

Waveguide Components

6

6.1 INTRODUCTION

In all practical applications, similar to the radar, the waveguide system carries the power from the transmitter end to the input of an antenna. However, the entire waveguide system cannot be formed into one piece. Therefore, it should be constructed using many individual sections (or components), and these sections have to be connected with waveguide bends, joints, and so on. In addition to this, a variety of signal control components such as directional couplers, circulators, isolators, attenuators, phase shifters, and multiport junctions have to be used in the microwave system. For example, waveguide attenuators and phase shifters are used to control the amplitude and to shift the phase, respectively. The isolators and circulators allow the microwave signals to travel in only one direction, whereas the directional couplers sample the power flowing in one direction and are used for determining the frequency and the level of power. This chapter describes the principle of operation of various waveguide components, coupling methods, and signal control components. The waveguide components are characterized by the network parameters known as *scattering parameters*, which are explained in detail in Chapter 7.

6.2 COUPLING MECHANISMS

The microwave signal to be carried from one point to other is introduced into the waveguide with an antenna like probes or loops. The probe is coupled to the waveguide parallel to the point where the electric field is maximum, and the loops are coupled at a point where the magnetic field strength is maximum.

6.2.1 Probes

The probe is defined as a $\lambda/4$ vertical antenna that is inserted in the waveguide at a distance of $\lambda/4$ from the closed end and the center of the broader dimension of the waveguide (as shown in Figure 6.1). It is inserted at that particular point, because there the electric field is maximum. The probe will now act as an antenna that is polarized in the plane parallel to that of an electric field.

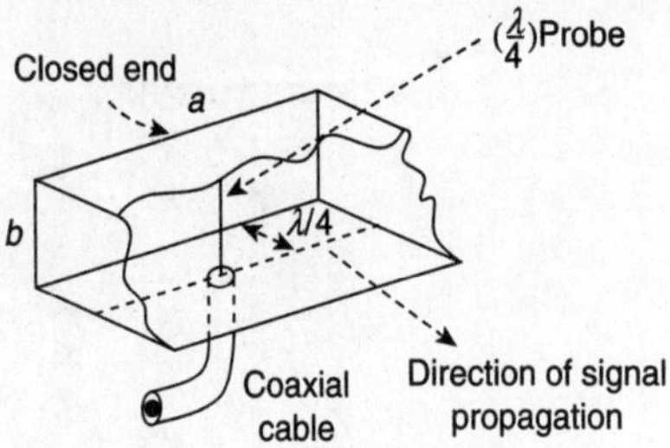

Figure 6.1 *Coupling probe*

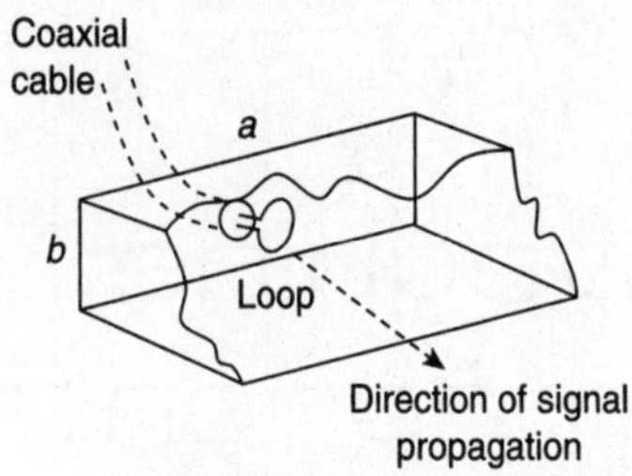

Figure 6.2 *Coupling loop*

6.2.2 Loops

A loop can be used to introduce a magnetic field into a waveguide. It is generally mounted at a distance of $\lambda/2$ (or an integer multiple of $\lambda/2$)from the closed end of the waveguide. It can also be placed at the middle of the top wall or bottom wall (Figure 6.2). Microwave energy applied through a short piece of coaxial cable causes a magnetic field to be set up in the loop. The magnetic field also establishes an electric field, which is then propagated down the waveguide.

It is often desirable to couple a coaxial cable to a waveguide or a cavity resonator by means of a coupling loop rather than by a coupling probe.

6.2.3 Coupling to a Cavity Resonator

There are three ways of coupling a field into a cavity or from it. The coupling mechanisms are mainly loop, probe, and hole (Figure 6.3). For efficient coupling, the prerequisite is that the field of the resonance mode should have some common components with the fields of the coupling element. A loop at the maximum of the magnetic field perpendicular to the field or a probe at the maximum of the electric field along the field works as a good coupling element.

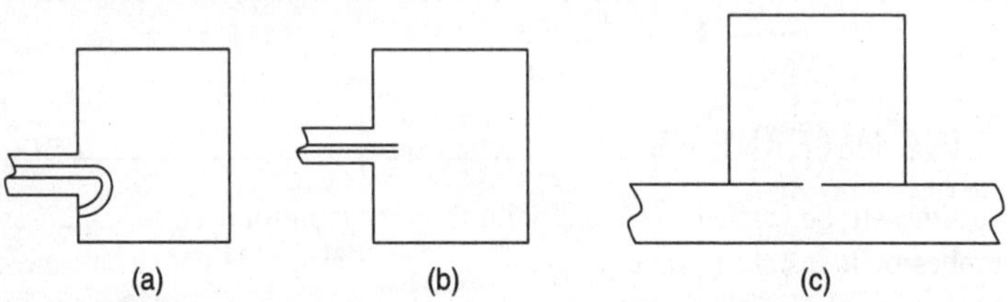

Figure 6.3 *Couplings to a cavity resonator (a) loop; (b) probe; (c) hole*

For efficient hole coupling, the fields of the waveguide and cavity should have some common components at the coupling hole.

6.3 WAVEGUIDE DISCONTINUITIES

Any interruption in the uniformity of a transmission line leads to *impedance mismatch* and is known as *impedance discontinuity* or *discontinuity*. In a waveguide system, when there is a mismatch, reflections will occur. In transmission lines, in order to overcome this mismatch, lumped impedances or stubs of required values are placed at the pre-calculated points. In waveguides also, some discontinuities are used for matching purposes.

6.3.1 Waveguide Irises

Fixed or adjustable projections from the walls of waveguides are used for impedance matching purposes, and these are known as *windows* or *irises*. An iris is a metal plate that contains an opening through which the waves may pass. It is located in the transverse plane of either a magnetic or an electric field. Irises are classified according to the sign of the imaginary part of the impedance. If the reactance of the impedance is positive or if the susceptance of the admittance is negative, we have an inductive iris. If the reactance is negative or if the susceptance is positive, we have a capacitive iris.

Inductive Iris

Usually inductive irises are used as coupling networks between half-wavelength cavities in rectangular waveguides. Generally an inductive iris is placed where either magnetic field is strong or electric field is weak. The plane of polarization of the electric field becomes parallel to the plane of inductive iris. This causes a current flow which sets up a magnetic field. Then the energy is stored in the magnetic field. Hence, inductance will increase at that point of the waveguide.

Capacitive Iris

A capacitive iris is also known as *capacitive window* (as given in Figure 6.4 (b)). It extends from the top and bottom walls into the waveguide. The capacitive iris has to be placed in strong electric field. This capacitive iris creates the effect of capacitive susceptance which is in parallel to that point of waveguide where the electric field is strong.

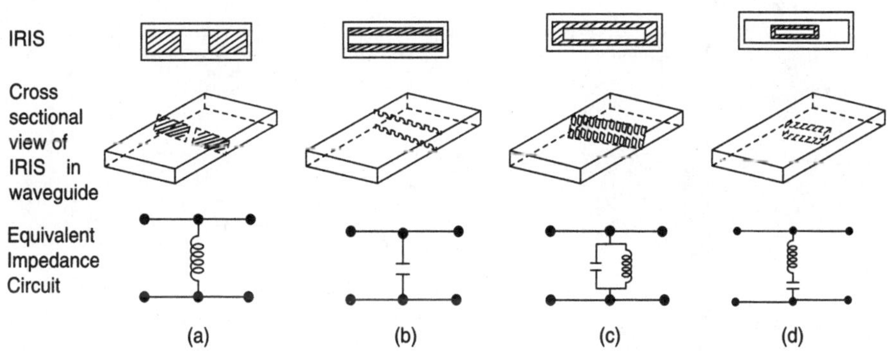

Figure 6.4 *Waveguide irises*

If the inductive and capacitive irises are combined suitably (correctly shaped and positioned), the inductive and capacitive reactances introduced will be equal, and the iris will become a parallel resonant circuit (Figure 6.4 (c)). For the dominant mode, the iris presents a high impedance, and the shunting effect of this mode will be negligible. Other modes are completely attenuated, and the resonant iris acts as a band-pass filter to suppress unwanted modes. Figure 6.4 (d) shows the series resonant iris that is supported by a non-metallic material and is transparent to the flow of microwave energy.

6.3.2 Tuning Screws and Posts

Posts and screws made from conductive material can be used for impedance-changing devices in waveguides. A post or screw can also serve as a reactive element. The only significant difference between posts and screws is that posts are fixed and screws are adjustable. A post (or screw) that only penetrates partially into the waveguide acts as a shunt capacitive reactance. When a post extends completely through the waveguide, making contact with the top and bottom walls, it acts as an inductive reactance. The screw acts similar to an LC-tuned circuit in such cases.

Screws

A screw is generally inserted into the top or bottom walls of the waveguide, parallel to the electric-field lines. It can give a variable amount of susceptance depending on the depth of penetration. A screw with an insertion distance (screw depth) less than $\lambda/4$ produces capacitive susceptance. When the distance is equal to $\lambda/4$, we have series resonance. When the distance is greater than $\lambda/4$, it produces inductive susceptance, as shown in Figure 6.5.

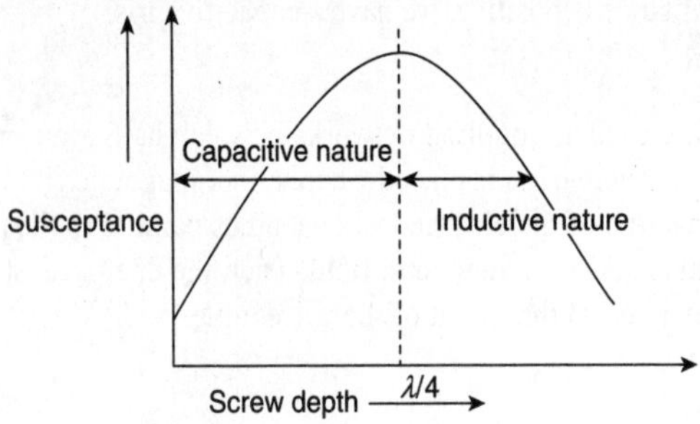

Figure 6.5 *Susceptance nature at different screw depths*

The adjustable waveguide screw for capacitive setting is shown in Figure 6.6 (a) and for inductive setting is shown in Figure 6.6 (b).

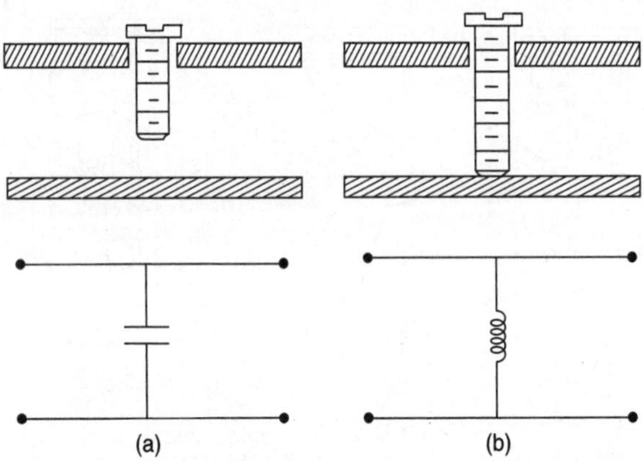

Figure 6.6 *Adjustable waveguide components: (a) capacitive setting; (b) inductive setting*

The most direct method of impedance matching with a matched screw involves using a single screw that is adjustable in both length and position along the waveguide. However, it requires a slot in the waveguide.

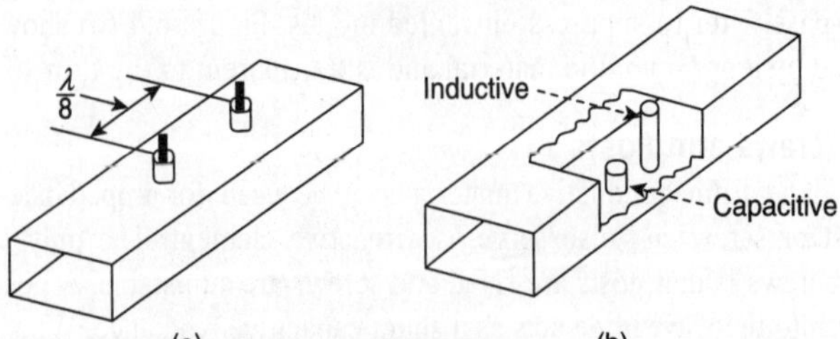

Figure 6.7 *(a) Two-screw matcher; (b) Waveguide post*

An alternative arrangement is to use double or triple screw units with a spacing of $\lambda/8$ or $\lambda/4$. A two-screw matcher is shown in Figure 6.7 (a).

Posts

A cylindrical post is introduced into the broader side of the waveguide; it produces a similar effect as an iris in providing lumped capacitive/inductive reactance at that point. When a metalpost extends completely across the waveguide, parallel to an electric field, it adds an inductive susceptance that is parallel to the waveguide. A post extending across the waveguide at right angles to the electric field produces an effective capacitive susceptance that is in shunt with the waveguide at the position of the post. The waveguide post is shown in Figure 6.7 (b). The advantage of such posts over irises is the flexibility they provide, which results in ease of matching.

6.3.3 Matched Loads

The most commonly used waveguide terminations are the matched loads. Whenever the load impedance and characteristic impedance of the transmission line are not matched/equal, reflections exist. These reflections would cause frequency instability to the source. Matched loads are used for minimizing the reflections by placing a material in the waveguide parallel to the electric field to absorb the incident power completely.

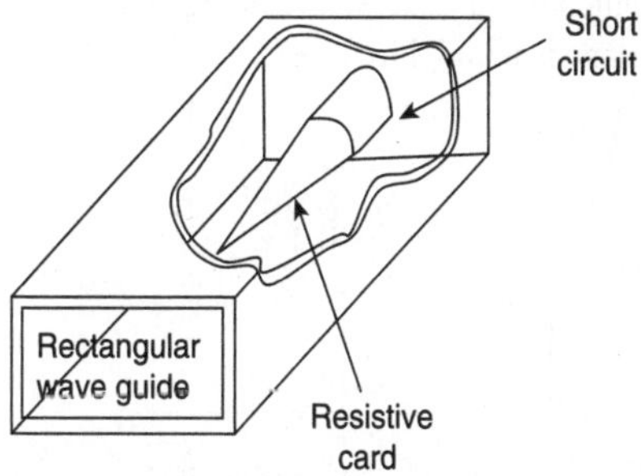

Figure 6.8 *Matched load with resistive card*

One of the methods involved in the matched load is to place a resistive card in the waveguide parallel to the electric field as shown in Figure 6.8. The front portion of the card is tapered to avoid discontinuity of the signal, and it almost absorbs the incident field.

6.4 WAVEGUIDE ATTENUATORS

An attenuator is a passive device that is used to reduce the strength or amplitude of a signal. At microwave frequencies, the attenuators were not only meant to do this, but also meant to maintain the characteristic impedance (Z_0) of the system. If the Z_0 of the transmission line is not maintained, the attenuator would be seen as impedance discontinuity, which causes reflections. Usually, a microwave attenuator controls the flow of microwave power by absorbing it.

Attenuation in dB of a device is ten times the logarithmic ratio of power flowing into the device (P_i) to the power flowing out of the device (P_o) when both the input and output circuits are matched.

$$\text{Attenuation in dB} = 10 \log P_i / P_o \tag{6.1}$$

Principle

In a microwave transmission system, the microwave power transferring from one section to another section can be controlled by a device known as *microwave attenuator*. These attenuators operate on the principle of interfering with electric or magnetic or both the fields. A resistive material placed in parallel to electric field lines (of field current) will induce a current in the material,which will result in I^2R loss. Thus, attenuation occurs by heating of the resistive element.

Attenuators may be of three types:

- Fixed
- Mechanically or electronically variable
- Series of fixed steps

6.4.1 Fixed Attenuators

Fixed attenuators are used where a fixed amount of attenuation is needed. They also called *pads*. In this type of attenuator tapering is provided by placing a short section of a waveguide with an attached tapered plug of absorbing material at the end. The purpose of tapering is for the gradual transition of microwave power from the waveguide medium to the absorbing medium. Because of the absorbing medium, reflections at the media interface will be minimized. In a fixed attenuator (Figure 6.9), plug is nothing but a dielectric slab which has a glass slab with aquadog or a carbon film coating. The pad is placed in such a way that the plane is parallel to the electric field. For this, two thin metal rods are used.

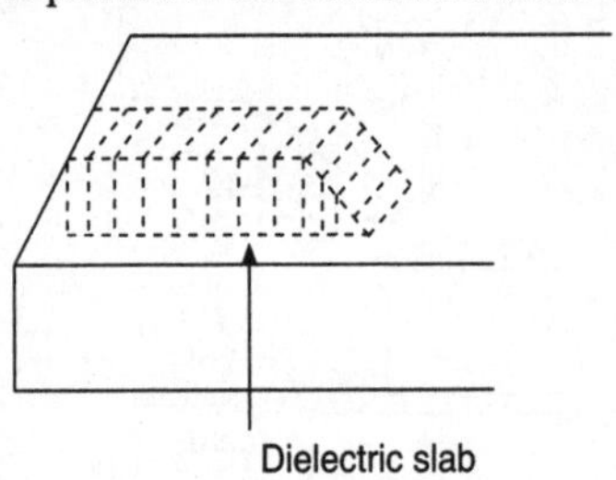

Figure 6.9 *Fixed attenuator*

The amount of attenuation provided by the fixed attenuator depends on the strength of the dielectric material, the location and area of the pad, type of material used for pad within waveguide and the frequency of operation.

6.4.2 Variable Attenuators

For providing continuous or stepwise attenuation variable attenuators are used. The provided attenuation depends on the insertion depth of the absorbing plate into the waveguide. The maximum attenuation will be achieved when the pad extends totally into the waveguide. This type of variable attenuation is provided by knob and gear assembly which can be properly calibrated. The power transmitted to the load can be varied manually or electronically from nearly the full power of the source to as little as a millionth of a percent of the source power depending on the frequency of operation. The types of variable attenuators are

1. Flap or resistive card-type attenuators
2. Slide vane attenuators
3. Rotary vane attenuators

6.4.2.1 Resistive card (flap type) and slide vane attenuators

Mechanically, variable attenuators are stepwise variable attenuators. Examples are *flap type, slide vane type attenuators* (shown in Figure 6.10). In contrast, electronically variable attenuators are continuously variable attenuators. They are used for applications requiring automatic signal leveling and control, amplitude modulation, remote signal control, and so on. A simple form of an attenuator consists of a thin, tapered resistive card, whose depth of penetration into the waveguide is adjustable (Figure 6.10 (a)). The card is inserted into the waveguide through a longitudinal slot cut in the center of the broad wall of a rectangular waveguide.

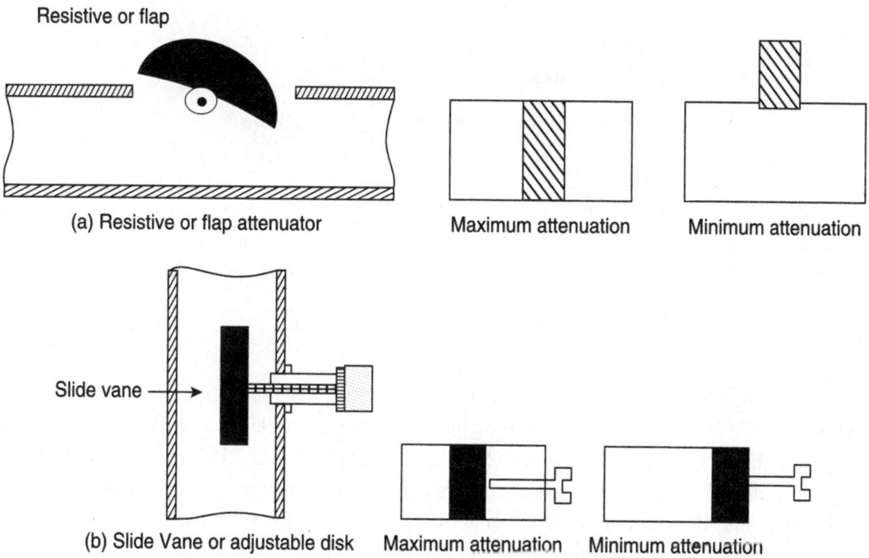

Figure 6.10 *Adjustable or variable attenuators*

In Figure 6.10 (b), the vane is positioned at the center of the waveguide and can be moved laterally from the center, where it provides maximum attenuation to the edges. However, the attenuation is reduced at the edges, as the electric field lines are always concentrated at the center of the waveguide. The vane is tapered at both the ends for matching the attenuator with the waveguide. An adequate match is obtained if the taper length is made equal to $\lambda/2$. The biggest disadvantage with these attenuators is that their attenuation is frequency sensitive, and also, the phase of the output signal is a function of attenuation.

6.4.2.2 Rotary Vane Attenuators

The most satisfactory precision attenuator is the rotary vane attenuator. The structure of this attenuator is shown in Figure 6.11. It consists of two rectangular to circular waveguide tapered transitions, along with an intermediate section of a circular waveguide that is free to rotate. All the three sections contain thin resistive cards.

The input signal passes the first card with a negligible attenuation, because the electric field of the TE_{10} wave mode is perpendicular to the card. Then, the wave enters through a transition to the circular waveguide. The attenuation is adjusted by rotating the circular waveguide section and the resistive card within it. The field of the TE_{11} wave mode can be divided into two components: one perpendicular to the card and the other parallel to it. The latter component is absorbed by the card; the former component enters the output of the waveguide, in which again its component parallel to the resistive card is absorbed.

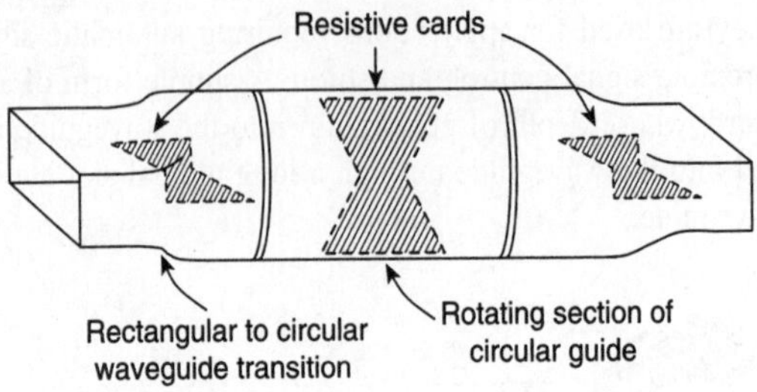

Figure 6.11 *Rotary vane attenuators*

The plates are usually thin with $\varepsilon_r > 1$, $\mu_r = 1$, and conductivity σ of a finite nonzero value. The plates attenuate the wave that is travelling, and the amount of attenuation is dependent on the properties of the material from which the plate is made, the dimensions of the slab, and the angle between the electric field at the input and the plane of the resistive card in the circular section. The attenuation in decibels is given by

$$\text{Attenuation in dB} = -40 \log(\cos\theta) \text{ dB}$$

where θ is the angle between the electric field at the input and the plane of the resistive card in the circular section. Hence, the attenuation is controlled by the rotation of the center section. Minimum attenuation at $\theta = 0°$, and maximum attenuation at $\theta = 90°$.The attenuation provided by this device depends only on the rotation angle θ but not on the frequency. This device is very accurate, and is, hence, being used as a calibration standard. Its accuracy is limited only by imperfect matching and by misalignment of the resistive cards.

6.5 WAVEGUIDE PHASE SHIFTERS

A phase shifter is a two-port component that provides a *fixed* or *variable* change in the phase of the traveling wave. An ideal phase shifter is lossless and matched. It only shifts the phase of the output wave. Example: Phase shifters are used in phased antenna arrays. A structure resembling the attenuator in Figure 6.11 also operates as a phase shifter when the resistive cards are replaced with dielectric cards having proper lengths. Electrically controlled phase shifters are much faster than mechanical phase shifters. They are often based on p-i-n diodes or field effect transistors (FETs).

6.5.1 Fixed Phase Shifters

Fixed phase shifters are usually extra transmission-line sections of a certain length that are meant to shift the phase with regard to the reference line. Therefore, depending on the bias current, the wave traveling along the transmission line will have an additional traveling path. Since these devices are binary switches, only discrete phase shifts are possible.

6.5.2 Variable Phase Shifters

The variable phase shifters use mechanical or electronic means to achieve a dynamic range of phase difference. The mechanically tuned phase shifter usually consists of variable short circuits that are used with hybrids or, in the case of waveguide components, a dielectric slab with a variable position in the guide. Step motors move the slab across the guide (from its center toward the outer walls), there by accomplishing a maximum or minimum phase shift. Another method for obtaining the desired mechanically tuned phase shift involves combining variable short circuits and hybrid circuits. The movement of the short circuit along a transmission line results in the phase shift, thus making it appear shorter or longer.

6.5.2.1 Dielectric phase shifters

Variable phase shifters in rectangular waveguides are shown in Figure 6.12 (a). The variable type of dielectric phase shifters employs a low-loss dielectric insertion in the air-filled guide at a point of the maximum electric field to increase its effective dielectric constant. This causes the guide wavelength, λ_g to decrease (as shown in Figure 6.12 (a)). Thus, the insertion of the dielectric increases the phase shift in the wave passing through the fixed length of the waveguide section. Tapering of the dielectric slab is resorted in order to reduce the reflections.

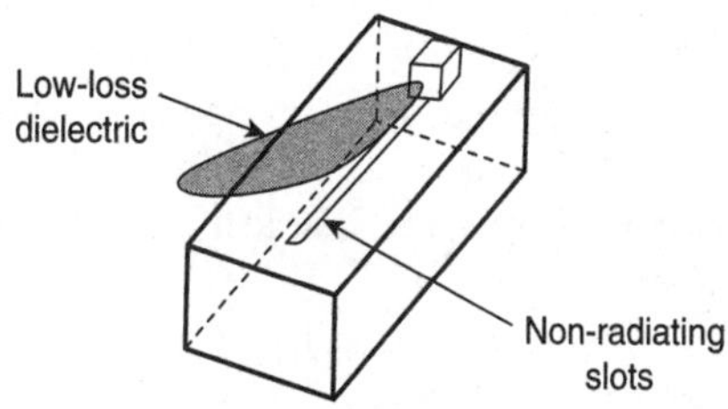

Figure 6.12 (a) *Variable phase shifters in rectangular waveguides*

6.5.2.2 Rotary phase shifters

The precision phase shifter can be realized by a rotary phase shifter, which is useful in microwave measurement. The essential parts of this phase shifter are three waveguide sections: two fixed and one rotary. The fixed sections consist of quarter-wave plates, and the rotary section consists of half-wave plates; all the plates are of dielectric type. The center section is rotatable to provide the required phase shift. The structure of the rotary vane attenuator is shown in Figure 6.12 (b).

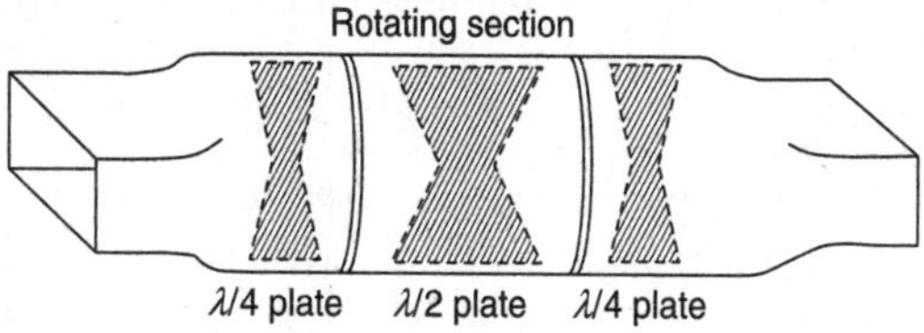

Figure 6.12 (b) *Rotary phase shifter*

The two fixed quarter-wave sections are identical in all respects, and the rotatable half-wave section is just the double of a quarter-wave section. Each of the two fixed sections, attached to a transition, consists of a piece of a circular waveguide with a dielectric plate, making an angle of 45° with the horizontal. The dielectric plate is usually thin with $\varepsilon_r > 1$, $\mu_r = 1$. The output remains vertically polarized, which means that the phase shifter is lossless and reflection less for any position of the rotary section. It is used as a calibration standard due to its high accuracy.

6.6 WAVEGUIDE MULTIPORT JUNCTIONS

Microwave multiport junctions are the devices that are used to split or combine microwave power. The important parts of microwave junctions are ports, arms, and junction regions. Ports are openings to which the source or load is connected. Arms are pieces of the transmission lines or waveguides with which the junction device is fabricated. The junction region is the common space where all the arms of the device meet each other.

6.6.1 Microwave or Waveguide Junctions

A microwave circuit is a combination of several microwave devices that are connected in a way so as to achieve the desired transmission of microwave signals. In general, a microwave junction is an interconnection of two or more microwave components as shown in Figure 6.13.

A port is said to be perfectly matched to the junction if nothing out of the power incident at the port is reflected back to the port by the junction. Two ports are said to be perfectly isolated if nothing out of the power incident at one port appears at the other port. When an input from a microwave source is fed to port 1, it spreads all of its power into ports 2, 3, 4, and some of the power is reflected back to port 1 due to the existence of a mismatch between the port and that junction. Low-frequency circuits can be described by Z, Y, h, and ABCD parameters, and these parameters are related in terms of currents and voltages. However, at microwave frequencies, instead of currents and voltages, we talk of traveling waves with their associated powers.

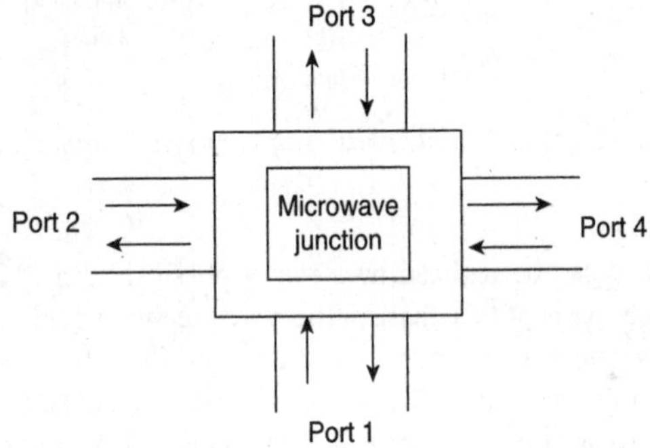

Figure 6.13 *Microwave or waveguide junctions*

6.6.2 Microwave TEE Junctions

Waveguide Tees are used for the purpose of connecting a branch or a section of a waveguide in a series or parallel to the main waveguide. The intersection of waveguides in the shape of the English capital letter "T" is called a T junction. E-plane Tee and H-plane Tee are examples of three-port waveguide T junctions. Normal reciprocal three-port junctions has one drawback, that is, lack of isolation between the output ports. This results in dependence of the power consumed at one port on the termination at the other output port. This lack of isolation between the output ports limits the usefulness of the three port junctions, particularly in power monitoring and divider applications.

6.6.2.1 E-plane Tees

E-plane Tee is a voltage or series junction. A side arm is attached to a waveguide by cutting a rectangular slot along the broader dimension of waveguide as shown in Figure 6.14 (a). If the E-plane junction is completely symmetrical and if waves enter through the side arm, the waves that leave the main arms are equal in magnitude and opposite in phase. Since the electric field lines change their direction when they come out of ports 1 and 2, it is called an *E-plane Tee*. Any signal that is to be split or any two signals which are to be combined will be fed to the E arm.

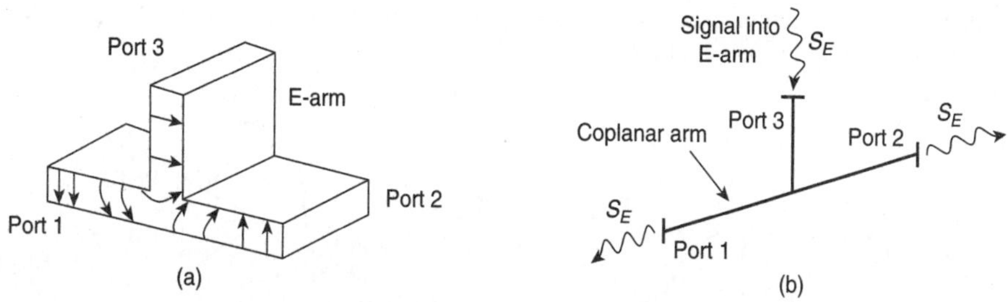

Figure 6.14 *(a) E-plane waveguide Tee junction; (b) Transmission-line equivalent circuit of E-plane Tee*

Figure 6.14 (c) *Schematic of E-plane Tee*

As power divider

- If the amplitude of the input wave at port 3 is A, the amplitudes of the waves at ports 1 and 2 are same and equal to $A/\sqrt{2}$. They are out of phase when its collinear arms lengths are same.
- When the power incident at port 3 is P, the powers that appear at ports 1 and 2 are $P/2$ each.

  i.e., $P_3 = P, \quad P_2 = P/2, \quad P_1 = P/2 \Rightarrow P_3 = P_1 + P_2 = 2P_1 = 2P_2$

  The amount of power coming out of ports 1 and 2 in decibels is

$$= 10\log_{10}\frac{P_1}{P_3} = 10\log_{10}\frac{P_1}{2P_1} = 10\log_{10}\frac{P_2}{2P_2} = 10\log_{10}\left(\frac{1}{2}\right) = -3dB$$

That is why it is called a *3dB splitter*.

As power combiner

- When equal input signals are given at both the collinear ports, the output signal appears at the side arm port whose power is the sum of the powers of the input signals provided the collinear arm lengths are same and the sources are out of phase.
- The output power is zero. When the sources are equal, in the phase and collinear arms lengths are same.

6.6.2.2 H-plane Tee

H-plane Tee is a current, shunt, or parallel junction. Since the axis of the side arm is parallel to the plane of the H field of the main waveguide, it is called a *H-plane Tee*. A rectangular slot is cut along the narrow dimension of a long waveguide, and a side arm is attached as shown in Figure 6.15 (a). If the H-plane junction is completely symmetrical and waves enter through the side arm, the waves that leave the main arms are equal in magnitude and phase.

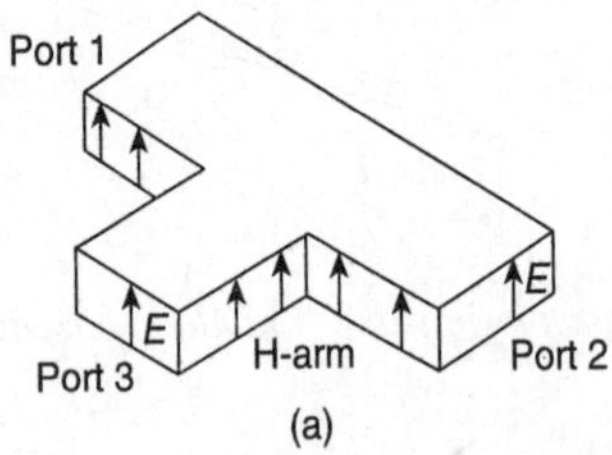

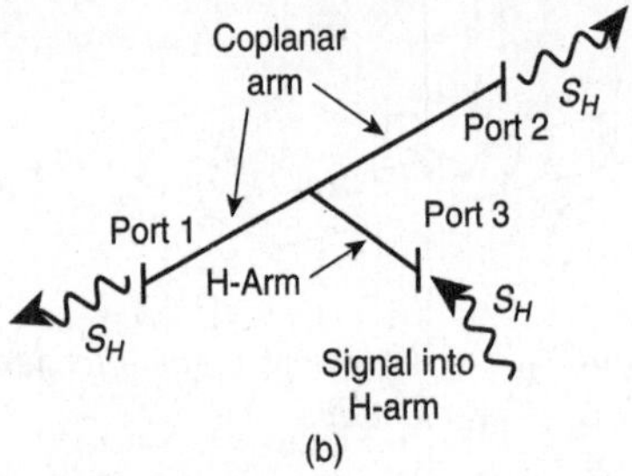

Figure 6.15 *(a) H-plane Tee junction; (b) Transmission-line Equivalent circuit of H-plane Tee*

Figure 6.15 (c) *Schematic of H-plane Tee*

As power divider

- They are inphase when its collinear arm lengths are same. If the amplitude of the input wave at port 3 is A, the amplitudes of the waves at ports 1 and 2 are same and equal to $A/\sqrt{2}$.
- It is called a *3db splitter*, because when the power incident at port 3 is *P*, the powers that appear at ports 1 and 2 are *P*/2 each.

As power combiner

- When equal input signals are given at both the collinear ports, the output signal appears at the side arm whose power is the sum of the powers of the input signals provided the collinear arm lengths are same and the sources are in phase.
- The output power is zero. When the sources are equal, in the out of phase and collinear arms the lengths are same.

Applications:

Rectangular waveguide Tees are used

- As tuners by placing a short circuit in the symmetrical arm
- As power dividers and adders
- In the duplexer assemblies of radar installations

6.6.2.3 *Comparison of E-plane Tee with H-plane Tee*

Table 6.1 Comparison of E- and H-plane Tees

E-plane Tee	H-plane Tee
1. In an E-plane Tee, the axis of its side arm is parallel to the electricfield of the main waveguide. 2. The E-plane Tee is also called a *series Tee.* 3. When the power is fed at port (3), that is, at the side arm, the resulting power is equally divided between port (1) and (2), but a phase shift of 180° is introduced between the two outputs.	1. In an H-plane Tee, the axis of its side arm is parallel to the magneticfield or shunting the electricfield of the main waveguide. 2. An H-plane Tee is also called a *parallel* or *shunt Tee.* 3. When the power is fed at port (3),that is, at the side arm, the resulting power is equally divided between port (1) and port (2) within phase.
4. When the equal input power is fed to both ports (1) and (2), no output is obtained at port (3).	4. When the equal input power is fed to both ports (1) and (2), the maximum power (i.e., addition of two inputs) is obtained at port (3).
5. When the input signal is applied at any one of the collinear ports i.e., port (1) or port (2), the resulting power is obtained at port (3).	5. When the input is applied at any one of the collinear ports i.e., port (1) or port (2), the resulting power is obtained at port (3).

6.6.2.4 *Magic Tee*

The combination of an E-plane Tee and an H-plane Tee is called as Magic Tee. A Magic Tee can be formed by attaching arms to the slots made in the broad and narrow walls of a waveguide. It is also called as hybrid tee in which the power distributes equally between the output ports. The outputs may have 0° or 180° phase difference. Magic Tee is a 3db hybrid coupler which is also called as an anti-symmetric coupler. If one of the coplanar arm is terminated, then the power delivered to another coplanar arm is independent of terminated port. The hybrid (Magic) Tee Junction is shown in Figure 6.16.

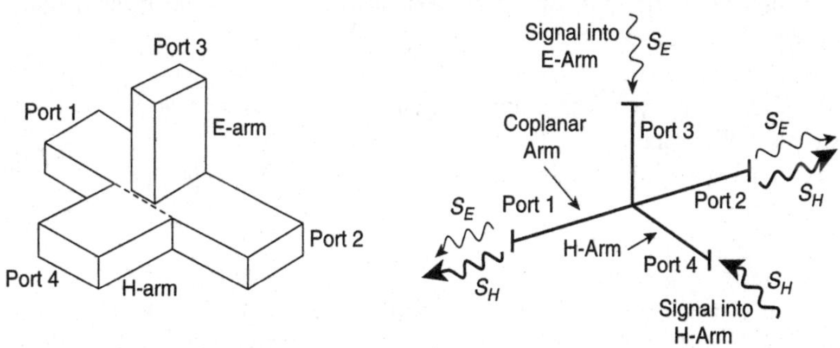

Figure 6.16 (a) *Hybrid(Magic) Tee junction and its equivalent circuit*

Figure 6.16 (b) *Schematic of Magic Tee*

Figure 6.16 (b) represents the schematic of the Magic Tee. Its characteristics are as follows:

- Let us consider two waves whose magnitude and phase are equal. If we feed these waves to ports 1 and 2, the outputs at port 3 and port 4 will be zero and additive respectively.
- The power distributes equally at ports 1 and 2 if a wave is incident at port 4 and no power wil appear at port 3.
- The power at ports 1 and 2 appears with an equal magnitude and the opposite phase if a wave is incident at port 3 and no power will appear at port 4.
- If a wave is incident on any one of the coplanar arms i.e., port 1 or port 2, then no power wil appear at other coplanar arm i.e., port 2 or port 1. This is due to the occurring of phase delay and phase advance in E arm and H arm respectively.
- In Magic Tee the imaginary plane bisects arms 3 and 4 symmetrically. There will not be any reflec tions in the junction, if ports 1 and 2 are terminated with matched loads.
- Since all the ports are the collinear arm ports in Magic Tee, they are perfectly matched to the junc tion, and the E and H arm ports are decoupled individually
- The signal distributes equally between the E and H arms, if signal is incident on collinear arm and output signal is given as $P_{out} = P_{in}/2$ and $A_{out} = A_{in}/\sqrt{2}$.
- A signal into the H arm splits equally between the collinear arms, the outputs being in phase equidistant from the junction.
- A signal into the E arm splits equally between the collinear arms, the outputs being out of phase equidistant from the junction.
- For signals into both collinear arms.
    1. The signal output from the E arm is equal to $1/\sqrt{2}$ times the phasor difference of the input signals. (Difference arm)
    2. The signal output from the H arm is equal to $1/\sqrt{2}$ times the phasor sum of the input signals (Sum arm)

Advantages of magic tee

1. Due to the decoupling property of output ports, the power delivered to one of the output port becomes independent of the termination at the other output port.
2. In the E- or H-plane Tee, the power division between ports depends on terminations existing at the respective output ports; but in Magic Tee (in which all the ports are perfectly matched), power division between the ports is independent of terminations.

Disadvantages of magic tee

There is an impedance mismatch at the junctions, when a signal is applied to any arm of the Magic Tee. Because of this impedance mismatch the flow of energy in the output arms is affected by reflections. These reflections cause the following two disadvantages of Magic Tee:

1. When all the energy that is fed into the junction does not reach the load due the reflections, it results in power loss.
2. The standing waves that are produced due to reflections can result in internal arcing. Thus, it results in reduction of the maximum power that a Magic Tee can handle.

6.6.2.4.1 Applications of magic tee

Depending on the above explained properties, a magic tee has many applications as follows

1. As an isolator
2. As a matching device
3. As a phase shifter
4. As a duplexer
5. As a mixer
6. As a measurement of impedance

Measurement of Impedance

A magic tee is used for measuring impedance in the form of a bridge as shown in the Figure 6.17 (a).

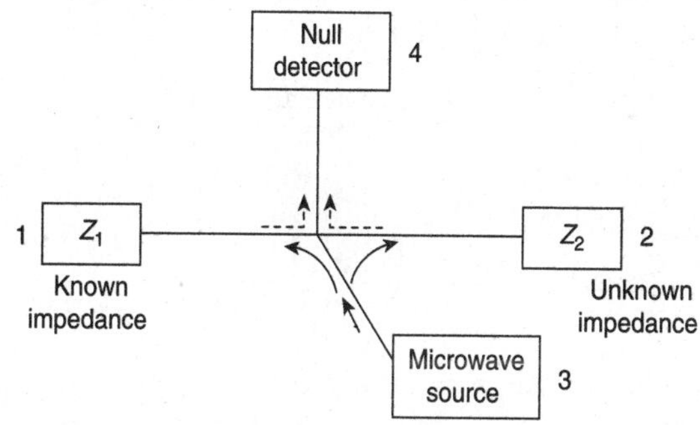

Figure 6.17 (a) *Magic Tee for measurement of impedance*

Arm (3) is connected to a Microwave source. Arm (4) is connected to a null detector. Arm (2) is connected to the unknown impedance. Arm (1) is connected to a standard variable known as impedance. The power from the microwave source in arm (3) gets divided equally to the unknown impedance and standard variable impedance between arms (1) and (2), by the properties of the Magic Tee. There will be reflections from arms (1) and (2), since these impedances are not equal to the characteristic impedance Z_0.

The powers enter the Magic Tee junction from the arms (1) and (2) are given by $\frac{\rho_1 a_3}{\sqrt{2}}$ and $\frac{\rho_2 a_3}{\sqrt{2}}$ respectively, if the reflection coefficients are ρ_1 and ρ_2 respectively. The net wave reaching the null detector which is also the resultant wave enters arm (4). It is also equal to zero if the bridge is balanced. For perfect balancing of the bridge (null detection), $\rho_1 - \rho_2 = 0$ (or) $\rho_1 = \rho_2 \therefore Z_1 = Z_2$.

Thus, by adjusting the standard variable impedance till the bridge is balanced and both the impedances become equal, the unknown impedance can be measured.

Magic Tee as a Duplexer

The receiver is connected to port 1 and the transmitter is connected to port 2. The antenna is in the E arm (or port 4) and port 3 of the Magic Tee are terminated with a matched load, as shown in Figure 6.17 (b). Since Ports 1 & 2, 3 & 4 isolated ports, the power does not reaches the receiver; that means, the power transmitted gets divided equally during the transmission and only half of the power is radiated into space. The other half of the power is absorbed without reflections in the matched load. As ports (1) and (2) are isolated ports in the Magic Tee during reception, the power received by the antenna divides equally, and half of the power reaches the receiver.

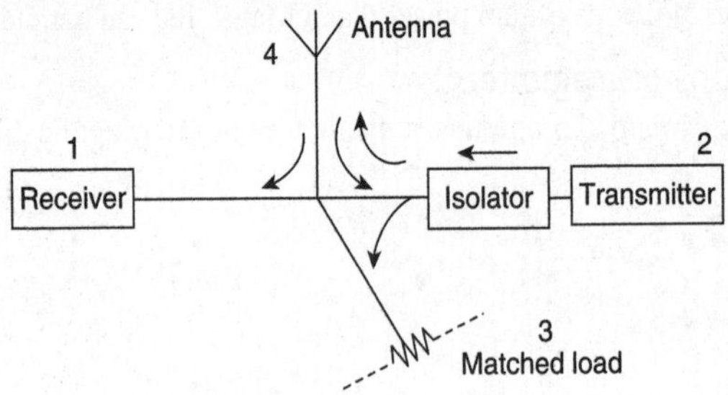

Figure 6.17 (b) *Magic Tee as a duplexer*

Magic Tee as Mixer

A Magic Tee is also used as a mixer in a microwave receiver where the signal and local oscillator are fed into the E and H arms as seen in Figure 6.17 (c).

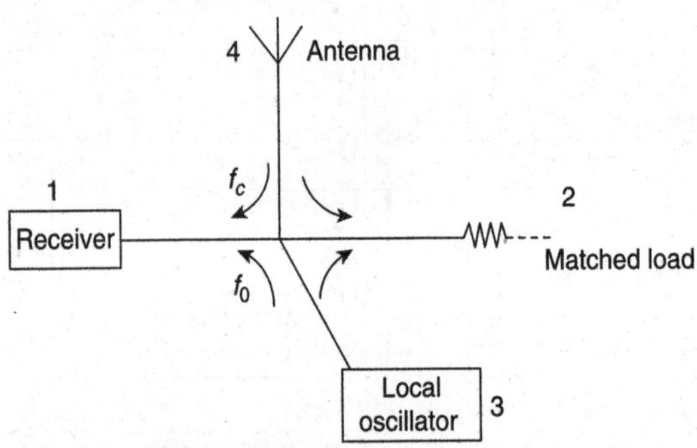

Figure 6.17 (c) *Magic Tee mixer*

Since the isolated ports are ports 3 and 4, the oscillator power does not reach the antenna. Also, the signal f_c received by the antenna will not reach the local oscillator. Half of the local oscillator signal f and half of the power f_c go to the mixer and gets mixed to generate the IF frequency which is given by,

$$\text{IF} = f_c - f_0$$

Magic Tee is also used as microwave bridge and microwave discriminator.

EXAMPLE PROBLEM 6.1

Consider a radar system that consists of two transmitters (Tx_1,Tx_2) and an antenna. Explain how the antenna gets twice more the power at its output than a single transmitter can usually deliver.

Solution

We can make use of the Magic Tee to couple the two transmitters to an antenna in such a way that the transmitters do not load each other. In Figure 6.17 (d), two transmitters (Tx_1,Tx_2) are connected to ports 3 and 4, respectively.

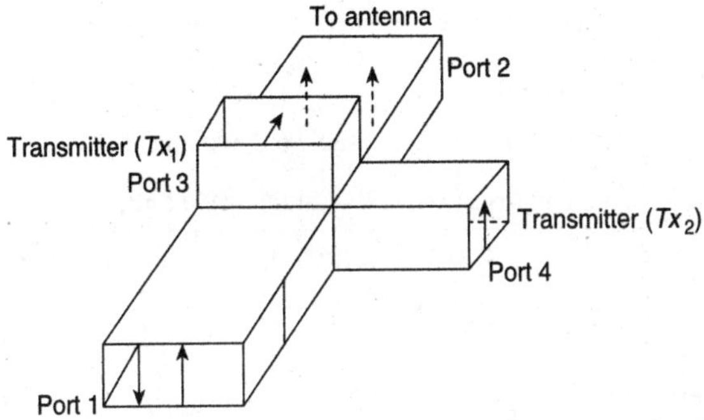

Figure 6.17 (d) *Magic Tee–coupling of transmitters with antenna*

The Tx_1 connected to port 3 causes a wave to emanate from ports 1 and 2. These waves are equal in magnitude and opposite in phase. Similarly, Tx_2 connected to port 4 gives rise to a wave at ports 1 and 2, both of which are equal in magnitude and in phase. At port 1, the two opposite waves cancel each other, and at port 2, two in-phase waves are added together. Therefore, the double output power is obtained by the antenna at port 2. ■

6.6.3 Hybrid Ring (Rat Race Junction)

The characteristics of the hybrid ring are similar to those of the MagicTee, but with a different construction. The rat race ring is shown in Figure 6.18. The arrangement shown consists of a piece of a rectangular waveguide bent in the E plane to form a complete loop whose median circumference is $1\frac{1}{2}\lambda$. It has four openings from each of which a waveguide emerges, forming parallel junctions. Ports 1, 2, 3, and 4 are with a spacing of $\lambda/4$,and the spacing between ports 1 and 4 is $3\lambda/4$.

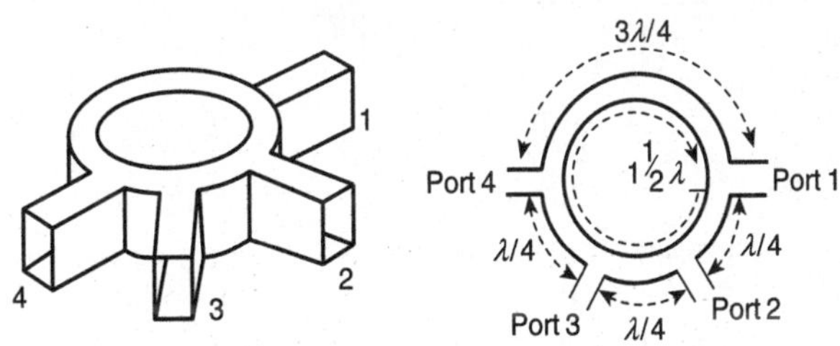

Figure 6.18 *Rat race ring*

Applications of Rat Race Ring

1. **As-in phase power divider**
   When a wave is fed to port 2, it splits equally (i.e., clockwise and anti-clockwise) into ports 1 and 3, and nothing appears at port 4. Since it is an *in phase* divider, and the difference of the phase shifts for

the waves traveling from port 2 to 1 (also from port 2 to 3) in clockwise and anti-clockwise direction is 0^0 (i.e path difference is λ), it gets divided equally into ports 1 and 3. However, at port 4, the difference of phase shifts for the waves traveling from port 2 to 4 in clockwise and anti-clockwise direction is 180^0 (i.e. cancellation occurs at port 4 due to a path difference of $\lambda/2$). Similarly, when the input is applied at port 3,it gets divided equally into ports 2 and 4, and the output at port 1 will be zero.

Similarly, the rat race ring can also be used as an *out of phase* power divider.

2. **As-in phase power combiner**

When power is applied at ports 1 and 3, it gets added at port 2, and the output at port 4 is zero (We can, therefore, consider the port 4 as isolated). Table 6.2 explains this.

Table 6.2 Path lengths associated with the wave direction

Wave direction from	Clockwise (path lengths)	Anticlockwise (path lengths)
Port 1 to 2	$\lambda/4$	$5\lambda/4$
Port 1 to 4	$3\lambda/4$	$3\lambda/4$
Port 3 to 2	$5\lambda/4$	$\lambda/4$
Port 3 to 4	$\lambda/4$	$5\lambda/4$

The difference between the path lengths of the wave traveling from 1 to 2 and from 3 to 2 is λ (i.e. $(5\lambda/4-\lambda/4)= \lambda$). Therefore, the phase difference at port 2 is zero, and the difference between the path lengths of the wave traveling from 1 to 4 and from 3 to 4 is $\lambda/2$ (i.e.$(3\lambda/4-\lambda/4)= \lambda/2$). Therefore, the phase difference at port 4 is 180°. Since it is an inphase combiner, powers coming from ports 1 and 3 get added at port 2 (as phase difference at port 2 is zero), and no power appears at port 4 (as phase difference at port 4 is 180°).

When we use an out of phase combiner, we can get power at port 4. In such a case, the power at port 2 is zero.

Hybrid ring vs Hybrid Tee

The rat race and Magic tee may be used interchangeably, but

- Inspite of being less bulky the hybrid ring requires internal matching, which does not require a hybrid ring if the thickness is properly chosen.
- At higher frequencies, the dimensions of the hybrid ring are not significant. Hence it is preferable at higher frequencies.

6.7 DIRECTIONAL COUPLERS

In some applications such as radar, very often we need to check the exact frequency/power applied to the antenna or that is radiated into space. Directional couplers allow us to sample or monitor the frequency level and/or power level of a given signal as it goes from one point to another. The directional coupler is a 4–port reciprocal device. Direction couplers consist of two transmission lines and a mechanism for coupling signals between them. Let us understand the meaning of the two terms (viz.coupler and directional) in the directional coupler.

Coupler: A coupler is a device that consists of two waveguides which are placed very close to each other(as shown in Figure 6.19 (a)). Thus, a portion of energy traveling in waveguide A will be coupled on waveguide B.

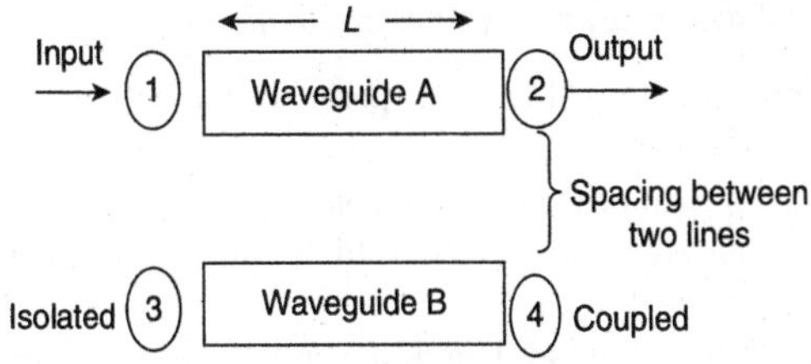

Figure 6.19 (a) *Coupler*

We can make this coupler directional by using a specific length (L) of the transmission line.

Directional: The term *directional* means the energy is passed in one direction only, and no energy passes in the reverse direction. The directional property is obtained by using a specific length (L) of a transmission line, that is, a quarter wavelength ($\lambda/4$). A $\lambda/4$ transmission line offers high impedance at one end and low impedance at the other end. Therefore, the specific length ($L = \lambda/4$ or $(2n+1)\ \lambda/4$) makes a coupler directional over a certain band of frequency (as shown in Figure 6.19 (b)).

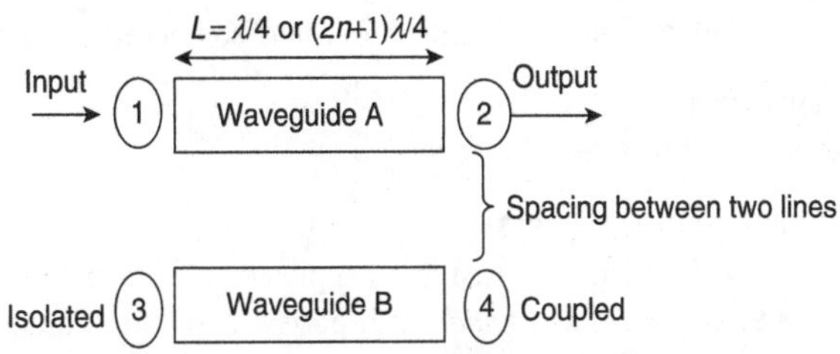

Figure 6.19 (b) *Coupler with a specific length of $\lambda/4$*

Power flow in a directional coupler:

The power incident at port 1(input) is split between two other ports (port 4 (coupled) and port 2 (output)), and no power appears from port 3 (isolated). Power flow in a directional coupler is shown in Figure 6.19 (c).

P_i or P_1 = power incident at port 1

P_f or P_2 = forward power or output power at port 2

P_b or P_3 = reflected power at isolated port 3 in secondary waveguide

P_{fc} or P_4 =forward coupled power in the secondary waveguide, that is, at port 4

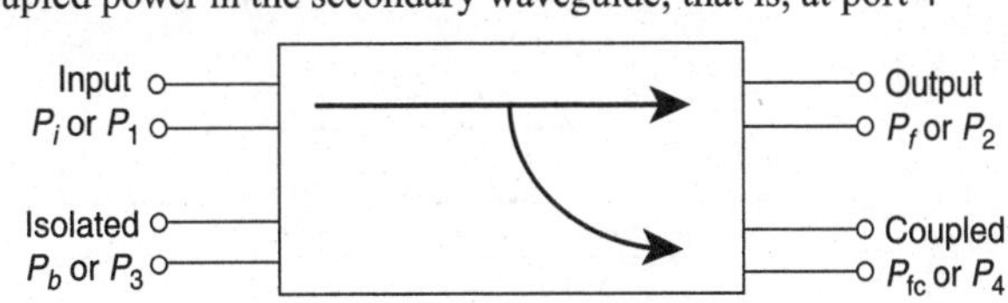

Figure 6.19 (c) *Four-port directional coupler*

Figure 6.19 (d) *Schematic of directional coupler*

The properties of an ideal directional coupler are as follows:

1. In an ideal directional coupler, all the four ports are perfectly matched and also ports1,3 and ports 2, 4 are perfectly isolated.
2. A portion of the wave is coupled to port 4 but not coupled to port 3 which is traveling from port 1 to port 2. Similarly, a portion of the wave travelling from port 2 to port 1 is coupled to port 3 but not to port 4.
3. Likewise, the portion of the wave traveling from port 4 to port 3 is coupled to port 1 but not to port 2. Similarly, a portion of the wave traveling from port 3 to port 4 is coupled to port 2 but not to port 1
4. The coupling between port 1 and port 4 is similar to that between port 2 and port 3, and the degree of coupling depends on the structure of the coupler.
5. The outputs are always in phase quadrature; that is, they exhibit a phase difference of 90^0. For this reason, a directional coupler is called a *quadrature-type hybrid.*

Figure 6.19 (d) represents the schematic of directional coupler.

Types of directional couplers:
There are two types of directional couplers; both are four-port components and are reciprocal.

1. Two-hole directional coupler
2. Single-hole or Bethe-hole directional coupler

6.7.1 Two-Hole Directional Couplers

The two-hole directional coupler is mostly used in all applications. The directional coupler consists of two waveguides referred to as a *main waveguide* with ports 1 and 2 and an auxiliary waveguide with ports 3 and 4. When a power is applied at port 1 of the main waveguide, the output is taken at port 2 of the main waveguide. A fraction of the power is coupled into port 4 of the auxiliary waveguide, and no power flows in port 3 of the auxiliary waveguide. Since the device is reciprocal, the power incident in port 3 of the auxiliary waveguide flows in port 4, a fraction of the power couples in port 2, and no power flows in port 1 of the main waveguide.

Functional Operation of 2 Hole Directional Coupler

To have the directional property of a coupler, the spacing between the centers of two holes should be

$$L = (2n+1)\lambda/4, \text{ where } n \text{ is any positive integer.}$$

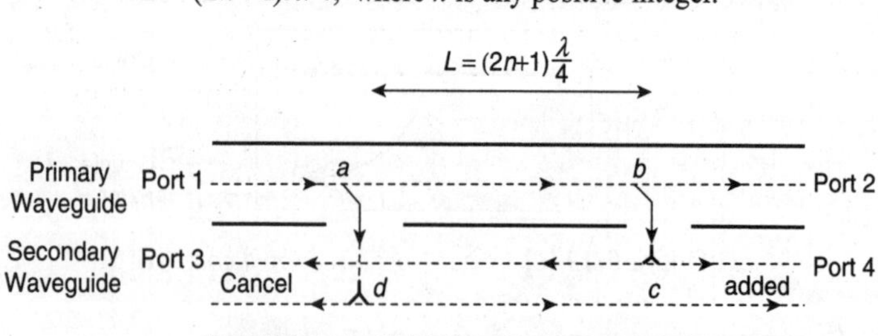

Figure 6.20 *Two-hole directional coupler*

The hole acts as a slot antenna. A portion of the wave energy entering into port 1 passes through holes and radiates into the secondary guide.

- Forward waves in the secondary guide are added at port 4 and are in a similar phase. Waves travelling from $a \to b \to c$ and from $a \to d \to c$ have similar path lengths.

- Backward waves in the secondary waveguide are out of phase $\left(180° \text{ or} \left(\frac{2L}{\lambda}\right)2\pi\, rad\right)$ and are cancelled at port 3.

Waves travelling from paths $a \rightarrow d$ & $a \rightarrow b \rightarrow c \rightarrow d$ have a difference of $2\times(2n+1)\frac{\lambda}{4}=(2n+1)\frac{\lambda}{2}$.

(i.e.,Two wave components, one coupled out immediately from a and the other from b, are 180° out of phase at d; therefore, waves traveling toward port 3 vanish.

Parameters that Characterize the Directional Coupler

A directional coupler is characterized by 3 parameters:

- Coupling factor
- Directivity
- Isolation

1. Coupling factor(C): It indicates the fraction of input power coupled to the coupled port.

$$C = 10\log_{10}\left(\frac{P_1}{P_4}\right) dB$$

where P_1 is the power incident at port 1, and P_4 is the power coupled at port 4.

2. Directivity(D): Directivity is the ability to isolate coupled (port 4) and backward(port 3) ports.

$$D = 10\log_{10}\left(\frac{P_4}{P_3}\right) dB$$

3. Isolation(I): *Isolation* is defined as the ratio of power incident to the power coupled in the isolated port and is expressed in dB.

$$I = 10\log_{10}\left(\frac{P_1}{P_3}\right) dB$$

In an ideal directional coupler, D and I are infinite (as $P_3 = 0$), and C is of the order of 10 dB.

6.7.2 Bethe-hole Directional Couplers

This is the simplest form of a waveguide directional coupler. In the Bethe-hole coupler, two waveguides are placed one above the other. A hole is located at the center of a common broad wall of two waveguides. The two waveguides are placed at an angle, θ as shown in Figure 6.21 (a).

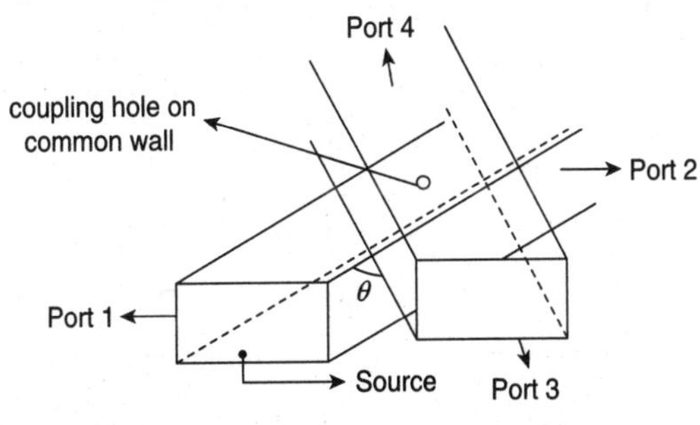

Figure 6.21 (a) *Bethe-hole directional couplers (waveguides placed at an angle θ)*

The input is incident at port 1 of the main waveguide (i.e. lower waveguide). The mode of propagation is the TE_{10} mode. If the hole (or aperture) is small compared with the propagating signal wavelength (λ), the hole acts similar to an electric dipole that is normal to the aperture plane. This dipole moment is a function of the normal component of the electricfield in the main waveguide and the tangential component of the exciting magnetic field at the aperture. Due to radiation from this dipole, coupling to the auxiliary guide is achieved.

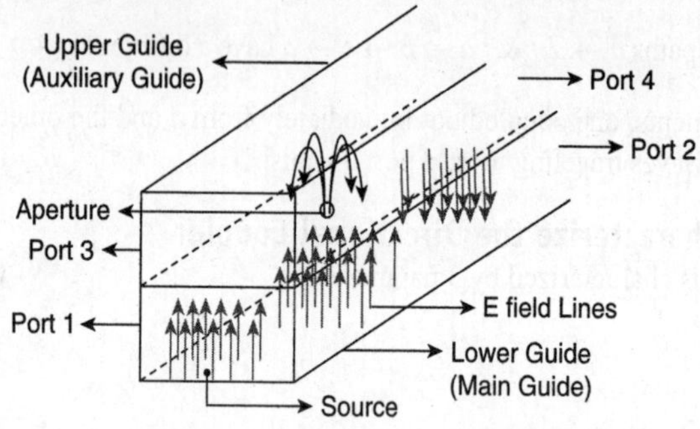

Figure 6.21 (b) *TE_{10} mode Electric field configuration in Bethe-hole coupler*

The electric dipole radiates symmetrically in both directions longitudinally as shown in Figure 6.21 (b). However, the magnetic field dipole radiates asymmetrically in longitudinal directions. In the auxiliary waveguide, both H_y and H_z components are present in the direction of propagation (port 4) as shown in Figure 6.22. The H_y and H_z fields are in the opposite direction and have different magnitudes; whereas the H_y component will be present in port 3(coupled port).

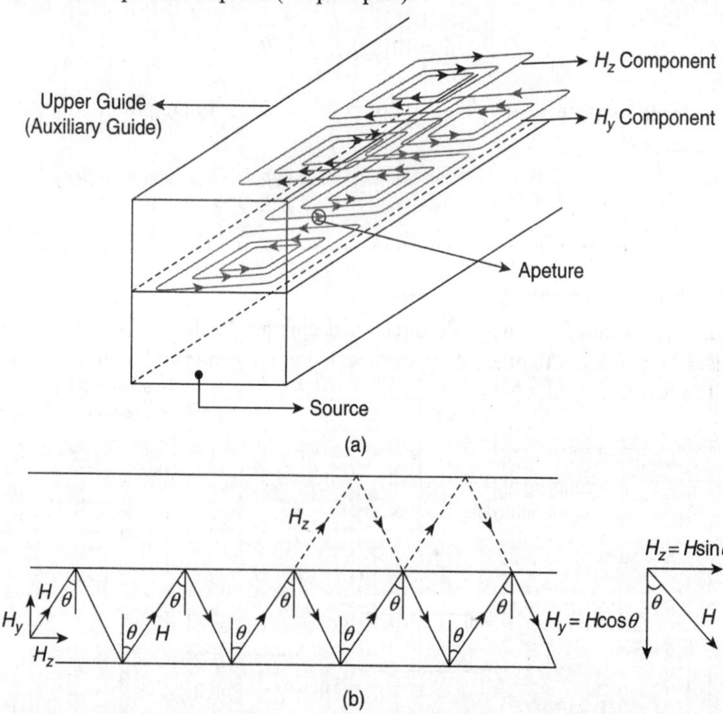

Figure 6.22 *(a) Direction of magnetic field component H_z in auxiliary waveguide;*
(b) Direction of magnetic field components H_y and H_z in auxiliary waveguide

By varying the angle between the waveguides, the magnitudes of H_y and H_z components at port 4 can be made equal. This leads to the zero magnetic field at the output port of the auxiliary waveguide, and power is coupled only at port 3 (coupled port).

6.7.3 Applications of Directional Couplers

Directional couplers are extensively used in systems that measure the amplitude and phase of traveling waves. The major applications are as follows:

1. Power monitoring and source leveling
2. SWR measurements
3. In unidirectional power measurements
4. In reflectometers
5. Unidirectional wave launching
6. Isolation of signal sources

EXAMPLE PROBLEM 6.2

The incident power is 500 mW for a directional coupler. If the coupling factor is 20 dB, calculate the power in the main and auxiliary arm.

Solution

Given incident power $P_1 = 500$ mW and coupling factor $C = 20$ dB.
Coupling factor = $10 \log P_1/P_4$

$$20 = 10 \log 500 \times 10^{-3}/P_4 \Rightarrow 500 \times 10^{-3}/P_4 = 100$$

Therefore $P_4 = 5$ mW
Power in auxiliary arm = 5 mW
Power in main arm = Output power
Input power = Output power + Auxiliary power
Output power = Input power –Auxiliary power

$$= P_1 - P_4 = (500 \times 10^{-3}) - (5 \times 10^{-3}) = 495 \text{ mW}$$

EXAMPLE PROBLEM 6.3

The incident power is 100 W for a directional coupler. It has a coupling factor of 25 dB and a directivity of 40 dB. Find coupled and isolated port powers.

Solution

Given $P_1 = 100$ W, $C = 25$ dB and $D = 40$ dB

Coupling factor is given by $C = 10 \log \dfrac{P_1}{P_4} = 10 \log \dfrac{100}{P_4} \Rightarrow \dfrac{100}{P_4} = 316.23$

$$P_4 = 0.316 \text{ W i.e. Coupled port power} = 0.316 \text{ W}$$

Directivity is given by,

$$D = 10 \log \frac{P_4}{P_3} \Rightarrow 40 = 10 \log \frac{0.316}{P_3} \Rightarrow \frac{0.316}{P_3} = 10000$$

$P_3 = 31.623$ μW i.e. Isolated port power = 31.623 μW

EXAMPLE PROBLEM 6.4

The incident and reflected power in a waveguide can be sampled using a 40 dB directional coupler. The value of VSWR is 3, and the coupler sampling power value is 4 mW. Find the value of reflected power?

Solution

Voltage standing wave ratio is given by $VSWR = \frac{1+\Gamma_L}{1-\Gamma_L} = 3$
Where Γ_L is the reflection coefficient.
Therefore, $(1+\Gamma_L) = 3(1-\Gamma_L) \Rightarrow 2-4\Gamma_L = 0$

$$\Rightarrow \Gamma_L = \frac{1}{2} = 0.5$$

We know that,

$$C = 10 \log \frac{P_1}{P_4} \Rightarrow \frac{P_1}{P_4} = 10^4 \Rightarrow P_4 = \frac{P_1}{10^4} \Rightarrow \frac{P_1}{10000} = 4 \text{ mW}$$

Therefore $P_1 = 40$ W

$\Gamma_L = \sqrt{\frac{P_3}{P_1}}$ ($\because$ Reflection coefficient, Γ_L = ratio of reflected voltage to incident voltage)

$$P_3 = \Gamma_L^2 \times P_1 = (0.5)^2 \times 40 = 10 \text{ W} \Rightarrow \text{Reflected power} = 10 \text{ W}$$

■

EXAMPLE PROBLEM 6.5

The directivity of a 10 dB coupler is 20 dB. Find its isolation

Solution

Given Coupling factor $C = 10$ dB, Directivity $D = 20$ dB
Isolation can be written as I = coupling factor(C) + Directivity (D)

$$= 10 + 20 = 30 \text{ dB}$$

$\therefore$ Isolation, $I = 30$ dB

■

EXAMPLE PROBLEM 6.6

A 20 mV signal is fed to the series arm of a lossless Magic Tee junction. Calculate the power delivered through each port when other ports are terminated with a matched load.

Solution

For a given lossless Magic Tee shown in Figure 6.16 (a), power is applied at the series arm (E arm), that is, at port 3. Therefore, $a_3 = 20$ mV. It behaves similar to an E-plane Tee. Therefore, the resulting power is equally divided between port (1) and port (2) with a phase shift of 180° between them; no power is obtained at port (4), that is, at shunt arm. Therefore, from the principle of operation of Magic Tee, power through port (1) is

$$= \frac{a_3}{2} = \frac{20 \text{ mV}}{2} = 10 \text{ mV}$$

Voltage through port (1) is

$$= \frac{a_3}{\sqrt{2}} = \frac{20 \text{ mV}}{\sqrt{2}} = 14.142 \text{ mV}$$

Power through port (2) is

$$= -\frac{a_3}{2} = -\frac{20 \text{ mV}}{2} = -10 \text{ mV}$$

Voltage through port (2) is

$$= -\frac{a_3}{\sqrt{2}} = -\frac{20 \text{ mV}}{\sqrt{2}} = -14.142 \text{ mV}$$

Power obtained at port (4), that is, at shunt arm, is zero. ■

EXAMPLE PROBLEM 6.7

A 20 dB-directional coupler is found to be given 3 dBm as output power through a coupled port. If the isolation is specified as 55 dB, find the power available at the isolated port.

Solution

Given the coupling factor, C = 20 dB
Power through the coupling port, $P_4 = 3$ dBm
Isolation, I = 55 dB
The directivity can be written as
Directivity (dB) = Isolation (dB) – Coupling factor (dB)

$$D = I - C = 55 - 20 = 35 \text{ dB}$$

$$D = 10 \log \frac{P_4}{P_3} \Rightarrow 35 = 10 \log \frac{P_4}{P_3}$$

Since

$$\log \frac{P_4}{P_3} = 3.5 \Rightarrow \frac{P_4}{P_3} = 10^{3.5}$$

where $P_4 = 3$ dBm

$$10 \log\left(\frac{P_4}{10^{-3}}\right) = 3$$

$$P_4 = 10^{-3} \times 10^{0.3} = 1.99526 \times 10^{-3} \text{ W}$$

$$P_3 = \frac{1.99526 \times 10^{-3}}{10^{3.5}} = 6.3 \times 10^{-4} \text{ watts}$$

∴ Power available at the isolated port, $P_3 = 0.63$ mW ■

EXAMPLE PROBLEM 6.8

A directional coupler (20 dB) is used in a guide to sample the incident and reflected powers. The incident and isolated part powers are 300 mW and 10 mW, respectively. What is the value of VSWR in the main waveguide?

Solution

Given incident power $P_1 = 300$ mW and isolated part power $P_3 = 10$ mW,

$$\text{Reflection coefficient,} \quad \Gamma_L = \sqrt{\frac{P_3}{P_1}}$$

$$= \sqrt{\frac{10 \text{ mW}}{300 \text{ mW}}} = \sqrt{0.033} = 0.1816$$

$$\text{Therefore, VSWR} = \frac{1+\Gamma_L}{1-\Gamma_L} = \frac{1+0.1816}{1-0.1816} = 1.44$$

■

EXAMPLE PROBLEM 6.9

The incident and reflected powers can be sampled using a directional coupler in a waveguide. The output of the two couplers is found to be 2 mW and 0.1 mW. Calculate the value of VSWR in the waveguide.

Solution

We know that, the reflection coefficient, $\Gamma_L = \sqrt{\frac{P_3}{P_1}} = \sqrt{\frac{0.1}{2}} = \sqrt{0.05} = 0.224$

$$\text{Therefore, VSWR} = \frac{1+\Gamma_L}{1-\Gamma_L} = \frac{1+0.224}{1-0.224} = 1.577$$

■

6.8 FERRITES

A device that is composed of material which has useful magnetic properties and, simultaneously, it provides high resistance to current flow is a ferrite. The electron movement within the atoms of the material results in the magnetic property of that material. There are two types of motions of Electrons: (1) Orbital movement of the electrons around the nucleus of the atom; (2) Movement of the electron about its own axis, called *electron spin*. The different types of electron movement are shown in Figure 6.23 (a). Movement of the electrons within the atom causes the current to flow. Therefore, the magnetic field is generated. Under the influence of the applied external magnetic field, the electron spin axes within some materials, such as iron or nickel, can be caused to align. Therefore, magnetic fields get added.

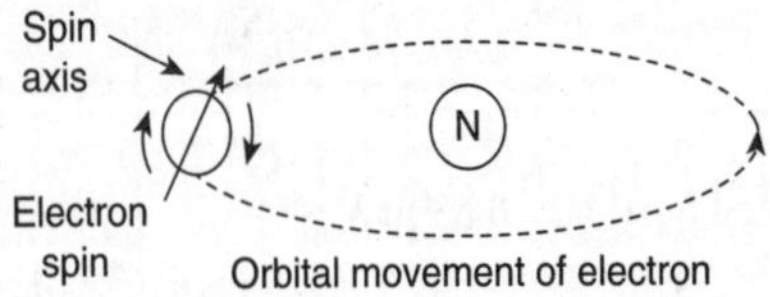

Figure 6.23 (a) *Two types of electron movement*

In the case of ferrites, electrons try to balance between two forces. They are as follows: (1) A force that holds the atoms together (i.e. orbital motion of the electrons about the nucleus); (2) An external static magnetic field. Interaction of these two forces causes the electrons to wobble on their axis (as shown in Figure 6.23 (b)).

Ferrite action depends on the behaviour of electrons due to the influence of the external field. This result is wobble frequency. Electrons that wobble also have natural resonant wobble frequency. It varies with the strength of the applied field.

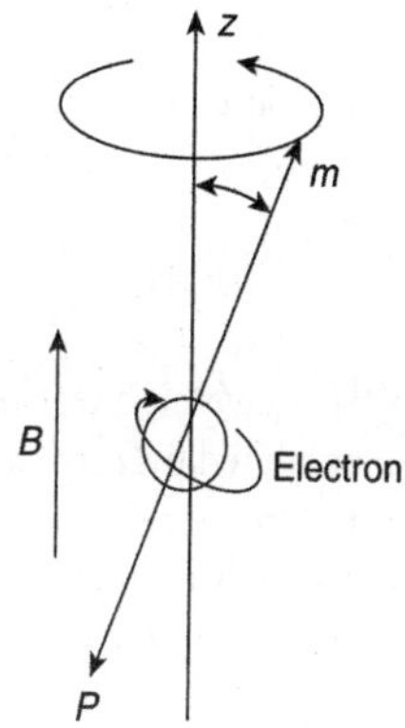

Figure 6.23 (b) *Electron wobbles in a magnetic field*

6.8.1 Faraday Rotation Principle

If a linearly polarized wave is made to pass through a ferrite rod and if it is influenced by the magnetic field, the axis of polarization gets tilted in clockwise direction. This is because the frequency of the microwave energy is much greater than the electron wobble frequency (Figure 6.24). This is known as the *Faraday rotation effect.* The strength of the magnetic field and the geometry of ferrite is the basis for the amount of tilt. The direction of the Faraday rotation depends on whether the signal frequency is smaller or larger than the resonance frequency.

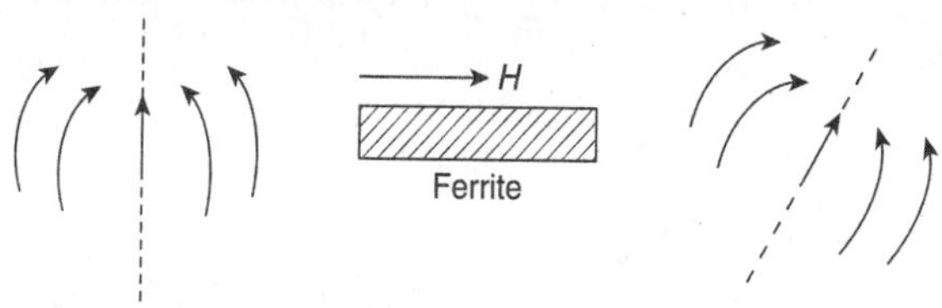

Figure 6.24 *Rotation of signal due to ferrite*

The phase shift of the resultant wave is given by $\left(\beta^{+}-\beta^{-}\right)\frac{l}{2}$, and the tilt angle is given by

$$\theta = \arctan\left(E_y / E_x\right) = -\left(\beta^{+}-\beta^{-}\right)\frac{l}{2}$$

Where β^{+} and β^{-} are phase constants of the components E_x and E_y

Hence, as the wave propagates to the distance of 'l' in a ferrite, the tilt angle of the polarization vector changes. This is called Faraday rotation. A typical change is 100° per centimeter at 10 GHz.

The tilt angle θ rotates in the same direction with respect to the coordinate system, if the direction of propagation is reversed. Thus the tilt angle does not return from θ to 0°, but its value becomes twice the tilt angle. Therefore, the Faraday rotation is a non-reciprocal phenomenon

6.8.2 Composition and Characteristics of Ferrites

Ferrites are non-metallic materials with resisitivities and dielectrics. They provide high resistance to current flow.

Characteristics of Ferrites:

- Ferrite materials are a mixture of metallic oxide and ferric oxide ($MeOFe_2O_3$) where Me is any divalent such as Mn^{+2}, Zn^{+2}, Cd^{+2}, and Ni^{+2}
- Ferrites have strong magnetic properties.
- In microwave devices ferrites are most suitable to reduce the reflected power, for modulation purposes, and in switching circuits.
- Ferrites are used up to 100 GHz as they have high resistivity.
- The non-reciprocal property is also exhibited by Ferrites.
- Their resistivities are around 10^{14} times greater than metals.
- The dielectric constant of ferrite materials is around 10 to 15.
- These materials have relative permeabilities of the order of 1000.

One widely used ferromagnetic material is Yttrium-Iron-Garnet [Y_3 Fe_2 $(FeO_4)_3$] or YIG (Yttrium iron garnet) in short.

6.9 FERRITE COMPONENTS

Microwave gyrator, isolator, and circulator use the principle of Faraday rotation. So, these are ferrite components.

6.9.1 Faraday rotation-based Gyrator

A gyrator is a two-port non-reciprocal ferrite device having a relative phase difference of 180° when wave is transmitting from port 1 to port 2 and a 0° phase shift when wave is transmitting from port 2 to port 1 (Figure 6.25 (a)).

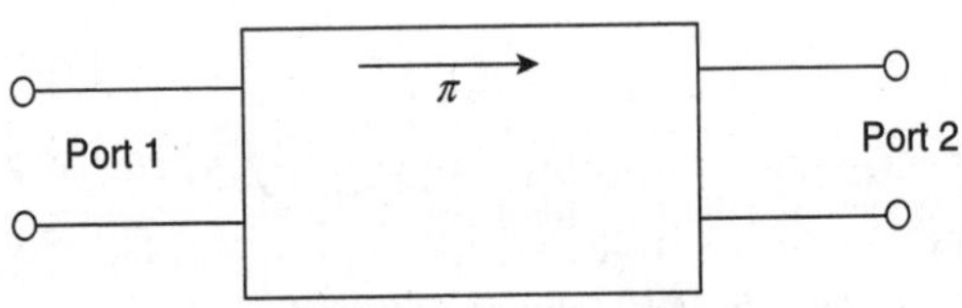

Figure 6.25 (a) *Symbol of gyrator*

It has a circular waveguide propagating the dominant mode (TE_{11}), which changes over to a rectangular waveguide with the dominant mode (TE_{10}) at both ends. The circular waveguide consists of a thin, circular ferrite rod which is tapered at both the ends to reduce the attenuation and is supported by polyfoam. This also helps for smooth rotation of the polarized wave. A dc magnetic field is generated by permanent magnet which is placed around the waveguide for appropriate operation of ferrites (as shown in Figure 6.25 (b)). To this waveguide's input end a 90° twisted rectangular waveguide is attached.

Operation

The plane of polarization of incident wave rotates by 90° when it enters port 1. This is because of waveguide's twist. The wave again experiences a faraday rotation of 90° due to ferrite rod. So the wave

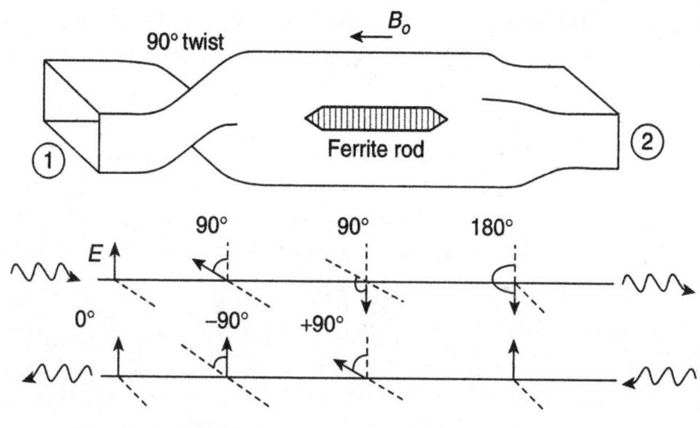

Figure 6.25 (b) *Faraday rotation-based gyrator*

coming out of port 2 will have a total phase shift of 180° with respect to the input wave entered the port 1. In the same way when TE_{10} mode signal is incident at port 2 it experiences a faraday rotation of 90° in anti-clock wise direction. It again rotates back by 90° because of twist in the waveguide. So the resultant phase shift when wave comes out of port 1 is 0°. Hence we can conclude that, the wave enters port 1 experiences a phase shift of 180° but the same wave when incident on port 2 does not undergo any change in the phase shift.

6.9.2 Faraday Rotation-Based Isolator

An isolator is a unilateral, two-port nonreciprocal transmission device. It is used to isolate one component from reflections of other components in the transmission line. The flow of power can be from input to output, but cannot be the other way. Hence, the bad effects of changing load impedance can be reduced by the use of the isolator on a signal source. Ferrites are used as the main material in isolators. The function of an isolator is shown in Figure 6.26.

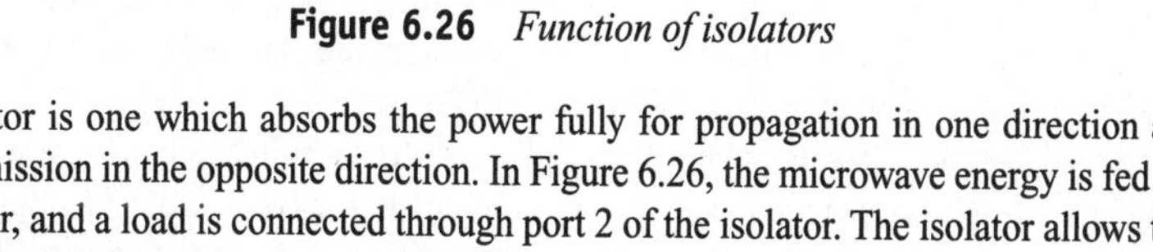

Figure 6.26 *Function of isolators*

An ideal isolator is one which absorbs the power fully for propagation in one direction and provides lossless transmission in the opposite direction. In Figure 6.26, the microwave energy is fed through port 1 of the isolator, and a load is connected through port 2 of the isolator. The isolator allows the energy to travel through it and to reach the load with minimum attenuation and provides maximum attenuation to the energy traveling from load to source. Therefore, isolators are used to improve the frequency stability of the microwave generators, such as klystrons and magnetrons, in which the reflection from the load affects the generating frequency.

Construction of Faraday Rotation based isolator

Figure 6.27 shows the Faraday rotation isolator. The isolator consists of a piece of circular waveguide supporting the dominant TE_{11} mode with transitions to a standard rectangular guide supporting the

TE_{10} mode at both ends. A thin pencil-shaped ferrite is located inside the circular guide, supported by polyfoam, and the waveguide is surrounded by a permanent magnet that generates a magnetic field in the ferrite core.

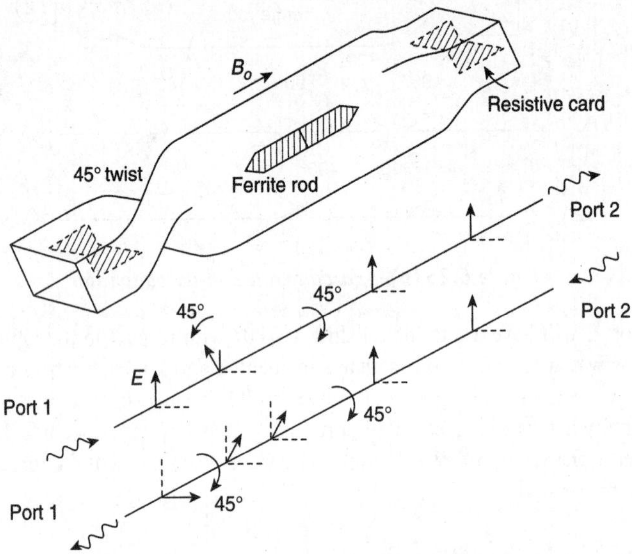

Figure 6.27 *Faraday rotation-based isolator*

Two resistive plates are placed in x-z plane at the ends of rectangular waveguide as shown in Figure 6.27. The transition from rectangular to circular waveguide results in 45° phase shift. The plane of polarization of the wave can be rotated by 45° by the DC magnetic field, which is applied longitudinally to the ferrite rod. The degree of rotation depends on the applied dc magnetic field and on the length and diameter of the ferrite rod. If TE_{10} wave is incident on the isolator's left end which is perpendicular to the input resistive card, then the wave passes through the ferrite rod without attenuation. The operation of isolator based on Faraday rotation is explained below.

A TE_{10} wave passes from port 1 through the resistive card without attenuation. The wave is shifted by 45° due to twist in the anti-clockwise direction after coming out of the card. Then, because of the ferrite rod, there is a shift of another 45° in the clock-wise direction. Hence, the polarization of the wave at port 2 will be same as at port 1 where there is no attenuation. As the plane of polarization of the wave is perpendicular to the plane of the resistive card, when the TE_{10} wave is fed from port 2, it passes from the resistive card placed near port 2. This wave suffers a phase shift of 45° in clock-wise direction due to the ferrite rod and again rotates by 45° in the same direction due to the twist. Now, the input card absorbs the wave as the plane of polarization of the wave is parallel to the input resistive card. Therefore, zero output will appear at port 1. In reverse transmission, the typical performance of these isolators is about 20 to 30 dB isolation and in forward transmission is about 1 dB insertion loss.

6.9.2.1 Applications of isolators

Figure 6.28 show the applications of isolators. Figure 6.28 (a) is designed to keep the high level of a local oscillator in a mixer circuit and not to have it radiate out through the incoming antenna.

Since the local oscillator power level is much higher than the RF signal that is coming in from the antenna, there is the possibility of this signal leaking back to the input circuit. That circuit may be attached to an antenna as shown in Figure 6.28 (a), and the local oscillator signal will radiate back out into the air. This is possible, because the RF and local oscillator signals are not that far apart in frequency, and the antenna will usually very easily transmit that signal out into the air. With the isolator in the circuit as shown, the signal coming from the mixer will hit the isolator and be dropped instantly into the termination. This keeps the signal in the mixer, where it can do the job it was intended to do.

The application shown in Figure 6.28 (b) is designed to supply a constant load to an oscillator circuit. If the load attached to an oscillator varies in value, there is a good possibility that the oscillator can be pulled off frequency and have a different output level than it was designed to produce. With the isolator in the circuit, variations in the load will be sent back to the isolator and end up in the termination, when the oscillators never see them. This application is used many times when you have a transmission system in which the carrier frequency and level are to be held constant. The isolator is a small, relatively inexpensive way of ensuring that these properties are preserved.

The application in Figure 6.28 (c) is similar to the second one. In this application, the isolator is placed between the generator and a test setup. Generally, when a circuit or system is being tested, there may be variations in the test setup that you do not want the generator to see. If the isolator is used, the variations are sent to the termination of the isolator and never get to the generator.

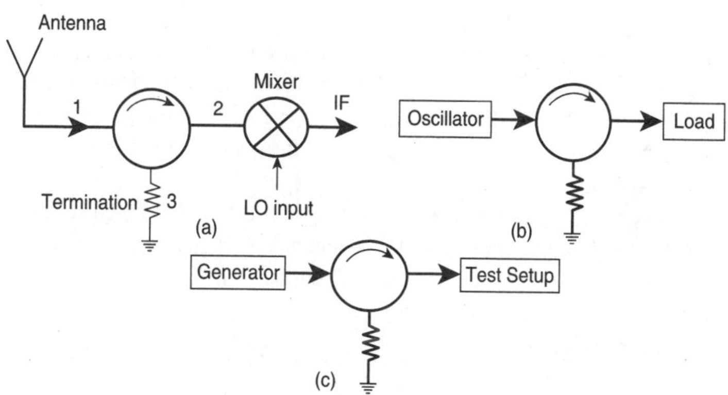

Figure 6.28 *Isolator applications (a) local oscillator re-radiation reduction; (b) oscillator pulling reduction; (c) generator oscillation*

6.9.3 Faraday Rotation-Based Circulator

The most important ferrite component is a circulator. A circulator is conceptually similar to the isolator, except that it is a multiport device. The circulator is also a unilateral device; i.e., power flows in only one direction. The main application of the circulator is in connection with multiple isolation in radars, parametric amplifiers, and so on.

It is a nonreciprocal device in which the ports are arranged in such a way that the electromagnetic energy which is entering a certain port is coupled to an adjacent port and not coupled to the other ports. The three-port symmetrical devices are commonly used as circulators. The Figure 6.29 represents a three-port circulator's circuit symbol.

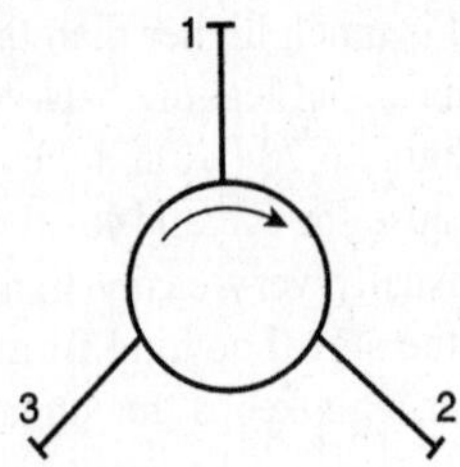

Figure 6.29 *Symbol of a circulator*

A signal applied to port 1 emerges from port 2 with a loss if all ports of a circulator are matched. This loss is called the insertion loss, which is given in decibels.

$$IL(dB) = 10\log_{10}\left(\frac{P_{in1}}{P_{out2}}\right)$$

A small part of the input signal emerges from port 3. Assuming that port 2 is terminated by a matched load Isolation can also be defined as the ratio of that emerging signal to the input signal. Isolation is given in decibels as below.

$$I(dB) = 10\log_{10}\left(\frac{P_{out3}}{P_{inp1}}\right)$$

The circulator is a three-port network that can be used to prevent reflection at the antenna from returning to the source. Figure 6.30 (a) and (b) show applications of circulators. In Figure 6.30 (a), we see a three-port circulator used as a two-port isolator. In this case, power flows from port 1 to port 2, while port 3 is terminated in the characteristic impedance. Figure 6.30 (b) shows a circulator used as a duplexer to connect a receiver and a transmitter to a common antenna.

The transmitter is connected at port 1, the antenna at port 2, and the receiver at port 3. When the transmitter sends out a signal, the signal goes to the antenna with great ease and does not enter into the receiver because of the isolation of the circulator. When the signal comes back to the antenna, it goes directly to the receiver and not to the transmitter, because of the circulator operation. It should be evident that to have the required isolation in the circulator, the transmitter, the receiver, and the antenna should be well matched to the circulator.

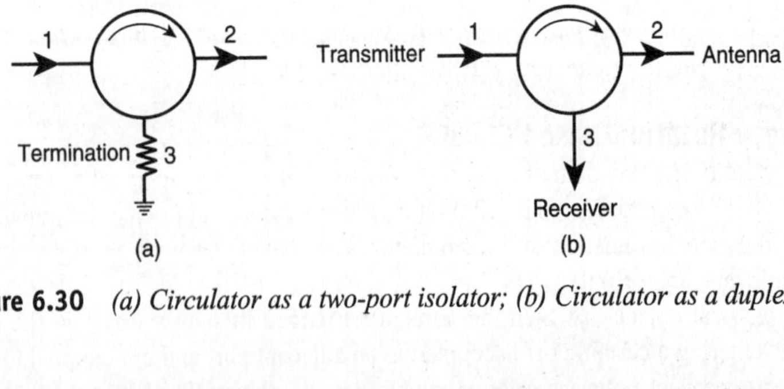

Figure 6.30 *(a) Circulator as a two-port isolator; (b) Circulator as a duplexer*

Y-Circulator

Three-port circulator is a symmetrical Y-type junctions of three identical waveguides with an axially magnetized ferrite placed at the center. Three-port circulator shown in Figure 6.31. The static magnetic field B_0 along the axis magnetizes the ferrite post. The necessary non-reciprocal property is provided by

magnet. Suitable tuning elements are placed in each arm to match the junctions. The input and output in a negative resistance amplifier can be isolated using this essential component. A transmitter can be coupled to various receivers by using three-port circulators.

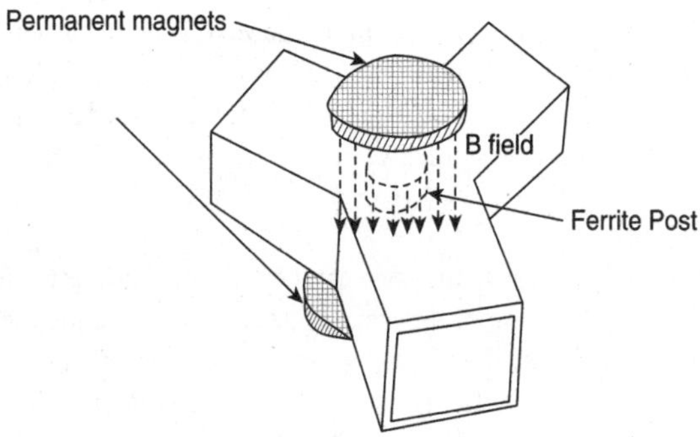

Figure 6.31 *Y circulator*

Construction of Circulator using two Magic Tees and Phase Shifters

Circulators are designed in many ways; however, their principle of operation remains the same. Figure 6.32 shows a four-port circulator constructed using two Magic Tees and a gyrator.

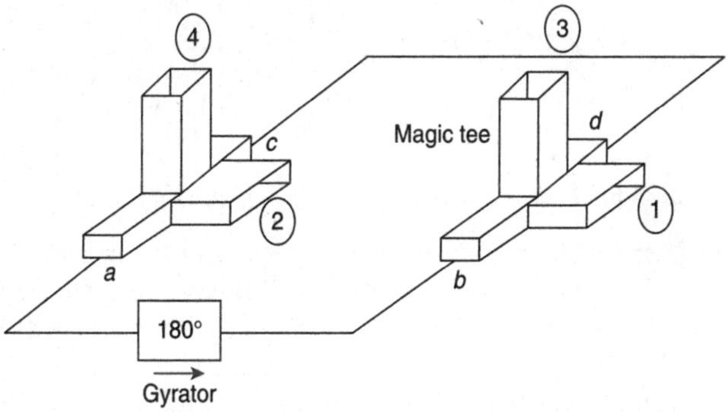

Figure 6.32 *Four-port circulator*

An additional phase shift of 180° is produced by the gyrator for propagation in the direction from *a* to *b*. The electrical path lengths are equal for propagation in all the directions i.e., from *a* to *b* and from *b* to *a* and also from *c* to *d* and from *d* to *c*.

Consider a wave incident at port 1. This wave is split into phase waves propagating in the side arms of the hybrid junction. It is also split into two equal amplitudes. These waves enter at points *a* and *c* from *b* and *d* and emerge from port 2, as they are in phase. Due to the presence of the gyrator a wave incident at port 2 is split into two waves: One at point *d* with phase φ, and the other at point *b* with phase φ+180°. As these waves have the right phase relationship, they combine and emerge from port 3 in the hybrid junction. A wave incident at port 3 is split into two equal amplitude waves with a phase difference of 180°. Hence, they will arrive at the other hybrid junction with the correct phase to combine and emerge from port 4. The wave incident at port 4 will split into two equal waves and 180° apart in the phase. The gyrator will restore the phase quality, so that the waves combine and emerge from port 1. Thus, the sequence of power flow is from 1→2→3→4→1.

6.10 WAVEGUIDE BENDS AND JOINTS

For the smooth flow of signal energy from one end of a waveguide to the other end without reflections, throughout its entire length waveguide should have constant dielectric material, size and shape. Reflections and a loss in the overall efficiency can be caused due to an abrupt change in the waveguide size or shape causes. Therefore, the change should be very gradual unless special devices are used. However, in some applications such as radar, a change in the waveguide size and shape is necessary. To change the shape or direction of a waveguide, we use bends, joints, or twists.

6.10.1 Waveguide Bends

Bents can be made in waveguides in several ways that they do not cause reflections. A bend can be made in either the narrow or wide dimension of the waveguide without changing the mode of operation. The waveguide bends are of 3 types. They are twisted, gradual, and sharp bends.

Twisted bends: The electromagnetic field should be rotated so that the antenna is polarized properly, when a waveguide is terminated with an antenna. This is achieved by twisting the waveguide, as shown in Figure 6.33. The reflections can be prevented by the twist which is gradual and can be extended over two or more wavelengths.

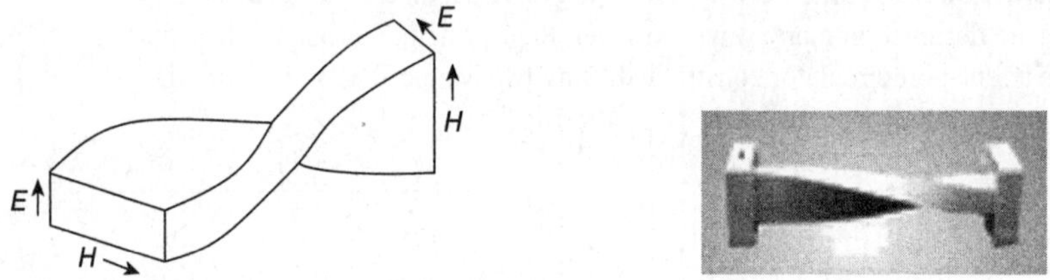

Source: globalspec.com

Figure 6.33 *Twisted section of a waveguide and it's schematic*

Gradual bends: In some applications, it is necessary to change the direction of the waveguide. For this purpose, a gradual bend is used. There are two types of bends: E bend and H bend. Neither the E bend in the wider dimension nor the H bend in the narrow dimension changes the normal mode of operation.

Gradual E bend: In E bends (Figure 6.34 (a)), to minimize the reflections, we have to ensure that the radius of the bend is greater than 2λ. This gradual bend is known as *E bend,* because it distorts the *E* fields (as shown in Figure 6.34 (b) and (c)).

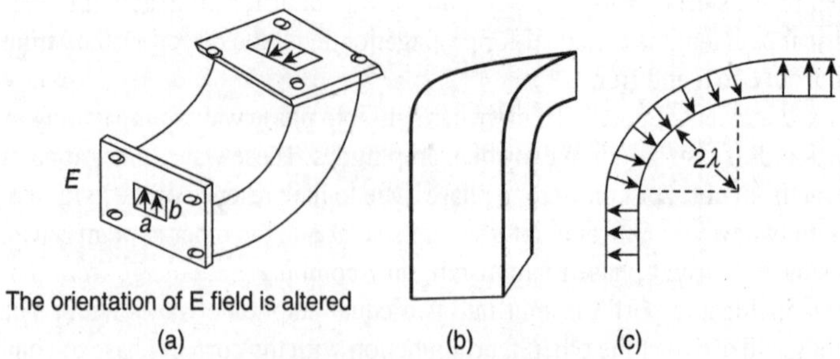

Figure 6.34 *Gradual E bend*

Gradual H bend: It is shown in Figure 6.35 (a). When a waveguide is bent in this way, the H fields will be distorted (as shown in Figure 6.35 (b) and (c)). Again, to prevent reflections the radius of the bend should be greater than 2λ.

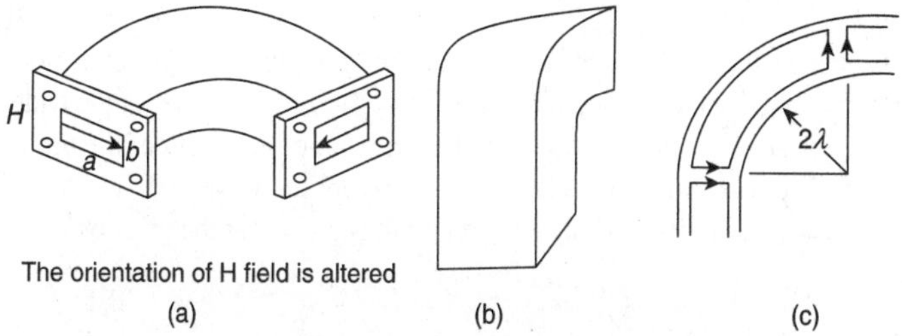

Figure 6.35 *Gradual H bend*

Figure 6.36 represents the schematic of E-plane bend and H-plane bend.

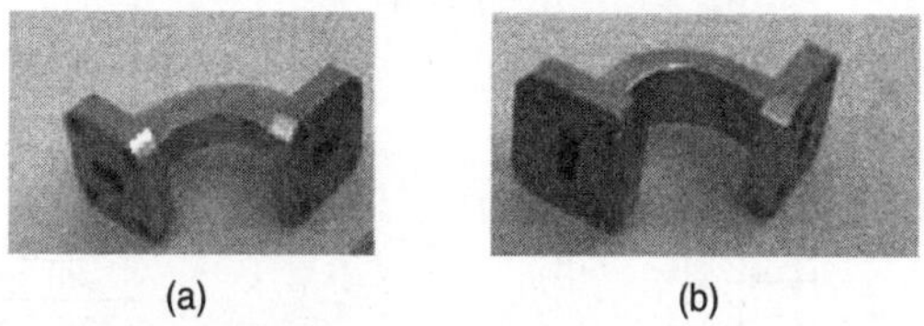

Source: microwaves101.com

Figure 6.36 *Schematic of E bend and H bend*

Sharp bend

If certain requirements are met, a sharp bend in any dimension can be used. The two 45-degree bends are shown in Figure 6.37. The bends are $\lambda/4$ a part. The reflections occur at the 45-degree bends. However, the combination of direct reflection at one bend and the inverted reflection at the other bend will cancel each other, leaving the fields as though no reflections had occurred.

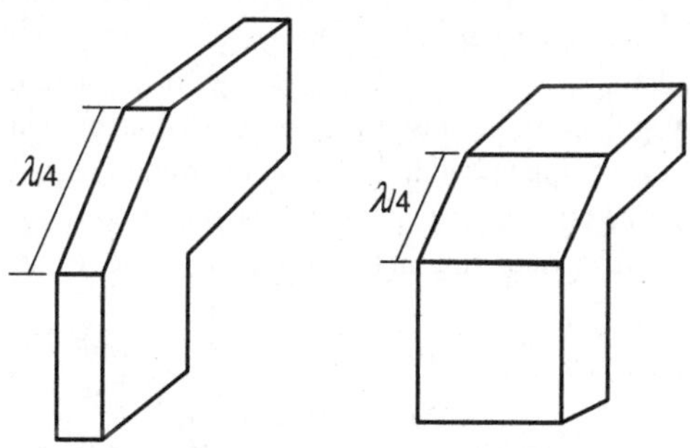

Figure 6.37 *Sharp bends*

6.10.2 Waveguide Joints(Flanges)

In applications such as radar, the waveguide system carries the power from the transmitter end to the input of the rotating antenna. However, the entire waveguide system cannot possibly be molded into one piece. The waveguide system should be constructed in sections which are connected with the joints.

There are three basic types of waveguide joints. They are permanent, semi-permanent, and rotating. The permanent joint is a factory-welded joint for which maintenance is not required. Hence, only the semi permanent and rotating joints will be discussed.

Semi-permanent joint: A cross-sectional view of the choke joint is shown in Figure 6.38. It is made up of a flat flange and a slotted flange. The slotted flange shown in Figure 6.38 (a) has a slot that is one-quarter wave deep. This slot is at a distance of one-quarter wave ($\lambda/4$) from the center of the wider side of the waveguide. Note that the distance of one-half wavelength is the sum of the depth of the groove and the distance from the waveguide. The half wave now reflects a short where the waveguide walls are joined together as bottom of the groove is shorted. Electrically a short circuit is created at the junction of the two waveguides.

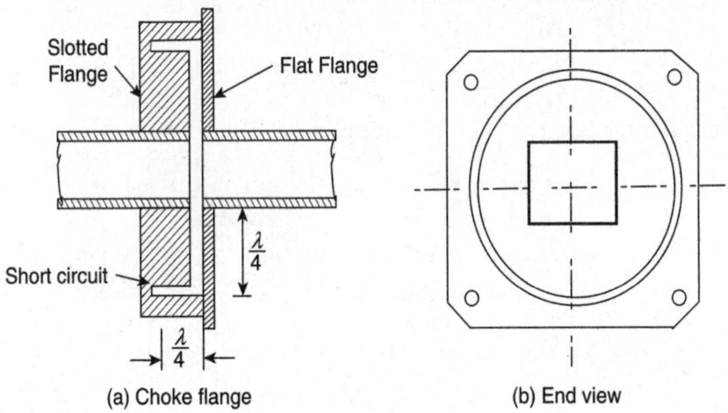

Figure 6.38 *Choke joint*

The operation of choke joint is similar to an RF choke in a power supply. The choke joint keeps the electromagnetic energy in the waveguide and the RF choke keeps RF energy in the circuit where it belongs.

Rotating joints

A rotating joint should be used whenever a stationary rectangular waveguide running from the transmitter is connected to a rotating antenna. A circular waveguide is generally used in a rotating joint as a rotating rectangular waveguide would cause field pattern distortion. The electrical connection with the stationary section can be completed with the use of the rotating section of the joint which uses a choke joint (Figure 6.39). The circular waveguide is designed to operate in the TM_{01} mode. To prevent the circular waveguide from operating in the wrong mode the rectangular sections are attached.

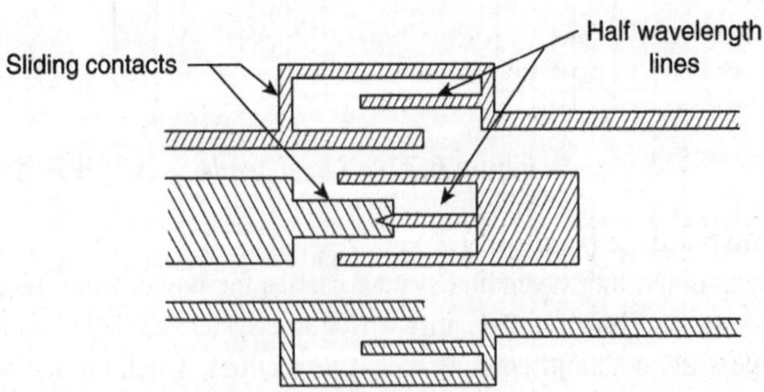

Figure 6.39 *Rotary joint*

SUMMARY

1. Probes and loops are used to couple coaxial line to waveguides or resonators. They are used for waveguide excitation in the desired mode (TE or TM).
2. In microwave circuits, a waveguide or coaxial junction with three independent ports is referred to as a *Tee junction.*
3. H-plane or E-plane Tee junctions split power equally, but because of the different field configurations at the junction, the electric fields at the output arms are in phase for the H-plane tee and are anti-phase for the E-plane tee. E- and H-plane Tee junctions act as 3-dB splitters.
4. Magic Tee is a combination of the E-plane tee and H-plane Tee junctions. It is a 4-port hybrid circuit and is also known as *hybrid Tee.*
5. Ferrites are ceramic-like materials. These are made by sintering a mixture of metallic oxides (MnO, Fe_2O_3). Ferrites possess strong magnetic properties and high resistivity. Non-reciprocal property of ferrites is useful in various applications. Examples of ferrite devices are isolators, circulators, phase shifters, modulators, and power limiters.
6. A gyrator is a two-port device that provides a relative phase shift of 180 degree for transmission from port 1 to port 2 and a zero-degree phase shift for transmission from port 2 to port 1.
7. An ideal isolator completely absorbs the power for propagation in one direction and provides lossless transmission in the opposite direction. The isolator protects the microwave source from the reflections.
8. The directional coupler is a device that samples the microwave incident power through the main waveguide.

$$\text{Coupling Factor, } C = 10\log_{10}\frac{P_1}{P_4}\text{ dB}, \text{ Directivity, } D = 10\log_{10}\frac{P_4}{P_3}\text{ dB}$$

$$\text{Isolation, } I = 10\log_{10}\frac{P_1}{P_3}$$

Directivity is a measure of how well the coupler distinguishes between forward and reverse traveling waves.

OBJECTIVE-TYPE QUESTIONS

1. A direction coupler is

   (a) four-port device (b) a three-port device (c) a two-port device (d) a one-port device

2. Application of Magic Tee:

   (a) mixer (b) duplexer (c) (a) and (b) (d) none

3. Isolated ports in Magic Tee:

   (a) E and H arms (b) collinear arms (c) (a) and (b) (d) none

4. Which device is based on Faraday rotation?
   (a) Magic Tee (b) H-plane T junction (c) E-plane T junction (d) isolator
5. Rat race junction:
   (a) Hybrid ring junction (b) Magic Tee (c) H junction (d) Ejunction
6. A circulator is
   (a) two-port device (b) a three-port device (c) a one-port device (d) a multi-port device
7. A matched load .... incident power
   (a) reflects (b) transmits (c) absorbs (d) none
8. Which has high permeability values?
   (a) ferrite (b) steel (c) copper (d) resistive card
9. How much phase shift does a gyrator produce in forward direction?
   (a) 0 (b) 90 (c) 320 (d) 180
10. How much phase shift does an isolator produce in forward direction?
   (a) 0 (b) 90 (c) 320 (d) 180
11. An isolator is
   (a) two-port device (b) a three-port device (c) a one-port device (d) none
12. Which device minimizes the reflections?
   (a) gyrator (b) circulator (c) Magic Tee (d) isolator
13. Coupling factor of a directional coupler is
   (a) $10\log_{10}\frac{P_i}{P_c}$ (b) $10\log_{10}\frac{P_i}{P_f}$ (c) $10\log_{10}\frac{P_b}{P_c}$ (d) $10\log_{10}\frac{P_b}{P_f}$
14. When the screw length is $\lambda_g/4$ in the waveguide, its nature is
   (a) capacitive (b) inductive (c) resistive (d) none
15. When the screw length is $> \lambda_g/4$ in the waveguide, its nature is
   (a) capacitive (b) inductive (c) resistive (d) none

ANSWERS TO OBJECTIVE-TYPE QUESTIONS

1. (a) 2. (c) 3. (c) 4. (d) 5. (a) 6. (d) 7. (c) 8. (a) 9. (d)
10. (a) 11. (a) 12. (d) 13. (a) 14. (a) 15. (b)

REVIEW QUESTIONS

1. Explain coupling probes and coupling loops.
2. Explain the functioning of rotary vane attenuators.
3. What is a phase shifter? Explain its principles of operation with a neat sketch. Give its applications.
4. What is Magic associated with a Magic Tee? Draw a neat sketch of a Magic Tee and list out its applications and properties.
5. Explain the operations of a directional coupler with the help of a sketch.
6. Explain the working of a two-hole directional coupler with a neat diagram and derive the expression for the coupling and directivity of a two-hole directional coupler.
7. Explain the characteristics of ferrites materials.
8. Explain Faraday rotation with a neat diagram. Explain the working of a ferrite isolator.
9. Write short notes on "inductive and capacitive posts".
10. Write short notes on the following:

    (a) Bethe-hole directional coupler
    (b) Rotary phase shifter
11. Incident power to a directional coupler is 80 watts. The direction coupler has coupling factor of 20 dB, directivity of 30 dB and insertion loss of 0.5 dB. Find the output power at (a) main arm, (b) coupled and (c) isolated ports.
12. A 30 dB directional coupler is used to sample incident and reflected power in a waveguide. The value of VSWR is 2.5 and coupler sampling power is 4 mW. What is the value of reflected power.
13. Calculate coupling factor of a directional coupler when incident power is 400 mW and power in auxiliary waveguide is 200 μ watts.
14. For a directional coupler the incident power is 500 mW. Calculate the power in the main arm and auxiliary arm. The coupling factor is 20 dB.

Scattering Matrix for Waveguide Components

7

7.1 INTRODUCTION

The analysis of networks or devices at low frequencies commonly uses circuit parameters such as *Z* parameters, *Y* parameters, and *ABCD* parameters. These parameters are derived based on terminal voltage and current measurements. They require the application of short- and open-circuit terminations for evaluation. A brief description of these parameters is given in Section 7.8.

However, it is very difficult to measure voltages and currents at microwave frequencies because of the following reasons:

- They are distributed values and vary with their position in microwave structures.
- Active devices become unstable when they are terminated with open or short circuits at high frequencies.
- The effect of the interconnecting leads between the test equipment and the device under test also becomes critical as the frequencies are increased.

One method of describing the behavior of a network at microwave frequencies is in terms of scattering parameters, which are commonly referred to as *S parameters*.

Scattering is a general term that refers to transmitting in all directions and reflecting back to the source. The derivation of *S* parameters is based on forward and backward travelling waves on terminal transmission lines. That's why they are most commonly used in microwave engineering. The forward and backward traveling waves are also referred to as *incident* and *reflected waves*, respectively. The *S* parameters are defined using the amplitudes of incident waves (i.e. voltage waves entering the ports) and reflected waves (i.e. voltage waves leaving the ports).

Scattering matrix: The directly measurable quantities at high frequencies are amplitudes and phase angles of reflected waves (or scattered waves). The scattered wave amplitudes are linearly related to the incident wave amplitudes. The matrix describing the relationship between the voltage waves incident at the ports and those reflected from the ports is called a *scattering matrix* or an *S matrix*.

Normalized incident and reflected voltage waves (a_i and b_i): The incident wave is defined as that component which would exist if the port under consideration were conjugately matched to the normalizing impedance at that port. Therefore, *S* parameters describe the interrelationships of a new set of variables, a_i and b_i. The variables a_i and b_i are defined in terms of the terminal current I_i, terminal voltage V_i and arbitrary reference impedance Z_i. They are the normalized complex voltage waves which are incident on and reflected from the ith port of the network.

The *Z*, *Y*, *h*, *ABCD*, and scattering parameters are introduced in this chapter to characterize the low- and high-frequency circuits, respectively. Scattering matrix calculations are also presented for one-port, two-port (isolator, gyrator, attenuator, etc), three-port (circulator, *E*-plane and *H*-plane Tees), and four-port (directional coupler, rat race, and Magic Tee) junction devices.

7.2 SIGNIFICANCE OF SCATTERING PARAMETERS (*S* PARAMETERS)

S parameters play a major role in network analysis at microwave frequencies. It is difficult to get the required short and open circuit tests for higher frequencies by using the *Z*, *Y*, *h*, and *ABCD* parameters at higher frequencies. Thus, the *S* parameters gained importance at higher frequencies.

In order to use *Z*, *Y*, *h*, and *ABCD* parameters at high frequency, let us suppose we short the circuit with a wire; the wire itself possesses an inductance that is of large magnitude. Likewise for open circuit, capacitive loading at the terminal will be of large magnitude.

The *S* parameters are measured with 50 Ω network analyzers (Figure 7.1); hence they are easiest parameters to be measured at frequencies over and above a few tens of MHz. The scattering parameters (magnitude and phase) of a one- or two-port microwave network from 0.05 GHz to 26.5 GHz are measured using this test instrument. This analyzer can also performs a Fast Fourier transform of the frequency domain data to provide a time-domain response of the network under test.

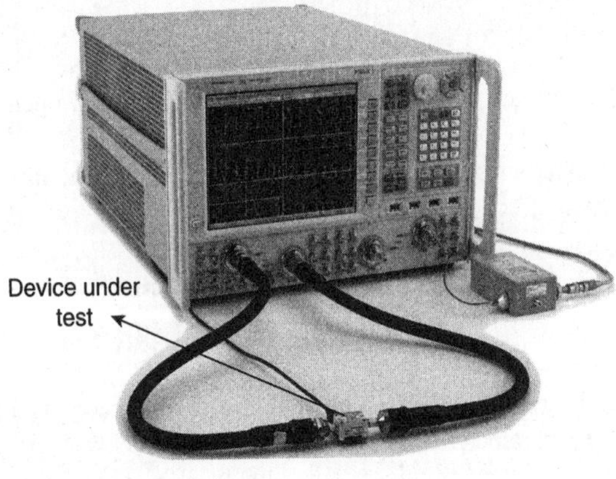

Source: www.to.gstatistic.com

Figure 7.1 *Network analyzer*

In microwave engineering, multiport networks are characterized using scattering matrices at high frequencies. They are used to represent microwave devices, such as isolators, circulators, directional couplers, amplifiers, *E* planes, *H* planes, Magic Tees, and hybrid rings and are easily related to concepts of gain, loss, and reflection.

Advantages/Disadvantages of *S* parameters

Advantages:

- The matched loads are used for terminating the ports while using *S* parameters rather than open and short the circuits as in (*Z*, *Y*, *h* and *ABCD*) parameters. So the capacitance and inductance effects do not affect the network. Most familiar measurements required to determine the *S* parameters are phase, attenuation (gain) and reflection coefficient.

- Easy to measure
  - Power at high frequencies than the current and voltage measured at the short- and open-circuit terminals
  - Termination is resistive and is more likely to be stable.
  - Different devices can be measured on the same setup.
- The terminal voltages and currents vary in magnitude at points along a lossless transmission line but S parameters are basically traveling waves, they do not vary along the transmission line. For example, if the measuring device and the transducers are located at a distance from each other and are connected by low-loss transmission lines, S parameters can be measured.

Disadvantages: They are difficult to understand and interpret measurements.

7.3 FORMULATION OF *S* MATRIX

Linear networks can be completely characterized by parameters that are measured at the network ports without knowing the content of the networks. Networks can have any number of ports. An analysis of a 2-port network (Figure7.2) is taken into consideration to explain the theory.

At each port i, an entering voltage and current waves are defined as V_i^+ and I_i^+ and the leaving voltage and a current waves are defined as V_i^- and I_i^- (as shown in Figure 7.2). These voltage and current waves are defined in such a way that the voltage is proportional to the transverse electric field and current is proportional to the magnetic fields of the wave. V_i^+, I_i^+, V_i^-, and I_i^- are the complex amplitudes of sinusoidal excitations. The power of the entering or leaving wave is given by the product of the voltage and current. The characteristic impedance (Z_0) of the port is given by the ratio of the voltage and current.

The total voltage and the total current at port i are as follows:

$$V_i = V_i^+ + V_i^-$$

$$I_i = I_i^+ + I_i^-$$

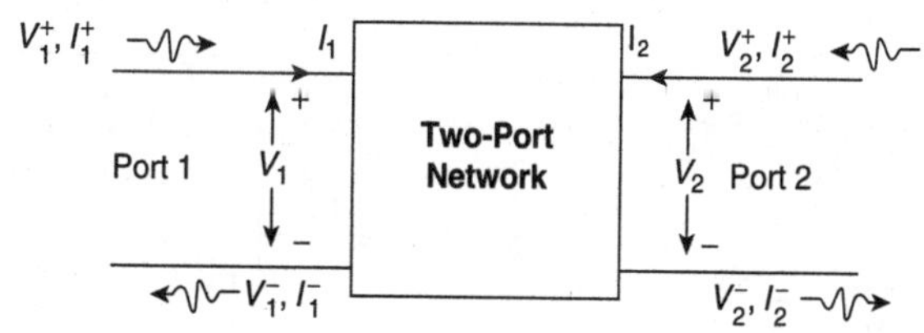

Figure 7.2 *Two-port network with incident and reflected waves*

When a voltage wave (V_i^+) is incident at one port, some fraction of the signal bounces back out of that port, some of it scatters and exits from other ports, and some of it disappears as heat or electromagnetic radiation. Usually, all the microwave network ports have similar connectors (coaxial connectors or waveguide flanges) with an impedance of 50 Ω, and the characteristic impedances (Z_0) of the ports have a similar value. However, in a general case, the characteristic impedances (Z_0) may have different values. For example, the ports of a coaxial-to-waveguide adapter have different characteristic impedances. Then, the voltage waves should be normalized.

Normalization of Voltage waves

We can convert the power toward a two-port network or a power away from the two-port network into normalized voltages (a_i) and (b_i), respectively (Figure 7.3). They can be interpreted in terms of voltage amplitude by normalizing the power to characteristic impedance (Z_0) as shown in Table 7.1.

Table 7.1 Power domain to Voltage domain conversion

Power domain		Voltage domain
Starting with Power (P)	Normalised to characteristic impedance (Z_0)	Normalised amplitude for voltage and current
$P = VI^*$	$= \dfrac{V^*V}{Z_0} = \dfrac{\lvert V\rvert^2}{Z_0}$	$\sqrt{P} = \dfrac{V}{\sqrt{Z_0}} = I\sqrt{Z_0}$

a_i represents the square root of the power wave injected into port i.

$$a_i = \sqrt{P} = \frac{V_i^+}{\sqrt{Z_0}} \tag{7.1}$$

b_i represents the square root of the power wave leaving port i

$$b_i = \frac{V_i^-}{\sqrt{Z_0}} \tag{7.2}$$

where V^+ and V^- are the incident and reflected wave voltages, respectively, at port i.

Since, a_i and b_i represent the square root of powers, $|a_1|^2$ is the power incident into port 1of the network, $|b_1|^2$ is the power coming out from port1.

S-parameter representation of a 2-port network

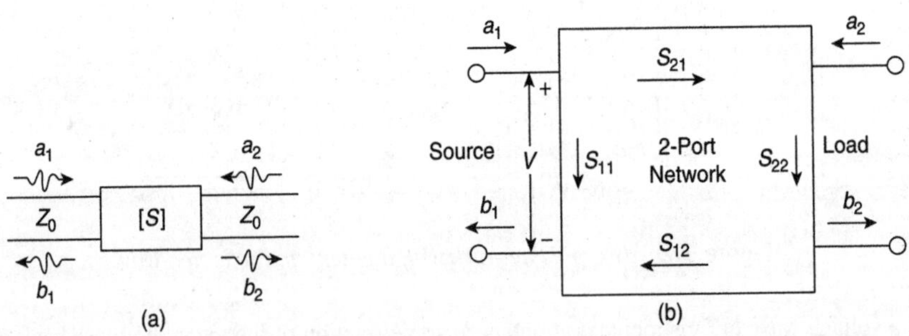

Figure 7.3 *(a) S-matrix representation; (b) Detailed S-matrix representation of a 2-port network*

The behavior of the network in terms of the injected and reflected power waves can be described by a set of linear equations (Figure 7.3). For the 2-port case, the outputs can be related to the inputs by

$$b_1 = S_{11}a_1 + S_{12}a_2 \tag{7.3}$$

$$b_2 = S_{21}a_1 + S_{22}a_2 \tag{7.4}$$

where b_1 is the wave traveling away from port 1 (i.e. toward source)
= wave incident on port 1 times the reflection coefficient ($S_{11}a_1$) + wave incident on port 2 times transmission coefficient from port 2 to port 1 ($S_{12}a_2$).

Similarly, b_2 is the wave traveling away from port 2 = $S_{21}a_1 + S_{22}a_2$

From Eqs. 7.3 and 7.4, we can say that each signal coming out of a two-port network (i.e. b_1 or b_2) will have two components: some signal reflected from the same port and some signal transferred from the other port.

We can interpret S_{ij} as the power measured at port i due to the incident power at port j.

The term S_{ij} can be computed directly by the following formula:

$$S_{ij} = \left.\frac{b_i}{a_j}\right|_{a_k=0\forall k\neq j} \tag{7.5}$$

where k is 1 to n (here, $k = 1$ or 2)

Here, n represents the number of ports.

In matrix form, the above equations can be written as

$$\begin{bmatrix} b_1 \\ b_2 \end{bmatrix} = \begin{bmatrix} S_{11} & S_{12} \\ S_{21} & S_{22} \end{bmatrix}\begin{bmatrix} a_1 \\ a_2 \end{bmatrix} \tag{7.6}$$

The outgoing waves are expressed in terms of the incoming waves by the matrix equation

$$[b] = [S]\,[a] \tag{7.7}$$

where S is an $n \times n$ square matrix of complex numbers called the *scattering matrix*.

$$[S] = \begin{bmatrix} S_{11} & S_{12} \\ S_{21} & S_{22} \end{bmatrix} \tag{7.8}$$

The behavior of the network is determined by the S matrix ($[S]$) and all the elements of this matrix are called *S parameters*. They are frequency dependent.

S parameters for two-port networks are given as follows:

$$S_{11} = \left.\frac{b_1}{a_1}\right|_{a_2=0} \qquad S_{22} = \left.\frac{b_2}{a_2}\right|_{a_1=0}$$

$$S_{21} = \left.\frac{b_2}{a_1}\right|_{a_2=0} \qquad S_{12} = \left.\frac{b_1}{a_2}\right|_{a_1=0}$$

Require proper termination on Port 2	Require proper termination on Port 1

$$S_{11} = \left.\frac{b_1}{a_1}\right|_{a_2=0} = \frac{\text{reflected power wave at port 1}}{\text{incident power wave at port 1}}$$

$$S_{21} = \left.\frac{b_2}{a_1}\right|_{a_2=0} = \frac{\text{transmitted power wave at port 2}}{\text{incident power wave at port 1}}$$

$$S_{22} = \left.\frac{b_2}{a_2}\right|_{a_1=0} = \frac{\text{reflected power wave at port 2}}{\text{incident power wave at port 2}}$$

$$S_{12} = \left.\frac{b_1}{a_2}\right|_{a_1=0} = \frac{\text{transmitted power wave at port 1}}{\text{incident power wave at port 2}}$$

7.3.1 S-Parameter Evaluation

S parameters can only be determined under conditions of perfect matching at the input or output port.

Determination of S_{11}, S_{21} parameters: S_{11}, S_{21} parameters can be computed if port 2 is terminated with a matched load (Z_0). Then, the incident wave applied at port 2 (i.e. a_2) becomes zero ($a_2 = 0$), and the wave leaving port 2 (i.e. b_2) is presented (Figure 7.4).

The b_2 is the wave traveling away from port 2 and is due to the wave incident at port 1 (i.e. a_1) times the transmission factor (S_{21}) from port 1 to port 2. Similarly, b_1 is the wave traveling away from port 1 toward the source and is equivalent to the wave incident at port 1 (i.e. a_1) times the reflection coefficient (S_{11}).

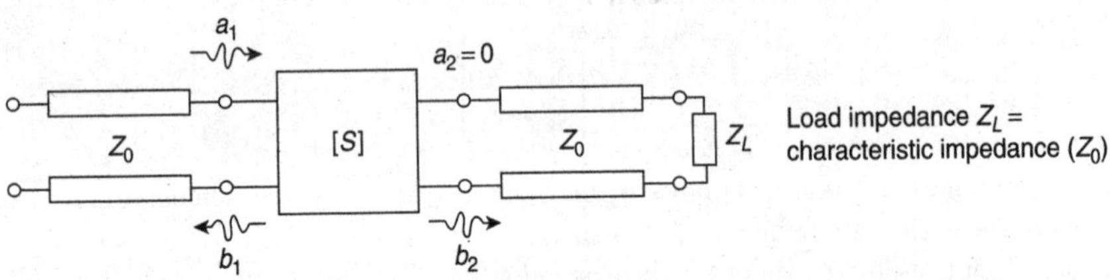

Figure 7.4 *Determination of S_{11}, S_{21} parameters with matched load at port 2*

$$S_{11} = \left.\frac{b_1}{a_1}\right|_{a_2=0}, \quad S_{11} = \left.\frac{b_1}{a_1}\right|_{Z_L=Z_0}$$

By defining Z_{in} as the input impedance given $Z_l = Z_0$ (i.e; $Z_{in}|_{Z_L=Z_0}$),

$$S_{11} = \frac{Z_{in}|_{Z_L=Z_0} - Z_0}{Z_{in}|_{Z_L=Z_0} + Z_0} \tag{7.9}$$

From the above equation, we can say that the reflection coefficient (S_{11}) is the method of specifying the input impedance Z_{in}.

$$\begin{Bmatrix} S_{11} \\ Z_{in} \end{Bmatrix} \text{ is the input } \begin{Bmatrix} \text{reflection coefficient} \\ \text{impedance} \end{Bmatrix} \text{ given that the load } \begin{Bmatrix} \text{reflection coefficient} \\ \text{impedence } (Z_L) \end{Bmatrix} \text{ is } \begin{Bmatrix} 0 \\ Z_0 \end{Bmatrix}$$

$S_{11} = b_1/a_1$ that is, the input reflection coefficient (when $a_2 = 0$)

$S_{21} = b_2/a_1$ that is, the forward transmission gain/loss (when $a_2 = 0$)

Determination of S_{22}, S_{12} parameters: S_{22}, S_{12} parameters can be computed, if port 1 is terminated with a matched load (Z_0). Then, the incident wave at port 1 (i.e. a_1) becomes zero ($a_1 = 0$), and the wave leaving port 1 (i.e. b_1) is presented (Figure7.5).

The b_1 is the wave traveling away from port 1 and is due to the wave incident at port 2 (i.e. a_2) times the transmission factor (S_{12}) from port 2 to port 1. Similarly, b_2 is the wave traveling away from port 2 toward the source and is due to the wave incident at port 2 (i.e. a_2) times the reflection coefficient (S_{22}).

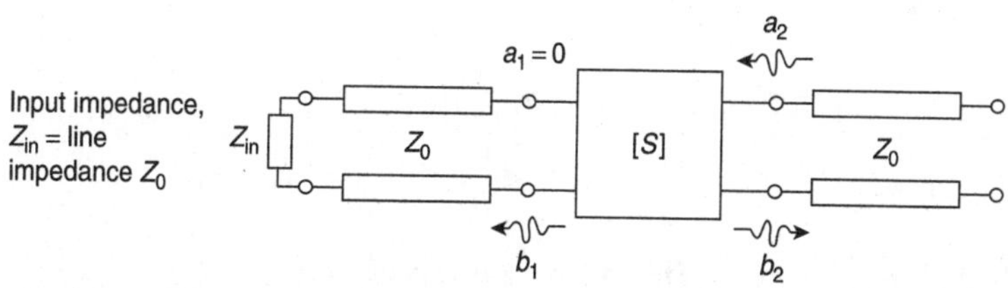

Figure 7.5 *Determination of S_{22}, S_{12} parameters with matched load at port 2*

Mathematically, the above statements can be written, as the same analysis and comments clearly apply to S_{22} as Eq. (7.9):

By symmetry,

$$S_{22} = \frac{Z_{out}\big|_{Z_{in}=Z_0} - Z_0}{Z_{out}\big|_{Z_{in}=Z_0} + Z_0} \tag{7.10}$$

From the above equation, we can say that the reflection coefficient (S_{22}) is the method of specifying the output impedance Z_0:

$$\begin{Bmatrix} S_{22} \\ Z_{out} \end{Bmatrix} \text{ is the output } \begin{Bmatrix} \text{reflection coefficient} \\ \text{impedance} \end{Bmatrix} \text{ given that the source } \begin{Bmatrix} \text{reflection coefficient} \\ \text{impedence } (Z_{in}) \end{Bmatrix} \text{ is } \begin{Bmatrix} 0 \\ Z_0 \end{Bmatrix}$$

$S_{22} = b_2/a_2$; that is, the output reflection coefficient (when $a_1 = 0$)

$S_{12} = b_1/a_2$; that is, the reverse transmission gain/loss (when $a_1 = 0$)

a_1 and b_1 are rms voltages normalized by $\sqrt{Z_0}$. S_{11} and S_{22} are the reflection coefficients with the opposite port terminated at Z_0 (usually 50 Ω). S_{21} and S_{12} are the forward and reverse 50 Ω transducer gains, respectively.

S Parameters for 3 Ports

A 3-port network is shown in Figure 7.6. The matrix equations for *a* 3-port network are as follows:

$$\begin{aligned} b_1 &= S_{11}\, a_1 + S_{12}\, a_2 + S_{13}\, a_3 \\ b_2 &= S_{21}\, a_1 + S_{22}\, a_2 + S_{23}\, a_3 \\ b_3 &= S_{31}\, a_1 + S_{32}\, a_2 + S_{33}\, a_3 \end{aligned} \tag{7.12}$$

In matrix form,

$$\begin{bmatrix} b_1 \\ b_2 \\ b_3 \end{bmatrix} = \begin{bmatrix} S_{11} & S_{12} & S_{13} \\ S_{21} & S_{22} & S_{23} \\ S_{31} & S_{32} & S_{33} \end{bmatrix} \begin{bmatrix} a_1 \\ a_2 \\ a_3 \end{bmatrix} \tag{7.13}$$

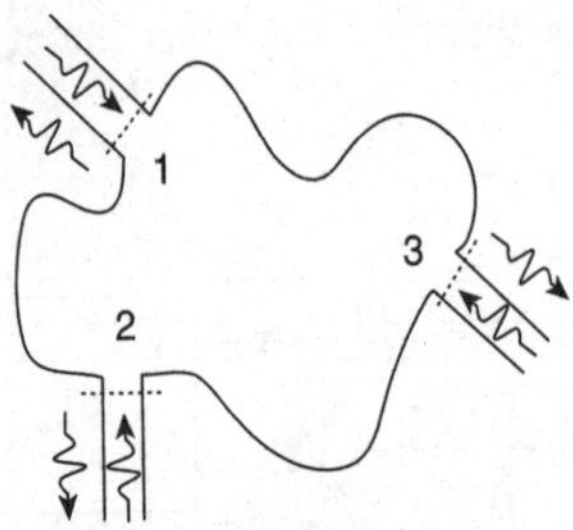

Figure 7.6 *3-port network*

7.3.2 S Parameters for *n* Ports

It is common to consider microwave networks of three and four ports, as in power dividers and directional couplers, respectively. The extension to the example of 3-port network equations and an equivalent matrix should support the concept of equivalence:

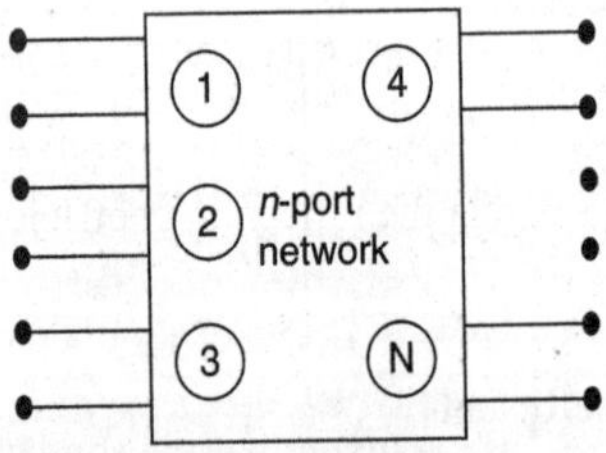

Figure 7.7 *N-port networks*

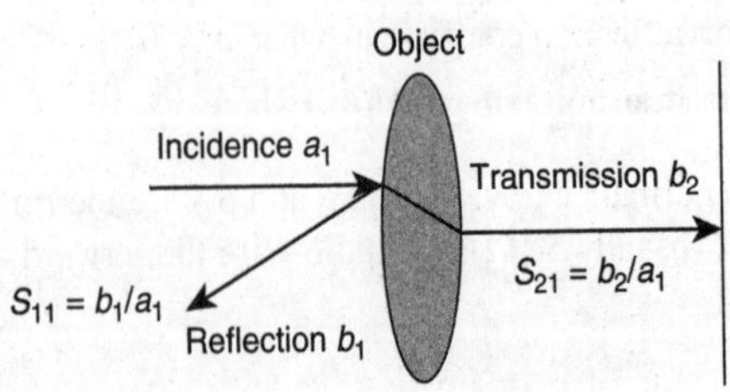

Figure 7.8 *Conceptual diagram of S parameters (optical analogy)*

Consider an *n*-port network/device as shown in Figure 7.7. Let us take two cases, one where the incident wave is applied at port1 only and another where incident waves are applied at *n* ports.

Case (1) If the incident wave (a_1) is applied to port 1 only, $a_2, a_3, \ldots a_n = 0$ (Figure 7.8) $b_1 = S_{11}a_1$ where S_{11} is the reflection coefficient

However, the wave (a_1) will also be scattered out of other ports, and we will have $b_n = S_{n1}a_1$ where $n = 2, 3, 4, \ldots, n$

Case (2) If waves are incident at all ports and the input and output ports of a network/device are numbered, we formulate the following:

- *S* parameter corresponding to a wave incident at port *j* and detected at port *i* is described as S_{ij}.
- Reflection is represented as $i = j$.
- Transmission is described as $i \neq j$.

The equations of an *n*-port network are as follows:

$$\begin{aligned} b_1 &= S_{11}a_1 + S_{12}a_2 + \cdots + S_{1n}a_n \\ b_2 &= S_{21}a_1 + S_{22}a_2 + \cdots + S_{2n}a_n \\ &\vdots \\ b_n &= S_{n1}a_1 + S_{n2}a_2 + \cdots + S_{nn}a_n \end{aligned} \tag{7.14}$$

where

$$a_n = \frac{V_n^+}{\sqrt{Z_{0n}}} \quad b_n = \frac{V_n^-}{\sqrt{Z_{0n}}}$$

$$S_{ij} = \left.\frac{b_i}{a_j}\right|_{a_{k=0,k\neq j}} = \left.\frac{V_i^- / \sqrt{Z_{0i}}}{V_j^+ / \sqrt{Z_{0j}}}\right|_{V^+_{k=0,k\neq j}} \tag{7.15}$$

V_n^+ and V_n^- are amplitudes of forward and reverse traveling waves, respectively. When these *S* parameters are aligned in a matrix form, they are referred to as an *S matrix* (Eq. 7.16). The *S* matrix for an *n-port* network contains n^2 coefficients (*S parameters), with* each one representing a possible input–output path. The number of rows and columns in an *S* matrix is equal to the number of ports.

$$\begin{bmatrix} b_1 \\ b_2 \\ \vdots \\ b_n \end{bmatrix} = \begin{bmatrix} S_{11} & S_{12} & \cdots & S_{1N} \\ S_{21} & & & \vdots \\ \vdots & & & \\ S_{n1} & \cdots & & S_{nn} \end{bmatrix} \begin{bmatrix} a_1 \\ a_2 \\ \vdots \\ a_n \end{bmatrix} \tag{7.16}$$

The *S* parameter is a non-dimensional parameter (no unit) since it is described by a ratio of transmitted power and reflected power. However, the magnitude of the *S* parameter, the unit *dB* is used with a common logarithm.

7.4 PROPERTIES OF A SCATTERING MATRIX

Properties of the *S* matrix for an *n*-port network are as follows:

(i) Scattering matrix [*S*] is always the square matrix of the order $n \times n$.
(ii) Under perfect matched conditions, diagonal elements of [*S*] are equal to "0".

i.e. $$S_{ii} = 0$$

(iii) $[S]$ is symmetric for all reciprocal networks, that is, $[S] = [S]^T$ where $[S]^T$ is the transpose of $[S]$.
In other words, $[S]$is symmetric about the main diagonal.
i.e $$S_{ij} = S_{ji}$$
(iv) If a device (or network) is lossless, then the S matrix is unitary $[S]\,[S]^* = [I]$ where $[S]^*$ is the complex conjugate of S.
$[I]$ is the unit matrix defined as

$$\begin{bmatrix} 1 & & 0 \\ & \cdot & \\ & & \cdot \\ 0 & & 1 \end{bmatrix}$$

For a lossless network:

(a) The dot product of any column (or row) of $[S]$ with the conjugate of that same column (or row) is equal to 1

$$\sum_{i=1}^{N} S_{ij} S_{ij}^* = \sum_{i=1}^{N} \left| S_{ij} \right|^2 = 1 \tag{7.17}$$

(b) The dot product of any column of $[S]$ with the conjugate of another column is equal to 0

$$\sum_{i=1}^{N} S_{ij} S_{ik}^* = 0 \tag{7.18}$$

Detailed Description of Reciprocal, Lossless Networks and Matched Termination

The $[S]_{n \times n}$ matrix for a given network has n^2 parameters, and these parameters need to be determined for analyzing the network. For example, *a* 2-port network has a 2×2 scattering matrix, and, in this case, the network is determined by 8 real numbers (each scattering parameter is complex). Fortunately, in many cases, it is possible to reduce the number of unknown coefficients by knowing some of the properties of the network, such as reciprocity, lossless, and symmetry. So, an understanding of these parameters helps in solving problems.

Reciprocity

If the power transfer and the phase do not change when the input and output ports are interchanged, then the network is said to be reciprocal. A network can be reciprocal, only if it is linear, time invariant, made of reciprocal materials. A reciprocal network cannot have any dependent voltage or current sources in the network.

For example, a two-port network is said to be reciprocal, if the transmission characteristics are same in both directions (i.e. $S_{21} = S_{12}$). Exchanging the incident voltage and current waves results in an equivalent definition of reciprocity. In general, reciprocity is a property of passive circuits (circuits with no active devices or ferrites). Usually, a network is non reciprocal if it contains active components such as generators.

Mathematically, for a reciprocal network, the S matrix is equal to its transpose.

$$[S] = [S]^T \tag{7.19}$$

$$\begin{bmatrix} S_{11} & S_{12} \\ S_{21} & S_{22} \end{bmatrix} = \begin{bmatrix} S_{11} & S_{21} \\ S_{12} & S_{22} \end{bmatrix}$$

∴ The condition for reciprocity is $S_{12} = S_{21}$

The scattering matrix of a reciprocal circuit is symmetrical: $S_{ij} = S_{ji}$, that is, $[S] = [S]^T$

Lossless networks

All the energy entering into the lossless network can be expressed in terms of reflection or scattering. It means, the energy conservation is the most important property of the lossless network.

A lossless network does not contain any resistive elements or other power dissipative elements, and there is no attenuation of the signal. In terms of scattering parameters, a network is lossless if

$$[S][S^*] = [I] \tag{7.20}$$

where * represents conjugation, $[I]$ is a unitary matrix, and

$$[S^*] = [S]^{-1} \tag{7.21}$$

Modulus of det $[S]$ is equal to 1.

For a losslessness network, sum of the $S_{m\times n}S^*_{m\times n}$ of any column should be unity.

Reciprocal and lossless condition

If the network is reciprocal and lossless, $\text{Power}_{\text{in}} = \text{Power}_{\text{out}}$.

So, one condition for the two-port network to be lossless is

$$|S_{11}|^2 + |S_{21}|^2 = 1 \tag{7.22}$$

If a signal is applied at port 2 rather than port 1 same condition as above is applied rather the parameters are changed; so, the second condition for a two-port network to be lossless is that

$$|S_{12}|^2 + |S_{22}|^2 = 1 \tag{7.23}$$

$$S_{11}S_{12}^* + S_{21}S_{22}^* = 0 \tag{7.24}$$

In addition, a reciprocal and lossless network satisfies $[S] = [S]^T$ and $[S][S]^* = [I]$ (7.25)

EXAMPLE PROBLEM 7.1

Determine the S parameters for a reciprocal and lossless network, a perfectly matched 2-port network.

Solution

For a reciprocal network, the S matrix should be symmetrical.

$$S_{12} = S_{21}$$

For a perfectly matched network,

$$S_{11} = S_{22} = 0$$

For a lossless network,

$$[S][S^*]=[I]$$

So, we get

$$|S_{12}|^2=1,\ |S_{21}|^2=1$$

$$[S]=\begin{bmatrix}0 & 1\\ 1 & 0\end{bmatrix}$$

■

Network ports terminated with matched loads

If all the ports are terminated with matched loads, the reflection coefficient (Γ) for port i is computed as

$$\Gamma_i = S_{ii} = b_i/a_i \tag{7.26}$$

and the transmission power gain from port j to port i is

$$G_{ji} = |S_{ij}|^2 = \left|\frac{b_i}{a_j}\right|^2 \tag{7.27}$$

Return Loss and Insertion Loss

Two-port networks are described by their return loss and insertion loss. The return loss (RL) at the ith port of a network is defined as

$$RL_i = -20\log V^-/V^+ = -20\log\frac{b}{a} = -20\log|\Gamma_i| \tag{7.28}$$

The S-parameter equivalent to the return loss is S_{11}.

$$S_{11} = \left.\frac{b_1}{a_1}\right|_{a_2=0} = \frac{V_1^-/\sqrt{Z_0}}{V_1^+/\sqrt{Z_0}} = \frac{V_1^-}{V_1^+} = \frac{V_{\text{reflected}}}{V_{\text{incident}}}$$

In general,

$$S_{ii} = \left.\frac{b_i}{a_i}\right|_{a_{j=0},\, j\neq 0} = \Gamma_i \tag{7.29}$$

The amount of signal lost when it goes from j^{th} port to an ith port is the Insertion Loss. It is defined as the measure of the attenuation as a result of insertion of a network between a source and a load.

The S- parameter equivalent to the insertion loss is S_{21}.

$$S_{21} = \left.\frac{b_2}{a_1}\right|_{a_2=0} = \frac{V_2^-/\sqrt{Z_o}}{V_1^+/\sqrt{Z_o}} = \frac{V_2^-}{V_1^+} = \frac{V_{\text{transmitted}}}{V_{\text{incident}}} \tag{7.30}$$

$$IL_{ij} = -20\log\left|\frac{V_i^-}{V_j^+}\right| \tag{7.31}$$

EXAMPLE PROBLEM 7.2

The [S] matrix of a two-port network is given below:

$$[S] = \begin{bmatrix} 0.1\angle 0^\circ & 0.8\angle 90^\circ \\ 0.8\angle 90^\circ & 0.2\angle 0^\circ \end{bmatrix}$$

(a) Determine whether the network is reciprocal or lossless.
(b) If a short circuit is placed at port 2, what will be the resulting return loss at port 1?

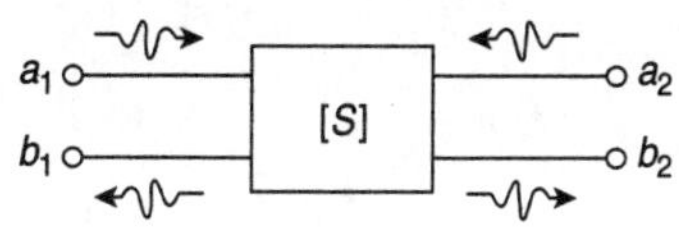

Short at port 2
means $-a_2 = b_2$

Figure (a) *A two-port network with a short at port 2*

Solution

(a) The given network is reciprocal, because $S_{12} = S_{21}$, which says that [S] is symmetry. To be lossless, the S parameters should satisfy

$$\sum_{i=1}^{n} S_{ij} S_{ik}^{*} = \begin{matrix} 1 & for\ j = k \\ 0 & for\ j \neq k \end{matrix} \Bigg\}$$

For $j = k$ (as per Eq. (7.12)) $|S_{11}|^2 + |S_{12}|^2 = (0.1)^2 + (0.8)^2 = 0.65$
Since the above summation is not equal to 1, the given network is not a lossless network.

(b) Reflected power at port 1, when port 2 is shorted, can be calculated as follows:
Since $a_2 = -b_2$ as port 2 is short circuited, thus

$$b_1 = S_{11}a_1 + S_{12}a_2 = S_{11}a_1 - S_{12}b_2 \tag{1}$$

$$b_2 = S_{21}a_1 + S_{22}a_2 = S_{21}a_1 - S_{22}b_2 \tag{2}$$

From (2), we have

$$b_2 = \frac{S_{21}}{1 + S_{22}} a_1 \tag{3}$$

Dividing (1) by a_1 and substituting the result in (3), we have

$$\Gamma = \frac{b_1}{a_1} = S_{11} - S_{12}\frac{b_2}{a_1} = S_{11} - \frac{S_{12}S_{21}}{1 + S_{22}} = 0.1 - \frac{(j0.8)(j0.8)}{1 + 0.2} = 0.633$$

$$\text{Returnloss} = -20\log\Gamma = -20\log(0.633) = 3.97\,\text{dB}$$

■

Examples of S Matrices

1-port S matrix

Consider 1-Port as simple lumped elements and cavities with one test port, long transmission lines or antennas. 1-port S-matrix is a matrix consisting of a single element, and the single element is the scattering parameter or reflection coefficient. It can be a 1×1 matrix; one row and one column.

- 1-port is characterized by their reflection coefficient (Γ), or in terms of S parameters, S_{11}.
- Ideal short for 1-port is given by $S_{11} = -1$
- Ideal termination is given by $S_{11} = 0$
- Active termination or reflection amplifier is given by $|S_{11}| > 1$

2-port S Matrix

An isolator is a 2-port device. Both the ports are perfectly matched to the junction, that is, the diagonal elements are zero. $S_{ii} = 0$, that is, $S_{11} = S_{22} = 0$, and the junction is non reciprocal ($S_{ij} \neq S_{ji}$). Hence, if the source is at port (1), its purpose is to prevent the reflected wave from port (2), that is, $S_{12} = 0$

The S matrix for an ideal isolator is $[S] = \begin{bmatrix} 0 & 0 \\ 1 & 0 \end{bmatrix}$

3-port S matrix

For a 3-port circulator, all the three ports are perfectly matched to the junction.
$\therefore S_{11} = S_{22} = S_{33} = 0$, and the junction is non reciprocal ($S_{ij} \neq S_{ji}$), $\therefore S_{12} = S_{23} = S_{31} = 0$

Hence, for *a* 3-port circulator, *a* 3×3 S matrix is $[S] = \begin{bmatrix} 0 & 0 & 1 \\ 1 & 0 & 0 \\ 0 & 1 & 0 \end{bmatrix}$ ■

EXAMPLE PROBLEM 7.3

Prove that the S matrix of a 3-port network cannot be matched, lossless, and reciprocal at the same time. Hint: Assume all three conditions are fulfilled in a 3×3 S matrix, and show that this will lead to a contradiction (Use proof by contradiction).

Solution

Assuming 3-port network as matched and reciprocal, the matrix is given by

$$S = \begin{bmatrix} 0 & S_{12} & S_{13} \\ S_{12} & 0 & S_{23} \\ S_{13} & S_{23} & 0 \end{bmatrix}$$

The diagonal elements are zero, and this matrix is symmetry. By the unitary condition (for a lossless 3-port network):

$$S.(S)^* = \begin{bmatrix} 1 & 0 & 0 \\ 0 & 1 & 0 \\ 0 & 0 & 1 \end{bmatrix}$$

$$\begin{bmatrix} 0 & S_{12} & S_{13} \\ S_{12} & 0 & S_{23} \\ S_{13} & S_{23} & 0 \end{bmatrix} \begin{bmatrix} 0 & S_{12}^* & S_{13}^* \\ S_{12}^* & 0 & S_{23}^* \\ S_{13}^* & S_{23}^* & 0 \end{bmatrix} = \begin{bmatrix} 1 & 0 & 0 \\ 0 & 1 & 0 \\ 0 & 0 & 1 \end{bmatrix}$$

Expanding the above equation,

$$|S_{12}|^2 + |S_{13}|^2 = 1 \quad \text{.......................................1}$$

$$|S_{12}|^2 + |S_{23}|^2 = 1 \quad \text{.......................................2}$$

$$|S_{13}|^2 + |S_{23}|^2 = 1 \quad \text{.......................................3}$$

$$S_{13}S_{23}^* = S_{12}S_{23}^* = S_{12}S_{23}^* = 0 \quad \text{........................4}$$

To fulfil Eq. (4) for arbitrary S parameters, S_{12}, S_{13}, and S_{23} should be 0. Eq. (3) cannot be fulfilled, by substituting this result in Eq. (3). This shows that a 3-port network with matched, reciprocal, and lossless conditions is wrong.

■

7.5 SCATTERING MATRIX CALCULATIONS FOR 3-PORT JUNCTION

A 3-port network S matrix contains 9 elements in a 3×3 arrangement. There are two types of 3-port junctions: E-plane tee and H-plane tee.

7.5.1 *E*-Plane Tee

There is a change of structure in the E plane (that is the side-arm port is in the E plane) so it is called an *E-plane tee*. It is also called as a *series junction* or *voltage junction*. In a long wave guide, a rectangular slot is cut along the broader dimension and a side arm is attached to it as shown in Figure 7.9. The *coplanar arms* are the two arms that are in line, while the other arm attached to it is called *sidearm* or *E arm* or *series arm*.

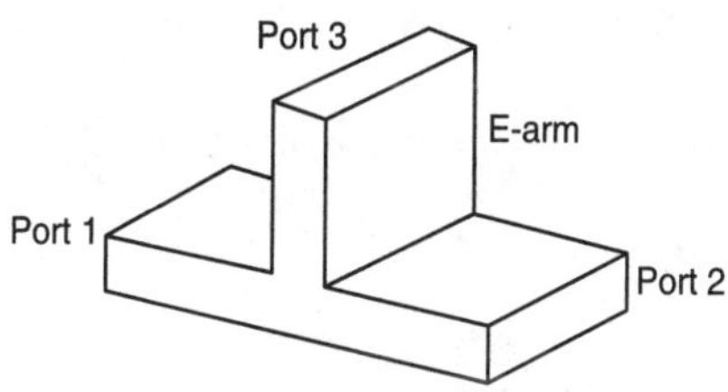

Figure 7.9 *E-plane Tee*

The scattering matrix can be derived as follows:

- Since the E-plane tee is a three-port junction, it should be a 3×3 square matrix as given below

$$[S] = \begin{bmatrix} S_{11} & S_{12} & S_{13} \\ S_{21} & S_{22} & S_{23} \\ S_{31} & S_{32} & S_{33} \end{bmatrix} \tag{7.32}$$

- Due to the plane of symmetry of junction, scattering coefficient outputs at port (1) and (2) are 180° out of phase with the input at port (3), $S_{23} = -S_{13}$.
- Symmetry property $S_{ij} = S_{ji}$

$$S_{12} = S_{21}, S_{13} = S_{31}, S_{23} = S_{32}$$

- If port (3) is perfectly matched to the junction, $\therefore S_{33} = 0$

- With the above properties, $[S]$ can be written as

$$[S]=\begin{bmatrix} S_{11} & S_{12} & S_{13} \\ S_{12} & S_{22} & -S_{13} \\ S_{13} & -S_{13} & 0 \end{bmatrix} \tag{7.33}$$

- That is, we have four unknown parameters as per Eq. (7.33).
- From the unitary property, $[S][S]^* = I$

$$\begin{bmatrix} S_{11} & S_{12} & S_{13} \\ S_{12} & S_{22} & -S_{13} \\ S_{13} & -S_{13} & 0 \end{bmatrix}\begin{bmatrix} S_{11}^* & S_{12}^* & S_{13}^* \\ S_{12}^* & S_{22}^* & -S_{13}^* \\ S_{13}^* & -S_{13}^* & 0 \end{bmatrix}=\begin{bmatrix} 1 & 0 & 0 \\ 0 & 1 & 0 \\ 0 & 0 & 1 \end{bmatrix}$$

Briefly, from the unitary property, we can say that
The sum of products of elements of the column of the S matrix is equal to 1.

$$R_1C_1 : S_{11}S_{11}^* + S_{12}S_{12}^* + S_{13}S_{13}^* = 1$$
$$|S_{11}|^2 + |S_{12}|^2 + |S_{13}|^2 = 1 \tag{7.34}$$

$$R_2C_2 : S_{12}S_{12}^* + S_{22}S_{22}^* + S_{13}S_{13}^* = 1$$
$$|S_{12}|^2 + |S_{22}|^2 + |S_{13}|^2 = 1 \tag{7.35}$$

$$R_3C_3 : S_{13}S_{13}^* + S_{13}S_{13}^* = 1$$
$$|S_{13}|^2 + |S_{13}|^2 = 1 \tag{7.36}$$

$$S_{13} = \frac{1}{\sqrt{2}} \tag{7.37}$$

Comparing Eqs. 7.34 and 7.35, we get $S_{11} = S_{22}$

$$R_3C_1 : S_{13}S_{11}^* - S_{13}S_{12}^* = 0$$
$$S_{13}\left(S_{11}^* - S_{12}^*\right) = 0$$
$$S_{13} \neq 0 \left(S_{11}^* - S_{12}^*\right) = 0$$
$$\therefore S_{11} = S_{12} = S_{22} \tag{7.38}$$

By substituting Eqs. (7.37) and (7.38) in Eq. (7.34), we get

$$|S_{11}|^2 + |S_{11}|^2 + \frac{1}{2} = 1 \quad \therefore S_{11} = \frac{1}{2}$$

Substituting the values of S_{11} S_{12} S_{13} S_{22} from the above equations, we get the resultant S matrix for the E-plane tee, which is

$$\therefore [S] = \begin{bmatrix} \frac{1}{2} & \frac{1}{2} & \frac{1}{\sqrt{2}} \\ \frac{1}{2} & \frac{1}{2} & \frac{-1}{\sqrt{2}} \\ \frac{1}{\sqrt{2}} & \frac{-1}{\sqrt{2}} & 0 \end{bmatrix} \quad (7.39)$$

7.5.2 *H*-Plane Tee

Since the side-arm port is in the *H* plane, it is called an *H-plane tee*. It is also called a *current junction*, *shunt junction*, or *parallel junction*. A rectangular slot is cut along the width (along *b*) of a long wave guide, and a side arm is attached as shown in Figure 7.10. The *coplanar arms* are the two arms that are in line, while the other arm which is attached to side is called a *side arm* or *H arm* or *shunt arm*.

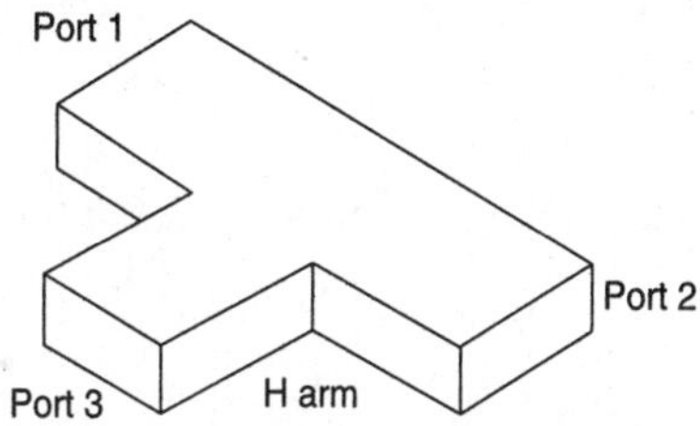

Figure 7.10 *H-plane Tee*

- The *H*-plane tee is a three-port junction. So, its *S* matrix should be a 3 × 3 matrix. Let it be

$$[S] = \begin{bmatrix} S_{11} & S_{12} & S_{13} \\ S_{21} & S_{22} & S_{23} \\ S_{31} & S_{32} & S_{33} \end{bmatrix} \quad (7.40)$$

- Since the plane of symmetry of junction $S_{13} = S_{23}$
- Symmetry property $S_{ij} = S_{ji}$

$$S_{12} = S_{21}, S_{13} = S_{31}, S_{23} = S_{32}$$

- Port (3) is perfectly matched to junction. $\therefore S_{33} = 0$
- With the above properties, [*S*] can be written as

$$[S] = \begin{bmatrix} S_{11} & S_{12} & S_{13} \\ S_{12} & S_{22} & S_{13} \\ S_{13} & S_{13} & 0 \end{bmatrix} \quad (7.41)$$

- That is, we have four unknown parameters as per Eq. (7.41).
- From unitary property, $[S][S]^* = I$

$$\begin{bmatrix} S_{11} & S_{12} & S_{13} \\ S_{12} & S_{22} & S_{13} \\ S_{13} & S_{13} & 0 \end{bmatrix}\begin{bmatrix} S_{11}^* & S_{12}^* & S_{13}^* \\ S_{12}^* & S_{22}^* & S_{13}^* \\ S_{13}^* & S_{13}^* & 0 \end{bmatrix} = \begin{bmatrix} 1 & 0 & 0 \\ 0 & 1 & 0 \\ 0 & 0 & 1 \end{bmatrix}$$

$$R_1C_1 \quad S_{11}S_{11}^* + S_{12}S_{12}^* + S_{13}S_{13}^* = 1$$

$$|S_{11}|^2 + |S_{12}|^2 + |S_{13}|^2 = 1 \tag{7.42}$$

$$R_2C_2 \quad S_{12}S_{12}^* + S_{22}S_{22}^* + S_{13}S_{13}^* = 1$$

$$|S_{12}|^2 + |S_{22}|^2 + |S_{13}|^2 = 1 \tag{7.43}$$

$$R_3C_3 \quad S_{13}S_{13}^* + S_{13}S_{13}^* = 1 \tag{7.44}$$

$$S_{13} = \frac{1}{\sqrt{2}} \tag{7.45}$$

Comparing Eqs. (7.44) and (7.45), we get $S_{11} = S_{22}$.

$$R_3C_1 \quad S_{13}S_{11}^* + S_{13}S_{12}^* = 0 \tag{7.46}$$

$$S_{13}\left(S_{11}^* + S_{12}^*\right) = 0 \quad S_{13} \neq 0$$

$$\therefore \quad S_{11} = -S_{12}$$

$$S_{12} = -S_{11}$$

Using these in Eq. (7.44), $\quad |S_{11}|^2 + |S_{11}|^2 + \frac{1}{2} = 1 \qquad \therefore S_{11} = \frac{1}{2}$

Substituting the values of S_{11} S_{12} S_{13} S_{22} from the above equations, we get the resultant S matrix for the H-plane tee, which is

$$\therefore [S] = \begin{bmatrix} \frac{1}{2} & \frac{-1}{2} & \frac{1}{\sqrt{2}} \\ \frac{-1}{2} & \frac{1}{2} & \frac{1}{\sqrt{2}} \\ \frac{1}{\sqrt{2}} & \frac{1}{\sqrt{2}} & 0 \end{bmatrix} \tag{7.47}$$

7.6 SCATTERING MATRIX CALCULATIONS FOR 4-PORT JUNCTION

The S matrix of a 4-port network consists of 16 elements in 4×4 square matrix. The matrix is symmetric and unitary. Compared to 3-port network, a 4-port network can be lossless, reciprocal, and matched at all ports simultaneously; i.e., the S matrix has the following form

7.6.1 Magic Tee

By connecting the sidewalls to the slots cut in the narrow wall and the broad wall of a piece of a waveguide, a magic tee is formed. At structure, it is a combination of the E-plane tee and the H-plane tee.

It is a kind of hybrid where the power is divided equally among the output ports. There can be a phase difference of 0° or a 180° in the outputs. One of the major benefits of the magic tee is that the power delivered to one port is not affected by the termination at the other output port given that the other port is match terminated.

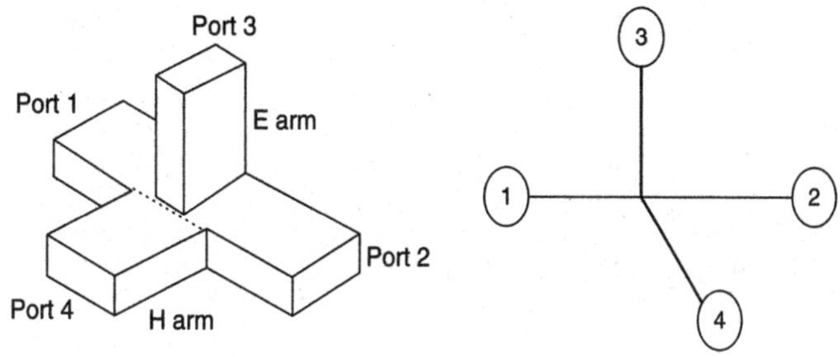

Figure 7.11 *Magic Tee*

Let the magic tee with the port designations as shown in the diagram (Figure 7.11).

- Since Magic tee is a four port junction, its s-matrix is a 4 × 4 square matrix. Let the matrix be

$$[S]=\begin{bmatrix} S_{11} & S_{12} & S_{13} & S_{14} \\ S_{21} & S_{22} & S_{23} & S_{24} \\ S_{31} & S_{32} & S_{33} & S_{34} \\ S_{41} & S_{42} & S_{43} & S_{44} \end{bmatrix} \tag{7.48}$$

- All the four ports are perfectly matched to the junction, so $S_{11} = S_{22} = S_{33} = S_{44} = 0$.
- The ports (1) (2) and (3) (4) are perfectly isolated; that is, $S_{12} = S_{21} = S_{34} = S_{43} = 0$, as the junction is ideal and it is reciprocal.
- So, the matrix is symmetrical; that is, $S_{ij} = S_{ji}$ for $i \neq j$; that is,

$$S_{23} = S_{32},\ S_{13} = S_{31},\ S_{24} = S_{42},\ S_{34} = S_{43},\ S_{41} = S_{14}\ \textit{and}\ S_{12} = S_{21}$$

- Due to the E-plane tee junction, $S_{23} = -S_{13}$, as scattering coefficient outputs at ports 1 and 2 are 180° out of phase with the input at port (3); the H-plane tee at port 4 is asymmetrical with ports 1 and 2, so $S_{24} = S_{14}$
- Incorporating the above aspects,

$$[S]=\begin{bmatrix} 0 & 0 & S_{13} & S_{14} \\ 0 & 0 & -S_{13} & S_{14} \\ S_{13} & -S_{13} & 0 & 0 \\ S_{14} & S_{14} & 0 & 0 \end{bmatrix} \tag{7.49}$$

- Since the junction is ideal, it should be lossless and its S matrix is unitary.

$$[S][S]^{*} = I$$

$$\begin{bmatrix} 0 & 0 & S_{13} & S_{14} \\ 0 & 0 & -S_{13} & S_{14} \\ S_{13} & -S_{13} & 0 & 0 \\ S_{14} & S_{14} & 0 & 0 \end{bmatrix} \begin{bmatrix} 0 & 0 & S_{13}^* & S_{14}^* \\ 0 & 0 & -S_{13}^* & S_{14}^* \\ S_{13}^* & -S_{13}^* & 0 & 0 \\ S_{14}^* & S_{14}^* & 0 & 0 \end{bmatrix} = \begin{bmatrix} 1 & 0 & 0 & 0 \\ 0 & 1 & 0 & 0 \\ 0 & 0 & 1 & 0 \\ 0 & 0 & 0 & 1 \end{bmatrix}$$

- $R_3C_3 : |S_{13}|^2 + |S_{13}|^2 = 1$, resulting in $S_{13} = \frac{1}{\sqrt{2}}$.

  $R_4C_4 : |S_{14}|^2 + |S_{14}|^2 = 1$, resulting in $S_{14} = \frac{1}{\sqrt{2}}$.

Then, the resultant S matrix for the magic tee is

$$[S] = \begin{bmatrix} 0 & 0 & \frac{1}{\sqrt{2}} & \frac{1}{\sqrt{2}} \\ 0 & 0 & -\frac{1}{\sqrt{2}} & \frac{1}{\sqrt{2}} \\ \frac{1}{\sqrt{2}} & -\frac{1}{\sqrt{2}} & 0 & 0 \\ \frac{1}{\sqrt{2}} & \frac{1}{\sqrt{2}} & 0 & 0 \end{bmatrix} \tag{7.50}$$

7.6.2 Directional Couplers

The directional coupler is a passive, reciprocal four-port network. In a four-port coupler, the power incident at one port (the input) is split between two other ports (the coupled and through ports), and little or no power emerges from the third port (isolated port). The schematic of the directional coupler is shown in Figure 7.12.

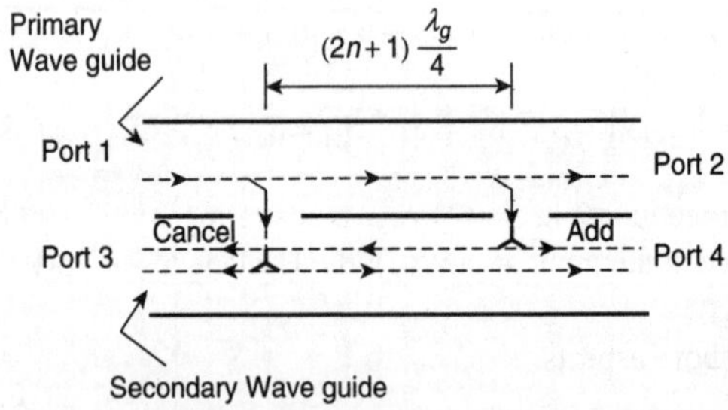

Figure 7.12 (a) *Schematic of a directional coupler*

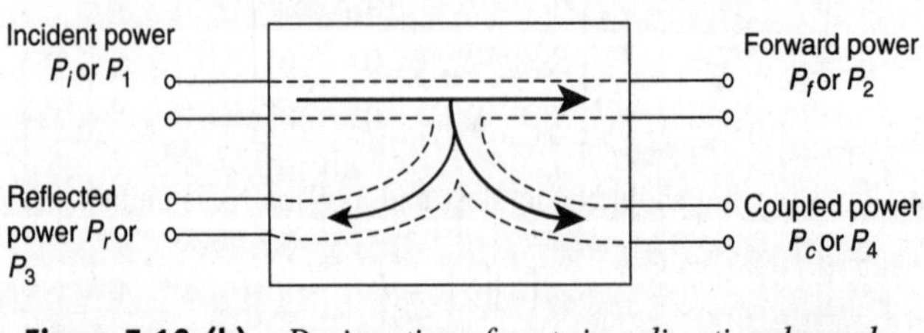

Figure 7.12 (b) *Designation of ports in a directional coupler*

The power incident at port1 is divided to through the port (port 2) and coupled port (port 4) but not to the isolated port (port 3).

Similarly,

power incident at port (2) is divided to (1) and (3) but not to (4)
power incident at port (3) is divided to (4) and (2) but not to (1)
power incident at port (4) is divided to (3) and (1) but not to (2)

- As the directional coupler is a four port junction its s-matrix is a 4×4 square matrix.

$$[S]=\begin{bmatrix} S_{11} & S_{12} & S_{13} & S_{14} \\ S_{21} & S_{22} & S_{23} & S_{24} \\ S_{31} & S_{32} & S_{33} & S_{34} \\ S_{41} & S_{42} & S_{43} & S_{44} \end{bmatrix} \tag{7.51}$$

- All four ports are perfectly matched to the junction; that is, all diagonal elements are zero.

$$S_{ii} = 0$$

- From the symmetry property, $S_{ij} = S_{ji}$

That is, $S_{23} = S_{32}, S_{13} = S_{31}, S_{24} = S_{42}, S_{34} = S_{43}, S_{41} = S_{14}$ *and* $S_{12} = S_{21}$

Ideally, there is no back power ($P_b = 0$); ports 1 and 3, ports 2 and 4 are decoupled (isolated)

$$S_{13} = S_{31} = 0$$
$$S_{24} = S_{42} = 0$$

$$\therefore [S]=\begin{bmatrix} 0 & S_{12} & 0 & S_{14} \\ S_{12} & 0 & S_{23} & 0 \\ 0 & S_{23} & 0 & S_{34} \\ S_{14} & 0 & S_{34} & 0 \end{bmatrix} \tag{7.52}$$

- $[S][S]^* = I$

$$\begin{bmatrix} 0 & S_{12} & 0 & S_{14} \\ S_{12} & 0 & S_{23} & 0 \\ 0 & S_{23} & 0 & S_{34} \\ S_{14} & 0 & S_{34} & 0 \end{bmatrix}\begin{bmatrix} 0 & S_{12}^* & 0 & S_{14}^* \\ S_{12}^* & 0 & S_{23}^* & 0 \\ 0 & S_{23}^* & 0 & S_{34}^* \\ S_{14}^* & 0 & S_{34}^* & 0 \end{bmatrix} = \begin{bmatrix} 1 & 0 & 0 & 0 \\ 0 & 1 & 0 & 0 \\ 0 & 0 & 1 & 0 \\ 0 & 0 & 0 & 1 \end{bmatrix}$$

$$R_1C_1 \qquad |S_{12}|^2 + |S_{14}|^2 = 1 \tag{7.53a}$$

$$R_2C_2 \qquad |S_{12}|^2 + |S_{23}|^2 = 1 \tag{7.53b}$$

$$R_3C_3 \qquad |S_{23}|^2 + |S_{34}|^2 = 1 \tag{7.53c}$$

$$R_1C_3 \qquad S_{12}S_{23}^* + S_{14}S_{34}^* = 0 \tag{7.53d}$$

Comparing Eqs. 7.53 *a* and 7.53 *b*, $S_{14} = S_{23}$
7.53 *b* and 7.53 *c*, $S_{12} = S_{34}$
Let us assume S_{12} is real and positive = '*p*'

$$S_{12} = S_{34} = p = S_{34}^*$$

From Eq. (7.53 *d*), $p\left(S_{23}^* + S_{23}\right) = 0$, S_{23} should be imaginary

$$\because p \neq 0, S_{23}^* + S_{23} = 0$$

or
$$S_{23}^* = -S_{23} = -jq \text{ or } S_{23} = jq$$

where *q* is real and positive.

$$S_{23} = jq = S_{14}$$

$$S_{12} = S_{34} = p$$

Substituting the values of unknown parameters from the above equations, we get the resultant *S* matrix for the directional coupler, which is

$$[S] = \begin{bmatrix} 0 & p & 0 & jq \\ p & 0 & jq & 0 \\ 0 & jq & 0 & p \\ jq & 0 & p & 0 \end{bmatrix} \tag{7.54}$$

EXAMPLE PROBLEM 7.4

Find the *S* parameters for a lossless 10 dB directional coupler. The directivity is 30 dB, and the VSWR at each port is 1.0 under matched condition.

Solution

Given D = 30 dB VSWR = 1.0

The coupling factor $C = -10\log\left(\frac{P_1}{P_4}\right) \Rightarrow 10 = -10\log\left(\frac{P_1}{P_4}\right)$

$$10^{-1} = \frac{P_1}{P_4} = \left|S_{41}^2\right|$$

$$S_{41} = \sqrt{0.1} \qquad \therefore S_{41} = 0.3162$$

For the lossless directional coupler under matched condition, $S_{41} = S_{14} = 0.3162$

Directivity
$$D = 10\log\left(\frac{P_4}{P_3}\right)$$

$$30 = 10\log\frac{|S_{41}^2|}{|S_{31}^2|}$$

$$\therefore 10^3 = \frac{|S_{41}^2|}{|S_{31}^2|}$$

$$|S_{31}^2| = \frac{|S_{41}^2|}{10^3} = \frac{(0.3162)^2}{10^3}$$

$$\therefore |S_{31}^2| = 10\times10^{-5} \quad \therefore S_{31} = 0.01$$

Hence,
$$S_{31} = S_{13} = 0.01$$

$$S_{11} = \frac{VSWR-1}{VSWR+1} = \frac{1-1}{1+1} = 0 \quad \therefore S_{11} = 0$$

Hence,
$$\therefore S_{11} = S_{22} = S_{33} = S_{44} = 0$$

The obtained S parameters are put in the S matrix given in Eq. (7.53):

$$\therefore [S] = \begin{bmatrix} 0 & S_{12} & 0.01 & 0.3162 \\ S_{12} & 0 & S_{23} & S_{24} \\ 0.01 & S_{23} & 0 & S_{34} \\ 0.3162 & S_{24} & S_{34} & 0 \end{bmatrix}$$

If the input power is provided to port 1, then

$$P_1 = P_2 + P_3 + P_4 \Rightarrow \frac{P_2}{P_1} + \frac{P_3}{P_1} + \frac{P_4}{P_1} = 1$$

$$\Rightarrow |S_{21}|^2 + |S_{31}|^2 + |S_{41}|^2 = 1$$

$$\Rightarrow |S_{21}|^2 = 1 - |S_{31}|^2 - |S_{41}|^2$$

$$\Rightarrow |S_{21}|^2 = 1 - 0.0001 - 0.1$$

$$|S_{21}|^2 = 0.8999 \; S_{21} = 0.9486$$

From the $[S][S]^* = I$, we have

$$S_{12}^2 + S_{23}^2 + S_{24}^2 = 1 \tag{A}$$

$$S_{23}^2 + S_{34}^2 + 10^{-4} = 1 \tag{B}$$

$$S_{24}^2 + S_{34}^2 + 10^{-1} = 1 \quad \text{(C)}$$

$$S_{12}^2 + S_{24}^2 - 10^{-2} = S_{34}^2 \quad \text{(A-B)}$$

$$S_{24}^2 + 10^{-1} = 1 - S_{34}^2$$

Substitute C in (A-B)

$$S_{12}^2 - 10^{-2} - 10^{-4} = 2S_{34}^2 - 1$$

$$\Rightarrow 2S_{34}^2 = 1 + S_{12}^2 - 10^{-1} - 10^{-4}$$

$$S_{34}^2 = 0.8999$$

$$S_{34} = S_{43} = 0.9486$$

From Eq. (1), we have

$$S_{23}^2 = 1 - 10^{-4} - S_{34}^2$$

$$= 1 - 10^{-4} - 0.8999$$

$$S_{23}^2 = 0.1$$

$$S_{23} = S_{32} = 0.3162$$

Also,

$$S_{24}^2 = 1 - 0.1 - S_{34}^2$$

$$= 1 - 0.1 - 0.8999$$

$$S_{24}^2 = 0.0001$$

$$S_{24} = S_{42} = 0.01$$

Hence, for the obtained S parameter, the S matrix is

$$S = \begin{bmatrix} 0 & 0.9486 & 0.01 & 0.3162 \\ 0.9486 & 0 & 0.3162 & 0.01 \\ 0.01 & 0.3162 & 0 & 0.9486 \\ 0.3162 & 0.01 & 0.9486 & 0 \end{bmatrix}$$

■

7.6.3 Rat-Race Coupler or Hybrid Ring Coupler

The hybrid ring coupler, also known as the *rat-race coupler*, is a four-port 3-dB directional coupler. The rat-race tee is shown in Figure 7.13.

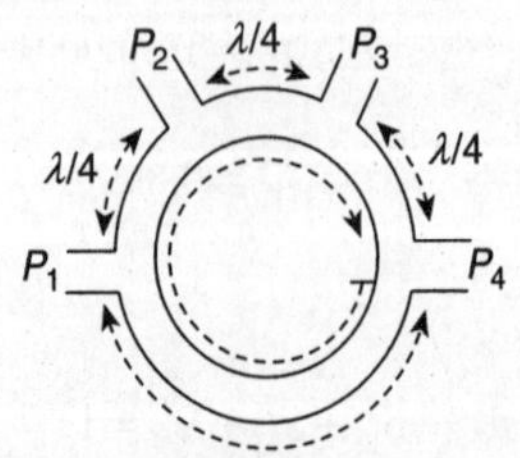

Figure 7.13 *Rat-race tee*

All the four ports of the rat-race coupler are placed one-quarter wavelength away from each other around the top half of the ring. However, the bottom half of the ring is three-quarter wavelengths in length. The characteristic impedance of the ring has factor $\sqrt{2}$ compared with port impedance.

A signal entering port 1 will be split between port 2 and port 4, and port 3 will be isolated. The scattering matrix for an ideal 3dB rat race is

$$S = \frac{-i}{\sqrt{2}}\begin{bmatrix} 0 & 1 & 0 & -1 \\ 1 & 0 & 1 & 0 \\ 0 & 1 & 0 & 1 \\ -1 & 0 & 1 & 0 \end{bmatrix} \tag{7.55}$$

Hybrid ring couplers are used to sum two in-phase signals without any loss. Similarly, they can be used to equally split an input signal without any phase difference between outputs and inputs. It is also possible to configure the coupler as a 180-degree phase-shifted output divider or to sum up two 180-degree phase-shifted signals with almost no loss.

7.7 SCATTERING MATRIX CALCULATIONS FOR FERRITE COMPONENTS

Unique transmission paths are provided by circulators and isolators. This makes the *RF* energy to pass in one direction with little (insertion) loss, and with high loss (isolation) in the other direction. An isolator can be a Strip line, Coaxial, Micro strip, or Waveguide type. The isolators are rated for powers from mill watts ranging upto megawatts. An important application of a circulator is in transmitting and receiving signals simultaneously as in duplexer, connecting a transmitter to port 1, antenna to port 2, and receiver to port 3.

7.7.1 Gyrators

A gyrator is a passive, linear, lossless, two-port network element. This device shifts the signal by 180 degrees.

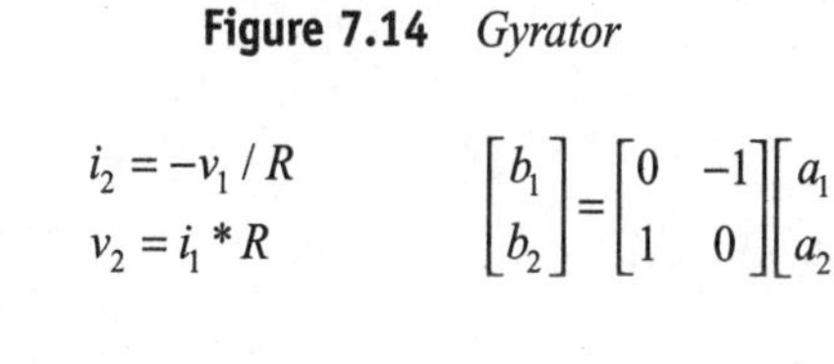

Figure 7.14 *Gyrator*

$$i_2 = -v_1 / R \qquad \begin{bmatrix} b_1 \\ b_2 \end{bmatrix} = \begin{bmatrix} 0 & -1 \\ 1 & 0 \end{bmatrix}\begin{bmatrix} a_1 \\ a_2 \end{bmatrix}$$
$$v_2 = i_1 * R$$

The input signal in a gyrator is shifted by 180° (π) at port 1. This shifted input signal is transmitted to the output at port 2. This transformation is given by the trans-impedance relations:

$$V_1/I_1 = R_1 + jX_1, -V_2/I_1 = R_2 + jX_2$$

The scattering matrix for the gyrator whose gyration resistance is chosen to be equal to the characteristic impedance of the two ports or to their geometric mean if these are not the same, is as below and thus the signal incident on port 2 will not be shifted

$$[S]=\begin{bmatrix}0 & -1\\ 1 & 0\end{bmatrix} \tag{7.56}$$

which is also asymmetric.

- It is a vital two-port, non-reciprocal component and has an 180° differential phase shift.
- Since a gyrator is a two-port component, the S matrix should be a 2×2 matrix.

$$[S]=\begin{bmatrix}S_{11} & S_{12}\\ S_{21} & S_{22}\end{bmatrix}$$

Two ports are perfectly matched to the junction; that is, the diagonal elements are zero $S_{ii}=0$,

i.e., $$S_{11}=S_{22}=0$$

- A gyrator is non reciprocal ($S_{ij} \neq S_{ji}$), and ports 1 and 2 have a 180° differential phase shift.

$$S_{21}=-S_{12}$$

- The S matrix is $[S]=\begin{bmatrix}0 & 1\\ -1 & 0\end{bmatrix}$.

7.7.2 Circulators

A microwave circulator has three or more ports and it is a non-reciprocal ferrite device. The input given at port n will come out at port $n+1$ but could not come at any other port. A *Y-junction circulator* is the widely used device which is a three-port ferrite junction circulator. It is a passive device. However it shows some characteristics that make it behave almost like a active device.

Clockwise circulator

A 3-port clockwise circulator is shown in Figure 7.15 (a).

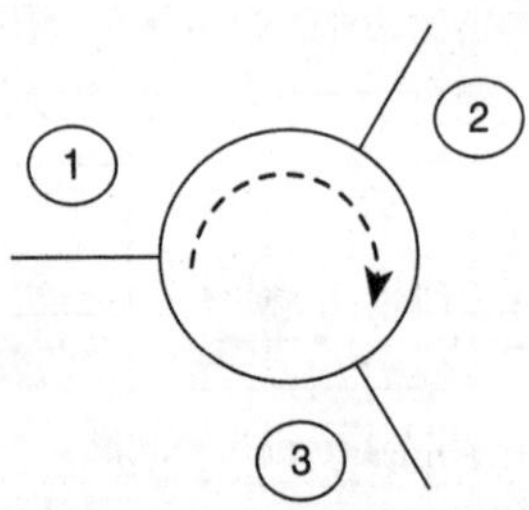

Figure 7.15 (a) *3-port circulator*

- For a 3-port circulator, the S matrix is a 3×3 matrix.

$$[S]=\begin{bmatrix}S_{11} & S_{12} & S_{13}\\ S_{21} & S_{22} & S_{23}\\ S_{31} & S_{32} & S_{33}\end{bmatrix} \tag{7.57}$$

All the three ports are perfectly matched to the junction. $\therefore S_{11}=S_{22}=S_{33}=0$

- A circulator obeys the non-reciprocity and lossless condition, since it has ferrite material at the junction.

- From the unitary property, $[S][S]^* = I$

$$\begin{bmatrix} 0 & S_{12} & S_{13} \\ S_{21} & 0 & S_{23} \\ S_{31} & S_{32} & 0 \end{bmatrix}\begin{bmatrix} 0 & S_{12}^* & S_{13}^* \\ S_{21}^* & 0 & S_{23}^* \\ S_{31}^* & S_{32}^* & 0 \end{bmatrix} = \begin{bmatrix} 1 & 0 & 0 \\ 0 & 1 & 0 \\ 0 & 0 & 1 \end{bmatrix}$$

$$\begin{aligned} S_{12}S_{12}^* + S_{13}S_{13}^* &= 1 \\ S_{21}S_{21}^* + S_{23}S_{23}^* &= 1 \\ S_{31}S_{31}^* + S_{32}S_{32}^* &= 1 \\ S_{13}S_{23}^* = S_{12}S_{32}^* = S_{21}S_{31}^* &= 0 \end{aligned} \tag{7.58}$$

Let us consider $S_{21} \neq 0$; from the above Eq. (7.58), we get

$$S_{13} = S_{21} = S_{32} = 1$$
$$S_{31} = S_{12} = S_{23} = 0$$

Substituting the values of unknown parameters from the above equations, we get the resultant S matrix for the circulator, which is

$$[S] = \begin{bmatrix} 0 & 0 & 1 \\ 1 & 0 & 0 \\ 0 & 1 & 0 \end{bmatrix} \tag{7.59}$$

Anti-clockwise circulator

A 3-port anti-clockwise circulator is shown in Figure 7.15 (b).

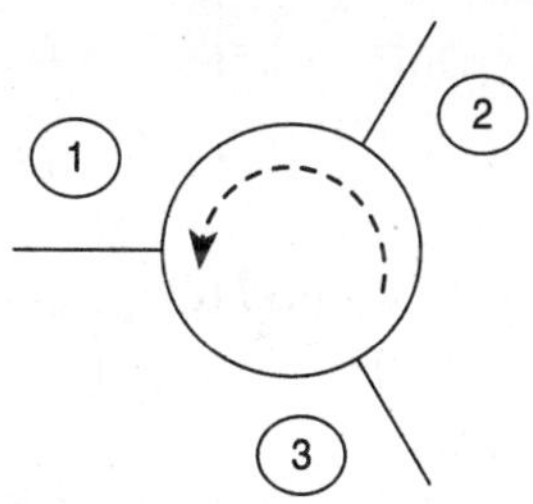

Figure 7.15 (b) *Anti-clockwise circulator*

For an anti-clockwise circulator from non-reciprocal

$$S_{13} = S_{21} = S_{32} = 0,$$
$$S_{31} = S_{12} = S_{23} = 1$$

We get the resultant S matrix for the anti-clockwise circulator, which is

$$[S] = \begin{bmatrix} 0 & 1 & 0 \\ 0 & 0 & 1 \\ 1 & 0 & 0 \end{bmatrix} \tag{7.60}$$

EXAMPLE PROBLEM 7.5

Find the S matrix for a three-port port circulator with an insertion loss of 0.5 dB, an isolation of 15 dB, and a VSWR of 1.5.

Solution

For a three port circulator the S matrix is of the form explained above,

$$[S] = \begin{bmatrix} S_{11} & S_{12} & S_{13} \\ S_{21} & S_{22} & S_{23} \\ S_{31} & S_{32} & S_{33} \end{bmatrix}$$

Insertion loss = $-20\log|S_{21}|$

$$0.5 = -20\log|S_{21}|$$

$$S_{21} = 10^{\frac{-1}{40}}\ S_{21} = 0.944$$

Similarly, $$|S_{21}| = |S_{13}| = |S_{32}| = 0.944$$

Isolation = $-20\log|S_{12}|$

$$15 = -20\log|S_{12}|$$

$$|S_{12}| = 10^{\frac{-15}{20}}\ \therefore S_{12} = 0.1778$$

Similarly, $$|S_{31}| = |S_{23}| = |S_{12}| = 0.1778$$

$$\rho = \frac{s-1}{s+1} = \frac{1.5-1}{1.5+1} = 0.2$$

$$|S_{11}| = |S_{22}| = |S_{33}| = 0.2$$

Hence, the scattering matrix is $$[S] = \begin{bmatrix} 0\cdot2 & 0\cdot1778 & 0\cdot944 \\ 0\cdot944 & 0\cdot2 & 0\cdot1778 \\ 0\cdot1778 & 0\cdot944 & 0\cdot2 \end{bmatrix}$$

■

7.7.3 Isolators

An isolator is a two-port device with input and output ports. It is nothing but a circulator with third port being terminated. An isolator is shown in Figure 7.16.

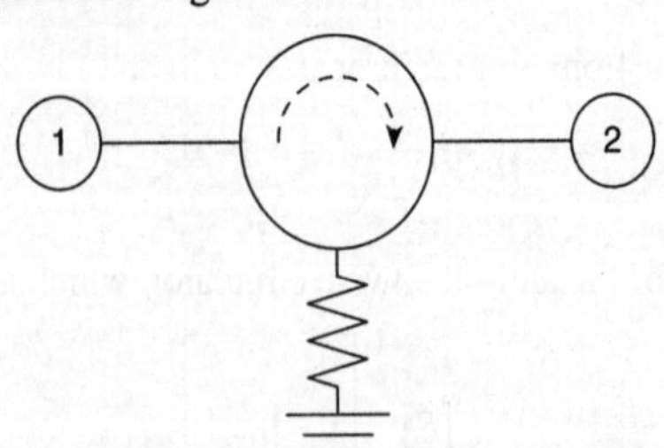

Figure 7.16 *Isolator*

- Since an isolator is a two-port component, the S matrix should be a 2×2 matrix.

$$[S] = \begin{bmatrix} S_{11} & S_{12} \\ S_{21} & S_{22} \end{bmatrix}$$

- Two ports are perfectly matched to the junction; that is, the diagonal elements are zero.

$$S_{ii} = 0$$

i.e., $$S_{11} = S_{22} = 0$$

- Isolators usually have ferrite material at the junction to cause the non-reciprocity condition. When the S matrix is non reciprocal ($S_{ij} \neq S_{ji}$), the conditions of port match and losslessness apply.
- In microwave applications, an isolator is used between a high power source and a load which prevents possible reflections that may damage the source. Isolators always follow the source. If the source is at port 1, its purpose is to prevent the reflected wave from port 2; that is, $S_{12} = 0$

  The S matrix for an ideal isolator is $[S] = \begin{bmatrix} 0 & 0 \\ 1 & 0 \end{bmatrix}$

- If the source is at port 2, its purpose is to prevent the reflected wave from port 1; that is, $S_{21} = 0$

  The S matrix for an ideal isolator is $[S] = \begin{bmatrix} 0 & 1 \\ 0 & 0 \end{bmatrix}$ (7.61)

EXAMPLE PROBLEM 7.6

Find the S matrix for a matched isolator having an insertion loss of 0.5 dB and an isolation of 25 dB.

Solution

The S matrix for the isolator is of the form explained above:

$$[S] = \begin{bmatrix} S_{11} & S_{12} \\ S_{21} & S_{22} \end{bmatrix}$$

Insertion loss $$= -20 \log |S_{21}|$$

$$0.5 = -20 \log |S_{21}|$$

$$S_{21} = 10^{\frac{-0.51}{20}} \therefore S_{21} = 0.944$$

Similarly, isolation $$= -20 \log |S_{12}|$$

$$25 = -20 \log |S_{12}|$$

$$|S_{12}| = 10^{\frac{-25}{20}} \therefore S_{12} = 0.0562$$

and $$|S_{11}| = |S_{22}| = 0.$$

Hence, the scattering matrix for the given isolator is

$$[S] = \begin{bmatrix} 0 & 0.0562 \\ 0.944 & 0 \end{bmatrix}$$

■

7.7.4 S Matrix for an Ideal Attenuator

Ideal reciprocal attenuator with the attenuation α in Neper/m:

$$S = \begin{pmatrix} 0 & e^{-\alpha} \\ e^{-\alpha} & 0 \end{pmatrix}$$

The attenuation in Decibel is given by $A = -20\log_{10}(S_{21})$, 1 Neper = 8.686 dB. An attenuator can be realized with three resistors in a T circuit or with resistive material in a waveguide.

7.7.5 S Matrix for an Ideal Amplifier

Ideal amplifier $$S = \begin{pmatrix} 0 & 0 \\ G & 0 \end{pmatrix}$$

with the voltage gain $G > 1$. Please note the similarity between an ideal amplifier and an ideal isolator.

7.7.6 S Matrix for an Ideal Transmission Line of Length L

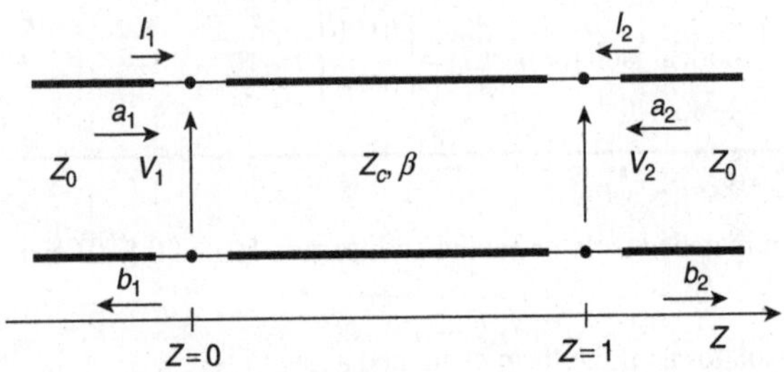

Figure 7.17 *Ideal transmission line*

For an ideal transmission line of length l can be regarded as a ideal two port network with incident waves a_1, a_2 and outgoing waves b_1, b_2.

Since there is no reflections at the two ports

$$b_2 = a_1 e^{-\gamma l}$$

$$b_1 = a_2 e^{-\gamma l}$$

where $\gamma = \alpha + j\beta$ is the complex propagation constant, α is line attenuation in [Neper/m] and $\beta = 2\pi/\lambda$ with the wavelength λ.

For a perfectly matched $s_{11} = s_{22} = 0$

$$s_{12} = \frac{b_2}{a_1} = e^{-\gamma l}$$

$$s_{21} = \frac{b_1}{a_2} = e^{-\gamma l}$$

$$S = \begin{pmatrix} 0 & e^{-\gamma l} \\ e^{-\gamma l} & 0 \end{pmatrix}$$

Since there is no attenuation $\alpha = 0$

then
$$S = \begin{pmatrix} 0 & e^{-j\beta l} \\ e^{-j\beta l} & 0 \end{pmatrix} \tag{7.62}$$

7.7.7 S Matrix for an Ideal Phase Shifter

Ideal phase shifter

$$S = \begin{pmatrix} 0 & e^{-j\phi_{12}} \\ e^{-j\phi_{21}} & 0 \end{pmatrix} \tag{7.63}$$

In the gyrator, the phases are related as $\varphi_{12} = \varphi_{21} + \pi$ but in a reciprocal phase shifter they are equal as given by $\varphi_{12} = \varphi_{21}$. The difference between gyrator and an ideal gyrator is that the ideal gyrator is lossless ($S^*S = 1$), but it is not reciprocal. Gyrators are implemented using active electronic components and magnetically saturated ferrite elements are used to implement passive gyrators in the microwave range.

7.8 CHARACTERIZING THE NETWORK USING *Z, Y, h* AND ABCD PARAMETERS

There are several ways to characterize the network. In this section we discuss the *Z*, *Y*, *h* and ABCD parameters that characterize the network at low frequencies. Figure 7.18 shows a general two-port network. I_1 and I_2 are input and output currents respectively. V_1 and V_2 are input and output voltages respectively.

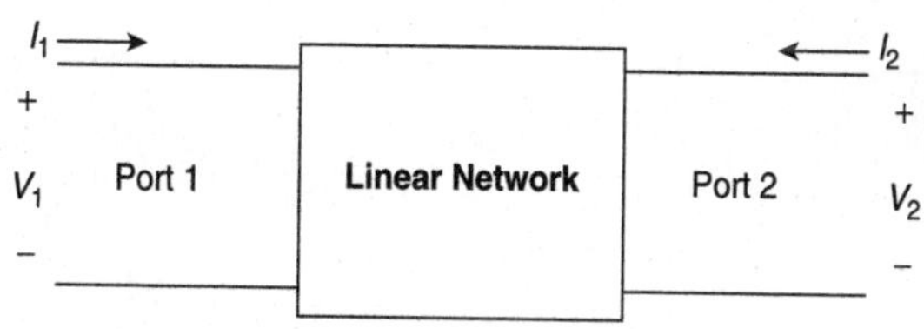

Figure 7.18 *Two-port network*

7.8.1 *Z* Parameters

Z parameters are also known as *impedance parameters*. Consider the two-port network shown in Figure 7.18. Since the network is linear, the superposition principle can be applied. Assuming that it contains no independent sources, voltage V_1 at port 1 can be expressed in terms of two currents as follows:

$$V_1 = Z_{11}I_1 + Z_{12}I_2$$

Since V_1 is in volts and I_1 and I_2 are in amperes, parameters Z_{11} and Z_{12} should be in ohms. Therefore, these are called the *impedance parameters*. Similarly, we can write V_2 in terms of I_1 and I_2 as follows:

$$V_2 = Z_{21}I_1 + Z_{22}I_2$$

Using the matrix representation, we can write

$$\begin{bmatrix} V_1 \\ V_2 \end{bmatrix} = \begin{bmatrix} Z_{11} & Z_{12} \\ Z_{21} & Z_{22} \end{bmatrix} \begin{bmatrix} I_1 \\ I_2 \end{bmatrix}$$

or

$$[V]=[Z][I]$$

where $[Z]$ is called the *impedance matrix* of the two-port network.

The current is the independent variable that is set to zero by using open-circuit terminations. These parameters are, therefore, called the *open-circuit impedance parameters*, and they are shown in the following equations:

$$Z_{11}=\frac{V_1}{I_1}(I_2=0)$$

$$Z_{12}=\frac{V_1}{I_2}(I_1=0)$$

$$Z_{21}=\frac{V_2}{I_1}(I_2=0)$$

$$Z_{22}=\frac{V_2}{I_2}(I_1=0)$$

Z_{11} is the input impedance with the output port terminated in an open circuit ($I_2 = 0$). This may be measured, for example, by placing a voltage V_1 across port 1 and measuring I_1. Similarly, Z_{22} is the output impedance with the input terminals open circuited. We can obtain Z_{21} (the forward transfer impedance) when the output terminal is open circuited. Similarly we can obtain Z_{12} (the reverse transfer impedance) when the input port is open.

The Z parameters are also called as open-circuit parameters since we need to open the ports to calculate Z_{11}, Z_{12}, Z_{21} and Z_{22}. These parameters obtain the relation between output currents and their input voltages.

EXAMPLE PROBLEM 7.7

Find the impedance parameters for the two-port network shown in the figure.

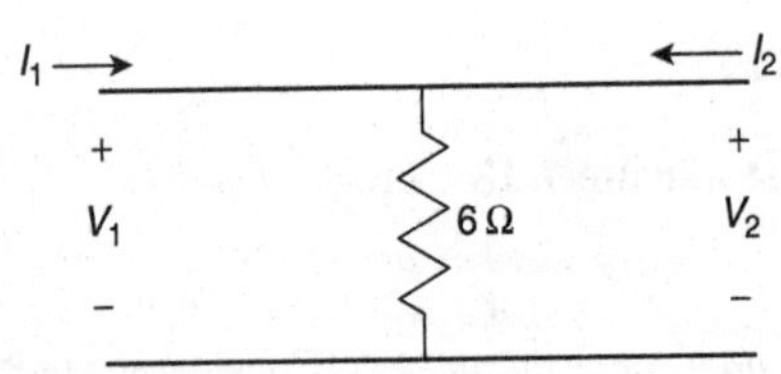

Solution

From the above figure, the impedance parameters are

$$V_2=V_1=6I_1$$

$$Z_{11}=\left.\frac{V_1}{I_1}\right|_{I_2=0}=\frac{6I_1}{I_1}=6\ \Omega$$

$$Z_{21}=\left.\frac{V_2}{I_1}\right|_{I_2=0}=\frac{6I_1}{I_1}=6\ \Omega$$

nd $$V_2 = V_1 = 6I_2$$

$$Z_{12} = \left.\frac{V_1}{I_2}\right|_{I_1=0} = \frac{6I_2}{I_2} = 6\ \Omega$$

$$Z_{22} = \left.\frac{V_2}{I_2}\right|_{I_1=0} = \frac{6I_2}{I_2} = 6\ \Omega$$

herefore, the impedance matrix is

$$\begin{bmatrix} Z_{11} & Z_{12} \\ Z_{21} & Z_{22} \end{bmatrix} = \begin{bmatrix} 6 & 6 \\ 6 & 6 \end{bmatrix}$$

■

EXAMPLE PROBLEM 7.8

ind the impedance parameters for the two-port network shown in the figure.

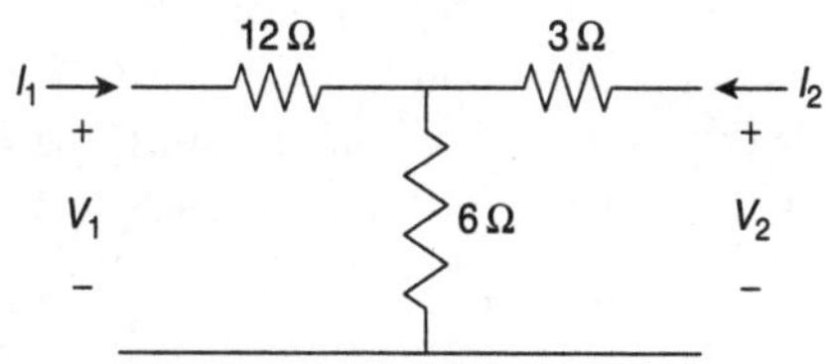

olution

rom the above figure, the impedance parameters are

$$Z_{11} = \left.\frac{V_1}{I_1}\right|_{I_2=0} = \frac{18I_1}{I_1} = 18\ \Omega \quad Z_{21} = \left.\frac{V_2}{I_1}\right|_{I_2=0} = \frac{6I_1}{I_1} = 6\ \Omega$$

$$V_2 = (6+3)I_2 = 9I_2 \text{ and } V_1 = 6I_2$$

$$Z_{12} = \left.\frac{V_1}{I_2}\right|_{I_1=0} = \frac{6I_2}{I_2} = 6\ \Omega \quad Z_{22} = \left.\frac{V_2}{I_2}\right|_{I_1=0} = \frac{9I_2}{I_2} = 9\ \Omega$$

herefore, the impedance matrix is

$$\begin{bmatrix} Z_{11} & Z_{12} \\ Z_{21} & Z_{22} \end{bmatrix} = \begin{bmatrix} 18 & 6 \\ 6 & 9 \end{bmatrix}$$

■

.8.2 *Y* Parameters

parameters are also called as *admittance parameters*. These are also called as *short-circuit Y parameters*, ince V_1 and V_2 are independent variables (as shown in Figure 7.18). They are good for reverse-biased ollector base junctions, but not so good for forward-biased base emitter junctions so they are useful or measuring higher impedance circuits. Capacitors should be used as the load for active circuits. The -parameter matrix for a two-port network is, therefore,

$$I_1 = y_{11}V_1 + y_{12}V_2$$

$$I_2 = y_{21}V_1 + y_{22}V_2$$

Using the matrix representation, we can write

$$\begin{bmatrix} I_1 \\ I_2 \end{bmatrix} = \begin{bmatrix} y_{11} & y_{12} \\ y_{21} & y_{22} \end{bmatrix} \begin{bmatrix} V_1 \\ V_2 \end{bmatrix}$$

or

$$[I] = [Y][V]$$

where [*Y*] is called the *admittance matrix* of the two-port network.

The voltage is the independent variable that is set to zero by using *S/C* terminations. These parameters are therefore, called the *S/C admittance parameters*. These parameters are shown in the following equations:

$$y_{11} = \left.\frac{I_1}{V_1}\right|_{V_2=0} \qquad y_{22} = \left.\frac{I_2}{V_2}\right|_{V_1=0}$$

$$y_{21} = \left.\frac{I_2}{V_1}\right|_{V_2=0} \qquad y_{12} = \left.\frac{I_1}{V_2}\right|_{V_1=0}$$

y_{11} is the input admittance with the output port terminated at *S/C* ($V_2 = 0$). This may be measured for example, by placing a current I_1 across port 1 and by measuring V_1. Similarly, y_{22} is the output admittance with the input terminals that are short circuited. y_{21} is the forward transfer admittance with the output terminal that is short circuited, and y_{12} is the reverse transfer admittance with the input port terminated at *S/C*.

The *Y* parameters are also called as short-circuit parameters because the output port is shorted to calculate Y_{11}, Y_{12}, Y_{21} and Y_{22}. These parameters obtain the relation between output voltages and their input currents.

EXAMPLE PROBLEM 7.9

Find the admittance parameters for the two-port network shown in the figure.

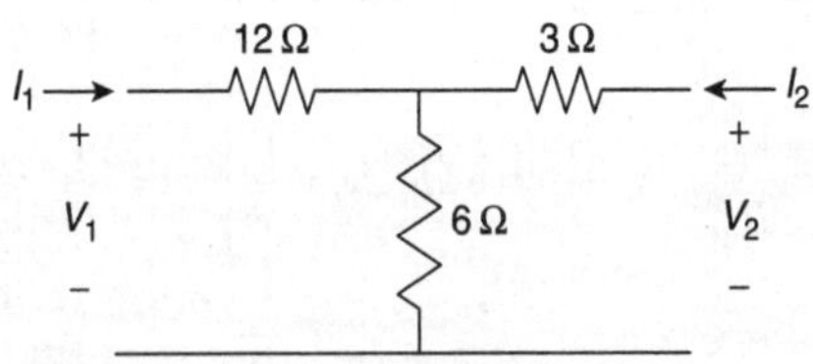

Solution

From problem 7.9, we know that the impedance matrix is

$$\begin{bmatrix} Z_{11} & Z_{12} \\ Z_{21} & Z_{22} \end{bmatrix} = \begin{bmatrix} 18 & 6 \\ 6 & 9 \end{bmatrix}$$

$$|Z| = Z_{11}Z_{22} - Z_{12}Z_{21} = 126$$

$$y_{11} = \frac{Z_{22}}{|Z|} = \frac{9}{126} = 0.072$$

$$y_{12} = \frac{-Z_{12}}{|Z|} = -0.048$$

$$y_{21} = \frac{-Z_{21}}{|Z|} = -0.048$$

$$y_{22} = \frac{Z_{11}}{|Z|} = 0.143$$

■

7.8.3 *h*-Parameters

The *h*-parameters are also called *hybrid parameters*. A two-port network can be represented using the h-parameters. If the circuit to be measured has a fairly low input impedance and a fairly high output impedance as in the case of common emitter or common base configurations, we require the following for the greatest accuracy of measurement: *A S/C* at the output, so V_2 is the independent variable and an open circuit on the input, so I_1 is the independent variable.

The describing equations for the *h* parameters are

$$V_1 = h_{11}I_1 + h_{12}V_2$$
$$I_2 = h_{21}I_1 + h_{22}V_2$$

where

I_1 and V_2 are independent variables, and
V_1 and I_2 are dependent variables.

Using the matrix representation, we can write

$$\begin{bmatrix} V_1 \\ I_2 \end{bmatrix} = \begin{bmatrix} h_{11} & h_{12} \\ h_{21} & h_{22} \end{bmatrix} \begin{bmatrix} I_1 \\ V_2 \end{bmatrix}$$

The *h* parameters can be found as follows:

$$h_{11} = \left.\frac{V_1}{I_1}\right|_{V_2=0} \quad h_{12} = \left.\frac{V_1}{V_2}\right|_{I_1=0}$$

$$h_{21} = \left.\frac{I_2}{I_1}\right|_{V_2=0} \quad h_{22} = \left.\frac{I_2}{V_2}\right|_{I_1=0}$$

The *h* parameters contain both open-circuit parameters ($I_1 = 0$) and short-circuit parameters ($V_2 = 0$), so they are also called as *hybrid parameters*.

EXAMPLE PROBLEM 7.10

Find the *h* parameters for the two-port network shown in the figure.

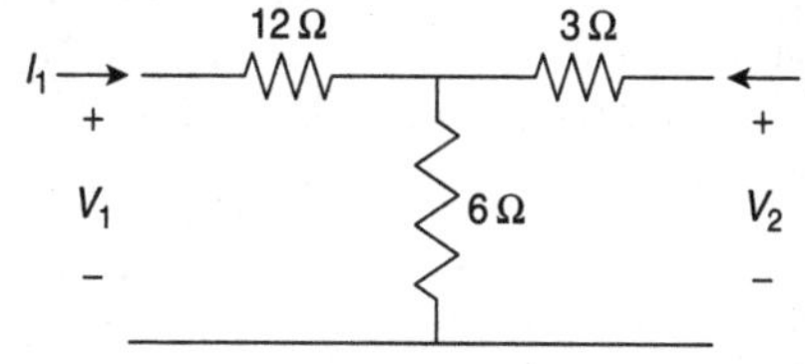

Solution

$$h_{11} = \left.\frac{V_1}{I_1}\right|_{V_2=0} = 14$$

$$h_{12} = \left.\frac{V_1}{V_2}\right|_{I_1=0} = 0.67$$

$$h_{21} = \left.\frac{I_2}{I_1}\right|_{V_2=0} = 0.67$$

$$h_{22} = \left.\frac{I_2}{V_2}\right|_{I_1=0} = 0.11$$

$$\begin{bmatrix} h_{11} & h_{12} \\ h_{21} & h_{22} \end{bmatrix} = \begin{bmatrix} 14 & 0.67 \\ 0.67 & 0.11 \end{bmatrix}$$

7.8.4 ABCD Parameters

A two-port network can be described by transmission parameters. The transmission matrix describes the network in terms of both voltage and current waves (similar to a Thevinen Equivalent). The describing equations are

$$V_1 = AV_2 - BI_2$$
$$I_1 = CV_2 - DI_2$$

where
V_2 and I_2 are independent variables, and
V_1 and I_1 are dependent variables.
In matrix form, the above two equations can be written as

$$\begin{bmatrix} V_1 \\ I_1 \end{bmatrix} = \begin{bmatrix} A & B \\ C & D \end{bmatrix} \begin{bmatrix} V_2 \\ -I_2 \end{bmatrix}$$

The transmission parameters can be found as

$$A = \left.\frac{V_1}{V_2}\right|_{I_2=0} \qquad B = -\left.\frac{V_1}{I_2}\right|_{V_2=0}$$

$$C = \left.\frac{I_1}{V_2}\right|_{I_2=0} \qquad D = -\left.\frac{I_1}{I_2}\right|_{V_2=0}$$

In the transmission parameters, the primary variables which are at the sending end, V_1 and I_1 are given in terms of the secondary variables V_2 and $-I_2$ which are at the receiving end. The minus sign indicates that the current is entering the load at the receiving end. ABCD matrix is used here to represent the ports in terms of currents and voltages because it is the most appropriate for cascading elements

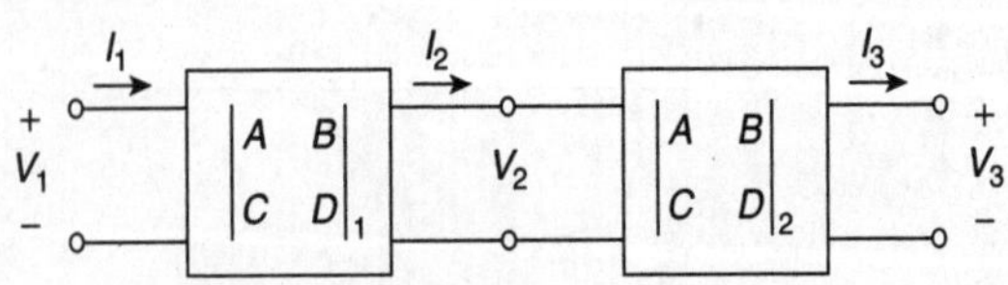

Figure 7.19

- Mathematically, the matrices are cascaded by multiplication:

$$\begin{bmatrix} V_1 \\ I_1 \end{bmatrix} = \begin{bmatrix} A & B \\ C & D \end{bmatrix}_1 \cdot \begin{bmatrix} V_2 \\ I_2 \end{bmatrix}$$
$$\begin{bmatrix} V_2 \\ I_2 \end{bmatrix} = \begin{bmatrix} A & B \\ C & D \end{bmatrix}_2 \cdot \begin{bmatrix} V_3 \\ I_3 \end{bmatrix} \Longrightarrow \begin{bmatrix} V_1 \\ I_1 \end{bmatrix} = \begin{bmatrix} A & B \\ C & D \end{bmatrix}_1 \cdot \begin{bmatrix} A & B \\ C & D \end{bmatrix}_2 \cdot \begin{bmatrix} V_3 \\ I_3 \end{bmatrix}$$

- In the frequency domain, this is the best method to cascade elements.
- It is accurate and easy to use.

EXAMPLE PROBLEM 7.11

Find the transmission parameters for the two-port network shown in the figure.

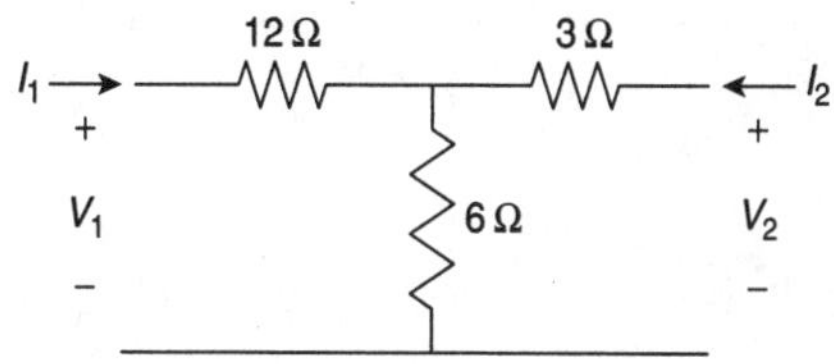

Solution

From problem 7.8, we know that the impedance matrix is

$$\begin{bmatrix} Z_{11} & Z_{12} \\ Z_{21} & Z_{22} \end{bmatrix} = \begin{bmatrix} 18 & 6 \\ 6 & 9 \end{bmatrix}$$

So, the transmission parameters are

$$|Z| = Z_{11}Z_{22} - Z_{12}Z_{21} = 126$$

$$A = \frac{Z_{11}}{|Z_{21}|} = \frac{18}{6} = 3$$

$$B = \frac{|Z|}{Z_{21}} = \frac{126}{6} = 21$$

$$C = \frac{1}{Z_{21}} = 0.167$$

$$D = \frac{Z_{22}}{Z_{21}} = 1.5$$

■

7.8.5 Parameter Conversion

The two-port network parameters can be inter-converted from one to another. Table 7.2 gives the relationship between the *Z*, *Y*, ABCD, and *S* parameters.

Table 7.2 Conversion for Z, Y, ABCD, and S parameters.

	$[Z]$	$[Y]$	[ABCD]	[S]
Z_{11}	Z_{11}	$\frac{y_{22}}{\lvert Y \rvert}$	$\frac{A}{C}$	$Z_0 \frac{(1+S_{11})(1-S_{22})+S_{12}S_{21}}{(1-S_{11})(1-S_{22})-S_{12}S_{21}}$
Z_{12}	Z_{12}	$\frac{-y_{12}}{\lvert Y \rvert}$	$\frac{AD-BC}{C}$	$Z_0 \frac{2S_{12}}{(1-S_{11})(1-S_{22})-S_{12}S_{21}}$
Z_{21}	Z_{21}	$\frac{-y_{21}}{\lvert Y \rvert}$	$\frac{1}{C}$	$Z_0 \frac{2S_{21}}{(1-S_{11})(1-S_{22})-S_{12}S_{21}}$
Z_{22}	Z_{22}	$\frac{y_{11}}{\lvert Y \rvert}$	$\frac{D}{C}$	$Z_0 \frac{(1-S_{11})(1+S_{22})+S_{12}S_{21}}{(1-S_{11})(1-S_{22})-S_{12}S_{21}}$
y_{11}	$\frac{Z_{22}}{\lvert Z \rvert}$	y_{11}	$\frac{D}{B}$	$Y_0 \frac{(1-S_{11})(1+S_{22})+S_{12}S_{21}}{(1+S_{11})(1+S_{22})-S_{12}S_{21}}$
y_{12}	$\frac{-Z_{12}}{\lvert Z \rvert}$	y_{12}	$\frac{BD-AC}{B}$	$Y_0 \frac{-S_{12}}{(1+S_{11})(1+S_{22})-S_{12}S_{21}}$
y_{21}	$\frac{-Z_{21}}{\lvert Z \rvert}$	y_{21}	$\frac{-1}{B}$	$Y_0 \frac{-2S_{21}}{(1+S_{11})(1+S_{22})-S_{12}S_{21}}$
y_{22}	$\frac{Z_{11}}{\lvert Z \rvert}$	y_{22}	$\frac{A}{B}$	$Y_0 \frac{(1+S_{11})(1-S_{22})+S_{12}S_{21}}{(1+S_{11})(1+S_{22})-S_{12}S_{21}}$
A	$\frac{Z_{11}}{Z\ldots}$ [illegible]	$-\frac{y_{22}}{\ldots}$ [illegible]	A	$Y_0 \frac{(1+S_{11})(1-S_{22})+S_{12}S_{21}}{2S\ldots}$ [illegible]

Table 7.2 (Continued)

B	$\dfrac{\lvert Z\rvert}{Z_{21}}$	$-\dfrac{1}{y_{21}}$	B	$Z_0\dfrac{(1+S_{11})(1+S_{22})-S_{12}S_{21}}{2S_{21}}$
C	$\dfrac{1}{Z_{21}}$	$-\dfrac{\lvert Y\rvert}{y_{21}}$	C	$Z_0\dfrac{(1-S_{11})(1-S_{22})-S_{12}S_{21}}{2S_{21}}$
D	$\dfrac{Z_{22}}{Z_{21}}$	$-\dfrac{y_{11}}{y_{21}}$	D	$Z_0\dfrac{(1-S_{11})(1+S_{22})+S_{12}S_{21}}{2S_{21}}$
S_{11}	$\dfrac{(Z_{11}-Z_0)(Z_{22}+Z_0)-Z_{12}Z_{21}}{\Delta Z}$	$\dfrac{(Y_0-y_{11})(Y_0+y_{22})+y_{12}y_{21}}{\Delta Y}$	$\dfrac{A+\dfrac{B}{Z_0}-CZ_0-D}{A+\dfrac{B}{Z_0}+CZ_0+D}$	S_{11}
S_{12}	$\dfrac{2Z_{12}Z_0}{\Delta Z}$	$\dfrac{-2y_{12}Y_0}{\Delta Y}$	$\dfrac{2(AD-BC)}{A+\dfrac{B}{Z_0}+CZ_0+D}$	S_{12}
S_{21}	$\dfrac{2Z_{21}Z_0}{\Delta Z}$	$\dfrac{-2y_{21}Y_0}{\Delta Y}$	$\dfrac{2}{A+\dfrac{B}{Z_0}+CZ_0+D}$	S_{21}
S_{22}	$\dfrac{(Z_{11}+Z_0)(Z_{22}-Z_0)-Z_{12}Z_{21}}{\Delta Z}$	$\dfrac{(Y_0+y_{11})(Y_0-y_{22})+y_{12}y_{21}}{\Delta Y}$	$\dfrac{-A+\dfrac{B}{Z_0}-CZ_0-D}{A+\dfrac{B}{Z_0}+CZ_0+D}$	S_{22}

EXAMPLE PROBLEM 7.12

For the given scattering parameters for a two-port network, calculate the equivalent impedance parameters if the characteristic impedance is 50 Ω.

$$S_{11} = 0.4 + j\,0.7$$
$$S_{12} = S_{21} = j\,0.6$$
$$S_{22} = 0.3 - j\,0.8$$

Solution

From the Table 7.2 the impedance parameters in terms of S parameters are,

$$Z_{11} = Z_0\left[\frac{(1+S_{11})(1-S_{22})+S_{12}S_{21}}{(1-S_{11})(1-S_{22})-S_{12}S_{21}}\right]$$

$$Z_{11} = Z_0\left[\frac{(1+0.4+j0.7)(1-0.3-j0.8)+j0.6\times j0.6}{(1-(1+0.4+j0.7))(1-0.3-j0.8)-j0.6\times j0.6}\right]$$

$$Z_{11} = (0.031 + j1.31)50 = 1.55 + j65.5$$

$$Z_{12} = Z_{21} = Z_0\left[\frac{2S_{21}}{(1-S_{11})(1-S_{22})-S_{12}S_{21}}\right]$$

$$Z_{12} = Z_{21} = Z_0\left[\frac{2S_{21}}{(1-(1+0.4+j0.7))(1-0.3-j0.8)-j0.6\times j0.6}\right]$$

$$Z_{12} = Z_{21} = 50(j0.657) = j32.38$$

$$Z_{22} = Z_0\left[\frac{(1-S_{11})(1+S_{22})+S_{12}S_{21}}{(1-S_{11})(1-S_{22})-S_{12}S_{21}}\right]$$

$$Z_{22} = Z_0\left[\frac{(1-0.4+j0.7)(1+0.3-j0.8)+j0.6\times j0.6}{(1-(1+0.4+j0.7))(1-0.3-j0.8)-j0.6\times j0.6}\right]$$

$$Z_{22} = 50(0.716 - j0.915) = 35.8 - j45.75$$

$$Z = \begin{bmatrix} 1.55 + j65.5 & j32.38 \\ j32.38 & 35.8 - j45.75 \end{bmatrix}$$

SUMMARY

1. The need for open- and short-circuit terminations to measure Z, Y, and ABCD parameters make them unsuitable for microwave frequencies. Since it is very difficult to achieve low inductanc short circuits and low capacitance open circuits at these frequencies.

2. For high frequencies, it is easier to describe a given network in terms of ratios between the forwar and reverse waves rather than between voltages and currents.

3. The S matrix is defined to relate the incident and reflected wave amplitudes at the ports of the network. An S-parameter representation of the network treats the AC signal as a voltage wave for a reference impedance (50 Ω usually).

4. S parameters describe the network characteristics using the degree of scattering when an AC signal is considered a wave.

5. The waves a_n, b_n are defined such that $|a_1|^2 = P_{\text{incident}}$ at port n or $a_n = \frac{v_{\text{inc}}}{\sqrt{z_{0n}}} |b_1|^2 = P_{\text{reflected}}$ at port n or $b_n = \frac{v_{\text{refl}}}{\sqrt{z_{0n}}}$

6. a_1 and b_1 are rms voltages normalized by $\sqrt{Z_0}$. S_{11} and S_{22} are the reflection coefficients with the opposite port terminated with Z_0. S_{21} and S_{12} are the forward and reverse 50 Ω transducer gains, respectively.

7. The S matrix of a 3-port network cannot be matched, lossless, and reciprocal at the same time.

8. The S matrix for an n-port network contains n^2 coefficients, with each one representing a possible input–output path. The number of rows and columns in an S-parameter matrix is equal to the number of ports. For the S-parameter subscripts "ij," "j" is the port that is excited, and "i" is the port where the power is measured.

9. A network is reciprocal if $[S] = [S]^T$. For a 2-port device, *network symmetry* means that $S_{12} = S_{21}$, so that port 1 behaves similar to port 2 if a signal is the input to one of the ports and the other port is terminated with a matched load.

10. Unlike a 3-port network, a 4-port network can be lossless, reciprocal, and matched at all ports simultaneously.

11. Unique transmission paths are provided by circulators and isolators. This makes the *RF* energy to pass in one direction with little (insertion) loss, and with high loss (isolation) in the other direction.

OBJECTIVE-TYPE QUESTIONS

1. When the co-planar arm lengths are not the same, then outputs in the H-plane tee can be
   (a) in phase (b) out of phase
   (c) with phase difference (d) either

2. When the co-planar arm lengths are not the same, then outputs in the E-plane tee can be
   (a) in phase (b) out of phase
   (c) with phase difference (d) either

3. The outputs of directional couplers have a phase difference of
   (a) 90° (b) 45°
   (c) 180° (d) none

4. Co-planar arm ports in 3-port junctions can be
   (a) matched to junction (b) mismatched
   (c) either (d) none

5. For its S matrix to be unitary, the circuit should be
   (a) reciprocal (b) lossless
   (c) both (d) none

6. Chain S parameters are similar to
   (a) ABCB parameters (b) transmission-line parameters
   (c) both (d) none

7. An E-plane tee is
   (a) a voltage junction (b) a series junction
   (c) both (d) none

8. An H-plane tee is
   (a) a voltage junction (b) a shunt Junction
   (c) both (d) none

ANSWERS TO OBJECTIVE-TYPE QUESTIONS

1. (c) 2. (c) 3. (a) 4. (a) 5. (b) 6. (c) 7. (b) 8. (b)

REVIEW QUESTIONS

1. Why are S parameters used in a microwave network analysis?
2. What is a scattering matrix? Derive the scattering matrix formulation for an n-port network.
3. What are the advantages of S parameters compared with Y or Z parameters?
4. Write down the S matrix of a two-port network.
5. Derive the S matrix for an ideal 3 dB directional coupler.
6. Obtain the S matrix for the magic tee.
7. Derive the S matrix for a 3-port circulator.
8. Explain the S-matrix representation of a multiport microwave network and its significance.
9. For a 2-port network, define the S parameters involved, and obtain the relations for insertion loss, reflection loss, and return loss in terms of S parameters.
10. What are the properties of the S matrix?
11. What is an ABCD matrix? Give the ABCD matrix for a two-port network.

12. Give the S matrix of a uniform transmission line. Ans: $\begin{bmatrix} 0 & e^{-j\beta l} \\ e^{-j\beta l} & 0 \end{bmatrix}$

13. Define a scattering matrix.

14. Why are the S parameters used in microwaves?

15. Explain briefly why it is difficult to define and measure the voltage and current in distributed circuits.

16. Write the properties of the [S] matrix.

17. Define one port circuit. Give two examples.

18. Prove that it is impossible to construct a 3-port lossless reciprocal network.

19. Prove that any matched, lossless three-port network is non-reciprocal.

Microwave Tubes

8

8.1 INTRODUCTION

Microwave tubes operate at frequencies from 300 MHz to 300 GHz with output power ranging from 10 mW to a few hundred watts. The operating principle of microwave tubes is based on velocity modulation of electrons. The velocity-modulation concept is used to prevent the problems caused due to transit-time effects that are encountered in conventional tubes. A long transit time is used very effectively in microwave tubes in the conversion of dc power to RF power. By using the resonant cavities in the microwave tubes, and the velocity modulation of electrons, the power can be interchanged.

In conventional vacuum tubes, the modulation of electron beam can be done by altering the number of electrons. The microwave tubes are also called *velocity-modulated tubes*. In velocity-modulated tubes, the electron beam is formed by varying the velocity of electrons, in order to enable some electrons to move slowly and others to move rapidly through the inter-electrode space. Bunching of electrons occurs when the fast moving electrons overtake the slowly moving electrons (which arrive before the fast moving electrons).

The microwave tubes are mainly classified into linear beam tubes (O-type) and crossed-field tubes (M-type). In linear beam tubes, the electron beam is parallel to both the electric and magnetic fields. However, in crossed-field tubes, the electric beam is perpendicular to both the electric and magnetic fields. This chapter begins with the limitations of conventional vacuum tubes at microwave frequencies, the common operating principles of many microwave tubes and finally the detailed descriptions of Klystron, Reflex Klystron, TWT, BWO, magnetrons and CFA.

8.2 LIMITATIONS OF CONVENTIONAL TUBES AT MICROWAVE FREQUENCIES

Due to the following reasons, at microwave frequencies above 1 GHz, the conventional tubes (triodes, tetrodes, and pentodes) are less useful signal sources. They are

- Effects of inter electrode capacitance
- Effects due to lead inductance
- Effects due to transit time
- Limitation of gain bandwidth product.

The details of each limitation are discussed in the following sections:

8.2.1 Inter-electrode Capacitance Effect

The inter-electrode capacitances in a vacuum tube produce capacitive reactance. At low and medium frequencies, the tube operation is not affected, even though the reactance value is very large. However, as the frequency increases, the reactance, $X_c = 1/2\pi fc$, decreases and the output voltage decreases due to shunting effect. Since at higher frequencies, X_c becomes almost a short, C_{gp} (grid-to-plate capacitance), C_{gc} (grid-to-cathode capacitance) and C_{pc} (plate-to-cathode capacitance) are the inter-electrode capacitance (IECs) that come into effect and are shown in Figure 8.1.

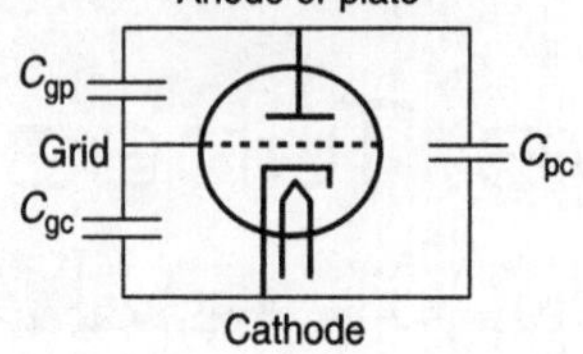

Figure 8.1 *Inter-electrode capacitance*

The inter-electrode capacitance can be reduced by decreasing the area of the electrodes, that is, by using smaller electrodes or by increasing the distance between electrodes. The relation for the capacitance is

$$C = \varepsilon_0 \varepsilon_r A/d \tag{8.1}$$

where A = area of the electrode
d = distance between electrodes
ε_0 = free space permittivity
ε_r = relative permittivity

8.2.2 Lead Inductance Effect

The lead inductance that appears at the connecting leads of the vacuum tube elements (e.g. cathode, plate and grid) is another limiting factor at higher frequencies. The three lead inductances that limit the performance of vacuum tube are: inductance at cathode (L_c), inductance at plate (L_p), and inductance at grid (L_g) as shown in Figure. 8.2. The main effects of the lead inductance at high frequencies are:

(i) they form unwanted tuned circuit by combining with the capacitance, which produce the parasitic oscillations
(ii) they create an input impedance matching problem due to increase in the inductive reactance.

The inductive reactance ($X_L = 2\pi fL$) in the connecting leads (wires or base pins) increases with the increase in frequency. Therefore, the input voltage drops across the lead inductance and only a fractional part of the applied input voltage reaches the terminals (e.g. grid), which decreases the gain of tube amplifier

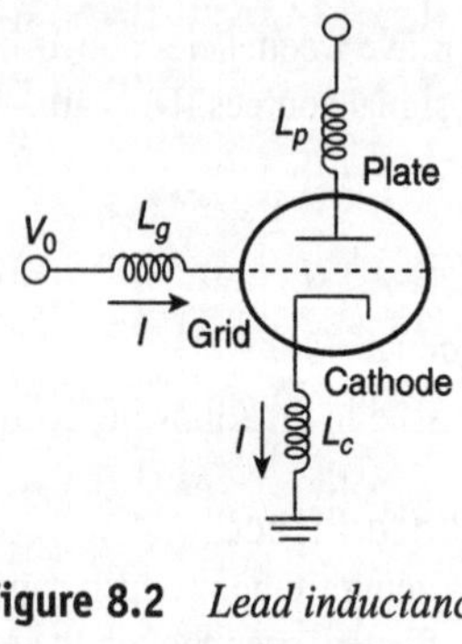

Figure 8.2 *Lead inductance*

Inductance is calculated by using the relation $$L = \frac{l}{\mu_0 \mu_r A} \quad (8.2)$$

where A = area of the electrode

l = length of the coil

μ_0 = permeability of free space = $4\pi \times 10^{-7}$ H/m

μ_r = relative permeability

The effect of lead inductance can be minimized by decreasing L, that is, by using larger-sized short leads without base pins, and by increasing A and decreasing l. Due to this the power handling capability is reduced.

8.2.3 Transit-time Effect

The time required for the electrons to travel from cathode to anode plate is called as transit time (τ). It is the main limitation caused at higher frequencies only. At low frequencies, transit time is not important and usually it is not considered as an affecting factor because the time period of the signal (T) is much greater than the transit time ($T >> \tau$). On the other hand, at high frequencies, it is a considerable portion of the signal cycle and obstructs the efficiency because of the time period of the signal (T) is much smaller than the transit time ($T << \tau$). In general, transit time may cause the reduction of efficiency if it exceeds 0.1 of the signal cycle.

> Transit time causes a phase shift between the plate current and the grid voltage resulting in reduction of efficiency.

For example, at a frequency of 1 MHz , the transit time of 1 nsec is only 0.001 of the signal period ($T = 1/f = 10^{-6}$ sec). So, in this case the transit time is much smaller than the time period of the signal and its effect is insignificant. If the frequency is increased to 100 GHz, with the same transit time of 1 nsec, then the transit time becomes equal to 100 times the signal period ($T = 1/f = 10^{-11}$ sec). In this case the transit time is greater than the time period of the signal and thereforc its effect is significant.

Transit time is the time taken by the electron to travel from cathode to anode.

$$\tau = \frac{d}{v_0} \quad (8.3)$$

Under the equilibrium condition, static energy is equal to kinetic energy, that is,

$$eV_0 = \frac{1}{2} m v_0^2 \quad (8.4)$$

Therefore, $$\tau = \frac{d}{\sqrt{\frac{2eV_0}{m}}} \quad (8.5)$$

where V_0 = dc voltage

d = distance between anode and cathode

v_0 = velocity of an electron

e = charge of an electron

m = mass of an electron

To reduce the transit time, the separation between electrodes, "d" can be decreased (but this increases IEC), and the anode to cathode voltage can be increased (this cannot be increased indefinitely). Therefore, a trade-off between IEC and transit time is a must.

8.2.4 Gain Bandwidth Product Limitation

Using a resonant circuit (LC tank circuit) across the load resistance (R) of an ordinary vacuum tube maximum gain is achieved at a particular frequency as shown in Figure 8.3. The gain bandwidth product is constant for all the vacuum tube amplifiers that employ resonant circuits and is given by

$$A_m \times BW = \frac{g_m}{C} \tag{8.6}$$

where A_m = maximum voltage gain at resonance
BW = bandwidth
g_m = transconductance
C = the capacitance of the tank circuit

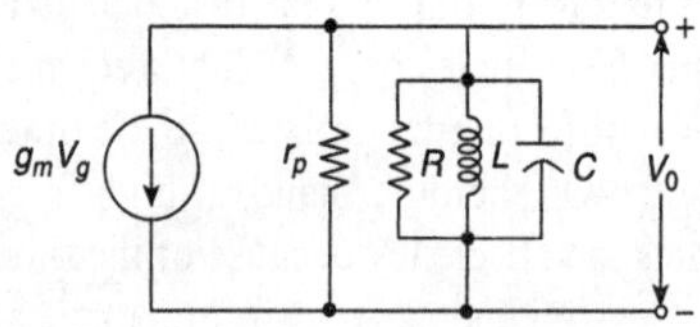

Figure 8.3 *Output tuned circuit of a pentode*

We note that the gain bandwidth product is independent of frequency. It depends on the tube parameter (g_m) and the external capacitance (C) of the resonator. These resonant circuits can be used at low frequencies to achieve a higher gain, but at a narrower bandwidth. However, higher gain at broader bandwidth is desired at higher frequencies. This limitation can be overcome by using re-entrant cavities or slow-wave structures in place of the resonant circuits in microwave tubes.

8.3 RE-ENTRANT CAVITIES

The re-entrant cavities (also known as *irregular-shaped resonators*) are used in place of tuned circuits at microwave frequencies. These devices are easily incorporated into the microwave device structure. Below the microwave frequencies, the cavity resonator can be represented by lumped-constant resonant circuit. In order to maintain resonance at high operating frequency, the values of inductance and capacitance are decreased to a minimum by using a short wire. Hence, for use in klystron and other microwave tubes, the re-entrant cavities are designed. One more advantage of the re-entrant cavity is we can easily couple and take out the signal from these devices. In these devices the metallic boundaries are extended into interior of the cavity.

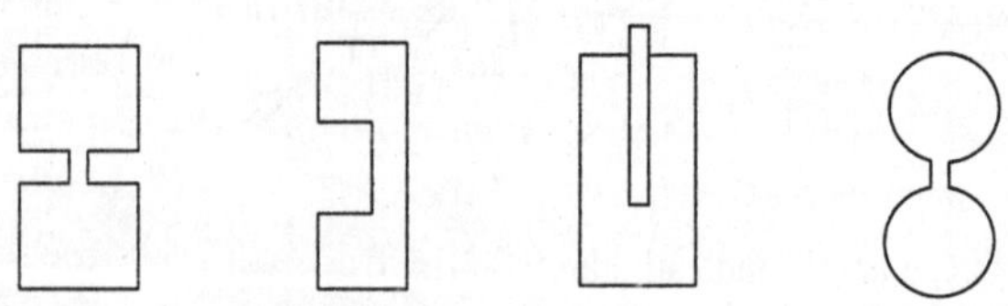

(a) Coaxial cavity (b) Radial cavity (c) Tunable cavity (d) Toroidal cavity

Figure 8.4 *Re-entrant cavities*

Several types of re-entrant cavities are shown in Figure 8.4. One of the commonly used re-entrant cavities is the coaxial cavity shown in Figure 8.5 (a). From the Figure 8.5(a), it can be observed that the inductance as well as the resistance losses is reduced, and the radiation losses are also prevented by

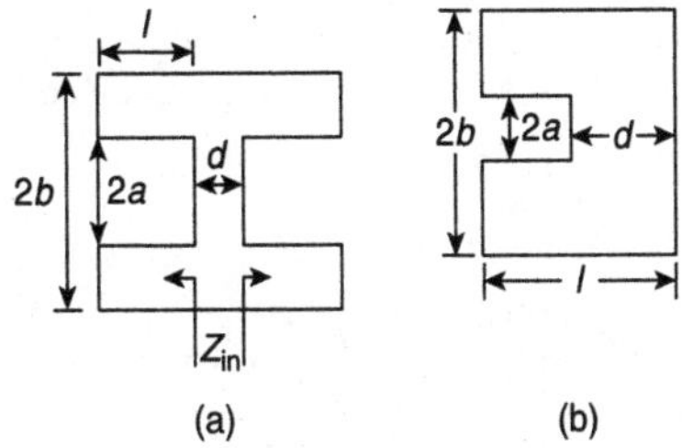

Figure 8.5 *(a) Coaxial cavity; (b) Radial cavity*

shelf-shielding enclosures. Calculation of the resonant frequency of the coaxial cavity is very difficult. Using transmission line theory an approximation can be made. The characteristic impedance of the coaxial line is given by

$$Z_0 = \frac{1}{2\pi}\sqrt{\frac{\mu}{\varepsilon}}\ln\frac{b}{a} \quad \text{ohms} \tag{8.7}$$

When two coaxial lines are shorted at both the ends and are joined at the centre by a capacitor, then a structure is formed which is similar to a coaxial cavity. For each shorted coaxial line the input impedance is given as

$$Z_{in} = jZ_0 \tan(\beta l) \tag{8.8}$$

where l = the length of the coaxial line

Substituting Eq. 8.7 in Eq. 8.8 results in

$$Z_{in} = j\frac{1}{2\pi}\sqrt{\frac{\mu}{\varepsilon}}\ln\frac{b}{a}\tan(\beta l) \tag{8.9}$$

The inductance of the cavity is given by

$$L = \frac{2X_{in}}{\omega} = \frac{1}{\pi\omega}\sqrt{\frac{\mu}{\varepsilon}}\ln\frac{b}{a}\tan(\beta l) \tag{8.10}$$

and the capacitance of the gap is given by

$$C_g = \frac{\varepsilon\pi a^2}{d} \tag{8.11}$$

When the circuit is in resonance, the inductive reactance of the two shorted coaxial lines which is in series has same magnitude to that of capacitive reactance of the gap.

That is $\omega L = \frac{1}{\omega C_g}$. Thus $\tan(\beta l) = \frac{dv}{\omega a^2 \ln(b/a)}$ (8.12)

where, $v = \frac{1}{\sqrt{\mu\varepsilon}}$ is the phase velocity in any medium.

The resonant frequency of the coaxial cavity is obtained by solving this equation. The Eq. 8.12 has infinite number of resonant modes, because it bears the tangent function. Each resonant mode corresponds to a particular value of resonant frequency and because of which each re-entrant cavity has an innumerable resonant frequencies. The mode which possesses the lowest resonant frequency is known as the *dominant mode*. It can be seen that a shorted coaxial-line cavity stores more magnetic energy than electric energy. The balance of the stored electric energy appears in the gap. At resonance, the magnetic and electric energies stored are equal.

Another commonly used re-entrant resonator is the radial re-entrant cavity shown in Figure 8.5 (b). The inductance and capacitance of a radial re-entrant cavity are expressed by

$$L = \frac{\mu l}{2\pi}\ln\frac{b}{a} \tag{8.13}$$

and

$$C = \varepsilon_0\left[\frac{\pi a^2}{d} - 4a\ln\frac{0.765}{\sqrt{(l^2+(b-a)^2)}}\right] \tag{8.14}$$

The resonant frequency is given by

$$f_r = \frac{c}{2\pi\sqrt{\varepsilon_r}}\left\{al\left[\frac{a}{2d} - \frac{2}{l}\ln\frac{0.765}{\sqrt{(l^2+(b-a)^2}}\right]\ln\frac{b}{a}\right\}^{-\frac{1}{2}} \tag{8.15}$$

where $c = 3\times10^8$ m/s is the velocity of light

8.4 CLASSIFICATION OF MICROWAVE TUBES

Microwave tubes are constructed so as to overcome the limitations of conventional and UHF tubes. The basic operating principle of microwave tubes involves the transfer of power from source of the dc voltage to the source of the ac voltage by means of a current density modulated electron beam. The same is achieved by accelerating electrons in a static field and retarding them in an ac field.The microwave tubes are mainly classified into linear beam tubes and crossed field tubes, and sub classified as shown in Figure 8.6.

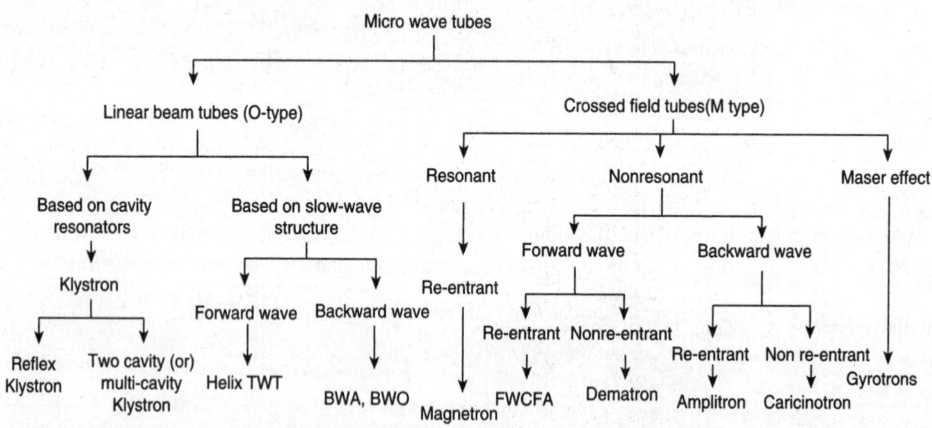

Figure 8.6 *Classification of microwave tubes*

Linear beam tubes are often called *O-type*. In linear beam tubes, the electron beam travels along a straight path between the cathode and the collector. This electron beam is parallel to both the electric and magnetic fields. However, in crossed field tubes, the electron beam is perpendicular to both the electric and magnetic fields. Crossed-field tubes deals with the propagation of magnetic field waves so they are known as *M-type devices.*

8.5 LINEAR BEAM (O TYPE) TUBES

At present, the most impoprtant microwave tubes are the linear beam tubes. There are two basic types of linear beam tubes. One type of tube uses electromagnetic cavities, and the other type of tubes use slow-wave structure. Both types of tubes use an electron beam. As the name implies, in a linear beam tube, the electron beam and the circuit elements with which it interacts are arranged linearly. A simple schematic of a linear beam tube is shown in Figure 8.7.

The electrons which are accelerated by the anode voltage possesses kinetic energy and this kinetic energy is calculated by using anode voltage. The applied RF input when interacts with the electron beam, a small part of kinetic energy that is present in it gets converted into microwave energy. At the RF output port the microwave energy is extracted and the remaining portion of electron beam is dissipated in the form of heat from the tube or it is sent back to the collector. As the electrons in the electron beam tends to repel each other, a magnetic field, is focused on the electrons as a beam going from the cathode to the collector. The magnetic field is generated either by an electromagnet or permanent magnets.

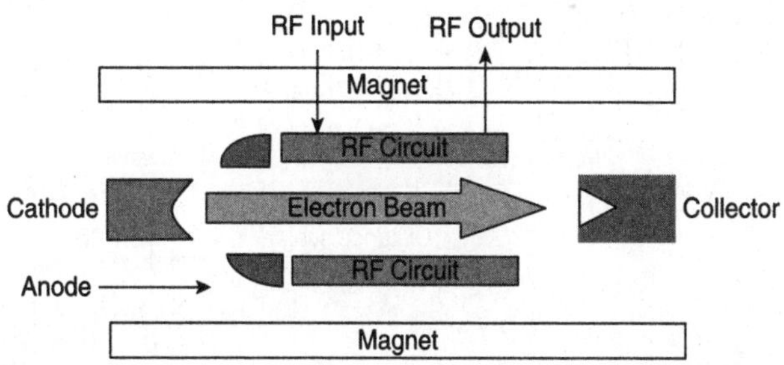

Figure 8.7 *Schematic diagram of a generic linear beam tube*

Klystrons: Klystrons have high peak power, high average power, good efficiency, high gain, and low spurious signals. The working principle of klystron is based on velocity and density modulation. The resonant cavities are used to produce the bunching effect. The klystron is basically a vacuum electron device and it may be an oscillator or an amplifier since it is used for transforming DC energy into RF energy. In a klystron, the signal and the electron beam interacts at a very short range. Therefore, a strong electrostatic field is required for efficient operation of the klystron. When the space charge in the beam decreases, then the klystron's efficiency increases. By changing the velocity of an electron beam, the transit-time effect is utilized by the klystrons.

1. Two-cavity or multi-cavity klystron: It is used as a low-power microwave amplifier.
2. Reflex klystron: It is used as a low-power microwave oscillator.

8.6 TWO-CAVITY KLYSTRON AMPLIFIER

The two-cavity klystron is a widely used microwave amplifier that is operated by the principles of velocity and current modulation. In a two cavity-klystron amplifier, a high-velocity electron beam is produced, focused, and made to travel along a glass tube, a field-free drift space, and an output cavity (catcher) to a collector electrode/anode. The anode is maintained at a positive potential with respect to the cathode. Electrons from the cathode start drifting with constant velocity after being accelerated by a DC voltage. When the electrons pass through a pair of narrowly spaced grids, their velocity is modulated by a sinusoidal RF signal.

8.6.1 Structure of Two-Cavity Klystron

The construction and essential components of a two-cavity klystron are shown in Figure 8.8. It consists of a (i) cathode-anode configuration (i.e. electron gun) to produce an accelerated electron beam; (ii) a buncher (input) cavity; (iii) a catcher (output) cavity; (iv) field-free drift space between input and output cavities; and (v) a collector.

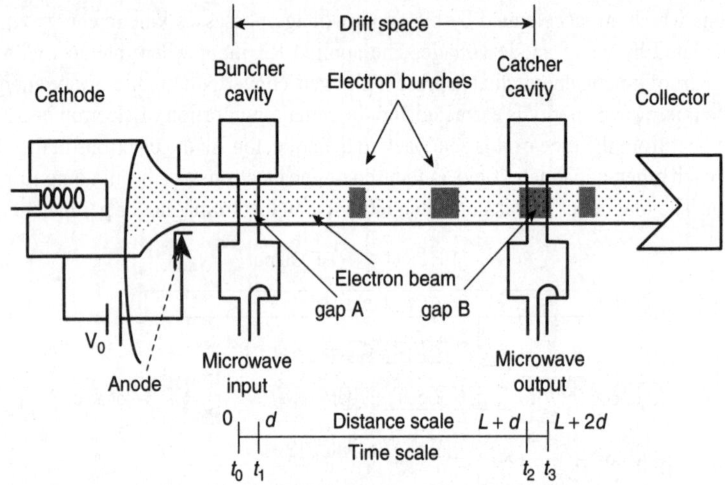

Figure 8.8 *Functional and schematic diagram of a two-cavity klystron*

Electrons emitted by the cathode are focused by one or more focusing electrodes placed in front of the cathode. The electron beam then passes through the buncher and catcher cavities and finally reaches the collector. By using coupling loop, the RF power can be coupled through the buncher and catcher cavities.

8.6.2 Velocity-modulation Process and Applegate Diagram

The velocity-modulation principle and operation of the klystron is explained by using the Applegate diagram as shown in Figure 8.9. Assume a condition when there is no voltage across the gap A, and the electrons that pass through gap A are unaffected and continue to the collector with the same velocities which they had before approaching the gap. If input is given to the buncher cavity, then the electrons which pass through this cavity are affected by the voltage depending on the variations of the voltage. Consider the instant when the electron passes through the gap A when the voltage across gap A is zero and is going positive. At this instant, the electric field across gap A is zero, and an electron that passes through gap A is unaffected by the input signal. Let this electron be called the *reference electron* (r_e), which travels with an unchanged velocity $v_0 = \sqrt{2eV_0 / m}$.

As shown in Figure 8.9, a late electron l_e passing the gap A slightly later than r_e is accelerated by the new positive voltage whose velocity (v) is greater than (v_0) across gap A. This electron will catch up with the reference electron r_e and the early electron e_e that was retarded by the negative voltage whose velocity (v) is less than (v_0).

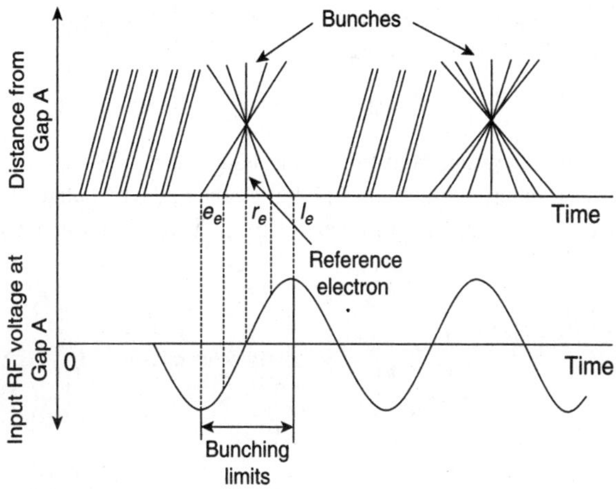

Figure 8.9 *Applegate diagram of a two-cavity klystron*

As a result of these actions, the average velocity of electrons with which they leave the grid gap is similar to that with which they enter. While traveling through the drift region, the beam under goes density modulation. The density of electrons passing the gap B vary cyclically with time, that is, the electron beam contains an ac current and is current modulated. The drift space converts the velocity modulation into current modulation. Bunching occurs only once for the cycle centered around the reference electron. With a proper design, a little RF power applied to the buncher cavity results in large beam currents at the catcher cavity with a considerable power gain.

Equation of Velocity Modulation

The equation of velocity modulation can be given by

$$v(t_1) = v_0\left[1+\frac{\beta_1 V_1}{2V_0}\sin\left(\omega t_0+\frac{\theta_g}{2}\right)\right]$$

where v_0 = initial velocity of electrons
β_1 = beam coupling coefficient
V_1 = amplitude of the input signal applied at the buncher cavity
V_0 = anode to cathode voltage

Derivation of Velocity-Modulation Equation

As explained earlier, the velocity of electrons before entering the buncher grid is uniform and is given by

$$v_0 = \sqrt{\frac{2eV_0}{m}} = 0.593\times10^6\sqrt{V_0}\,\text{m/s} \tag{8.16}$$

where V_0 = applied beam voltage
e = charge of electrons
m = mass of electrons

From Eq. 8.16 it is inferred that the electrons leave the cathode with zero velocity. Consider that the RF signal fed to the input buncher grid is specified as

$$V_s = V_1 \sin \omega t \tag{8.17}$$

where, V_1 is the amplitude of the signal and $V_1 << V_0$ is assumed.

By considering either the entering time t_0 or the exiting time t_1, the modulated velocity in the buncher cavity can be determined. The average microwave voltage in the buncher gap needs to be determined as shown in Figure 8.10.

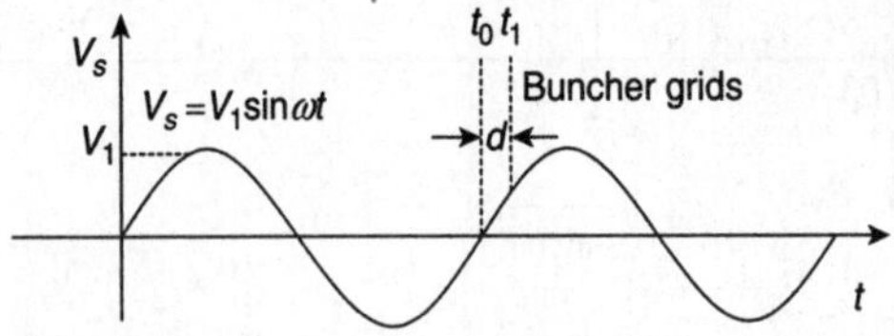

Figure 8.10 *Signal voltages in the buncher gap*

As $V_1 << V_0$, the average transit time all the way through the buncher gap of distance d is

$$\tau \approx \frac{d}{v_0} = t_1 - t_0 \tag{8.18}$$

The phase delay caused during transit time across the gap is referred to as gap transit angle (θ_g) and can be given as

$$\theta_g = \omega\tau = \omega(t_1 - t_0) = \frac{\omega d}{v_0} \tag{8.19}$$

Eventually, the average microwave voltage in the buncher gap can be given as

$$V_s = \frac{1}{\tau}\int_{t_0}^{t_1} V_1 \sin(\omega t)dt = \frac{-V_1}{\omega\tau}[\cos(\omega t_1) - \cos(\omega t_0)] \tag{8.20}$$

$$V_s = \frac{V_1}{\omega\tau}\left[\cos(\omega t_0) - \cos\left(\omega t_0 + \frac{\omega d}{v_0}\right)\right] \tag{8.21}$$

Let $\quad \omega t_0 + \dfrac{\omega d}{2v_0} = \omega t_0 + \dfrac{\theta_g}{2} = A$ and $\dfrac{\omega d}{2v_0} = \dfrac{\theta_g}{2} = B$

By using trigonometric relations, i.e., $\cos(A - B) - \cos(A + B) = 2\sin A \sin B$, Eq. 8.21 can be written as

$$V_s = V_1 \frac{\sin[\omega d / 2v_0]}{\omega d / 2v_0} \sin\left(\omega t_0 + \frac{\omega d}{2v_0}\right) \tag{8.22}$$

$$V_s = V_1 \beta_1 \sin\left(\omega t_0 + \frac{\theta_g}{2}\right) \tag{8.23}$$

where β_1 the beam coupling coefficient of the input cavity gap and is given as

$$\beta_1 = \frac{\sin[\omega d/2v_0]}{\omega d/2v_0} = \frac{\sin(\theta_g/2)}{\theta_g/2} \tag{8.24}$$

We can observe that when the gap transit angle increases the coupling between the electron beam and buncher cavity reduces which means for a given microwave signal the velocity modulation decreases. The exit velocity from the buncher gap after velocity modulation, can be instantly calculated as

$$v(t_1) = \sqrt{\frac{2e}{m}(V_0 + V_s)} \tag{8.25}$$

substituting Eq. 8.23 in Eq. 8.25,

$$v(t_1) = \sqrt{\frac{2e}{m}V_0\left[1 + \frac{\beta_1 V_1}{V_0}\sin\left(\omega t_0 + \frac{\theta_g}{2}\right)\right]} \tag{8.26}$$

where, the factor $\beta_1 V_1 / V_0$ is called the depth of velocity modulation

$$v(t_1) = v_0\sqrt{1 + \frac{\beta_1 V_1}{V_0}\sin\left(\omega t_0 + \frac{\theta_g}{2}\right)} \tag{8.27}$$

Assuming that $\beta_1 V_1 << V_0$ and by means of binomial expansion the Eq. 8.27 is modified as

$$v(t_1) = v_0\left[1 + \frac{\beta_1 V_1}{2V_0}\sin\left(\omega t_0 + \frac{\theta_g}{2}\right)\right] \tag{8.28}$$

This is called the velocity modulation equation, this equation can also be written as,

$$v(t_1) = v_0\left[1 + \frac{\beta_1 V_1}{2V_0}\sin\left(\omega t_1 - \frac{\theta_g}{2}\right)\right] \tag{8.29}$$

8.6.3 Bunching Process and Small Signal Theory

The electrons drift immediately after leaving the buncher cavity, with a velocity analogous to that shown in Eq. 8.28 (or Eq. 8.29) along the field-free space between the two cavities. This effect of velocity modulation creates bunching of the electron beam or current modulation.

The bunching parameter of the two-cavity klystron is given by

$$X = \frac{\beta_1 V_1}{2V_0}\theta_0$$

where β_1 = beam coupling coefficient

V_1 = amplitude of the input signal applied at buncher cavity

θ_0 = dc transit angle

V_0 = anode to cathode voltage

Derivation of Bunching Parameter of Two-Cavity Klystron

The electrons form the bunching centre when they pass through the buncher at $V_s = 0$ with an unchanged velocity v_0. During the positive half cycles of the microwave input voltage V_s, the electron passes the gap faster compared to the electrons that pass the gap at $V_s = 0$. The electrons that enter buncher cavity during the negative half cycle of V_s are slow compared to the electrons that pass the gap at $V_s = 0$. The beam electrons drift into dense clusters at a distance of ΔL all along the beam from the buncher cavity.

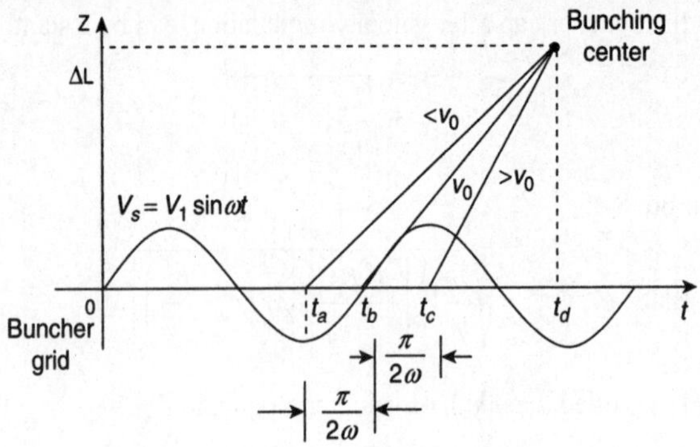

Figure 8.11 *Bunching process*

Figure 8.11 shows the trajectories of minimum, zero, and maximum electron acceleration ($-\pi/2$ to 0 then 0 to $\pi/2$). The location of dense electron bunching for the electron at t_b is at distance ΔL from the bunching grid and it is given as

$$\Delta L = v_0(t_d - t_b) \tag{8.30}$$

Similarly, the distances for the electrons at t_a and t_c are

$$\Delta L = v_{\min}(t_d - t_a) = v_{\min}\left(t_d - t_b + \frac{\pi}{2w}\right) \tag{8.31}$$

$$\Delta L = v_{\max}(t_d - t_c) = v_{\max}\left(t_d - t_b - \frac{\pi}{2w}\right) \tag{8.32}$$

From Eq. 8.28 or 8.29, the minimum and maximum velocities are as follows:
Maximum velocity occurs at $+\pi/2$, so that

$$v_{\max} = v_0\left(1 + \frac{\beta_1 V_1}{2V_0}\right) \tag{8.33}$$

Minimum velocity occurs at $-\pi/2$, so that

$$v_{\min} = v_0\left(1 - \frac{\beta_1 V_1}{2V_0}\right) \tag{8.34}$$

Substituting Eqs. 8.34 and 8.33 in Eqs. 8.31 and 8.32, respectively,

$$\Delta L = v_0(t_d - t_b) + v_0\left[\frac{\pi}{2\omega} - \frac{\beta_1 V_1}{2V_0}(t_d - t_b) - \frac{\beta_1 V_1}{2V_0}\frac{\pi}{2\omega}\right] \tag{8.35}$$

and $$\Delta L = v_0(t_d - t_b) + v_0\left[-\frac{\pi}{2\omega} + \frac{\beta_1 V_1}{2V_0}(t_d - t_b) + \frac{\beta_1 V_1}{2V_0}\frac{\pi}{2\omega}\right] \tag{8.36}$$

The necessary condition for those electrons at t_a, t_b, and t_c to meet at the same distance ΔL is

$$\frac{\pi}{2\omega} - \frac{\beta_1 V_1}{2V_0}(t_d - t_b) - \frac{\beta_1 V_1}{2V_0}\frac{\pi}{2\omega} = 0 \tag{8.37}$$

and $$-\frac{\pi}{2\omega} + \frac{\beta_1 V_1}{2V_0}(t_d - t_b) + \frac{\beta_1 V_1}{2V_0}\frac{\pi}{2\omega} = 0 \tag{8.38}$$

Consequently,

$$t_d - t_b \approx \frac{\pi V_0}{\omega \beta_1 V_1} \tag{8.39}$$

and $$\Delta L = v_0\left[\frac{\pi V_0}{\omega \beta_1 V_1}\right] \tag{8.40}$$

The transit time for velocity-modulated electrons to travel at a distance L is given by (From Figure 8.8)

$$T = (t_2 - t_1) = \frac{L}{v(t_1)} = \frac{L}{v_0\left[1 + \frac{\beta_1 V_1}{2V_0}\sin\left(\omega t_0 + \frac{\theta_g}{2}\right)\right]}$$

$$= \frac{L}{v_0}\left[1 + \frac{\beta_1 V_1}{2V_0}\sin\left(\omega t_0 + \frac{\theta_g}{2}\right)\right]^{-1}$$

As for the Binomial expansion,

$$= \frac{L}{v_0}\left[1 - \frac{\beta_1 V_1}{2V_0}\sin\left(\omega t_0 + \frac{\theta_g}{2}\right)\right] \tag{8.41}$$

Multiplying by ω on both sides of the above equation, we get

$$\omega T = \omega t_2 - \omega t_1 = \frac{\omega L}{v_0}\left[1 - \frac{\beta_1 V_1}{2V_0}\sin\left(\omega t_0 + \frac{\theta_g}{2}\right)\right] \tag{8.42}$$

In the above equation, $L/v_0 = T_0$ is the dc transit time.

$$\omega T = \omega(t_2 - t_1) = \theta_0\left[1 - \frac{\beta_1 V_1}{2V_0}\sin\left(\omega t_0 + \frac{\theta_g}{2}\right)\right] \tag{8.43}$$

$$\theta_0 = \frac{\omega L}{v_0} = 2\pi N \tag{8.44}$$

where θ_0 = dc transit angle between cavities
N = number of electron transit cycles in the drift space

By expanding Eq. 8.43, we get the value of the bunching parameter

$$\omega T = \omega(t_2 - t_1) = \theta_0 - \theta_0 \frac{\beta_1 V_1}{2V_0} \sin\left(\omega t_0 + \frac{\theta_g}{2}\right)$$

where
$$X = \frac{\beta_1 V_1}{2V_0}\theta_0 \tag{8.45}$$

is defined as the *bunching parameter* of the klystron.

Substituting Eq. 8.44 in Eq. 8.45, we get

$$X = \frac{\beta_1 V_1}{2V_0}\frac{\omega L}{v_0} \Rightarrow L = \frac{2v_0 V_0}{\omega \beta_1 V_1} X \tag{8.46}$$

The beam current at the catcher cavity is a periodic waveform of period $\frac{2\pi}{\omega}$ about the dc current. Therefore, the current i_2 can be expanded in a Fourier series and so,

$$i_2 = a_0 + \sum_{n=1}^{\infty}\left[a_n \cos(n\omega t_2) + b_n \sin(n\omega t_2)\right] \tag{8.47}$$

By using Bessel's function ($J_n(nX)$) and trigonometric functions the beam current i_2 is

$$i_2 = I_0 + \sum_{n=1}^{\infty} 2I_0 J_n(nX)\cos(n\omega(t_2 - \tau - T_0)) \tag{8.48}$$

The magnitude of the fundamental component of the beam current at the catcher cavity is

$$I_f = 2I_0 J_n(nX) \tag{8.49}$$

It has maximum amplitude at $X = 1.841$. From the Eqs. 8.44 and 8.45 the maximum distance L at which the optimum fundamental component of current occurs can be calculated as

$$L_{\max} = 3.682\frac{v_0 V_0}{\omega V_1 \beta_1} \tag{8.50}$$

The distance mentioned in Eq. 8.40 is lesser by 15% than result of Eq. 8.50. This inconsistency is due to the approximations made in Eq. 8.40.

8.6.4 Expressions for Output Power and Efficiency

The beam coupling coefficients β_1 and β_0 can be equal, only when the buncher and catcher cavities are identical. The current induced by the electron beam (which is induced in the walls of the catcher cavity) is directly proportional to the amplitude of the microwave input voltage V_1. Eventually the fundamental component of the induced microwave current in the catcher is given by

$$i_{2ind} = \beta_0 2I_0 J_1(X)\cos\left[\omega(t_2 - \tau - T_0)\right] \tag{8.51}$$

Its magnitude is given by

$$I_{2ind} = \beta_0 I_2 = \beta_0 2I_0 J_1(X) \tag{8.52}$$

Figure 8.12 shows an output equivalent circuit of two cavity klystron amplifier. It comprises of wall resistance (R_c) taken in parallel combination with the catcher cavity, beam loading resistance (R_b) and an external load resistance R_L.

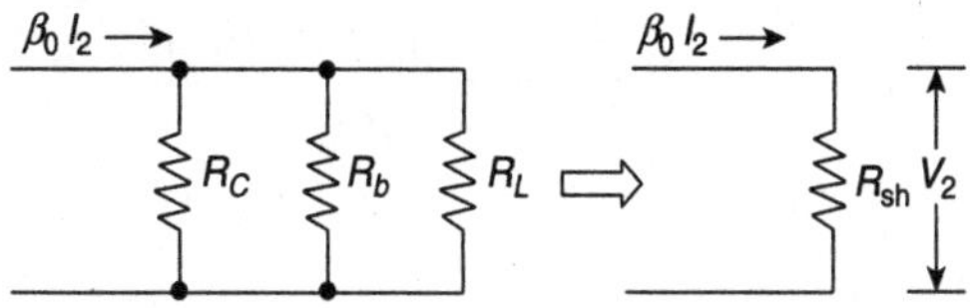

Figure 8.12 *Output equivalent circuit of two-cavity klystron amplifier*

The output power delivered to the catcher cavity and the load is given as

$$P_{out} = \frac{(\beta_0 I_2)^2}{2} R_{sh} = \frac{\beta_0 I_2 V_2}{2} \tag{8.53}$$

where R_{sh} = total equivalent shunt resistance of the catcher circuit, including the load
V_2 = fundamental component of the catcher gap voltage

Input power (P_{in})

The input power is basically the dc input and is given by

$$P_{in} = I_0 V_0 \tag{8.54}$$

Efficiency of klystron

For a klystron amplifier the electronic efficiency can be defined as the ratio of output power to the input power. From Eqs. 8.53 and 8.54 we get

$$\text{Efficiency} \equiv \frac{P_{out}}{P_{in}} = \frac{\beta_0 I_2 V_2}{2 I_0 V_0} \tag{8.55}$$

Here we also include the power loss due to the beam loading and cavity walls.

Under the perfect coupling conditions, the maximum electronic efficiency is about 58%, where $\beta_0 = 1$, the maximum beam current reaches to $I_{2\max} = 2I_0(0.582)$, and the voltage V_2 is equal to V_0. Because the efficiency is a function of the catcher gap transit angle θ_g, practically, efficiency of a klystron amplifier is in the range of 15 to 30%.

Voltage gain of a klystron amplifier

The equivalent mutual conductance of the klystron amplifier can be defined as the ratio of the induced output current to the input voltage. That is,

$$|G_m| \equiv \frac{i_{2ind}}{V_1} = \frac{2\beta_0 I_0 J_1(X)}{V_1} \tag{8.56}$$

From Eq. 8.45, the input voltage V_1 can be expressed in terms of the bunching parameter X as

$$V_1 = \frac{2V_0}{\beta_0 \theta_0} X \tag{8.57}$$

In Eq. 8.57, it is assumed that $\beta_0 = \beta_1$. Substitution of Eq. 8.57 in Eq. 8.56 yields the normalized mutual conductance as

$$\frac{|G_m|}{G_0} = \beta_0^{\,2}\theta_0 \frac{J_1(X)}{X} \tag{8.58}$$

where $G_0 = I_0/V_0$ is the dc beam conductance.

The voltage gain of a klystron amplifier is defined as $A_v \equiv \frac{|V_2|}{|V_1|}$.

By substituting Eq. 8.57 and $V_2 = \beta_0 I_2 R_{sh}$ the voltage gain is given as

$$A_v = \frac{\beta_0 I_2 R_{\text{sh}}}{V_1} = \frac{\beta_0^{\,2}\theta_0}{R_0}\frac{J_1(X)}{X} R_{\text{sh}} \tag{8.59}$$

where $R_0 = V_0/I_0$ is the dc beam resistance.
By substituting Eqs. 8.52 and 8.57 in Eq. 8.59, we get

$$A_v = G_m R_{\text{sh}}$$

Performance characteristics of a two-cavity klystron amplifier:

- Frequency : 250 MHz to 100 GHz
- Power : 10 kW – 500 kW (CW), 30 MW (Pulsed)
- Power gain : 15 dB – 20 dB
- Bandwidth : 10 – 60 MHz generally used in fixed frequency applications
- Noise figure : 15 – 20 dB
- Theoretical efficiency : 58% (50 – 60%)

Applications of a two-cavity klystron amplifier: Klystrons are widely used in particle accelerators, UHF television transmitters and communication system uplinks. The following are its applications:

- It is used as a power amplifier.
- It is used as a frequency multiplier.
- As power output tubes
  a) In UHF TV transmitters
  b) In troposphere scatter transmitters
  c) At satellite communication ground stations

EXAMPLE PROBLEM 8.1

A two-cavity klystron operates at 10 GHz with $I_o = 3.5$ mA, $V_o = 10$ kV. The drift space length is 3 cm, and the output cavity total shunt conductance is $G_{\text{sh}} = 20$ μ mho with beam coupling coefficient $\beta_0 = 0.92$. Find the maximum voltage gain.

Solution

Maximum voltage gain $\quad A = \dfrac{\beta_0^2 \theta_0 I_0 J_1(X)_{\max}}{X V_0 G_{sh}}$

The dc beam velocity $\quad v_0 = 0.593\times10^6\sqrt{V_0} = 0.593\times10^8\ \text{m/s}$

Transit angle in drift space $\quad \theta_0 = \dfrac{\omega L}{v_0} = \dfrac{2\pi\times10\times10^9\times3\times10^{-2}}{0.593\times10^8} = 31.786\ \text{rad}$

$$A_{max} = \frac{\beta_0^2\theta_0 I_0 J_1(X)_{max}}{XV_0G_{sh}} = \frac{0.92\times0.92\times31.78\times3.5\times10^{-3}\times0.582}{1.841\times10\times10^3\times20\times10^{-6}}$$

$$= 2.715\times10^{-3}\ (X_{max} = 1.841,\ J_1(X_{max}) = 0.582)$$

■

EXAMPLE PROBLEM 8.2

A two-cavity Klystron amplifier has the following parameters: DC voltage for accelerations of electron = 1000 V, DC beam resistance = 50 kΩ, DC current = 25 mA, f = 3 GHz, gap spacing in either cavity = 1 mm, spacing between the two cavities = 4 cm, effective shunt impedance including the beam loading = 30 kΩ, $J_1(X)$ is maximum at 1.841, and J_1 (1.841) = 0.582.

(a) Find the input gap voltage to give maximum output voltage.
(b) Find voltage gain, neglecting the beam loading in the output cavity.

Solution

(a) The electron velocity just leaving the cathode is

$$v_0 = \sqrt{\frac{2eV_0}{m}} = 0.593\times10^6\sqrt{V_{dc}} = 0.593\times10^6\sqrt{10^3} = 1.88\times10^7\,\text{m}\backslash\text{s}$$

The gap transit angle is

$$\theta_g = \omega.\frac{d}{v_0} = 2\pi\times\left(3\times10^9\right)\times\frac{10^{-3}}{1.88\times10^7} = 1\ \text{rad}$$

The dc transit angle between the cavities is

$$\theta_0 = \frac{\omega.L}{v_0} = 2\pi\times\left(3\times10^9\right)\times\frac{4\times10^{-2}}{1.88\times10^7} = 40\ \text{rad}$$

The beam coupling coefficient is

$$\beta_1 = \beta_{dc} = \frac{\sin\left(\theta_g/2\right)}{\theta_g/2} = \frac{\sin(1/2)}{1/2} = 0.952$$

The maximum input voltage is

$$V_{1max} = \frac{2V_{dc}X}{\beta_1\theta_0} = \frac{2\times10^3\times1.841}{0.952\times40} = 96.5\ \text{V}$$

(b) The voltage gain is found as

$$A_v = \frac{\beta_1^2\theta_0}{R_0}.\frac{J_1(X)}{X}R_{sh} = \frac{(0.952)^2\times40\times0.582\left(30\times10^3\right)}{5\times10^4\times1.841} = 6.876$$

■

EXAMPLE PROBLEM 8.3

A two-cavity klystron amplifier has the following parameters: $V_0 = 1200$ V, $I_0 = 25$ mA, $R_0 = 48$ KΩ, $f = 10$ GHz, $d = 1$ mm, $L = 4$ cm, and $R_{Sh} = 30$ KΩ.

Calculate (i) the input voltage for maximum output voltage
(ii) the voltage gain in decibels

Solution

Given that for a two-cavity klystron amplifier,

$$V_0 = 1200 \text{ V} \quad I_0 = 25 \text{ mA} \quad R_0 = 48 \text{ K}\Omega$$
$$f = 10 \text{ GHz} \quad d = 1 \text{ mm} \quad L = 4 \text{ cm}$$
$$R_{Sh} = 30 \text{ K}\Omega$$

(i) The input voltage applied at a two-cavity klystron is

$$X = \frac{\beta_1 V_1}{2V_0}\theta_0 \Rightarrow V_1 = \frac{2V_0}{\beta_1\theta_0}X$$

For maximum output power, $X = 1.84$

$$V_{1(\max)} = \frac{3.68\,V_0}{\beta_1\theta_0}$$

$$\theta_0 = \frac{\omega L}{v_0} = \frac{2\pi f \times L}{0.593\times10^6\sqrt{V_0}} = \frac{2\pi\times10\times10^9\times4\times10^{-2}}{0.593\times10^6\sqrt{1200}} = 122.347 \text{ rad}$$

$$\text{Average transit angle } \theta_g = \frac{\omega d}{v_0} = \frac{2\pi\times10\times10^9\times10^{-3}}{0.593\times10^6\sqrt{1200}} = 3.059 \text{ rad}$$

$$\beta_1 = \frac{\sin(\theta_g/2)}{\theta_g/2} = \frac{\sin(3.0591/2)}{3.058/2} = 0.653$$

The input voltage for maximum output voltage is

$$V_{1(\max)} = \frac{3.68\times1200}{0.653\times122.35} = 55.275 \text{ V}$$

(ii) The voltage gain A_V is given by

$$A_v = \frac{V_2}{V_{1(\max)}}$$

$$V_2 = \beta_0 I_2 R_{sh}$$
$$I_2 = 2 I_{dc} J_1(X)$$

For $X = 1.84$, $J_1(X) = 0.582$ (from Bessel function table)

$$I_2 = 2 \times 25 \times 10^{-3} \times 0.582 = 29.1 \text{ mA}$$
$$V_2 = \beta_0 I_2 R_{sh} = 0.653 \times 29.1 \times 10^{-3} \times 30 \times 10^3 = 570.069 \text{ V}$$

The voltage is

$$A_v = \frac{V_2}{V_{1(max)}} = \frac{570.069}{55.273} = 10.314$$

$$A_v\,(\text{dB}) = 20\log(10.314) = 20.269\text{ dB}$$

■

EXAMPLE PROBLEM 8.4

The operating frequency of a two-cavity klystron is 5 GHz. For a input RF voltage of 40 KV, the magnitude of the gap voltage is 100 volts and the capacity gap is 4 mm. Calculate the following:

(i) the transit time at the cavity gap,
(ii) the transit angle
(iii) the velocity of the electrons from the gap.

Solution

DC beam velocity $v_0 = 0.593\times10^6\sqrt{V_0}$

$$v_0 = 0.593\times10^6\sqrt{(40\times10^3)}\text{ m/s}$$
$$= 1.186\times10^8\text{ m/s}$$

Gap transmit time $\tau_g = \dfrac{d}{v_o} = \dfrac{4\times10^{-3}}{1.186\times10^8} = 33.7\times10^{-12}$ sec

The gap transmit angle $\theta_g = \omega\tau_g = 2\pi\times5\times10^9\times33.7\times10^{-12} = 1.059\text{ rad} = 60.7\text{ deg}$

The beam coupling coefficient $\beta_1 = \dfrac{\sin[\omega d/2v_0]}{\omega d/2v_0} = \dfrac{\sin(\theta_g/2)}{\theta_g/2}$

$$= 0.505/0.5295 = 0.95337$$

The velocity of electrons leaving the input cavity gap is changing sinusoidally at the input cycle and is given by

$$v(t) = v_0\left[1+\frac{\beta_1V_1}{2V_0}\sin(\omega t_1)\right]$$

$$= 1.186\times10^8\left[1+(0.954\times100)/(2\times40\times10^3)\sin(\omega t_1)\right]$$
$$= 1.186\times10^8\left[1+0.001192\sin(\omega t_1)\right]$$

The maximum velocity

$$v(t)\max = v_0(1+\alpha/2) = 1.186\times10^8(1+0.001192)$$
$$= 1.1874\times10^8\text{ m/s}$$

The minimum velocity

$$v(t)\min = v_0(1-\alpha/2) = 1.186\times10^8(1-0.001192)$$
$$= 0.999\times10^8\text{ m/s}$$

■

EXAMPLE PROBLEM 8.5

The operating frequency of a two-cavity klystron is 10 GHz. The current and voltage when the *RF* input not applied to the klystron is $I_0 = 14.4$ mA, $V_0 = 40$ kV. Find the maximum voltage and power gain when the drift space length is 1cms and the output cavity total shunt conductance is $R_{sh} = 20\ \mu$ mho and beam coupling coefficient $\beta = 0.92$.

Solution

Maximum voltage gain

$$A = \frac{\beta^2 \theta_0 I_0 J_1(X)\max}{X V_0 G_{sh}}$$

$$\text{dc beam voltage} = V_0 = 0.593\times10^6\sqrt{V_0}$$

$$= 0.593\times10^6\sqrt{40\times10^3} = 1.186\times10^8 \text{ m/s}$$

Transit angle in drift place

$$\theta_0 = \frac{\omega L}{v_0} = \frac{2\pi\times10\times10^9\times1\times10^{-2}}{1.186\times10^8} = 21.19 \text{ rad}$$

$$A_{\max} = \frac{0.92\times0.92\times21.19\times14.4\times10^{-3}\times0.582\times10^{-2}}{1.841\times40\times10^3\times20\times10^{-6}} = 1.002\times10^{-3}$$

■

EXAMPLE PROBLEM 8.6

The operating frequency of an identical two-cavity klystron is 4 GHz. The current and voltage when the RF input not applied to the klystron is $I_0 = 22$ mA, $V_0 = 4$ kV. The gap in the cavity is 2 mm, and the gap between the cavities is of 6 cms. If the dc beam conductance and catcher cavity total equivalent conductance are 0.25×10^{-4} mhos and 0.3×10^{-14} mhos, respectively, calculate the following:

(a) the beam coupling coefficient, the dc transit angle in the drift space, and the input cavity voltage magnitude for maximum output voltage
(b) voltage gain neglecting the beam loading

Solution

DC beam velocity, $v_0 = 0.593\times10^6\sqrt{V_0} = 0.59\times10^6\sqrt{4\times10^3} = 3.76\times10^7$ m/s

Gap transit angle, $\theta_g = \dfrac{\omega d}{v_0} = \dfrac{2\pi\times4\times10^9\times2\times10^{-3}}{3.76\times10^7} = 1.337 \text{ rad} = 76.6 \text{ deg}$

The beam coupling coefficient,

$$\beta_1 = \beta_0 = \frac{\sin\theta_g/2}{\theta_g/2} = \sin 38.3^\circ/0.6685 = 0.927$$

DC transit angle in the drift space,

$$\theta_0 = \frac{\omega L}{v_0} = \frac{2\pi \times 4\times 10^9 \times 6\times 10^{-2}}{3.76\times 10^7}\text{ rad } = 40.11 \text{ rad}$$

For maximum output voltage, $X = 1.84$, $J_1(X) = 0.582$, so that the input cavity gap voltage magnitude,

$$V_1 = \frac{2V_0 X}{\beta_1\theta_0} = \frac{2\times 4\times 10^3 \times 4\times 1.841}{0.927\times 40.11}\text{volts}$$

(b) voltage gain

$$A_v = \frac{\beta^2\theta_0 J_1(X) I_0}{XV_0 R_{sh}} = \frac{0.927\times 0.927\times 40.11\times 0.582\times 88\times 10^{-3}}{1.841\times 0.55\times 10^{-4}\times 4\times 10^3} = 4.36 = 12.8 \text{ dB}$$

Catcher voltage, $V_2 = A_v \times V_1 = 4.36 \times 99 = 431.64$ V ■

8.7 MULTI-CAVITY KLYSTRON

Extra cavities help to modulate the electron beam's velocity and increase the output energy. Hence, intermediate cavities are added between the input and output cavities of a klystron amplifier. This will improve the klystron parameters like amplification, efficiency and power output to a great extent. Two-cavity klystron tubes generally have a gain of 10dB-20 dB. Higher gains can be obtained by cascading more two-cavity klystron tubes by connecting the output of a tube to the input of the following tube.

A four-cavity klystron is illustrated in Figure 8.13; with four cavities, power gains of around 50 dB can be easily achieved. Each intermediate cavity increases power gain by 15 to 20 dB. The intermediate cavities are kept at a distance of the bunching parameter X of 1.841, away from the previous cavity, which acts as buncher cavities. These intermediate cavities pass the electron beam and induce a higher RF voltage than the previous cavity. This in turn, results in increased velocity modulation. This increasing beam voltage V_0 could be used in the subsequent cavities. Therefore, more output power can be achieved at the output cavity.

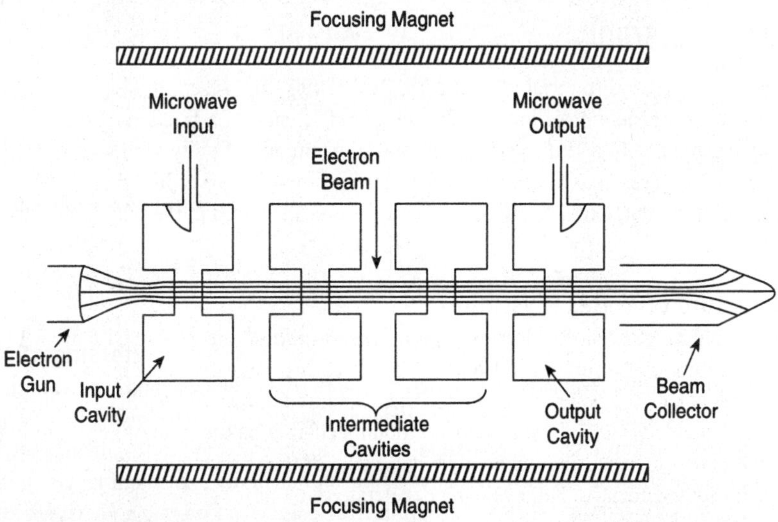

Figure 8.13 *Schematic diagram of a four-cavity klystron*

For a four-cavity klystron,

The output voltage and the output power are given by

$$|V_4| = \frac{1}{8}\left(\frac{I_0\omega}{V_0\omega_q}\right)^3 \beta_0^6 |V_1| R_{sh}^2 R_{shl}$$

$$P_{out} = |I_4|^2 R_{shl} = \frac{1}{64}\left(\frac{I_0\omega}{V_0\omega_q}\right)^6 \beta_0^{12} |V_1|^2 R_{sh}^4 R_{shl} \quad (8.60)$$

where ω = angular frequency

ω_q = reduced plasma frequency(the minimum frequency at which the electron will oscillate in the electron beam)

β_0 = beam coupling coefficient

R_{sh} = total shunt resistance of the input cavity

R_{shl} = total shunt resistance of the output cavity, including external load

I_4 = output current at 4th (or output) cavity

V_1 = magnitude of the input signal

Two types of tuning are possible in multi cavity klystron. They are: *synchronous tuning and staggered tuning. Synchronous tuning*, when all the cavities are tuned to the same frequency, result in high gain and narrow bandwidth. *Staggered tuning*, in which the cavities are tuned differently, result in an increase in the bandwidth of about 800 MHz but results in a decrease of gain of about 45 dB.

Performance Characteristics of a Multi-Cavity Klystron Amplifier:

- Power gain: 40 to 50 dB
- Bandwidth: several percent
- Frequency: 0.5 GHz to 14 GHz
- Power range: 25 kW to 40 MW

8.8 REFLEX KLYSTRON

For applications which require variable frequency, Reflex klystron is used. It is a single cavity variable frequency microwave generator of low power and low efficiency. If a fraction of the output power is fed back to the input cavity and if the loop gain has a magnitude of unity with a phase shift which is a multiple of 2π, the klystron will oscillate. It produces an output power in the range of 10–500 mW and a frequency in the range of 1 to 25 GHz. This type is widely used in the laboratory for microwave measurements.

8.8.1 Structure of Reflex Klystron

The schematic diagram of a simple Reflex klystron is shown in Figure 8.14 (b). Figure 8.14 (a) shows the detailed structure of the reflex klystron. An electron gun, a filament surrounded by a cathode, an accelerating grid at cathode potential, and a repeller constitute the setup. The electron beam is accelerated towards the anode cavity, which is velocity modulated after passing through the gap in the cavity. All the electrons that cross during the positive half cycle of the gap voltage get accelerated and those which cross during the negative half cycle get decelerated.

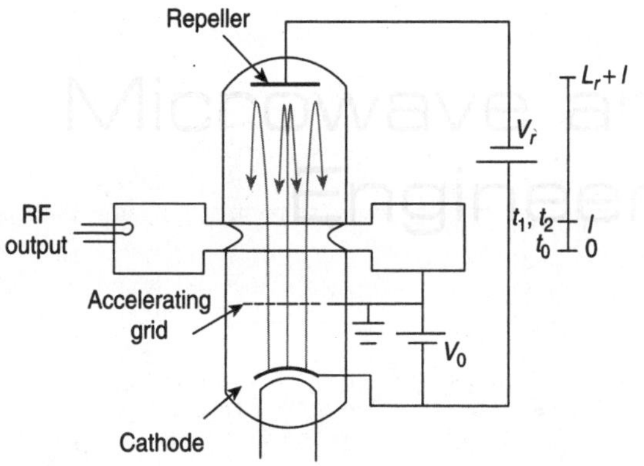

t_0 = time for electron entering cavity gap at $x = 0$
t_1 = the same electron leaving cavity gap at $x = l$
t_2 = time for the same electron returned by retarding field $x = l$ and collected on walls of the cavity.

(a)

Source: electrapk.com

(b)

Figure 8.14 *(a) Constructional details of a reflex klystron; (b) Schematic diagram of a simple reflex klystron*

The velocity modulated electrons travel towards a repeller electrode which is at a high negative potential. The electrons never reach this electrode because of the negative field and are returned back towards the gap. Under suitable conditions, the electron gives more energy to the gap than they took from the gap on their forward journey and oscillations are sustained.

8.8.2 Applegate Diagram and Principle of Working

The electrons, which are reflected due to negative potential at the repeller, enter the cavity which acts as a catcher for these reflected electrons. At the same time, the cavity acts as a buncher for the new electrons. Some of the electrons come together, forming a bunch after spending a different amount of time in the repeller region. This bunching process can be clearly explained in the Applegate diagram as shown in Figure 8.15

In Figure 8.15, the paths of electrons e_e, r_e, and l_e are shown. Let r_e be a reference electron. It passes the gap towards the reflector, there is no effect of the gap voltage, and it is returned to the anode without reaching the reflector. Now consider an electron e_e which passes the gap slightly before r_e. If there is no gap voltage, early electron e_e returns before r_e. However, RF voltage modulates the velocity of electrons. The electron e_e is accelerated by the influence of the positive voltage, and it moves closer to the reflector than r_e.

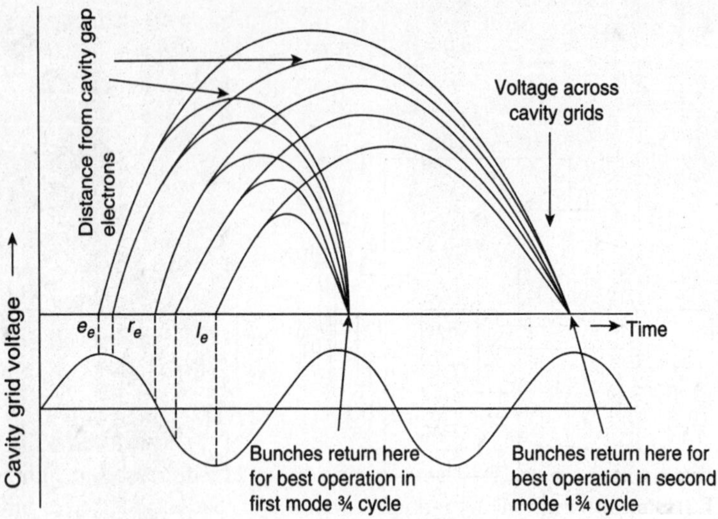

Figure 8.15 *Applegate diagram of reflex klystron*

However, it is possible for r_e to catch up with the electron e_e as it returns into the gap. In a similar manner, the late electron l_e leaving the gap slightly after the electron r_e is under the influence of the negative field, and, hence, it does not reach the reflector as close as the electron r_e. The electron l_e also catches up with e_e and r_e electrons forming a bunch. Once the electrons leave the buncher and catcher gap, the velocity modulation is converted into current modulation. A reference electron is used to form one bunch per cycle of oscillations. For sustained oscillations, the energy is given by these bunches to the gap. An optimum value for time taken by the electrons for the round trip between the repeller space and the gap should be ensured for sustained oscillations.

The transit time (τ) corresponding to oscillation, $\tau = \left(n + \frac{3}{4}\right) T_p$

Where T_p is Time period of arriving electron bunches (or time period of RF signal); n is any integer (0, 1, .., n) that depends on repeller and anode voltages.

Halfway between the catcher grids, maximum bunching occurs. The bunched electron beam should go through the retarding phase. In this retarding phase, the kinetic energy of the bunched electron beam is transferred to the field of the catcher cavity. The electrons finally come out from the catcher grids with a reduced velocity and are accumulated at the collector.

8.8.3 Mathematical Theory of Bunching

The analysis of a reflex klystron is similar to that of a two-cavity klystron. The bunching parameter of the reflex klystron oscillator is given by

$$X' = \frac{\beta_1 V_1}{2V_0}\theta_0$$

where β_1 = beam coupling coefficient

V_1 = amplitude of the input signal applied at the buncher cavity

θ_0 = dc transit angle
V_0 = anode to cathode voltage

Derivation of Bunching Parameter of Reflex Klystron

For simplicity, the effect of space-charge forces on the electron motion will again be neglected. Suppose the oscillator is lying along the x axis with its grid walls at $x = 0$ and $x = l$, let t_0 be the instant at which the reference electron enters the cavity gap at $x = 0$, t_1 be the instant at which the electron leaves cavity gap at $x = l$ and t_2 be the instant at which the same electron returned to the gap by the retarding field at $x = l$ and collected by the walls of the cavity. Let v_0 is the velocity with which the electrons enter the cavity gap after getting accelerated by the potential V_0. The electron entering the cavity gap from the cathode at $x = 0$ and the time t_o are assumed to have uniform velocity

$$v_0 = 0.593 \times 10^6 \sqrt{V_0}$$

The same electron leaves the cavity gap at $x = l$ at time t_1 with velocity

$$v(t_1) = v_0\left[1 + \frac{\beta_1 V_1}{2V_0}\sin\left(\omega t_1 - \frac{\theta_g}{2}\right)\right]$$

Problems are identical to those of the two cavity klystron amplifier as observed up to this point. The retarding electric field E forces the same electron to the cavity at $x = l$ and the time t_2, which is given by

$$E = \frac{V_r + V_0 + V_1 \sin(\omega t)}{L_r} \tag{8.62}$$

where V_r = repeller voltage

$V_1 \sin \omega t$ = RF voltage at cavity gap
V_0 = anode to cathode voltage
L_r = distance between cavity gap and repeller

The x component of retarding field E is taken as constant and the force equation for one electron in the repeller region is

$$m\frac{d^2x}{dt^2} = -eE = -e\frac{V_r + V_0}{L_r} \tag{8.63}$$

where $E = -\nabla V$ is used in the x direction only, V_r is the magnitude of the repeller voltage, and $|V_1 \sin \omega t| << (V_r + V_0)$ is assumed.

Integration of Eq. 8.63 twice yields

$$\frac{dx}{dt} = \frac{-e(V_r + V_0)}{mL_r}\int_{t_1}^{t} dt = \frac{-e(V_r + V_0)}{mL_r}(t - t_1) + K_1$$

At $t = t_1$, $dx/dt = v(t_1) = K_1$; then,

$$x = \frac{-e(V_r + V_0)}{mL_r}\int_{t_1}^{t}(t - t_1)dt + v(t_1)\int_{t_1}^{t} dt$$

$$x = \frac{-e(V_r + V_0)}{2mL_r}(t - t_1)^2 + v(t_1)(t - t_1) + K_2$$

At $t = t_1, x = l = K_2$; then,

$$x = \frac{-e(V_r + V_0)}{2mL_r}(t - t_1)^2 + v(t_1)(t - t_1) + l \tag{8.64}$$

On the assumption that the electron leaves the cavity gap at $x = l$ and time t_1 with a velocity of $v(t_1)$ and returns to the gap at $x = l$ and time t_2, then at $t = t_2, x = l$

$$0 = \frac{-e(V_r + V_0)}{2mL_r}(t_2 - t_1)^2 + v(t_1)(t_2 - t_1)$$

The round-trip transit time in the repeller region is given by

$$T' = t_2 - t_1 = \frac{2mL_r}{e(V_r + V_0)}v(t_1) = T_0'\left[1 + \frac{\beta_1 V_1}{2V_0}\sin\left(\omega t_1 - \frac{\theta_g}{2}\right)\right] \tag{8.65}$$

where

$$T'_0 = \frac{2mL_r v_0}{e(V_r + V_0)} \tag{8.66}$$

Equation 8.66 is the round-trip dc transit time of the center-of-the-bunch electron. Multiplication of Eq. 8.65 through by a radian frequency results in

$$\omega(t_2 - t_1) = \theta_0 + X'\sin\left(\omega t_1 - \frac{\theta_g}{2}\right) \tag{8.67}$$

$$\theta_0 = \omega T_0' \tag{8.68}$$

Equation 8.68 is the round-trip dc transit angle of the center-of-the-bunch electron and

$$X' = \frac{\beta_1 V_1}{2V_0}\theta_0 \tag{8.69}$$

Equation 8.69 is the bunching parameter of the reflex klystron oscillator.

8.8.4 Power Output and Efficiency

The returning electron beam in the klystron should reach the cavity gap when the *RF* field in the cavity gap is maximum retarding so that oscillations can be produced with maximum energy. In this manner, there will be a maximum amount of kinetic energy transfer from the returning electrons to the cavity walls. Figure 8.15 shows that when the energy transfer is maximum, the round trip transit angle is given by

$$\theta_0 = \omega(t_2 - t_1) = \omega T_0' = \left(n - \frac{1}{4}\right)2\pi = N2\pi = 2\pi n - \frac{\pi}{2} \tag{8.70}$$

This transit angle is with reference to the centre of bunch.

Assuming $V_1 << V_0$, n = number of cycles (positive integer), and $N = (n - 1/4)$ is the number of modes.

When the electron beam enters the cavity again from the repeller region, its current modification can be determined in the similar manner as in a two-cavity klystron amplifier. The bunching parameters of a reflex klystron oscillator (X') and of a two-cavity klystron amplifier (X) are of opposite sign. The beam current also flows in negative Z-direction. Therefore we can write the beam current of reflex oscillator as

$$i_2 = -I_0 - \sum_{n=1}^{\infty} 2I_0 J_n(nX')\cos[n(\omega t_2 - \theta_0 - \theta_g)] \tag{8.71}$$

where $\theta_g << \theta_0$, hence θ_g is neglected.

The fundamental component of the current induced in the cavity by the modulated electron beam is given by

$$i_2 = -\beta_1 I_2 = 2I_0\beta_1 J_1(X')\cos(\omega t_2 - \theta_0) \tag{8.72}$$

The magnitude of the fundamental component is

$$I_2 = 2I_0\beta_1 J_1(X') \tag{8.73}$$

The dc power supplied by the beam voltage V_0 is

$$P_{dc} = V_0 I_0 \tag{8.74}$$

and the ac power delivered to the load is given by

$$P_{ac} = \frac{V_1 I_2}{2} = V_1 I_0 \beta_1 J_1(X') \tag{8.75}$$

From Eqs. 8.69 and 8.70, the ratio of V_1 over V_0 is expressed by

$$\frac{V_1}{V_0} = \frac{2X'}{\beta_1(2\pi n - \pi/2)} \tag{8.76}$$

Substituting Eq. 8.76 in Eq. 8.75 gives the power output as

$$P_{ac} = \frac{2V_0 I_0 X' J_1(X')}{2\pi n - \pi/2} \tag{8.77}$$

From Eq. 8.74 and Eq. 8.77, the electronic efficiency of a reflex klystron oscillator can be given as

$$\text{Efficiency} = \frac{P_{ac}}{P_{dc}} = \frac{2X' J_1(X')}{2\pi n - \pi/2} \tag{8.78}$$

The factor $X'J_1(X')$ reaches a maximum value of 1.25 at $X' = 2.408$ and $J_1(X') = 0.52$. Actually, the mode of $n = 2$ has the most power output. If $n = 2$ or $1^3/_4$ mode, the maximum electronic efficiency becomes

$$\text{Efficiency}_{\max} = \frac{2(2.408)J_1(2.408)}{2\pi(2) - \pi/2} = 22.7\% \tag{8.79}$$

The maximum theoretical efficiency of a reflex klystron oscillator ranges from 20 to 30%.

Relationship between reflector voltage and accelerating voltage

The relationship between the repeller voltage and the cycle number required for oscillation is

$$\frac{V_0}{(V_r+V_0)^2}=\frac{1}{8}\frac{1}{\omega^2 L_r^2}\frac{e}{m}\left(2\pi n-\frac{\pi}{2}\right)^2$$

where V_0 = anode to cathode voltage

V_r = reflector voltage

ω = angular frequency

L_r = distance from cavity grid to repeller electrode

n = mode number

Derivation of Relation Between Reflector Voltage and Accelerating Voltage

When $V_1 << V_0$, the presence of RF voltage does not effect the bunch of electrons at the center.

$$\omega t_2=\omega t_1+\theta_g$$

In order to transfer maximum amount of energy, mode 2 is used where the repeller voltage produces an electron transit of 1 ¾ cycles. Then, for maximum energy transfer, the optimum value of θ_g is

$$\theta_g'=\left(2\pi n-\frac{\pi}{2}\right)$$

Since each cycle consists of $2\pi\left(n-\frac{1}{4}\right)$ when $\left(n-\frac{1}{4}\right)=\frac{3}{4},1\frac{3}{4}$ and so on,

However,

$$\theta_g'=\frac{2mL_r\omega}{e(V_r+V_0)}v_0 \tag{8.80}$$

$$v_0=\frac{e(V_r+V_0)}{2mL_r\omega}\theta_g' \tag{8.81}$$

From the mass and voltage relationship of electrons:

$$\frac{1}{2}mv_0^2=eV_0 \text{ or } V_0=\frac{m}{2e}v_0^2$$

$$\therefore V_0=\frac{m}{2e}\frac{e^2(V_r+V_0)^2\theta_g'^2}{4\omega^2 m^2 L_r^2}$$

$$\frac{V_0}{(V_r+V_0)^2}=\frac{m}{2e}\frac{e^2}{4\omega^2 m^2 L_r^2}\left(2\pi n-\frac{\pi}{2}\right)^2$$

For a given beam voltage V_0, the relationship between the repeller voltage and the cycle number required for oscillation is

$$\frac{V_0}{(V_r+V_0)^2}=\frac{1}{8}\frac{1}{\omega^2 L_r^2}\frac{e}{m}\left(2\pi n-\frac{\pi}{2}\right)^2 \tag{8.82}$$

The power output can be expressed in terms of the repeller voltage V_r. That is,

$$P_{ac} = \frac{V_0 I_0 X' J_1(X')(V_r + V_0)}{\omega L_r}\sqrt{\frac{e}{2mV_0}} \tag{8.83}$$

It can be seen from Eq. 8.82 that, we can determine the center repeller voltage V_r in terms of centre frequency, if the beam voltage V_0 and cycle number n or mode number N are given.

From Eq. 8.83, the power output can be determined. When the frequency changes from the centre frequency and the repeller voltage from the centre voltage, the power output will also be changed, and a bell shape is formed (Figure 8.16).

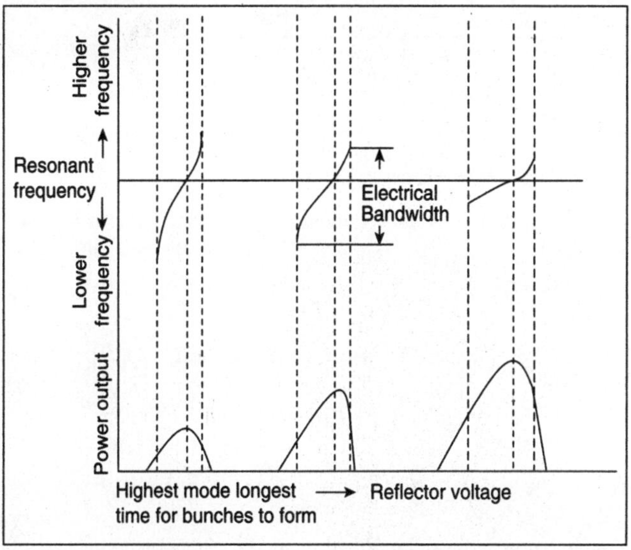

Figure 8.16 *Power output and frequency characteristics of a reflex klystron*

8.8.5 Electronic Admittance

The electronic admittance is a function of the dc beam admittance, the transit angle, and the second transit of the electron beam through the cavity gap.

From Eq. 8.72, the induced current can be written in phasor form as

$$i_2 = 2I_0 \beta_i J_1(X') e^{-j\theta_0} \tag{8.84}$$

The voltage across the gap at time t_2 can also be written in phasor form:

$$V_2 = V_1 e^{-j\pi/2} \tag{8.85}$$

The ratio of i_2 to $V_2$is defined as the electronic admittance of the reflex klystron.That is,

$$Y_e = \frac{I_0}{V_0}\frac{\beta_i^2 \theta_0}{2}\frac{2J_1(X')}{X'} e^{j(\pi/2 - \theta_0)} \tag{8.86}$$

From Eq. 8.86, it is evident that the electron admittance is non linear, as it is proportionl to the factor $2J_1(X')/X'$, and X' is proportional to the signal voltage. When the signal voltage reaches zero, the factor approaches unity.

The equivalent circuit of a reflex klystron is shown in Figure 8.17. It consists of a parallel combination of L and C, with both representing the energy storage elements of the cavity. The three conductances G_c, G_b, and G_l represent copper losses, beam loading, and load conductance, respectively.

The condition required for the oscillation is that, the total conductance of the cavity circuit should not be greater than the magnitude of the negative real part of the electronic admittance.

$$|-G_e| \geq G \tag{8.87}$$

where $G = G_c + G_b + G_l = \dfrac{1}{R_{sh}}$ and R_{sh} is the effective shunt resistance.

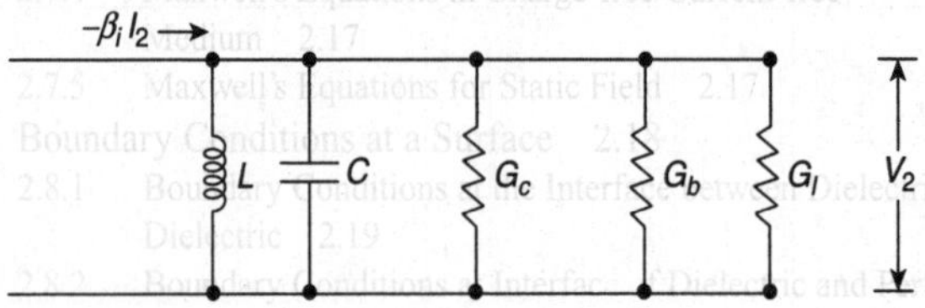

Figure. 8.17 *Equivalent circuit of a reflex klystron*

Equation 8.86 can be rewritten in rectangular form:

$$Y_e = G_e + jB_e \tag{8.88}$$

The electronic admittance shown in Eq. 8.86 is in exponential form; its phase is $\pi/2$ when θ_0 is zero. The rectangular plot of the electronic admittance Y_e is a spiral (see Figure 8.18). Any value of θ_0 for which the spiral lies in the area to the left of line ($-G - jB$) will yield oscillation. That is,

$$\theta_0 = \left(n - \frac{1}{4}\right) 2\pi = N 2\pi \tag{8.89}$$

where N is the mode number as indicated in the plot, and the phenomenon verifies the early analysis.

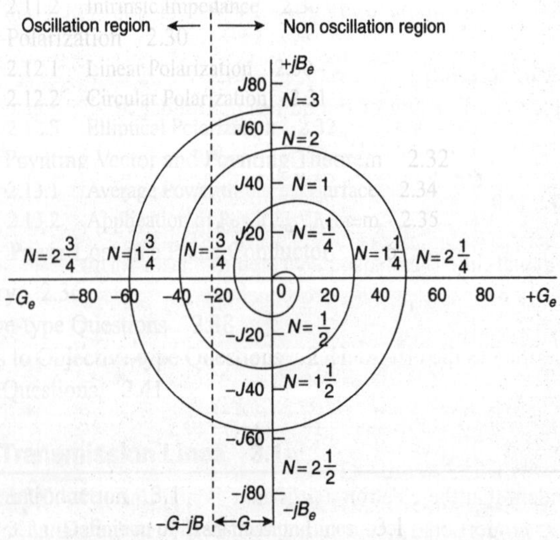

Figure 8.18 *Modes of reflex klystron*

8.8.6 Oscillating Modes and Output Characteristics

Mode curves

The output frequency and the output power vary with the changes in repeller voltage for different modes as shown in Figure 8.16. These modes are called mode curves. The oscillation frequency is determined by the frequency of resonance of the output cavity. This is called as electronic tuning range of reflex klystron. Therefore the reflex klystron can be used as frequency modulated oscillator (voltage tunes oscillator).

Operating characteristics

The adjustment of repeller and anode voltage is in such a way that the bunch appears exactly at any of the positive maximum voltage of the RF signal, which is necessary for reflex klystron to undergo oscillations. The oscillations can be achieved only for some combination of anode and repeller voltages. The voltage or operating characteristics of reflex klystron are shown in the Figure 8.19 where the repeller and beam voltage combinations are represented in shaded portion. For a fixed frequency these diagrams are drawn. The pattern remains same for other frequencies, but there is a shift in the positions of regions because of variations in n. Large output power is obtained at lower modes which require high values of repeller and beam voltages. The modes corresponding to $n = 2$ or $n = 3$ are preferred usually.

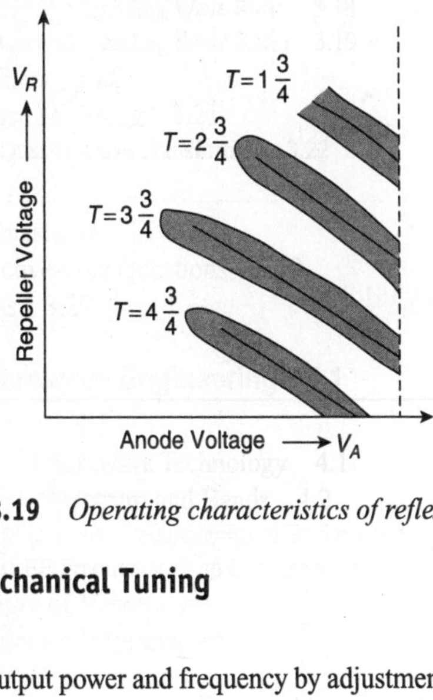

Figure 8.19 *Operating characteristics of reflex klystron*

8.8.7 Electronic and Mechanical Tuning

Electronic tuning

The nature of the variation of output power and frequency by adjustment of the repeller voltage is called the *electronic tuning*. It can be measured by electronic tuning sensitivity (ETS). This can be determined by considering the slope of the frequency characteristics of the modes. The tuning range is about ±8 MHz at the X-band and ±80 MHz for sub-millimeter klystron.

We know the equation

$$(V_r + V_0)^2 = \frac{8mL_r^2V_0}{\left(2\pi n - \frac{\pi}{2}\right)^2 .e} .\omega^2 \quad (8.90)$$

Differentiating V_r with regard to ω, we get

$$\frac{dV_r}{d\omega} = \frac{8mL_r^2V_0}{e.\left(2\pi n - \frac{\pi}{2}\right)^2} \cdot \frac{\omega}{(V_r + V_0)} \tag{8.91}$$

$$\frac{dV_r}{d\omega} = \frac{8mL_r^2V_0\omega}{e.\left(2\pi n - \frac{\pi}{2}\right)^2} \cdot \frac{1}{\sqrt{\frac{8mL_r^2V_0}{\left(2\pi n - \frac{\pi}{2}\right)^2 .e} .\omega^2}}$$

$$= \sqrt{\frac{8mL_r^2V_0}{e}} \cdot \frac{1}{\left(2\pi n - \frac{\pi}{2}\right)} \tag{8.92}$$

$$\frac{dV_r}{df} = \sqrt{\frac{8mV_0}{e}} \cdot \frac{2\pi L_r}{\left(2\pi n - \frac{\pi}{2}\right)} \tag{8.93}$$

This is a very useful relationship for the electronic tuning of reflex klystron. If the repeller voltage varies by even 2%, frequency will vary considerably.

EXAMPLE PROBLEM 8.7

Typically, if $V_r = 2000$ V, $V_0 = 500$ V, drift space = 2 cm, mode $n = 1$, $f = 2$ GHz, and variation in V_r is 2%.

Then,
$$df = \frac{dVr}{\sqrt{\frac{8mV_0}{e}} . \frac{2\pi L_r}{\left(2\pi n - \frac{\pi}{2}\right)}}$$

$$= \frac{\frac{2}{100} \times 2000}{\left[2\pi \times \frac{2}{100} \times \frac{1}{2\pi - \frac{\pi}{2}} \sqrt{\frac{8 \times 9 \times 10^{-31} \times 500}{1.6 \times 10^{-19}}}\right]} = 10 \text{ MHz}$$

Variation in frequency is quite sensitive to repeller voltage adjustments, draws large currents, and gets overheated. The precaution to be taken is that, the application of repeller voltage should be before the application of anode voltage and the connection of a protective diode across the klystron, so that the repeller can never become positive. This device is very easy to modulate frequency, simply by the application of modulating voltage to the repeller. ■

Mechanical tuning

In mechanical tuning, the effective capacitance and the resonant frequency of the klystron changes by changing the width of the cavity. But the output power remains unchanged inspite of the tuning. Mechanical tuning of reflex klystron may give a frequency variation that ranges from ± 20 MHz at the *X*-band to ± 4 GHz at 200 GHz.

The frequency of resonance is mechanically adjusted by the following methods:

- By mechanically adjustable screws or the most popular method called *post*
- The *Q* of the cavity depends on its tuning by a screw or sliding piston. The current flows through the tuning elements. Because of the presence of these elements, the area becomes larger which decrease the *Q*. The resonant frequency is varied because of the introduction of dielectric materials.
- In another method there will be a wall which can be moved in or out slightly by an automatic screw, and it tightens or loosens small bellows. Sometimes this method can be used with the permanent cavities in reflex klystron which act as a form of limited frequency shifting. Thus, mechanical tuning can also be used to change the resonant frequency.

Performance characteristics of Reflex klystron:

- Frequency range : 4 to 200 GHz
- Output power : 1.0 mW to 2.5 W
- Theoretical efficiency : 22.78%
- Practical efficiency : 10% to 20%
- Tuning range : 5 GHz at 2 watts to 30 GHz at 10 mW

Applications of Reflex klystron:

- In radar receivers
- Local oscillator in microwave receivers
- Signal source in microwave generators of variable frequency
- Portable microwave links
- Pump oscillators in parametric amplifiers

EXAMPLE PROBLEM 8.8

A reflex klystron operates under the following conditions: $V_0 = 600$ V, $L_r = 1$ mm, $R_{sh} = 15$ kΩ, e/m = 1.759×10^{11} (MKS system), and $f_r = 9$ GHz. The tube is oscillating at f_r at the peak of the $n = 2$ mode or $1\frac{3}{4}$ mode. Assume that the transit time through the gap and beam loading can be neglected.

(a) Find the value of the repeller voltage V_r.

(b) Find the direct current that is necessary to give a microwave gap voltage of 200 V.

(c) What is the electronic efficiency under this condition?

Solution

(a) We know that

$$\frac{V_0}{(V_r + V_o)^2} = \left(\frac{e}{m}\right)\frac{(2\pi n - \pi/2)^2}{8\omega^2 L_r^2}$$

$$= (1.759 \times 10^{11}) \frac{(2\pi \times 2 - \pi/2)^2}{8(2\pi \times 9 \times 10^9)^2 (10^{-3})^2} = 0.832 \times 10^{-3}$$

$$(V_r + V_0)^2 = \frac{600}{0.832 \times 10^{-3}} = 0.721 \times 10^6$$

$$V_r = 250 \text{ V}$$

(b) Assume that $\beta_0 = 1$. as $V_2 = I_2 R_{sh} = 2I_0 J_1(X') R_{sh}$
The direct current I_0 is

$$I_0 = \frac{V_2}{2J_1(X')R_{sh}} = \frac{200}{2(0.582)(15 \times 10^3)} = 11.45 \text{ mA}$$

(c) The equation for electronic efficiency is given by

$$\text{Efficiency} = \frac{2X' J_1(X')}{2\pi n - \pi/2} = \frac{2(1.84)(0.582)}{2\pi(2) - \pi/2} = 19.49\%$$

■

EXAMPLE PROBLEM 8.9

The reflex klystron operates at $V_0 = 1200$ V, $f = 10$ GHz, and $L_r = 8$ cm. Calculate repeller voltage.

Solution

We know that in order to calculate the repeller voltage, V_r. We have the formula

$$\frac{V_0}{(V_r + V_0)^2} = \frac{1}{8} \frac{1}{\omega^2 L_r^2} \frac{e}{m} \left[2\pi n - \frac{\pi}{2} \right]^2$$

$$\frac{1200}{(V_r + 1200)^2} = \frac{1}{8} \frac{(1.759 \times 10^{11}) \left(2\pi \times 2 - \frac{\pi}{2} \right)^2}{(2\pi \times 10 \times 10^9)^2 (8 \times 10^{-2})^2}$$

Therefore, repeller voltage $V_r = 108$ kV (approx.)

■

EXAMPLE PROBLEM 8.10

The beam voltage $V_0 = 250$ V, beam current $I_0 = 15$ mA, and the signal voltage $V_m = 35$ V are the parameters of a reflex klystron which operates at the mode $n = 2$. Find the input voltage and electronic efficiency.

Solution

Input power,
$$P_{dc} = V_0 I_0$$
$$= 250 \times 15 \times 10^{-3}$$
$$= 3.75 \text{ watts}$$

In order to calculate the efficiency, we have to calculate the output power

$$\text{Output power, } P_{ac} = \frac{2V_0 I_0 X' J_1(X')}{2n\pi - \frac{\pi}{2}} = \frac{2 \times 250 \times 15 \times 10^{-3} \times 2.408 \times 0.52}{4\pi - \frac{\pi}{2}}$$

$$= 0.85 \text{ watts}$$

Now, efficiency $\eta = \frac{P_{ac}}{P_{dc}} \times 100 = \frac{0.85}{3.75} \times 100$

Therefore, efficiency is $\eta = 22.7\%$ ■

EXAMPLE PROBLEM 8.11

The operating frequency of a reflex klystron is 10 GHz, It has a DC beam voltage of 200 V, a repeller spacing of 0.1cm for 1¾ mode. Determine the maximum value of power and the corresponding repeller voltage for a beam current of 60 mA.

Solution

Maximum power,

$$P_{max} = \frac{1.25V_0 I_0 (V_R + V_0)}{\omega L_r} \times \sqrt{\frac{e}{2mV_0}}$$

Repeller voltage must be calculated first in order to calculate maximum power

Now, $$\frac{V_0}{(V_R + V_0)^2} = \frac{1}{8}\frac{1}{\omega^2 L_r^2}\frac{e}{m}\left(2\pi n - \frac{\pi}{2}\right)^2$$

$$\frac{200}{(V_R + 200)^2} = \frac{1}{8}\frac{\left(1.759\times10^{11}\right)\left(2\pi\times2 - \frac{\pi}{2}\right)^2}{\left(2\pi\times10\times10^9\right)^2\left(0.1\times10^{-2}\right)^2} = 6.71\times10^{-4}$$

Therefore, repeller voltage = 372 V and maximum power is

$$P_{max} = \frac{1.25V_0 I_0 (V_R + V_0)}{\omega L_r} \times \sqrt{\frac{e}{2mV_0}}$$

$$= \frac{1.25\times200\times60\times10^{-3}\times(346 + 200)}{2\pi\times10\times10^9\times0.1\times10^{-2}} \times \sqrt{\frac{1.759\times10^{11}}{2\times200}} = 2.73 \text{ watts}$$

Therefore, maximum value of power is equal to 2.73 W ■

8.9 TRAVELING-WAVE TUBE

Traveling-wave tube (TWT) is a broadband device. In this device the propagating speed of the wave is same as that of the electrons in the beam, and the microwave circuit is non resonant. The weak electric fields which are associated with the travelling wave produces a small amount of velocity modulation that initially effects the beam. Later on this velocity modulation can be translated into current modulation same as in the klystron, where the RF current is induced in the circuit, causing the amplification. For better understanding of TWT, a comparison is made between the TWT and klystron (Table 8.1).

Table 8.1 Comparison of TWT and klystron

TWT	Klystron
1. RF signal travels along with the beam.	1. RF signal is stationary and only the beam travels.
2. The interaction of electron beam and RF field in the TWT is continuous over the entire length of the circuit.	2. The interaction of electrons in the klystron occurs only at the gaps of a few resonant cavities.
3. The TWT circuit is nonresonant.	3. The klystron circuit is resonant type.
4. The wave in TWT is a propagating wave.	4. The wave in klystron is not a propagating wave.
5. TWT uses slow-wave structures for input and output.	5. Klystron uses cavities for input and output.
6. High-power output	6. Low-power output
7. Long-life period	7. Short-life period

8.9.1 Significance of TWT

Travelling wave tubes (TWTs) have gains of 40 dB and above, with bandwidths more than an octave. A bandwidth of 1 octave is one in which the upper frequency is twice the lower frequency. TWTs have high-gain, low-noise, and wide bandwidth and this make them ideal for RF amplifier. The frequency range of TWTs is from 300 MHz to 50 GHz. Voltage amplification is the main application of TWTs (even if, same characteristics are developed for the high power TWTs and the power klystron).

TWT is a broadband slow-wave device. Its operation is based on the interaction between the traveling wave structure and the electron beam. For extending the interaction of electron beam and RF field, it is compulsory to make sure that both of them travel in the same direction having virtually same velocity. TWT is a linear beam tube in which the interaction between the electron beam and the RF field is continuous over the full length of the tube.

8.9.2 Types and Characteristics of Slow-wave Structures

The traveling-wave tubes (TWTs) are commonly employed where a high power is required. The ordinary resonators, which are used in klystrons, cannot generate a large output, because the gain-bandwidth product is limited by the resonant circuit.The TWT uses slow-wave structures in its construction for obtaining large gain over a wide bandwidth.The special features of slow-wave structures can reduce the RF wave velocity in a certain direction so that the electron beam and signal wave can interact over a length.

The phase velocity of a wave in ordinary waveguides is greater than the velocity of light in vacuum. In the operation of TWT, the electron beam should keep in step with the microwave signal. Since the electron beam can be accelerated only to velocities that are about a fraction of the velocity of light, a slow-wave structure should be incorporated in the microwave devices so that the phase velocity of the micro wave signal can keep pace with that of the electron beam.

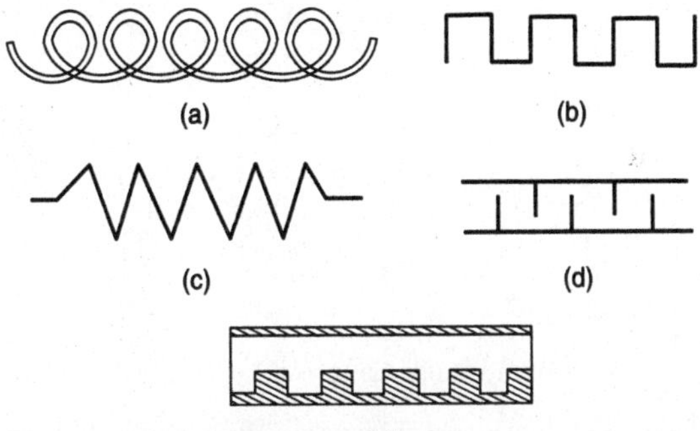

Figure 8.20 *Types of slow-wave structures: (a) Helical line; (b) Folded back line; (c) Zigzag line; (d) Inter-digital line; (e) Corrugated waveguide*

Helix: Different types of slow-wave structures are shown in Figure 8.20. A helix is the most commonly used slow-wave structure. It consists of a thin ribbon of metal that is wound into a helical structure. A helix is also constructed by the use of a round wire that acts as a slow-wave structure.

From the Figure 8.21, the ratio of the phase velocity v_p (phase velocity along the pitch) and c (phase velocity along the coil) is given as

$$\frac{v_p}{c} = \frac{p}{\sqrt{p^2 + (\pi d)^2}} = \sin\psi$$

where $c = 3 \times 10^8$ m/s is the velocity of light in free space
p = helix pitch
d = diameter of the helix
ψ = pitch angle

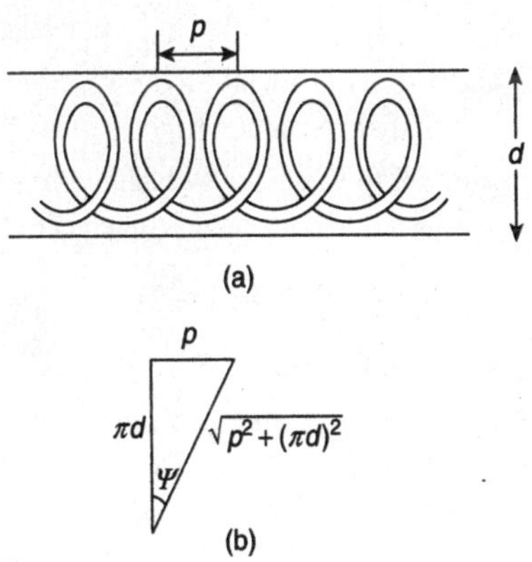

Figure 8.21 *(a) Helical coil; (b) One turn of the helix*

Mostly, the helix is surrounded by a dielectric filled cylinder. In the axial direction, the phase velocity can be given as

$$v_{pe} = \frac{p}{\sqrt{\mu\varepsilon\left[p^2 + (\pi d)^2\right]}} \tag{8.94}$$

Care has to be taken about the dielectric constant, such that it is not too large, because the slow wave structure causes a sizeable loss to the microwave devices and thus the efficiency is reduced. If we consider the case of small pitch angle, the phase velocity along the coil in free space is given by

$$v_p \approx \frac{pc}{\pi d} = \frac{\omega}{\beta} \tag{8.95}$$

The ω-β (or Brillouin) diagram as shown in Figure 8.22 is very useful in designing a helix slow-wave structure. Once β is found, v_p can be computed from Eq. 8.95. Furthermore, the group velocity of the wave is merely the slope of the curve and is given by

$$v_g = \frac{\partial\omega}{\partial\beta} \tag{8.96}$$

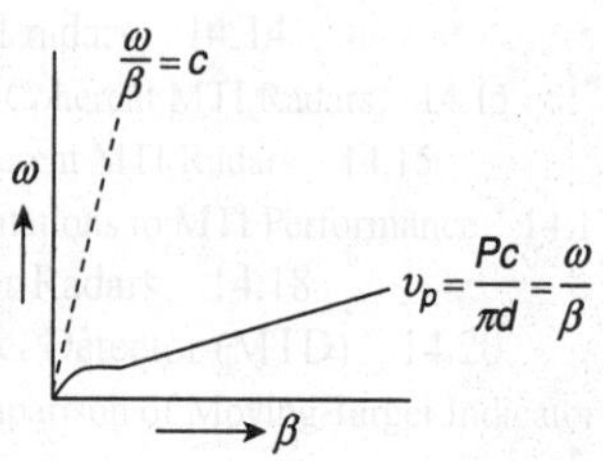

Figure 8.22 *ω-β diagram for a helical structure*

In order to maintain a slow-wave structure, it should have the property of periodicity in the axial direction. The Fourier analysis of wave guide is used to obtain the phase velocity of some of the spatial harmonics in the axial direction, which may be smaller than the velocity of light. If we move a distance of one pitch length either in forward direction or backward direction along the helical slow wave structure, we obtain identical structure again. Hence, the period of helical slow-wave structure can be taken as its pitch.

The helix periodic structure can be expanded as an infinite series of waves with a period 'L', all at the same frequency but with different phase velocities, and is given by

$$v_{pn} = \frac{\omega}{\beta_n} \equiv \frac{\omega}{\beta_0 + (2\pi n / L)} \tag{8.97}$$

The group velocity that can be calculated from Eq.8.96 is

$$v_g = \left[\frac{d(\beta_0 + 2\pi n / L)}{d\omega}\right]^{-1} = \frac{\partial\omega}{\partial\beta_0} \tag{8.98}$$

where β_0 is the phase constant of the average electron velocity

L = period of the helix

n = any integer value

From the above Eq. 8.97, we can observe that, for higher values of β_0 and positive n, the phase velocity in the axial direction decreases. Thus, for suitable values of n, the phase velocity of the wave is less than the velocity of light. At that time, there is the possibility of communication between the electron beam and the microwave signal, and also the amplification of microwave devices can be achieved.

The ω - β diagram for a helical slow-wave structure with several spatial harmonics is shown in Figure 8.23. The second quadrant of the ω - β diagram indicates the negative phase velocity that corresponds to the negative *n*. It is clearly understood that the electron beam is moving in positive *z*-direction and the beam velocity matches with the negative spatial harmonic's phase velocity. The shaded area shown in figure is the area where the propagation of the wave is not allowed, because if the axial phase velocity of any spatial harmonic is more than the velocity of light, at that moment the structure will start radiating energy, which is not desirable.

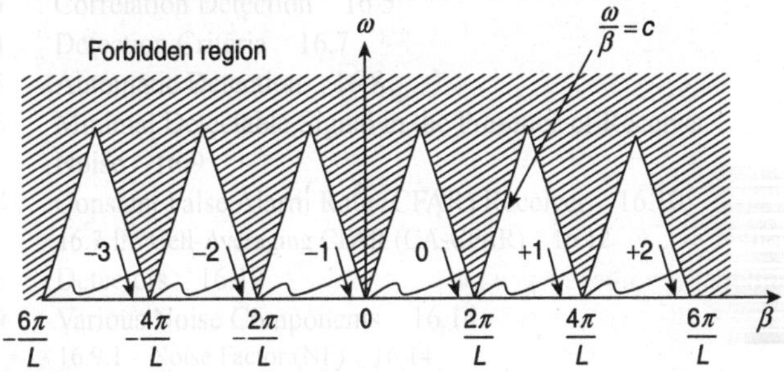

Figure 8.23 *ω -β diagram for spatial harmonics of a helical structure*

8.9.3 Structure of TWT and Amplification Process

The schematic diagram of a typical TWT is shown in Figure 8.24. The TWT consists of an electron gun that is used to produce a narrow constant velocity electron beam. This electron beam is, in turn, passed through the center of a long axial helix. Hence we use a magnetic field of high focusing capacity to avoid spreading and it will guide the wave through the centre of the helix. A helix is a loosely wound, thin conducting helical wire that acts as a slow-wave structure. The signal to be amplified is applied to the end of the helix that is adjacent to the electron gun. The amplified signal appears at the output or the other end of the helix under appropriate conditions.

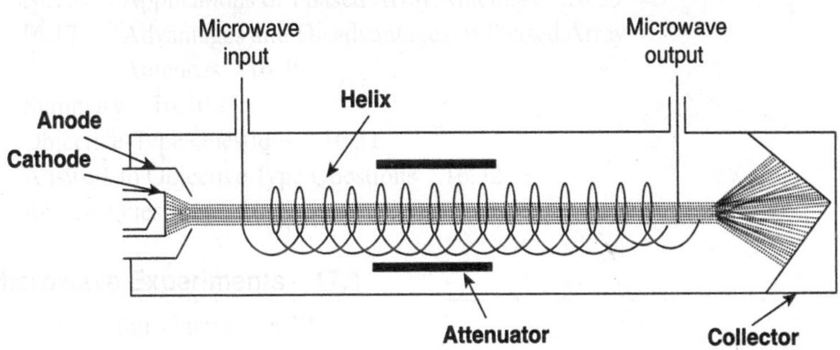

Figure 8.24 *Schematic diagram of a* Traveling-Wave Tube

Amplification process

According to the ω - β diagram (Figure 8.23), the phase shift per period *L* of the fundamental wave on the helix slow-wave structure is given by

$$\theta_1 = \beta_0 L \tag{8.99}$$

where $\beta_0 = \omega / v_0$ is the phase constant of the average beam velocity.

Since the dc transit time of an electron is given by

$$T_0 = \frac{L}{v_0} \tag{8.100}$$

The phase constant of the nth space harmonic is

$$\beta_n = \frac{\omega}{v_0} = \frac{\theta_1 + 2\pi n}{v_0 T_0} = \beta_0 + \frac{2\pi n}{L} \tag{8.101}$$

In Eq. 8.101, for the interaction between electronic beam and electric field, it is assumed that the axial space harmonic velocity should be synchronized with beam velocity.

$$\upsilon_{pn} = v_0 \tag{8.102}$$

When the signal voltage is coupled into the helix, a force is acted upon the electrons due to the axial electric field. The amount of force is given by

$$F = -eE$$
$$E = -\nabla V$$

When we give an RF signal as input to the helix, part of RF signal's electric field is in parallel with the direction of the electron beam and this causes an interaction between RF signal and the electronic beam. Bunching occurs due to interaction of electron beam and RF signal when the electrons in the beam are accelerated and travel faster than the RF signals.

The interaction between fields formed by these bunches and field from the RF signal produces amplification of RF signal. Each newly formed electron bunch adds a small amount of energy to the RF signal which is travelling in the helix as shown in the Figure 8.25. Now this merely amplified RF signal interacts with a dense electron bunch which again interacts with a denser electron bunch and gives additional energy to the RF signal. These types of interactions occur continuously over the full length of the helix. This energy is then coupled from the helix to the output side.

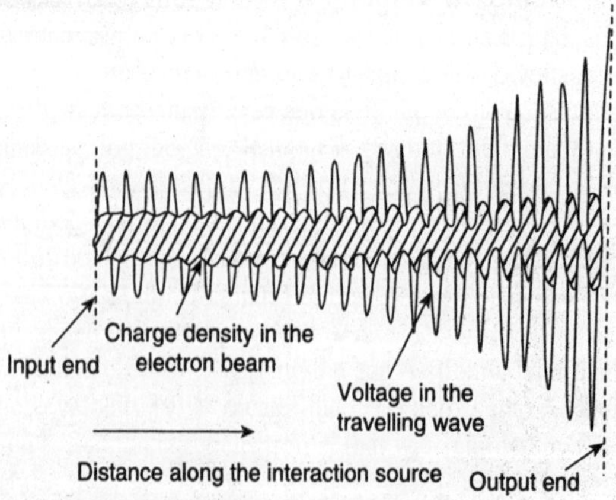

Figure 8.25 *Growth of signal and bunching along traveling-wave tube*

Near the centre of the helix an attenuator is placed and it reduces the waves travelling by the side of the helix to zero. Therefore the reflected waves from the mismatched loads are prohibited from reaching the input and causing oscillation. A new electric field having same frequency can be induced by the bunched electrons from the attenuator.

A new amplified microwave signal gets induced on the helix from this field. The analysis of motion of electrons in the helix type travelling wave tube can be done in terms of axial electric field.

Mathematical analysis of the physical phenomenon taking place in TWT amplifier

If the direction of the propagation of the travelling wave is in z-direction then the z-component of electric field can be given as

$$E_z = E_1 \sin(\omega t - \beta_p z) \tag{8.103}$$

where E_1 is the magnitude of the electric field in the z direction.

The electric field is assumed to be maximum, when $t = t_o$ at $z = 0$. $\beta_p = \dfrac{\omega}{v_p}$ is the axial phase constant of the microwave, and the axial phase velocity of the wave is v_p. The equation for electron motion is given as

$$m\frac{dv}{dt} = -eE_1 \sin(\omega t - \beta_p z) \tag{8.104}$$

Assume that the velocity of the electron is

$$v = v_0 + v_e \cos(\omega_e t + \theta_e)$$

$$\Rightarrow \frac{dv}{dt} = -v_e \omega_e \sin(\omega_e t + \theta_e) \tag{8.105}$$

where v_0 = dc electron velocity

v_e = magnitude of velocity fluctuation in the velocity-modulated electron beam

ω_e = angular frequency of velocity fluctuation

θ_e = phase angle of the fluctuation

Substituting Eq. 8.105 in Eq. 8.104,

$$mv_e\omega_e \sin(\omega_e t + \theta_e) = eE_1 \sin(\omega t - \beta_p z) \tag{8.106}$$

The velocity of the velocity modulated electron beam should be equal to dc electron velocity for the interaction between the electrons and the electric field. This is

$$v \approx v_0$$

Hence, the distance z traveled by the electrons is

$$z = v_0(t - t_0)$$

and

$$mv_e\omega_e \sin(\omega_e t + \theta_e) = eE_1 \sin\left(\omega t - \beta_p v_0 (t - t_0)\right) \tag{8.107}$$

A comparison of the left-hand and right-hand sides of Eq. (8.107) shows that

$$v_e = \frac{eE_1}{m\omega_e}$$

$$\omega_e = \beta_p(v_p - v_0)$$

$$\theta_e = \beta_p v_0 t_0 \tag{8.108}$$

From the above relationship we can say that there is a directly proportional relation between the magnitude of velocity fluctuation and magnitude of the axial electric field.

- For determining the relationship between the circuit and electron beam quantities, two terms should be calculated. They are
- The convection current of the axial electric field
- The axial electric field

Convection current of Axial Field: Considering the space charge effect, the electron velocity, the charge density, the current density, and the axial electric field will perturbate depending on their averages or dc values. These quantities can be mathematically written as below:

Electron velocity = $v = v_0 + v_1 e^{(j\omega t - \gamma z)}$

Charge density = $P = p_0 + p_1 e^{(j\omega t - \gamma z)}$

Current density = $J = -J_0 + J_1 e^{(j\omega t - \gamma z)}$

(minus sign indicates that J_0 may be positive in the negative z direction)

Axial electric field $E_z = E_1 e^{j\omega t - \gamma z}$

where $\gamma = \alpha + j\beta$ is propagation constant of the axial waves
The convention current in electron beam is given by

$$i = j\frac{\beta_e I_0}{2V_0(j\beta_e - \gamma)^2} E_1 \tag{8.109}$$

where $\qquad \beta_e = \dfrac{\omega}{v_0}$

$v_0 = \sqrt{\dfrac{2eV_0}{m}}$ has been used.

Axial Electric Field: In the slow wave circuit, an electric field gets induced into the electron beam by the convection current. This induced electric field gets added to the field which already exists in the circuit and causes the circuit power to increase with distance.

$$V = -\frac{\gamma\gamma_0 Z_0}{\gamma^2 - \gamma_0^{\,2}} i$$

Since $E_z = -\nabla V = -(\partial V / \partial z) = \gamma V$, the axial electric field is given by

$$E = -\frac{\gamma^2 \gamma_0 Z_0}{\gamma^2 - \gamma_0^{\,2}} i \tag{8.110}$$

where

$$\gamma_0 = j\omega\sqrt{LC}$$

$$\text{Characteristic impedance } (Z_0) = \sqrt{\frac{L}{C}}$$

Equation 8.110 is known as a *circuit equation*, and it determines the effect of the spatial ac electron beam current on the axial electric field of the slow-wave helix.

8.9.4 Suppression of Oscillations

In order to prevent oscillations from being spontaneously generated in a traveling-wave tube, it is necessary to prevent internal feedback arising from reflections due to slight impedance mismatches at the output terminal. The energy reflected at the output terminal will travel back to the gun end of the TWT, and on reflection, it provides a feedback signal that is further amplified along the desired signal.

It is necessary to prevent backward-wave oscillations from being generated in TWT. This situation is controlled by introducing an attenuator which is placed near the input end of the TWT that absorbs any wave propagated along the helix. Aquadag is used to coat the glass wall of TWT. It acts as an attenuator and attenuates the parasitic signals and spurious signals. The attenuator does not affect the bunching of the electrons.

8.9.5 Nature of the Four Propagation Constants

By solving the electronic and circuit equations at the same time the wave modes of helix type travelling wave tube are determined. Thus, the values of the four propagation constants γ are given by

$$\gamma_1 = -\beta_e C\frac{\sqrt{3}}{2} + j\beta_e\left(1+\frac{C}{2}\right)$$

$$\gamma_2 = \beta_e C\frac{\sqrt{3}}{2} + j\beta_e\left(1+\frac{C}{2}\right)$$

$$\gamma_3 = j\beta_e(1-C)$$

$$\gamma_4 = -j\beta_e\left(1-\frac{C^3}{4}\right)$$

Derivation of Expression for Four Propagation Constants of TWT

From Eqs. 8.109 and 8.110, it can be observed that there are four different solutions for the propagation constants. It implies that there are four modes of travelling waves in the O-type travelling-wave tube. Substituting Eq. 8.109 in Eq. 8.110 gives

$$(\gamma^2 - \gamma_0^{\ 2})(j\beta_e - \gamma)^2 = -j\frac{\gamma^2\gamma_0 Z_0\beta_e I_0}{2V_0} \tag{8.111}$$

It can be seen that the above equation is of fourth order in γ and therefore it has four roots. By numerical methods and digital computer, exact solutions can be obtained. On the other hand , by equating the dc electron beam velocity to the axial phase velocity of the travelling wave , we can get the approximate solutions, which is comparable to

$$\gamma_0 = j\beta_e$$

Then, Eq. 8.111 is reduced to

$$(\gamma - j\beta_e)^3(\gamma + j\beta_e) = 2C^3\beta_e^2\gamma^2 \tag{8.112}$$

where C is the travelling-wave tube gain parameter and is given as

$$C = \left(\frac{I_0 Z_0}{4V_0}\right)^{1/3} \tag{8.113}$$

From Eq. 8.112, it can be observed that there are three travelling waves equivalent to $e^{-j\beta_e Z}$ and one backward travelling wave which is equivalent to $e^{+j\beta_e Z}$. For the three forward travelling waves, the propagation constant is given by

$$\gamma = j\beta_e - \beta_e C\delta \tag{8.114}$$

where it is assumed that $C\delta << 1$

Substitution of Eq. 8.114 in Eq. 8.112 results in

$$(-\beta_e C\delta)^3(j2\beta_e - \beta_e C\delta) = 2C^3\beta_e^2(-\beta_e^2 - 2j\beta_e^2 C\delta + \beta_e^2 C^2\delta^2) \tag{8.115}$$

Since $C\delta << 1$, Eq. 8.115 is reduced to

$$\delta = (-j)^{1/3} \tag{8.116}$$

From the theory of complex variables, the three roots of $(-j)$ can be plotted in Figure 8.26.
Equation 8.116 can be written in exponential form as

$$\delta = (-j)^{1/3} = e^{-j[(\pi/2 + 2n\pi)/3]} \quad (n = 0, 1, 2)$$

The first root δ_1 at $n = 0$ is $\delta_1 = e^{-j\pi/6} = \dfrac{\sqrt{3}}{2} - j\dfrac{1}{2}$

The second root δ_2 at $n = 1$ is $\delta_2 = e^{-j5\pi/6} = -\dfrac{\sqrt{3}}{2} - j\dfrac{1}{2}$

The third root δ_3 at $n = 2$ is $\delta_3 = e^{-j3\pi/6} = j$

The fourth root δ_4 corresponding to the backward traveling wave can be obtained by the setting

$$\gamma = -j\beta_e - \beta_e C\delta_4 \quad \text{Similarly, } \delta_4 = -j\frac{C^2}{4}$$

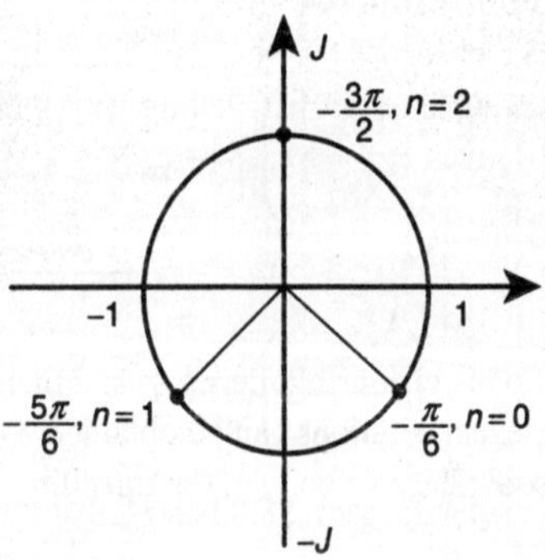

Figure 8.26 *The roots of (−j)*

Thus, the values of the four propagation constants γ are given by

$$\gamma_1 = -\beta_e C\frac{\sqrt{3}}{2} + j\beta_e\left(1+\frac{C}{2}\right)$$

$$\gamma_2 = \beta_e C\frac{\sqrt{3}}{2} + j\beta_e\left(1+\frac{C}{2}\right)$$

$$\gamma_3 = j\beta_e(1-C)$$

$$\gamma_4 = -j\beta_e\left(1-\frac{C^3}{4}\right) \tag{8.117}$$

The above four equations represent four different modes of wave propagation in the O-type helical travelling-wave tube.

Therefore the waves related to γ_1, γ_2, and γ_3 are forward waves but their amplitudes increase exponentially with respect to the distance; decay with distance; and remain constant. Fourth wave is backward wave which is corresponding to γ_4, and the amplitude is not changed with distance.

The growing wave is propagated at phase velocity which is little less than the electron beam velocity, and the energy flows from the electron beam to the wave. Whereas the decaying wave is propagated with the growing wave velocity, but the energy flows in reverse direction. The velocity of the constant-amplitude wave is slightly greater than the velocity of electron beam velocity, but there won't be any net energy exchange between the wave and electron beam. The backward wave is propagated in the negative *z*-direction with the velocity slightly more than the velocity of the electron beam.

8.9.6 Gain Considerations

The output gain of TWT in decibels is defined as

$$A_p \equiv 10\log\left|\frac{V(l)}{V(0)}\right|^2 = -9.54 + 47.3\ \mathrm{N}_l C\ \mathrm{dB}$$

where $V(l)$ is the output signal voltage

$V(0)$ is the input signal voltage

N_l = Circuit length

C = gain parameter

Derivation of Expression for Gain of TWT

Assume that the structure is perfectly matched, then there is no backward travelling wave. An attenuator is placed around the centre of the tube which controls the reflected wave and reduces it to zero level. Hence, the total circuit voltage is equal to the sum of three forward voltages corresponding to the three forward travelling waves. It is given as

$$V(z) = V_1e^{-\gamma_1 z} + V_2e^{-\gamma_2 z} + V_3e^{-\gamma_3 z} = \sum_{n=1}^{3} V_n e^{-\gamma_n z} \tag{8.118}$$

The input current can be found from Eq. 8.113 as

$$i(z) = -\sum_{n=1}^{3} \frac{I_0}{2V_0C^2} \frac{V_n}{\delta_n^{\,2}} e^{-\gamma_n z} \tag{8.119}$$

in which $C\delta << 1$, $E_1 = \gamma V$, and $\gamma = j\beta_e(1 - C\delta)$ have been used.

The input fluctuating component of velocity of the total wave may be found as

$$v_1(z) = \sum_{n=1}^{3} j \frac{v_0}{2V_0C} \frac{V_n}{\delta_n} e^{-\gamma_n z} \tag{8.120}$$

where $E_1 = \gamma V$, $C\delta << 1$, $\beta_e v_0 = \omega$, and $v_0 = \sqrt{(2e/m)V_0}$ have been used.

In order to determine the amplification of the growing wave, the input reference point is set at $z = 0$, and the output reference point at $z = l$. The voltage, current and velocity at the input point at $z = 0$ is given as

$$V(0) = V_1 + V_2 + V_3 \tag{8.121}$$

$$i(0) = -\frac{I_0}{2V_0C^2}\left(\frac{V_1}{\delta_1^{\,2}} + \frac{V_2}{\delta_2^{\,2}} + \frac{V_3}{\delta_3^{\,2}}\right) \tag{8.122}$$

$$v_1(0) = -j\frac{v_0}{2V_0C}\left(\frac{V_1}{\delta_1} + \frac{V_2}{\delta_2} + \frac{V_3}{\delta_3}\right) \tag{8.123}$$

The simultaneous solution of Eqs. 8.121, 8.122, and 8.123 with $i(0) = 0$ and $v_1(0) = 0$ is

$$V_1 = V_2 = V_3 = \frac{V(0)}{3} \tag{8.124}$$

As the growing wave is exponentially increased with respect to distance, it will dominate the total voltage beside the circuit. The output voltage is almost equal to voltage of growing wave, when the length l of the slow-wave structure is large. By substituting the Eqs 8.117 and 8.124 in 8.118, the output voltage can be expressed as

$$V(l) \approx \frac{V(0)}{3} \exp\left(\frac{\sqrt{3}}{2}\beta_e Cl\right) \exp\left[-j\beta_e\left(1 + \frac{C}{2}\right)l\right] \tag{8.125}$$

where $\beta_e l = 2\pi N_l$, N_l is circuit length and it can be written as

$$N_l = \frac{l}{\lambda_e} \text{ and } \beta_e = \frac{2\pi}{\lambda_e}$$

The amplitude of the output voltage is then given by

$$V(l) = \frac{V(0)}{3} \exp(\sqrt{3}\pi N_l\, C)$$

The output gain in decibels is defined as

$$A_p \equiv 10 \log \left|\frac{V(l)}{V(0)}\right|^2 = -9.54 + 47.3 N_l C \text{ dB} \tag{8.126}$$

The above equation represents the output power gain which indicates an initial loss at the circuit input of 9.54 dB. This loss occurs because of the fact that the input wave is divided into three waves of equal magnitude, and the growing wave voltage is one third of the total input voltage. From the above equation it can also be observed that the power gain is proportional to the length N1 in electronic wavelength of the slow-wave structure and the gain parameter C in circuit.

Performance characteristics of TWT

- Frequency of operation : 0.5 GHz to 95 GHz
- Power outputs : 5 mW (10 – 40 GHz) (Low-power TWT)
  250 KW (CW) at 3 GHz (High-power TWT)
  10 MW (pulsed) at 3 GHz
- Efficiency : 5 to 20% (30% with depressed collector)
- Noise Figure : 4 – 6 dB (Low-power TWT 0.5 to 16 GHz)
  25 dB (High-power TWT at 40 GHz)

Applications of TWT

- Low-noise RF amplifier in broadband microwave receivers
- Repeater amplifier in wide band communication links and coaxial cables
- Due to long tube life, TWT is used as a power output tube in communication satellites.
- Continuous-wave high-power TWTs are used in troposcatter links.
- TWTA transmitters are extensively used in radars, particularly in airborne fire-control radar systems, and in electronic warfare and self-protection systems.
- Another major use of TWTs is in the electromagnetic compatibility (EMC) testing industry for immunity testing of electronic devices.

Advantages of TWT

- Bandwidth is large.
- High reliability
- High gain
- Constant performance in space
- Higher duty cycle

EXAMPLE PROBLEM 8.12

The traveling-wave tube (TWT) is operated at a frequency f=10 GHz with Voltage $V_0 = 3$ kV and Beam current $I_0 = 30$ mA. If the Circuit length $N_l = 50$ and Characteristic impedance of helix $Z_0 = 10\ \Omega$ then determine the following: (a) The gain parameter (b) The output power gain A_p in decibels (c) All four Propagation constants

Solution

(a) The gain parameter is

$$C \equiv \left(\frac{I_0 Z_0}{4V_0}\right)^{1/3} \equiv \left(\frac{30\times10^{-3}\times10}{4\times3\times10^{3}}\right)^{1/3} \equiv 2.92\times10^{-2}$$

(b) The output power gain is

$$A_p = -9.54 + 47.3\,N_l C \text{ dB}$$
$$= -9.54 + 47.3 \times 50 \times 2.92 \times 10^{-2} = 59.52 \text{ dB}$$

(c) The four propagation constants are

$$\beta_e = \frac{\omega}{v_0} = \frac{2\pi \times 10^{10}}{0.593 \times 10^6 \sqrt{3 \times 10^3}} = 1.93 \times 10^3 \text{ rad/m}$$

$$\gamma_1 = -\beta_e C \frac{\sqrt{3}}{2} + j\beta_e \left(1 + \frac{C}{2}\right)$$
$$= -1.93 \times 10^3 \times 2.92 \times 10^{-2} \times 0.87 + j1.93 \times 10^3 \left(1 + \frac{2.92 \times 10^{-2}}{2}\right) = -49.03 + j1952$$

$$\gamma_2 = \beta_e C \frac{\sqrt{3}}{2} + j\beta_e \left(1 + \frac{C}{2}\right) = 49.03 + j1952$$

$$\gamma_3 = j\beta_e (1 - C) = j(1.93) \times 10^3 (1 - 2.92 \times 10^{-2}) = j(1872.25)$$

$$\gamma_4 = -j\beta_e \left(1 - \frac{C^3}{4}\right) = -j(1.93) \times 10^3 \left(1 - \frac{(2.92 \times 10^{-2})^3}{4}\right) = -j1930$$

■

EXAMPLE PROBLEM 8.13

A TWT is operated at 10 GHz with a beam voltage and a beam current of 3 kV and 30 mA, respectively. If the slow-wave structure has a characteristic impedance Z_0 of 10 Ω, the electronic circuit length N_l = 50. Find out

(i) The gain parameter C
(ii) The output power gain A_p (in dB)
(iii) v_0

Solution

(i) The gain parameter is

$$C = \left(\frac{I_0 Z_0}{4V_0}\right)^{1/3} = \left(\frac{30 \times 10^{-3} \times 10}{4 \times 3 \times 10^3}\right)^{1/3} = \left(25 \times 10^{-6}\right)^{1/3} = 3 \times 10^{-2}$$

(ii) The output power gain in decibels is defined as

$$A_p = -9.54 + 47.3 N_l C$$
$$= -9.54 + 47.3 \times 50 \times 3 \times 10^{-2} = 61.41 \text{ dB}$$

(iii) Since $\beta_e v_0 = \omega$ implies $\beta_e = \dfrac{\omega}{v_0}$

$$v_0 = 0.593 \times 10^6 \sqrt{V_0}$$

$$\beta_e = \frac{2\times3.14\times10\times10^9}{0.593\times10^6\sqrt{3\times10^3}} = 2 \times 10^3 \text{ rad/m}$$

■

EXAMPLE PROBLEM 8.14

An O-type TWT operates at 2 GHZ. The slow-wave structure has a pitch angle of 4.4° and an attenuation constant of 2 Np/m. Determine the propagation constant of the traveling wave in the tube.

Solution

Given data,
For an O-type traveling-wave tube,
Operating frequency, $f = 2$ GHZ
Pitch angle of the slow-wave structure, $\phi = 4.4°$
Attenuation constant $\alpha = 2$ Np/m
Propagation constant $\gamma = ?$
Then, the phase velocity along pitch of the TWT is given by

$$v_p = \text{c} \sin\phi = 3\times10^8 \times \sin(4.4^0) = 23.016\times10^6 \text{ m/sec}$$

The phase constant, $\beta = \dfrac{\omega}{v_p} = \dfrac{2\pi f}{v_p} = \dfrac{2\pi\times2\times10^9}{23.016\times10^6} = 545.984$ rad/m

Then, the attenuation constant of the traveling-wave tube is given by

$$\gamma = \alpha + j\beta = 2 + j545.984 = 45.988 < 89.79^0 \text{ m}^{-1}$$

■

EXAMPLE PROBLEM 8.15

A helix TWT is operated at a frequency of 20 GHz with beam current of 600 mA, a beam voltage of 10 kV, and a characteristic impedance of 20 Ω. Find the length of the helix to produce an output power gain of 50 dB.

Solution

The given parameters of helix travelling tube are Beam current, $I_0 = 600$ mA, Beam voltage, $V_0 = 10$ kV, Characteristic impedance, $Z_0 = 20\ \Omega$, Operating frequency, $f = 20$ GHz, Output power gain, $A_p = 50$ dB. We have to determine the length of the helix, l which is given as $l = N\lambda$

So the gain parameter is,

$$C = \left(\frac{I_0 Z_0}{4V_0}\right)^{1/3} = \left(\frac{600\times10^{-3}\times20}{4\times10\times10^3}\right)^{1/3} = 0.067$$

And output power gain is given by,

$$A_p = -9.54 + 47.3N_l C \text{ dB}$$

$$50 = -9.54 + 47.3\times N_l \times 0.067$$

$$N_l = 18.788$$

Then, the length of the helix is given by

$$l = N_l \lambda = N_l \frac{c}{f} = \frac{18.788 \times 3 \times 10^8}{20 \times 10^9} = 28.2 \text{ cm}$$

■

EXAMPLE PROBLEM 8.16

Calculate gain parameter of a TWT when it operates at a frequency of $f = 10$ GHz with beam current, $I_0 = 5$ mA, beam voltage, $V_0 = 5$ kV, characteristic impedance, $Z_0 = 30\ \Omega$, and $N_l = 50$.

Solution

Gain parameter

$$C = \left(\frac{I_0 Z_0}{4V_0}\right)^{1/3} = \left(\frac{5 \times 10^{-3} \times 30}{4 \times 5 \times 10^3}\right)^{1/3} = 1.95 \times 10^{-3}$$

■

8.10 BACKWARD-WAVE OSCILLATORS

The backward oscillator does not contain an attenuator as in case of TWT, therefore the RF signals which are reflected towards the cathode are not attenuated. But the helix is terminated with matched impedance since the output is taken near electron gun from the end of the helix. By inserting the matched impedance for terminating, it causes the dissipation of RF signal which is travelling in forward direction toward the collector. This operation does not occur in the TWT. The schematic diagram of BWO is shown in Figure 8.27.

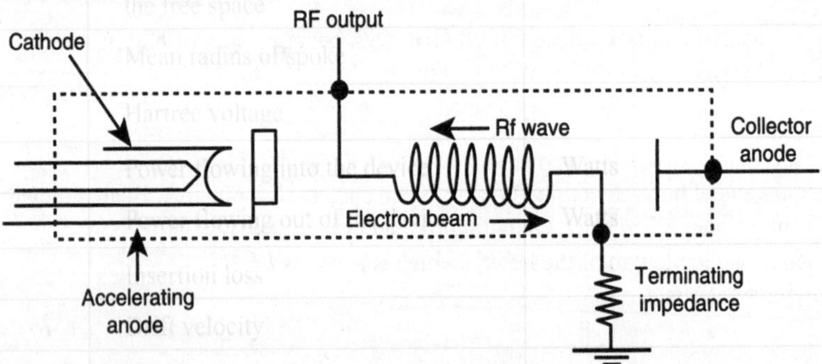

Figure 8.27 *Backward-wave oscillator*

The electron gun emits a stream of electrons and are accelerated towards the collector. They generate noise by shot and thermal effects. The noise signal is arbitrary in frequency, and almost all frequencies in range of 0–109 Hz are present. A wave travelling on the helix is developed from all these frequencies. This developed wave is travelled in the reverse direction towards the electron gun at the end of the tube. There will be interaction between the electron beam and the signal of helix, when the velocity of electron beam is slightly greater than the velocity of the signal on helix and this causes the electron beam to give up energy to the RF wave. By taking the energy from the electron beam, RF signal amplitude on the helix increases as it nears the gun end of the tube. The transfer of energy takes place due to the formation of bunches of the electron beam.

The bunched electron beam transfers energy to RF signal which is fed to the terminating impedance at end of helix. This phenomenon represents a travelling wave moving towards the cathode. This new wave will again produces bunching in the electron beam and the energy which has been taken from the electron beam will produce amplification and oscillations at a new frequency. As we knew electron beam can assume only one velocity at a given time, to maintain this phenomenon the beam will give energy to only one of the backward waves on the helix. Depending upon the velocity of the beam an appropriate frequency is being selected. For this purpose the potential difference between the cathode and anode are taken as a dependent variable. The change in the potential difference in turn causes an appropriate change in the frequency of oscillation.

Performance characteristics of BWO:

- Frequency range : 1 GHz to 1000 GHz
- Power output : 10 mW to 150 mW (CW)
  20 W (at high frequencies)
  250 kW (pulsed) with duty cycle < 1 sec.
- Tuning range : up to about 40 GHz

Application of BWO:

- A BWO can be used as a source of signals in microwave instruments and transmitting devices.
- It can be used as a broad band source with noise with an application of creating confusion to the enemy radar about its characteristics.
- It can be used as noise less oscillator in the frequency range 3–9 GHz with a desirable bandwidth.
- It is used as a continuous-wave generator.
- It is used to generate a wide range of frequencies.

8.11 M-TYPE TUBES

Crossed-field tubes are referred to as *M-Type tubes*, which deal with the propagation of waves in a magnetic field. In crossed field tubes both static electric and magnetic fields are present and they are perpendicular to each other. The electron motion takes place in area where the fields are perpendicular to each other. These fields affect the RF behavior of the electrons under RF fields. The magnetron is the most commonly used resonant crossed-field tube that is used in microwave circuits.

8.11.1 Crossed-Field Effects

If both electric and magnetic fields are present, motion of electrons depends on the orientation of electric and magnetic fields.

(a) If electric and magnetic fields are in the same direction or the opposite direction, the magnetic field exerts no force on electrons. Therefore, electron motion depends only on the electric field as shown in Figure 8.28 (b). Example: linear beam tubes.

(b) If electric and magnetic fields perpendicular to each other, electron motion depends on both electric and magnetic fields, this type of field is called *cross field*.

In crossed-field tubes, the electrons emitted by the cathode are accelerated by the electric field, and the motion of electrons is perpendicular to both fields as is indicated in Figure 8.28 (a). Analyzing the operation of crossed field tube, consider that an RF field is applied to the anode circuit. During the retarding field, electrons which enter the field at this point are decelerated by the field and loose some of their energy to the RF field. Hence the electron velocity decreases. Now these electrons with less velocity will travel into the dc electric field, a far enough distance, they will retain their earlier velocity at the end of the field.

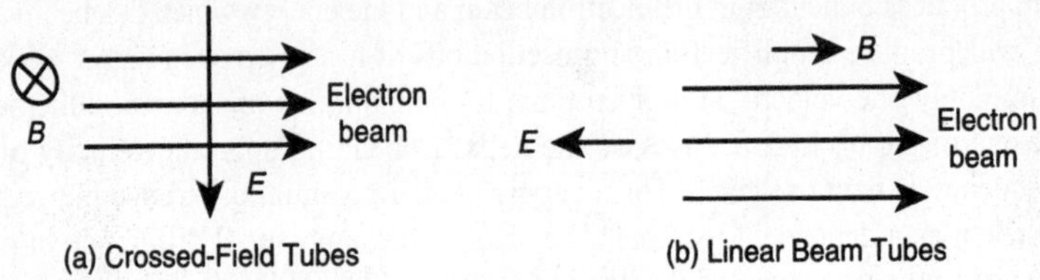

Figure 8.28 *Comparison of field configurations in crossed-field and linear beam tubes*

The presence of cross field interactions makes the electrons to give up some of its energy to the RF field. Only those electrons which have given sufficient energy to the RF field can only be eligible to travel to the anode end. Hence this phenomenon makes M-type devices relatively efficient devices. On the other hand when an electron enters into the circuit at the time of accelerating fields, it receives energy from RF fields which are in turn are accelerated, or returned towards cathode. This phenomenon produces heat in the cathode and in turn reduces efficiency of the device.

8.12 MAGNETRONS

The magnetron is a crossed field device, in which electric field and magnetic field are produced in a direction perpendicular to each other, in a way to cross each other. Therefore, the flow of electrons is perpendicular to both the fields. In magnetrons anode and cathode are concentric and cylindrical type structures. The magnetic field causes the electrons that are emitted from the cathode to move in curved paths. Magnetrons use various shapes of cavities to build oscillations and power.

8.12.1 Types of Magnetrons

There are three basic types of magnetrons:

- Cyclotron-frequency magnetrons
- Negative-resistance (split-anode) magnetrons
- Cavity-type magnetrons

Cyclotron-frequency magnetrons: Its principle of working is based on the synchronization between orbiting electrons in a magnetic field and a resonant circuit that is tuned to the cyclotron frequency. In this magnetron the ac component of electric field and the oscillations of electrons are parallel to the field.

Negative-resistance (split-resistance) magnetrons: It uses the static negative resistance between two anode segments. In this operation when both segments are at the same potential, the magnetic field effects can only be sufficient to keep flow of electrons to reach anode. By connecting a resonant circuit between the two anode segments, we can obtain sustained oscillations.

Traveling-wave magnetrons: These magnetrons provide oscillations of high peak power and peak power capability that is increased by about an order of magnitude to 100 kW. The operating frequency of negative resistance magnetron is generally below the microwave region. In the case of cyclotron frequency magnetron, it operates at microwave frequency range, but it has low power output and efficiency ($\gamma = 10\%$ – split anode type, 1% single anode type). Since the efficiency is very low in the first two types, they are not dealt in this chapter. In general, travelling wave magnetrons uses cavity resonators.

8.12.2 8-cavity Cylindrical Magnetron

Cavity magnetron is a high power microwave oscillator with high efficiency. The operating principle of this device is interaction of electrons with the perpendicularly oriented electric and magnetic fields. An 8-cavity cylindrical magnetron is shown in Figure 8.29 (a). It is a diode with eight re-entrant cavities and is concentric with an oxide-coated cathode. A permanent magnet was used for applying a magnetic field that is parallel to the cathode surface as shown in Figure 8.29 (c). A cavity magnetron is usually of a cylindrical configuration with a thick cylindrical cathode at the center and a co-axial cylindrical block of copper as anode as shown in Figure 8.29 (b).

In the anode block, a number of holes and slots act as resonant anode cavities. The electric field due to DC voltage applied between anode and cathode is radial, whereas the magnetic field produced by a permanent magnet is axial. When DC voltage and magnetic field are adjusted properly, due to the magnetic field, the electrons follow curved cyclodial paths in the cathode to anode interaction space.

The electric field in this resonant oscillator can be resolved into two components i.e., alternating current field, direct current fields. In ac fields, it is undesirable effect for the electrons to take the energy from the ac fields. The RF oscillations which are induced at the anode block (cavity tank circuits) due to noise transients affects the dc field to extend radially between adjacent anode segments. Cloud of electrons will be formed around the cathode due to DC voltage (or thermonic emission). The energy has to be given to the inputs in the correct phase, in order to not to disturb the sustained oscillations in the resonant circuit, that is, for this, the anode DC voltage should be adjusted so that the average rotational velocity of electrons coincides with ϕ of gap voltage at various gaps.

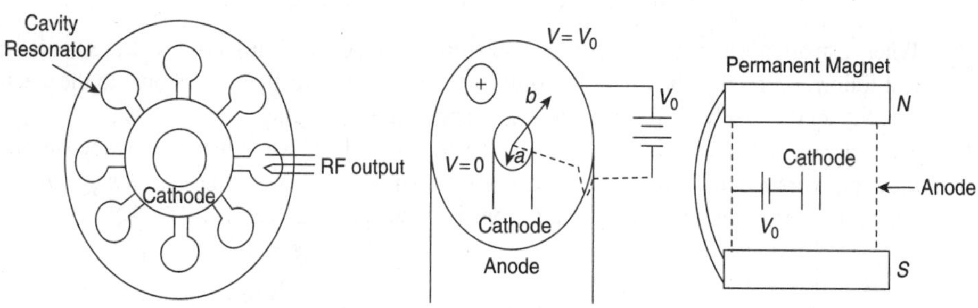

Figure 8.29 *(a) Structure of a cavity magnetron; (b) Cylindrical configuration; (c) Magnetic field*

The heated cathode is a source of electrons in a magnetron. The cavity magnetron consists of 8 cavities that are tightly coupled to each other. If we consider the case of N cavity magnetron it will have N modes of operation. Each mode of operation will be distinctively characterised by, "frequency and phase of oscillation", which are relative to next cavities. Sufficient care has to be taken while considering these combinations for a given mode so that it is self consistent. That is the total phase shift produced by this system is $2n\pi$, n is an integer. The minimum phase shift should be 45°(45° × 8 = 360°). The relative phase change, ϕ_v of the electric field across the adjacent cavities can be given as

$$\phi_v = \frac{2\pi n}{N}$$

where $$n = 0, \pm 1, \pm 2, \ldots, \pm\left(\frac{N}{2} - 1\right), \pm\frac{N}{2}$$

That is, $N/2$ mode of resonance can exist only in resonator systems that have an even number of resonators. If $n = N/2$, $\phi_v = \pi$. Since the phase angle of π radians is in the $N/2$ mode, this mode of resonance is called the *π-mode*. If $n = 0$, $\phi_v = 0$, this mode is the zero mode; that is, there will be no RF electric field between the anode and cathode (called the *fringing field*) and it will be of no use in magnetron operations.

To have an in-depth analysis of operation of cavity magnetron in which the RF field is applied in presence of perpendicularly oriented electric and magnetic fields, let us clearly understand that the incidence of electron in the EMF field.

Electron trajectories at various magnetic fields: Comparing the magnitude of electric and magnetic fields, we can understand the trajectory of an electron coming from cathode, moving towards anode takes different paths through the interaction space. Electron trajectories at various magnetic fields V_0 are presented.

(a) If $B = 0$, electrons emitted from the cathode move along the radial direction

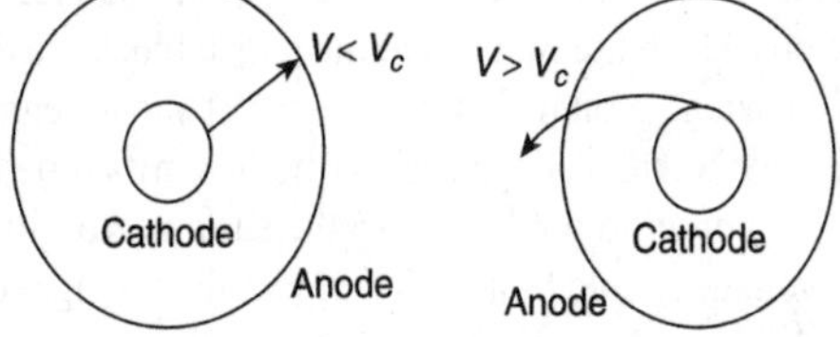

(a) No magnetic field (b) Small magnetic field

(b) When a small B is applied (at a perpendicular to radial electric field), electron trajectories bend and follow a curved path. The radius of the curved path is directly proportional to the electron velocity and inversely proportional to the magnetic field strength.

(c) The magnetic field required to return electrons to the cathode while just grazing the surface of the anode is called the *critical magnetic field* (B_c) and is also known as the *cut-off magnetic field*. Under this condition, the motion of electrons is shown in Figure 8.30.

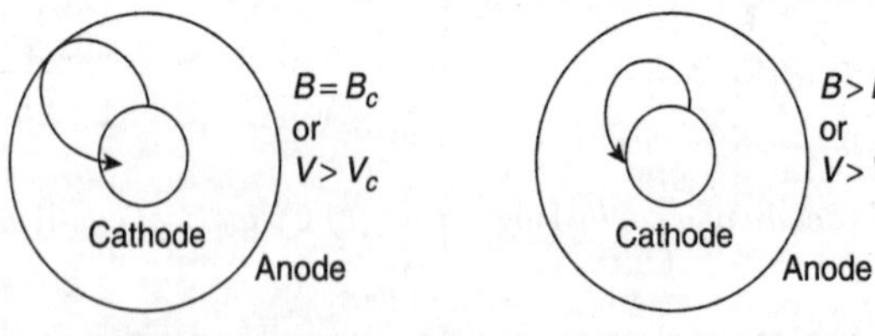

(a) Magnetic field = B_c (b) Magnetic field > B_c

Figure 8.30 *Electron trajectories in cavity magnetrons*

(d) If the magnetic field is made larger than the critical field ($B > B_c$), the electrons travel with a greater velocity and may return to the cathode quite faster; these electrons may cause back heating of the cathode. To disallow this back heating of cathode we have to switch off the heater supply after oscillation.

The equation of the cut-off magnetic field is given by (the derivation of this equation will be explained in sec.8.12.4)

$$B_c = \frac{(8V_0 m / e)^{1/2}}{b\left(1-\frac{a^2}{b^2}\right)}$$

Conversely, the cutoff voltage is given by

$$V_c = \frac{e}{8m} B^2 b^2 \left(1-\frac{a^2}{b^2}\right)^2$$

8.12.3 Modes of Resonance and π Mode Operation

We have discussed the effect of electric and magnetic fields in the previous section when no RF field is applied. Let us assume RF oscillations are initiated and are maintained sustainably and assume that these oscillations are created by some noise which is transient in the magnetrons. The device is having high Q- cavity resonators. Now, we proceed to explain the mechanism by which the oscillations are sustained. Best results are obtained when $n = 4$, that is, the phase difference between adjacent cavities is π radians, and then there is a π mode of operation which is shown in Figure 8.31(a).

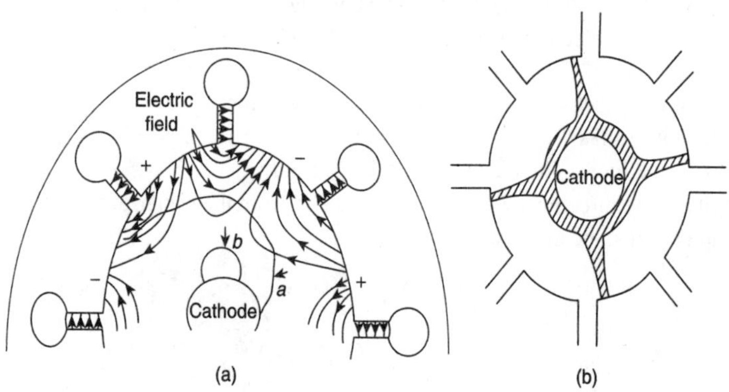

Figure 8.31 *(a) Magnetron operation in π mode; (b) Electron cloud showing spokes*

The electron "a" that is entering the interaction space during the decelerating field gives some of its energy to the RF field; therefore, its velocity decreases and it spends more time in interaction space during its long journey. In the same way, the electrons that are emitted a little later to be in the correct position move faster and try to catch up with electron "*a*". The electrons that are emitted a little earlier than "*a*" slow down, and they fall back in step with electron "*a*".

All these favored electrons come together, form electron bunches, and are confined to spokes or electron clouds. The process is called *phase-focusing effect*. The spokes so formed in the π mode rotate with an angular velocity that corresponds to two poles per cycle. The RF oscillations are sustained due to phase processing effect of the forward electrons which give enough energy to RF oscillations.

The electron '*b*' which is introduced during accelerated RF field takes energy from the oscillations. This results in increased velocity of electrons. Since the velocity is increased, the trajectory path of an electron bends more sharply and it stays short time in the interaction space. Hence they are returned to the cathode early. These electrons are unfavoured electrons as they do not participate in bunching process, rather, they are harmful in the sense that they cause back heating.

8.12.4 Hull Cut-Off Voltage Equation

A cavity cylindrical magnetron is the most commonly used magnetron, because for a cross-field device the electric and magnetic fields are perpendicular to each other, and the path of the electrons in the presence of this cross-field is naturally parabolic. The equation for the hull cut-off voltage is given by

$$V_c = \frac{e}{8m} B^2 b^2 \left(1 - \frac{a^2}{b^2}\right)^2$$

where B = magnetic flux density
a = cathode radius
b = anode radius
e = charge of the electron
m = mass of the electron

Derivation of Hull Cut-off Voltage Equation

The Hull cut-off condition is obtained, under the condition that there is no RF field, which in turn defines anode voltage is a function of magnetic field. The magnetic field tends to prevent the flow of electrons to the anode. On the other hand, under right circumstances, the electrons leave the hub after getting interacted with the RF wave that is rotating about the cathode, and flow to the anode. It happens when the rotating speed of electrons is more than the RF wave. By interacting with the RF wave, the electrons speed is reduced to RF rotation rate. In this process the electrons amplify the wave and losses the energy. Here, we will discuss the Hull cut-off voltage equation:

Force acting on the electron is $F = Bev$

In the direction of ϕ, the force component is given by $F_\phi = eBv_\rho$

where v_ρ = velocity in the direction of the radial distance ρ, from the center of the cathode cylinder.

Torque in direction of ϕ can be given as

$$T_\phi = \rho F_\phi = e.\rho.v_\rho.B \tag{8.127}$$

$$\text{Angular momentum} = \text{angular velocity} \times \text{moment of inertia} = \frac{d\phi}{dt} \times m\rho^2 \tag{8.128}$$

$$\text{Time rate of angular momentum} = \frac{d}{dt}\left(\frac{d\phi}{dt} \times m\rho^2\right) \tag{8.129}$$

This gives the torque in ϕ direction. Equating Eqs. 8.129 and 8.127 (the two values of torque in ϕ direction),

$$\frac{d}{dt}\left(\frac{d\phi}{dt} \times m\rho^2\right) = e.\rho.v_\rho.B \text{ That is, } 2m\rho\frac{d\phi}{dt} + m\rho^2\frac{d^2\phi}{dt^2} = e.\rho.v_\rho.B \tag{8.130}$$

We know that $v_\rho = \dfrac{d\rho}{dt}$

Therefore, Eq. 8.130 becomes

$$2m\rho\frac{d\phi}{dt} + m\rho^2\frac{d^2\phi}{dt^2} = eB.\rho.\frac{d\rho}{dt} \tag{8.131}$$

Integrating Eq. 8.131 with regard to "t," we will get

$$2m\rho.\phi + m\rho^2 \frac{d\phi}{dt} = eB\frac{\rho^2}{2}$$

For a particular direction, $m\rho.\phi$ can be considered a constant.

$$m\rho^2 \frac{d\phi}{dt} + C = eB.\frac{\rho^2}{2} \tag{8.132}$$

The value of C can be determined by applying boundary conditions (i.e., at surface of the cathode $\rho = a$ and $\frac{d\phi}{dt} = 0$ being the zero angular velocity at emission):

$$0 + C = \frac{e.B.a^2}{2} \quad \text{or} \quad C = \frac{eBa^2}{2}$$

Substituting the above value of C in Eq. 8.132, we get

$$m\rho^2 \frac{d\phi}{dt} = \frac{eB}{2}(\rho^2 - a^2)$$

or

$$\frac{d\phi}{dt} = \frac{eB}{2m}\left(1 - \frac{a^2}{\rho^2}\right) \tag{8.133}$$

When ρ = a (i.e., at cathode), $\frac{d\phi}{dt}$ approaches 0.

When $\rho >> a$, $\frac{d\phi}{dt}$ approaches $(\omega)_{\max}$ (maximum angular velocity).

i.e.,

$$\left(\frac{d\phi}{dt}\right)_{\max} = (\omega)_{\max} = \frac{eB}{2\text{m}} = \frac{eB_c}{2\text{m}} \tag{8.134}$$

where $B = B_c$ is the cut-off magnetic flux density.

We know that the potential energy of electron = kinetic energy of electrons

That is,

$$eV_0 = \frac{1}{2}mv^2$$

$$eV_0 = \frac{m}{2}(v_\rho^2 + v_\phi^2) \tag{8.135}$$

where $v_\rho = \frac{d\rho}{dt}$ and $v_\phi = \rho\frac{d\phi}{dt}$

Rewriting the equation (substituting for v_ρ and v_ϕ), Eq. 8.135 becomes

$$eV_0 = \frac{m}{2}\left[\left(\frac{d\rho}{dt}\right)^2 + \rho^2\left(\frac{d\phi}{dt}\right)^2\right]$$

From Eqs. 8.133 and 8.134,

$$\left(\frac{d\phi}{dt}\right) = (\omega)_{\max}\left(1-\frac{a^2}{\rho^2}\right)$$

$$\therefore eV_0 = \frac{m}{2}\left[\left(\frac{d\rho}{dt}\right)^2 + \rho^2(\omega)^2_{\max}\left(1-\frac{a^2}{\rho^2}\right)^2\right]$$

At anode $\rho = b$, $\frac{d\rho}{dt} = 0$, substituting these boundary conditions in the above equation,

$$\frac{m}{2}\left[b^2(\omega)^2_{\max}\left(1-\frac{a^2}{b^2}\right)^2\right] = eV_0 \tag{8.136}$$

Substituting Eq. 8.134 in Eq. 8.136, we get

$$\frac{m}{2}b^2\left(\frac{eB_c}{2m}\right)^2 \times \left(1-\frac{a^2}{b^2}\right)^2 = eV_0$$

i.e.,

$$\frac{e^2B_c^2b^2}{8m}\left(1-\frac{a^2}{b^2}\right)^2 = eV_0$$

or

$$B_c = \frac{(8V_0 m/e)^{1/2}}{b\left(1-\frac{a^2}{b^2}\right)} \tag{8.137}$$

i.e., for a given V_0, the electrons will not reach at anode, if $B > B_c$.
On the other hand, the cut-off voltage is given by

$$V_c = \frac{e}{8m}B^2b^2\left(1-\frac{a^2}{b^2}\right)^2 \tag{8.138}$$

It can be observed that for a given B, the electrons will not reach at anode, if $V_0 < V_c$. Equation 8.138 is called the *Hull cut-off voltage equation.*

8.12.5 Hartree Condition

The Hull cut-off condition is obtained, under the condition that there is no RF field, which in turn defines anode voltage is a function of magnetic field. The magnetic field tend to prevent the flow of electrons to the anode. On the other hand, under right circumstances, the electrons leave the hub after getting interacted with the RF wave that is rotating about the cathode, and flow to the anode. It happens when the rotating speed of electrons is more than the RF wave. By interacting with the RF wave, the electrons speed is reduced to RF rotation rate. In this process the electrons amplify the wave and losses the energy. This happens when the anode voltage is such that the electrons are rotating faster than the RF wave. Therefore, velocity of electrons is reduced by giving up energy to the RF wave.

The rate of rotation of electrons is reduced if the anode voltage is kept below the Hull cut-off voltage. Due to this, the electrons transfer less energy to the rotating wave. At a critical anode voltage, the rate of rotation of electrons and wave becomes equal. At this point the magnetron stops functioning as the electrons can no longer give up energy to the wave. The magnetron cannot work below that critical voltage as the rotation of the electrons is much slower than the RF wave. The critical voltage at which the magnetron stops functioning is called *Hartree Voltage.*

The Hartree anode voltage equation is a function of the magnetic flux density and the spacing between the cathode and anode.

$$V_H = Bdv_m - \frac{1}{2\eta}v_a^2$$

where B = magnetic flux density
d = spacing between anode and cathode
η = circuit efficiency
v_m = mean spoke velocity
v_a =Velocity of spoke at anode.

Derivation of Expression for Hartree Condition

Let us assume, in a magnetron an RF wave is rotating at an angular rate ϕ' and an electron is rotating at an angular rate θ' as indicated in Figure 8.32(a). The electron can transfer energy to the wave as long as $\theta' > \phi'$. The energy, W, which can be transferred from the electron to the RF wave is

$$W = \frac{1}{2}mr^2(\theta' - \phi')^2$$

$$W = \frac{1}{2}mr^2\left(\theta'^2 - 2\theta'\phi' + \phi'^2\right) \tag{8.139}$$

where $$\theta' = \frac{d\theta}{dt} \text{ and } \phi' = \frac{d\phi}{dt}$$

where r is the radius of the electron path.

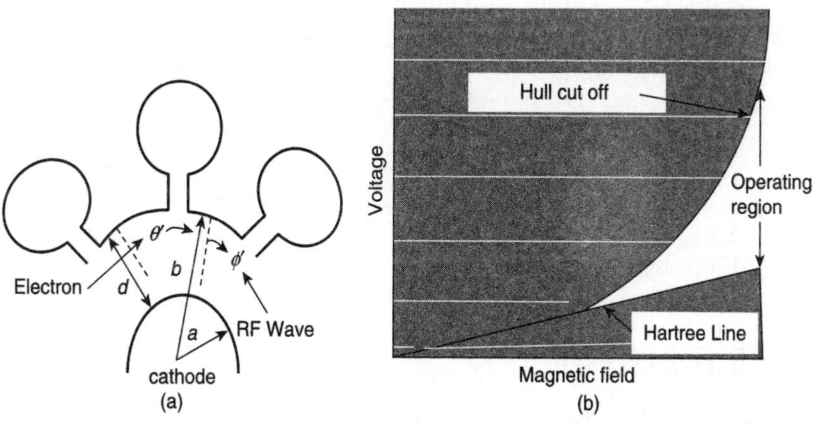

Figure 8.32 *(a) Electron and wave rotation in a magnetron; (b) Operating region of magnetrons*

Now, the condition for transfer of energy is zero, and from the electron spoke to the RF wave is

$$\frac{1}{2}mr^2\theta'^2 = \frac{1}{2}mr^2\left(2\theta'\phi' - \phi'^2\right)$$

$$\theta'^2 = \left(2\theta'\phi' - \phi'^2\right) \quad 8.140)$$

If we assume that the amplitude of the RF wave is small, $\frac{1}{2}mr^2\theta'^2 = eV_r$ (8.141)

where V_r is the voltage that causes the electron to rotate at θ'. This means that the voltage for which the energy transfer goes to zero (i.e Hartree voltage from Eqs. 8.141 and 8.140) is

$$V_H = \frac{1}{2}\frac{m}{e}r^2\left(2\theta'\phi' - \phi'^2\right) \quad (8.142)$$

where

$$\theta' = \frac{\eta B}{2}\left(1 - \frac{a^2}{b^2}\right) \quad (8.143)$$

For the normal π mode of magnetron operation, there are $N/2$ cycles of the RF wave around the anode, where N is the number of cavities. This means that the rate of rotation of the wave should be the operating frequency of the magnetron divided by the number of cycles

That is,

$$\phi' = \frac{2\omega}{N} \quad (8.144)$$

Substituting Eqs. 8.143 and 8.144 in Eq. 8.142, the Hartree voltage becomes

$$V_H = \frac{1}{2}B(b^2 - a^2)\frac{2\omega}{N} - \frac{1}{2\eta}b^2\left(\frac{2\omega}{N}\right)^2$$

By making the following substitutions, this can be rewritten in a form that leads to a straight forward physical interpretation

Spacing between anode and cathode, $d = b - a$
Mean radius of spoke, $r_m = (b - a)/2$
Mean spoke velocity, $v_m = r_m\theta'$
Velocity of spoke at anode, $v_a = r_a\phi'$

$$V_H = Bdv_m - \frac{1}{2\eta}v_a^2 \quad (8.145)$$

From the above Eq. 8.145, the Hartree voltage varies linearly with the magnetic field and so, the Hartree voltage and the Hull cut-off voltage can be plotted as shown in Figure 8.32(b). The significant voltages for a magnetron are between the Hartree and Hull cut-off voltages.

EXAMPLE PROBLEM 8.17

A normal circular magnetron has the following parameters: inner radius 0.15 m, outer radius 0.45 m, and magnetic flux density 1.2 milli weber/m^2.

(a) Determine Hull cut-off voltage
(b) Determine the Hull cut-off magnetic flux density if the beam voltage is 6000 V.

Solution

Given $a = 0.15$ m, $b = 0.45$ m, and $B = 1.2$ mWb/m^2

(a) Hull cut-off voltage

$$V_C = \frac{e}{8m} B^2 b^2 \left(1 - \frac{a^2}{b^2}\right)^2 = \frac{1.759\times10^{11}}{8} \left(1.2\times10^{-3}\right)^2 (0.45)^2 \left(1 - \frac{0.15^2}{0.45^2}\right)^2$$

$$= 5.699 \text{ kV}$$

(b) Hull cut-off magnetic flux density

$$B_C = \frac{\left(8V_0 \frac{m}{e}\right)^{1/2}}{b\left(1 - \frac{a^2}{b^2}\right)} = \frac{\left(8\times6000\times\frac{1}{1.759\times10^{11}}\right)^{1/2}}{0.45\left(1 - \frac{0.15^2}{0.45^2}\right)} = 1.3 \text{ m weber / m}^2$$

■

EXAMPLE PROBLEM 8.18

The magnetic flux density of a normal circular magnetron is 0.2 Wb/m^2 and find the cut-off magnetic flux density if $V_0 = 20$ kV. If the cathode radius = 2 mm and anode radius = 4 mm, then determine the Hull cut-off voltage.

Solution

Given $a = 2$ mm $= 2\times10^{-3}$ m, $b = 4$ mm $= 4\times10^{-3}$m, and $B_0 = 0.2$ Wb/m^2

Hull cut-off voltage $V_C = \frac{e}{8\text{m}} B^2 b^2 \left(1 - \frac{a^2}{b^2}\right) = \frac{1.759\times10^{11}}{8} (0.2)^2 \left(4\times10^{-3}\right)^2 \left(1 - \frac{2^2}{4^2}\right)$

$$= 10.55 \text{ kV}$$

Cut-off magnetic flux density

$$B_C = \frac{\left(8V_0 \frac{m}{e}\right)^{1/2}}{b\left(1 - \frac{a^2}{b^2}\right)} = \frac{\left(8\times20\times10^3 \frac{1}{1.759\times10^{11}}\right)^{1/2}}{4\times10^{-3}\left(1 - \frac{2^2}{4^2}\right)}$$

$$= 317.911 \text{ mWb/m}^2$$

■

8.12.6 Separation of π Mode

Modes of operation: The resonant circuit that is used in cavity resonators acts similar to an LC tank circuit. If two resonant circuits are coupled, they produce two different resonant frequencies. In general, if resonant circuits are coupled together, they produce "*n*" different and distinct resonant frequencies. However, the difference in frequency value is very small. The resonant modes of magnetrons are very close to each other. As a result, there is every possibility that one resonant frequency (or mode of operation) gets shifted easily to another and is called *mode jumping*. To avoid this problem, resonant frequencies should be separated as widely as possible. The best desired mode is the π mode, where adjacent blocks of the anode become positive and negative, respectively. This can be done by strapping.

Strapping: Keeping magnetron operations in the π mode is difficult; unless special means are employed, strapping is one method that is used. *Strapping* means to connect alternate anode plates with two conducting rings of heavy gauge touching the anode's poles at the dots as shown in Figure 8.33. This is done in order to make the anode poles together. One ring is strapped to the even blocks, and the other is strapped to the odd blocks; that is, it keeps alternate anode blocks at the same potential and keeps two rings at the opposite (positive and negative) potential. By using strapping we can achieve only the dominant mode in magnetrons.

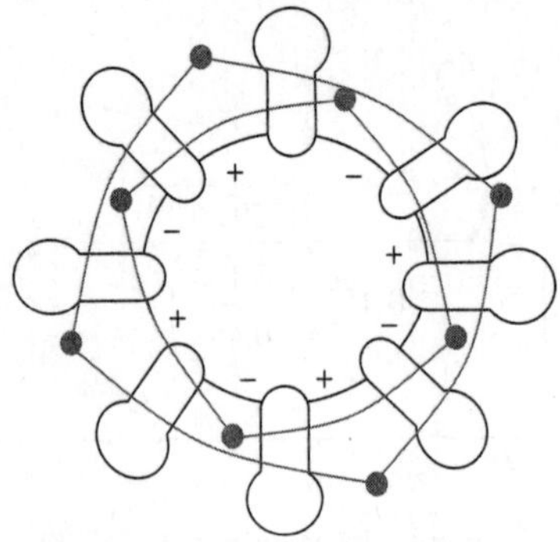

Figure 8.33 *Strapping of magnetrons*

Disadvantages of strapping

- Strapping may cause power loss in the conducting rings.
- Strapped resonators are very difficult.
- As the number of cavities increase (16 or 32), strapping has no effect on mode jumping.
- At higher frequencies, it will be difficult to maintain the RF field within the interaction space and it may introduce stray effects.

A magnetron that needs no strapping is the rising sun magnetron and is shown in Figure 8.34 (a). Here, the anode cavities are designed to be dissimilar, and only the dominant mode with 2π phase will be effective. The adjacent cavities oscillate at widely different frequencies, and, hence, separation will be quite effective. The other types of resonators that are used in the magnetron structure are slot and vane resonators and are shown in Figure 8.34 (b) and (c), respectively.

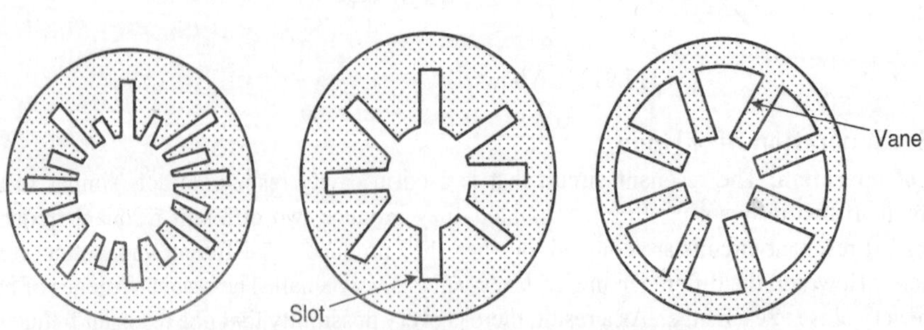

Figure 8.34 *Traveling-wave magnetron resonators*

Frequency pushing and pulling: Similar to reflex klystron, it is possible to change the resonant frequency of magnetrons by changing the anode voltage, which results in a change in the orbital velocity of electrons. This process is called *frequency pushing*. This alters the rate at which the energy is transferred to anode resonators and results in a change of oscillation of frequency.

Magnetrons are also susceptible to frequency variation due to changes in load impedance. This takes place regardless of whether these load variations are purely resistive or reactive variations. However, magnetron frequency variations are more severe for reactive variations. These frequency variations are known as *frequency pulling*. To prevent frequency pushing, a stabilized power supply is employed. Frequency pulling is prevented by using a circulator, which does not allow backward flow of electromagnetic energy. It is placed before the waveguide connection at the output of the magnetism.

8.12.7 Sustained Oscillations in Magnetrons

The cavities of magnetrons have(possesses) high quality factor (Q). The transient switching in the cavities is good (ample) enough to start oscillations. The electrons from the cathode are deflected(repelled) towards the anode as soon as the anode voltage is applied. The axial magnetic field acts on them as these electrons gain velocity. The electrons are deflected with a tangential velocity due to magnetic field and the radial directions (due to radial electric field) of electrons are now changed (at this instant).Due to the (because of the presence of) RF field in the interaction space, the tangential velocity of the electrons experiences a drag or opposition, thus slowing them in the process (the electrons). In losing velocity, the electrons part off their energy to the RF field. Since the velocity of the electrons is less, the deflection force of the magnetic field on them also is reduced, and, hence, these electrons (favourable electrons only) move toward the anode instead of curving back to the cathode in spite of $B > B_c$.

The spacing between the anodes is adjusted such that it is equal to half cycle of the RF frequency, so when the electrons reach the second anode the polarity of RF is reversed. The electrons continue to decelerate, as the energy acquired by them in falling through the dc anode to the cathode voltage is delivered to the RF oscillating wave. After delivering the kinetic energy got by anode to cathode, the electrons at last reaches the anode after slowing down to dead spot.

The outward centrifugal force should be equal to the magnetic force on the electron which is assumed to be a favourable electron in presence of cross-field.

i.e.,

$$\frac{mv^2}{r} = evB$$

where r = radius of cycloidal path

v = tangential velocity of electrons

Angular velocity, $\omega = \frac{v}{r} = \frac{eB}{m}$

Period of one revolution, $T = \frac{2\pi}{\omega} = \frac{2\pi m}{eB}$

The feedback should be in phase or integral multiples of 2π radians, for oscillations. If there are N cavities, the phase should be

$$\phi = \frac{2\pi n}{N}$$

where n = integer of the nth mode oscillation.

From the above discussions, magnetron oscillators are operated in the π mode therefore $\phi = \pi$. It can be observed from the Figure 8.31(a) that in this mode the successive cavities of RF fields are in anti-phase. The angular velocity in the interaction space of the RF field is given by

$$v_{ph} = \frac{d\phi}{dt} = \frac{\omega}{\beta}$$

Maximum amount of energy is transferred from electrons to the RF field when the angular velocity and cyclotron frequency of electron of RF wave are equal.

That is, $$\frac{\omega}{\beta} = \frac{d\phi}{dt} \text{ or } \omega = \beta\frac{d\phi}{dt} = \frac{eB}{m}$$

Comparison of magnetron and reflex klystron tubes

Magnetron	**Reflex klystron**
(1) Magnetron is a cross-field device. These are generally referred to as *M-type tubes.*	(1) Reflex klystron is a linear beam device. These are generally referred to as *O-type tubes.*
(2) The electrons carrying energy are made to interact with the RF field for a long duration.	(2) The electrons carrying energy are in contact with the RF field for a short duration.
(3) The range of frequencies over which magnetrons work properly is 500 MHz to 12 GHz.	(3) The range of frequencies over which reflex klystron works properly is 1-25 GHz.
(4) The efficiency provided by magnetrons is in the range of 40 to 70%.	(4) The efficiency provided by reflex klystron is very less than magnetrons and is in the range of 10 to 20%.
(5) For these tubes, the output power is in the range of 2 mW to 250 kW.	(5) For these tubes, the output power is in the range of 1 mW to 2.5 W.
(6) A permanent magnet is used that generates the magnetic field; it is so placed that the magnetic field is perpendicular to the electric field inside the magnetron.	(6) No permanent magnet is used here, but a repeller electrode is used at the end of the tube that bounces back the electron beam.
(7) Magnetrons are used for industrial heating and microwave ovens.	(7) Reflex klystron is used as a signal generator in microwave radars or receivers.

Performance characteristics

- Power output : In excess of 250 kW (pulsed mode)
  10 mW (UHF band) 2 mW (X band)
  8 kW (at 95 GHz)
- Frequency : 500 MHz to 12 GHz
- Duty cycle : 0.1 %
- Efficiency : 40% to 70%

Applications of Magnetrons

The most important application having large pulse power is pulsed radar.

- Dielectric heating on industrial scale
- In telemetry sweep oscillators, Voltage tunable magnetrons (VTMs) are used.
- In missiles applications also voltage tunable magnetrons are used (200 MHz to X band with CW, powers up to 500 W, and efficiency of 70 %)
- In industrial heating and microwave ovens, Fixed-frequency CW magnetrons are used. (500 MHz – 2.5 GHz frequency range, 300 W to 10 kW power outputs, and efficiency of 50%).

EXAMPLE PROBLEM 8.19

A linear magnetron has the following operating parameters: anode voltage $V_0 = 15$ kV, cathode current $I_0 = 1.2$ A, operating frequency $f = 8$ GHz, magnetic flux density $B = 0.015$ Wb/m^2, hub thickness $h =$ 2.77 cm, and distance between anode and cathode, $d = 5$ cm.

Calculate

(a) The electron velocity at the hub surface
(b) The phase velocity for synchronism
(c) The Hartee anode voltage

Solution

(a) The electron velocity is

$$v = \frac{eB}{m}r = 1.759 \times 10^{11} \times 0.015 \times 2.77 \times 10^{-2} = 0.73 \times 10^{3} \text{ m/s}$$

(b) The phase velocity is

$$v_{ph} = \frac{\omega}{\beta} = 0.73 \text{ x } 10^{8} \text{ m/s}$$

(c) The Hartee anode voltage is

$$V_H = 0.73 \times 10^{8} \times 0.015 \times 5 \times 10^{-2} - 1/(2 \times 1.759 \times 10^{11}) \times (0.73 \times 10^{8})^2$$

$$= 5.475 \times 10^{4} - 1.515 \times 10^{4} = 39.60 \text{ kV}$$

■

EXAMPLE PROBLEM 8.20

A magnetron has a cathode radius of 2.5 mm and an anode radius of 5 mm. What is the cut-off potential if a 0.27-Wb/m^2 magnetic field is applied?

Solution

Hull potential in volts

a = cathode radius in meters = 0.0025 m
b = anode radius in meters = 0.005 m
$e = 1.6 \times 10^{-19}$ coulombs
$m = 9.1 \times 10^{-31}$ kg
$B = 0.27$ Wb / m^2

$$V_C = \frac{e}{8\text{m}} B^2 b^2 \left(1 - \frac{a^2}{b^2}\right)^2 = 10.018 \text{ kV}$$

■

EXAMPLE PROBLEM 8.21

A magnetron has the following parameters: Anode voltage $V_0 = 25$ kV, cathode current $I_0 = 25$ A, operating frequency $f = 8$ GHz, Magnetic flux density $B = 0.34$ Wb/m^2, cathode radius $a = 4$ cm, Radius of the vane edge center $b = 8$ cm. Calculate the cyclotron frequency and cutoff voltage

Solution

The expression for Cyclotron frequency of a magnetron is given by

$$\omega = \frac{eB}{m} = \frac{1.6\times10^{-19}\times0.34}{9.1\times10^{-31}} = 0.0598\times10^{12}\ \text{rad/sec}$$

$$f = \frac{0.0598\times10^{12}}{2\pi} = 9.52\times10^{9}\ \text{Hz}$$

The expression for cutoff voltage of a magnetron is given by

$$V_C = \frac{e}{8\text{m}}B^2b^2\left(1-\frac{a^2}{b^2}\right)^2 = \frac{1.759\times10^{11}}{8}(0.34)^2\left(8\times10^{-3}\right)^2\left(1-\frac{4^2}{8^2}\right)^2 = 91503.18\ \text{V}$$

■

EXAMPLE PROBLEM 8.22

Find the angular frequency, cut-off voltage, cut-off magnetic flux density of a pulsed cylindrical magnetron when it has beam current =30 A, Magnetic Flux density = 0.34 Wb/m^2 and other parameters are $V_0 = 30$ KV, $a = 5$ cm, $b = 10$ cm.

Solution

Angular frequency = $\omega = \frac{e}{m}B = 1.759\times10^{11}\times0.34 = 0.5981\times10^{11}$ rad/sec

$$\text{Cut-off voltage} = V_C = \frac{e}{8\text{m}}B^2b^2\left(1-\frac{a^2}{b^2}\right)^2 = \frac{1}{8}\times1.759\times10^{11}\times(0.34)^2\times(10\times10^{-2})^2\times\left(1-\frac{5^2}{10^2}\right)^2$$

$$= 14297\ \text{kV}$$

$$\text{Cut-off magnetic flux density } B_C = \frac{\left(8V_0\dfrac{m}{e}\right)^{1/2}}{b\left(1-\dfrac{a^2}{b^2}\right)} = \frac{\left(8\times30\times10^3\dfrac{1}{1.759\times10^{11}}\right)^{1/2}}{10\times10^{-2}\left(1-\dfrac{5^2}{10^2}\right)} = 155.73\ \text{mWb/m}^2$$

■

EXAMPLE PROBLEM 8.23

A normal circular magnetron has the following parameters: inner radius, $a = 0.15$m, outer radius, $b = 0.45$ m and magnetic flux density, B = 1.2 m Wb/m^2. Determine the cyclotron frequency in GHz.

Solution

Cyclotron frequency is given by the following equation $\omega_c = \frac{e}{m}B$

where $e = 1.6 \times 10^{-19}$ c

$m = 9.1 \times 10^{-31}$ kg

$\omega_c = 1.759 \times 10^{11} \times 1.2 \times 10^{-3}$ rad/sec

$f_c = 0.0336$ GHz ■

8.13 CROSSED-FIELD AMPLIFIERS

A crossed-field amplifier (CFA) is a broadband microwave power amplifier. CFA is similar to magnetron in structure. The RF and dc interacts in the region of crossed electric and magnetic fields. It has high efficiency in providing moderately large amounts of power. There are two general types of CFAs based on their electron stream source:

- injected-beam CFA and
- distributed-emission CFA

These CFAs describe the method by which electrons reach the interaction region and how they are controlled. In injected-beam CFAs, the electrons are injected into the interaction region by an electron gun; whereas in distributed-emission CFAs, electrons are emitted by the cathode or sole. The injected-beam CFA is generally not suited for high powers. Therefore, distributed-emission CFAs are mostly preferred. Two types of formats can be used in construction of distributed-emission CFA. They are the circular format and the linear format. In the circular format, electrons from the output may be isolated from the input, forming the non-reentrant configuration. In the re-entrant configuration, the feedback electrons may be bunched, forming RF feedback, or the electrons may be de-bunched, eliminating the RF feedback. The linear-format tubes are of a non re-entrant type.

CFAs are also classified by their mode of operation in to two types:

- Forward-wave CFAs
- Backward-wave CFAs

The forward wave and backward wave CFAs are mainly troubled with the direction of the phase and group velocity of the energy on the microwave circuit. The behaviour of the phase velocity with frequency is of primary concern because of electron stream reacts to the RF field forces. For the forward waves CFA, the helix slow wave structure is selected as a microwave circuit and for the backward waves CFA the strapped bar line is used.

Basic Principle of Operation of CFA

The CFAs have an odd number of resonant cavities which are coupled with each other and they act as slow-wave structures. These resonant cavities produce oscillations and the oscillating resonant cavity excites the next cavity and this cavity excites the next one and so on. The actual oscillation starts from the input wave guide and they continue until the output waveguide. Due to the influence of electric field, (anode voltage) and the magnetic field (a strong permanent magnet), all the electrons will move evenly from cathode to anode in a cycloid path without applying any input signal. The vanes of the resonator get a voltage difference synchronously to the oscillation when the input of the wave guide introduces an RF oscillation into the first resonator. The electron bunches (which are formed due to acceleration and deceleration of the electrons) are formed due to the difference in the speed of electrons. These bunches of electrons rotate alike to a "Space-Charge Wheel" that is similar to the magnetron operation. But the Space Charge Wheel will be interrupted due to the odd number of cavities causing an opposite phase in the last

odd cavity that is the cavity between input and output waveguide. The last odd cavity may exist as a block containing graphite to decouple input and output to avoid a negative feedback in this resonant cavity.
In the input cavity of the CFA, the oscillation is weak. As the electron bunches hit the vanes of the other cavities, their energy is distributed synchronously to the oscillation. The oscillations get stronger from cavity to cavity since the electrons can alternatively accelerate and slow down near the next cavity because of the alternating magnetic field. These electrons finally hit the anode cavity, causing amplification, and the anode current is coupled through the output waveguide as shown in Figure 8.35. CFA also requires strapping in its construction similar to a magnetron to avoid ineffective modes of operation.

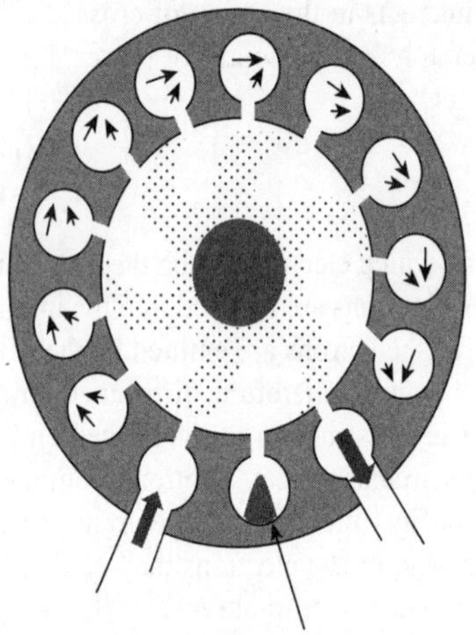

Block containing graphite

Figure 8.35 *The interrupted "Space-Charge Wheel" in a Crossed-Field Amplifier*

CFA has many advantageous characteristics such as wide bandwidth, high efficiency, and the ability to handle large amounts of power . So now a days CFA are used in many microwave electronic system applications. The CFA is of less cost, particularly when a relatively simple power supply/ modulator system is used for operating it.

From the below Table 8.2 we can compare various types of tubes.

Table 8.2 Comparison of characteristics of microwave tubes

Tube type	Frequency Bandwidth	Power out (Typical)	Advantages	Drawbacks
Klystron	0.1 – 300 Hz 5 – 10%	10 KW CW 10 MW Pulse	High Power 40–60%Efficient Low noise	Narrow Bandwidth
Traveling – Wave Tube (Helix)	1 – 90 GHz Wide Bandwidth 2-3 Octaves	20 W CW 20 KW Pulse.	Broad Bandwidth	Power – Handling limitations, Low Efficiency

(Continued)

Table 8.2 (Continued)

Tube type	Frequency / Bandwidth	Power out (Typical)	Advantages	Drawbacks
Magnetron	1 – 90 GHz / N/A	100 W CW 10 MW Pulse	Simple-Inexpen-sive Rugged	Noisy
Crossed-Field Amplifier	1–30 GHz / 10 – 20%	1000 W CW 5 MW Pulse	Compact Size 30 – 40% Efficient	Complex and expensive slow-wave structure

EXAMPLE PROBLEM 8.24

A reflex klystron has following parameters, beam voltage $V_0 = 200$ V. Beam current $I_0 = 20$ mA, signal voltage $V_1 = 40$ V. Find (i) the input power in watts (ii) output power in watts (iii) efficiency, when it operates at the peak mode of $n = 2$

Solution

(i) P_{dc} (input power) $= V_0 I_0 = 4$ W.

(ii) P_{ac} (output power) $= \dfrac{2V_0 I_0 X' J_1(X')}{2n\pi - \dfrac{\pi}{2}} = 0.906$ W.

(iii) $\eta = \dfrac{P_{ac}}{P_{dc}} \times 100\% = 22.6\%$.

■

EXAMPLE PROBLEM 8.25

A reflex klystron tube is oscillating at a frequency $f_r = 9$ GHz at the peak on $n = 2$ mode or 1 ¾ mode under the following conditions: $R_{\text{sh}} = 30$ kΩ, $V_0 = 500$ V. The spacing between the repeller and the cavity is $L = 1$ mm. Assume that the transit time through the gap and through beam loading effect can be neglected. Find the value of (a) repeller voltage V_r (b) Find the dc necessary to give a microwave gap of voltage of 100 V (c) Calculate the electronic efficiency.

Solution

(a) The relationship between repeller voltage and accelerating voltage is given by

$$\frac{V_0}{(V_R + V_0)^2} = \frac{1}{8}\frac{1}{\omega^2 L_r^2}\frac{e}{m}\left(2\pi n - \frac{\pi}{2}\right)^2$$

By substituting the given values in the above formula we get $V_r = 503.39$ V.

(b) Assuming $\beta_0 = 1$,

$$I_0 = \frac{V_1}{2J_1(X')R_{sh}} = 2.863 \text{ mA}$$

(c) Electronic efficiency $\eta = \dfrac{2X'J_1(X')}{2\pi n - \dfrac{\pi}{2}} = 15.28\,\%$ ■

EXAMPLE PROBLEM 8.26

Determine the following when the reflex klystron operates at the peak of $n = 1$ or ¾ mode and has dc power input of 50 mW, and the ratio of V_1 over V_0 is 0.178

(a) The efficiency of the reflex klystron.
(b) The total power output in mW.
(c) Power delivered to the load, if 20% of the power delivered by the electron beam is dissipated in the cavity walls

Solution

(a) Efficiency, $\eta = \dfrac{2X'J_1(X')}{2n\pi - \dfrac{\pi}{2}} = 0.655$

(b) $P_{out} = 4.455$ mW

(c) Power delivered to load = 4.455*0.8 = 3.564 mW

■

SUMMARY

1. Microwave tubes perform the same functions of generation and amplification in the microwave portion of the frequency spectrum that vacuum tubes perform at lower frequencies.

2. At microwave frequencies, the size of electronic devices required for generation of microwave energy becomes smaller and smaller. This results in lesser power-handling capability and increased noise levels.

3. To produce resonance a circuit would require a parallel connection of a inductor and a capacitor at low frequencies. At microwave frequencies, this is achieved by using a cavity, which may be constructed of brass, copper, or aluminum.

4. Conventional vacuum triodes, tetrodes, and pentodes are less useful signal sources at frequencies above 1GHz because of lead-inductance and inter-electrode-capacitance effects, transit-time effects, and gain-bandwidth product limitations.

5. Microwave tubes are constructed so as to overcome the limitations of conventional and UHF tubes. The basic operating principle of microwave tubes involves transfer of power from a source

of the dc voltage to a source of the ac voltage by means of a current density modulated electron beam. The same is achieved by accelerating electrons in a static field and retarding them in an ac field.

6. The linear beam tubes are the most important microwave tubes which are currently in use.
7. In linear beam tubes, the electron beam travels along a straight path between the cathode and the collector. It is parallel to both the electric and magnetic fields.
8. The two-cavity klystron is a microwave amplifier that is operated by the principles of velocity and current modulation. Extra cavities help to modulate the electron beam's velocity and increase the output energy. Hence, intermediate cavities are added between the input and output cavities of a klystron amplifier. This will improve the klystron parameters like amplification, efficiency and power output to a great extent.
9. For applications which require variable frequency, Reflex klystron is used. It is a single cavity variable frequency microwave generator of low power and low efficiency. It is a low-power, low-efficiency microwave oscillator that is used as a signal source in microwave generators, as a local oscillator in microwave receivers, as a pump oscillator in parametric amplifiers, and as frequency-modulated oscillators in portable microwave links.
10. Klystrons are essentially narrow band devices, as they utilize cavity resonators to velocity modulate the electron beam over a narrow gap; whereas TWTs are broadband devices in which there are no cavity resonators.
11. The Backward-Wave Oscillator (BWO) is a slow-wave device that operates on the principle of velocity modulation. BWO is a self-oscillating TWT that is capable of delivering microwave power over a wide range of frequency.
12. In crossed-field tubes, the electric and magnetic fields are perpendicular to each other. Crossed-field tubes are also known as *M-type devices*, as they deal with propagation of waves in a magnetic field.
13. The magnetron is a crossed-field device. This means that the flow of electrons, the electric field, and the magnetic field are mutually perpendicular to each other. Magnetrons use various shapes of cavities to build oscillations and power.
14. A cavity magnetron is a high-power, high-efficiency microwave oscillator that depends on the interaction of electrons with a traveling electromagnetic wave for its operation.
15. The Hull cut-off condition determines the anode voltage or magnetic field that is necessary to obtain non zero anode current as a function of the magnetic field or anode voltage in the absence of an electromagnetic field.
16. Hartree anode voltage equation is a function of the magnetic flux density and the spacing between the cathode and anode. Magnetrons using identical cavities in the anode block employ strapping to prevent mode jumping.
17. A Crossed-field Amplifier (CFA) is a broadband microwave power amplifier where RF-dc interaction region is a region of crossed electric and magnetic fields.

18. CFAs are classified based on their electron stream source as injected-beam CFAs and distributed-emission CFAs, and by their mode of operation, they are classified as forward-wave CFAs or backward-wave CFAs.

OBJECTIVE-TYPE QUESTIONS

1. Both axial magnetic field and radial electric fields are used in the following vacuum tube

(a) magnetron (b) a reflex klystron
(c) Klystron (d) traveling-wave tube

2. The following vacuum tube can be used as an oscillator and an amplifier?

(a) klystron (b) BWO (c) TWT (d) magnetron

3. The transit time can be reduced in microwave tubes,

(a) if electrodes are brought closer together
(b) if a higher anode current is used
(c) if multiple or coaxial leads are used
(d) none

4. The modes in a reflex klystron

(a) give the same frequency but different transit time
(b) result from excessive transit time across the resonator gap
(c) are caused by spurious frequency modulation
(d) are just for theoretical considerations

5. Vacuum tubes fail at microwave frequencies, because

(a) noise figure increases (b) shunt capacitive reactances become too large
(c) transit time becomes too short (d) series inductive reactances become too small

6. For use as a local oscillator for frequency measurements, the most suitable microwave source would be

(a) TWT (b) double-cavity klystron
(c) reflex klystron (d) magnetron

7. The main advantage of TWT over a multi-cavity klystron is:

(a) greater bandwidth (b) more efficient
(c) higher number of modes (d) higher output power

8. In a travelling-wave tube, the purpose of helix structure is

(a) to make-sure broadband operation

(b) to minimise the noise figure

(c) to minimse the RF field's axial velocity

(d) none

9. The purpose of attenuator in a travelling-wave tube is

(a) to prevent saturation (b) to increase gain

(c) to prevent oscillation (d) to help bunching

10. The oscillating frequencies of different modes (or cavity resonators) of magnetrons are not same and are quite close to each other, which results in

(a) helping focusing (b) providing attenuation

(c) improving bunching (d) mode jumping

11. The following is not used as a TWT slow-wave structure

(a) coupled cavity (b) helix

(c) ring bars (d) periodic permanent magnet

12. The phase velocity of RF field's axial component in the TWT slow-wave structure is

(a) equal to the velocity of the electrons

(b) slightly less than the velocity of the electrons

(c) slightly greater than the velocity of the electrons

(d) equal to the velocity of light in vacuum

13. In the following microwave tubes, RF energy travels at nearly the same speed as the electrons that are traveling from the cathode to the collector:

(a) magnetron (b) TWT

(c) CFA (d) Klystron

14. A backward-wave oscillator is based on

(a) cross-field amplifier (b) coaxial magnetron

(c) traveling-wave tube (d) rising sun magnetron

15. Which one of the following can be used for the amplification of microwave energy?

(a) traveling-wave tube (b) magnetron

(c) reflex klystron (d) Gunn diode

16. Strapping is used in magnetrons to

(a) prevent mode jumping (b) ensure bunching

(c) improve the phase-focusing effect (d) prevent cathode back heating

17. The purpose of the slow-wave structure used in TWT amplifiers is

(a) to increase wave velocity

(b) to reduce spurious oscillations

(c) to reduce wave velocity so that the electron beam and the signal wave can interact

(d) None of the above

18. The following cavity structure is preferred for use in magnetron to overcome problems with strapping at high frequencies is

(a) slot (b) rising sun

(c) Vane (d) all

19. The time taken by the electron to travel into the repeller space and back to the gap in a reflex klystron is referred to as

(a) transit time, $T = n + 1/4$ (b) bunching time, $T = n + 1/4$

(c) transit time, $T = n + 3/4$ (d) bunching time, $T = n + 3/4$

ANSWERS TO OBJECTIVE-TYPE QUESTIONS

1. (a)	2. (d)	3. (a)	4. (a)
5. (b)	6. (c)	7. (a)	8. (c)
9. (c)	10. (d)	11. (d)	12. (b)
13. (b)	14. (c)	15. (a)	16. (a)
17. (c)	18. (d)	19. (c)	

REVIEW QUESTIONS

1. What are the limitations of conventional vacuum tubes at microwave frequencies?

2. Explain clearly the classification of microwave sources.

3. Explain the principle of operation of a two-cavity klystron with a neat diagram.

4. Find out the expression for efficiency of a two cavity Klystron amplifier.

5. Derive the equation of velocity modulation for a two-cavity Klystron amplifier.

6. Explain in detail bunching process & obtain expression for bunching parameter in a two cavity klystron.

7. Explain the construction and working of a multi-cavity klystron.

8. Explain the construction and operation of Reflex Klystron Oscillator.

9. Derive the relationship between accelerating voltage and repeller voltage.

10. Draw the equivalent circuit of reflex klystron and explain about the electronic admittance.

11. Explain about electronic and mechanical tuning of reflex klystron.

12. What is slow wave structure? Explain how a helical TWT achieves amplification.

13. Explain the principle of working of Travelling Wave Tube.

14. Explain why there are four propagation constants in TWT & derive equations to these propagation constants.

15. How oscillations are prevented in a Travelling Wave Tube.

16. Explain the principle of working of Backward Wave Oscillator.

17. Explain about the different types of magnetron.

18. Explain the working principle of 8-Cavity Cylindrical Magnetron.

19. Derive an expression for the Hull cutoff equation for cylindrical magnetron.

20. Derive the Hartree anode Voltage equation for linear magnetron.

21. Explain the π-mode operation of magnetron.

22. A two-cavity klystron amplifier has the following parameters: $V_0 = 1000$ V, $R_0 = 40$ K ohm, $I_0 = 25$ mA, frequency = 3 GHz, gap spacing (d) = 1 mm, cavity spacing = 4 cm, effective shunt impedance, and excluding beam loading = 30 K ohm.

    (a) Find the input gap voltage to give maximum voltage V_2.

    (b) Find the voltage gain and efficiency of the amplifier neglecting the beam loading.

23. The parameters of a two-cavity klystron are given by $V_b = 900$ V, $f = 3.2$ GHz, and $d = 10^{-3}$ m. Determine electron velocity, transit angle, and beam coupling coefficient.

24. A reflex klystron operates at the peak mode of $n = 2$ with beam voltage $V_0 = 300$ V, beam current $I_0 = 20$ mA, and signal voltage $V_1 = 40$ V. Determine

    i. Input power in watts
    ii. Output power in watts
    iii. The efficiency

25. A reflex klystron having an accelerated field of 300 V oscillates at a frequency of 10 GHZ with a retarding field of 500 V. If its cavity is returned to 9 GHz, what should be the new value of the retarding field for oscillations in the same mode to take place?

26. A reflex klystron has the following parameters: V_0 = 800 V, L = 1.5 mm, R_{sh} = 15 kΩ, and f = 9 GHZ. Calculate

    i. The repeller voltage for which the tube can oscillate in 1 ¾ mode
    ii. The direct current necessary to give a microwave gap voltage of 200 V
    iii. Electron efficiency

27. A reflex klystron is operated at 56 Hz with an anode voltage of 1000 V and a cavity gap of 2 mm. Calculate the gap transit angle. Find optimum length of the drift region. Assume n =1 ¾ and V_r = –500 V.

28. A reflex klystron operates at 8 GHz with dc beam voltage 300 V, repeller space = 1 mm for 1 ¾ mode. Calculate $P_{RF\,max}$ and corresponding repeller voltage for a beam current of 18 mA.

29. A TWT operates under following parameters: beam voltage V_o = 3 KV, beam current I_o = 20 mA, characteristic impedance of helix Z_o = 10, circuit length N_l = 50, and frequency f = 10 GHz. Determine

    i. Gain parameter
    ii. Output power gain in dB and all
    iii. Four propagation constants

30. A TWT has the following parameters: V_o = 3 KV, I_o = 4 mA, f = 9 GHz, Z = 20, and N = 50. Calculate the

    i. Gain parameter
    ii. Power gain in db

31. A linear magnetron has the following operating parameters: V_o = 15 KV, I_o = 1.2 A, f = 8 GHZ, B_o = 0.015 wb/m^2, d = 5 cm, and h = 2.77 cm. Calculate

    i. Electron velocity at hub surface
    ii. Phase velocity for synchronism
    iii. Hartree anode voltage.

32. A magnetron operates with the following parameters: V_o = 25 KV, I_o = 1.25 A, B_o = 0.4 wb/m^2, diameter of the cathode = 8 cm, and radius of vane edge to center = 8 cm. Find the cyclotron frequency and cut-off voltage.

33. A normal circular magnetron has the following parameters: inner radius of 0.15 m, outer radius of 0.45 m, and magnetic flux density of 1.6 milli weber/m^2.

    (a) Determine Hull cut-off voltage.
    (b) Determine the Hull cut-off magnetic flux density if the beam voltage is 4000 V.

34. A normal circular magnetron has the following parameters: Cathode radius = 2 mm and anode radius = 4 mm. Determine the Hull cut-off voltage if the magnetic flux density is 0.3 Wb/m^2 and the cut-off magnetic flux density if V_o = 15 KV.

35. A magnetron is operating in the π mode and has the following specifications:

$$N = 10, f = 3 \text{ MHz}, a = 0.4 \text{ cm}, b = 0.9 \text{ cm},$$
$$L = 2.5 \text{ cm}, V_o = 18 \text{ KV}, \text{B} = 0.2 \text{ Wb/m}^2,$$

and $$m = 9.1 \times 10^{-31} \text{ kg}.$$

Determine angular velocity of electrons.

36. For a magnetron, $a = 0.6$ m, $b = 0.8$ m, $N = 16$, $B = 0.06$ T, $f = 3$ GHz, and $V_0 = 1.6$ KV. Calculate the average drift velocity for electrons in the region between the cathode and anode.

Microwave Solid-state Devices

9

9.1 INTRODUCTION

Before the 1960s, vacuum tube technology was widely used in electronics, which increased the size and cost of the entire system. Later, the development in semiconductor technology led to the growth of solid-state devices, which were more reliable and had a longer life. These devices generally require low power and are very compact and lighter in weight.

In the recent past, tremendous research activities have taken place for the development of better, low-noise, high-frequency, and greater-bandwidth components to achieve lesser switching time and better performance characteristics. In this chapter, several microwave solid-state devices, including two terminal devices such as transferred electron devices (Gunn diodes), avalanche transit time devices (IMPATT, TRAPATT, and BARITT diodes), tunnel diodes, and three terminal devices such as bipolar junction transistors and field effect transistors (MESFETs and HEMTs) are discussed.This chapter also discusses the PIN diode, Schottky diode, Varactor diode, parametric amplifier, step-recovery diode, and crystal diode in detail.

9.2 NEGATIVE RESISTANCE PHENOMENON

There are various microwave solid state devices which show negative resistance characteristics. The real part of their impedance is negative over a range of frequencies. In a positive resistance, the current through the resistor and the voltage across it are in phase. The voltage drop across positive resistance is positive, and the power that is given by the product of voltage (V) and current (I) is dissipated in the resistor. In a negative resistance, the current and voltage are out of phase by 180°. Therefore, the voltage drop across it is negative, and a negative power is generated by the power supply associated with the negative resistance. In other words, negative resistance generates or supplies power (e.g. active solid state devices) to the external circuits and positive resistance absorbs power (e.g. passive solid state devices).

However, the reason for exhibiting negative resistance varies from one device to another. For instance, in tunnel diode the negative resistance is due to heavy doping where the majority carries tunnel through the thin barrier. In an IMPATT, it is due to delay, which causes the current to lag behind the voltage; whereas in TEDs, it is due to differential negative electron mobility characteristic in materials such as GaAs and InP.

9.3 CLASSIFICATION OF SOLID-STATE DEVICES

Solid-state devices can be classified into two categories depending on the number of terminals these device have. They are diodes and transistors. Depending on their principle of operation, the two terminal diodes are further classified as shown in Figure 9.1(a).Transferred Electron Devices (TED), tunnel diodes, and Avalanche Transit Time Devices (ATD) show negative resistance. The three terminal devices are also further divided into two types as shown in Figure 9.1(b).Microwave BJTs are bipolar devices, whereas microwave FETs are unipolar devices.

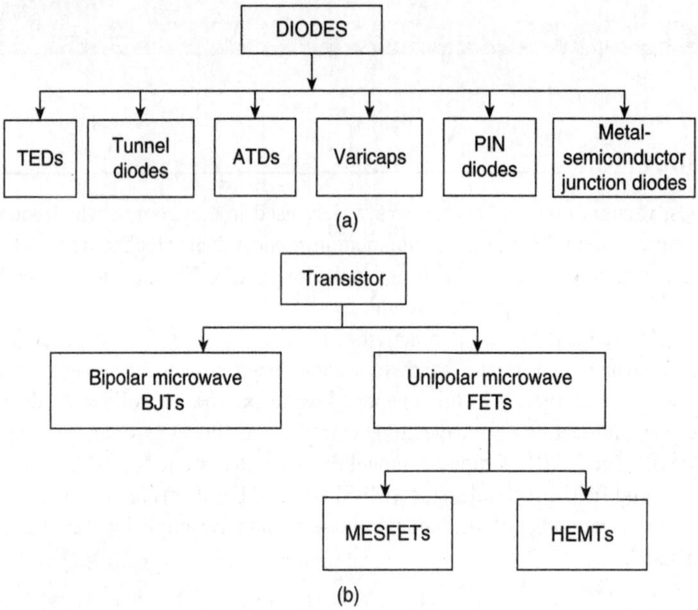

Figure 9.1 *(a) Two terminal devices; (b) Three terminal devices*

9.4 APPLICATIONS OF SOLID-STATE DEVICES

Microwave solid-state devices are becoming more important. These devices have been invented for various applications in microwave frequency regions such as frequency multiplication, signal detection, attenuation, generation of oscillation, switching, phase shifting, amplitude limiting, and amplification. Some of their applications are as follows:

- As microwave generators
- As amplifiers in satellite communications and also in other space applications
- As transmitters for millimeter communication systems
- In radio transmitters, such as CW Doppler radar
- In broadband linear amplifiers and in low-power amplifiers
- As a pumping source for parametric amplifiers
- In transponders
- In both combinational and sequential logic circuits
- In microwave receivers

9.5 TRANSFERRED ELECTRON DEVICES (TEDS)

Transferred Electron Devices (TEDs) are one of the important microwave devices. They are bulk devices that have no junction or gates as compared with microwave transistors, which operate with either junction or gates. Transferred electron devices are fabricated with compound semiconductor materials (e.g. gallium arsenide (GaAs), cadmium telluride (CdTe) and indium phosphide (InP)). They have two energy regions or valleys in the conduction band. The TEDs show a transferred electron effect in which transfer of electrons takes place from lower valley to upper valley in the conduction band. Most of the electrons will be in lower valley at low electric fields. When the electric field strength is increased to higher values, most of the electrons will be transferred into high-energy bands. In the higher-energy bands, the effective mass of electrons is larger than in lower-energy bands. Since the conductivity is directly proportional to mobility, the higher-energy band has lower conductivity than the lower-energy band. Hence, conductivity decreases with an increase in electric field strength. Thus, the current decreases with an increase in voltage, showing negative resistance. Gunn diode is an example of TEDs, and its theory is better explained in the next section.

A few differences between microwave transistors and TEDs are as follows:

- First, TEDs do not have junctions or gates as in the case of transistors.
- TEDs with smaller physical dimensions have a limited power output. In order to get a reasonable power output, the physical dimensions are to be made large as compared with a microwave transistor.
- TEDs are made of compound semiconductors from group III–V and II–VI elements (of periodic table) such as gallium arsenide (GaAs), indium phosphide (InP), and cadmium telluride (CdTe); whereas the majority of transistors are made from elements such as germanium or silicon.

9.6 GUNN DIODE

In 1963, Gunn observed a periodic variation of current passing through an N-type GaAs semiconductor. He found the following when a DC bias voltage is applied to the contacts of N-type GaAs or InP:

- Current first rises linearly from zero.
- Then, it begins to oscillate when a certain threshold is reached.
- The time period of oscillation is equal to the travel time of electrons from cathode to anode.

This is known as *Gunn effect* or *bulk effect*. The device that shows Gunn effect is known as *Gunn diode*. Gunn diodes are usually fabricated using N-type semiconductor materials (eg. GaAs, and InP); so, they should be associated with electrons rather than with holes. Gunn diode operations do not depend on junction properties. Even though it has no junction, it is called a *diode*, because it has two terminals (anode and cathode) attached to it. It is an active two-terminal solid-state device and is mainly used as a local oscillator in the microwave frequency range of 1 to 100 GHz. Gunn effect can be explained on the basis of domain formation, two-valley theory of Ridley–Watkins–Hilsum (RWH), or the transfer electron mechanism.

9.6.1 Operation and Characteristics of Gunn Diode

The Gunn diode is made up of a single piece of N-doped semiconductor material (e.g. GaAs, InP) with two thinner N^+-doped layer contacts on opposite ends. The two N^+ layers are required to connect the anode and cathode leads.

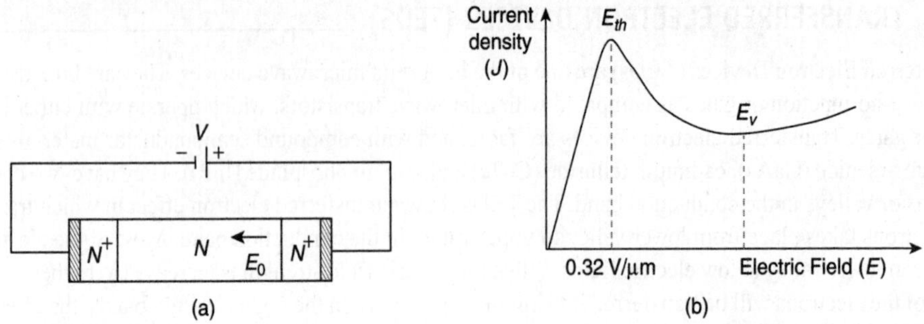

Figure 9.2 *(a) Three-layer structure of Gunn diode; (b) J/E curve of a Gunn diode*

When a DC voltage (V) is applied to the two terminals, an electric field (E_o) will be established across the piece of GaAs (just as in a resistor) (Figure 9.2(a)). Figure 9.2(b) shows how the current density ($\vec{J}$) through the material varies with the electric field (E_o) across it. We know that the drift velocity (v_d), current density, and electric field have the following relations:

$$v_d = \mu \vec{E}$$
$$\vec{J} = nq\mu \vec{E}$$

Initially, when the electric field is increased, from the above relations it can be seen that the drift velocity and current density increase. Thus, the current density increases with an increase in the electric field, resulting in a positive resistance. This continues till the electric field reaches a value known as *threshold value* E_{th} (corresponding threshold voltage V_{th}). When the electric field is increased beyond the threshold value E_{th}, the Gunn effect takes place and the current density decreases, causing the device to exhibit negative resistance. This behavior is due to domain formation, which will be explained in the next section. This will continue till the field reaches a value known as E_v (corresponding valley voltage). When the voltage is increased beyond E_v, the current density increases. Thus, the device again exhibits positive resistance.

Negative resistance region: In common the current first rises linearly from zero with increasing electric field, however there is a region between the threshold electric field and valley electric field, where the current decreases as the electric field is increased. This is called the *negative resistance region*. The dynamic resistance, r, in this voltage range is given by

$$r = dV/dI, r < 0.$$

9.6.2 Domain Formation

When bias voltage (V) is applied across the diode terminals, an electric field (E_0) is established across the GaAs piece. As E_0 increases, electron drift velocity increases up to a certain potential (E_{th}). This is because the velocity (v_d)is proportional to the applied electric field. However, with a further increase in voltage, the mobility decreases, which causes the velocity to decrease, thereby causing the electrons to slow down. This slowing down results in a traffic jam, and more electrons pileup to form a charge layer or domain. This charge layer produces an electric field (E) such that it decreases the original field (E_0) on the left and increases the original field (E_0) to the right (Figure 9.3). As a result, the charge bunch gets pushed toward the anode on the right, forming a current pulse.

When this pulse moves to the anode, E_0 goes back to the original value, resulting in another traffic jam and another pulse formation, which causes oscillation.

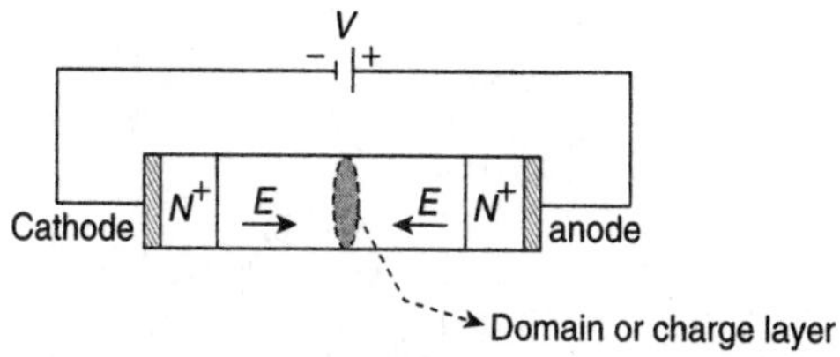

Figure 9.3 *Domain formation in a Gunn diode*

In this procedure, N-type GaAs releases the power that it has acquired during the domain formation in the direction of the anode. If conditions are appropriate, this release of power at anode will be utilized by the noise energy of suitable frequency existing in the Gunn diode configuration and gets amplified. The amplification of noise energy results in sustained oscillations. The frequency of oscillation depends on the length of the GaAs piece and concentration of electrons.

9.6.3 RWH Theory or Two-valley Theory of Gunn Diode

The Gunn diode, which is made of *n*-doped semiconductor material (e.g. GaAs or InP), is characterized by having two valleys in their conduction bands with different mobility. The two-valley model is also called the *Ridley–Watkins–Hilsum(RWH) theory*. There are two regions in the conduction band of the N-type GaAs. These conduction band regions are known as *upper valley* and *lower valley*. The energy band and transfer of electrons between the two valleys are shown in Figures 9.4 and 9.5, respectively. The following can be noted about GaAs:

- In N-type GaAs, the valence band is filled with electrons, and the conduction band is partly filled.
- The forbidden energy gap between the valence band and the conduction band is about 1.43 eV.
- Electrons in the lower valley (n_1) exhibit a small effective mass (m_1) and very high mobility, μ_1.
- Electrons in the upper valley (n_2) exhibit a large effective mass (m_2) and very low mobility, μ_2.
- The two valleys are separated by a small energy gap, ΔE, of approximately 0.36 eV.
- At a low electric field, electrons remain in the lower valley and material behaves ohmically. When the electric field reaches a certain threshold value E_{th}, the electrons will be swept from the lower valley to the upper valley. If the rate of transfer of electrons from the lower to the upper valley is very high, the current will decrease with an increase in voltage. This leads to a decrease in the average electron mobility, μ with an increase in the field in a bulk semiconductor, thus resulting in equivalent negative resistance. This manifests itself as a bulk negative differential resistance.
- The average electron mobility, μ is given by

$$\mu = (n_1\mu_1 + n_2\mu_2)/(n_1 + n_2) \tag{9.1}$$

where n_1 = electron density in lower valley

n_2 = electron density in upper valley

μ_1 = mobility of electron in lower valley

μ_2 = mobility of electron in upper valley

Total carrier concentration = $n_1 + n_2$

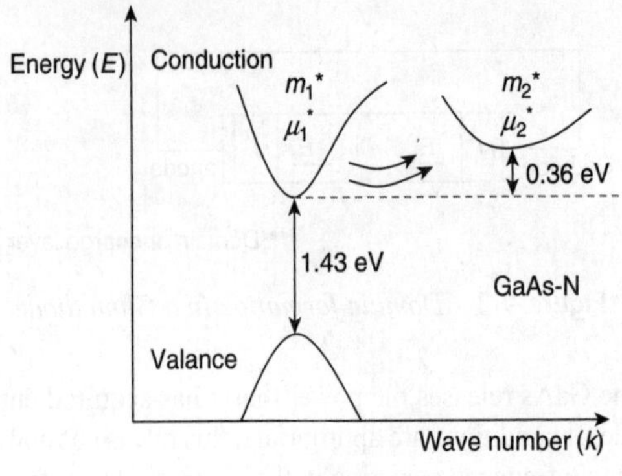

Figure 9.4 *Energy band diagram for GaAs*

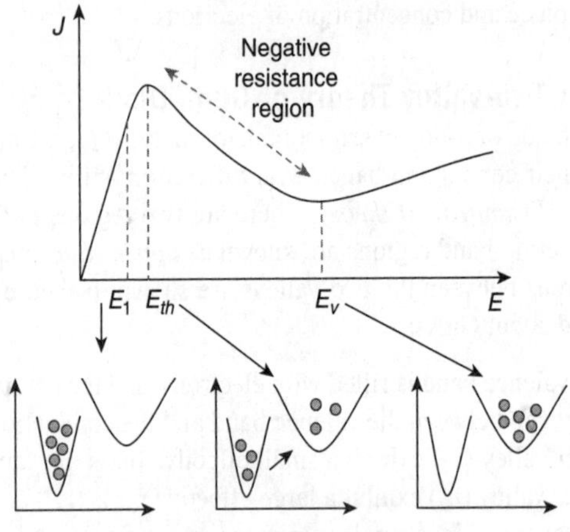

Figure 9.5 *Inter-valley transfer*

where m_1^* = effective mass of electron in lower valley
m_2^* = effective mass of electron in upper valley

The inter-valley transfer of electrons with regard to the E_{th} and E is illustrated in Figure. 9.5. Three different ranges of electric field strength may be considered.The current (I_1) and the current density ($\overrightarrow{J_1}$) corresponding to the electric field < ($\vec{E}_{th}$) where all the electrons occupy the lower-energy valley, is given by

$$\overrightarrow{J_1} = \mu_1 e n_1 \vec{E},\ I_1 = eA\mu_1 n_1 \vec{E} \left(\therefore \vec{J} = \frac{I}{A} \right)$$

The current (I_2) and the current density ($\overrightarrow{J_2}$) corresponding to the electric field $\vec{E}_{th} < \vec{E} < \vec{E}_v$, where the total number of electrons are distributed between the lower and upper valleys, is given by

$$\overrightarrow{J_2} = e(n_1\mu_1 + n_2\mu_2)\vec{E},\ I_2 = e(n_1\mu_1 + n_2\mu_2)A\vec{E}$$

The current (I_3) and the current density ($\overrightarrow{J_3}$) corresponding to the electric field $\vec{E} = \vec{E}_v$, where all electrons occupy the upper-energy valley, is given by

$$\overrightarrow{J_3} = e\mu_2 n_2 \vec{E},\ I_3 = eA\mu_2 n_2 \vec{E}$$

where $(\vec{J})$ = current density, $(\vec{E})$ = electric field, A = area of cross-section of device, and e = charge of electrons.

Performance characteristics: Gunn diode operates at 10-12 V of typical biased voltage, 300 mA of bias current, and the power output is 100 mW in the *X* band. The typical frequency range that it can operate is from 4 GHz to 94 GHz, efficiency is 10-20 %, CW power output is 250 mW at 15 GHz, and pulsed power is 5 W at 12 GHz.

9.6.4 Equivalent Circuit of Gunn Diode

The scale of the parasitic impedances accredited to the package element decreases with the package dimension. The equivalent circuit of Gunn diode is shown in Figure 9.6(a).

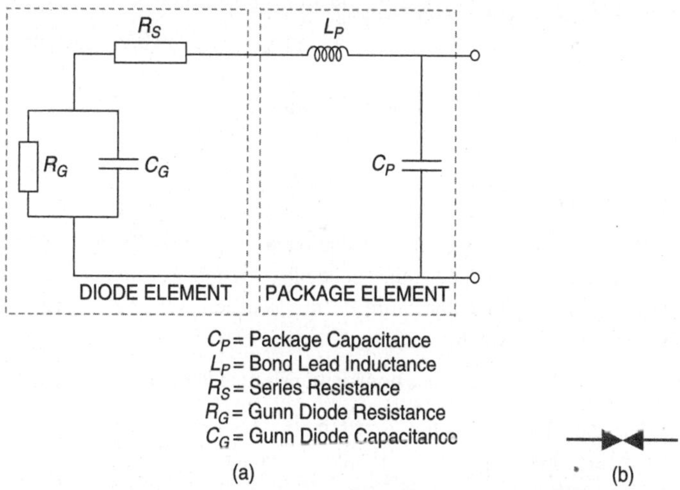

Figure 9.6 *(a) Equivalent circuit Symbol; (b) Symbol*

As a result, at lower microwave frequencies the stronger package styles are normally precise and for the higher microwave frequency the smaller parasitic impedance packages are suggested. Its symbol is as shown in Figure 9.6(b).

9.6.5 Basic Modes of Operation

Depending on the device characteristics and external circuitry, the Gunn diode can be made to oscillate in any one of the following four frequency modes:

- Gunn oscillation mode
- Limited space charge accumulation (LSA) mode
- Stable amplification mode
- Bias circuit oscillation mode

The modes of operations are classified based on the condition

$$n_o l > \frac{\varepsilon_s v_d}{e|\mu_n|}$$

v_d = electron drift velocity
ε_s = semiconductor dielectric permittivity
μ_n = electron mobility
e = electron charge
n_o = doping concentration
l = device length

9.6.5.1 *Gunn oscillation modes*

This mode has the following features:

- The electric field is greater than the threshold.
- The product of frequency and length is about 10^7cm/s.
- The product of doping and length is more than 10^{12} /cm^2.
- Due to the cyclic arrangement of either accumulation layer or the high voltage domain the Gunn diode is unstable.
- There are three types of Gunn oscillation mode. They are
    - Transit time domain mode
    - Delayed domain mode
    - Quenched domain mode.

Transit time domain mode: Transit time domain oscillation is the basic mode and is not depend on the external circuit. When a domain is quenched at the anode the current peaks are obtained. Then, another is nucleated near the cathode. At any time, the entire electric field across the device is above the threshold electric field. The frequency is given by

$$f = 1/\tau = v_d / l \text{ since domain transit time is } \tau = l/v_d$$

where v_d is the domain or drift velocity $= fl = 1 \times 10^7$ cm/s.
l is the effective length, and fl is the product of frequency and length
The oscillation time (τ_o) is equal to the transit time (τ).It is a low-power (< 2 W), low-efficiency mode, and the operating frequency is between 1 GHz and 18 GHz.

Delayed domain mode: This is also called an *inhibited mode*. In this mode, the domain is collected when the dc bias is less than the threshold electric field (E_{th}). The next domain can only be formed when the field again reaches the threshold. The oscillation period is greater than the transit time of the critical field (E_{th}), that is, $\tau_o > \tau$. The frequency of oscillation is determined by the resonant circuit. The efficiency of this mode is limited to 20%. The drift velocity (v_d) or ($f \times l$)lies between 10^6 cm/s and 10^7 cm/s.

Quenched domain mode: In this mode, the domain collapses before it reaches the anode; that is, it is quenched before it is collected, hence the name *quenched mode*. As shown in Figure 9.7, the bias electric field reduces below the sustaining electric field (E_s) in the negative half cycle. When the bias electric field swings back again more than the threshold electric field, another domain is nucleated and the procedure repeats. The operating frequencies are higher than the transit time frequency (Figure 9.7). The maximum efficiency of this mode is 13%. The drift velocity (v_d) or ($f \times l$) = 10^7 cm/s.

9.6.5.2 *Limited Space charge Accumulation (LSA) mode*

This mode operates by using a high-Q resonant cavity with current pulses from the Gunn diode. The Gunn diode is placed in a resonant cavity, which is tuned to a frequency of the LSA mode (f_0), so that the circuit operates like a negative resistance device and the domains do not have enough time to form. In this mode, the device can be biased to several times higher than E_{th}. When the input field increase

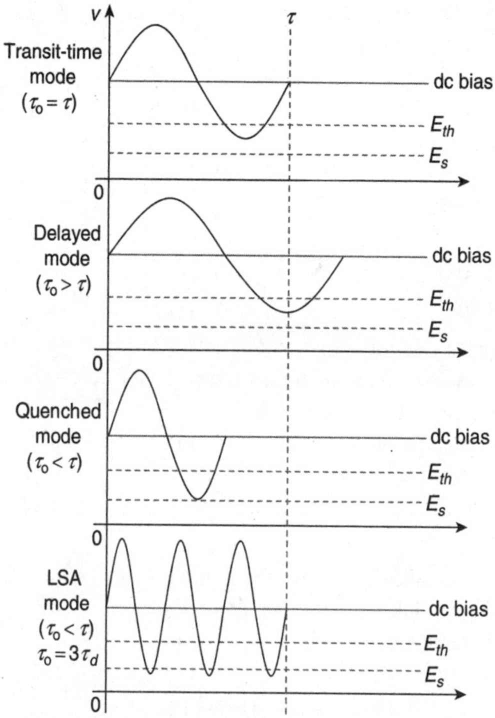

Figure 9.7 *Various modes of Gunn diode*

more than the threshold electric field, the device remains in the negative resistance region. Oscillation time is obtained by the external circuit, which is given by

$$\tau_0 = 2\pi\sqrt{LC} + \frac{L}{R(V_b / V_{th})} \tag{9.2}$$

where τ_0 is period in seconds

L is inductance in Henry

C is capacitance in farad

R is resistance in ohm

V_{th} is threshold potential

V_b is bias voltage

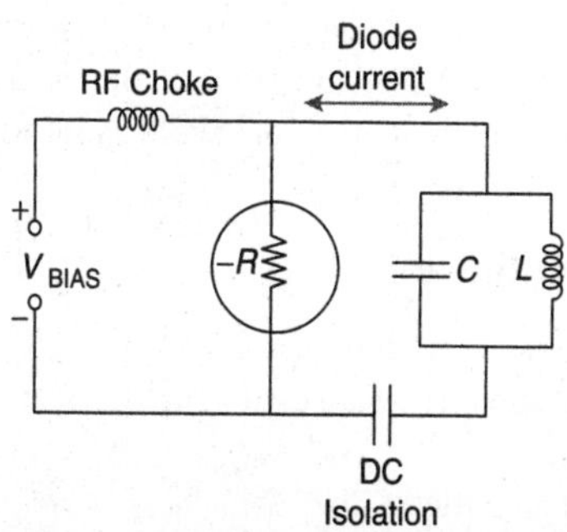

Figure 9.8 *Gunn diode in LSA mode*

Figure 9.8 shows the LSA mode Gunn oscillator simplified circuit. Gunn oscillator with LSA mode consists of Gunn diode which is shunted by either an LC tank circuit or a tuned cavity that behaves like a tank circuit. The condition for resonant oscillation is given by

$$(-G) = 1/-R \geq \text{conductance represented by circuit losses } (G_0).$$

Where G = Negative conductance

$$-G \geq G_0 \tag{9.3}$$

Output power of LSA mode $P_{out} = \eta I_0 V_0$

where η = conversion parameter of material

I_0 = operating current

V_0 = operating voltage

Advantages of LSA mode are as follows:

- Operating frequency is not limited by transit time effect. Therefore, device length can be made larger and the device can sustain higher voltage.
- This mode gives high-pulsed output power (> 100 W) with a high frequency (100 GHz). However, the power decreases with frequency.

EXAMPLE PROBLEM 9.1

A Gunn diode with 3.5 V critical threshold potential and 25 ohms resistance is connected to 3 V dc. If the diode is connected to a tank circuit where L = 0.1 nH and C = 2.5 pH, what is the period of oscillation? Also find the frequency of oscillation.

Solution

Given V_b = 3 V, V_{th} = 3.5 V, R = 25 ohms, L = 0.1 nH, and C = 2.5 pH

The period of oscillation is $\tau_o = 2\pi\sqrt{LC} + \dfrac{L}{R(V_b/V_{th})}$

$$= 2\pi\sqrt{2.5\times 0.1\times 10^{-21}} + \frac{0.1\times 10^{-9}}{25(3/3.5)} = 9.9\times 10^{-11} + 0.466\times 10^{-11}\ \text{s} = 1.366 \text{ picoseconds}$$

The frequency of oscillation is $f = 1/\tau_o = \dfrac{1}{1.366}\times 10^{11}\,\text{Hz} = 7.32\times 10^{12}\,\text{Hz}$

9.6.5.3 *Stable amplification mode*

This mode is defined as the region where the frequency times length is about 10^7cm/s and the doping time length is between 10^{11}cm^{-2} and 10^{12} cm^{-2}.

9.6.5.4 *Bias circulation mode*

In this mode, the product of frequency and length is very less. This mode occurs along with Gunn or LSA oscillations. When a bulk diode is biased to the threshold, the average current suddenly drops as Gunn oscillations begin. The drop in current at the threshold can lead to oscillations in the bias circuit that are typically from 1 kHz to 100 MHz.

9.6.6 Applications of Gunn Diode

1. It is used in Radar transmitters as low power oscillator (e.g. CW Doppler Radar, Police Radar).
2. Gunn diode oscillators with pulsed signal are used in transponders in industry telemetry systems and for air traffic control.
3. It is used in broadband linear amplifier.
4. It is used in sequential logic circuits and fast combinational circuits

5. It is used in microwave receivers as low and medium power oscillator.
6. It is used in parametric amplifier as pump sources.

EXAMPLE PROBLEM 9.2

A Gunn diode has the following parameters: electron drift velocity, $v_d = 2.5 \times 10^5$m/s, negative electron mobility, $|\mu_n| = 0.015$ m^2/Vs, relative dielectric constant, and $\varepsilon_r = 13.1$. Determine the criterion for classifying the modes of operation.

Solution

The data given are $v_d = 2.5 \times 10^5$ m/s, $|\mu_n| = 0.015$ m^2/Vs, and $\varepsilon_r = 13.1$

We have $n_o l > \dfrac{\varepsilon_s v_d}{e|\mu_n|} \Rightarrow n_0 l > \dfrac{\varepsilon_0 \varepsilon_r v_d}{e|\mu_n|} (\because \varepsilon = \varepsilon_0 \varepsilon_r)$

$$> \frac{8.85 \times 10^{-12} \times 13.1 \times 2.5 \times 10^5}{1.6 \times 10^{-19} \times 0.015} > \frac{2.9 \times 10^{-5}}{2.4 \times 10^{-21}} > 1.2 \times 10^{16}$$

This means that the product of the doping concentration and the device length should be

$$n_o l > 1.2 \times 10^{16}/\text{m}^2$$

■

EXAMPLE PROBLEM 9.3

Determine the conductivity of *n*-type GaAs Gunn diode if
Electron density $n = 10^{18}$ cm^{-3}, electron density at lower valley $n_l = 10^{10}$ cm^{-3}
Electron density at upper valley $n_u = 10^8$ cm^{-3}, temperature $T = 300°$ K

Solution

Given that electron density, $n = 10^{18}$ cm^{-3}, $n_l = 10^{10}$ cm^{-3},$n_u = 10^8$ cm^{-3}, and temperature T = 300° K.
We know that

$$\mu_l = 8000 \text{ cm}^2/\text{v.sec} = 8000 \times 10^{-4} \text{ m/v.s}, \ \mu_u = 180 \text{ cm}^2/\text{v.sec} = 180 \times 10^{-4} \text{ m/v.s}$$

We have conductivity,

$$\sigma = e(n_l \mu_l + n_u \mu_u) = 1.6 \times 10^{-19}(8000 \times 10^{-4} \times 10^{16} + 180 \times 10^{-4} \times 10^{14}) = 1.28 \times 10^{-3} \mho$$

■

EXAMPLE PROBLEM 9.4

A GaAs Gunn diode has an active region of 10 micrometers. If the electron drift velocity is 10^5 m/sec, calculate the natural frequency and the threshold voltage. The critical electric field is 3 kv/cm.

Solution

The given data are $v_d = 10^5$ m/sec $= 10^7$ cm/sec, $l = 10$ μm $= 10 \times 10^{-4}$ cm, and $E_c = 3$ kv/cm $f = ?, V = ?$

We have natural frequency, $f = \dfrac{v_d}{l} \quad \therefore f = \dfrac{10^7}{10 \times 10^{-4}} = 10^{10} \quad \therefore f = 10$ GHz

Critical voltage, $\quad V = l \times E_c$

i.e., $\quad$ Critical voltage = Active length × Critical electric field

$$\therefore V = 10 \times 10^{-4} \times 3 \times 10^3 = 3 \text{ V}$$

∴ Critical voltage, $\quad V = 3$ V

■

9.7 TUNNEL DIODES

In a conventional PN junction diode, the concentration of impurities is in the order of one part in 10^8. When the impurity concentration is increased to 1 part in 10^3, the conventional PN diode becomes the tunnel diode.

Principle of Operation: In a tunnel diode, the impurity concentration is greatly increased as compared with the PN diode. This reduces the width of the depletion layer to the order of hundred angstroms. According to quantum mechanics, if the barrier is less than $3A^0$, there is a large probability that particles will tunnel through the potential barrier. This is known as *tunneling*.

The operation can be best explained by considering energy band diagram of P-type and N-type materials. Due to high impurity concentration, there are many holes in valence band of P-type material and many electrons in conduction band of N-type material. Now, if no voltage is applied, the alignment of valence band and conduction band of P-type and N-type is as shown in Figure 9.9 (a). Hence, there is no flow of current. Now, if some forward voltage is applied the energy levels of N-region move upward and that of P-region move downward relatively. Therefore, the electrons on the right side of the potential barrier are precisely opposite to the holes on the left side and the corresponding energy band diagram is as shown in Figure 9.9 (b). This results in flow of current, whose magnitude gradually increase as forward voltage increase till $V = V_p$ which is known as *peak voltage*.

At the peak, as shown in Figure 9.10, the voltage energy levels on either side are in perfect alignment. As the voltage is increased beyond V_p, the energy levels of the N region move further upward and those of the *P* region move further downward, causing a condition as shown in Figure 9.9 (c). In this condition, some of the N-region electron energy levels are opposite to the band gap and some are opposite to the holes (or empty states). This results in a decrease of current flow; this condition continues till $V = V_v$, called *valley voltage*. At this voltage, current ($I = I_v$) is minimum. If the voltage is further increased as shown in Figure 9.9 (d), then all electrons in the N-region are opposite to the band gap and no currents flows. The corresponding point in the Figure 9.10 is the V_v. If the voltage continues to increase further as shown in Figure 9.9 (e), the diode shows characteristics of a normal PN diode. Thus, between peak voltage and valley voltage, the tunnel diode shows negative resistance characteristics. The volt-ampere characteristic curve of a tunnel diode is as shown in Figure 9.10 The *V-I* characteristic of a tunnel diode exhibits the current as multi-valued and the voltage as a single-valued.

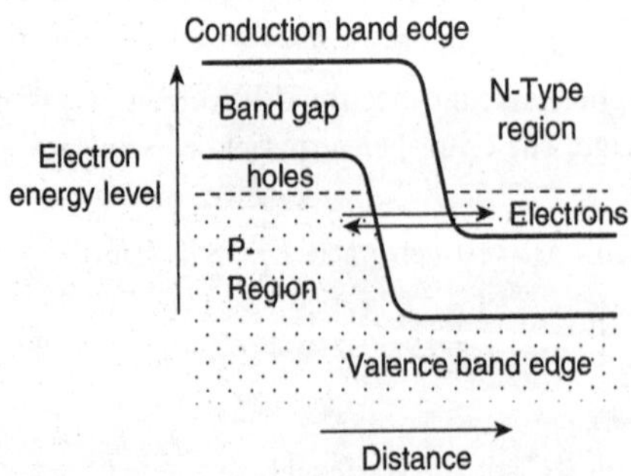

Figure 9.9 (a) *Energy bands in unbiased tunnel diode*

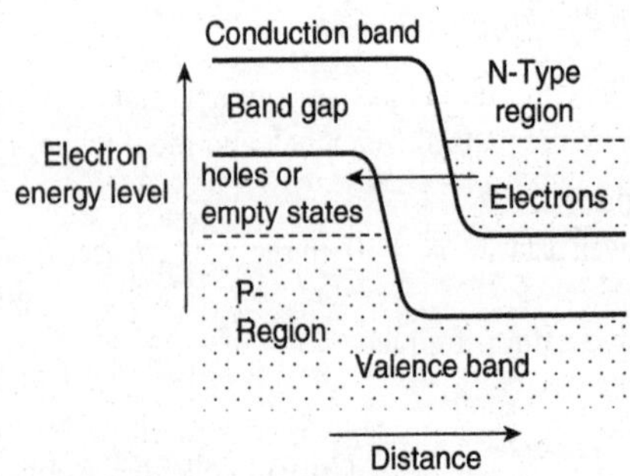

Figure 9.9 (b) *Electron energy levels of N-region are equal to the empty states level on P region.*

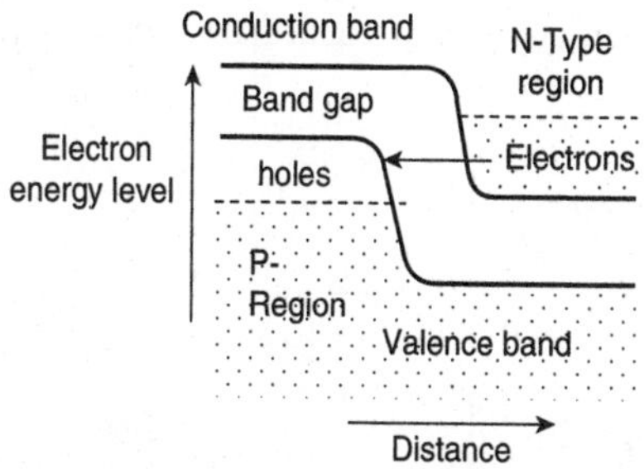

Figure 9.9 (c) *Some N-region electrons energy level are opposite to "band gap," and some are opposite to holes*

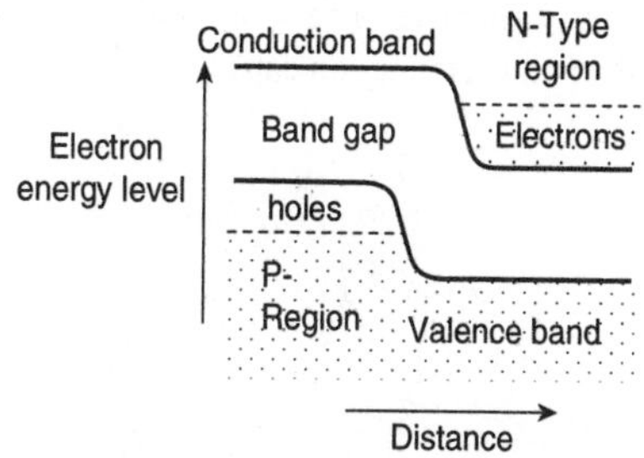

Figure 9.9 (d) *All N-region electrons energy level are opposite to band gap*

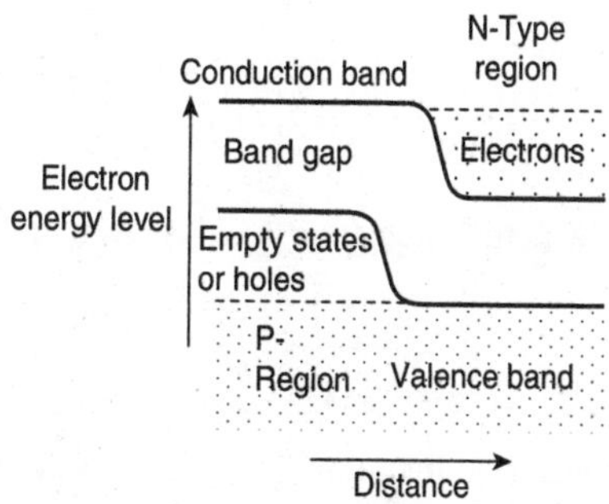

Figure 9.9 (e) *Tunnel diode functions as normal diode*

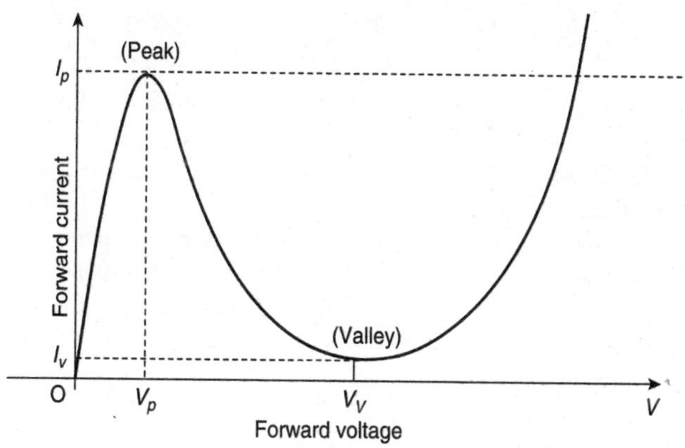

Figure 9.10 *VI characteristics of a tunnel diode*

The equivalent circuit of a tunnel diode is as shown in Figure 9.11. R_n is negative resistance. From the volt-ampere characteristics,

$$-g = \frac{\partial i}{\partial v} = -\frac{1}{R_n} \tag{9.4}$$

R_s is series ohmic resistance; L_s is series inductance that depends on lead length and packaging circuit of the diode; and C is junction capacitance.

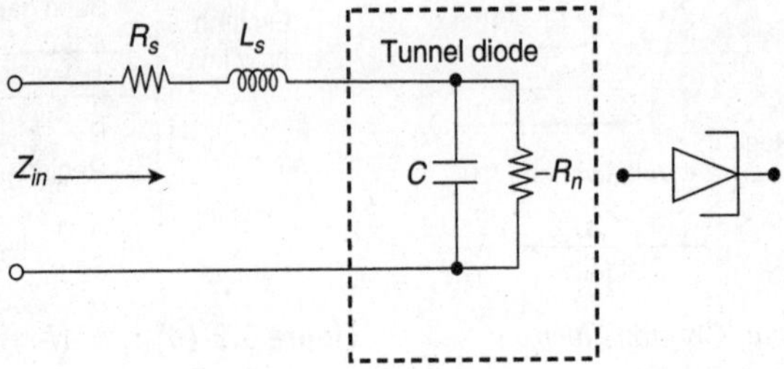

Figure 9.11 *Equivalent circuit of tunnel diode and its symbol*

The input impedance of the circuit is

$$Z_{\text{in}} = R_s - \frac{R_n}{(1+(\omega R_n C)^2)} + j\left[\omega L_s - \frac{\omega R_n^2 C}{(1+(\omega R_n C)^2)}\right] \tag{9.5}$$

The cutoff frequency is obtained by putting Re (Z_{in}) = 0 and is given by

$$f_c = \frac{1}{2\pi R_n C}\sqrt{\frac{R_n}{R_s} - 1} \tag{9.6}$$

Self-resonance frequency is obtained by putting Img (Z_{in}) = 0 and is given by

$$f_r = \frac{1}{2\pi R_n C}\sqrt{\frac{R_n^2 C}{L_s} - 1} \tag{9.7}$$

Applications

- It is used as a high-speed switch. The switching times of the order of nanoseconds are obtained.
- As a logic memory storage device
- As a microwave oscillator
- In a relaxation oscillator
- The tunnel diode can be used in bistable, a stable, and monostable circuits. If the operating point is made to vary over the entire range of the voltage, the circuit is in bistable mode. If the operating point is between V_p and V_v, the circuit is a stable. If the operating point is between 0 and V_p, the circuit is monostable.

Some important characteristics of a tunnel diode are as follows:

- Doping is very high, usually 1000 times of normal diodes.
- It has zero breakdown voltage.
- It has a dynamic negative resistance region, because when we increase its voltage, the current will decrease beyond the peak voltage.
- For an ideal tunnel diode, V_p or I_p should be very large.
- It behaves similar to both an amplifier and an oscillator.
- In reverse bias, it behaves similar to a good conductor.

9.8 AVALANCHE TRANSIT TIME DEVICES

It is possible to realize a microwave diode that exhibits negative resistance, by providing a delay between voltage and current through the material. In 1958 Read had proposed that there must be a phase delay of more than 90° between RF input voltage and avalanching current if the total voltage go above *breakdown* voltage in a diode due to the RF input voltage. Such devices are called avalanche transit time devices. At microwave frequencies, avalanche transit time devices produce negative resistance by using carrier impact ionization and transit time in the high voltage region of a semiconductor junction. There are distinct modes of avalanche oscillators, out of which the following three are discussed:

1. IMPATT: Impact Ionization Avalanche Transit Time diode
2. TRAPATT: Trapped Plasma Avalanche Triggered Transit diode
3. BARITT: Barrier injection transit time diode

9.9 IMPATT DIODE

IMPATT diode is abbreviated as impact-ionization avalanche transit time diode. It is an active solid state device that operates by a reverse bias adequate to cause avalanche breakdown. This is a high-power diode and a very powerful microwave source that is used in high-frequency electronics and microwave devices. They may be operated at frequencies up to about 350 GHz when manufactured with silicon. The IMPATT diode exhibits a dynamic negative resistance that is required for microwave oscillation and amplification applications. This is due to the following two reasons:

Impact Ionization avalanche effect: This causes the carrier current to lag behind the ac voltage by 90 degrees.

Transit time effect: This causes a further time delay and causes the external current to lag behind the ac voltage by a further 90 degrees.

The summation of delay involved in generating avalanche current multiplication along with delay due to transit time through drift space provides the necessary 180° phase difference between the applied voltage and the resulting current in an IMPATT diode.

These devices can be classified as follows:

- **Single drift devices:** Devices such as P^+NN^+, N^+PP^+, P^+NIN^+, and N^+PIP^+ come under this category. Consider the P^+NN^+ device. In this device, when the P^+N junction is reverse biased, it causes an avalanche breakdown. This causes the P^+ region to inject electrons into the NN^+ region. These electrons move with a saturated velocity. However, the holes injected from the NN^+ region do not drift. Hence, these are called *single drift devices*.
- **Double drift devices:** The example of a double drift device is P^+PNN^+. In this device, when the PN junction is biased near an avalanche breakdown, electrons drift along the NN^+ region and holes drift along the PP^+ region. Hence, they are called *double drift devices*.

9.9.1 Principle of Operation of IMPATT Diode

To understand the operation of an IMPATT diode, here we consider the N^+PIP^+ diode. Let V_b be the reverse bias breakdown voltage that is applied to the IMPATT device. Assume that a sinusoidal waveform $V_1 \sin\omega t$ is superimposed on V_{dc}, resulting in a total device voltage $V(t) = (V_{dc} + V_1 \sin\omega t)$ as shown in Figure 9.12.

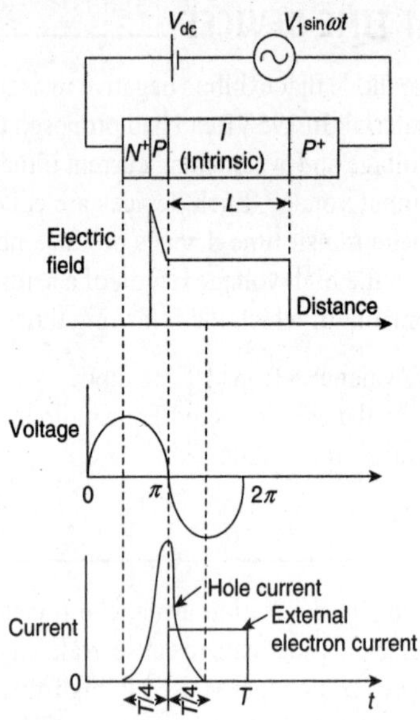

Figure 9.12 *IMPATT diode operation*

Initially, the device contains a few thermally generated free electrons. When $V(t) > V_b$ breakdown occurs at the N^+P junction, these electrons gain energy from the applied voltage and knock off electrons in the valence band to the conduction band. As a result, a new electron hole pair is created. An electron-hole pair generated because of such impact ionization is called a *secondary electron-hole pair*. These secondary electrons again pick up sufficient energy and generate more secondary electron-hole pairs. Therefore, as long as $V(t) > V_b$, the number of carriers increases exponentially, even beyond the voltage maximum irrespective of magnitude of $V(t)$. This is because of sufficient number of secondary electron-hole pairs presence. This exponential increase continues until the sine wave crosses zero and then drops exponentially until the sine wave reaches its negative peak. This avalanche current (generated holes) is injected into the I-region and drifts toward P^+ region with saturated velocity along the depletion region. The electrons move toward the positive terminal. In this way, this current will have a one-quarter period ($T/4$) delay or a 90° phase shift with regard to the applied signal voltage.

To achieve the desired 180° phase shift between input voltage and external current, additional T/4 delay is essential. This is made available by the hole drift along the depletion region. It is the property of semiconductor materials that the drift velocity tends to be constant at high field strengths. Since the holes move at the constant velocity v_d, the device length may be chosen to provide the necessary delay for a 180° phase shift between the device voltage and current, which is given by

$$l = v_d \frac{T}{4} \tag{9.8}$$

EXAMPLE PROBLEM 9.5

The carrier drift velocity in silicon semiconductors is $v_d = 105$ m/s. What should be the length of silicon IMPATT in order to obtain a negative resistance at about 12 GHz?

Solution

In order to obtain a negative resistance, the carrier drifting through the device should give rise to a phase shift of $\pi/4$, to be added to the avalanche phase shift of $\pi/4$. The drift time should, therefore, be $T/4$. The device length is given by

$$l = v_d \frac{T}{4}$$

$$T = \frac{1}{12\times 10^9} = 8.33\times 10^{-11}\,\text{s}$$

Therefore,

$$l = 105\times \frac{8.33\times 10^{-11}}{4}$$

$$l = 2\ \mu\text{m}$$

■

Equivalent Circuit: The IMPATT diode equivalent circuit is as shown in Figure 9.13. It is composed of two parts that are the avalanche and drift regions, and a loss resistance (R_s). One part of the equivalent circuit is the avalanche region it consists of a resonant circuit, with an avalanche inductance (L_a) and a capacitance (C_a). The avalanche capacitance is given by:

$$C_a = \frac{\varepsilon_s S}{w_a} \tag{9.9}$$

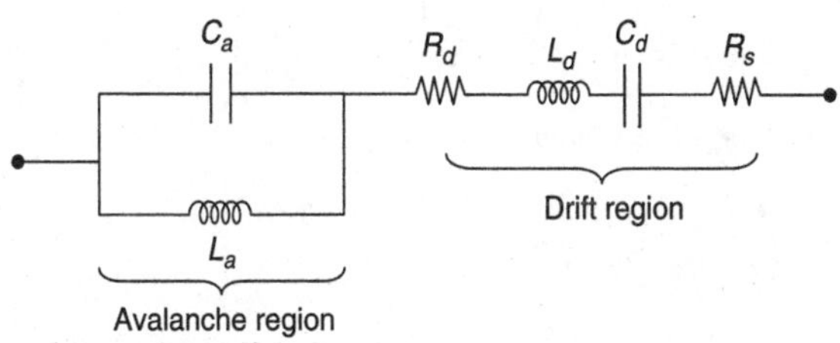

Figure 9.13 *Equivalent circuit of IMPATT diode*

where, w_a is the width of the avalanche region,
ε_s = permittivity of the dielectric, and
A is the area.

The IMPATT diode exhibits negative resistance for frequencies higher than the avalanche resonant frequency (f_a), and is given by

$$f_a = 1/\sqrt{L_a C_a}$$

The second part of the equivalent circuit is the drift region, it consist of a series resonant circuit, with drift capacitance (C_d), and is given by

$$C_d = \frac{\varepsilon_s S}{w - w_a} \tag{9.10}$$

where w is total device width.

The operating frequency (f) of the IMPATT diode is given by

$$f = l/(2\tau) = v_d/2L$$

where v_d = Carrier drift velocity

L = Length of the drift space charge region

For the condition, $f > f_a$ the drift resistance (R_d) is negative. At frequencies more than avalanche frequency, the avalanche sub-circuit behaves like capacitor. The drift region jointly with the avalanche capacitance, account for the phase shift. At a specified frequency, the maximum power output of a simple diode is limited by semiconductor material and the achievable impedance level in microwave circuits. The most commonly used semiconductors are GaAs and Si. Though GaAs is costly and difficult to fabricate, it is the most preferred semiconductor because it gives less noise, high efficiency and maximum operating frequency.

Derivation for the output power and efficiency of IMPATT diode

The maximum voltage applied across the diode is given by

$$V_m = E_m l$$

where $E_m \rightarrow$ Maximum electric field

$l \rightarrow$ Depletion length

The breakdown voltage limits the applied voltage V_m; as a result of this, the maximum current is also limited, that is, the maximum current is given by

$$I_m = J_m A = \sigma E_m A = \frac{\varepsilon_s}{\tau} E_m A = \frac{v_d \varepsilon_s E_m A}{l}$$

$$\therefore I_m = J_m A = \frac{v_d \varepsilon_s E_m A}{l} \tag{9.11}$$

The upper limit of the input power is obtained as

$$P_m = V_m I_m = (E_m l) \times \frac{v_d \varepsilon_s E_m A}{l}$$

$$\therefore P_m = v_d \varepsilon_s E_m^2 A \tag{9.12}$$

We know that $$C = \frac{\varepsilon_s A}{l} \tag{9.13}$$

Substituting Eq. 9.13 in Eq. 9.12,

$$P_m = v_d E_m^2 Cl$$

Multiplying and dividing the R.H.S with $2\pi f$, we get

$$P_m = \frac{v_d E_m^2 Cl \times 2\pi f}{2\pi f} \Rightarrow P_m = \frac{E_m^2 v_d l}{2\pi f X_c} \left[\because X_c = \frac{1}{2\pi f C}\right] \tag{9.14}$$

Using $2\pi f \tau = 1$, we have

$$f = \frac{1}{2\pi\tau} = \frac{v_d}{2\pi l}\left[\because \tau = \frac{l}{v_d}\right]$$

$$\because l = \frac{v_d}{2\pi f} \tag{9.15}$$

Substituting Eq. 9.15 in Eq. 9.14, we get

$$P_m = \frac{E_m^2 v_d^2}{4\pi^2 f^2 X_c} \quad \therefore P_m f^2 = \frac{E_m^2 v_d^2}{4\pi^2 X_c}$$

$$\text{Efficiency, } \eta = \frac{P_{ac}}{P_{dc}} = \left[\frac{V_a \times I_a}{V_d \times I_d}\right] \tag{9.16}$$

For practical IMPATT diodes, the efficiency is less than 30%.

9.9.2 Characteristics of IMPATT Diode

- IMPATT diode operates in reverse bias. It exhibits negative resistance region due to the impact avalanche and transit time effects.
- The phase difference between voltage and current is 180°. Here 90° phase delay is due to avalanche effect, and the remaining 90° is due to transit time effect.
- It is a narrow-band amplifier that provides output power in the millimeter-wave frequency range.
- At low frequencies, their power output is inversely proportional to frequency. At high frequencies, their power output is inversely proportional to the square of frequency.
- They are often used in the design of oscillators and amplifiers when a high output power is required. They provide higher output power than Gunn diodes.
- They are manufactured in Si, GaAs, and InP. They can be operated up to 350 GHz when manufactured in Si.
- These diodes are of low cost, reliable, and compact. They are moderately efficient milliwatt power sources.
- These are noisier than Gunn diodes. Therefore, they are rarely used for local oscillators in receivers.

Performance characteristics

- Theoretical, $\eta = 30\%$ ($< 30\%$ in practice) and 15% for Si, 23% for GaAs
- Frequency: 1 to 300 GHz
- Maximum output power for a single diode: 5W in X band to 6.5 W at 30 GHz
- Several diodes combined: 40 W in X band
- Pulsed powers = 4 kW

Disadvantages of IMPATT diode

- In terms of noise figure an IMPATT diode is not good as in comparison with the TWT amplifier or Gunn diode oscillator or klystron tube. Because the avalanche is a high noise process, so the IMPATT is very noisy diode, the value of noise figure is 30 dB.
- In IMPATT diode matching is difficult because of the low value of their negative resistance.
- It is sensitive to operational conditions.
- It has large electronic reactance, which can cause detuning or burn out the device if proper care is not taken.

Applications

- IMPATT diodes are used as microwave oscillators in microwave generators, in modulated output oscillators.
- They are used in microwave links, continuous wave radars, and electronic counter measures.
- IMPATT diodes are also used as amplification with negative resistance. In police radars, low power transmitters, and intrusion alarm devices are used the high-Q IMPATT diodes. In frequency modulated telecommunication transmitters and continuous wave Doppler radar transmitters are used the low-Q IMPATT diodes.

EXAMPLE PROBLEM 9.6

Determine transit time of the carriers and operating frequency of the IMPATT diode. If an IMPATT diode has drift length of 4 μm and $v_d = 10^6$ m/sec.

Solution

Given that $l = 4$ μm, we have $f = \frac{1}{2\tau} = \frac{v_d}{2l}$

where v_d = Carrier drift velocity

l = Drift length

$$\text{Transit time, } \tau = \frac{l}{v_d} = \frac{4\times10^{-6}}{10^6} [\because v_d = 10^6 \text{ m/sec}] \therefore \tau = 4\times10^{-12} \text{ Sec}$$

$$\text{Operating frequency, } f = \frac{1}{2\tau} = \frac{1}{2\times4\times10^{-12}} \therefore f = 0.25\times10^{12} \text{ Hz} = 125 \text{ GHz}$$

■

EXAMPLE PROBLEM 9.7

An IMPATT has the following parameters: pulse operating current is 1A and a pulse operating voltage is 200 V. The efficiency is 15%. If the pulse width is 0.02 ns and the frequency is 20 GHz, determine the power output and the duty cycle.

Solution

Given that, $V_o = 200$ V, $I_o = 1$ A.

Output power, $P = \eta \times P_{dc}$

where $P_{dc} = V_0 \times I_0 = 200 \times 1 = 200$ W

$$P = 0.15 \times 200 \text{ W} = 30 \text{ W}$$

Pulse width = 0.02 ns = 0.02×10^{-9} s, Frequency = 20 GHz = 20×10^9 Hz

$$\text{Time, } T = \frac{1}{f} = \frac{1}{20\times10^9} = 5\times10^{-11} \text{s}$$

$$\text{Duty cycle} = \text{pulse width/total time} = \frac{0.02\times10^{-9}}{5\times10^{-11}} = 0.4$$

■

9.10 TRAPATT DIODE

IMPATT and Gunn diodes cannot operate at lower frequencies and they operate at frequencies of 3 GHz or above. In these diodes as it is difficult to increase transit time. A slightly modified structure of the IMPATT diode that can be used at low frequencies is the TRAPATT diode, where TRAPATT stands for *trapped plasma avalanche triggered transit mode*. In 1967,the first TRAPATT diode was produced, which has an efficiency of 25% and produces 400 W at 1000 MHz. Now a days, efficiencies of 60% to 75% are obtained.

Principle of operation: The TRAPATT diode depicted in Figure 9.14 (a) is a P^+NN^+ diode that is driven by a large repetitive pulse of current. Breakdown will occur at one of the junctions. Since the current drive is very large, a large collection of disassociated electrons and holes known as *plasma* is generated. These carriers do not easily recombine. The violent breakdown creates a high electric field shock front that moves across the N-type drift region. After the passage of shock front, the plasma is located in a low field region and is trapped because it takes a long time to clear the drift region of the carriers. When the plasma has been cleared from the drift region, the cycle will be repeated. The formation of plasma in the active region increases the transit time that is required for low-frequency operation. Initially, the device starts oscillating in the IMPATT mode. When the amplitude of oscillations increases, the TRAPATT mode of oscillation is established.

The operation can be better explained with help of Figure 9.14 (a).When a square current drive is applied to the diode, the output will be as shown in Figure 9.14 (b). To better explain the electric field variations, let it be divided into six regions as shown in Figure 9.14 (a).

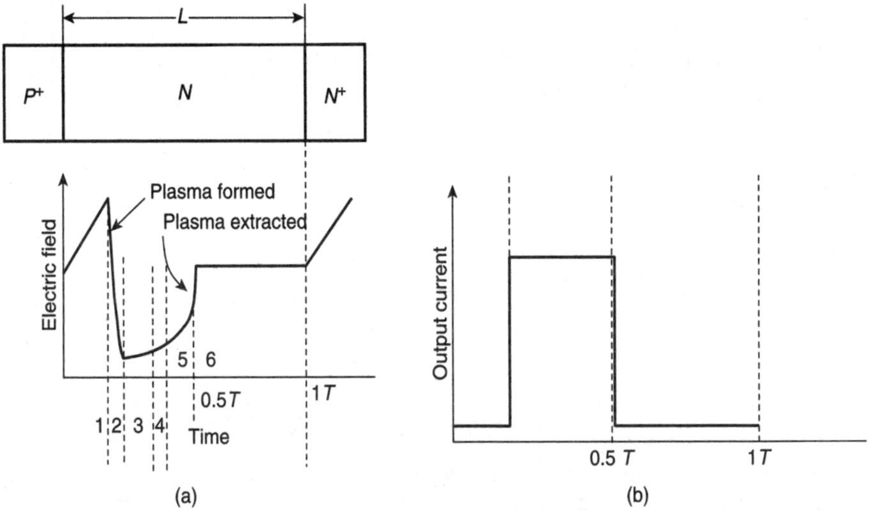

Figure 9.14 *(a) Structure of TRAPATT; (b) Output of TRAPATT*

Region 1 (Charging): Initially, only thermally generated minority charge carriers are present, and, hence, the current pulse charges up the depletion-layer capacitance. Therefore, the output voltage ramps up as shown in Figure 9.14 (a).

Region 2 (Plasma formation): When the electric field becomes more than what is required to cause a breakdown, a large number of carriers is generated. Hence, in region 2, internal current exceeds external current. This results in the formation of a plasma region of a large number of disassociated holes and electrons that are hard to be recombined. As a result of this, conductivity of the medium increases, voltage is reduced to lower values, and the electric field at the depletion region decreases.

Region 3 (Plasma extraction): As the voltage drops significantly, the plasma starts being removed. However, the process takes a lot of time. As the plasma gets extracted, the current starts decreasing. As the conductivity is decreasing, the voltage starts building up. In this region, the entire plasma is removed.

Region 4 (Residual Plasma extraction): A few charge carriers remain on either side of the depletion layer, resulting in a residual electric field in region 4, which gradually decreases.

Region 5 (Charging): In this region, the diode again behaves similar to a capacitor, resulting in the electric pattern.

Region 6: The voltage remains constant as the current drops to zero.

A sharp rise time current pulse, harmonic rich output of TRAPATT diode is shown in Figure 9.14 (b). This pulse must be applied to a low-pass filter which is connected to the TRAPATT diode at the input of the waveguide or transmission line. Here the harmonics are not accepted with the filter and thus are reflected back to the TRAPATT diode to activate the next current pulse.

Performance characteristics of TRAPATT diode

- It can be operated at comparatively low frequencies.
- With five diodes connected in a series, the highest pulse power of 1.2 kW is obtained at 1.1 GHz.
- Operating voltage: 60–150 V
- Frequency: 3 to 50 GHz
- Efficiency: 15 to 40% (8 GHz to 0.5 GHz)
- Continuous power: 1–3 W between 8 GHz and 0.5 GHz
- Noise figure: greater than 30 dB

Advantages

- It is more efficient than the IMPATT diode
- Low power dissipation

Drawbacks

- High noise figure
- Strong harmonics due to short current pulse

Applications

- They are used in low-power Doppler radars, microwave beacon landing systems, phased array radars, and so on.

Comparison of IMPATT and TRAPATT diodes

IMPATT diode	TRAPATT diode
1. It stands for Impact Avalanche and Transit Time	1. It stands for Trapped Plasma Avalanche Triggered Transit
2. Efficiency of operation is 30%	2. Efficiency is in between 15% and 40%
3. Frequency = 1 to 300 GHz	3. Frequency = 3 to 50 GHz
4. Pulsed power = 4 kW	4. Pulsed powers =1.2 kW at 1.1 GHz
5. It finds applications in microwave oscillators	5. They are used in low-power Doppler radars, phased array radars, radio altimeters, and so on

9.11 BARITT DIODE

BARITT diode is an acronym for Barrier Injection Transit Time diode. These are improved versions of IMPATT diodes. The ionization technique of IMPATT makes it very noisy; therefore it is avoided in BARITT devices. It is replaced by barrier-injected minority carriers generated from forward-biased junctions instead of being generated from impact ionization of the avalanche region as in the IMPATT diode. Therefore, these devices are less noisy than IMPATT. The negative resistance is due to the drift of injected holes to the P region. It consists of three regions of semiconductor material, as shown in Figure 9.15.

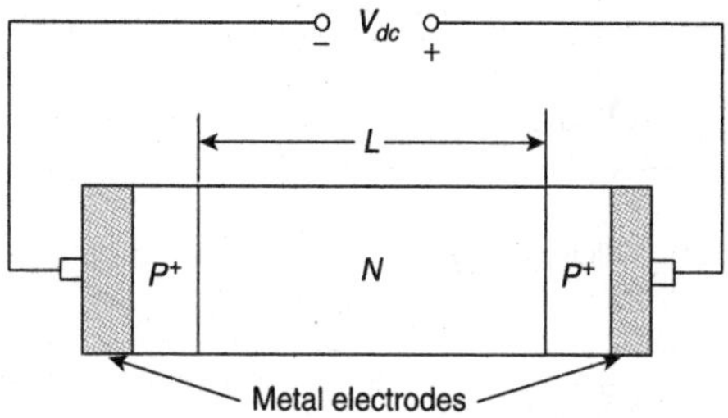

Figure 9.15 *Construction of BARITT*

It has a pair of abrupt PN junctions as shown in Figure 9.15. The P^+N region is reverse biased, and the NP^+ is forward biased. Initially, when small magnitudes are applied, the depletion region is formed at both junctions but no current flows. As the biasing voltage increases, the depletion region that is formed in the N region meets from either side and causes the depletion zone to cover the entire N region. These situations are called *punch through* or *reach through*. The forward bias causes the holes to enter the depletion region and then to drift across this region with saturation velocity, forming a current pulse. The current increase is due to the magnitude of critical voltage (V_c) and its negative temperature coefficient, not due to avalanche multiplication.

$$V_c = \frac{qNL^2}{2\varepsilon_s}$$

Where N = doping concentration
L = semiconductor thickness
ε_s = semiconductor dielectric permittivity
q = charge of carriers

The BARITT operation should be satisfied.

- The voltage across the BARITT must be large enough to send charge carriers to space charge region on the saturation velocity.

Performance characteristics

- Frequency: 4–10 GHz
- Power: 50 mW at 4.9 GHz
- Efficiency: 1.8%
- Noise figure: 9 dB at 6.35 GHz with a 15–dB gain

Applications

- Used as an amplifier in the microwave region

9.12 PIN DIODE

It is a variation of the conventional PN junction diode with a small layer between P and N layers. It consists of a layer of intrinsic semiconductors between two heavily doped P- and N-type semiconductors (Figure 9.16 (a)). The width of the intrinsic region may vary from 10 to 200 micro meters. This long intrinsic region makes the device capable of withstanding high breakdown voltages. The equivalent symbol of the PIN diode is shown in Figure 9.16 (b).

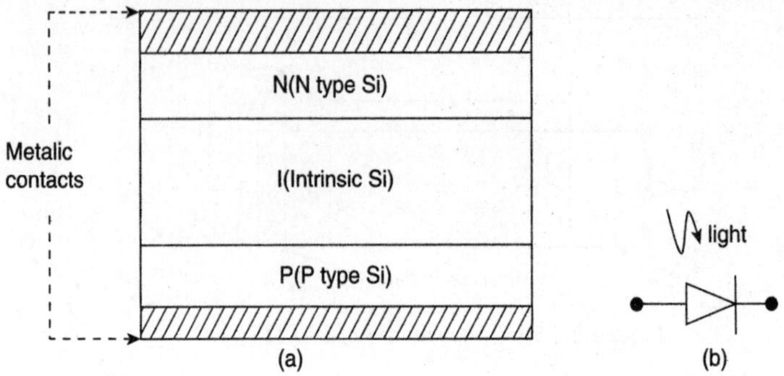

Figure 9.16 *(a) PIN diode; (b) PIN diode symbol*

Principle of operation Under the no-bias condition, the I-region of a PIN diode has a very few charge carriers. Holes from the P region and electrons from the N region diffuse into the intrinsic region. The width of the space charge region is inversely proportional to impurity concentration. Therefore, the space charge regions of P and N are very narrow.

When a reverse bias is applied, thermally generated electrons move toward the N-region, and thermally generated holes move toward the P-region; as a result of this, a small electron-hole current flows through the device. The space charge region extends into the intrinsic region and almost covers it such that the depletion width (w_d) and the width of the intrinsic layer (w_I) are equal. The junction capacitance is

$$C_j = \frac{\varepsilon_s A}{w_I} \tag{9.17}$$

where A is the effective area, and ε_s is the permittivity of the semiconductor. Therefore, the large width of the intrinsic region makes the junction capacitance less, making it suitable for switching applications.

When a forward bias is applied, a large number of holes from the P-region and electrons from the N-region enter the intrinsic region. If the width of the intrinsic region is smaller than the diffusion length of electrons, injected carriers do not combine immediately but they remain in the intrinsic region. This reduces the resistance of the intrinsic layer. If the forward bias is increased further, more and more carriers are injected into the I-region, further reducing its resistance. This is provided by the doping which is sufficient enough to supply the carriers. Hence, the resistance of the I layer can be determined by controlling the forward bias. This is known as *conductivity modulation.*

The PIN diode can be used as a microwave device only above 200 MHz.At low frequencies, the diffusion length of carriers is small. Therefore, the carriers recombine in the intrinsic layer, and the diode behaves similar to an ordinary PN diode.

Specifications

The following are the few important parameters that are associated with a PIN diode:

- **Breakdown voltage:** It is the maximum RF voltage applied; when applied beyond this, it causes rupture of the intrinsic layer.
- **Junction capacitance:** The capacitance associated with the charge variation in the depletion layer is called the *junction capacitance.*

- **Series resistance:** It is the total resistance of the diode when a certain amount of current flows through the diode. In data sheets, it is mentioned along with a value of current.
- **Carrier lifetime:** It is the average time taken by the minority carriers to recombine. It is a measure of the ability of the PIN diode to store electrical charge.

Applications

- It is used as a limiter.
- As a modulator
- It is used as a dc-operate attenuator in TV tuners, antenna distribution amplifiers, and so on.
- Provides isolation in certain applications
- Used as a switch
- Used as a phase shifter
- If the intrinsic layer is thick enough, it can be used as a rectifier.

Performance characteristics

Power: Up to 200 kW peak and 200 W average
Switching times: 40 ns to 1 μs

9.13 SCHOTTKY DIODE

This diode is formed by the junction of a semiconductor and a metal. This junction has unipolar properties. Aluminum is generally used.

Working: The Schottkydiode is as shown in Figure 9.17. It contains a junction between an *n*-type semiconductor and a metal. Initially, when no bias is applied, electrons from the conduction band of the N-type semiconductor reach the metal, leaving a region known as the *depletion layer* that has no free electrons. This results in the buildup of a positive space charge region in the semiconductor, because each donor atom loses an electron and, hence, becomes positive. This results in the buildup of an electric field that opposes the further flow of electrons to the metal. Since the metal contains a greater number of free electrons, the depletion width inside it is small.

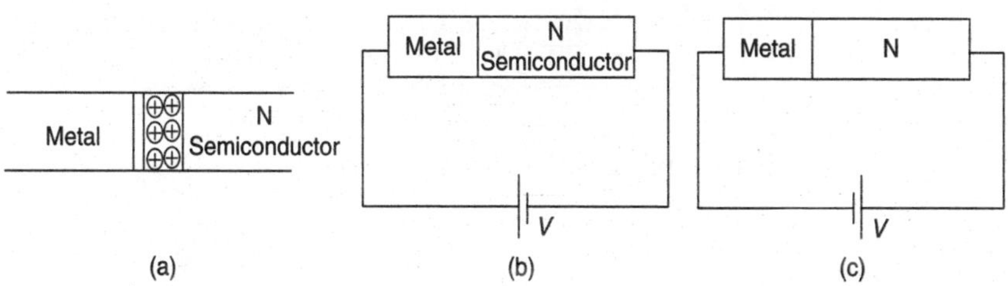

Figure 9.17 *(a) Schottky diode; (b) Forward bias; (c) Reverse bias*

In the forward-biased condition, the metal is biased positively with regard to the semiconductor. Then, a few electrons from the N-type semiconductor whose thermal energy happens to be many times the average thermal energy will fall down from the potential barrier and move into the metal. Therefore, these electrons are known as *hot electrons*. Merely those electrons are considered for the forward currents from the semiconductor into the metal. These will have higher energy and velocity till they come into equilibrium with the free electrons of the metal. This raises the temperature of the diode and is, therefore, called a *hot carrier diode.*

An important feature of this diode is that there is no flow of holes from the metal into the semiconductor. Therefore, there is no hole-electron recombination (that takes place in a conventional PN diode). Therefore, if the forward voltage is removed, the time taken by the current to stop is in the order of a few picoseconds, and reverse voltage can be established in this time. As in junction diodes there is no delay effect due to charge storage.

Equivalent circuits

The equivalent circuit of a Schottky diode is as shown in Figure 9.18 and the electronic symbol of Schottky diode is shown in Figure 9.19.

These parameters are as follows:

- The series inductance, L_s, is the inductance of the bonding wires. It has typical values of 0.4 to 0.9 nH.
- The series resistance in the Schottky diode is the total resistance in the diode, including that of the semiconductor and the substrate on which the diode is mounted. This value is particularly important when we are choosing a diode for a detector or a mixer application. Its value ranges from 4 to 6 ohm.
- The junction capacitance is the capacitance that is present across the actual junction between the semiconductor and the metal. It ranges from 0.3 to 0.5 pF.
- R_j is the junction resistance, that is, the resistance of the area where the semiconductor and the metal come together.
- The overlay capacitance, C_o, is the value of capacitance produced from the Schottky junction to the metal contact of the opposite lead of the diode.

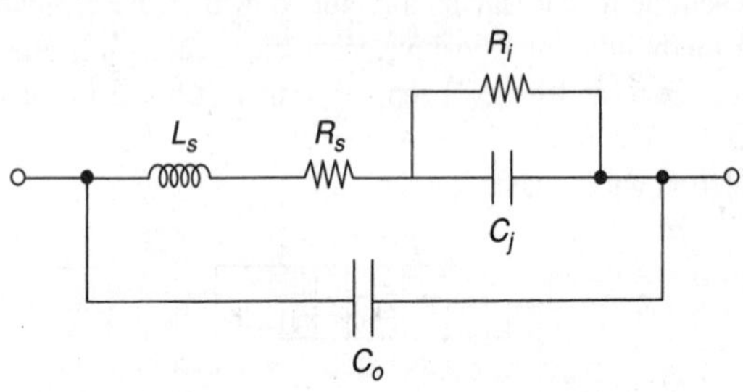

Figure 9.18 *Equivalent circuit of Schottky diode*

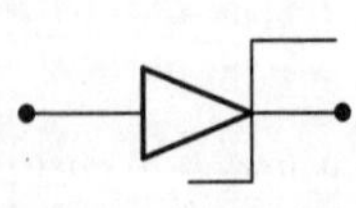

Figure 9.19 *Electronic symbol of Schottky diode*

Applications

- They are used in fast-switching applications such as in microwave mixers, where the diode should switch conductance states at the rate of the frequency of a microwave local oscillator.

- A Schottky diode is used as a square-law detector when bias is zero, whose power output is proportional power to input.
- They are used in detectors.

9.14 VARACTOR DIODE

A varactor is a variable capacitance junction diode. These two terminal devices are also called *varicaps* or *voltacaps*. This is a special type of PN junction that is designed to operate in a microwave range. It works on the principle of voltage variable nature of the depletion capacitance.

Working: In a PN junction, due to the density gradient, holes diffuse to the N region and electrons diffuse to the P region. This causes a few ions on either side of the junction to be depleted of mobile charges as shown in Figure 9.20. This region is known as the *depletion region* or *space charge region*.

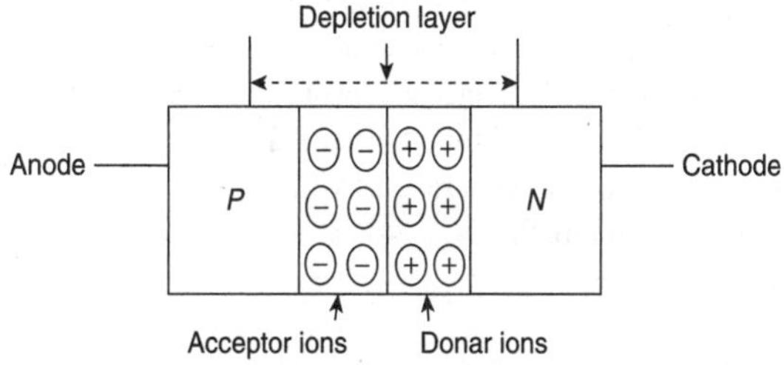

Figure 9.20 *PN junction*

If a forward bias is applied, the carriers move toward the junction as shown in Figure 9.21, which reduces the depletion width.

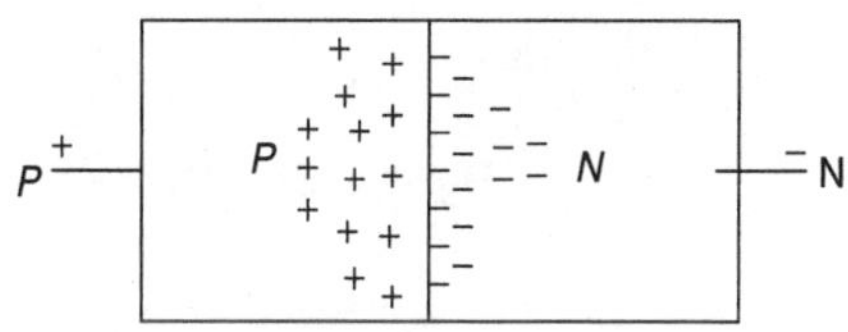

Figure 9.21 *Forward-biased PN junction*

If a reverse bias is applied, the carriers move away from the junction; as a result, the depletion width increases as shown in Figure 9.22.

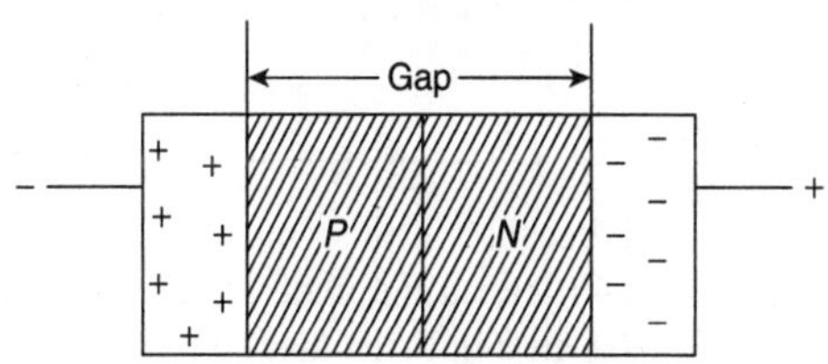

Figure 9.22 *Reverse-biased PN junction*

The variation of width with voltage may be considered a capacitive effect with the depletion region as a dielectric, and P and N regions as parallel plates. This capacitance is known as *transition capacitance* or *junction capacitance* and is given by

$$C_j = \frac{\varepsilon_s A}{W} \tag{9.18}$$

where A is area of junction
W is depletion width
ε_s is permittivity

Therefore, this capacitance is not constant, but depends on the magnitude of the applied bias. It is inversely proportional to the applied voltage. For example, if the reverse bias is 4 volts, the capacitance is 20 pf. If the reverse voltage is increased to 6 volts, the capacitance becomes 10 pf. Thus, the voltage to capacitance ratio is 1:5. Thus, a 1 volt increase in bias causes the capacitance to decrease by 5 pf. This ratio can be as high as 1:10.

Equivalent circuit: The equivalent circuit is as shown in Figure 9.23,where R_j is the junction resistance that has a high value. If the junction area is made small, this resistance can further be increased and, hence, can be neglected. R_s is the series resistance and is due to doped semiconductors, the contact resistance, and lead resistance. It can be reduced by increasing the doping density. Typical values of C_j and R_s are 20 pF and 8.5 ohm, respectively, when a reverse voltage of 4 V is applied. The electronic symbol of a varactor diode is as shown in Figure 9.24.

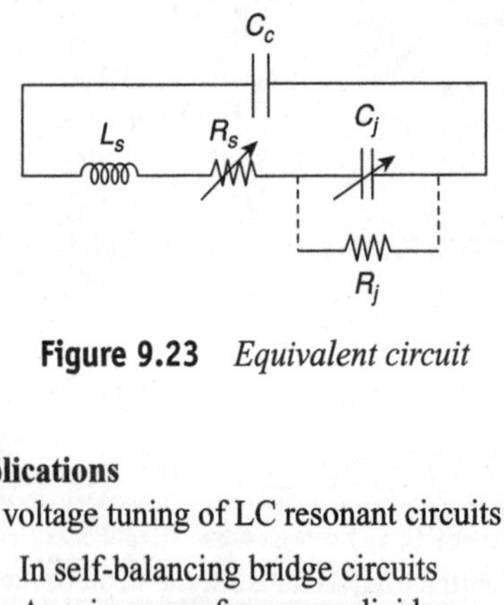

Figure 9.23 *Equivalent circuit*

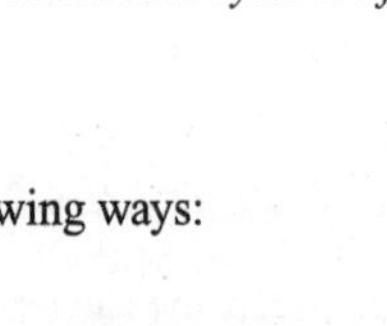

Figure 9.24 *Electronic symbol of varactor diode*

Applications

The voltage tuning of LC resonant circuits is used in the following ways:

- In self-balancing bridge circuits
- As microwave frequency dividers
- In construction of a low-noise amplifier known as a *parametric amplifier*

9.15 PARAMETRIC AMPLIFIERS

The parametric amplifier is so called, because in this amplifier a parameter is made to vary with time. It is also called a *reactance amplifier*, as the underlying principle of operation is based on reactance. A variable resistance is used in amplifier that consumes energy from a dc source and increases ac energy. On the other hand, a nonlinear variable reactance is used in the parametric amplifier to supply energy from an ac source to a load. The advantage of reactance-based amplification is that it does not add thermal noise to a circuit and therefore, parametric amplifiers have low-noise characteristics than most conventional amplifiers. Therefore, they are used as a front end in a microwave receiver.

The operation of the parametric amplifier depends on a capacitance that varies with time. To understand its working better, let us consider a simple series circuit as shown in Figure 9.25. The capacitor charges to value (Q), when the switch is closed. If the switch is opened, the capacitor will have a voltage across the plate that is determined by the charge Q divided by the capacitance C.

$$V = \frac{Q}{C} \tag{9.19}$$

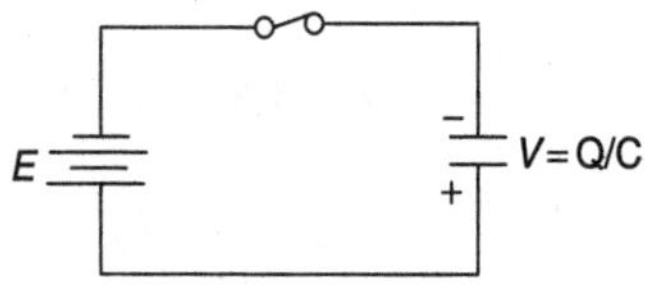

Figure 9.25

The decrease of capacitance causes an increase in the voltage across the plates. Therefore, if the capacitance can be decreased by some means, voltage amplification takes place. A signal known as a *pump signal* is used to electronically vary the capacitance. In a parametric amplifier, three frequencies are involved. The first frequency is one at which the signal is applied (f_s);the second is the frequency of the pump signal (f_p); and the third one is idle frequency (f_i) (or) output frequency (f_0),which is the difference between pump frequency and signal frequency. Therefore,

$$f_i = f_p - f_s \tag{9.20}$$

Manley–Rowe relations: MR power relations are general power relations that are useful in predicting whether power gain is possible in any non-linear reactance. They represent conservation of energy. They considered the circuit as shown in Figure 9.26. It consists of resistive loads in series with band

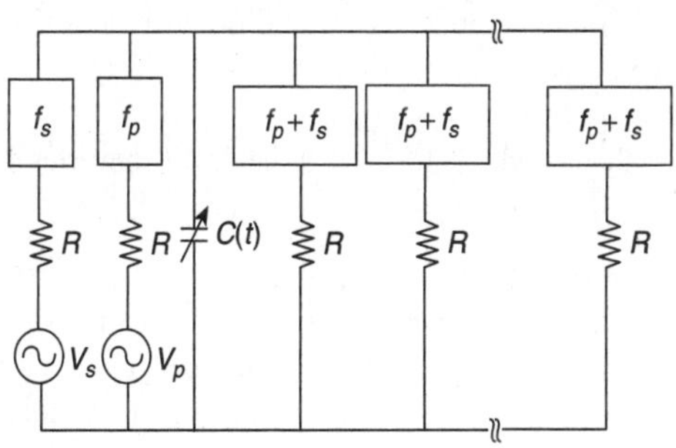

Figure 9.26 *Equivalent circuit of Manley–Rowe derivation*

pass filters connected in parallel with a lossless nonlinear capacitance. These filters reject power at all frequencies other than their respective signal frequencies. A signal generator voltage (v_s) and a pump generator voltage (v_p), at their respective frequencies, are connected as shown. The non-linear capacitance generates frequencies at harmonics of f_s and f_p ($mf_s \pm nf_p$), where m and n are integers.

Manley and Rowe related the input power at frequencies, f_s and f_p, to the output power at frequencies, $mf_s \pm nf_p$.

The voltage across the capacitance is the sum of signal and pump voltages and is given by

$$v = v_p + v_s = V_p \cos\omega_p t + V_s \cos\omega_s t = V_p \frac{(e^{j\omega_p t} + e^{-j\omega_p t})}{2} + V_s \frac{(e^{j\omega_s t} + e^{-j\omega_s t})}{2} \tag{9.21}$$

The standard forms of Manley-Rowe power relations are given below (for more details refer Appendix E):

$$\sum_{m=0}^{\infty}\sum_{n=-\infty}^{\infty} \frac{mP_{m,n}}{m\omega_s + n\omega_p} = 0 \tag{9.22}$$

$$\sum_{m=-\infty}^{\infty}\sum_{n=0}^{\infty} \frac{nP_{m,n}}{m\omega_s + n\omega_p} = 0 \tag{9.23}$$

Depending on these frequencies, the parametric amplifiers can be classified into two types:

- Non-degenerate parametric amplifier
- Degenerate parametric amplifier

Non-Degenerate Parametric amplifier: Here, the signal and idle frequencies are not the same but are clearly separated. The pump frequency need not be a multiple of the signal frequency. There are two types of non-degenerate parametric amplifiers. They are upconverter and down converter.

Upconverter: In this, the pump frequency is a few times (usually 5 to 10) that of the signal frequency. The output frequency is the sum of pump frequency and signal frequency ($f_s + f_p$). Thus, the output frequency is more than the signal frequency. That is why it is known as an *up converter*.

Up convertor is a unilateral stable device with high bandwidth and low gain. The power gain of the up convertor is not dependent of the change in source impedance. It is useful as a modulator.

Down converter: In this, the output frequency is the difference between signal frequency and pump frequency ($f_s - f_p$). Thus, the output frequency is lower than the signal frequency, and it is known as *down converter*.

Since $f_0 < f_s$ gain is less than 1, which means that the down converter causes loss. Down converters are used as demodulators.

Degenerate Parametric amplifier: In these amplifiers, pump frequency is twice that of signal frequency. Therefore, signal and idle frequencies are almost equal. At the signal frequency, it will present a negative resistance to any external circuit. The bandwidth and gain are same as in the up converter. It is a simple device, uses a relatively low pump frequency, and has a low noise figure. Its major drawback is that it is phase sensitive.

Equivalent circuit: The Parametric amp's equivalent circuit is as shown in Figure 9.27. The signal circuit, idle circuit, and pump circuit are clearly shown. The output frequency in the idle circuit will be a linear combination of pump and signal frequencies, which is given by

$$f_0 = mf_p \pm nf_s$$

where *m* and *n* range from zero to infinity.

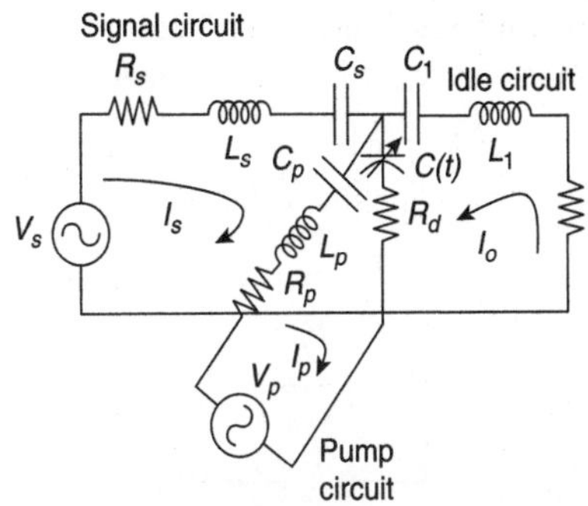

Figure 9.27 *Equivalent circuit of parametric amplifier*

Performance: The noise figure of parametric amplifiers is very low, as they use variable reactance for amplification instead of variable resistance. The only noise source is base resistance. Their performance can further be improved by cooling. The noise figure decreases with an increase in temperature.

Power gain: 10 to 60 dB
Frequency range: 3 to 60 GHz
Bandwidth: up to 700 MHz
Noise figures: 0.2 to 6.0 dB

Applications

- Used in the front end of microwave receivers, which require extremely low noise figures
- Used in satellite and space probe tracking stations
- Used in communication satellites, radar receivers
- Used in radio telemetry

9.16 STEP-RECOVERY DIODE

This is a PN diode whose construction is similar to that of a varactor diode. These are also known as *snap-off diodes.* These are usually made up of silicon. These operate under the forward-biased condition. They also use PIN configuration.

Working: In a P-N junction, when the bias changes from forward condition to reverse condition, the recovery of the diode has two stages:

1. In storage stage, the minority carriers near the junction are removed. However, the carrier concentration does not reduce evenly along the junction. There will be a quicker decrease close to the junctions. After a time t_a, the concentration level at the junctions becomes zero. However, there will be stored charge at the junction region that is to be removed.
2. A *transition stage* is one in which the junction transition capacitance is charged due to the left-out minority charges. In this stage, minority carriers that are located at some distance from the junction are removed with a certain time constant t_b.

In step-recovery diodes, the doping of the diodes is controlled in such a way that the minority carriers are restricted to a very narrow region in the immediate vicinity of the junction. This has an effect of increasing t_a. Since most of the minority carriers are removed in t_a, a few minority carriers are left away from the junction.

This has an effect of decreasing t_b. When t_b is less than t_a, the output will be a pulsed response. When using a sinusoidal input voltage, this pulsed current can be used for harmonic generation.

Equivalent circuit: This is as shown in Figure 9.28. C_d is the diffusion capacitance, R_d is the junction resistance under the forward-biased condition, R_s is the series resistance, and L_s is the lead inductance.

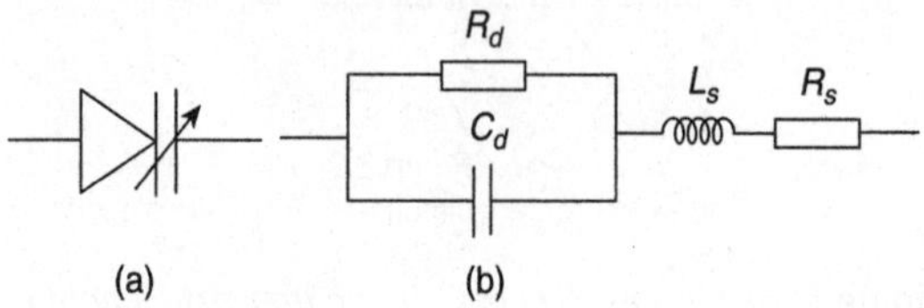

Figure 9.28 *(a) Symbol; (b) Equivalent circuit of step-recovery diode*

Applications: They are commonly employed in the design of frequency multipliers of a high order.

9.17 CRYSTAL DIODE

Similar to a Schottky diode, the crystal diode has a metal semiconductor junction. It is a unipolar device and is, hence, free from ill effects of minority carrier storage that are present in the conventional PN diode. It depends on the pressure of contact between a semiconductor crystal and a whisker made of gold-plated tungsten.

Construction and Operation: The diode consists of a small rectangular crystal of N-type silicon. The other part of the diode has a gold-plated tungsten wire that is also called *cat's whisker.* This whisker is alloyed onto silicon wafer in the form of a pellet. To provide a protective covering, this pellet is immersed in wax, which also prevents moisture. The entire structure is then housed in a ceramic structure, with one metal contact at the top which acts as the cathode and another metal contact at the bottom that acts as the anode. During the manufacturing of the diode, a current of large magnitude is passed through the whisker. This results in the formation of a small P region around the crystal as shown in Figure 9.29. Thus, the device behaves similar to a normal P-N junction.

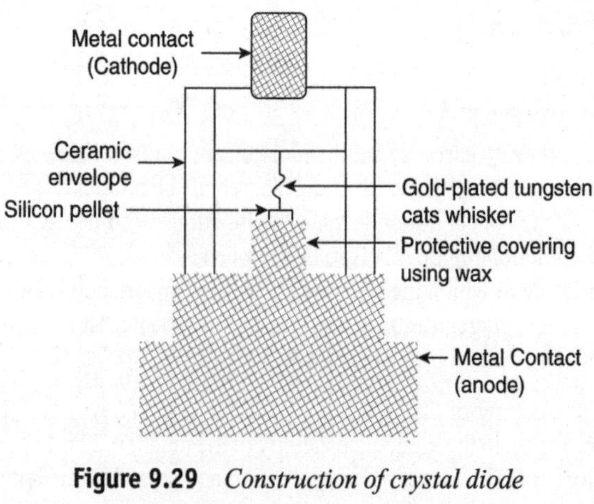

Figure 9.29 *Construction of crystal diode*

Applications

- In microwave mixer applications
- These are used to avoid mismatch in a microwave transmission path.

Disadvantages

- Reverse leakage current is very high.
- The device cannot handle currents more than a few milli amps.

9.18 MICROWAVE BJTS

A bipolar junction transistor operated in the microwave region is called a *microwave BJT*. The operation of the microwave BJT is similar to that of low-frequency transistors. However, the major differences are

- GaAs is to be used instead of silicon, because the mobility of electrons is more in GaAs than in Si.
- To minimize junction capacitance, the size of the device should be minimized.
- The emitter forward resistance is to be minimized by reducing the thickness of the emitter and also by applying maximum possible forward voltage at the emitter junction.
- The base transit time is to be reduced by minimizing its thickness and using a graded doping profile.

In the analysis and designing of a circuit with microwave BJT, the following points should be considered:

- A much more complex equivalent circuit model as shown in Figure 9.30 is considered. This includes package parasitic capacitance, lead inductance, and resistance.

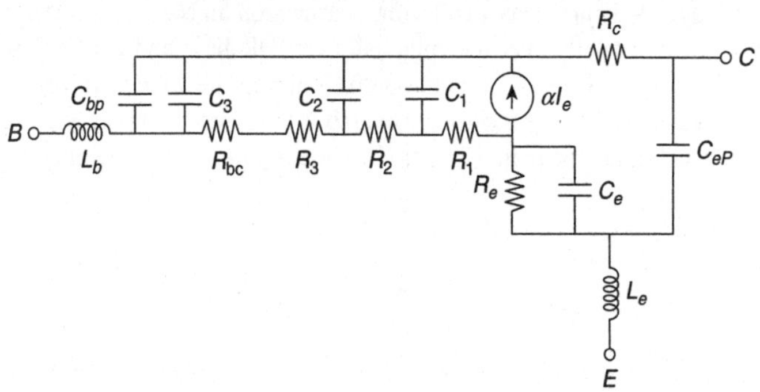

Figure 9.30 *Equivalent circuit of microwave BJT*

C_{bp}—base bond pad capacitance
C_{ep}—emitter bond pad capacitance
R_{bc}—base contact resistance
R_{ec}—emitter contact resistance
R_1, R_2, R_3—base distributed resistance
C_1, C_2, C_3—collector base distributed capacitance

R_e—dynamic emitter-base diode resistance
C_e—emitter-base diode junction capacitance
R_c—collector resistance
L_b, L_e— base and emitter bond wire inductances

- Care should be taken while designing input and output circuits to prevent oscillations that occur due to large capacitive feedback from the collector to the base.

Construction: These are usually N-P-N type, because the electron mobility is higher than the hole mobility. The structure of a conventional Si homojunction N-P-N BJT is as shown in Figure 9.31. N-P-N BJT is designed with planar shape as a double diffused epitaxial transistor.

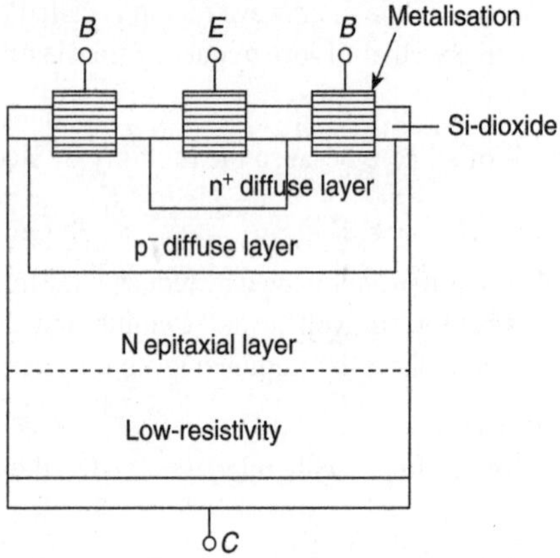

Figure 9.31 *N-P-N silicon double-diffused epitaxial transistor*

An N-type epitaxial layer, which acts as a collector, is grown on an N^+substrate. An insulating layer of silicon dioxide is grown thermally over the epitaxial layer. The base and emitter layer are formed by diffusion or ion implantation. Proper care is to be taken during the whole process of microwave BJT fabrication so as to reduce package parasites considerably. Contacts are provided by means of openings in the oxide, and connections are made in parallel. The surface geometry of the microwave transistor can be any one of the following:

(a) inter-digitized
(b) overlay
(c) matrix

All the three types are as shown in Figure 9.32. In an inter-digitized geometry, a large number of metallized emitter strips alternating with metallized base strips are present as shown in Figure 9.32. This type is used for small signal and power transistors. The second one is the overlay geometry that has a large number of segmented emitters overlying a number of wide metal strips. The third is the matrix or mesh geometry that has an emitter which forms the grid, with the base filling the meshes of this grid with a P^+contact area in the middle of each mesh. Overlay and matrix structures are useful for power transistors.

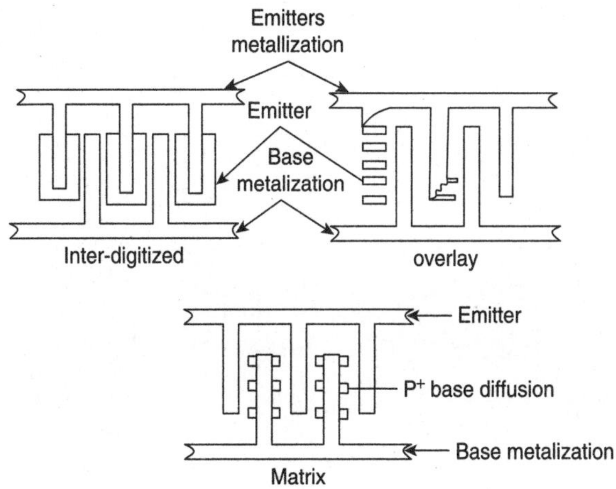

Figure 9.32 *Surface geometry of microwave transistor*

Operation of BJT: Microwave BJT behave very similarly to a low-frequency N-P-N bipolar transistor. In an N-P-N BJT, primarily, the collector-base and emitter-base junctions are reverse biased, consequent to the class C condition. The microwave frequency signal is applied between the emitter and base junction, this junction forward bias during the positive portion of the microwave signal. A pulse of current flows through the load connected in the collector circuit. In a proper-designed microwave BJT, the emitter current consists almost all of the electrons and is approximately equal to the collector current, while very small fraction of the emitter current is the base current.

The mode mentioned just now is known as an *active mode*. There are two other operation modes. One is the cut-off mode, in which the emitter and collector junctions are reverse biased and the collector current is negligible, and another mode is the saturation mode, in which both the emitter and collector junctions are forward biased, resulting in a large collector current.

Transistor biasing: The first consideration for designing a biasing circuit is that the operating point should be stable irrespective of temperature changes and device parameters. This condition can be obtained by incorporating the dc feedback circuit in the biasing circuit. Second, low-impedance capacitive elements are inserted in to shunt high-frequency currents, to obtain the condition where high-frequency signal should not flow in the dc biasing circuit. In addition, high-impedance, high-frequency circuit elements are inserted in series with DC components.

The transistor biasing circuit is as shown in Figure. 9.33(a). This biasing circuit is known as a *passive biasing circuit*. It provides a stable operating point. However, as the operating frequency increases, the parasitic capacitance associated with capacitor leads increases and causes difficulty.

Another biasing circuit known as an *active bias circuit* is as shown in Figure 9.33(b). The transistor Q_1 has a stable bias circuit by means of R_1, R_2, and R_3. It can be seen from Figure 9.33(b) that the base current of Q_2 is a collector current of Q_1. The collector current of Q_1 is stable, and, hence, the base current of Q_2 is stable, irrespective of transistor parameters.

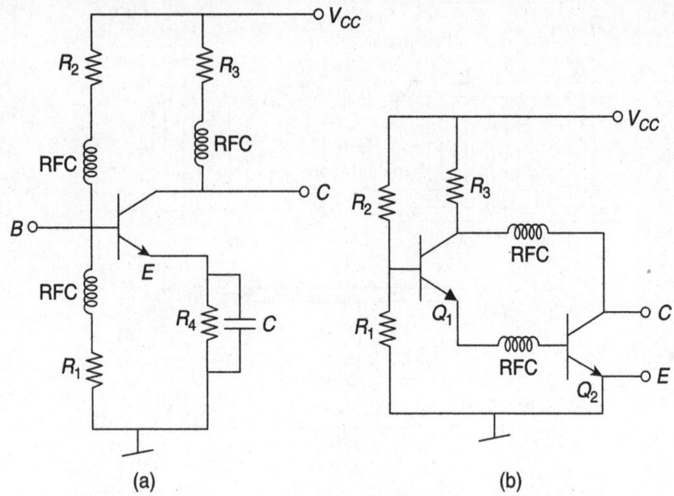

Figure 9.33 *(a) Passive biasing circuit; (b) Active biasing circuit*

Disadvantages: Microwave transistors also suffer from high-frequency limitations. The effects like transit time, inter electrode capacitance etc, come into view again at high frequencies. For example due to the effect of the inter electrode capacitance high frequency response is limited which causes increase the complexity of the transistor. Lead inductance cause undesirable effect. Ideally, these effects should be kept lowest with appropriate selection of geometry and packaging of BJTs. In addition, they operate at lower frequencies than microwave field effect transistors (FETs).

Differentiation between TEDs and transistors

Transferred Electron Devices	Transistors
1. The TEDs are bulky devices, and they do not have any junctions as in the case of conventional transistors.	1. Conventional transistors are simple in construction. They operate with junctions.
2. These are manufactured from compound semiconductor materials such as GaAs, InP, and CdTe.	2. These are manufactured from elemental semiconductors such as silicon and germanium.
3. TEDs operate with hot electrons.	3. Transistors operate with warm electrons.
4. The range of frequencies that these devices handle is in gigahertz range.	4. The range of frequencies that these transistors operate is in kilohertz range only.
5. The power handled by these is very high in the range of watts and kilowatts (kW).	5. The power handled by these is very low compared with TEDs.
6. Population inversion happens in these devices.	6. Population inversion does not happen in these devices.

9.19 MICROWAVE FETS

Microwave FET operates at higher frequencies than those of BJT. Basically, there are two kinds of microwave FETs. They are

- Metal Semiconductor FET(MESFET)
- Hetero structure FET(HFET)

All the above kinds have a source, a drain, and a channel between them. The electric field across the channel can be changed to vary the conductivity of the device. A detailed discussion of MESFETs is given next.

MESFET: Constructional features of a GaAs MESFET are as follows: Initially, a semi-insulating GaAs substrate is taken (To make it semi-insulating, GaAs is doped with chromium). Over that substrate, a thin active layer of N-type GaAs material is grown by ion implantation of donor atoms into the substrate. This active layer has a resistivity that is more than 10^7ohm-cm. Three metal contacts for source, gate, and drain are made as shown in Figure 9.34. To reduce the contact resistance, the regions beneath the source and drain contacts are diffused with highly doped materials.

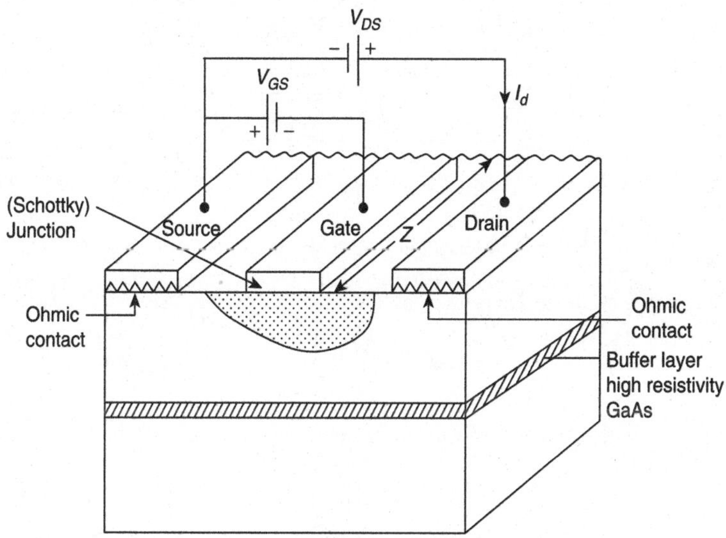

Figure 9.34 *MESFET*

The channel region is a thin epitaxial N-type GaAs grown on a chromium-doped insulating GaAs substrate. The gate is not a simple P-N junction, but has a Schottky barrier metal contact. The Schottky barrier voltage partially depletes the thin channel under the gate as long as channel doping is in the order of a few 10^{17}cm^{-3}. The positive drain to source voltage will let the current flow through the device. Negative V_{gs} further reduces the drain to the source current. As V_{GS} approaches the pinch-off voltage, I_{DS} remains constant. The cut-off frequency of the device is given by

$$f_c = \frac{1}{2\pi\tau}$$

where τ is the carrier transit time that is obtained by dividing the length of channel l with the saturation velocity of electrons. The expression for pinch-off voltage is given by

$$V_p = \frac{qN_d a^2}{2\varepsilon_s} \tag{9.24}$$

a = channel height
ε_s = semiconductor dielectric permittivity
N_d = Electron Concentration

The I-V characteristics of the device are as shown in Figure 9.35:

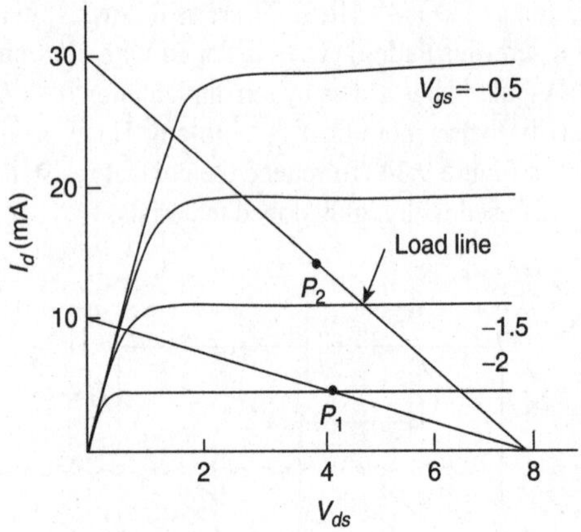

Figure 9.35 *Output characteristics of MESFET*

The output conductance of the device is given by

$$g_{DS} = \frac{1}{r_{DS}} = \left.\frac{dI_{DS}}{dV_{DS}}\right|_{V_{GS}=const} \tag{9.25}$$

The transconductance of the device is affected by its dimensions and material properties and is given by

$$g_m = \left.\frac{dI_{DS}}{dV_{GS}}\right|_{V_{DS}=const} \tag{9.26}$$

The gate-source capacitance of the device plays an important role in its applications, as it has a significant effect on frequency performance. It is given by

$$C_{GS} = \left.\frac{dQ_G}{dV_{GS}}\right|_{V_{DS}=const} \tag{9.27}$$

where Q_G is the depletion region charge in the gate.

The gate-drain capacitance is defined as

$$C_{GD} = \left.\frac{dQ_G}{dV_{DS}}\right|_{V_{GS}=const} \tag{9.28}$$

FET Biasing: The output characteristics of the GaAs MESFET with load line are as shown in Figure 9.35. For a dynamic-range operating point, the operating point should be at point P_1, where the dc voltage of the gate should be negative w.r.t. source voltage. The desired condition for FET biasing (Figure 9.36) can be obtained by using an RF choke between the ground and the gate. The desired bias voltage can be obtained from the voltage drop across the source resistance R_s, which is bypassed by means of a capacitor C_s (to ground any RF signals). The value of R_s is given by

$$R_S = \frac{-V_{GS}}{I_{DS}} \tag{9.7}$$

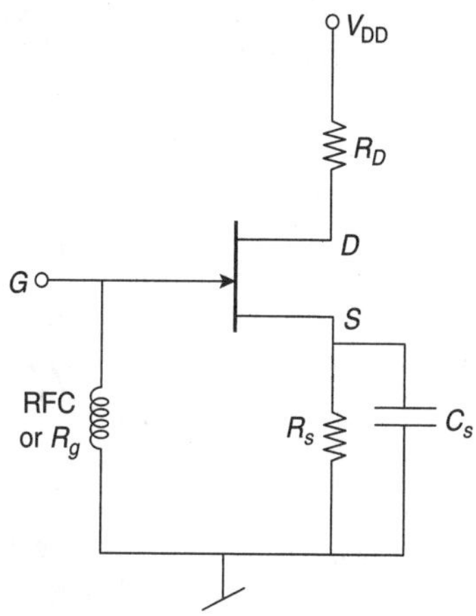

Figure 9.36 *FET biasing circuit*

Applications:

- They are used in various applications such as attenuators, mixers, oscillators, amplifiers, phase shifters, transfer switches, frequency multipliers, discriminators, and isolators.
- They are used as power amplifiers.

EXAMPLE PROBLEM 9.8

Calculate the pinch-off voltage for a GaAs MESFET that has a channel height of $0.1\ \mu\text{m}$, a relative dielectric constant of $\varepsilon_r = 11.8$, and an electron concentration of $N_d = 1.8\times10^{23}\ \text{cm}^{-3}$.

Solution

Given data $N_d = 8\times10^{23}\ \text{cm}^{-3}$, $a = 0.1\ \mu\text{m}$, and $\varepsilon_r = 11.8$

The pinch-off voltage for an Si MESFET is

$$V_p = \frac{qN_d a^2}{2\varepsilon_s} = \frac{1.6\times10^{-19}\times1.8\times10^{23}\times(0.1\times10^{-6})^2}{2\times8.854\times10^{-12}\times11.8} = 12.203\ \text{V}$$

■

HEMT Devices: *HEMT* stands for High-Electron Mobility Transistor. In these devices, a junction is formed between two different semiconductor materials having various band gaps. This results in the formation of low potential on one side of the junction. Electrons will concentrate in this low potential region, and they will travel through the un-doped material. This results in an increase of the mobility of carriers in the un-doped material. This results in fast response time of the device. The HEMT structure is a bit more complex than MESFET, which results in fabrication difficulties, added costs, and low yields. However, the main advantage of this device is its improved noise figure.

The structure of HEMT is as shown in Figure 9.37, in which an N-type AlGaAs is on an un-doped GaAs; N-Type AlGaAs has a wide band gap; and GaAs has a narrow band gap. The thickness of both un-doped GaAs and N-type AlGaAs is crucial in determining the device behavior.

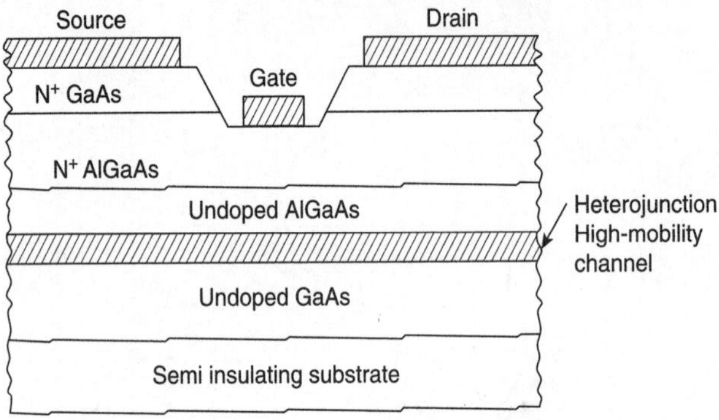

Figure 9.37 *HEMT structure*

Current voltage characteristics: The energy-band diagram of the device with zero gate bias is shown in Figure 9.38. A sharp dip in the conduction band results in a high carrier concentration of electrons along the GaAs side of the hetrojunction and is described as a two-dimensional electron gas. A heavily doped source, low-resistance and drain wells get in touch with the two dimensional electron gas wells. When a low voltage is applied between the source and drain, a current flows through the two-dimensional electron gas. As the voltage is further increased, the resulting current will increase up to a level and then saturate. The two-dimensional electron gas densities are controlled by the gate bias. The current-voltage characteristics are as shown in Figure 9.39.

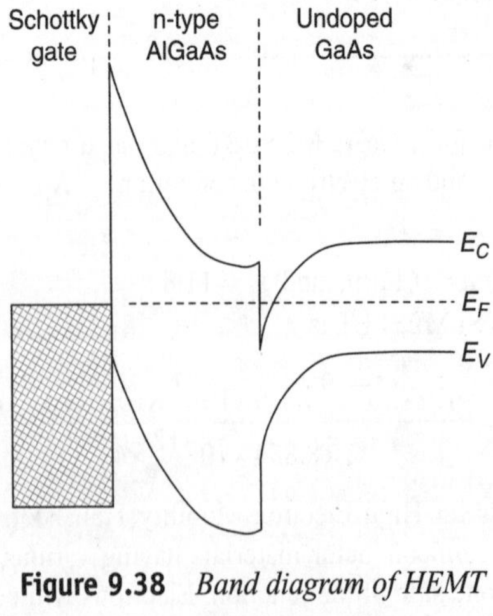

Figure 9.38 *Band diagram of HEMT*

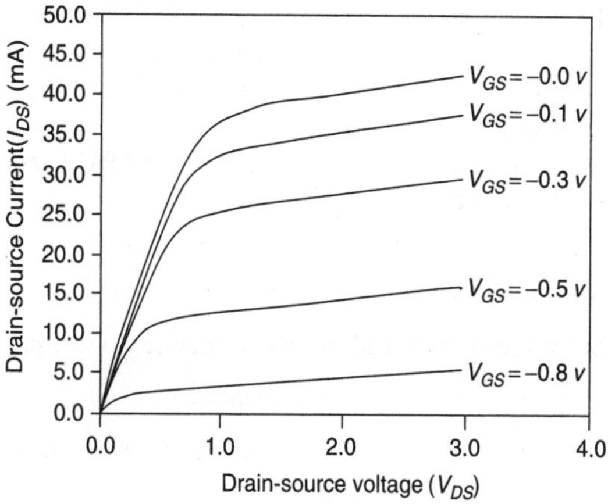

Figure 9.39 *I-V characteristics of HEMT*

Small Signal Model: The small-signal model of a FET is as shown in Figure 9.40. Each element in the equivalent circuit is briefly described as follows: L_S, L_D, and L_G are parasitic inductance. These are due to metal contacts on the device surface. C_{GS} and C_{GD} are capacitances due to depletion charge with regard to gate-source and gate-drain voltages. R_S and R_D are the resistances due to ohmic contacts. The gate resistance R_G is due to gate Schottky contact. All the resistances are of the order of 1ohm.

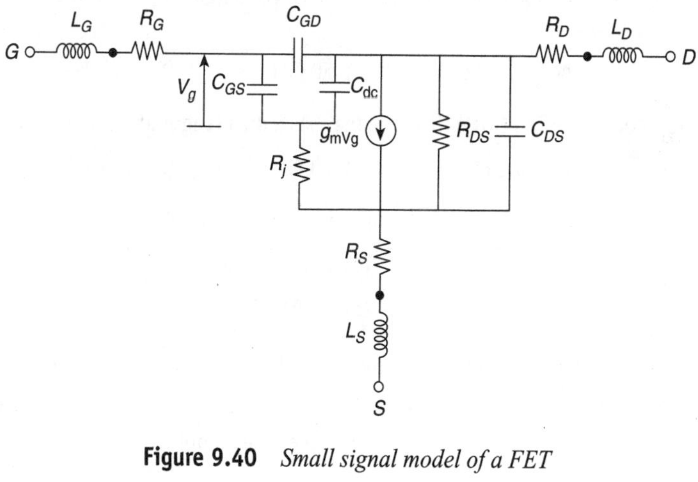

Figure 9.40 *Small signal model of a FET*

SUMMARY

1. The common characteristic of all active two-terminal devices (solidstate) is their negative resistance.
2. Gallium Arsenide (GaAs), Silicon (Si), and Germanium (Ge) are widely used for making different microwave solid-state devices.
3. Microwave solid-state devices are classified as microwave transistors, FETs, TEDs, and ATDs.

4. The microwave BJT has low cost, low power supply, high CW power output, and light weight and is reliable. BJT's operating frequency is less than that of FETs.
5. In a tunnel diode, the charge carriers cross the potential barrier through a process called *tunneling*.
6. TEDs operate on the principle of the RWH theory.
7. Avalanche transit time diodes are of three types: IMPATT, TRAPATT, and BARITT.
8. TRAPATT has more efficiency than the IMPATT diode.

OBJECTIVE-TYPE QUESTIONS

1. If the phase difference between voltage and current in a device is 180°, such a device is called

   (a) an inductor (b) a capacitor
   (c) a positive resistance device (d) a negative resistance device

2. Which of the following is a TED?

   (a) GUNN diode (b) TRAPATT (c) IMPATT (d) BARITT

3. Which of the following is a unipolar device?

   (a) SCHOTTKY diode (b) PN diode
   (c) PIN diode (d) Varactor diode

4. The natural frequency of a GUNN diode, having an active length of 2 micrometers and an electron drift velocity of 10^5 m/sec, is

   (a) 10 GHz (b) 20 GHz (c) 50 GHz (d) 1 GHz

5. The operating frequency of an IMPATT diode that has a drift length of 5 micrometers is

   (a) 10 GHz (b) 20 GHz (c) 50 GHz (d) 1 GHz

6. Which of the following diodes has high impurity ion concentration?

   (a) PN (b) tunnel diode
   (c) SCHOTTKY diode (d) IMPATT diode

7. In a parametric amplifier, if the output frequency is the difference between signal frequency and pump frequency, it is called

   (a) upconverter (b) degenerate amplifier
   (c) downconverter (d) none of the above

8. Which of the following has a metal semiconductor junction?

   (a) crystal diode (b) Schottky diode
   (c) step-recovery diode (d) a & b

ANSWERS TO OBJECTIVE-TYPE QUESTIONS

1. (d)	2. (a)	3. (a)	4. (c)
5. (b)	6. (b)	7. (c)	8. (d)

REVIEW QUESTIONS

1. Give the classification of solid-state microwave devices.
2. List out the applications of microwave solid-state devices.
3. Explain the construction of the GUNN diode using RWH theory.
4. Explain the Gunn effect using the two-valley theory. Also explain several modes of operation and applications of Gunn diodes.
5. Explain the principle of operation of an IMPATT diode.
6. What is Gunn effect?
7. Explain the V-I characteristics of a Gunn diode. What is the main advantage of using the Gunn diode compared with the IMPATT diode?

ANSWERS TO OBJECTIVE-TYPE QUESTIONS

1. (d) 2. [illegible] 3. [illegible] 4. (c)
5. (b) 6. (b) 7. (c) 8. (d)

REVIEW QUESTIONS

1. Give the classification of solid-state microwave devices.
2. List out the applications of microwave solid-state devices.
3. Explain the construction of the GUNN diode using RWH theory.
4. Explain the Gunn effect using the two-valley theory. Also explain several modes of operation and applications of Gunn diodes.
5. Explain the principle of operation of an IMPATT diode.
6. What is Gunn effect?
7. Explain the characteristics of a Gunn diode. What is the main advantage of using the Gunn diode compared with the IMPATT diode?

Monolithic Microwave Integrated Circuits

10

10.1 INTRODUCTION

Now a day any technology is moving towards smaller size, lighter weight, lower cost, and decreased complexity. The field of Microwave technology is also progressing in the direction of Microwave Integrated Circuits (MICs). An MMIC is an integrated circuit mounted on a monolithic substrate that operates at microwave frequencies. Bulky and expensive waveguide and coaxial components can be replaced by this MMIC technology with planar components which are small and inexpensive.

There are two types of MICs. Hybrid Microwave Integrated Circuits (HMIC) are made by placing the discrete components (active and passive) on a circuit board. A Monolithic Microwave Integrated Circuit (MMIC) is a more recent type of MIC in which the active and passive components are fabricated on the same semiconductor substrate. The frequency of operation of MMICs ranges from 3 GHz to 300 GHz. The word *monolithic* is taken from Greek words *monos* meaning single and *lithos* meaning stone. Thus, a monolithic IC is fabricated on a single crystal and it involves step by step process of epitaxial growth, masked impurity diffusion, oxidation growth, and oxidation etching.

The design of MMICs is quite different from the conventional VLSI design, which has a special range of applications. MMICs are smaller in size (from around 1 mm^2 to 10 mm^2) and can be produced in mass, which has allowed the creation of high-frequency devices such as cellular phones. MMICs perform functions such as microwave mixing, power amplification, low-noise amplification, and high-frequency switching. The purpose of this chapter is to provide a brief introduction to the types of MICs, general applications and technology of MMICs. The advantages and disadvantages of MMICs, comparison with hybrid MICs, materials used for MMICs, NMOS, and CMOS fabrication techniques, and the formation of thin film are described in this chapter.

10.2 MICROWAVE INTEGRATED CIRCUITS (MICs)

Microwave circuits exist in two different forms:

- Discrete circuits and
- Microwave integrated circuits (MICs)

A discrete circuit consists of packaged microwave active devices/components mounted in coaxial and waveguide assemblies. The components in discrete circuits can be removed from the assembly and replaced.

The main advantages of ICs are that they are smaller in size and usually require less power to operate. Similar to conventional ICs, microwave integrated circuits (MICs) can be fabricated in two ways. They are as follows:

- Hybrid MIC (HMIC)
- Monolithic MIC (MMIC)

Fabrication materials and component soldering affect electrical properties and overall performance of the circuit. For example, microwave circuits often use resistors, capacitors, and inductors, and these components are implemented differently in HMIC and MMICs. MMICs are more advantageous compared to HMICs based on their design differences and circuit performance capabilities. However, the design procedure of MMICs is more complex, because all components of entire circuit are fabricated on a single semiconductor substrate.

Hybrid MIC (HMIC): An HMIC consists of a number of discrete active devices (diodes/transistors) and passive components (resonators, capacitors, and circulators) that are fabricated separately and then integrated onto a common substrate using solder or conductive epoxy adhesive.

The passive elements are fabricated by using thick-film or thin-film technology. Despite being developed in 1960s, hybrid MIC still provides flexibility in circuit implementation and is also economical.

Monolithic MIC (MMIC): In MMIC, all active and passive components as well as transmission lines are built simultaneously on a single crystal substrate using various technologies such as ion implantation, diffusion, oxidation, epitaxial growth, masking, and etching.

Substrate of MMIC is defined as a substance on which electronic device are built. Materials used are GaAs, glass, rutile, Alumina, beryllium, ferrite.

The MMICs are quite different from conventional ICs. For example, the package density of conventional ICs is high; whereas that of MMICs is quite low and highly reliable. The main features of MMICs are as follows:

- Minimal mismatches and minimal signal delay.
- There are no wire bond reliability problems.
- We can fabricate up to thousands of devices simultaneously into a single MMIC.

10.3 ADVANTAGES AND DISADVANTAGES OF MMICs

Advantages:

MMICs offer the following advantages as compared with discrete circuits and hybrid ICs.

- These are economical when produced in large quantities.
- They are small in size and light in weight.
- They have very good reproducibility.
- They are highly reliable and suitable for space and military applications, because they can withstand adverse environmental conditions, compared with discrete and hybrid ICs.
- They offer more bandwidth.
- They exhibit improved performance.

Disadvantages:

- Development time is large compared with discrete components.
- It is expensive to produce a few ICs.
- Choice of equipment is very limited.
- They exhibit low power-handling capabilities.

10.4 COMPARISON OF MMICs WITH HMICs

The MMIC and HMIC technologies have been compared and presented in Table 10.1.

Table 10.1 Comparison of MMICs and HMICs

Hybrid MICs	Monolithic MICs
Manufacturing is laborious, because it contains individual components	It is easy to manufacture and fabricate on wafers in batches
Some of the interconnections among components are deposited, and some are wire bonded	All the interconnections are deposited
Cost per unit is more	Cost per unit is less
Has greater size and weight	Has lesser size and weight
Design flexibility is less	Basically, they use FET's geometry in different configurations to realize a wide variety of functions such as amplification, oscillation, and, hence, they have more design flexibility
Circuits can be tuned for optimum performance even after manufacturing; that is, circuit tweaking is possible	Almost impossible to make any changes after manufacturing; that is, circuit tweaking is not possible
Broadband performance is limited	Broadband performance is good
Reproducibility is good	Reproducibility is excellent

10.5 APPLICATIONS OF MMICs

The most important application of MMICs is electronically steered phased array antenna which is not practical with hybrid technology because of the size and mass of transceiver modules. Military and space applications are major driving force behind it. In military, MMICs are used in phased array radar, decoys, synthetic-aperture radar, remote sensing and instrumentation etc. In military technology, MMICs are used in synthetic-aperture radars, phased-array radars, decoys, remote sensing, and instrumentation. Their space applications include astronomy, low earth orbit mobile systems, radiometers, communication satellites, and so on. They are also used in many civil applications such as wireless LANs, medical systems, Internet systems, mobile phones, smart cards, and GPS.

Among the applications of MMICs, the electronically-steered phased-array antenna is important application. Because of its size and mass of the individual transmitter-receiver modules it is not practical to use hybrid technology.

MMICs are often used in mobile phones, GPS receivers, and WLAN systems that require small, inexpensive circuits which offer high speed and performance. These circuits operate in the 3-GHz to 300-GHz frequency band. The MMIC can easily be fabricated on an Si or GaAs chip with an area of 1 mm^2 and a thickness of 100 mm. In designing MMICs, commercially available software packages are used for both electrical and layout design. Figure 10.1 illustrates the GaAs-MMIC with the following typical elements:

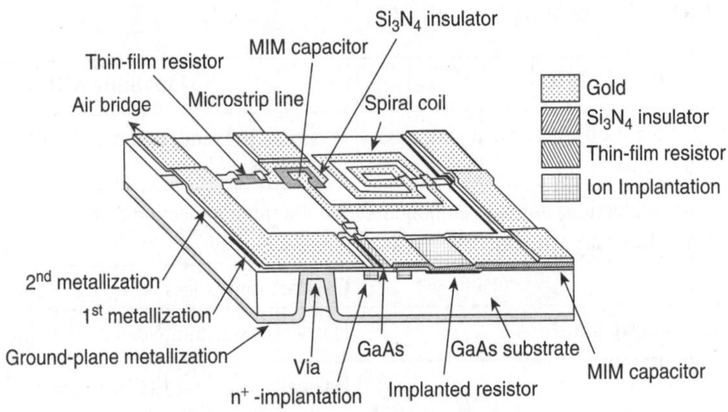

Figure 10.1 *Monolithic Microwave Integrated Circuit on GaAs*

(i) Transmission lines can be either coplanar lines or microstrip lines.
(ii) Resistors can be either thin metal films or ion planted directly in GaAs.
(iii) Inductors or coils possibly narrow microstrip lines in the form of a loop or a spiral.
(iv) Capacitors contain either a metal-insulator-metal (MIM) or an inter digital structure.
(v) The ground point can be realized by a metal through the substrate.

10.6 MATERIALS USED FOR MMICs

The basic materials used for fabricating MMICs, in general, are divided into four categories:

- Substrate material
- Conductor material
- Dielectric material
- Resistive films

10.6.1 Substrate Materials

A substrate of MMICs is a slice of a solid semiconductor substance on which active and passive electronic devices are built. Some of the commonly used substrates are silicon carbide, alumina, ferrite/garnet, silicon, glass borosilicate, quartz, GaAs, GaN, and so on. Properties of some of these materials are shown in Table 10.2.

Till recent times, very few semiconductors (e.g., GaAs) were often used for the realisation of MMIC components, as they are more suitable for operation at RF frequencies. Traditional silicon technology could not be used, because it leads to major issues associated with parasitic capacitance and can also influence the Q factor of the components. However, recent advancements in silicon bipolar transistor technology made it possible to realize silicon-based RF MMICs. These advanced silicon technologies use silicon wafer of diameters up to 300 mm.

Table 10.2 Properties of substrate materials

Property	Si	SiC	GaAs	InP	GaN
Semi-insulating	No	Yes	Yes	Yes	Yes
Resistivity (Ω–Cm)	$10^3 - 10^5$	$>10^{10}$	$10^7 - 10^9$	10^7	$>10^{10}$
Dielectric constant	11.7	9.7	12.9	14	8.9
Electron mobility (cm^2/Vs)	1450	500	8500	4000	800
Radiation hardness	Poor	Excellent	Very good	Good	Excellent
Density (g/cm^3)	2.3	3.1	5.3	4.8	6.1
Thermal conductivity (Wcm–°C)	1.45	3.5	0.46	0.68	1.3
Operating Temperature(°C)	250	>500	350	300	>500
Energy gap(eV)	1.12	2.86	1.42	1.34	3.39
Break down field(kV/cm)	~300	>2000	400	500	>5000

For low unit cost, high volume market sectors, the advanced silicon technologies are more attractive. The choice of substrate materials depends on various factors as in the case of high-power applications, where wide band-gap materials such as SiC or GaN are used.

10.6.2 Conductor Materials

Conductor materials are used in ICs for electrical connectivity. The ideal conductor materials used for MMICs should have high conductivity, good adhesion to the substrate, good etchability, and solderability. They should have a low temperature coefficient of resistance so that the properties do not vary widely with temperature, and they should be capable of being deposited by a number of methods.

Materials with good conducting properties and ideal for ICs are copper, gold, silver, aluminum, platinum, and so on. In microwave circuits some conductor materials are widely used to form both the conductor pattern and the bottom ground plane and the properties of those materials are shown in Table 10.3.

The conductor used should be thick enough to the current density. It can be seen in Table 10.3 that aluminium has relatively good conductivity and good adhesion but, generally, the higher the electrical conductivity, the poorer is the substrate adhesion, and vice versa. So, to achieve good adhesion with high-conductivity materials, a very thin film of one of the poorer conductors is used between the substrate and the good conductor; for example, Cr-Cu, Ta-Au. Aluminum has been used as interconnecting material in ICs. Copper is a better conductor than aluminum, but even a small trade off in the substrate will destroy its properties. Therefore, in the chips with a copper interconnect, a special protection layer between the substrate and copper is used.

Table 10.3 Properties of conductor materials

Material	Coefficient of thermal expansion (α_t/°c × 10^6)	Adherence to dielectric film or substrate	Surface resistivity (Ω/sq × $10^{-7}\sqrt{f}$)	Skin depth at 1 GHz
Ag	21	Poor	2.5	1.4
Au	15	Very poor	3.0	1.7
Cu	18	Very poor	2.6	1.5
Al	26	Very good	3.3	1.9
W	4.6	Good	4.7	2.6
Ta	6.6	Very good	7.2	4.0
Mo	6.0	Good	4.7	2.7
Cr	9.0	Good	4.7	2.7

10.6.3 Dielectric Materials

In MICs, the dielectric materials are used as insulators in capacitors, couple-line structure, as a protective layer for active devices, and as an insulating layer for passive circuits. Some of the commonly used dielectrics in microwave circuits are SiO, SiO_2, and Ta_2O_5. They perform the following functions:

- Provide insulation on a substrate that is especially important in a multi-level metallization system
- Act as a barrier to dopants during processing
- Provide surface passivation
- Isolate one device from another

The commonly used dielectric materials and their properties are listed in Table 10.4. The desirable properties of dielectric materials have the ability to withstand in high voltages i.e., high breakdown voltage; ability to undergo processes without developing pin holes; and a low tangent.

Table 10.4 Properties of dielectric materials

Material	Relative dielectric constant (ε_r)	Dielectric strength (V/cm)	Method of deposition
SiO	6–8	4×10^5	Evaporation
SiO_2	4	10^7	Deposition
Si_3N_4	7.6	10^7	Vapor phase, Sputtering
Ta_2O_5	22–25	6×10^6	Anodization, Sputtering
Al_2O_3	7–10	4×10^6	Anodization, evaporation

10.6.4 Resistive Materials

Resistive materials provide the isolation between the contacts in IC circuits. These are used in MMICs for realizing bias networks, terminators, and attenuators. Some of the thin-film resistive materials used in MICs are NiCr, Cr, Ta, Cr-SiO, and Ti. MMIC resistors are realized with materials having controlled thickness and resistivity, such as Ni–Cr, Ta, or the semiconductor itself with suitable doping. The properties required for a good microwave resistor are good stability, low temperature coefficient of resistance (TCR), adequate dissipation capability, and sheet resistivity.

The resistive materials and their properties are shown in Table 10.5. Depending on the process used, there are thick-film and thin-film resistors. Thin-film resistors have thickness of the order of 1μm or smaller, whereas the thickness of the thick film is of the order of 1 to 500 μm. Thin-film resistors usually perform better than thick-film ones, because their resistance is less dependent on the frequency, due to the lower influence of the skin effect. These thin-film resistors are used in applications that require high accuracy, high stability, or low noise.

Table 10.5 Properties of resistive materials

Material	Resistivity (Ω/square)	TCR (%/°C)	Method of deposition	Stability
Cr	10–1000	–0.100 – +0.10	Evaporation	Poor
NiCr	40–400	+0.002 – +0.10	Evaporation	Good
Ta	5–100	–0.010 – +0.01	Sputtering	Excellent
Cr-SiO	10–600	–0.005 – –0.02	Evaporation or cermets	Fair
Ti	5–2000	–0.100 – +0.10	Evaporation	Fair

10.7 GROWTH OF MMICs

MMICs can be made in monolithic form or hybrid form. In monolithic circuits, active components are grown on a semiconductor substrate, and passive components are either deposited on the substrate or grown on it.

10.7.1 Fabrication Techniques

The MMIC fabrication process includes different techniques, such as

- Ion implantation
- Diffusion
- Oxidation
- Film deposition
- Epitaxial growth
- Lithography
- Masking and etching
- Metallization

Ion implantation: It is a process in which controlled amounts of dopants are implanted in the substrate. Radiation of sufficient density of ions with required energy on substrate using ion beam is called ion implantation. Concentration of impurity ions is determined by the density, whereas the energy determines the depth of the implantation. Subsequent to implantation, the substrate is heated at high temperature. This process causes the substrate to melt and re-crystallize along with implants, thus activating the implant. In silicon based technology, this technique most widely used to form active regions of a device like base, emitter and collector regions.

Diffusion: It is a process in which a substrate is redistributed from an area of a relatively high concentration to an area of a relatively low concentration due to random thermal motion. Similar to ion implantation, this is also used for controlling amounts of dopants in MMIC fabrication. It consists of diffusing impurities into pure material to alter the basic pure characteristics of the material. Both processes are used to dope the semiconductor substrate selectively to produce either an n-type or a p-type layer.

Oxidation: It is the chemical process of joining oxygen with another element to form a compound of the material and oxygen. Oxidation is one of the most basic deposition technology used to form silicon dioxide on a silicon substrate. This is done by heating (800° C–1100° C) the substrate in an oxidized atmosphere; for example, in a chamber containing water vapor or oxygen as shown in Figure 10.2. This process consumes some of the substrate as it proceeds and is limited to materials that can be oxidized.

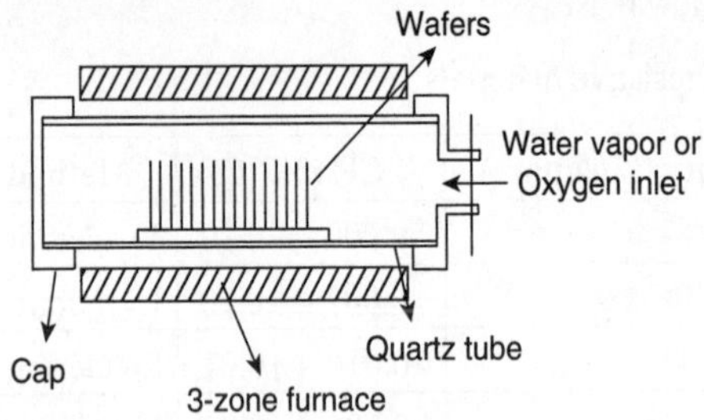

Figure 10.2 *Chamber used for oxidation*

Film deposition: Various types of thin films are used to fabricate integrated and discrete devices or circuits. Thin films are divided into four groups as follows:

- Thermal oxides
- Polycrystalline silicon
- Metal films
- Dielectric layer

  Deposition is the process in which the material is released from the source and transferred towards a substrate, forming a thin film. There are different methods of deposition. They are electron-beam evaporation, *DC* sputtering, and vacuum evaporation.
- In electron beam evaporation, a narrow beam of electrons is generated to heat the source material that is to be deposited on the substrate till a point where the material gets evaporated, as shown in Figure 10.3. Then, condensation on substrate takes place inside vacuum chamber. This enables the molecules to evaporate freely so that it condenses on surface.

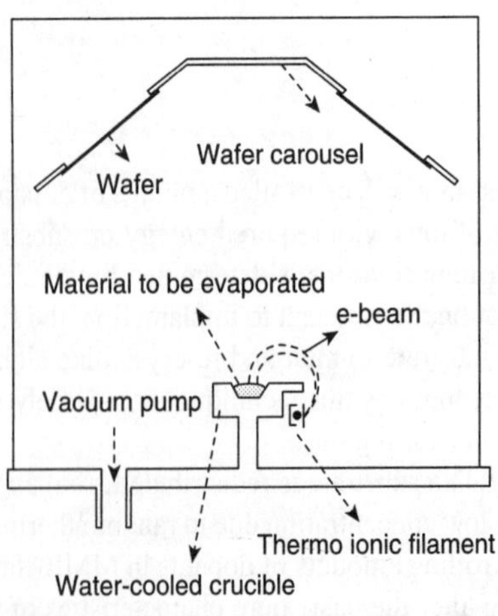

Figure 10.3 *Electron beam evaporation*

- DC (direct current) sputtering is one of the deposition methods that is used to coat substrate structures with thin films of different materials. In this process, source releases the material at temperature less than evaporation. The substrate is placed in a vacuum, which is maintained in a crucible. Inside it, the impurity is used as a cathode, and the substrate is used as an anode, as shown in Figure 10.4. A small amount of inert gas-like argon is introduced into the vacuum. A negative voltage of the order of kilovolts is applied between the cathode and anode, which results in acceleration of the positive argon ions toward the cathode, where they bombard the cathode and remove impurity atoms. Then, the impurities reach the substrate and adhere to it, creating a thin film.

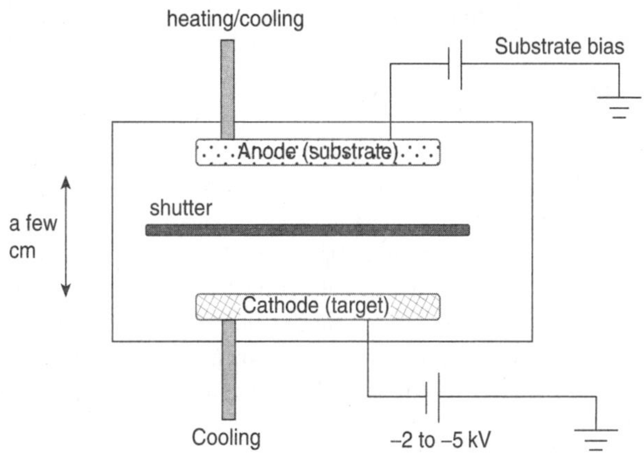

Figure 10.4 *DC sputtering*

Vacuum Evaporation: This process is similar to electron beam evaporation, but a thermal source is used in place of an electron beam. The advantages of this process are that high-purity films can be deposited, there is easy deposition rate monitoring and control, and it is less expensive.

Epitaxial Growth: *Epitaxial layer* means the formation of a single crystal layer on top of a single crystal substrate. The doping level of the epitaxial layer is different from the substrate on which the epitaxial layer is formed.

This process is used to form films of a thickness ranging from 1 μm to 100 μm. *Selective epitaxy* means the growth of a single-crystal semiconductor on a substrate patterned with oxide or nitride. This process is superior to the local oxidation process in providing isolation. Care should be taken to suppress the nucleation of silicon on dielectric. The growth process can be controlled by controlling the flow of gases or controlling the temperature. This process allows deposition of even thin layers. Depending on the phase of the material used to form the epitaxial layer, they are classified into Vapor Phase Epitaxy (VPE), Liquid Phase Epitaxy (LPE), and Molecular Beam Epitaxy (MBE).

- **Vapor Phase Epitaxy (VPE)**: In this process, the substrate is heated in an induction heated reactor to a point that is half the melting point of the material to be deposited. A number of gases are then made to enter the reactor as shown in Figure 10.5. The advantages of this process are very uniform layer formation, high growth rate of material, and conformal growth with very good step coverage.

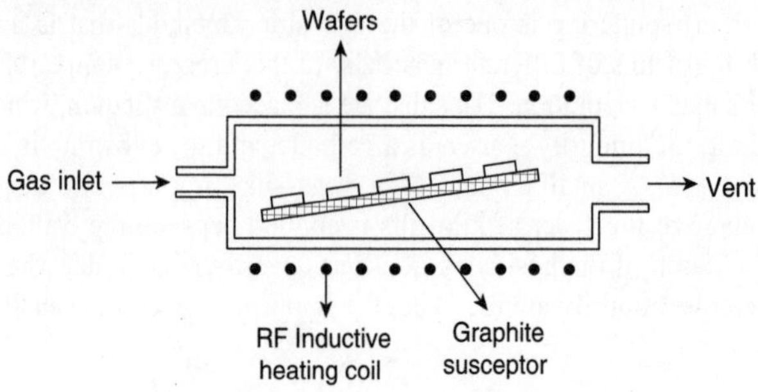

Figure 10.5 *Vapor phase epitaxy*

- **Liquid Phase Epitaxy (LPE):** It is a process in which the epitaxial layer is deposited on the substrate from a super-saturated solution. Figure 10.6 shows the LPE process. It contains a boat. Initially, a thin GaAs wafer is placed on a graphite slider. This slider can be moved under the boat. The boat consists of Ga along with a piece of GaAs. When the boat is heated to a temperature of 750–800° centigrade, the solution will be saturated. Required amounts of dopants can be added to this saturated solution to grow n-type or p-type layers.

  The advantages of this process are that the growth temperature can be low; the equipment is inexpensive, simple, and non-hazardous. In this process, very thin, uniform, and high-quality layers can be produced. However, the layers cannot be made uniform over large areas and also it is difficult to control the layer thickness. To overcome these, molecular beam epitaxy has been developed.

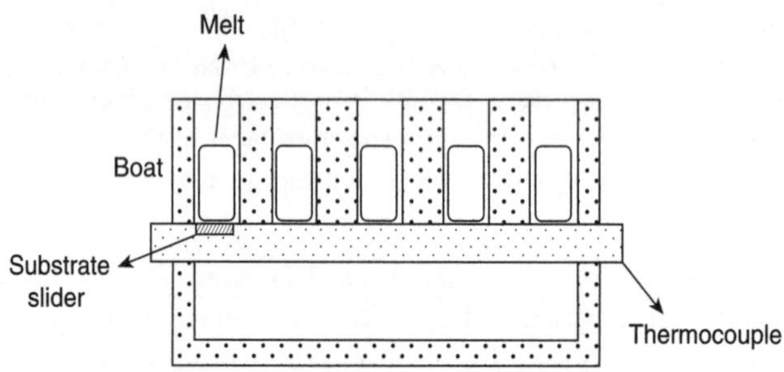

Figure 10.6 *Liquid phase epitaxy*

- **Molecular Beam Epitaxy (MBE):** It is a very refined high-vacuum evaporation technique, in which one or several molecular beams interact with the surface of a crystalline substrate.

  The set up contains a stainless steel bell jar. This process requires an ultra-high vacuum (10^{-11} torr). The required particles shown as sources in Figure 10.7 are made to evaporate by heating the tubes containing them. These tubes are generally made of inert materials such as tungsten. In addition, the substrate is heated to a temperature from 500°C to 700°C. As the required particles evaporate, they get deposited on the substrate and form the desired material. The MBE made it possible to develop key devices with dimensions less than 1 micro meter. It is a difficult and costly process, but it has following

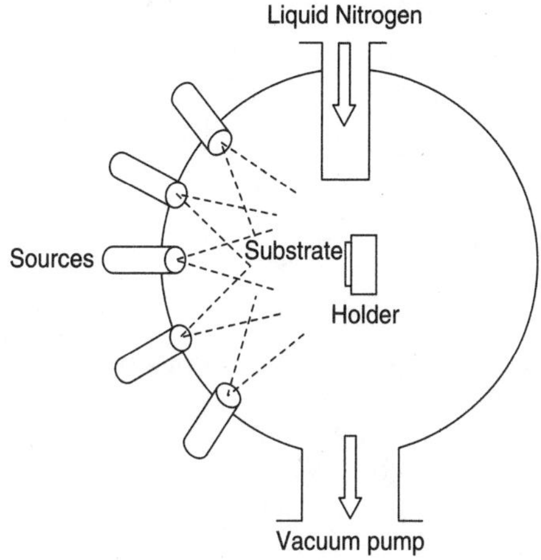

Figure 10.7 *Molecular beam epitaxy*

Advantages:

- An abrupt hetero junction can be grown.
- A complex hetero junction with many layers can be formed.
- Requires low temperature compared with vapor phase epitaxy
- Extremely sharp interfaces can be formed.
- Doping profiles in highly localized regions can also be controlled.

Lithography: The literal meaning of *lithography* is stone writing. The transfer of a pattern or an image from one medium to another, as from a mask to a wafer, is called *lithography*. In the context of VLSI fabrication, it means the patterning of shapes on a resist. Several lithography technologies are available, and some of them are as follows:

- *Electron-beam lithography*: In this process, an electron-sensitive resist film is applied over the substrate, and a beam of electrons is made to scan on the surface. This scanning deposits energy on the resist film in the required pattern. The advantages of this process are that it gives atomic-level resolution, and an infinite number of patterns can be formed. The disadvantages of this process are that it is slow, expensive, and complicated.
- *Ion-beam lithography*: It is similar to electron beam lithography, where ions are used in place of electrons. It gives more resolution than electron beam lithography.
- *Optical lithography*: It uses light to form shape patterns on the substrate. A light-sensitive material known as a *photoresist* is used in this process.
- *X-ray lithography*: In this process, X-rays are used to selectively form patterns on the substrate.

Masking and Etching: A transparent glass plate is covered by patterns of opaque areas that prevent light to pass through. Mask is used to mark areas, to be later etched on wafer, on a photoresist by using emulsion chrome iron oxide silicon to produce opaque areas.

Etching is the process of removing a material by a chemical reaction. During the IC fabrication, selective openings are required in silicon dioxide. These openings are required to diffuse the impurities. During this process, a uniform film of photosensitive emulsion is applied over the wafer. Kodak photoresist is one such photosensitive emulsion. Then, certain masks that are created from the layout information provided by the designer are used to expose the selected areas to UV light. Next, the exposed photoresist gets polymerized. After removal of mask, wafer is developed using chemicals such as trichloroethylene, as shown in Figure 10.8. This chemical dissolves the unpolymerized photoresist and leaves the required surface pattern.

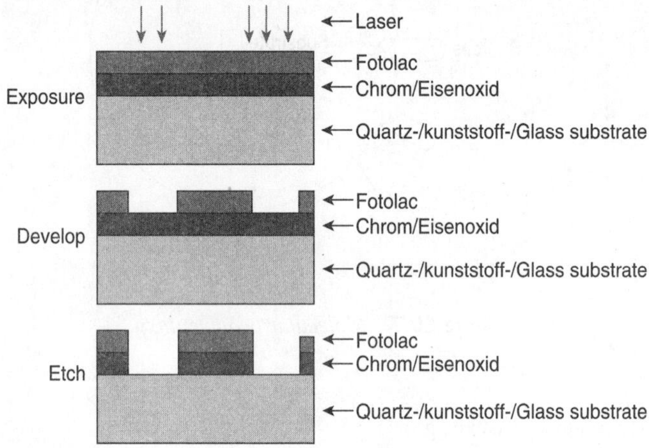

Figure 10.8 *Masking and etching*

Metallization: In this process, all the active and passive elements of the integrated circuit are interconnected using an appropriate metal. The desired pattern of interconnections is formed using the photoresist technique. The realization of an MMIC is a process in which the final circuit elements are constructed by a combination of substrate patterning followed by selective implantation, etching, and metallization.

10.8 MOSFET FABRICATION

Metal Oxide Semiconductor Field Effect Transistor (MOSFET) is one of the important devices in ICs. In the fabrication of any IC, initially, a large silicon wafer that acts as a substrate is considered. These substrates contain an equal number of free charge carriers (electrons and holes). Then, the substrate is doped with impurities to create concentrated regions of N (electron) and P (hole) regions. These regions are required, because the movement of electrons and holes is what enables transistors to work. Later steps in fabrication involve many processes such as oxidation, diffusion, and deposition. In this section, the MOSFET formation, fabrication procedure of NMOS and CMOS are discussed.

10.8.1 MOSFET Formation

The fabrication of MOSFET involves the following steps:

- Initially, an N-type or a P-type substrate of 100-mm to 200-mm diameter and 0.5-mm thickness is taken.
- Then, an insulation layer of silicon dioxide is grown on the substrate.
- The surface is covered with a photoresist material.
- The photoresist layer is then exposed to UV light through a mask. This mask will have windows in those regions where impurity (P^+ or N^+) diffusion takes place.
- The photoresist material along with underlying silicon dioxide is etched away.
- Again, a thin layer of SiO_2 is deposited over the surface.

- The oxide layer is covered by deposition of polysilicon.
- Then, using the photoetching process, two windows are opened in the thin oxide region, into which P^+ or N^+ impurities are diffused.
- Then, aluminum metal is deposited over its surface to a thickness of 1 micrometer.
- This metal layer is etched to form the required interconnection pattern.

10.8.2 NMOS Fabrication Process (or) Growth

The fabrication process of an NMOS involves the following steps:

1. Initially, a substrate is taken and is lightly doped with a P-type impurity such as boron. Impurity concentrations are in the order of $10^{15}/cm^3$, and they give a resistivity of an order from 25 ohm cm to 2 ohm cm.
2. Then, a thick oxide layer of thickness 1μm is deposited on the substrate which acts as a barrier to dopants and provides an insulating substrate to other layers may be deposited and patterned on it.
3. A silicon nitride layer (photoresist) is deposited on the oxide surface. Using the photo-etching process, regions in which diffusion takes place are defined.
4. UV rays are passed on to photo resist layer through a mask to mark regions for diffusion. Boron ions are implanted to prevent inversion under the field oxide. The areas that are exposed to ultra-violet radiation are polymerized, but the areas required for diffusion are shielded by the mask and remain unaffected.

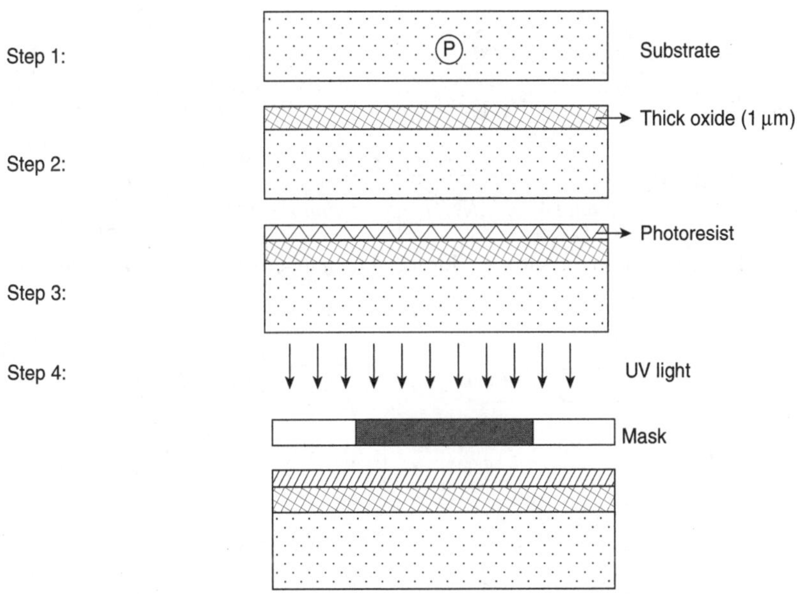

5. Then, the SiO_2–Si_3N_4 layers are cleaned, and the wafer surface is exposed in the window defined by the mask.

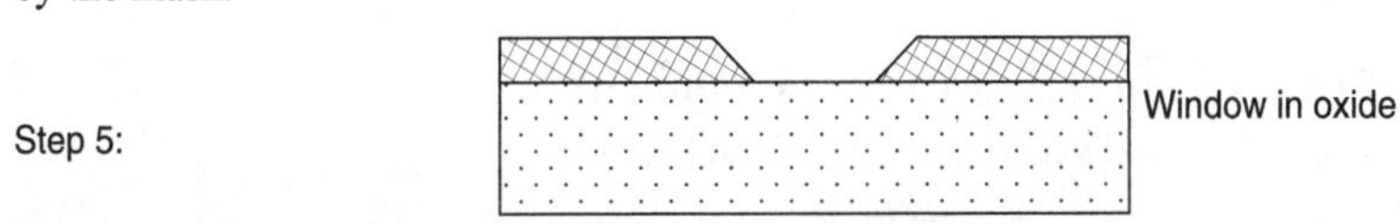

6. After removing the photoresist, a thin layer of SiO_2 of thickness 0.1μm is grown on the top of the surface. The polysilicon can be deposited by using chemical vapour deposition (CVD) method in

the selected area to form gate structure of fine pattern with control over thickness, impurity concentrations and resistivity should be taken care of.

7. More photoresist and masking allows patterning of polysilicon. After this thin oxide is removed to diffuse n–type impurities into exposed areas which leads to formation of source and drain. Diffusion is done by passing a gas with desired n impurity at high temperature over surface. The polysilicon with the underlying thin oxide acts as a mask during diffusion—this process is called *self-aligning*.

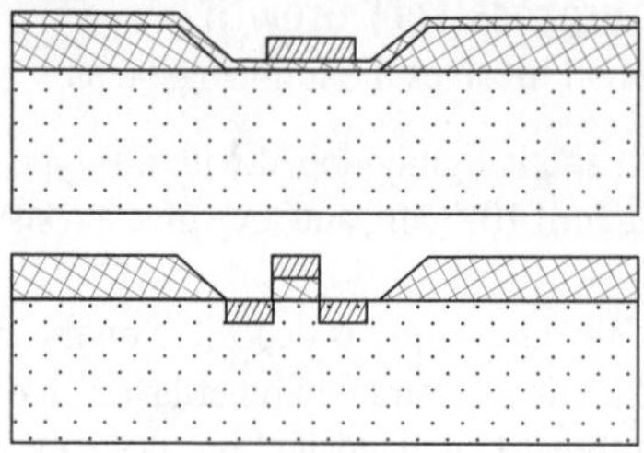

8. After growing thick oxide it is masked with photoresist and selected areas are etched for source, drain and some areas of polysilicon for connections.

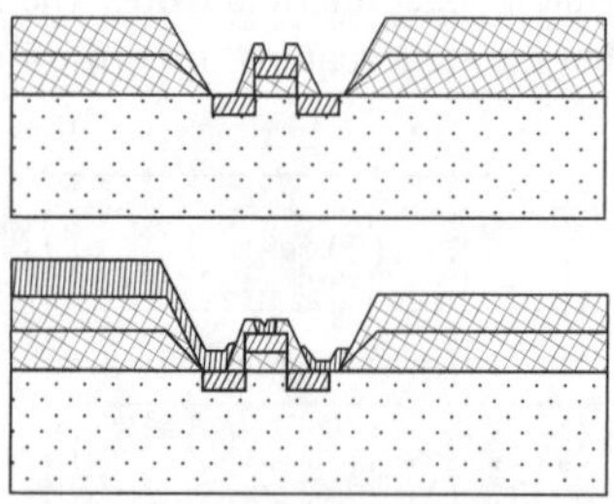

Figure 10.9 *NMOS fabrication steps*

9. Metallization is done all over the surface. This metal (aluminum) layer is then masked and etched to form the required interconnection pattern.
   Note:

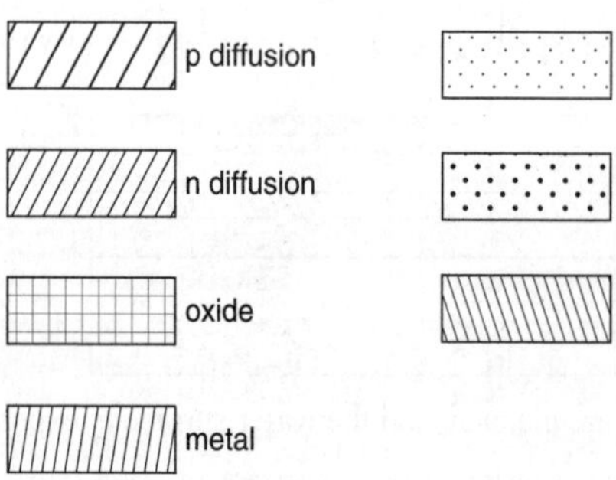

10.8.3 CMOS Fabrication Process (or) Development

The possible approaches for CMOS fabrication are listed below

- p well
- n well

- Twin tub
- Silicon on insulator processes

The basic processing steps are of a similar nature as those used for NMOS. The two processes *p* well and twin tub are described below:

***P*-well process**: Steps involved in this process are masking, patterning, and diffusion and are shown in Figure10.10 and how the *p*-well acts as inverter is shown in Figure 10.11.

Mask 1- Mark the areas for deep *p* well diffusion

Mask 2- After thick oxide is deposited this mask selects the areas for thin oxides on the thick oxides to grow the *p*-type and *n*-type transistors and wires.

Mask 3- This mask is used to deposit the polysilicon layer over the thin oxide regions

Mask 4- It uses a *p*-plus(+ve) mask to define all areas to diffuse *p* impurity in order to achieve *p* devices

Mask 5- It uses a *p*-plus (−ve) mask to define all areas to diffuse *n* impurity in order to achieve *n* devices in *p* well

Mask 6- Mark contact cuts.

Mask 7- Mark metal layer pattern

Mask 8- This masking is used to form the openings to access the bonding pads by applying an over-glass (passivation) layer.

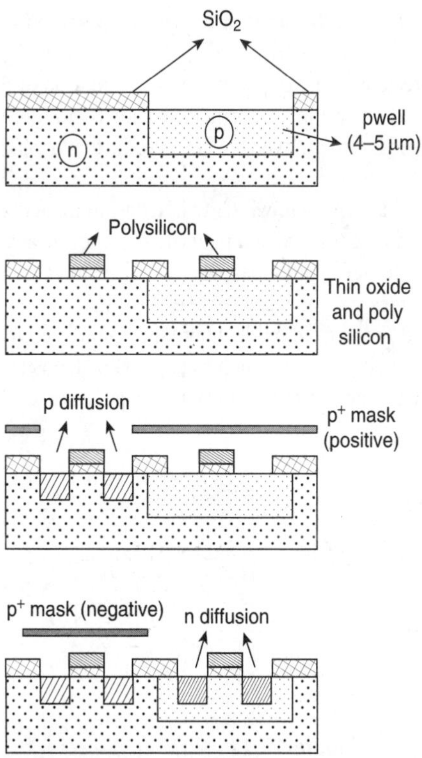

Figure 10.10 *CMOS p-well process*

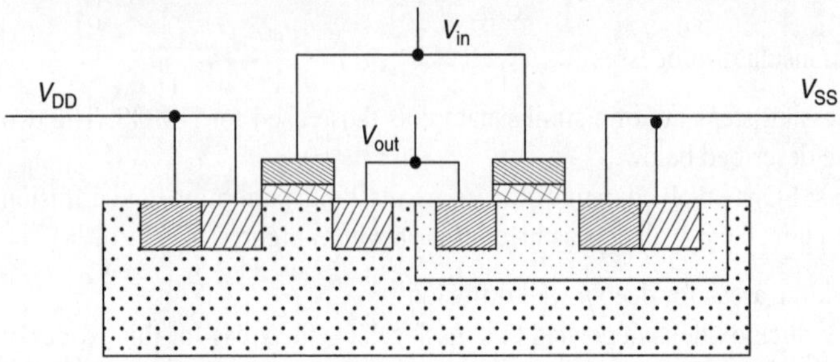

Figure 10.11 *CMOS p-well inverter showing V_{DD} and V_{SS} substrate connections*

Note: The p-well is used as the substrate for *n* devices within the *n*-parent substrate with voltage polarity restrictions so that there is a electrical isolation between two areas. However, since there are now, in effect, two substrates, two substrate connections (v_{DD} and v_{SS}) are required (as shown in Figure 10.11).

Twin tub process: Here, the twin tub process of CMOS fabrication is illustrated. This process is a logical extension of *p*-well and *n*-well fabrication processes. For separate optimization of the *n* and *p* transistors twin tub process is used.

- Initially, an *n*-type substrate is taken, and a thick layer of oxide is grown over the substrate.
- Using masks and a photoresist, the active areas (where *p* and n tubs are to be formed) are etched away.
- These active regions are selectively doped by ion implantation of dopant atoms into the material. In this way, one n tub and one p tub are formed, as shown in Figure 10.12.
- Then, a thick oxide layer is grown over the surface. In the areas directly over the substrate, this oxide is etched away.
- Then, a much thinner oxide is grown, which will act as an insulator for the transistor gates.
- Then, a layer of polysilicon is formed and patterned on the surface using a mask.
- Then, a P^+ mask is used to define all areas where *P* diffusion is to take place.
- A negative P^+ mask defines the areas of n-type diffusion.
- Contact cuts are defined in the field oxide where the connection to the substrate is desired.
- Aluminum (or any other suitable metal such as copper) is deposited over the surface in the appropriate contact cuts, and an interconnect pattern is defined.

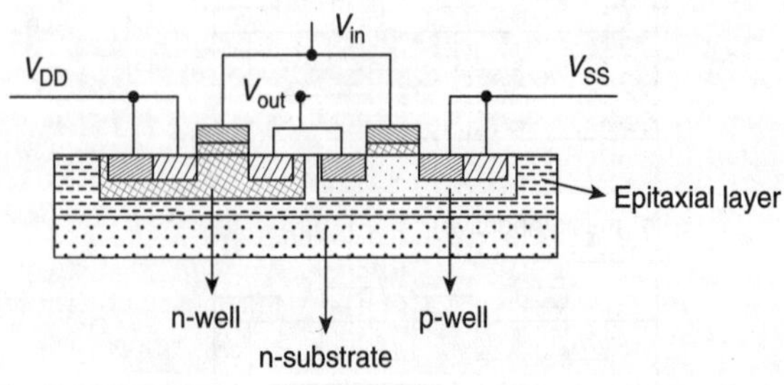

Figure 10.12 *Twin tub process*

10.9 THIN-FILM FORMATION

A thin-film formation is a thin solid layer formed on a substrate. The resistors and inductors developed using thin-film technology are lumped elements and are extremely useful in MMIC applications. These lumped elements can be used up to 20 GHz frequencies, and at these high frequencies, the lumped elements show a negligible phase shift.

The formation of a thin film basically consists of three stages. They are

- **Nucleation**: The molecules of whatever material is to be deposited on the substrate are absorbed by the wafer surface, and they become adatoms (isolated atoms sitting on a crystal surface).
- **Grain growth**: The adatoms from the previous stage combine to form clusters that are also called *island clusters*.
- **Continuous film**: These island clusters continue to grow and condense into a thin solid substance on the substrate.

Two important parameters in thin-film formation are step coverage and aspect ratio.

***Step coverage*:** This is used to measure the uniformity of the thin film. The uniformity of the film should be controlled to a high degree across the diameter of the wafer, because poor step coverage leads to high stress inside the device. It is defined as ratio of the minimum thickness deposited on the side of the step to deposition on the horizontal surface.

***Aspect ratio*:** It is ratio of depth to width. The gap between two metal lines is called a *contact hole*. A deep contact hole would have a large aspect ratio and is hard to fill.

Methods of thin-film deposition:

Thin-film deposition involves two methods. They are

- Chemical vapor deposition (CVD)
- Physical vapor deposition (PVD)

In chemical vapor deposition, gases are enabled to react, and the desired film is formed on the surface of the substrate. Physical vapor deposition uses physical processes to deposit the film. PVD techniques are more versatile than CVD methods.

Components formed using thin-film process: In MMICs, lumped resistors, lumped capacitors, and planar inductors are formed using thin-film methods.

10.9.1 Planar Resistor Films

A thin resistive film deposited on a semi insulating substrate forms a planar resistor. The resistive film materials with a resistivity range of 30–1k ohm-meter such as Al, Cu, Au, Ta, etc are mostly used and they are widely utilised for termination of power combiners, hybrid couplers, and bias voltage circuits. For the design of planar resistors, some factors are considered: (1) the material resistivity, (2) the thermal resistance of the material, (3) temperature coefficient of the resistive material, and (4) the frequency bandwidth.

Planar resistors are classified in to three groups depends on its fabrication process as follows (i) semiconductor films (ii) metal deposited films and (iii) cermets. There are two methods by which the semiconductor planar resistors are fabricated, (i) by forming a conducting epitaxial film on the substrate (ii) by implanting a semi-insulating, high resistivity region within a substrate as shown in Figure 10.13 (a). Metal film resistors are formed by evaporating a metal layer over the substrate and forming the desired pattern by photolithography.

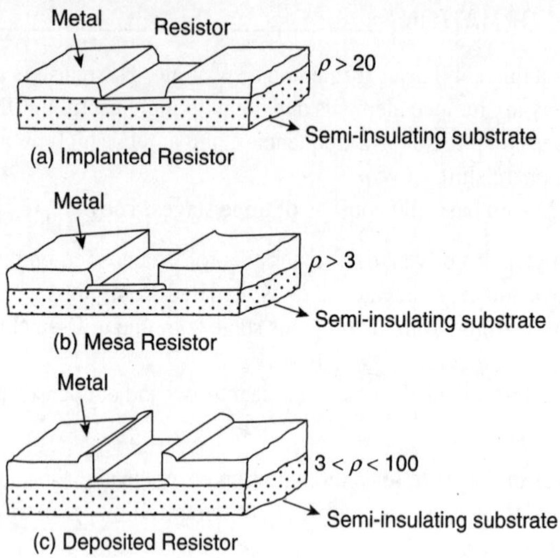

Figure 10.13 (a) *Planar configuration of plane resistor*

The resistance of a planar resistor film is given by the general formula

$$R = \frac{\rho l}{a} \tag{10.1}$$

$$R = \frac{\rho l}{wt} \tag{10.2}$$

where l is the length of the resistive film; ρ is the sheet resistivity of the film; and a is the area given by thickness of the film (t) multiplied by width of the film (w).

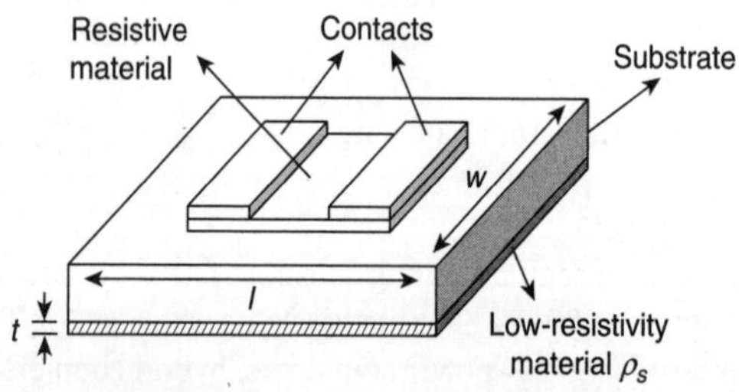

Figure 10.13 (b) *Thin-film resistor*

10.9.2 Planar Inductor Films

There are various configurations of a monolithic inductor. Figure 10.14 design procedures of some planar inductor shown in Figure 10.14. The inductors which provide inductances upto 0.5 nH are single microstrip line, meander, and single-loop inductors which are rarely used. Among these inductors, meander inductor is used to conserve chip area, but a more better approach is winding the microstrip line in a spiral which is shown in Figure 10.14 (d), (e), and (f). Because of layout simplicity rectangular spiral is used mostly. Thus, compared to circular layouts, higher losses will be obtained in rectangular layouts i.e., for equal inductance, the metallization loss results are as high as 10%. The outer dimension

OD, the strip width w, the number of turns n, spacing between lines s, and the gap between opposing groups of coupled lines G are its parameters.

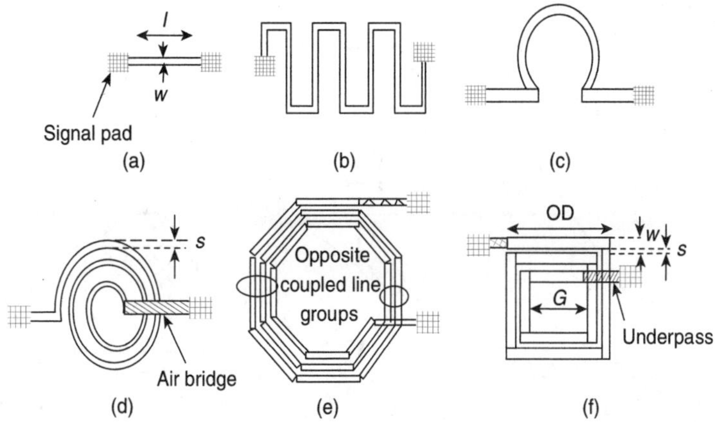

Figure 10.14 *Monolithic inductor configurations; (a) Microstrip line; (b) Meander line; (c) Single loop, (d) Circular spiral; (e) Octogonal spiral; (f) Rectangular spiral*

For the monolithic circuits, the distinctive inductance values ranges from 0.5 to 10 nH. The inductance of a circular spiral inductor is given by with n turns and outer diameter d_0 is (Figure 10.15)

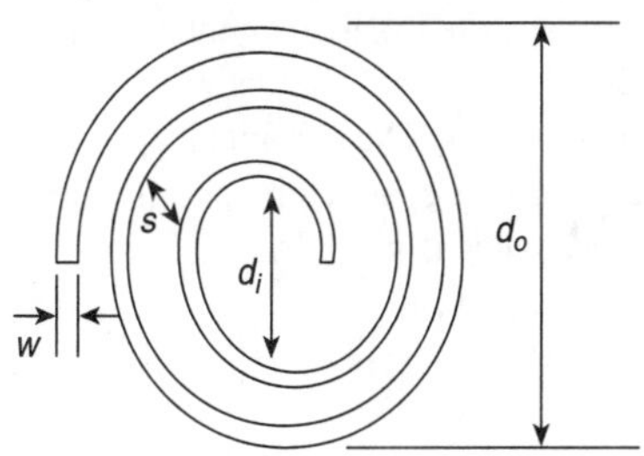

Figure 10.15 *Circular spiral inductor*

$$L = 0.03125 n^2 d_0 \text{ nH/mil} \tag{10.3}$$

where $d_0 = 5d_1 = 2.5\, n(w + s)$ in mils
n = number of turns
s = separation in mils
w = film width in mils

10.9.3 Planar Capacitor Films

In MMICs, two types of capacitors are generally used. They are the metal-oxide-metal capacitor and interdigitated capacitor.

***Metal-oxide-metal capacitor*:** It uses a dielectric film between two electrodes and forms an overlay structure. It is widely used for low impedance circuits, bypass, and dc blocking circuits with higher values. It has two electrodes at the top and bottom, and a dielectric layer is in between them as shown in Figure 10.16 (a). The capacitance of the dielectric metal-oxide-metal-capacitor can be expressed by

$$C = \varepsilon_0 \varepsilon_r \frac{lw}{h} \text{ Farad} \tag{10.4}$$

where $\varepsilon_0 = 8.854 \times 10^{-12}$ F/m,

ε_r = relative dielectric constant of dielectric material,

l = metal length,

w = metal width, and

h = height of the dielectric material

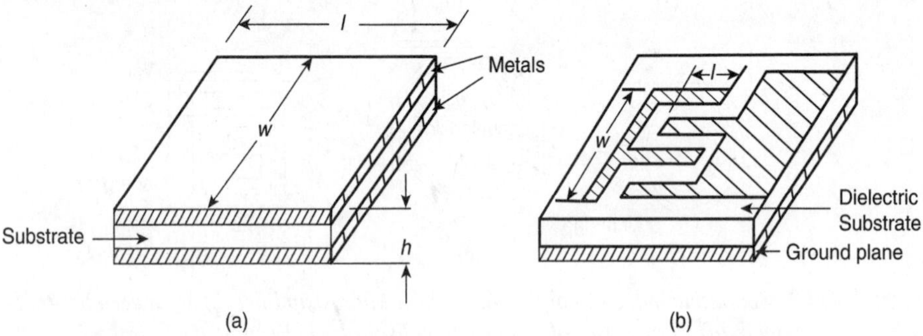

Figure 10.16 *(a) Metal-oxide-metal capacitor; (b) Interdigitated capacitor*

Interdigitated capacitor: The interaction area can be enhanced by incorporating a comb shaped electrode in between the strip; such a structure is called an interdigital capacitor structure, and it is shown in Figure 10.16 (b). These are suitable for applications where low values of capacitance are required, such as high impedance matching circuits. The capacitance value varies with some parameters. They are, w/l ratio, gap distance between two conductors, and the number of fingers N. It consists of a single-layer structure, and its capacitance is given by

$$C = \frac{\varepsilon_r + 1}{w} l\left[(N-3)K_1 + K_2\right] \text{ pF/cm} \tag{10.5}$$

where N is the number of fingers, $K_1 = 0.089$ pF/cm is the contribution of the interior finger for $h > w$, and $K_2 = 0.10$ pF/cm is the contribution of the two external fingers for $h > w$

w = finger base width in cm

l = finger length in cm

h = finger height in cm

EXAMPLE PROBLEM 10.1

Calculate the resistance of a planar resistor film with the following parameters: Film thickness is 0.2 μm, film length is 10 mm, film width is 10 mm, and sheet resistivity of gold film is 2.44×10^{-8} ohm-meter.

Solution

Given $\rho = 2.44 \times 10^{-8}$ ohm-meter, $l = 10$ mm, $w = 10$ mm, and $t = 2 \times 10^{-7}$m

Resistance of a planar resistor film is given by

$$R = \frac{\rho l}{wt} = 2.44 \times 10^{-8} \times 10/10 \times 2 \times 10^{-7} = 0.122 \text{ ohm}$$

■

EXAMPLE PROBLEM 10.2

Calculate the inductance of a circular spiral inductor with the following parameters: number of turns are 8, separation is 100 mils, and film width is 50 mils.

Solution

Given $n = 8$, $s = 100$ mils, and $w = 50$ mils
The inductance of a circular spiral inductor is given by

$$L = 0.03125n^2 d_0 \text{ nH/mil}$$

where $d_0 = 5d_1 = 2.5\, n\,(w + s)$ in mils
Therefore, inductance $L = 0.03125\,(8)^2 \times 2.5 \times (8)\,(50 + 100) = 6{,}000$ nH/mil ■

EXAMPLE PROBLEM 10.3

An inter digitised capacitor is fabricated from a GaAs substrate having the following parameters: relative dielectric constant of GaAs is 12.10, number of fingers are 10, substrate height is 0.2054 cm, finger length is 0.00310 cm, and finger base width is 0.050 cm. Calculate the capacitance of this capacitor.

Solution

Given $\varepsilon_r = 12.10$, $l = 0.00310$ cm, $N = 10$, and $w = 0.050$ cm.
We know that $K_1 = 0.089$ pF/cm, and $K_2 = 0.10$ pF/cm
The capacitance for the interdigitated capacitor is given by

$$C = \frac{\varepsilon_r + 1}{w} l\left[(N-3)K_1 + K_2\right] \text{ pF/cm}$$

Therefore,

$$C = \frac{12.10 + 1}{0.050}(0.00310)\left[(10-3)0.089 + 0.10\right] = 0.587 \text{ pF/cm}$$

■

SUMMARY

1. Microwave circuits are broadly classified into two categories: discrete and Microwave Integrated Circuits (MICs). In the discrete circuit, the circuit elements are separately manufactured and then interconnected by conducting wires. MICs are of two types: MMICs and HMICs.
2. MICs operate in a microwave frequency range (3GHz–300GHz).
3. The MMIC is a monolithic microwave integrated circuit in which the active and passive components are fabricated on the same semiconductor substrate.
4. HMICs comprise a number of discrete active devices and passive components such as diodes/transistors, capacitors, and circulators.
5. MMICs offer several advantages: small in size and weight, high reliability, improved reproducibility, and improved performance

6. Basic materials for MMICs are substrate materials, conductor materials, dielectric materials, and resistive materials.

7. MMIC fabrication techniques include ion implantation, diffusion, oxidation and film deposition, epitaxial growth, lithography, masking, and etching.

8. An MMIC whose elements are formed on an insulating substrate, such as glass or ceramic, is called a *film integrated circuit.*

9. Resistive materials are used in MMICs for bias networks, terminations, and attenuators.

10. Epi or epitaxy is the controlled growth of a layer of crystalline semiconductor material on a suitable substrate.

11. Etching is the process of removing material (such as oxides or other thin films) by chemical, electrolytic, or plasma (ion bombardment) means.

12. Dopants are the materials that are used to change the electrical characteristics of a semiconductor crystal, making it N or P type. Doping is usually accomplished through diffusion or ion-implantation processes.

13. Diffusion and ion implantation are the two processes that are used in controlling amounts of dopants in semiconductor fabrications.

14. Lithography is the process of transferring patterns of geometric shapes on a mask to a thin layer of radiation-sensitive material, which is known as resist, for covering the surface of a semiconductor wafer.

15. Chemical vapor deposition and physical vapor deposition are the two methods that are used for forming thin films

16. A photoresist is a light-sensitive liquid that is spread as a uniform thin film on a wafer or a substrate. After baking, the exposure of specific patterns is performed using a mask. The material remaining after development resists subsequent etching or implant operations.

17. Sputtering is a method of depositing a film of material on an IC wafer.

18. A wafer is a thin disk of semiconductor material (usually silicon) on which many separate chips can be fabricated.

OBJECTIVE-TYPE QUESTIONS

1. A film integrated circuit is
   (a) HMIC (b) MMIC
   (c) a discrete circuit (d) not useful at microwave frequencies

2. The property of dielectric material in MMIC is
   (a) low resistivity (b) reproducibility
   (c) low temperature coefficient (d) high RF dielectric loss

3. Lithography is a process of
   (a) deposition
   (b) evaporation
   (c) etching
   (d) transferring patterns of geometric shapes
4. Diffusion means
   (a) epitaxial growth
   (b) adding dopants
   (c) a method of lithography
   (d) same as evaporation
5. The conductor material used in MMICs is
   (a) alumina
   (b) GaAs
   (c) Ag
   (d) SiO
6. DC sputtering is
   (a) vacuum evaporation
   (b) vacuum deposition
   (c) etching
   (d) evaporation
7. Resistance of a planar resistor film is
   (a) $R = \dfrac{\rho l}{wt}$
   (b) $R = \dfrac{tl}{w\rho}$
   (c) $R = \dfrac{w\rho}{lt}$
   (d) $R = \dfrac{\rho t}{wl}$
8. Capacitance of a metal-oxide-metal capacitor is
   (a) $C = \varepsilon_0 \varepsilon_r \dfrac{w}{lh}$
   (b) $C = \varepsilon_0 \varepsilon_r \dfrac{l}{wh}$
   (c) $C = \varepsilon_0 \varepsilon_r \dfrac{l}{h}$
   (d) $C = \varepsilon_0 \varepsilon_r \dfrac{lw}{h}$

ANSWERS TO OBJECTIVE-TYPE QUESTIONS

1. (b) 2. (b) 3. (d) 4. (b) 5. (c) 6. (b) 7. (a) 8. (d)

REVIEW QUESTIONS

1. List the advantages of MMICs.
2. What are the applications of MMICs?
3. What are the basic materials used for MMICs?
4. What are the ideal properties of a substrate material?
5. What are the ideal properties of a conductor material?
6. What are the ideal properties of a resistor and dielectric materials?
7. Explain NMOS fabrication.
8. Explain CMOS fabrication.
9. Write a short note on thin films.

10. Write about different types of deposition.

11. Write a short note on epitaxial growth.

12. What are the advantages of MMICs compared with discrete circuits?

13. Differentiate between microstrips and substrates. What are the disadvantages of microstrips compared with waveguides?

14. Write the advantages and disadvantages of HMICs.

15. What are the different techniques used to fabricate MMICs?

16. Write the ideal characteristics of substrate material and conductor material.

17. Explain the comparison between diffusion and ion implantation.

18. What is the inductance of a round-wire inductor of 1m length and with 1mm diameter?

19. Calculate the resistance of a planar resistor made of aluminum with a thickness of 0.5 μm, a length of 10 mm, a width of 10 mm, and a resistivity of $2.82 \times 10^{-8}\ \Omega - \text{m}$.

20. Calculate the inductance for a planar circular spiral inductor with the following parameters: film width of 50 mils, separation of 100 mils, and number of turns equal to 10.
(Hint: 1 mil $= 2.54 \times 10^{-5}$ m)

Microwave Measurements

11

11.1 INTRODUCTION

The measurement parameters and techniques used for analyzing the circuits (or networks) are different at low and microwave frequencies. At low frequencies, the entire circuit is affected by the voltage and current waves at the same time. Thus the circuit elements are lumped. It is convenient to measure current and voltage and to use them to calculate power at low frequencies. However, at high frequencies, it is convenient to describe a given network in terms of voltage waves and current waves rather than voltage and current, because they vary with position in a transmission line (or waveguide). At microwave frequencies, most of the quantities measured are relative, not absolute as in low-frequency measurements. And it is also not necessary to know their absolute values at microwave frequencies. The parameters that can be conveniently measured at microwave frequencies are frequency, power, attenuation, VSWR, impedance, and Cavity *Q*. In this chapter, the basic microwave bench used to measure these parameters and the various techniques involved in measuring these parameters are described in detail.

11.2 DESCRIPTION OF MICROWAVE BENCH

The general setup for the measurement of parameters at microwave frequencies is usually done by a microwave bench and is operated in X band (8–12 GHz). The general microwave bench setup is shown in Figure 11.1.

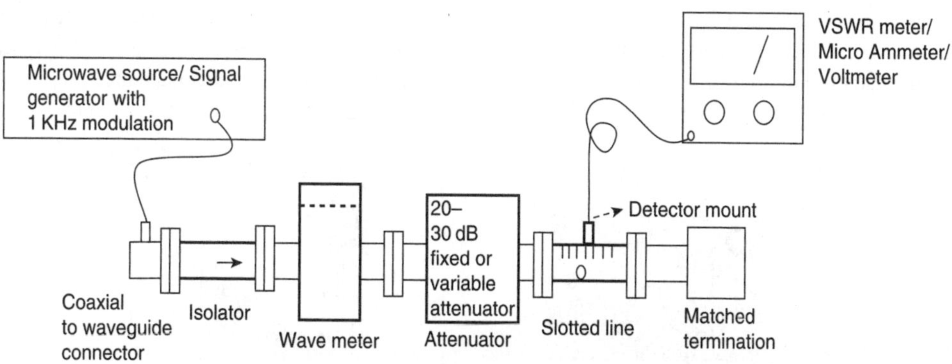

Figure 11.1 *Block diagram of microwave bench setup*

11.2.1 Description of Blocks of Microwave Bench and their Features

Signal generator

A microwave source is used to generate signal and the output is in the order of mill volts. Usually, it is a Reflex Klystron or a Gunn diode or a backward oscillator. The signal output from these sources is generally square-wave modulated at an audio frequency of 1 KHz. A typical Reflex Klystron tube setup and Gunn oscillator setup are shown in Figure 11.2 (a) and (b).

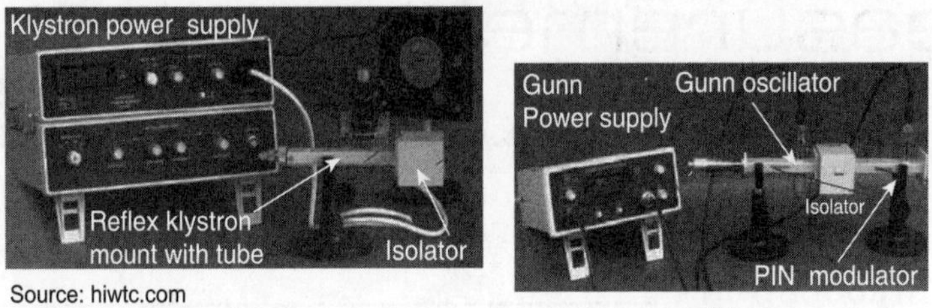

Source: hiwtc.com

(a) (b)

Figure 11.2 *(a) Reflex Klystron tube; (b) Gunn oscillator*

Isolator

An isolator is a two-port device that allows transmission in one direction only (i.e., from port 1 to port 2) and provides maximum attenuation for transmission from port 2 to port 1 (Figure 11.3). In all microwave bench setups, the microwave source is always followed by an isolator. This is to prevent the damage of the source due to load variations, which causes reflections in the circuit.

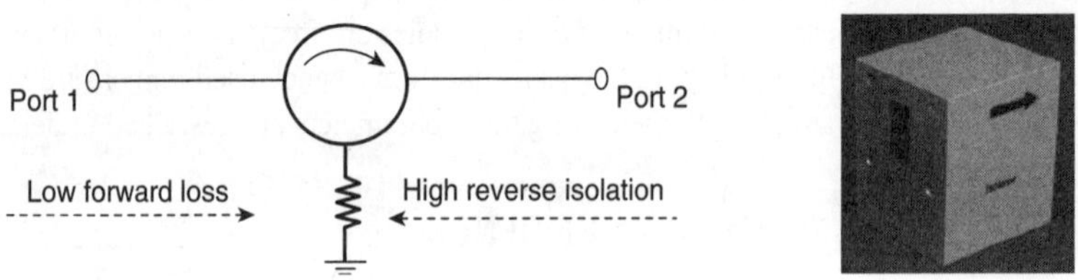

Source: hik-consulting.pl

Figure 11.3 *Isolator function and its schematic*

Precision attenuator

Attenuator is a device used to adjust the power or to reduce the power flowing in a waveguide. The power input can be reduced to a particular stage by using attenuators to prevent overloading. There are two types of attenuators: fixed and variable. Fixed attenuators are used where a fixed amount of attenuation is to be provided (Figure 11.4). Variable attenuators provide continuous or step-wise attenuation (Figure 11.5).

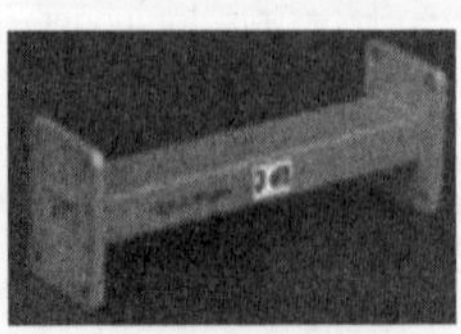

Source: hik-consulting.pl

Figure 11.4 *Fixed 3-dB attenuator*

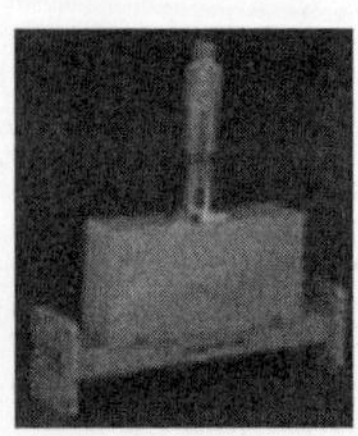

Source: hik-consulting.pl

Figure 11.5 *Variable attenuator*

Resonant cavity frequency meter/wave meter

It is used for a direct reading of frequency and it consists of a single cylindrical cavity. The length of the cavity is adjusted to resonance frequency and is slot coupled to the waveguide. The long scale length and numbered calibration marks provide high resolution (Figure 11.6).

Source: hik-consulting.pl

Figure 11.6 *Wave meter*

Slotted line

It is used for measuring the standing wave ratio (SWR). It consists of a slotted-line section of a waveguide and a traveling probe carriage facility. A slot is made in the center of the broader face of the waveguide, and a small probe (coaxial **E**- field probe) is inserted through the slot. A crystal detector is inserted in the probe, and it is used to adjust the modulated signal by sensing the relative field strength of the standing wave pattern in the waveguide (Figure11.7 (a) and (b)).

(a)

Source: 1.imimg.com

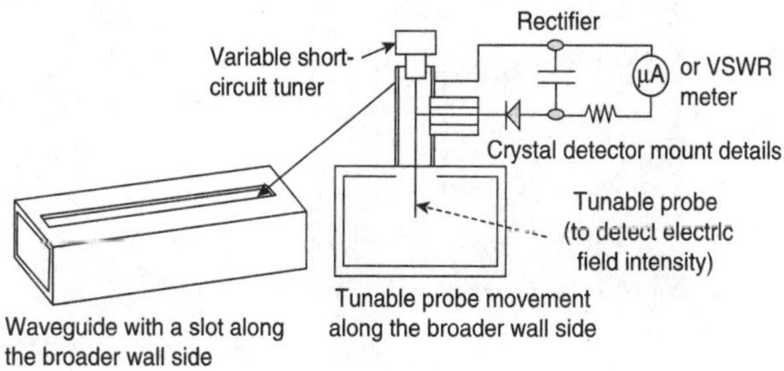

(b)

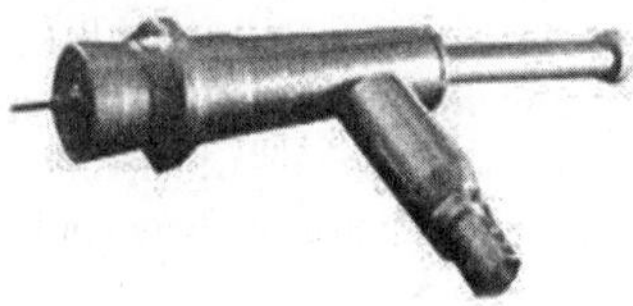

(c)

Source: 1.imimg.com

Figure 11.7 *(a) Slotted section of waveguide; (b) Slotted line section details; (c) Schematic of a tunable probe mount with a crystal detector inside*

Tunable probe and crystal detector

The low-frequency, square-wave-modulated microwave signal can be detected by a tunable probe. It senses the voltage (or current) at any point on the standing waves created inside the slotted waveguide due to unmatched load impedance. When the position of the probe is moved along the waveguide slot, it gives an output that is proportional to the standing wave inside the waveguide. The Tunable probe is connected to the crystal detector. At the position of the probe, the detector gives an output that is proportional to the square of the input voltage. (Figure 11.7 (c)).

Matched termination

The waveguide transmission line operating at low average power can be terminated using Matched termination. The loads are carefully designed in such a way that, all the applied power is absorbed avoiding the reflected power. The matched termination is shown in Figure 11.8.

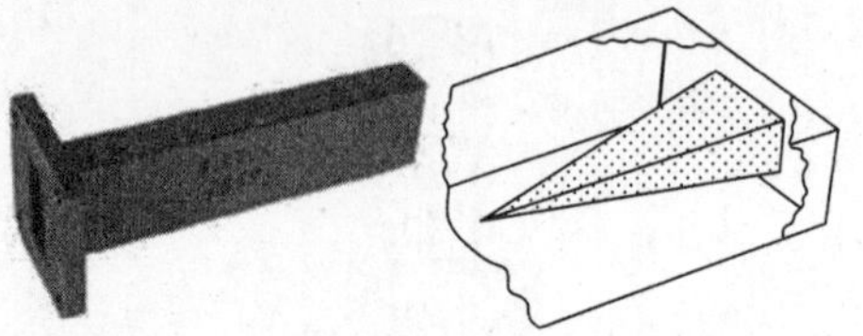

Figure 11.8 *Matched termination and its cross-section*

VSWR meter

A high-gain, high-Q, low-noise voltage amplifier that is tuned to a fixed frequency (1 KHz) is a VSWR meter. The VSWR meter in conjunction with a slotted waveguide line and detector carriage is used to the measure standing wave ratio. The scale on the VSWR meter is calibrated to read VSWR directly or its equivalent value in decibels. The VSWR scale is shown in Figure 11.9.

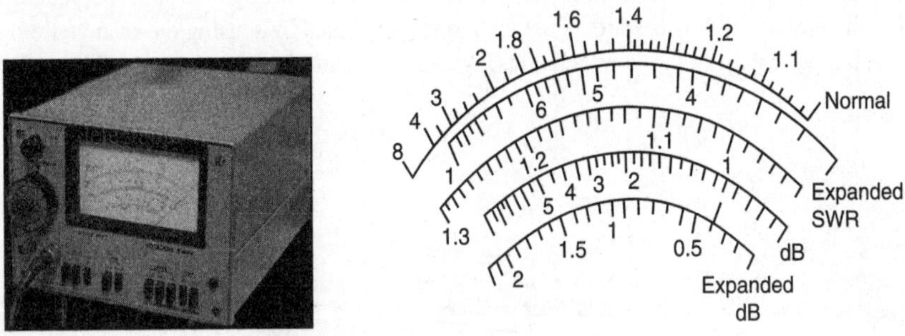

Figure 11.9 VSWR *meter and scale*

Power meter

The power meter is used to measure the amount of power in the microwave signals. It consists of a group of sensors such as a Schottky barrier diode, a bolometer, and the thermocouple that cover a range of frequencies. These sensors will convert microwave power into heat energy. This variation in temperature provides an output current in the low-frequency circuit, which indicates power. The power meter can be set to display power in mW or dBm. The power meter is shown in Figure 11.10.

Source: radiodan.com

Figure 11.10 *Power meter with matched load*

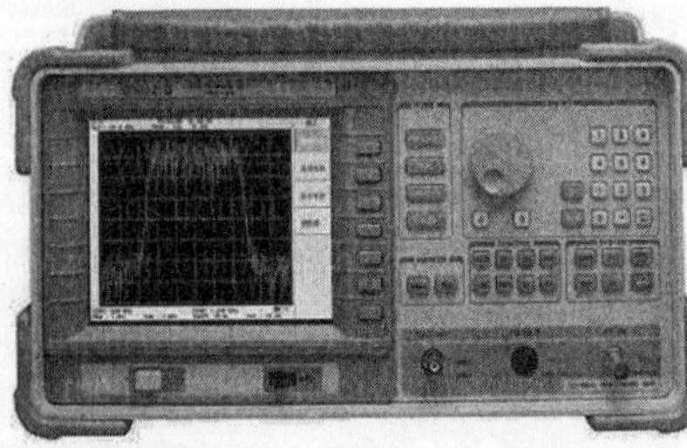

Source: spectrumanalyzerhq.com

Figure 11.11 *Spectrum analyzer*

Spectrum analyzer

A broadband super heterodyne receiver that can provide a plot of amplitude versus frequency of the received signal is the spectrum analyzer (Figure 11.11). It gives a frequency domain display of an input signal and allows the measurement of power of individual frequency components. This is especially useful when a signal contains components at several frequencies.

Network analyzer

The network analyzer is used to measure both amplitude and phase of a microwave signal over a wide range of frequencies. A network analyzer is shown in Figure 11.12.

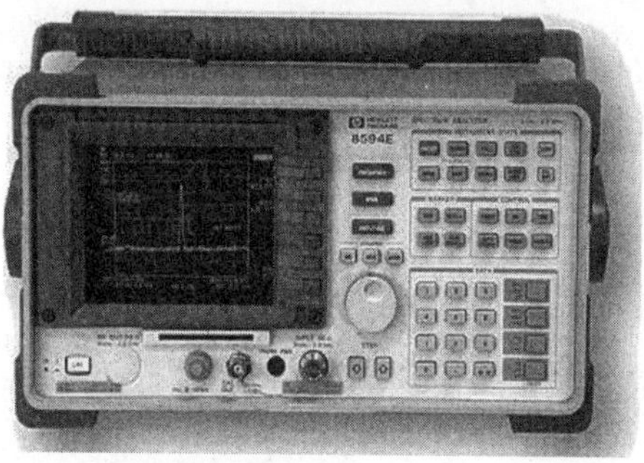

Source: teknetelectronics.com

Figure 11.12 *Network analyzer*

There are two types of network analyzers: (i) a vector network analyzer and (ii) a scalar network analyzer.

Vector network analyzer

The vector network analyzer can measure both magnitude and phase of a signal. The *S* parameters of a one-port or two-port network can be measured by a spectrum analyzer and these data can be converted to SWR, return loss, insertion loss, and phase.

Scalar network analyzer

The scalar network analyzer can measure only the magnitude of reflection or transmission.

11.2.2 Precautions

The following precautions should be taken while measuring the parameters at microwave frequencies.

- Connections among the components should be done tightly to avoid leakages.
- The source should be isolated from the load to prevent the damage of source due to reflected power.
- While using reflex klystron higher input, voltage should be avoided.
- Impedance matching has to be provided to avoid mismatching.
- The selected coaxial lines should be suitable for the desired frequency of operation and be of low loss.
- Never look into the open end of the waveguide that is connected to other equipment.
- The microwave power source should be turned off when assembling or dissembling components.

11.3 MICROWAVE POWER MEASUREMENT

Power at low frequencies is similar to the *Microwave power*; in the way that is, it is the product of rms voltage, rms current, and power factor.

$$P_{av} = I_{rms} V_{rms} \cos\phi \tag{11.1}$$

The power flow is the same at any point in the waveguide; that is, the microwave power inside a waveguide is invariant with the position of measurement, and the power measured is the average power. Depending on the power level, there are three different measuring techniques:

(a) Measurement of low power (0.10 mW–10 mW) –Bolometer technique
(b) Measurement of medium power (10 mW–1 W) – Calorimetric technique
(c) Measurement of high power (1 W–10 W) – Calorimetric watt meter

(a) Measurement of low power (Bolometer technique)

Devices that are capable of measuring low microwave powers are bolometers and thermocouples whose resistance changes with the applied power. Among these, bolometers are the most widely used. A bolometer is a device that converts RF power into heat, and this changes its resistance, from which power can be measured. There are two types of devices, namely barretters and thermistors, which are useful in the absolute measurement of power.

A barretter consists of a short length of very fine platinum wire that is suitably capsulated. These are positive temperature coefficient (PTC) components, and their resistance increases with an increase in temperature as shown in Figure 11.13 (a). Thermistors are semiconductor materials. These are negative temperature coefficient (NTC) components, and their resistance decreases with an increase in temperature as shown in Figure 11.13 (b).

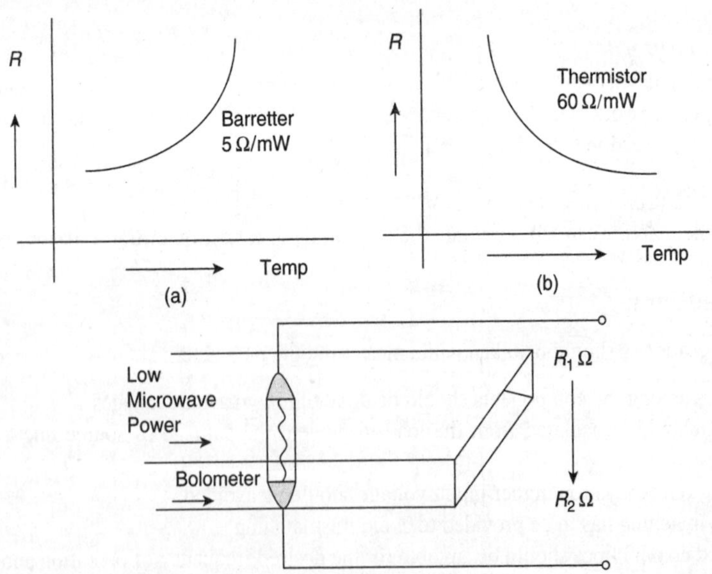

Figure 11.13 *(a) PTC of barretter; (b) NTC of thermistor; (c) Power measurement setup*

A bolometer is a square-law device, similar to a crystal diode. It produces a current that is proportional to the applied power, that is, the square of the applied voltage, rather than the applied voltage. A bolometer is placed inside the waveguide as shown in Figure 11.13 (c), where the bolometer itself acts as a load with R_1 Ω as operation resistance. Now, low microwave power is applied that is to be measured. Bolometer load absorbs some of the power applied and dissipates it as heat, and its resistance changes to R_2 Ω. This change in resistance (R_1~R_2) is proportional to the microwave power that can be measured using a bridge. Due to non-linear characteristics of the bolometer inaccuracy is introduced.

The bolometer itself becomes one of the arms of the bridge, in the balanced bolometer bridge technique and is as shown in Figure 11.14. At first, R_5 is adjusted so that the bridge is balanced. This changes the dc power applied to the bridge, and the bolometer element is brought to pre determined operating resistance before the microwave is applied. Let the battery voltage at balance be denoted by E_1.

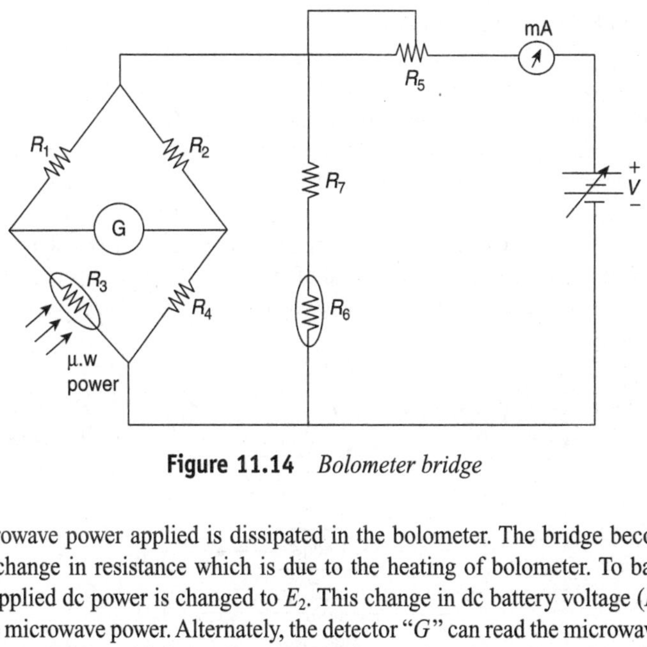

Figure 11.14 *Bolometer bridge*

The microwave power applied is dissipated in the bolometer. The bridge becomes unbalanced due to the change in resistance which is due to the heating of bolometer. To balance the bridge again, the applied dc power is changed to E_2. This change in dc battery voltage (E_1~E_2) is proportional to the microwave power. Alternately, the detector "*G*" can read the microwave power directly, by balancing the bridge. This balancing of the bridge can be achieved by calibrating the detector in terms of microwave power.

However some errors can occur which can be avoided by temperature compensation, since the bolometer is a temperature sensitive device. This can be achieved by using R_6 and R_7 resistors.

Limitations: Power-handling capability of barretters and thermistors are limited to about 10 mW. So, higher powers (> 10 mW) cannot be measured with them directly.

(b) Measurement of medium microwave power

The range of medium microwave power is from 10 mW to 1 W. The power in this range can be measured by calorimetric techniques. The principle of calorimetric technique is that a special load which is monitored, is proportional to the power responsible for the rise as shown in Figure 11.15.

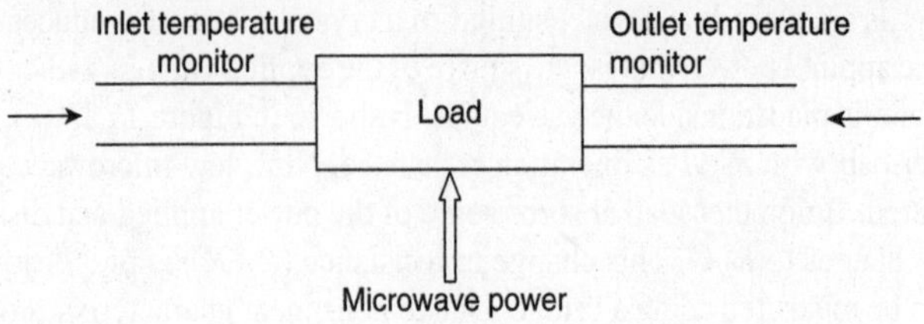

Figure 11.15 *Principle of medium power measurement*

The special load should have high specific heat. Generally water is a good load. The power can be measured by knowing mass, specific heat, and temperature rise at a fixed and known rate of fluid flow.

(c) Measurement of high microwave power

Power between 1 W and 10 W can be considered high microwave power. These are usually measured by calorimetric watt meters. In this method also, the temperature rise of a special load is monitored to measure the microwave power as shown in Figure 11.16.

A microwave calorimeter watt meter consists of two identical temperature-sensitive gauges, one in each arm, and an amplifier with high gain. The bridge is imbalanced as the input sensing resistor heats up. This heating up of the input sensing resistor is due the microwave power that is incident on the calorimeter. The circulating stream of either oil or ammonia gas is the dissipative load. By amplifying and applying the imbalance signal of the bridge to the comparison resistor and using the power delivered to it the bridge can be rebalanced. The watt meter keeps the record of the microwave power.

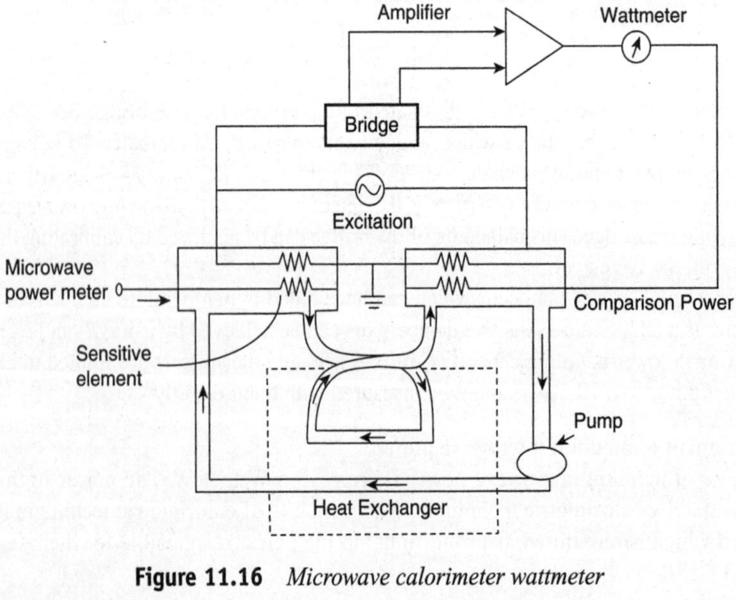

Figure 11.16 *Microwave calorimeter wattmeter*

11.4 MICROWAVE ATTENUATION MEASUREMENT

Attenuators are used to adjust the power level of the microwave signal. If the network is perfectly matched, the reflected power is zero and the insertion loss is similar to the attenuation provided by the microwave device or component. Attenuation is defined as the ratio of input power to the output power in dB.

$$\text{Attenuation in}(dB) = 10\log\left(\frac{P_1}{P_2}\right) \tag{11.2}$$

where P_1 = power detected by the load without the attenuator in the line
P_2 = power detected by the load with the attenuator in the line

Measurement of attenuation is done in two ways: (i) power ratio method; (ii) RF substitution method

11.4.1 Power Ratio Method

Power ratio method is a process of measuring the input and output power with the device (set up 1) as shown in Figure 11.17 (a) and without the device (or attenuator) in the circuit as shown in Figure 11.17 (b). The P_1 and P_2 are the powers measured in setup 1 and 2. The attenuation is the ratio of power (P_1/P_2) which is expressed in decibels.

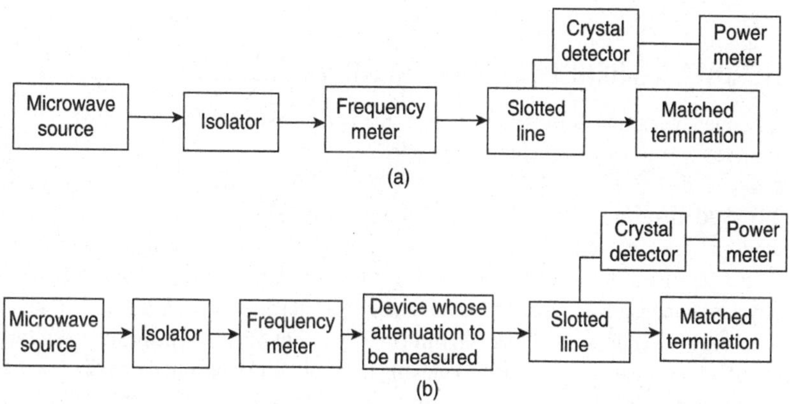

Figure 11.17 *(a) Setup 1 power ratio method without the device or attenuator; (b) Setup 2 power ratio method with the device or attenuator*

Disadvantage of the power ratio method

The attenuation calculated will not be accurate, particularly if it is large. Because the two powers measured (P_1 & P_2) is non-linear. It is also true for the networks with low input power. With this method, we can measure the attenuation up to 20 dB only.

11.4.2 RF Substitution Method

The attenuation through the device under test is compared with a standard microwave attenuator operating at the same frequency in this method. As shown in Figure 11.8 (a) the network whose attenuation is to be measured is included in the setup 1. Thus the output power "P" is measured by this method.

The drawbacks of the power ratio method can be overcome by this method as the attenuation is measured at a single power position.

A precision calibrated variable attenuator replaces the network in setup 2 as shown in Figure 11.18 (b). This attenuator is adjusted to get same output power "P," as in setup 1. Under this condition, attenuation of the device is to be measured in the precision attenuator.

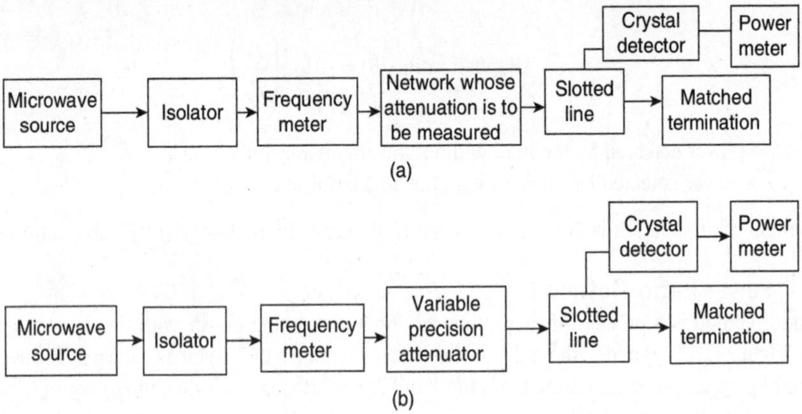

Figure 11.18 *(a) Setup 1 RF substitution method; (b) Setup 2 RF substitution method*

11.5 MICROWAVE FREQUENCY MEASUREMENTS

Microwave frequency can be measured by two methods: (i) slotted-line method; (ii) electronic technique.

11.5.1 Slotted-line Method (Mechanical Technique)

In this method, the measurement of wavelength in a waveguide will be made first and from that, frequency will be determined. A tunable resonator is required for this method, which has a known relation between a physical dimension and frequency, for example an absorption wave meter. The standing wave pattern appears only when the slotted line is terminated by a short circuit. The positions of two adjacent nulls are accurately positioned in two steps (i) moving the probe along a slotted line (ii) read the position of nulls in the vernier scale. The two positions are separated by half a guide wavelength $\lambda_g/2$.

The free space wavelength is given by $\lambda_0 = \dfrac{C}{f} \Rightarrow c = f\lambda_0 \Rightarrow f = \dfrac{C}{\lambda_0}$

The guided wavelength in the air-filled rectangular waveguide,

$$\lambda_g = \frac{\lambda_0}{\sqrt{1-(\lambda_0/\lambda_c)^2}} \tag{11.3}$$

and

$$\frac{1}{\lambda_0} = \sqrt{\left(\frac{1}{\lambda_g^2}+\frac{1}{\lambda_c^2}\right)} \tag{11.4}$$

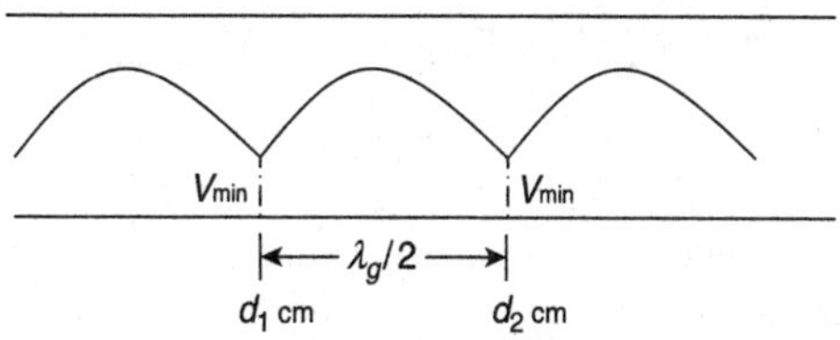

Figure 11.19 *Standing wave*

The distance between two successive voltage minima as shown in Figure 11.19 is given by

$$\lambda_g/2 = (d_2 - d_1).$$

The cutoff wavelength, $\lambda_c = 2a$ (for the dominant TE_{10} mode), where "a" is the broad dimension of the waveguide

Therefore,

$$f = \frac{c}{\lambda_0} = c\sqrt{\left(\frac{1}{\lambda_g^2} + \frac{1}{\lambda_c^2}\right)} \quad (11.5)$$

where c - speed of light in free space (3×10^8 m/s)

f – frequency, (Hz)
λ_0 – wavelength in free space, (m)
λ_g – wavelength of waveguide, (m)
λ_c – cutoff wavelength of waveguide (m)

In this method, the guided wavelength (λ_g) in a waveguide is measured by creating standing waves in a slotted-line section. The distance between a maxima and minima of the standing wave ($d_2 - d_1$) corresponds to ($\lambda_g/2$); hence, frequency can be determined from the measurement of (λ_g).

11.5.2 Electronic Technique

This method uses frequency heterodyne system. This system compares the unknown microwave frequency with a harmonic of the known standard frequency as shown in Figure 11.20. The unknown frequency f can be calculated as below from the output frequency f_0 and frequency nf_c.

$$f = nf_c - f_0 \quad (11.6)$$

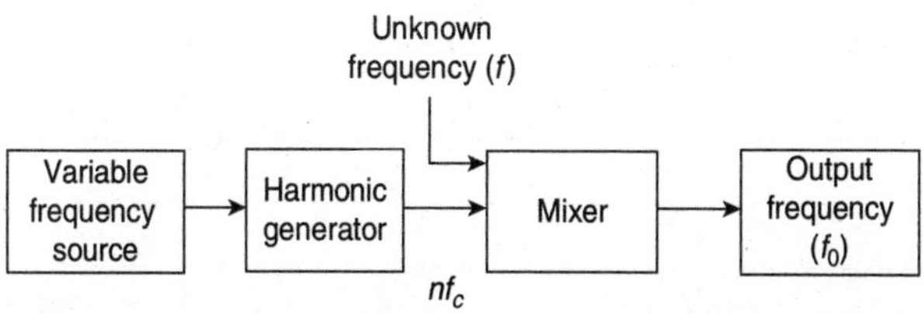

Figure 11.20 *Electronic technique for measuring frequency*

11.6 MICROWAVE VSWR MEASUREMENT

VSWR stands for voltage standing wave ratio. In a perfectly matched system, there is no variation in the field strength along the waveguide. A mismatch leads to reflected waves, thereby leading to standing waves along the length of the guide. Standing waves are the indication of the quality of the transmission. VSWR = 1 for a perfectly matched system. The ratio of the maximum to the minimum voltage gives the VSWR.

$$\text{VSWR} = \frac{V_{max}}{V_{min}} \tag{11.7}$$

$$\text{VSWR} = \frac{1+\Gamma}{1-\Gamma} \tag{11.8}$$

$$\Gamma = \frac{\text{VSWR-1}}{\text{VSWR+1}} \tag{11.9}$$

where Γ is the reflection coefficient.

A typical setup for VSWR measurement is shown in Figure 11.21.

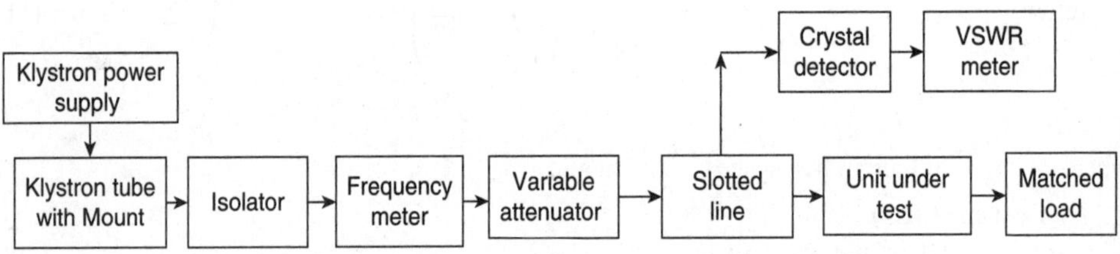

Figure 11.21 *Block diagram for measurement of VSWR*

11.6.1 Measurement of Low VSWR ($S < 10$)

VSWR values below 10 are very easily measured by this method. The VSWR meter directly displays these values. The setup is shown in Figure 11.21. The attenuator is adjusted to give a maximum reading on the meter. The attenuation has to be adjusted to get a full-scale reading. Then, the minimum reading on the meter is obtained by adjusting the probe on the slotted line. Thus the VSWR is defined by the ratio of the first reading to the second reading. The meter can be calibrated in terms of VSWR. Here, attenuator is adjusted so that the traveling probe gives a maximum deflection on the VSWR meter. The VSWR of 1 corresponds to full-scale deflection (i.e. 10 mV in the meter). By adjusting the traveling probe the minimum reading can be obtained on the meter. The variation in deflections and the corresponding VSWR values are tabulated in Table 11.1.

Table 11.1 VSWR for various deflections

Deflection in the meter	VSWR
5 mV	2
3.33 mV	3
2.5 mV	4
2 mV	5
1.69 mV	6

11.6.2 Measurement of High VSWR(S > 10)

VSWR greater than 10 can be measured by double minimum method. In this method, the probe is inserted to a depth where the minimum value can be read easily. Then the probe should be moved to a point where the power is twice the minimum ($P_{min} = 2V^2\text{min}/R_L$, i.e. $P_{min} = 2P$). Let d_1 be this position. Then again, the probe is moved to twice the power point on the other side of the minimum (say d_2) as shown in Figure 11.22.

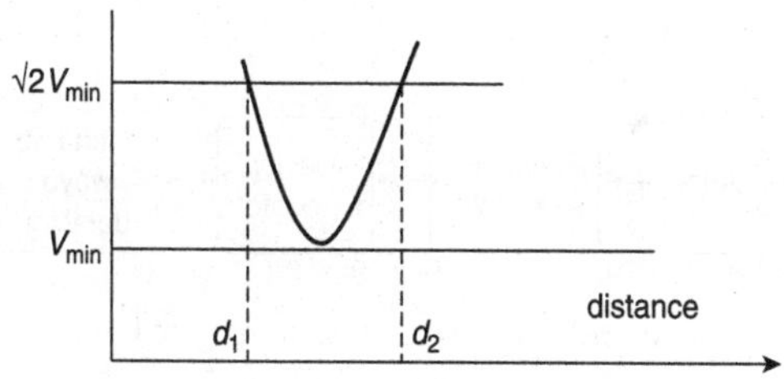

Figure 11.22 *Double minimum method*

For the dominant TE_{10} mode rectangular waveguide, λ_0, λ_g, and λ_c are related as below.

$$\frac{1}{\lambda_0^2} = \frac{1}{\lambda_g^2} + \frac{1}{\lambda_c^2} \tag{11.10}$$

where λ_0 is free space wave length
λ_g is guide wave length
λ_c is cutoff wave length

For the TE_{10} mode, $\lambda_c = 2a$ where "a" is the broad dimension of the waveguide

$$\lambda_0 = \frac{c}{f} \tag{11.11}$$

Where

$$\lambda_g = \frac{\lambda_0}{\sqrt{1-(\lambda_0/\lambda_c)^2}}$$

The VSWR can be calculated using the empirical relation

$$\text{VSWR} = \frac{\lambda_g}{\pi(d_2 - d_1)} \tag{11.12}$$

11.7 MEASUREMENT OF Q OF A CAVITY RESONATOR

A volume that is completely surrounded by a metallic surface is defined as a *cavity*. The important parameter of a cavity is quality factor. It is a measure of selectivity of the frequency resonant or anti-resonant circuit and is defined by the following equation:

$$Q = 2\pi \frac{\text{Maximum energy stored in resonant circuit}}{\text{Energy stored per cycle}} \tag{11.13}$$

The Q of a cavity resonator can be measured in three ways:

1. Transmission method

2. Impedance measurement
3. Transient delay or decrement method

In general, the transmission method is used among these three methods. The setup for the transmission method for measuring Q is shown in Figure 11.23.

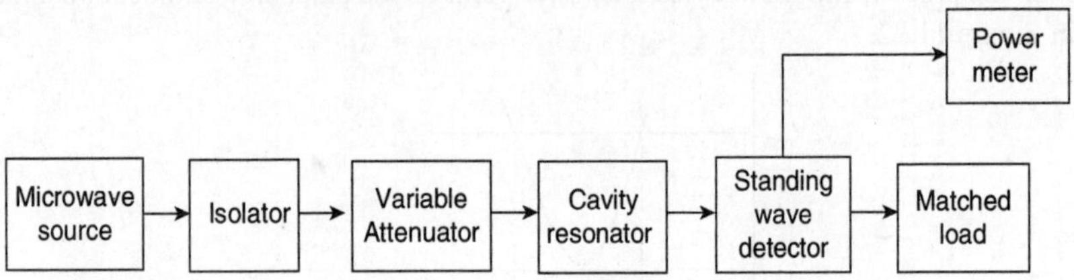

Figure 11.23 *Bench setup for measurement of Q by transmission method*

The cavity resonator is used as a transmission device, in this method, and the output power is measured as a function of the frequency resulting in the resonance curve (Figure 11.24).

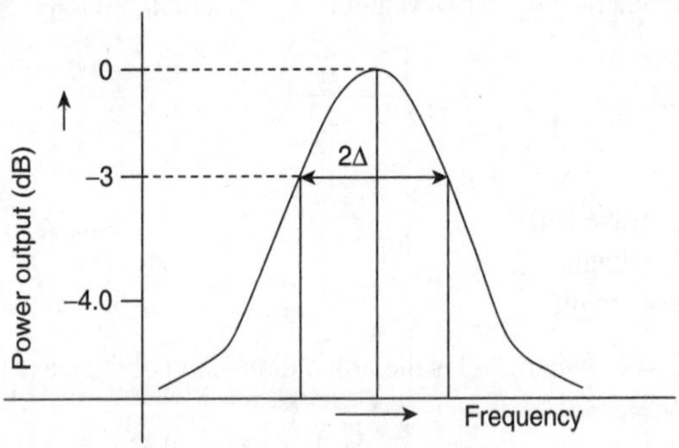

Figure 11.24 *Resonance curve*

The output power is to be measured by two methods

(i) Changing the frequency of the source of the microwave and by keeping the signal level constant.
(ii) Tuning the cavity and by keeping both signal level and frequency constant, the output power is measured. The half-power bandwidth can be calculated (2Δ) from the resonance curve in the Figure 11.24.

$$2\Delta = \pm\left(\frac{1}{Q_L}\right) \tag{11.14}$$

$$Q_L = \pm\left(\frac{1}{2\Delta}\right) = \pm\left(\frac{\omega}{2(\omega-\omega_0)}\right) \tag{11.15}$$

where Q_L = loaded value
Q_0 = unloaded Q

$Q_L = Q_0$ if the coupling between microwave source and cavity and that between detector and cavity are neglected.

11.8 IMPEDANCE MEASUREMENT

Impedance (Z) is defined as the ability to oppose the flow of an alternating current (AC) at a given frequency. It is represented as a complex quantity and is graphically shown on a vector plane. An impedance vector consists of a real part (resistance, R) and an imaginary part (reactance, X) as shown in Figure 11.25.

$$Z = R + jX = |Z|\angle\theta$$

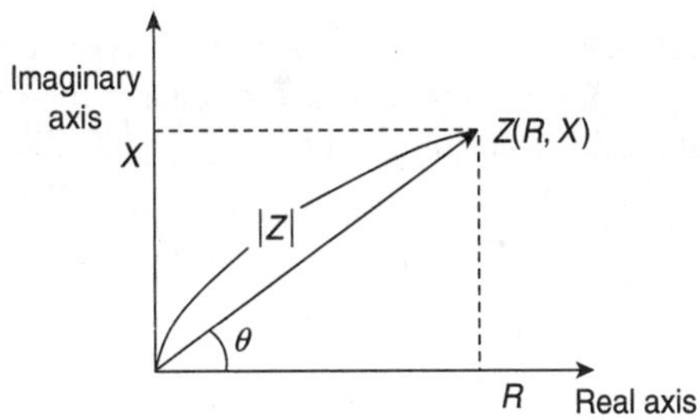

Figure 11.25 *Impedance (Z) consists of a real part (R) and an imaginary part (X)*

There are many methods for measuring impedance, each of which has advantages and disadvantages. The most commonly used method is the measurement of impedance using the slotted line.

11.8.1 Measurement of Impedance Using a Slotted Line

The end of the slotted line is connected to the unknown impedance as shown in Figure 11.26. From the other end of the coaxial line microwave power is fed. A part of the power is reflected by the unknown impedance. By evaluating the standing wave fields in the slotted line this reflection coefficient is measured. The reflection coefficient Γ, is given by

$$\Gamma = \frac{Z_L - Z_0}{Z_L + Z_0} \tag{11.16}$$

Where, the unknown impedance terminating a line of characteristic impedance Z_0 is Z_L. Thus, Z_L can be determined, if Γ is measured and Z_0 is known. As Z_L is complex, from the VSWR measurement both the magnitude and phase of Γ can be determined.

$$\Gamma = \frac{\text{VSWR} - 1}{\text{VSWR} + 1} \tag{11.17}$$

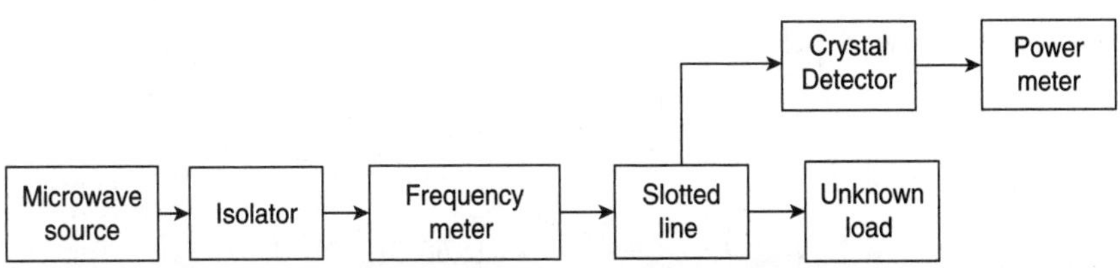

Figure 11.26 *Impedance measurement using slotted line*

VSWR and d_{min} (difference of two successive minima) are noted by moving the probe and observing the output on a VSWR meter. Knowing the VSWR and d_{min}, a graphical construction on Smith chart may be used to calculate the unknown impedance Z_L.

SOLVED PROBLEMS

EXAMPLE PROBLEM 11.1

Calculate the SWR of a transmission system operating at 8 GHz. Assume TE_{10} wave transmission inside a waveguide of dimensions $a = 3.5$ cm, $b = 2.1$ cm. The distance measured between twice minimum power points (successive minima) is 1 mm on a slotted line.

Solution

Given $f = 8$ GHz, $a = 3.5$ cm, and $b = 2.1$ cm. For TE_{10} mode, $\lambda_c = 2a = 2 \times 3.5 = 7$ cm

$$\lambda_0 = \frac{c}{f} = \frac{3\times10^{10}}{8\times10^{9}} = 3.75 \text{ cm}$$

Given that $d_2 - d_1 = 1$ mm

We know

$$\lambda_g = \frac{\lambda_0}{\sqrt{1-\left(\frac{\lambda_0}{\lambda_c}\right)^2}} = \frac{3.75}{\sqrt{1-\left(\frac{3.75}{7}\right)^2}} = 4.441 \text{ cm}$$

For the double minimum method, VSWR is given by

$$\text{VSWR} = \frac{\lambda_g}{\pi(d_2 - d_1)} = \frac{4.441}{\pi\times1\times10^{-1}} = 14.136$$

■

EXAMPLE PROBLEM 11.2

If the width of the waveguide is 3.5 cm and the distance between two successive minima is 1 cm of a standing wave pattern formed within the waveguide, calculate the frequency of the wave.

Solution

Given $a = 3.5$ cm, $d_2 - d_1 = 1$ cm

$\lambda_c = 2 \times 3.5 = 7$ cm; λ_c is cutoff wavelength

Guide wavelength, $\lambda_g = 2(d_2 - d_1) = 2 \times 1 = 2$ cm

We know that free space wavelength, $\lambda_0 = 1.923$ cm

$$\frac{1}{\lambda_0^2} = \frac{1}{\lambda_g^2} + \frac{1}{\lambda_c^2} = \frac{1}{2^2} + \frac{1}{7^2} = \frac{53}{196}$$

$$f = \frac{C}{\lambda_0} = \frac{3\times10^{10}}{1.923} = 15.60 \text{ GHz}$$

■

EXAMPLE PROBLEM 11.3

Calculate the VSWR of a transmission system operating at 15 GHz. The TE_{10} mode is propagating through the waveguide of dimensions 4.0 and 2.1 cm, respectively. The distance between two successive minima is 1.5 mm.

Solution

Given $f = 15$ GHz, $a = 4$ cm, $b = 2.1$ cm; $d_2 - d_1 = 1.5$ mm $= 0.15$ cm

Free space wavelength,
$$\lambda_0 = \frac{C}{f} = \frac{3\times 10^{10}}{15\times 10^9} = 2 \text{ cm}$$

Cutoff wavelength,
$$\lambda_c = 2a = 8 \text{ cm}$$

Guide wavelength,
$$\lambda_g = \frac{\lambda_0}{\sqrt{1-\left(\frac{\lambda_0}{\lambda_c}\right)^2}} = \frac{2.0}{\sqrt{1-\left(\frac{2.0}{8.0}\right)^2}} = 2.065 \text{ cm}$$

$$\text{VSWR} = \frac{\lambda_g}{\pi(d_2 - d_1)} = \frac{2.065}{\pi\times 0.15} = 4.382$$

EXAMPLE PROBLEM 11.4

The VSWR value on a waveguide can be determined by using the double minimum method. If 5 cm is the separation between two adjacent nulls and that between twice minimum power points is 3 mm, find the value of VSWR.

Solution

Given that $d_2 - d_1 = 3$ mm $= 0.3$ cm and $\lambda_g = 5$ cm

$$\text{VSWR} = \frac{\lambda_g}{\pi(d_2 - d_1)} = \frac{5}{\pi\times 0.3} = 5.308$$

EXAMPLE PROBLEM 11.5

The calibrated power from a generator as read at the power meter is 20 mW. When a 3 dB attenuator with a VSWR of 1.3/1 is inserted between the generator and the detector, what value should the power meter read?

Solution

Given $P_1 = 20$ mW, attenuation $\alpha = 3$ dB, and $\text{VSWR} = \frac{1.3}{1}$, $P_2 = ?$

Attenuation
$$\alpha = 10 \log\left(\frac{P_1}{P_2}\right)$$

$$3 = 10 \log\left(\frac{20}{P_2}\right) \Rightarrow \frac{20}{P_2} = 10^{0.3}$$

$$P_2 = 10.02 \text{ mW}$$

EXAMPLE PROBLEM 11.6

Calculate the VSWR of a rectangular guide of 2.5 cm × 1.0 cm operating at 10 GHz. The distance between twice minimum power points is 1mm.

Solution

Given $a = 2.5$ cm, $b = 1.0$ cm, $f = 10$ GHz, and $d_2 - d_1 = 0.1$ cm, VSWR = ?

$\lambda_c = 2a = 5$ cm;

$$\lambda_0 = \frac{C}{f} = \frac{3\times10^{10}}{10\times10^9} = 3 \text{ cm}$$

$$\lambda_g = \frac{\lambda_0}{\sqrt{1-\left(\frac{\lambda_0}{\lambda_c}\right)^2}} = \frac{3}{\sqrt{1-\left(\frac{3}{5}\right)^2}} = 3.75 \text{ cm}$$

$$\text{VSWR} = \frac{\lambda_g}{\pi(d_2-d_1)} = \frac{3.75}{\pi\times0.1} = 11.936$$

SUMMARY

1. The microwave Bench is operated in the X-band (8–12 GHz). The microwave signal source could be a reflex klystron or a Gunn diode oscillator or a backward wave oscillator.
2. An isolator is used to protect the microwave source from reflected power.
3. The slotted line is basically used for measuring the standing wave ratio.
4. The spectrum analyzer gives a frequency domain display of an input signal.
5. The network analyzer is used to measure both amplitude and phase of a microwave signal over a wide range of frequencies.
6. Bolometers and thermocouples are used to measure microwave power.
7. Attenuation is defined as the ratio of the powers measured at input and output.
8. Two types of methods are commonly used to measure the frequency:
   (a) Slotted-line method
   (b) Electronic techniques
9. The voltage standing wave ratio is defined as the ratio between the maximum and minimum field strength along the line. VSWR is denoted by S; if $S < 10$, it is known as low VSWR, and if $S > 10$, it is known as high VSWR.
10. The Q factor is a measure of selectivity of frequency resonant or anti-resonant circuit that can be measured by the transmission method.

OBJECTIVE-TYPE QUESTIONS

1. The microwave bench in the lab is designed to work in the band ________

(a) X band (b) L band
(c) S band (d) C band

2. The mode used in the laboratory bench is ________

(a) dominant mode (b) degenerate mode
(c) any one of the above two (d) none

3. The cutoff wave length of the bench is ________

(a) $2a$ (b) $2b$
(c) cf (d) none

4. The method used to measure high VSWR is ________

(a) slotted-line method (b) Double minimum method
(c) both (d) none

5. Low VSWR method can be used to measure VSWR up to ________

(a) ten (b) five
(c) three (d) none

6. The range of VSWR in dB indicated in the VSWR meter is ________

(a) 0–2 dB (b) 0–10 dB
(c) 0–5 dB (d) 0–100 dB

7. The temperature coefficient of the Thermistor is ________

(a) positive (b) negative
(c) zero (d) none

8. Barretters have ________

(a) positive temp efficient of resistance (b) negative temp coefficient of resistance
(c) both (d) none

9. Impedance of the line to the left of voltage minimum is ________

(a) inductive (b) capacitive
(c) high resistance (d) low resistance

10. Impedance of the line to the right of voltage minimum is ________

(a) inductive (b) capacitive
(c) high resistance (c) low resistance

11. The bolometer that has a negative temperature coefficient of resistivity is called

(a) barretter (b) varistor
(c) thermistors (d) calorimeter

12. For impedance measurement, the following oscillator is used:

(a) Reflex klystron tube oscillator (b) Gunn oscillator
(c) *a* and *b* (d) *a* or *b*

ANSWERS TO OBJECTIVE-TYPE QUESTIONS

1. (a) 2. (a) 3. (a) 4. (b) 5. (a) 6. (b) 7. (b) 8. (a) 9. (a) 10. (b) 11. (c) 12. (d)

REVIEW QUESTIONS

1. Draw a block diagram of the microwave bench setup and explain each block.
2. What is VSWR? Discuss the measurement of low and high VSWR.
3. Discuss methods for the measurement of low and high microwave power.
4. How do you measure microwave frequency?
5. Explain how the loaded Q of a cavity resonator can be measured.
6. Describe briefly the equipment that is used to measurc impedance using a slotted line.
7. What is a bolometer? Differentiate between a barretter and a thermistor.
8. What is a tunable detector?
9. What is a slotted section? Give the main purpose of a slotted section with a line carriage.
10. What is a VSWR meter? How will you determine the VSWR?

Introduction to Radars

12

12.1 INTRODUCTION

The word *RADAR* is an acronym from the words **Ra**dio **D**etection **A**nd **R**anging; that is, finding and positioning a target and determining the distance (range) between the target and the radar by using radio frequency. The basic radar equipment consists of a transmitter, a receiver, a duplexer, and an antenna. The basic principle behind the radar is simple: The transmitter sends out a very short duration pulse at a high power level. The pulse strikes an object (or a target), and energy will be reflected (known as *radar returns* or *echoes*) back to the radar receiver. The radar determines the distance (range) to the target by measuring the travel time of the radar pulse to the target and to come back from the target, and then divides that time by two.

For extracting the target information from the echo, the signal should be of sufficient magnitude. The radar equation is used to predict the echo power to assist in making the determination of whether or not the condition just mentioned is met. The radar receiver processes these echoes to determine the target information. These information can be a range, velocity, angular position, etc. The radar functions can be performed at any distance long or short and in any visibility levels. Radar can measure distances with high accuracy in all weather conditions.

This chapter describes the classification of radars, radar range equation, radar block diagram, pulse radar characteristics, radar cross-section of targets, and so on.

12.2 HISTORY OF RADARS, FREQUENCIES, AND APPLICATIONS OF RADARS

We can say that RADAR becomes 100 years old in 1988. It was Henrich Hertz, who with his underlying experiments in 1887/1888 was able to demonstrate that EM waves are reflected by metals and dielectric objects, which is the main principle that was used in radars. Doppler radar is named after Christian Andreas Doppler, an Austrian physicist. He first described in 1842, about his observation regarding the affect of the relative motion of the source and the detector on the frequency of light and sound waves. This phenomenon is called the *Doppler effect*. The earliest use of electromagnetic signals to detect targets was demonstrated in 1904 by Christian Hulsmeyer using a spark gap generator. Research contributed to further developments, with a significant acceleration during World War II. *Radar* is now a word in its own right, even though in 1941 the term *RADAR* was created.

Targets are the objects of interest such as an aircraft and *clutters are the* objects that can be confused with targets, such as reflections from the ground, buildings, or the sea. Radar designers are trying hard to filter out clutter returns from target returns. Targets on the land, on the sea, in the air, and outside the earth's atmosphere can be detected. The examples of such targets are aircraft, land vehicles, ships, and ballistic missiles. In a radar system, typically a high-gain antenna such as a parabolic antenna is used to transmit a radar signal, but always a high-gain antenna is used to receive the signal.

The radar equipment had been modified and improved progressively to meet the present-day standards. The following Table 12.1 gives a summary of the historic development of radars.

Table 12.1 The historic development of RADAR is summarized as below:

1860	Michael Faraday discovered electric and magnetic fields.
1864	James Clark Maxwell determined mathematical equations of electromagnetism.
1886	Heinrich Hertz tested the theories of Maxwell and the similarity between radio and light waves.
1888	Heinrich Hertz discovered EM waves laid down by Maxwell. He also proved that metallic and dielectric bodies reflect radio waves.
1903	Detection of reflected radio waves was experimented by Hulsmeyer.
1904	The detection of objects by radio's first patent was obtained by Christian Hulsmeyer.
1922	Albert Hoyt Taylor and Leo C. Young detected a wooden ship by using a CW radar.
1925	G. Breit and M. Truve used the first application of the pulse technique to measure distance.
1940	For long-range detection, microwaves were used.
1947	On February 14 in Washington D.C. the first weather radar was installed.
1950	First used for the detection and tracking of weather conditions such as thunderstorms and cyclones.
1990	With the Doppler radar a remarkable advance to radars came into existence.

Radar Frequencies

In the microwave region of the EM spectrum radar is operated, and the energy is emitted in the form of an EM wave into the atmosphere through an antenna. A great deal of information is provided with only a fraction of the energy returned. This entire process occurred at the speed of light. When the targets are struck by EM energy, the signals return from these targets which are called *radar echoes*. Figure 12.1 shows the different types of EM waves as a function of frequency, from EM telegraphy to gamma rays. Although all of them are EM waves, some of their characteristics are very different depending on their frequency. Radars operate in the range of 3 MHz to 300 GHz, though the large majority operate between about 300 MHz and 35 GHz.

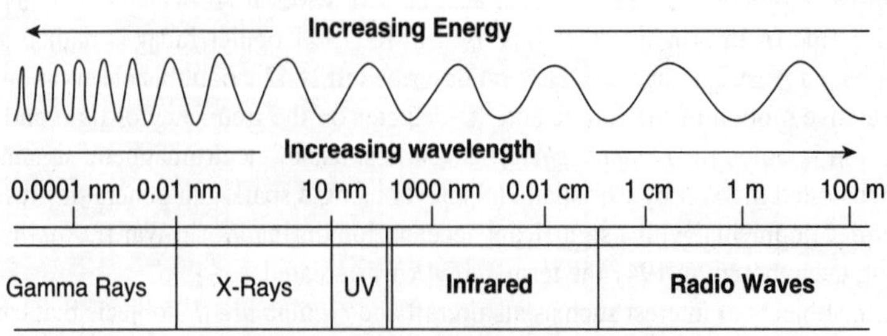

Figure 12.1 *Electromagnetic spectrum*

The International Telecommunication Union (ITU) assigned specific frequencies within the above ranges for radars are illustrated in Table 12.2. Various radar applications, their operating frequencies, and the corresponding band designations are given in Table 12.2.

Table 12.2 Band designation, nominal frequency, wavelengths, and various types of radars

Band designation		Nominal frequency	Nominal wavelength	Application	Types of radars
Radar bands	HF	3–30 MHz	100–10 m	Over-the-horizon radar, oceanographic mapping	Search Radars
	VHF	30–300 MHz	10–1 m	Oceanographic mapping, atmospheric monitoring, and long-range search	
	UHF	300–1000 MHz	1–0.3 m	Long-range surveillance, foliage penetration, ground penetration, and atmospheric monitoring	
	L	1–2 GHz	30–15 cm	Satellite imagery, mapping, long-range surveillance, and environmental monitoring	Search and Track Radars
	S	2–4 GHz	15–8 cm	Weather radar, air traffic control, surveillance, search, and IFF (identify, friend, or foe)	
	C	4–8 GHz	8–4 cm	Hydrologic radar, topography, fire control, and weather	
	X	8–12 GHz	4–2.5 cm	Cloud radar, air-to-air missile seeker, maritime, air turbulence, police radar, high-resolution imaging, and perimeter surveillance	Fire Control and Imaging Radars
	Ku	12–18 GHz	2.5–1.7 cm	Remote sensing, short-range fire control, perimeter surveillance, and pronounced "kay-you"	
	K	18–27 GHz	1.7–1.2 cm	Radar and satellite communications. Astronomical observations.	
	Ka	27–40 GHz	1.2–0.75 cm	Police radar, weapon guidance, remote sensing, perimeter surveillance, and pronounced "kay-a"	Missile Seekers
	V	40–75 GHz	0.75–0.40 cm	Perimeter surveillance, remote sensing, and weapon guidance	
	W	75–110 GHz	0.40–0.27 cm	Perimeter surveillance, remote sensing, and weapon guidance	
	Mm	110–300 GHz	0.27–0.1 cm	Radio astronomy, remote sensing. Wireless HD.	

Applications of Radars

By using RADAR technology, we can satisfy different applications such as search, detect, track, imaging, and remote sensing. Radars are employed everywhere on the ground, in the air, on the sea, and in space. Radars can be categorized based on the application, into

(i) Ground-based radars – For the detection, location, and tracking of aircrafts or spacecraft.
(ii) Ship-board radars – As navigational aid and safety devices to locate buoys, shore-lines, and other ships, etc.
(iii) Air-borne radars – For the detection of other aircrafts, ships, land mapping, weather indication, etc.
(iv) In space – For the guidance of spacecraft and remote sensing.

The radars are mainly used in the military, and also have important civilian applications especially for air and marine navigation. The following sections briefly describe the major radar applications.

Air traffic control: High resolution radars are used for safe controlling of the air traffic and ground vehicular traffic at large airports throughout the world.

Maritime navigation: In ship, radars are used to warn about the potential collision with other ships and for detecting navigation buoys, especially in poor visibility. Moderately high resolution shore-based radars are used for the surveillance of harbors as an aid to navigation.

Aircraft navigation: Weather forecasting is very essential in aircraft navigation. Pulse Doppler radar used for this purpose. A classic example of the above application is the weather avoidance radar installed on the nose of aircrafts which is used to outline the regions of precipitation to the pilot. For terrain avoidance and the terrain following, radars are used. Another application of radar is *radio altimeter* that measures the aircraft's height above the ground.

Military applications: In military, we have to identify and track the enemy aircraft, which can be done by using search and tracking radars and also by using over-horizon search radars. By using radars we can detect the missile activity at a very long range and we can guide and control the weapons for attacking enemy aircrafts.

Meteorological applications: Radars are used by the meteorological department to forecast approaching storms and to issue timely warning, thus avoiding the loss of life and property. There is a network of weather radars operating in our country, especially along the eastern coast.

Process control: Very short-range radars can be used to measure the fluid levels in enclosed tanks very accurately and to determine the dryness of products in a manufacturing process to provide feedback to the process controller.

Space applications: The advantage of Space-based radar systems is an unobstructed overhead view of the earth and objects on the earth surface. For rendezvous and docking of Space vehicles radars can be used. For the detection and tracking of satellites some of the largest ground-based radars are used. For remote sensing of earth resources, mapping of sea conditions, water resources, ice cover, agricultural and forest conditions, geological formations, and environmental pollution satellite borne radars can be used.

Law enforcement applications: Radars are widely used to measure the speed of automobile traffic by police on highways, thereby aiding them in enforcing road traffic regulations.

12.3 CLASSIFICATION OF RADARS

Radar systems can be classified into four categories as follows:

- Based on role of the targets during detection process
- Based on how transmitting antenna and receiving antenna are employed

- Based on waveforms used
- Based on services provided

We shall discuss each of these classifications in the following sections:

12.3.1 Classification of Radars Based on Role of Targets During Detection Process

Based on the role of targets during the detection, the radars are classified into two groups, namely,

- Primary radar
- Secondary radar

Primary radars

In primary radars, the cooperation of the target is not required to find the range, position, and relative velocity of the target. In other words, the role of the target is said to be passive and is limited only to reflect the radar signals back to the radar. Most of the radars used for the air traffic control, such as the approach surveillance radar (ASR) or airport surveillance radar, air-route surveillance radar (ARSR), and precision approach radar (PAR), belong to the group of primary radars.

Transmitter characteristics

The modern air traffic control (ATC) primary radars operate in the *S*-band frequency (2700 to 2900 MHz). Older systems tended to use vast amounts of power (several Megawatts), and employed vacuum tube devices such as magnetrons, klystrons, or traveling-wave tubes. More modern systems benefit from powerful signal processing techniques, and tend to require far less power than older types. In some new installations, solid-state transmitters are commonly used.

Advantages:

- It works independently, that is, the active cooperation of the target is not required.
- It engages several targets simultaneously and is not likely to get saturated.
- The electronic system is comparatively simpler, and requires only one set of transmitter and receiver.

Disadvantages:

- The efficiency of a primary radar is poor, because the echo signals depend on the target size, shape, material, and so on.
- The transmitter power has to be high, because the same energy has to return after getting reflected from the target.
- The receiver has to be highly sensitive, because the strength of echoes may be very weak.
- The critical alignment of the transmitter and receiver frequency is very essential.
- The selective response of targets is not possible.
- The echoes from fixed targets will cause disturbance in detecting moving targets.

Secondary radars

Here, the active cooperation of targets is very much required for finding the range and other details of the targets. Hence, the role of the targets is said to be active. The secondary radar system basically consists of two principal components, namely the interrogator, which is ground based and the transponder, which is carried out on the targets (aircrafts). Each of these components consists of a set of pulse transmitter and receiver. The interrogator radiates pulses, which when received by a corresponding transponder on a target, a reply will be initiated from that transponder. These replies are then collected by the interrogator to extract information about the targets. Examples of a secondary radar used in airports are as follows: Monopulse Secondary Surveillance Radar (MSSR).

The ground-based interrogator transmits a coded signal to a transponder on the aircraft, which then replies with a second code. The transmission from the interrogator to the transponder (known as the *uplink*) is on a frequency of 1030 MHz, and the replies (downlink) are on a frequency of 1090 MHz. Ground-based SSR antennas are often mounted directly on top of primary radar antennas (Figure 12.2).

Advantages:

- Considerable range increase is possible as the radar transmission has to travel the distance between the target and the radar only once.
- It allows low powers to be used to get a given performance.
- The echo is no longer dependent on the target size, material, and so on.
- Since there is a frequency difference between the transponder and the interrogator, received signals are totally free from permanent target echoes.
- By suitable coding, some useful information can be conveyed from the target to the ground station.

Disadvantages:

- It can be used for friendly targets only.
- The system operation depends on the equipment of the target remaining serviceable.
- All secondary radars are liable to bc saturated.

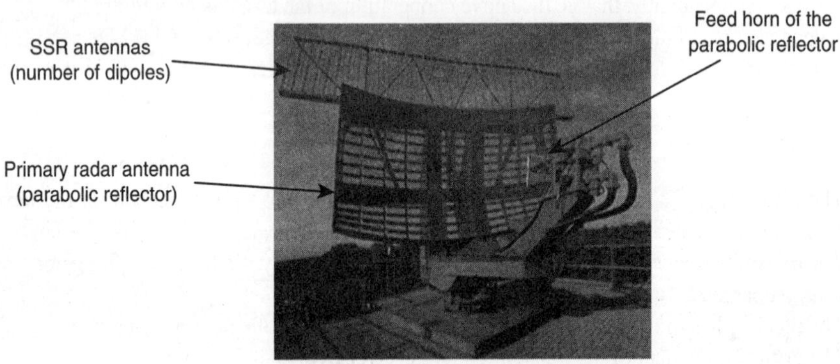

Source: upload.wikimedia.org

Figure 12.2 *Primary and secondary surveillance radar antennas*

12.3.2 Classification of Radars Based on How Transmitting and Receiving Antennas are Employed

If the same antenna is used to transmit and receive, the system is called a *monostatic radar* (see Figure 12.3 (a)). A radar with transmit and receive antennas at different sites is called a *bistatic radar* (shown in Figure 12.3 (b)).

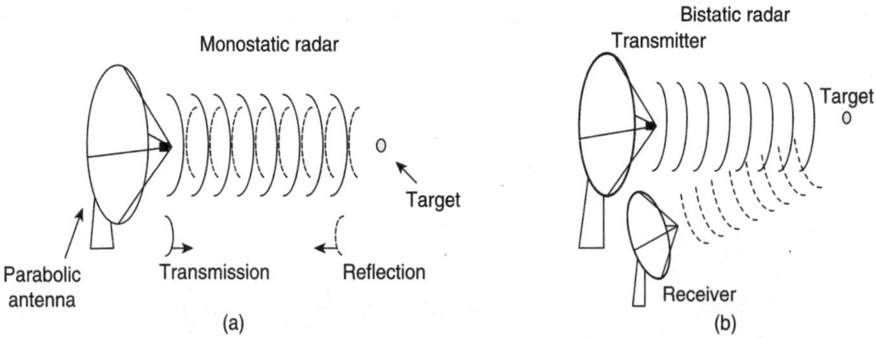

Figure 12.3 *(a) Monostatic radar; (b) Bistatic radar*

12.3.3 Classification of Radars Based on Waveforms Used

Radars can be classified into two groups depending upon the type of waveforms radiated.

- **CW radars** are of two types:
  – *CW radar (Doppler radar)* - uses unmodulated waveforms for radar transmission and
  – *FMCW radar* - uses frequency-modulated waveforms for radar transmission
- **Pulsed radar** uses pulse modulated waveforms for radar transmission.

Each type of waveform enjoys distinct advantages and has specific usage.

CW radars: The simplest type of radar is the CW radar, which uses a continuous waveform for its transmission (Figure 12.4 (a)). These radars are capable of detecting moving targets as well as their radial velocities. Their main disadvantages are their inability to find the range of the targets.

CW-FM radars: This radar is an improved version of the CW radar, and it can measure the range of targets by the use of frequency-modulated signals (Figure 12.4 (b)). They are usually used in radio altimeters for finding the height of an aircraft.

Pulsed radars: The most widely used radar is the pulsed radar, which uses pulse-modulated microwave signals for their transmission (Figure 12.4 (c)). The pulsed radars can be designed either for moving target detection or for radial velocity measurements.

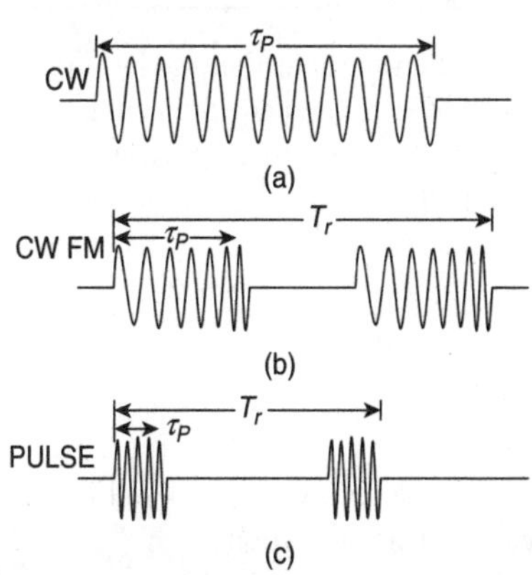

Figure 12.4 *Radar waveforms: (a) Continuous wave; (b) Frequency-modulated continuous wave; (c) Pulsed wave*

A detailed classification of radars based on waveforms radiated is given below (Figure 12.4 (d)).

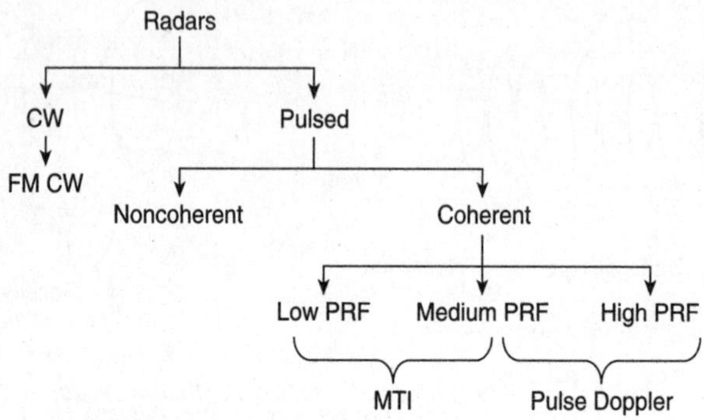

Figure 12.4 (d) *Radar classification*

12.3.4 Classification of Radars Based on Services Provided

A radar system is generally required to perform one of the two tasks. It should either search for targets or track them once they have been acquired. Accordingly, radars have been classified as

- SEARCH radar
- TRACK radar

Search radars: Also known as the SURVEILLANCE radars, these radars should acquire targets in a large volume of space regardless of whether their presence is known or not. These radars usually use an antenna system that is capable of continuous rotation. The most common application of these radars is for Air Traffic Control.

Tracking radars: Once the target has been acquired, it may then be tracked. Tracking radars usually locate the targets very accurately, perhaps in order to bring weapons to point toward it. Tracking radars can continuously give the angular position, the range, and the radial velocity of targets with precision. If radars are purely used for tracking, SEARCH radar should also be co-located.

12.4 BASIC RADARS

An electromagnetic system that is used for the detection and location of reflecting objects like vehicles, ships, aircraft, spacecraft, people and other natural environment is *Radar*.

Basically, pulses at a high power level are sent out by radar, where these pulses travel through the air or space until they meet an obstacle in their path. Some of the energy is reflected from the object when the pulse hits the object, while some energy is absorbed by the object and some is transmitted through the object. The amount of the energy reflected is dependent on the following parameters:

(i) The wavelength of the radar signal.
(ii) Composition of the object.
(iii) Size and shape of the object.

The amount of reflected energy to the radar is very small and is detected using a specially designed receiver as the reflected energy can go in any direction.

Basic Principle of Radars

The Figure 12.5 shows the basic principle of radar. An electromagnetic signal that is radiated into space by an antenna is generated by a transmitter. The target intercepts a part of the transmitted energy and reradiates it in many directions. The reradiated signal is collected by the radar antenna and sends it to a receiver. To locate the target's presence and to find its location it is processed further in the receiver. By measuring the time taken by the radar signal to travel to the target and back to the radar, the range or distance of a target is found. A directional antenna has a narrow beamwidth. Hence, it is used to find the target's location in angle to sense the angle of arrival of the received echo signal. If the target is in motion, due to the Doppler Effect there will be a shift in the frequency of the echo signal. In radars on the basis of differentiating the targeted moving object from fixed clutter echoes (unwanted) which are reflected from the natural environment such as land, sea, or rain, the Doppler frequency shift is broadly used. Nature of the target can also be provided by radars.

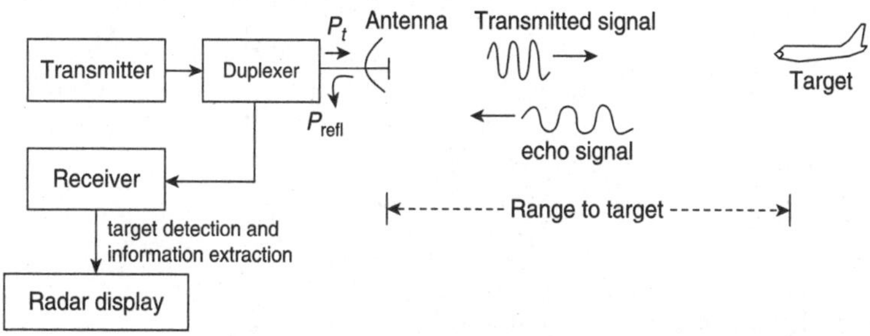

Figure 12.5 *Basic principle of radar*

12.4.1 Radar Range Equation

The transmitted power, the received power, the characteristics of the target and the radar itself is related by the radar equation. The equation is also helpful to assess the performance of the radar. The maximum radar range R_{max} is given by

$$R_{max} = \left[\frac{P_t G^2 \lambda^2 \sigma}{(4\pi)^3 S_{min}} \right]^{1/4} \tag{12.1}$$

where R_{max} is the maximum detection range between the radar and the target. It has the units of meters (m).

P_t is termed the *peak transmit power* and is the average power when the radar is transmitting a signal. It has the units of Watts.

λ is the radar wavelength. It has the units of meters (m).

G is the gain of the transmitter and receiver antenna.

σ is the target radar cross-section or RCS and has the units of square meters or m^2.

S_{min} is the minimum received power that the radar receiver can sense and is referred to as *Minimum Detectable Signal* (MDS). It has the units of Watts.

Derivation of radar range equation

The derivation of the radar equation involves the following three steps for an isolated target:

1. The power flux density which is given by radiated power per unit area incident on the target is determined.

2. Radar cross-section which is given by power flux density reflected back toward the radar is determined.
3. Antenna effective area which is given by the amount of power collected by the antenna is determined.

Consider radars with an omni-directional antenna that radiates energy equally in all directions. These kinds of antennas are isotropic radiators and have a spherical radiation pattern. Isotropic radiator is the one that radiates in all directions evenly. Assuming that the medium is a lossless propagation medium, the transmitted power divided by the surface area $4\pi R^2$ of an imaginary sphere of radius R determines the power density at range R away from the radar. Let the power transmitted by the antenna through an isotropic radiator be P_t.

$$\text{Power density from an isotropic antenna} = \frac{P_t}{4\pi R^2}\ \text{W/m}^2 \tag{12.2}$$

In order to increase the power density in particular direction radar systems utilizes directional antennas. There will be losses in the signal as it travels from the input port to the target point at which it is targeted into the atmosphere. Let G_t be the radar transmitter antenna gain; and then, the power density at range R away from the radar can be expressed as

$$P_d = \frac{P_t G_t}{4\pi R^2}\ \text{W/m}^2 \tag{12.3}$$

where P_t = peak transmitted power in Watts
P_d = power density in W/m^2
R = range between the antenna and the target
$4\pi R^2$ = surface area of a sphere of radius R

Power density P_d received by the target from a transmitter increases with antenna gain G_t. The impinging radiated power is scattered by the object and is dependent on shape, size, material, and orientation of the object. The measure of scattered power in direction of the radar is the scattering cross-section of the object. With this, from the target to the transmitter, the power S_R is scattered.

$$P_{\text{refl}} = P_d \cdot \sigma = \frac{P_t \cdot G_t \cdot \sigma}{(4\pi R^2)} \tag{12.4}$$

The signal reflected from the target propagates back toward the radar system over a distance R so that the power density back at the radar receiver antenna, S_R is

$$S_R = \frac{P_{\text{refl}}}{4\pi R^2} = \frac{P_t \cdot G_t \cdot \sigma}{(4\pi R^2)^2} \tag{12.5}$$

The receiving antenna has an effective area A_e, and it absorbs the power out of the power density S_R. The power density of the radio wave received back at the radar receiving antenna is given by

$$S = S_R A_e = \frac{P_t G_t \sigma A_e}{(4\pi R^2)^2}\ \text{W/m}^2 \tag{12.6}$$

Let the gain of the receiving antenna be G_r; and its effective area be A_e. The relation between G_r and A_e is given by

$$G_r = \frac{4\pi A_e}{\lambda^2} \tag{12.7}$$

The total received power S scattered by a target is then given by

$$S = \frac{P_t G_t G_r \lambda^2 \sigma}{(4\pi)^3 R^4} \tag{12.8}$$

The antenna does both transmitting and receiving functions, where $G_t = G_r = G$ in a mono-static radar. Therefore,

$$S = \frac{P_t G^2 \lambda^2 \sigma}{(4\pi)^3 R^4} \tag{12.9}$$

Basic Radar Range Equation by Considering Energy Losses

The total amount of power received will be reduced by the energy loss which is due to the circuit or equipment power dissipation. These losses are lumped together and are denoted by L, where $L < 1$. The total received power is given by

$$S = \frac{P_t G^2 \lambda^2 \sigma}{(4\pi)^3 R^4 L} \tag{12.10}$$

The above equation can also be given in terms of dB as

$$S(dB) = P_t(dB) + 2G(dB) + 2\lambda(dB) - 30\log_{10}(4\pi) + \sigma(dB) - 4R(dB) - L(dB) \tag{12.11}$$

Range of the target can be determined from the Equation (12.10) or (12.11) and is called the *radar range equation*. The range at which the received power equals the minimum detectable signal $S_{\min}$ is the maximum detection range $R_{\max}$. The maximum range can be calculated from Eq. (12.10) by substituting $R_{\max}$ and $S_{\min}$ and is as follows:

$$R_{\max} = \left[\frac{P_t G^2 \lambda^2 \sigma}{(4\pi)^3 S_{\min} L}\right]^{1/4} \tag{12.12}$$

All the above radar equation parameters are illustrated in the figures given below (Figure 12.6).

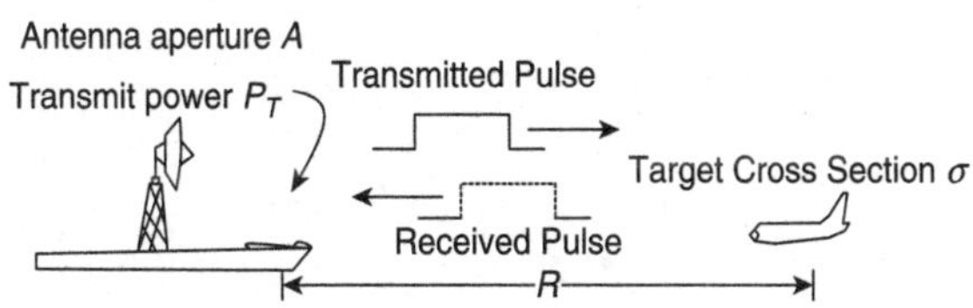

Figure 12.6 (a) *Radar equation parameters*

The general radar equation is derived for point targets. Point targets are objects whose dimensions D are small compared with the illumination (Range × Half power beam width (HPBW)) by the radar at the target site.

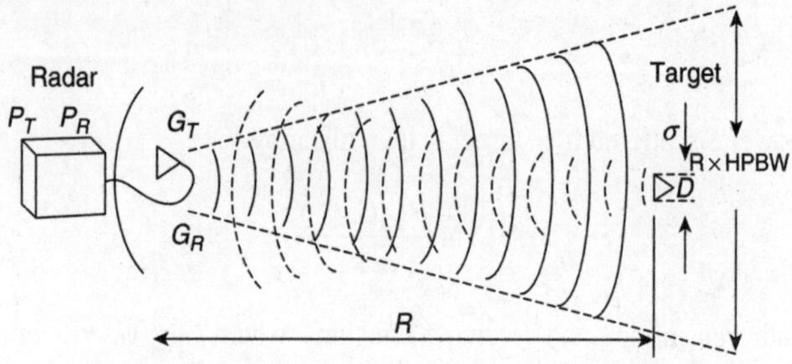

Figure 12.6 (b) *Illustration of principle of radar*

The power density of isotropic antenna (P_{ISO}) at range R from the transmitter is defined as the transmitted power (P_t) divided by the surface area $4\pi R^2$ of an imaginary sphere of radius R.

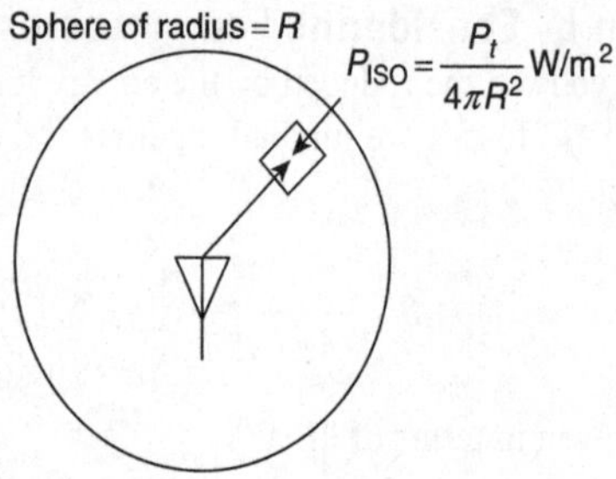

Figure 12.6 (c) *Power Flux Calculations – Isotropic Transmit Antenna*

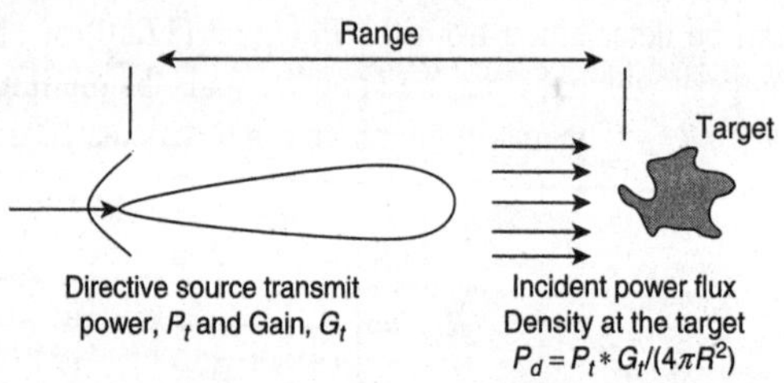

Fig.12.6 (d) *Power Flux Calculations – Directive Transmit Antenna*

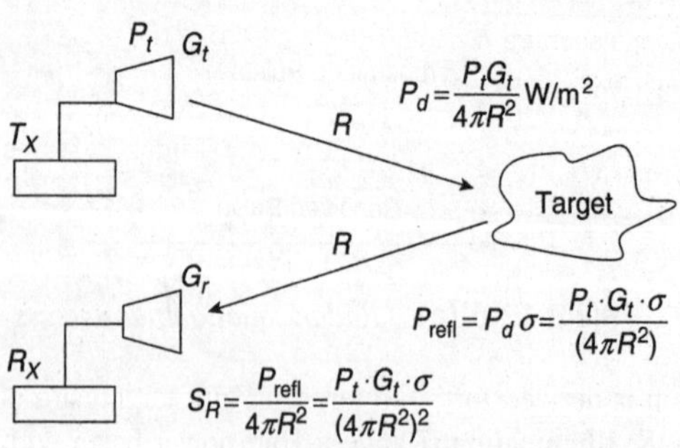

Figure 12.6 (e) *The power density relationship between the transmitter, the target, and the receiver*

Receiver thermal noise: The receiver bandwidth, B gives the range of frequencies for which the radar is subject to noise signals which affects the performance of the radar. Thus, the thermal noise power in the radar receiver, P_n will be directly proportional to B as given in the equation below

$$P_n = kT_sB = kT_0FB \tag{12.13}$$

where K is Boltzmann's constant (1.38×10^{-23} watt-sec/K)
T_0 is the standard temperature (290 K)
T_s is the system noise temperature ($T_s = T_0F$)
B is the instantaneous receiver bandwidth in Hz
F is the noise figure of the receiver subsystem (unit less)

Signal-to-noise ratio: The ratio of received signal power to the noise power is measured in dB.

$$\text{SNR} = \text{received signal/noise} \tag{12.14}$$

$$SNR = \frac{P_tG_tG_r\lambda^2\sigma}{(4\pi)^3 R^4 kT_0FB} \tag{12.15}$$

EXAMPLE PROBLEM 12.1

A pulsed radar operating at 8 GHz has an antenna with a gain of 10 dB and a transmitter power of 1 KW. If it is defined to detect a target with a cross-section of 12 square meters, and the minimum detectable signal is $S_{min} = -80$ dBm. What is the maximum range of the target?

Solution

Pulsed radar frequency = 8 GHz = 8×10^9 Hz

$$\lambda = \frac{c}{f} = \frac{3\times10^8}{8\times10^9} = 37.5\times10^{-3}\,\text{m}$$

Antenna gain G = 10 dB
Peak power 1kw = 1×10^3 W
Radar cross-section of target = σ = 12 sq.m
Minimum detectable signal (S_{min}) = – 80 dBm.

$$-80 = 10\log_{10}\frac{S_{min}}{1\text{mW}}$$

$$\frac{S_{min}}{1\text{mW}} = 1\times10^{-8}$$

$$S_{min} = 1\times10^{-8}\times10^{-3} = 1\times10^{-11}\,\text{W}$$

As per radar range equation,

$$R^4 = \frac{P_tG^2\lambda^2\sigma}{(4\pi)^3 S_{min}}$$

$$R^4 = \frac{1\times10^3\times10\times10\times37.5\times10^{-3}\times37.5\times10^{-3}\times12}{1981\times1\times10^{-11}} = 8.44\times10^{10}\,\text{m}$$

$$R = 539\text{ m}$$

■

EXAMPLE PROBLEM 12.2

Find the maximum range of a radar whose transmitted power is 200 kw. Cross-sectional area of the target is 10 sq.m. The minimum power received is 1 mw. The power gain of the antenna used is 2000, and the operating frequency is 3 GHz.

Solution

Given radar Transmitted power $P_t = 200 \text{ kW} = 200\times10^3 \text{ W}$
Radar cross-section area of target, $\sigma = 10 \text{ m}^2$
Minimum detectable signal $(S_{\min}) = 1 \text{ mW} = 1\times10^{-3} \text{ W}$
Power gain of antenna $G_t = 2000$
Transmitted frequency $f = 3 \text{ GHz} = 3\times10^9$

$$\lambda = \frac{c}{f} = \frac{3\times10^8}{3\times10^9} = 0.1 \text{ m} = 1\times10^{-1}\text{m}$$

$$R^4 = \frac{P_t G^2 \lambda^2 \sigma}{(4\pi)^3 S_{\min}} = \frac{200\times10^3\times2000\times2000\times10^{-1}\times10^{-1}\times10}{1981\times1\times10^{-3}}$$

$$= 10^5\times4\times10^6\times10^{-1} = 4\times10^{10} = 400\times10^8$$

$$\text{R} = 10^2\times(400)^{1/4} = 4.472\times10^2 = 447 \text{ m}$$

■

12.5 RADAR BLOCK DIAGRAM

Radar is an electromagnetic system that transmits radio frequency (RF) electromagnetic (EM) waves toward a region of interest and receives and detects EM waves when reflected from objects in that region.

Figure 12.7 (a) shows the major modules involved in the process of transmitting a radar signal, reflection of the signal from the target, receiving the reflected signals, and processing the received signals. The major subsystems of radar block diagram include the waveform generator, the transmitter, the duplexer, the receiver, the antenna, and a signal processor. The subsystem that generates the radiofrequency electromagnetic waves is the transmitter. The antenna is connected to the transmitter through a duplexer. The duplexer (circulator or transmitter/receiver (T/R) switch) is a circuit that switches the antenna from the transmitter to the receiver at the proper time so that the signal can be transmitted to perform its tasks without destroying the receiver in the process. At the same time, the switch allows the very low level reflected signal to the receiver but not into the transmitter.

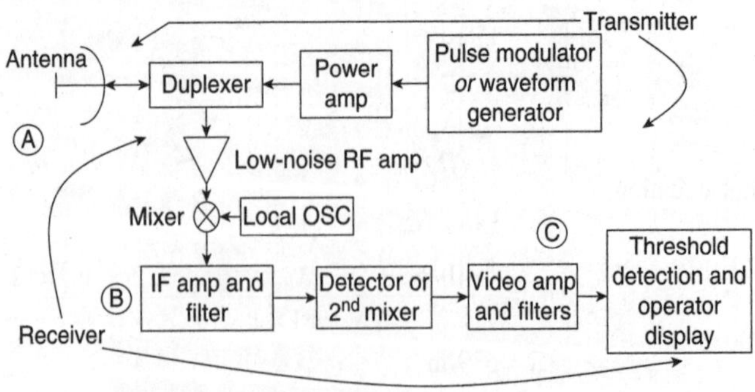

Figure 12.7 (a) *Basic radar block diagram*

The waveform generator can be a magnetron; the resultant signal peak can be turned ON/OFF by the modulator for generating pulses that are to be transmitted. The short bursts of radio energy generated by the modulator are amplified by the power amplifier and sent to the antenna through the duplexer. In radar systems, the receiver is usually of a superhetrodyne type in which the first device is a low-noise RF amplifier that may or may not be present always. By using a mixer and local oscillator, we can convert an RF signal to an IF signal. The IF amplifier should be designed such that the frequency response function of the matched filter should maximize the S/N ratio at the output. The pulse modulation done by the pulse modulator at the transmitter is extracted by the second detector after maximizing the S/N ratio in the IF amplifier. Then, by using the video amplifier, the extracted signal will be amplified to a level where it can be properly displayed.

The strength of the received signal depends on factors such as the shape, size, material, RCS of the target, propagation affect of atmosphere and earth, presence of interference, and noise in the propagation medium. In general, the noise with a similar frequency as that of the transmitted signal affects the ability of a radar receiver to detect a weak echo signal. The *minimum detectable signal* is the weakest signal that the receiver can detect.

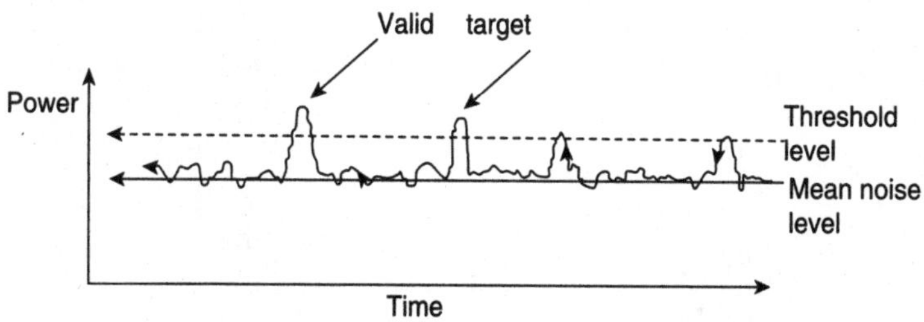

Figure 12.7 (b) *Schematic representation of received radar signals with noise*

The radar return is sampled at regular intervals with A/D (Analog to Digital) converters. The sampled returns may include the target of interest and noise. A threshold is used to reject noise. If the returned echo crosses the threshold, that echo is considered a valid target as shown in Figure 12.7 (b); if not, it will be rejected.

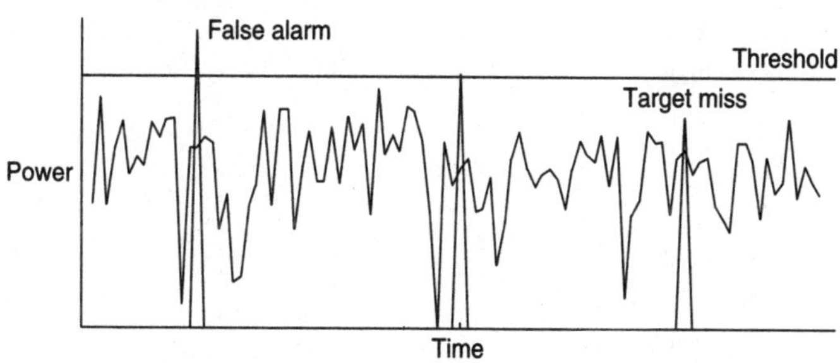

Figure 12.7 (c) *Detection of target echoes in noise*

In some cases, a strong interference spike can cross the threshold, which leads to a wrong inference about the presence of the target, which is known as a *false alarm*. Similarly, a weak target echo might not have enough power to cross the threshold, so that the target is not detected and this is known as *target miss*, as shown in Figure 12.7 (c).

12.5.1 Pulse Characteristics

The pulses of a short duration (τ_p), are transmitted by radar and then, reflected back from the target. The returning echo of that reflected pulse is received by a receiver. A sequence of identical pulses, transmitted at regular intervals is called a radar pulse train. The pulse train is formed by the radio frequency carrier wave which is amplitude modulated. The *pulse repetition rate* (PRR) or *pulse repetition frequency* (PRF) is defined as the number of pulses per second; while the *pulse repetition time* (PRT) is the time between the onset of successive pulses. The basic parameters of the pulse train are defined by referring to Figure 12.8 (a).

- τ_p is the pulse width, and it has units of time (usually expressed in μsec)
- PRT is the pulse repetition time. PRT has units of time and is commonly expressed in *m sec*. It is the interval between the start of one pulse and the start of another. PRT is also equal to the sum, PRT = τ_p + RT.
- PRF is the pulse repetition frequency; it has units of time^{-1} and is commonly expressed in Hz (1Hz = 1/sec) = 1/PRT.
- Duty cycle, D = pulse width/inter-pulse period (or PRT) = τ_p /PRT = $\tau_p \cdot$PRF

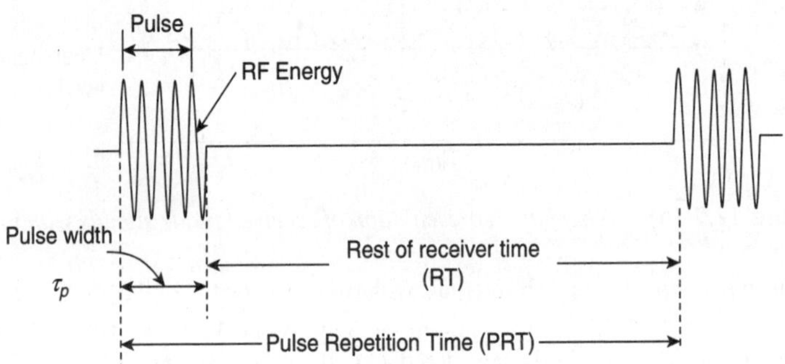

Figure 12.8 (a) *Pulse repetition time of radar*

The pulse propagates at a constant speed i.e., at the speed of light (c); hence, the range can be indirectly measured by measuring the time required for the echo to return (Δt) as shown in Figure 12.8 (b):

$$R = \frac{c \times \Delta t}{2}$$

where R = range to the target in meters

c = speed of light (3×10^8 m/s)

Δt = time taken for the signal to travel and return

Two as the factor in the equation, is resulted from the fact that the radar pulse should either travel to the target and return before the detection, or two times the range to the target, R

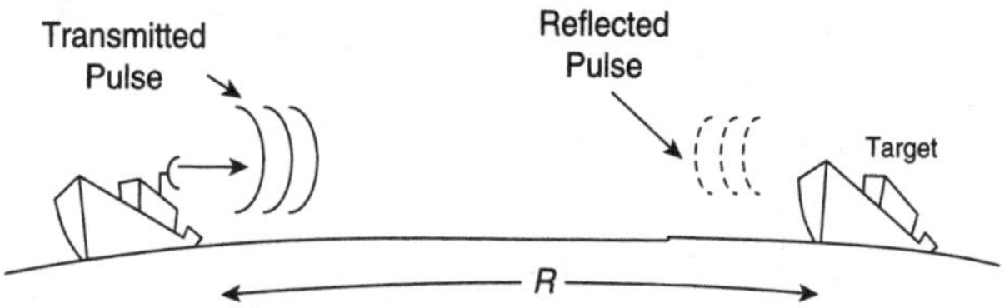

Figure 12.8 (b) *Range measurement*

The product of pulse width (τ_p) and operating frequency denotes the number of cycles in the pulses. For example, 4000 cycles of *RF* energy per pulse can be sent out by a 2 micro second pulse at a 2-GHz frequency. The peak power (P_p), the average power (P_a), and the duty cycle (D) should be known for radar pulses.

Peak power (P_p), the power averaged over the carrier cycle and that occurs at the maximum of the pulse power. The output power of the transmitter is usually measured in terms of peak power (P_p), and it is that peak power which we will use shortly to develop the radar range equation.

Average power (P_a), the average transmitted power over the pulse repetition time or period. The relationship between the average power (P_a), peak power (P_p), PRT, and PRF is as follows:

$$P_a = \frac{P_p \tau_p}{PRT} \tag{12.16}$$

$$P_a = P_p \tau_p (PRF) \tag{12.17}$$

where P_a = average power in watts
P_p = peak pulse power in watts
τ_p = pulse width in seconds
PRF = pulse repetition frequency in hertz
PRT = pulse repetition time in seconds

Duty cycle, ratio of average power to the peak power or pulse width to the pulse repetition time.
Duty cycle = P_a/P_p or τ_p/ PRT or Duty Cycle = $\tau_p \times$ PRF

Pulse repetition frequency (PRF): PRF is the rate at which pulses are transmitted (per second). The PRF is used primarily for knowing the maximum range at which targets are expected. If the period between successive pulses is very short (that is, one pulse is transmitted before the previous pulse has completed the roundtrip to the target and back), it is unclear as to which transmitted pulse originated which echo pulse. Such a condition is called *range ambiguity*. The maximum PRF that can be used for an unambiguous range R_{max} is given by

$$PRF \leq \frac{c}{2R_{max}} \text{ [Hz]} \tag{12.18}$$

Unambiguous range measurement: Usually, a train of pulses are transmitted and received by the pulsed radar as shown in Figure 12.9. The *Pulse Repetition Time* (PRT) is also known as Inter-Pulse Period (IPP). The pulse width is given by τ_p and the IPP is given by T_r. The pulse repetition frequency PRF is denoted by f_r which is the inverse of PRT.

$$f_r = 1/\text{PRT} = 1/T_r \tag{12.19}$$

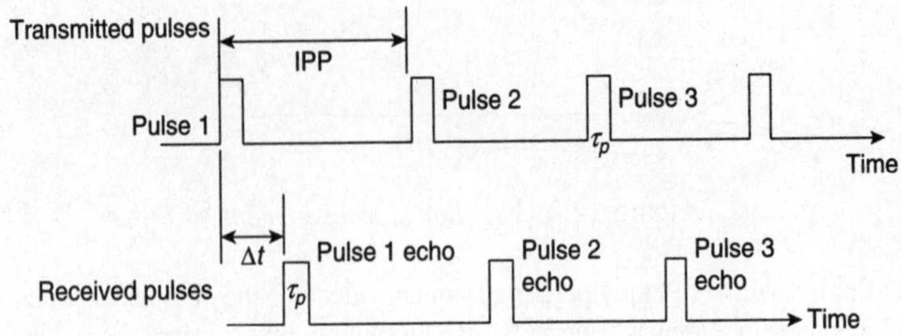

Figure 12.9 *Train of transmitted and received pulses*

During each PRT, the radar radiates energy only for seconds and listens for target returns for the rest of the PRT. The round-trip time for the radar wave is given by

$$\Delta t = \frac{2R}{c} \tag{12.20}$$

The *radar unambiguous range* R_{ua} is the range corresponding to the two-way time delay, Δt. Let us consider the below case in Figure 12.10. Here, the radar that returns from a target at a range of $R_1 = c\,\Delta t/2$ due to pulse 1 is represented by Echo 1. Echo 2 represents either the return from the same target caused by pulse 2, or it may be the return from a distant target caused by pulse 1 again. Thus,

$$R_2 = c\,\Delta t/2 \text{ or } R_2 = c\,(T_r + \Delta t)/2$$

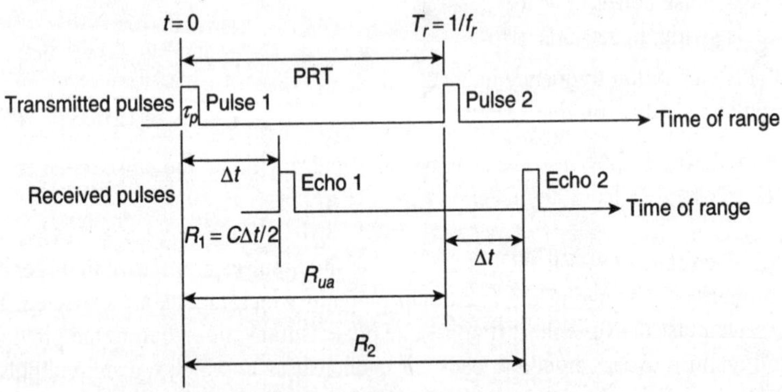

Figure 12.10 *Illustrating range ambiguity*

From the Figure 12.10 it is evident that echo 2 is in association with range ambiguity. So when a pulse is transmitted, the radar has to wait for a certain span of time for transmitting the next pulse. Thus the Maximum unambiguous range should be equal to the range corresponding to half of the PRT,

$$R_{ua} = \frac{c}{2PRF} \tag{12.21}$$

where R_{ua} is unambiguous range.

That is, range ambiguities can be avoided by ensuring that the inter-pulse period, PRT, is long enough or, equivalently, the pulse repetition frequency PRF is low enough. In general, the unambiguous range can be of two types, in which one minimum unambiguous range and one maximum unambiguous range are given by

$$\min(R_{ua}) = \frac{c\tau_p}{2} \tag{12.22}$$

$$\max(R_{ua}) = \frac{cT_r}{2} \tag{12.23}$$

EXAMPLE PROBLEM 12.3

A radar is required to have an unambiguous range of 500 km; what is the maximum PRF that may be used? If the pulse length is 5 ns, what is the duty cycle for the transmitter?

Solution

The maximum PRF can be found from

$$PRF = \frac{c}{2R_{max}} = \frac{3\times10^8}{2\times500\times10^3} = 300 \text{ Hz}$$

Inter-pulse Period (PRT) = 1/PRF = 3 ms

$$\text{Duty cycle} = \frac{\text{Pulse width}}{\text{inter pulse period}} = \frac{5\times10^{-9}}{3\times10^{-3}} = 1.67\ \mu\text{s}$$

■

Unambiguous doppler shift measurement: In a radar, Doppler shift is being sampled at the radar's PRF; thus, the maximum range of Doppler shift frequencies that can be unambiguously measured is

$$f_{d\max} = \pm PRF/2 \text{ or } PRF_{\min} = 2f_{d_{\max}} = \frac{4v_{r_{\max}}}{\lambda} \tag{12.24}$$

where $v_{r_{max}}$ is the radial velocity

Pulse shape: Pulse shape is important because the minimum range, range accuracy, and resolution is determined by the shape of the pulse. However, until the return pulse has enough energy to produce a measurable echo at the receiver, the pulse shape is not very important in target detection. Whereas for range, the shape of the pulse is very important. A major factor in the accuracy of range measurements is the steepness of the leading edge. The receiver can be remained off or without being aligned to antenna longer than required, thus reducing minimum range (R_{min}), performance, and target resolution by the slope of the leading edge. The shape of the pulse can be affected; the accuracy of range measurements can be influenced by the noise. A usual pulse shape, which is superimposed by a pulse degraded by noise, is illustrated in Figure 12.11. As the target appears closer, the time delay between transmitted pulses to received pulses will, in effect, be decreased by Δt.

Pulse width: The radar's maximum and minimum detection range, range resolution, can be determined by the width of the pulse. A narrow pulse width is necessary for a good range resolution which is defined as the capability to differentiate two targets at almost the same range. Figure 12.11 describes that when the distance between two targets is less than the pulse duration, the leading edge of the pulse will be hitting the farthest target; while the trailing edge of the pulse will be closing onto the closest target. The leading edge of pulse *B* is hidden in pulse *A*, when both the pulses return to the radar. Thus the second target is masked. From the above discussion, it is made clear that a narrow pulse is not desirable, in every case. A target should return a strong echo that is detectable on the scope or plan position indicator (PPI) for proper detection. The energy in the returned echo can be increased by two ways

(i) Increasing the peak transmitted power
(ii) Increasing the pulse width which is the widely practiced, because the former method will raise the weight, cost, and energy requirements of the radar.

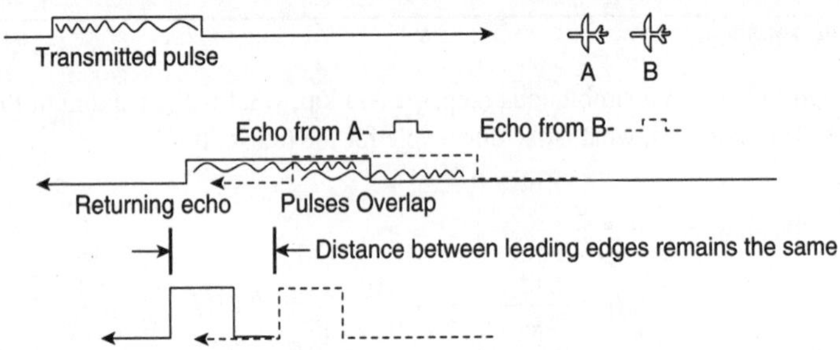

Figure 12.11 *Effects of pulse width on target resolution*

Carrier frequency: The required directivity and resolution, and the existing limitations on the design of the electronic equipment are the several factors on which the choice of a suitable carrier frequency depends. The higher frequency leads to shorter wavelength in quasi-optical antenna systems, and, therefore, smaller antennas are required. Because of the shorter wavelength, higher frequencies will improve target resolution and enable the detection of smaller sized targets. On the other hand, higher frequencies will increase the directivity for antennas of fixed dimensions. The greater propagation loss and natural difficulties of generating and amplifying the *RF* energy are the major disadvantages of high frequencies. The greater sea-clutter and backscatter can also be created by the greater resolution caused by the higher frequency. The main difficulty the operator finds is increasing clutter and false alarm problem at the same time since the higher-frequency radar will better couple with lower duct heights.

Receiver sensitivity: Sensitivity is defined as the minimum signal-to-noise ratio times the mean noise power. The minimum input signal (S_{min}) is the sensitivity in a receiver and is necessary for producing a specific output signal with a specific signal-to-noise (S/N) ratio. The minimum pulse width where the specified sensitivity applies should be mentioned; while stating the sensitivity of receivers that is to be used to detect and process the pulse signals.

$$S_{min} = (S/N)_{min}\, \text{kT B (NF)} \tag{12.25}$$

where $(S/N)_{min}$ = Minimum signal-to-noise ratio needed to process (vice just detect) a signal
NF = Noise figure/factor
K = Boltzmann's Constant = 1.38×10^{-23} Joule/EK
T_0 = Absolute temperature of the receiver input (°Kelvin) = 290° K
B = Receiver Bandwidth (Hz)

Signal-to-Noise (S/N) ratio: The ratio of the signal power in the receiver to the mean noise power of the receiver is known as the Signal-to-Noise Ratio (S/N) in a receiver. The prerequisite of all receivers is that the signal should exceed the noise by certain amount. The sum of the signal power and the noise power should exceed certain threshold value for detection of a signal. But the useful signal is not detectable, if the signal power is less than or is just equal to the noise power.

Pulse compression: Very short pulses will improve the range resolution of the radar *e* considerably. The normal modes of operation of the radar are affected, as the average transmitted power will be reduced due to these short pulses. It is always recommendable to increase the pulse width i.e., increase the average transmitted power and keeping sufficient range resolution, because the average transmitted power is related to the receiver SNR. This is achieved by using pulse compression radar. In this radar technique, a long pulse with pulse width of τ_p and peak power is transmitted. This pulse is coded by frequency or phase modulation to obtain a bandwidth B larger than the bandwidth of the uncoded pulse with the same duration. As a result the frequency of the wave is increased within the pulse; provide a large maximum range by packing sufficient power and allows good range resolution. For short pulsed transmission, this technique allows wide pulses to improve the detection and also maintains the range resolution at the same time.

12.5.2 Radar Cross-Section of Target (RCS)

The measure of reflective strength of the target is called RCS. The RCS is a product of its geometric cross-section, reflectivity, and directivity, and is denoted by σ. It is given by

$$\sigma = \text{geometric cross section} \times \text{reflectivity} \times \text{directivity}$$

Each of the function in the formula is defined as follows:

geometric cross-section is the target's size seen from the radar and *reflectivity* is the ratio of the power from the target to the radar power that light up the target and the remaining power is absorbed.

The ratio of power reflected back in the radar's direction to the amount of reflected power scattered in all directions is known as *directivity*.

Electromagnetic waves with a known polarization hitting on a target, are either scattered or diffracted in all directions. These scattered waves are splitted into two parts. The first part includes waves having same polarization as that of the receiving antenna. This polarization is called *Principle Polarization* (PP). This part well describes the target RCS. The other part includes the scattered waves having different polarization to that of the receiving antenna which will not be responded and is called *Orthogonal Polarization* (OP). Both are orthogonal.

The backscattered energy having the polarization of the radar's receiving antenna is used to determine the target RCS. By energizing the target with RF energy it acts as an antenna and possesses near and far fields. Usually the reflected waves which are measured in near field are spherical. On the other hand the wave fronts in far field are decomposed into linear combination of plane waves.

RCS vs. radar wavelength λ

The RCS (σ) of an object is given by

$$\sigma = \frac{4\pi A_p^{\,2}}{\lambda^2} \tag{12.26}$$

where λ is radar wavelength. Thus RCS is depends on the radar wavelength

A_p is the Flat area of the target

From the equation, following cases can be arrived:

Case 1: If $A_p >> \lambda$, then, the RCS is more or less similar to the size of the real area of the target. This is known as the *optical region*, as RCS nears the optical value.

Case 2: If $A_p \sim \lambda$, then, there will be wide changes in RCS with changes in wavelength, and it may be greater or smaller than the optical value. This is known as the *resonance or Mie region.*

Case 3: If $A_p << \lambda$, the RCS will become a λ^{-4}. This is known as the *Rayleigh region.*

The *RCS* of a few simple shapes are given in Table 12.3

Table 12.3 Radar cross-sections of various geometrical shapes

Object	RCS / Comment	
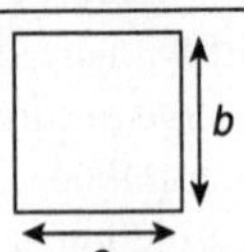 Flat plate	$\frac{4\pi a^2 b^2}{\lambda^2}\left[\frac{\sin((2\pi/\lambda)a\sin\theta)}{(2\pi/\lambda)a\sin\theta}\right]^2 \cos^2\theta$ Where θ= incident angle	Max RCS $= \frac{4\pi a^2 b^2}{\lambda^2}$
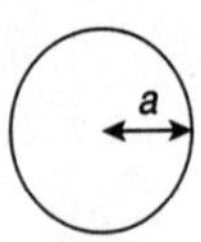 Sphere	For $\frac{2\pi a}{\lambda} > 1$ RCS $< \pi a^2$ and proportional to $\frac{1}{\lambda^4}$ (Rayleigh scattering region) For $1 > \frac{2\pi a}{\lambda} < 10$ Max RCS $\approx 4\pi a^2$ (Resonance or Mie region) For $\frac{2\pi a}{\lambda} > 10$, RCS $= \pi a^2$ (Optical region)	
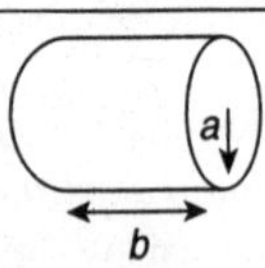	Right circular cylinder	Max RCS $= \frac{2ab^2}{\lambda}$
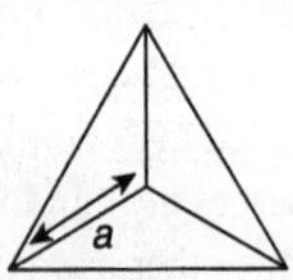	Triangular trihedral	Max RCS $= \frac{4\pi a^4}{3\lambda^2}$ Large RCS over wide range of angles

RCS vs. radar equation

It is the property of a scattering target. RCS in the radar equation represents the magnitude of the echo signal returned. Reradiated power density is given by

$$\frac{P_t G}{4\pi R^2} \times \frac{\sigma}{4\pi R^2} \tag{12.27}$$

The amount of reflected power from the target is

$$P_r = \sigma P_{\text{Di}} \tag{12.28}$$

Where P_{Di} is the power density of a wave incident on a target at range R
σ denotes the target cross-section.

P_{Dr} represents the power density of the scattered waves at the receiving antenna. It is given by

$$P_{Dr} = \frac{P_r}{(4\pi R^2)} \tag{12.29}$$

Substituting P_r in Eq. (12.29),

$$\sigma = 4\pi R^2 \left(\frac{P_{Dr}}{P_{Di}}\right) \tag{12.30}$$

From above we can say that the intensity of backscattered waves from a target is proportional to the ratio of the scope (size) of target to the wavelength of the incident wave. If the targets are very small compared to its operating wavelength then they cannot recognised by the radar. For example, if we operate weather radars in L-band frequency, the rain drops are not visible to the radar because rain drops are very small than the operating wavelength.

On the other hand the frequency region where the target size exceeds the radar operating wavelength, that region is considered as *optical region*. Generally most of the radar applications fall within the optical region.

12.5.3 Radar Antennas

A device which is used to convert electronic signals to EM waves or vice versa is known as an antenna and also radiates or receives radio waves. It also acts as the transitional structure between a guiding device and free space.

Antenna performance parameters

Antenna gain: The ability of an antenna to concentrate the transmitted energy in a certain direction is the gain of the antenna which is given by,

$$G = \frac{4\pi A_e}{\lambda^2}$$

where A_e is the effective area of the antenna.

Directivity: It is described by the antenna radiation pattern. An *antenna radiation* pattern is represented as the plots of the power gain and directivity, when normalized to unity.

$$D = \frac{4\pi}{\Omega_A}$$

where Ω_A is the beam solid angle.

Antenna aperture: It is a measure of how an antenna is effectively receiving the power of radio waves. The area, oriented perpendicular to the direction of an incoming radio wave, which would intercept the same amount of power from that wave as is produced by the antenna receiving is known as an aperture.

$$A_e = \varepsilon A$$

where ε is an efficiency factor, usually in the range of 0.4–0.9, for a parabolic dish antenna.

Antenna input impedance: It is defined as the ratio of the input voltage to the input current.

$$z_a = \frac{V_i}{I_i}$$

Radiation resistance (R_r): The imaginary resistance that scatters power which is equal to the radiated power is represented by the term R_r.

Frequency coverage: The range of frequency in which the antenna can transmit or receive signals and even appropriate parametric performance can be provided in this range.

Polarization: This deals with the orientation of the electric field of waves that are transmitted or received. This can be of any shape.

Beam width: The angular exposure of the antenna, given in degrees.

Efficiency: The percentage of signal power transmitted or received to the theoretical power of a part of a sphere covered by the antenna's beam.

Functions of an antenna

1. Radiation in the desired directions and suppression in the unwanted directions can be obtained by employing this device.
2. As a transducer: Converts electrical energy into EM energy if it is transmitter and EM energy into electrical energy if it is receiver.
3. Impedance matching: The transmitter and free space are matched if it is transmitter and free space and the receiver are matched if it is a receiver.
4. It acts as a radiator as well as a sensor of EM waves.
5. While using an antenna for transmitting and receiving purposes, it possesses the properties like identical impedance, identical directional characteristics, and identical effective wavelength. This in turn can be proved with the help of reciprocity theorem.

Antenna characteristics

1. It focuses the energy into a collimated beam so that energy density is focussed onto the target.
2. It listens selectively at the angle where the energy is coming back from the echo. So, it also serves as function of collecting very efficiently by having a well-collimated receiver beam from the echo.
3. It measures the angle where the targets are located; by having measurements in azimuth and elevation angle, we can resolve targets that are nearby in angle.
4. It tells us how much we have to revisit space; it tells us how much it takes to go through, transmit, and receive all the different areas of angle space that we want to invest in energy in search for targets.

Directional characteristics: These are also called *radiation characteristics* or *radiation patterns*. An antenna radiation pattern is a three-dimensional variation of the radiated field. They are of two types:

Field radiation pattern: It indicates the variation of the absolute value of field strength with θ.

Power pattern: It indicates the variation of radiated power with θ.

12.5.4 Information Available from Radars

If we want to perform any action by using radars, we have to know about the presence of the target and information about that target. By using radars, we can know about the below mentioned parameters:

Range: The most important feature of a radar is its ability to determine the range to a target by measuring the time it takes for a transmitted RF signal to propagate at the speed of light to the target and back to the radar and then divide that time in two. To measure range, a timing mark should be introduced on the radar signal; this timing mark can be a short pulse that can be a frequency-modulated or a pulse-modulated one. The accuracy of range measurement depends on the radar signal bandwidth; the wider the bandwidth, the greater will be the accuracy.

Radial velocity: Radial velocity is nothing but the rate of change of range over a period of time that can also be measured from the Doppler frequency shift.

Angular direction: The target's location in angle can be found from the direction in which the narrow-beam width radar antenna points when the received echo signal is of maximum amplitude, which requires an antenna of narrow beam width.

Size and shape: If the radar has sufficient resolution capability in range or angle, it can provide a measurement of the target extent in dimension and with sufficient resolution in both range and cross-range (which is nothing but the product of range and antenna beam width); not only the size in two orthogonal co-ordinates but also the target shape can be sometimes determined.

12.6 PULSE RADAR CHARACTERISTICS

Radar pulses travel at the speed of light (3×10^8 m/sec). Since the speed is known, distance at which the target is present can easily be calculated by just observing the time taken for the pulse to travel from transmission to its return. The target's range from the antenna can be determined by half the distance traveled by the pulse. The most widely used pulse shape for modulation is rectangular. There are several reasons that radar pulses are rectangular with vertical edges and a flat top.

Leading edge of the pulse: The leading edge of the transmitted pulse should be vertical to ensure that the leading edge of the received echo pulse is also close to vertical, so that accurate range measurements can be made possible. Otherwise, ambiguity will arise in knowing the precise instants of pulse transmission and reception and, therefore, inaccuracies will creep into the measurement of the target range. A flat top for the pulse is required, as this determines the anode voltage of the magnetron during the modulation. Since a magnetron requires a steady, regulated anode voltage for proper operation, the pulse driving the anode voltage should also have constant amplitude; that is, it should be flat. Otherwise, the magnetron frequency will be pushed.

Steep trailing edge of the pulse: A steep trailing edge for the pulse is needed for proper operation of the Duplexer; that is, the antenna is connected as soon as the transmission is over to the receiver. This will not happen if the pulse decays slowly and has a long time constant, as the duplexer can switch the antenna to the receiver only when the transmitted energy has dropped down to zero. This will result in lengthening the period of time for which the receiver remains disconnected from the antenna and, hence, has an effect on the minimum range of the radar.

Pulse length: The length measured from leading to trailing edge of pulse is called Pulse length (or pulse duration expressed in microseconds) and it efficiently indicates the amount of power present in the pulse. When radar emits long pulses the received echo power is obviously more which in turn provides better target information and data reliability. But the fine details which are enclosed within the pulse may be lost while receiving echoes of longer pulses.

12.6.1 Minimum Range

Pulse length provides radar's minimum detectable range or tells how close a target can get to the antenna without harmfully affecting its operation. Any distance that is greater than one-half of the pulse length is considered as *Minimum radar range*. If targets present in more than one-half pulse length from the antenna can be correctly processed, but a serious threat comes into picture when targets come closer. When target lies within the one-half pulse length or less from the antenna, the leading edge of the pulse strikes the target and returns before the radar reverts into its receive mode. During this process some part of the return energy is lost and due to this radar may get puzzled. The minimum detectable range is given by

$$R_{\min} = \frac{c \times \tau_p}{2} \tag{12.31}$$

12.6.2 Maximum Range

The peak power of the pulse tells us how far a pulse can travel to a target and return a usable echo. The smallest signal that can be detected by a receiver which can be processed and presented on an indicator is called a usable echo. The indicator time base which measures the returned echoes gets reset with the leading edge of transmitted pulse and eventually a new sweep appears on the screen. When a pulse is transmitted which is shorter than the time taken to detect the echo, then the target is indicated in the display at false range in a different sweep. The maximum detectable range is given by

$$R_{\max} = \left[\frac{P_t G^2 \lambda^2 \sigma}{(4\pi)^3 S_{\min}} \right]^{1/4} \tag{12.32}$$

12.6.3 Radar Resolution

The *target resolution* of radar is defined as the ability of radar to differentiate targets that are in very close range or bearing. Depending upon this, radar resolution is categorized into: range resolution and bearing resolution.

12.6.3.1 Range resolution

The *range resolution* is defined as the ability of radar to differentiate targets on the same azimuth but at different ranges.

The range resolution depends on several parameters such as transmitted pulse width (τ_p), size of the target, and efficiency of the receiver and radar display. The important parameter that decides the range resolution is the pulse width (τ_p). The range resolution in terms of pulse width is given by

$$\Delta R = \frac{c\tau_p}{2} \text{ meters} \tag{12.33}$$

For example, the range resolution of a radar with a transmitted pulse width of 1 microsecond is calculated as 150 meters using Eq. (12.33); that is, the targets on the same bearing would have to be separated by more than 150 meters to clearly distinguish them on the display.

The problem of range resolution will be discussed in later chapters in greater detail.

12.6.3.2 Azimuth resolution

Azimuth (or bearing) resolution is defined as the ability of radar to differentiate the target at the same range but at different bearings or azimuth. The radar beamwidth and the angular distance between (S_{AB}) the targets are the basis for the degree of bearing resolution. A radar beamwidth is given in terms of half-power points. If S_{AB} is more than the half-power beamwidth of the antenna, then the two point targets (*A* and *B*) at the same range can be distinguished by the relation(Figure 12.12.)

$$S_{AB} \geq 2R\sin\theta/2 \text{ meters} \tag{12.34}$$

where S_{AB} is the angular resolution as a distance between two targets,

R is the slant range,

θ is the antenna beam width.

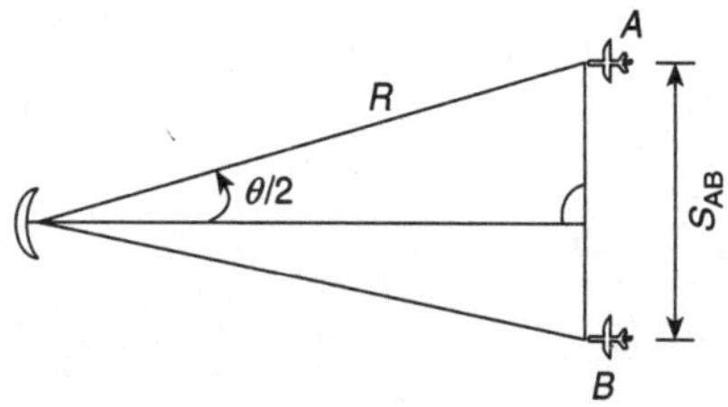

Figure 12.12 *Angular resolution*

12.7 RADOME

Antennas of ground-based radars are often subject to severe weather. So, some enclosure is needed for antennas to survive and to perform under adverse weather conditions. These enclosures are called *radome (radar dome)*.

A radome basically serves like a cover and protects the radar from accumulation of ice (especially freezing rain) onto the metal surface of the antenna as shown in Figure 12.13. The radome present in spinning radar dish antenna protects it from debris and rotational irregularities caused due to wind. This excessive accumulation of ice causes problems in antennas, in case of a stationary antenna this can detune the antenna to such an extent that its impedance at input frequency rises significantly, leading the voltage standing wave ratio to rise as well. This in turn makes the reflected power to go back to the transmitter where it causes overheating. To prevent this, a fold back circuit is used, which in turn causes the station's output power to drop severely reducing its range. So a radome plays a crucial role in protecting antenna's exposed parts with sturdy, weather-proof material, mostly a fiber glass which keeps the ice a safe distance from antenna prevents the occurrence of any serious issues.

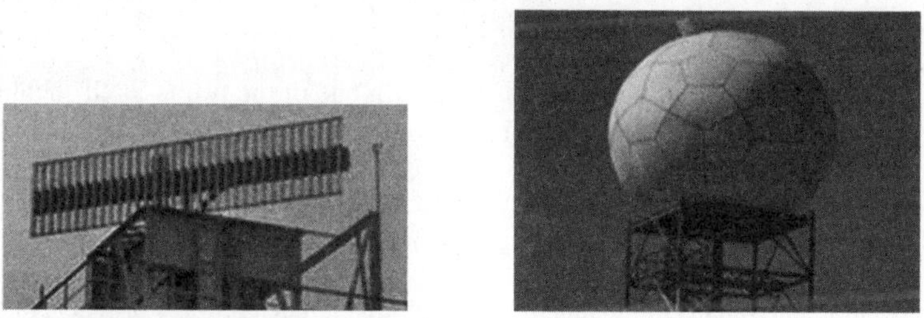

Source: upload.wikimedia.org

Figure 12.13 *Secondary surveillance radar antenna and radome*

Radome Characteristics

Though the radome plays an important role in protecting antenna but its presence affects the antenna gain, beam width, direction of the bore sight, side lobe level, and it may also change the VSWR and antenna noise temperature. For minimising the affects caused by radome the following characteristics are specified.

- One-way transmission loss for the dry and wet state

- Increase of side lobe level
- Reflected power
- Cross-polarization degradation
- Bore-sight error
- Beam width increase

These parameters are closely related to the antenna which is covered; therefore the radome performance can be estimated based on the antenna characteristics.

SUMMARY

1. RADAR is an abbreviation for Radio Detection and Ranging. Radar is used for detecting the existence and position of objects.

2. The radar is subjected to operate in the microwave region of the EM spectrum where it emits energy in the form of EM waves into the atmosphere through an antenna.

3. A radar is basically used for detecting and locating reflecting objects like aircraft, ships, vehicles, spacecraft, people and natural environment.

4. The range to a target can be known by measuring the time taken by the pulse to travel to and from the detected target (Δt) where the range to the target is given by $c(\Delta t)/2$.

5. Radar's resolution is meant to display various targets clearly and separately.

6. Even though there is increasing demand for radar application in civil and marine navigation, the major application of radars is in military.

7. The radars are divided into two groups: primary radars and secondary radars based on the role played by them when detecting the targets.

8. Radars can be classified depending on the type of waveforms they radiate into two groups: CW radars and pulsed radars.

9. A radar system is generally required to perform one of the two tasks. It should either search for targets or track them once they have been acquired.

10. The radar equation provides the relationship between the transmit power, the received power, the characteristics of the target, and the characteristics of the radar itself.

11. The major subsystems of the radar block diagram include the waveform generator, the transmitter, the duplexer, the receiver, the antenna, and a signal processor.

12. Pulse repetition frequency is the rate at which pulses are transmitted.

13. **Radar cross-section** is defined as the measure of reflective strength of the target RCS and is a function of its geometric cross-section, reflectivity and directivity of a target.

14. An antenna that transmits radiations equally in all directions is referred to as an isotropic antenna. The power flux density $(S_{\text{isotropic}}) = P_t/4\pi R^2$ W/m^2.

15. *Duplexer* is a switch that alternately connects the transmitter or the receiver to the antenna. It protects the receiver from the high power output of the transmitter. During the transmission of an outgoing pulse, the duplexer will be aligned to connect the transmitter to the antenna for the duration of the pulse.

OBJECTIVE-TYPE QUESTIONS

1. Increasing the pulse width in a pulse radar
   (a) increases resolution
   (b) decreases resolution
   (c) has no effect on resolution
   (d) increases the power gain

2. A high noise figure in a receiver means
   (a) a poor minimum detectable signal
   (b) a good detectable signal
   (c) receiver bandwidth is reduced
   (d) high power loss

3. The sensitivity of a radar receiver is ultimately set by
   (a) a high S/N ratio
   (b) a lower limit of signal input
   (c) overall noise temperature
   (d) a higher figure of merit

4. A RADAR that is used for measuring the height of an aircraft is known as
   (a) radar altimeter
   (b) radar elevator
   (c) radar speedometer
   (d) radar altitude

5. Second-time-around echoes are caused by
   (a) second-time reflection from target
   (b) echoes returning from targets beyond the cathode tube range
   (c) echoes that arrive after transmission of next pulse
   (d) extreme ends of bandwidth

6. The resolution of pulsed radars can be improved by
   (a) increasing the pulse width
   (b) decreasing pulse width
   (c) increasing the pulse amplitude
   (d) decreasing the pulse repetition frequency

7. In case the antenna diameter in a radar system is increased to four times, the maximum range will be increased by
   (a) 1/2 times
   (b) 2 times
   (c) 4 times
   (d) 8 times

8. The term *RADAR* stands for
   (a) radio direction and reflection
   (b) radio detection and ranging
   (c) radio waves dispatching and receiving
   (d) random detection and re radiation

9. The gain of a radar transmitting antenna is
   (a) less than that of a radar receiving antenna
   (b) almost equal to that of a radar receiving antenna
   (c) slightly higher than that of a radar receiving antenna
   (d) much higher than that of a radar receiving antenna

10. A large antenna is used in radars, because it
(a) gives higher gain
(b) gives lesser side lobes
(c) increases the beam width
(d) increases bandwidth

11. A radar cross-section is affected by
(a) target size and shape
(b) target aspect
(c) target material and carrier frequency
(d) all of the above

12. A radar range equation provides a relationship among
(a) received power
(b) target characteristics
(c) R_{max} of the radar, which is determined
(d) all of the above

13. A radome is a
(a) radar housed in a dome
(b) protective cover for the antenna
(c) dish-shaped antenna
(d) dome-shaped antenna

14. The range of a radar can be increased by
(a) increasing the peak transmitted pulse power
(b) increasing the diameter of the antenna
(c) increasing the frequency
(d) any one of the above

15. In order to double the radar range, the peak pulse power should be increased
(a) 2 times
(b) 4 times
(c) 8 times
(d) 16 times

16. If P_t is the peak transmitted pulse power, the maximum radar range will be proportional to
(a) P_t
(b) P_t^4
(c) $P_t^{1/2}$
(d) $P_t^{1/4}$

ANSWERS TO OBJECTIVE-TYPE QUESTIONS

1. (b) 2. (a) 3. (c) 4. (a) 5. (c) 6. (b) 7. (c) 8. (b) 9. (b) 10. (a) 11. (d) 12. (d) 13. (b) 14. (d) 15. (d) 16. (d)

REVIEW QUESTIONS

1. Draw the block diagram of the radar and explain each block.

2. Explain classification of radars.

3. Explain and derive the radar range equation.

4. What is the working principle of radars?

5. What is Doppler principle?

6. How do Doppler radars measure target velocity?

7. Explain the antenna performance parameters.
8. Describe various radar applications.
9. Explain radar performance factors.
10. Explain radar cross-section of targets.
11. Explain how system losses will affect the radar range.
12. What is the difference between pulse interval and PRF?
13. What is the maximum unambiguous range? How is it related with pulse repetition rate?
14. Calculate the range of a target, if the time taken by the signal to travel and return is 100 micro seconds.
15. If the peak power of a radar is 100 KW, PRF is 1000 Hz, and the pulse width is 1 micro second, calculate the average power in dB.
16. Determine the maximum unambiguous range and range resolution of a pulse radar having pulse width of 5 micro sec and PRF of 1 KHz.
17. With a maximum of 250 Km range, a radar is to be operated. Determine the maximum allowable PRF for unambiguous reception.
18. Calculate minimum receivable signal in a radar receiver that has an IF bandwidth of 1.5 MHz and a 9-dB noise figure.
19. Calculate maximum range of a radar system that operates at a 3-cm wavelength with a peak pulse power of 1 MW, if its minimum detectable signal is 4 mW, radar cross-section area is 15 m^2, and effective antenna aperture is 6 m^2.
20. Calculate the average transmitted power of a radar when the peak power is 200 KW, pulse width is 2 micro seconds, and rest time is 2000 sec.

CW Radar, FMCW Radar, and Pulse Radar

13

13.1 INTRODUCTION

The radar system main function is to detect the target first and then extract the parameters such as range and velocity of each target. The range and velocity estimation are mainly dependant on the accuracy of Doppler frequency and time delay measurement. A radar transmits either continuous waves or a sequence of pulses in most radar applications. This chapter describes the principle of operation and applications of CW radar, FMCW radar, and pulse radar.

13.2 CW RADAR

A continuous-wave (CW) radar system operating with a constant frequency can measure velocity, but not the range. A signal transmitted from an antenna which is having certain frequency is reflected back by the target with a slight change in frequency, i.e. Doppler frequency shift. Comparison of transmitted frequency with received frequency leads to determination of speed of target (but not its range). CW radars make use of sine wave of the form cos $2\pi f_0 t$, where the echo from clutter (i.e. stationary targets) is concentrated at f_0 in the frequency spectrum. The centre frequency extracted from echoes of moving targets, will be shifted by a frequency known as Doppler frequency, f_d. Thus by measuring the frequency shift ($f_0 \sim f_d$) target velocity can be determined accurately.

Doppler Effect

If either the transmitter or the receiver is in motion, resulting in an apparent shift in frequency, this is the Doppler effect and is the basis of the continuous-wave (CW) radar. Let us suppose the range of the target is R, and the wavelength is λ. By definition each wavelength (λ) corresponds to a phase change of 2π radians. The total wavelength for the two way propagation path (i.e. from radar to the target and its return to radar) is $2R/\lambda$.

Then the total phase change in the two-way path of the signal is given by,

$$\phi = 2\pi \times \frac{2R}{\lambda} \text{ rad}$$

$$\phi = \frac{4\pi R}{\lambda} \text{ rad} \tag{13.1}$$

Let us consider the target is in motion with respect to the radar, so when range (R) changes, the phase (ϕ) observed also changes. Then the rate of change of phase $\left(\frac{d\phi}{dt}\right)$ is referred to as angular frequency, ω_d. Therefore, the angular frequency can be obtained by differentiating Eq. (13.1) with respect to time.

$$\omega_d = 2\pi f_d = \frac{d\phi}{dt} = \frac{4\pi}{\lambda}\left(\frac{dR}{dt}\right) = \frac{4\pi v_r}{\lambda}$$

where, v_r = radial velocity = $\left(\frac{dR}{dt}\right)$ given in knots

Hence the Doppler frequency is given by

$$f_d = \frac{2v_r}{\lambda} \tag{13.2}$$

Usually, radial velocity is given in the units of knots (or kt), wavelength (λ) is in meters, then the Doppler frequency is measured in Hertz (Hz). Thus, Eq. (13.2) can be written as,

$$f_d = \frac{1.03 v_r(kt)}{\lambda(m)} \text{ Hz} \tag{13.3}$$

where, 1 knot = 1.852 Km/hr or 0.514 m/sec.

Figure 13.1 shows the block diagram of the simple CW radar. The CW radar uses the Doppler frequency shift principle to identify the moving target.

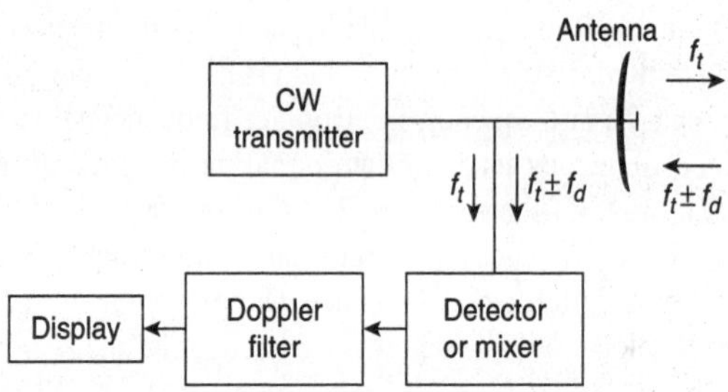

Figure 13.1 *The block diagram of a simple CW radar*

The main use of the Doppler filter is to filter and eliminate the echoes from the stationary targets, and also amplifies the weak Doppler echo signal so that it can be identified in the display device. These weak signals are filtered in frequency domain in the filter. The mixer mixes the echo signal at a frequency $f_t \pm f_d$ with the corresponding transmitted signal (or reference signal), f_t which is the frequency of continuous sinusoidal oscillations. While transmission, the antenna radiates a continuous sinusoidal oscillations at a frequency, f_t.

A portion of this radiated energy intercepted by the target and the re-radiated energy is collected by the receiver antenna. If we consider the target is moving with a velocity v_r, relative to the radar, then the signal received shifts in frequency from the transmitted frequency by $f_t \pm f_d$. If the target is moving towards the radar, then the frequency shift is given by $f_t + f_d$. Thus, the echo signal from a closer target has a higher frequency than that was transmitted. If the target is moving away from the radar, then the frequency shift is given by $f_t - f_d$. In order to use the Doppler frequency shift, the radar should be capable to distinguish the difference between received echo signal and transmitted signal. In the CW radar each target velocity produces a single distinctive Doppler frequency of the CW carrier. This results in clear Doppler measurement, which is a main advantage in CW radar. Whereas there is an uncertainty in the measurement of range using CW radar, because all the returned waveforms are continuous and hence the radar is not able to differentiate between the different echoes received. Most of the modern radars utilises a pulse waveform technology, in which single antenna is used for both transmitting and receiving functions.

Advantages of CW Doppler Radar

- It is simple, inexpensive, easy to maintain, and fully automated.
- It needs low power and is compact in size.
- Peak power is less, as duty cycle is unity.
- Stationary objects do not affect the performance of the radar.

Disadvantages of CW Doppler Radar

- There is a limitation by the power in the maximum range of the CW radar.
- Maximum power depends on the amount of isolation and the transmitter noise, which effects the receiver sensitivity.
- The target range cannot be obtained by the CW Doppler radar.
- When there are more number of targets there is a possibility of ambiguity.

Applications of CW Doppler Radar: CW Doppler radars are used to determine velocity information but not range; for instance,

- In cricket to measure ball speed
- In traffic to monitor traffic
- In police radar to catch cars exceeding speed limit
- In aircraft navigation to measure speed
- In planes as a rate-of-climb indicator for vertical takeoff planes

EXAMPLE PROBLEM 13.1

Calculate the Doppler frequency of an aircraft moving with a speed of 450 knots and when the CW radar is working with $\lambda = 9$ cms.

Solution

Given data $v_r = 450$ knots and $\lambda = 9$ cms

If we consider the radial velocity is in knots, Doppler frequency in Hertz and the radar wavelength in meters, f_d can be written as (as 1 knot is 1.852 Km/hr or 0.514 m/sec)

$$f_d = 1.03 v_r / \lambda$$

$$= 1.03 \times 450/(9\times10^{-2}) = 5150 \text{ CPS (CPS means cycles per second)}$$

■

EXAMPLE PROBLEM 13.2

Calculate f_d for a car traveling at a speed of 150 kmph with a CW radar of frequency 15 GHz.

Solution

Given data $v_r = 150$ kmph and $f = 15$ GHz

$$\lambda = \frac{3\times10^8}{15\times10^9} = 2 \text{ cm}$$

The Doppler frequency is given $\therefore f_d = 2v_r/\lambda$

$$f_d = \frac{2\times150\times1000}{60\times60}\times\frac{1}{2\times10^{-2}} = 4166.67 \text{ Hz}$$

■

EXAMPLE PROBLEM 13.3

A10 GHz police radar measures a Doppler frequency of 1800 Hz from a car approaching the stationary police vehicle in an 80 kmph speed limit zone. What should the police officer do?

Solution

Given f = 10 GHz, f_d = 1800 Hz

The police officer should calculate the relative velocity with which the car is approaching.

The Doppler frequency is given by $f_d = 2v_r/\lambda$

$$v_r = \frac{f_d.\lambda}{2}$$

$$\lambda = c/f = \frac{3\times10^8}{10\times10^9} = 0.03 \text{ m}$$

$$v_r = \frac{1800\times0.03}{2} = 27 \text{ m/sec}$$

$$v_r = \frac{27\times60\times60}{1000} = 97.2 \text{ kmph}$$

■

13.3 FMCW RADAR

The CW radar is usually limited in its transmitted power by interference between the transmitter and receiver, which should operate simultaneously. This limits their sensitivity and range. Since there is no timing reference, CW radar systems cannot measure range which is the main disadvantage. The target range may be measured by changing the transmitter frequency linearly with time; this is called a *frequency-modulated continuous-wave* (FMCW) radar. In this radar system beat signal is produced when a frequency-modulated signal is mixed with an echo. For a relative motion, CW radars receive a signal which is shifted by f_d but this is not helpful in determining the range from received echo signal. In FMCW radars, the transmitted signal is frequency modulated, so even if the target is stationary, the delayed signal is received with a different frequency. In FMCW radar the transmitted and received signal are mixed, which has information about the range and speed of the target.

The block diagram of FMCW Radar is shown in Figure 13.2.

A variety of modulations is possible; the transmitter frequency can slew up and down in the following manner:

- Sine wave
- Saw-tooth wave
- Triangle wave
- Square wave

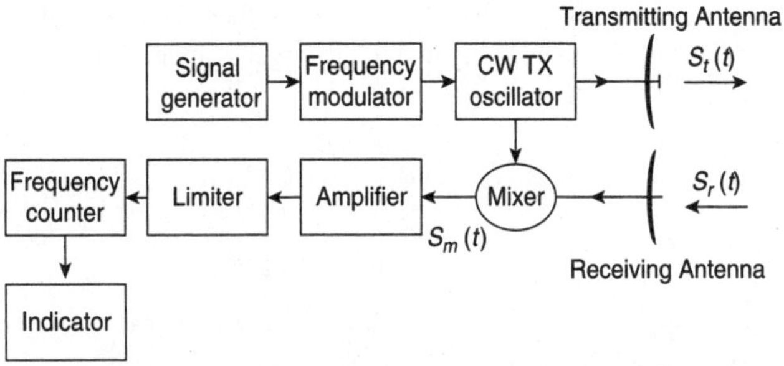

Figure 13.2 *Block diagram of FMCW radar*

In the FMCW radar, the transmitter frequency is varied with respect to time. If the transmitter frequency increases linearly with time, and the target is present at range R, an echo signal is returned as shown in Figure 13.3. The time difference between the transmitted and received signals is $\Delta t = 2R/c$; if the transmitted and received signals are multiplied within a mixer, filtering out the high-frequency term of the output will give a beat frequency f_b as shown in Figure 13.3. The beat frequency has to be amplified and limited for eliminating any amplitude fluctuations, this is done by using a amplifier and a limiter. This is measured by the cycle-counting frequency meter which is calibrated in distance.

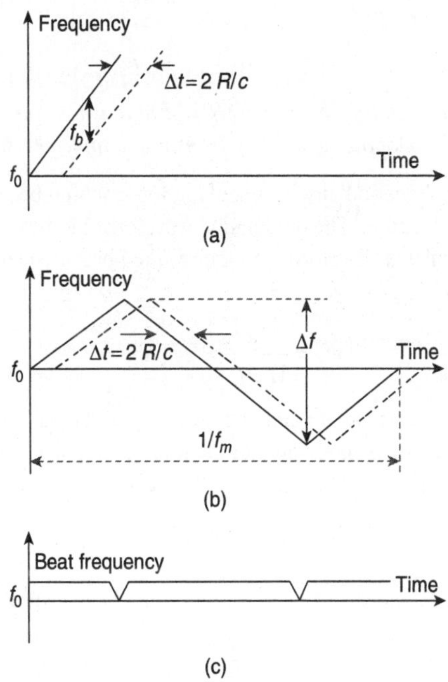

Figure 13.3 *Frequency–time relationship in FMCW radar; solid curve represents transmitted signal, dashed curve represents echo; (a) Linear frequency modulation; (b) Triangular modulation; (c) Beat frequency*

The beat frequency is a measure of the target's range when there is no Doppler shift in the signal i.e., $f_b = f_r$ (where f_r is the beat frequency only due to the target's range). If the slope of the frequency change in transmitted signal is m_f, then we get

$$f_b = \Delta t m_f = \frac{2R}{c} m_f \tag{13.4}$$

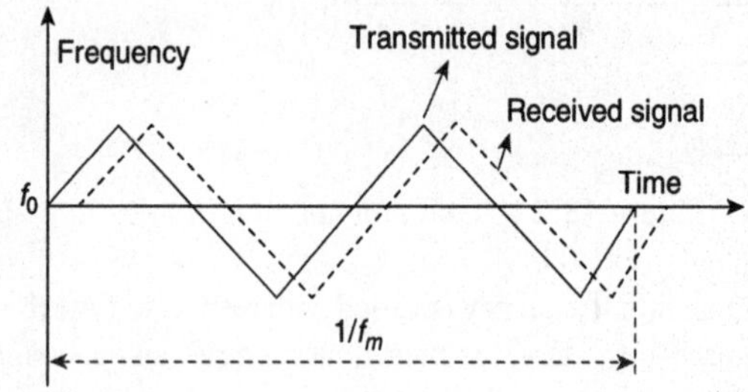

(a)

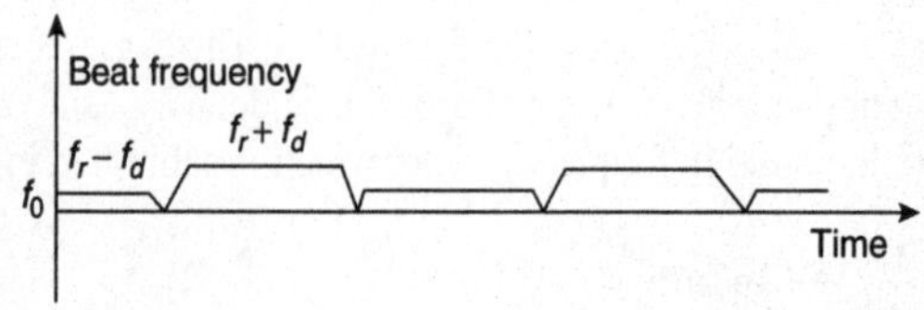

(b)

Figure 13.4 *Frequency–time relationships in FMCW radar when the received signal is shifted in frequency by the Doppler effect; (a) Transmitted and echo signal; (b) Beat frequency*

Basically in any CW radar, only periodicity in modulation is essential because the frequency cannot be continuously changed in one direction. The modulation waveform can be of any shape. Figure 13.4 show the triangular-frequency-modulated waveform and the resulted beat frequency using triangular modulation is shown in Fig 13.4(b). m_f is given as

$$m_f = \frac{\Delta f}{1/2f_m} = 2f_m \Delta f \tag{13.5}$$

The beat frequency is uniform everywhere and changes only at turn-around region. If the variation of frequency is Δf, and the frequency is modulated at the rate f_m, then the beat frequency where it is constant is arrived by substituting Eq. (13.5) in Eq. (13.4), then we get

$$f_b = \frac{4Rf_m \Delta f}{c} \tag{13.6}$$

From this Eq. (13.6), the range R can also be determined from the measurement of the beat frequency. If

$$K = \frac{4f_m \Delta f}{c}$$

then

$$f_b = KR \tag{13.7}$$

The frequency–time plot of the transmitted and echo signals for the moving target is given in Figure 13.4. The situation described in Figure 13.3 is the case when the target is stationary. If the target is moving, there will be a Doppler frequency shift superimposed to the beat frequency, and it should be considered in the demodulation. The Doppler frequency shifts the frequency–time plot of the echo (received) signal according to the relative direction of the target's velocity.

The Doppler shift increases the beat frequency in one portion and decreases the beat frequency in the other portion of the frequency modulation cycle exchanging it between f_{b1} and f_{b2} (Figure 13.4 (b)), where $f_{b1} = f_r - f_d$ and $f_{b2} = f_r + f_d$. By switching the frequency counter for every half-cycle, the beat frequencies f_{b1} (up) and f_{b2} (down) of the cycle can be measured separately. Average of the two beat frequencies gives the beat frequency, $f_r = (f_{b1} + f_{b2})/2$. Similarly Doppler frequency is given by, $f_d = (f_{b1} - f_{b2})/2$.

FMCW Altimeter (Main Application of FMCW Radar)

A typical altimeter utilises a transmitter power of about 1 to 2W and operates in C band. The target's range is calculated based on the measured delay Δt between the transmitted signal and the received signal, whereas the frequency offset Δf gives the velocity. From Δt and Δf, the height of the aircraft can be calculated as shown in Figure 13.5. The mixer output gives the frequency difference which is amplified and limited. This amplitude limited output is then fed to a frequency counter which in turn is fed to an indicator. The output thus obtained can be calibrated in feet or meters.

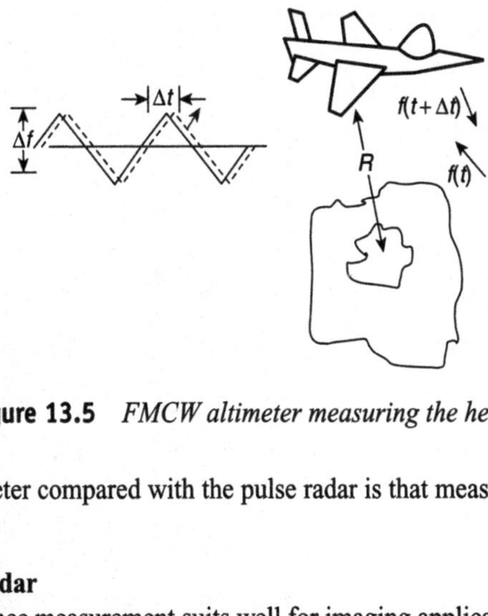

Figure 13.5 *FMCW altimeter measuring the height*

The advantage of the altimeter compared with the pulse radar is that measurement results are provided continuously.

Advantages of FMCW Radar

- High-resolution distance measurement suits well for imaging applications.
- Quick updating of measurement because of continuous transmitting signals.
- Functions well in all types of weather (rain, humidity, fog, and dusty) and atmospheric conditions, because electromagnetic radiations of short wavelength are used.
- Better in detecting tangential motion than Doppler-based systems.

Disadvantages of FMCW Radar

- More costlier than other competing technologies.
- Radar is mostly susceptible to interference from other radio devices.
- More computing power is required.
- It can be a disadvantage in defence applications because they can be easily blocked by electronic warfare systems.

Applications of FMCW Radar

- Used in high accuracy applications where repeatability and reliability is required, because FMCW radars provide accurate range measurement.
- In transportation, it is used as automotive collision avoidance radars and marine radars.
- Because of its resistance to dust, steam, heat, these are mostly used in the blast furnace of a steel mill.

EXAMPLE PROBLEM 13.4

An FMCW radar working at 4 GHz has frequency excursions of 60 MHz, a modulating frequency of 100 Hz, and a range of 3000 mts. Find the beat note frequency f_b.

Solution

Given data $\Delta f = 60$ MHz, $f_m = 100$ Hz, and $R = 3000$ mts

$$f_b = \frac{2R}{c} \times 2 f_m \times \Delta f = \frac{2\times 3000}{3\times 10^8} \times 2\times 100\times 60\times 10^6 = 240 \text{ KHz.}$$

■

EXAMPLE PROBLEM 13.5

Determine the range and Doppler velocity for an FMCW radar if the target is approaching the radar. Given the beat frequency f_b(up) = 15 KHz and f_b(down) = 25 KHz for the triangular modulation, the modulating frequency is 1 MHz and Δf is 1 KHz.

Solution

Given data $f_b(\text{up}) = 25$ KHz $\quad f_m = 1$ MHz

$f_b(\text{down}) = 15$ KHz $\quad \Delta f = 1$ KHz

$$f_r = \frac{1}{2}\left[f_b(\text{up}) + f_b(\text{down})\right] = 1/2\left[25+15\right] = 20 \text{ KHz}$$

$$f_d = \frac{1}{2}\left[f_b(\text{up}) - f_b(\text{down})\right] = 1/2\left[25-15\right] = 5 \text{ KHz}$$

$$f_r = \frac{4Rf_m\Delta f}{c}$$

$$\text{Range, } R = \frac{c\times f_r}{4 f_m \Delta f} = \frac{3\times 10^8 \times 20\times 10^3}{4\times 10^6 \times 10^3} = 300\times 20/4 = 1500 \text{ m}$$

$f_d = 2v_r/\lambda$ (Assuming radar operating frequency $f = 9.25$ GHz)

$$\lambda = c/f = \frac{3\times 10^8}{9.25\times 10^9} = 0.0324 \text{ m}$$

$$v_r = \frac{5\times 10^3 \times 0.0324}{2} = 81 \text{ m/sec}$$

■

13.4 PULSE RADAR

The pulse radar transmits signals in a sequence of pulse unlike CW radar which transmits continuous signal. Hence, it can be used to measure both the range and velocity of the target. Here the transmitted pulse consists of a burst of microwave signal with pulse duration of 100 ms to 50 ns, which is used to determine the target range based on the round-trip time delay of pulse. For a better resolution, shorter pulses are used and for a better signal-to-noise ratio, longer pulses are desired. To generate the pulses, either magnetrons or switched amplifiers are employed. In case of a non-coherent radar, there is no reference from the transmitter oscillator to the receiver; therefore, it cannot measure Doppler. Thus, only the distance can be measured but the velocity of a target cannot be determined.

The pulse radar makes use of pulse repetition frequencies (PRF) which ranges from 100 Hz to 100 KHz. In the higher range, PRF returns more pulses per unit time, thus enhancing the pulse radar performance. While lower range PRF prevents range ambiguities.

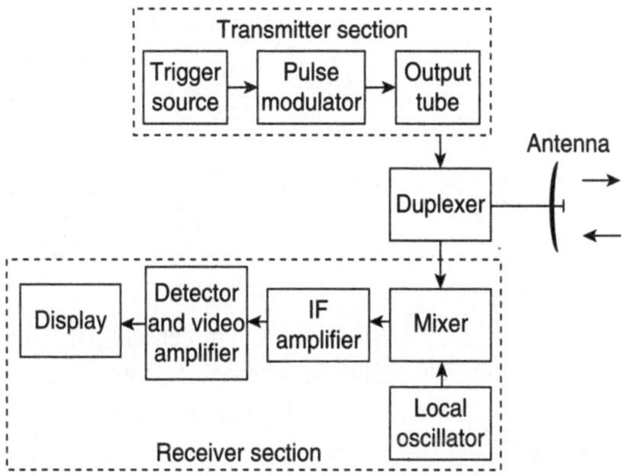

Figure 13.6 *Block diagram of pulse radar system*

As shown in the block diagram, pulses of suitable time and frequency are generated by the trigger source (Figure. 13.6). The output pulse from the trigger source is then sent to the modulator. Rectangular voltage pulses provided by the modulators are used as supply voltage for the output tube. The output tube can be a magnetron oscillator, a travelling wave tube, or klystron amplifier. The input rectangular pulses switches on and off in the output tube. The output of the tube is a strong pulse and is coupled to the antenna via duplexer and radiated by the antenna. If any target is encountered in the direction of propagation, an echo is produced. The echo signal in the direction of the radar will be collected by the radar antenna and directed to the receiving antenna via the duplexer. The echo signal which is received is amplified and demodulated. The antenna drive motor provides information regarding the azimuth (horizontal) and the elevation (vertical) of the target. The radar is capable of locating the exact target position by making use of this information. The time taken between the transmitting and receiving pulses is considered for measuring the distance of the target from the radar. Similar to the transmitter, the receiver also is connected to the antenna via duplexer.

A superheterodyne receiver is used, which consists of mixer, IF amplifier, detector, and video amplifier. The IF amplifier and the Mixer used here, must have very low noise values. So that the overall noise value of the receiver is not high. To achieve sufficient image frequency suppression, down conversion from microwave frequency to IF frequency is done in many stages. The IF output is fed into the detector which is a crystal diode. The crystal diode output is amplified by video amplifier which has the same bandwidth as that of IF amplifier. The output which is obtained is then fed to the display unit like CRT. The timing diagram for the pulse radar system is shown in Figure 13.7.

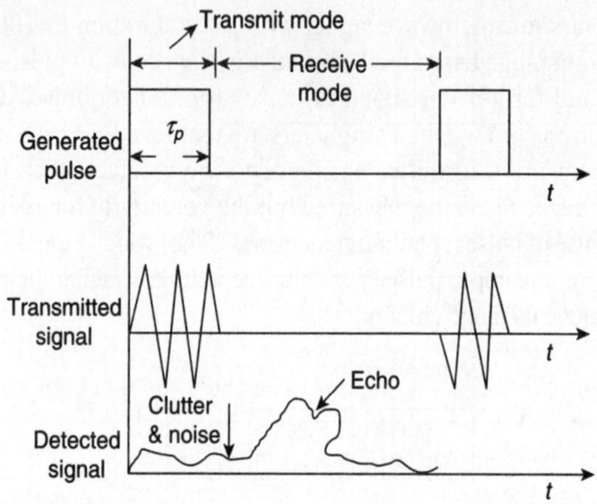

Figure 13.7 *Pulse radar timing diagram*

EXAMPLE PROBLEM 13.6

An FMCW radar operates at a frequency of 10 GHz. A symmetrical triangular modulating waveform is used, the magnitude of slope being 700 MHz/sec. The return from a moving target produces a beat frequency of 4 kHz over the positive slope and 3.5 kHz over the negative slope of the FM. Determine

(a) Target range
(b) Range rate
(c) Whether the target is moving toward or away from radar

Solution

Given $f_0 = 10$ GHz
Slope, $m_f = 700$ MHz/sec

(a) Target range

$$f_0 = \Delta t m_f = \frac{2R}{c} m_f$$

$$R = \frac{f_0 \times c}{2m_f} = \frac{10 \times 10^9 \times 3 \times 10^8}{2 \times 7 \times 10^8} = 2.1428 \times 10^9 \text{ meters}$$

(b) Range rate

$$f_d = 2v_r f_0 / c$$

$$f_d = \frac{1}{2}[f_b(\text{up}) - f_b(\text{down})] = \frac{1}{2}\left[4 \times 10^3 - 3.5 \times 10^3\right] = 250 \text{ Hz}$$

$$v_r = \frac{f_d \times c}{2f_0} = \frac{250\times3\times10^8}{2\times10\times10^9} = 3.75 \text{ m/sec}$$

(c) $f_r = \frac{1}{2}[f_b(\text{up}) + f_b(\text{down})] = \frac{1}{2}[4\times10^3 + 3.5\times10^3] = 3750 \text{ Hz}$

Here it is clear that $f_r > f_d$

Hence the target is moving towards the radar. ■

EXAMPLE PROBLEM 13.7

Determine the range and Doppler velocity of the target, if the target is moving away from FMCW radar. The beat frequency observed for triangular modulation as f_b (up) = 45 kHz and f_b (down) = 25 kHz. The modulating frequency is 3 MHz and Δf is 2 KHz.

Solution

Given that

$$f_b(\text{up}) = 45 \text{ kHz}$$
$$f_b(\text{down}) = 25 \text{ kHz}$$
$$\Delta f = 2 \text{ kHz}$$
$$\text{Modulating frequency } f_m = 3 \text{ MHz}$$
$$\text{Range } R = ?$$

Doppler velocity of the target, v_r =?

When the target is moving away from FMCW radar, the beat frequencies during increasing and decreasing portion of triangular wave are

$$f_b(\text{up}) = f_r + f_d$$
$$f_b(\text{down}) = f_r - f_d$$

where f_r is range frequency of radar

f_d is Doppler frequency

$$f_r = \frac{1}{2}[f_b(\text{up}) + f_b(\text{down})] = \frac{1}{2}[45 + 25] = 35 \text{ kHz}$$

$$f_d = \frac{1}{2}[f_b(\text{up}) - f_b(\text{down})] = \frac{1}{2}[45 - 25] = 10 \text{ kHz}$$

Then the range of FMCW radar can be measured by using the following relation

$$\text{Range, } R = \frac{c\times f_r}{4f_m\Delta f} = \frac{3\times10^8\times35\times10^3}{4\times3\times10^6\times2\times10^3} = 218.75 \text{ m}$$

and the Doppler velocity of the target can be obtained using (Assuming radar operating frequency $f = 9$ GHz)

$$v_r = \frac{f_d\times c}{2f_o} = \frac{10\times10^3\times3\times10^8}{2\times9\times10^9} = 166.66 \text{ m/sec}$$

■

EXAMPLE PROBLEM 13.8

For an unambiguous range of 60 nautical miles (1 nautical mile = 1852 m) in a two frequency CW radar. Determine f_2 and Δf when $f_1 = 3.2$ kHz. Derive the expression to solve this problem.

Solution

Derivation to find the unambiguous range in CW Radar

Consider a CW radar with the following waveform,

$$S(t) = A\sin(2\pi f_0 t)$$

The received signal from target at range 'R' is

$$S_r(t) = A_r \sin(2\pi f_0 t - \phi)$$

where, the phase ϕ is equal to,

$$\phi = 2\pi f_0 T$$

$$\Rightarrow \phi = 2\pi f_0 \frac{2R}{C} \qquad \left[\because T = \frac{2R}{C}\right]$$

where, C is velocity of light $= 3\times10^8$ m/s

Solving for R we obtain,

$$R = \frac{C\phi}{4\pi f_0} = \frac{\lambda}{4\pi}\phi$$

From the above equation we observe that, the maximum unambiguous range occurs when ϕ is maximum, i.e., $\phi = 2\pi$. Therefore, even for relatively large radar wavelengths, R is limited to impractical small values.

Now, consider a radar with two CW signals, denoted by

$$S_1(t) = A_1 \sin(2\pi f_1 t)$$

$$S_2(t) = A_2 \sin(2\pi f_2 t)$$

The received signals from target are,

$$S_{1r}(t) = A_{1r} \sin(2\pi f_1 t - \phi_1)$$

and

$$S_{2r}(t) = A_{2r} \sin(2\pi f_2 t - \phi_2)$$

where,

$$\phi_1 = 4\pi f_1 \frac{R}{C}$$

and

$$\phi_2 = 4\pi f_2 \frac{R}{C}.$$

After mixing with the carrier frequency, the phase difference between the two received signals is,

$$\phi_2 - \phi_1 = \Delta\phi = 4\pi \frac{R}{C}(f_2 - f_1) = 4\pi\Delta f \frac{R}{C}$$

Again R is maximum when $\Delta\phi = 2\pi$, substituting this in above equation, we get,

$$2\pi = 4\pi \frac{R}{C}\Delta f$$

$$\therefore R = \frac{C}{2\Delta f}$$

Problem

Given that,

Unambiguous range, $R = 60$ nautical miles = 111.12 km.

$$\text{Velocity, } C = 3\times10^8 \text{ m/s}$$
$$\text{Frequency, } f_1 = 3.2 \text{ kHz}$$
$$\Delta f = ?$$
$$f_2 = ?$$

The unambiguous range, R is given as,

$$R = \frac{C}{2\Delta f}$$
$$\Delta f = \frac{C}{2R}$$
$$\Delta f = \frac{3\times10^8}{2\times111.12\times10^3} = 1349.892$$
$$\therefore \Delta f = 1349.892 \text{ Hz}$$

f_2 is given as,

$$f_2 = \Delta f + f_1$$
$$\Rightarrow f_2 = 1.35\text{ k} + 3.2\text{ k} = 4.55 \text{ kHz}$$
$$\therefore f_2 = 4.55 \text{ kHz}$$

■

SUMMARY

1. The modern radar system basically detects intended targets besides estimating the position of target and its velocity.
2. A CW radar system which has constant frequency and is used to measure the velocity of the target in motion.
3. CW radars utilize CW waveforms, which may be considered a pure sine wave of the form Cos $2\pi f_0 t$.
4. Target range may be measured by changing the transmitter frequency linearly with time, which is called FMCW.
5. The CW radar is usually limited in its transmitted power by interference between the transmitter and the receiver, which should operate simultaneously.

6. FMCW radar is mainly used in aircraft altimeters to measure its height above the ground level.

7. The pulse radar transmits signals in a sequence of pulse unlike CW radar which transmits continuous signal. Hence, it can be used to measure both the range and velocity of the target.

8. The pulse radar uses pulse repetition frequencies (PRFs) ranging from 100 Hz to 100 kHz.

OBJECTIVE-TYPE QUESTIONS

1. The Doppler frequency shift produced by a moving target may be used in a pulse radar to
(a) combine moving targets from desired stationary objects
(b) determine the relative velocity of a target
(c) separate desired moving targets from desired stationary objects
(d) determine the displacement of a target

2. To operate with unambiguous Doppler, pulse repetition frequency is usually

(a) low (b) very low (c) high (d) very high

3. Increasing the pulse width in a pulse radar

(a) increases resolution (b) decreases resolution
(c) has no effect on resolution (d) increases the power gain

4. Which of the following is the biggest disadvantage of the CW radar?

(a) It does not give the target velocity. (b) It does not give the target position.
(c) A transponder is required at the target. (d) It does not give the target range.

5. The Doppler effect is used in

(a) MTI (b) pulse radar (c) FM (d) altimeter

6. The major advantage of the pulsed radar compared with the CW radar is that

(a) the pulsed radar readily gives the range of the target, whereas the CW radar cannot give the range information
(b) the pulsed radar can identify a target more easily than can the CW radar
(c) pulses get reflected from the target more efficiently as compared with CW waves
(d) pulses have both variation of magnitude and frequency

7. The minimum range of detection by the pulse radar depends on
(a) pulse width (b) average transmitter power
(c) beam width of the antenna (d) bandwidth of the antenna

ANSWERS TO OBJECTIVE-TYPE QUESTIONS

1. (b) 2. (d) 3. (b) 4. (d) 5. (d) 6. (a) 7. (a)

REVIEW QUESTIONS

1. Explain the CW radar with a neat block diagram.
2. Draw a block diagram of the FMCW radar and explain its operation.
3. Distinguish between CW radar and pulse radar.
4. What is Doppler principle?
5. Draw and explain the block diagram of a simple pulse radar system.
6. Give the advantages and disadvantages of the CW radar.
7. Write the applications of the CW radar.
8. What is the difference between a CW radar and a pulse radar?
9. What is the difference between an FMCW radar and a pulse radar?
10. Derive the expression for Doppler frequency in terms of radar velocity and wavelength.
11. Estimate the range of a FMCW radar, if its frequency is modulated at a rate f_m over a range Δf, given $\Delta f = 1.5$ kHz, $f_m = 100$ kHz and beat frequency is 40 Hz .
12. Determine the beat frequency and the quantization error if range = 100 m and the frequency excursion is 80 Hz and modulating frequency is 2 kHz (hint : range error $\delta R = \dfrac{c}{4\Delta f}$)
13. With a transmit (CW) frequency of 7 GHz, calculate the Doppler frequency seen by a stationary radar when the target radial velocity is 150 km/h.
14. A10 GHz police radar measures a Doppler frequency of 1600 Hz from a car approaching the stationary police vehicle in an 70 kmph speed limit zone. What should the police officer do?
15. Determine the range and Doppler velocity for FMCW radar if the target is approaching the radar. Given the beat frequency f_b (up) = 30 kHz and f_b(down) = 40 kHz for the triangular modulation, the modulating frequency is 1 MHz and Δf is 2 KHz.
16. An FMCW radar operates at a frequency of 9 GHz. A symmetrical triangular modulating waveform is used, the magnitude of slope being 700 MHz/sec. The return from a moving target produces a beat frequency of 4.85 kHz over the positive slope and 4.5 kHz over the negative slope of the FM. Determine
    (a) Target range
    (b) Range rate
    (c) Whether the target is moving toward or away from radar
17. Determine the range and Doppler velocity of the target, if the target is moving away from FMCW radar. The beat frequency observed for triangular modulation as f_b(up) = 60 kHz and f_b(down) = 25 kHz. The modulating frequency is 3MHz and Δf is 3 KHz.

MTI and Pulse Doppler Radars

14

14.1 INTRODUCTION

Radars usually receive the echoes scattered by the natural environment as well including land, sea and weather when detecting the targets in motion such as aircrafts or ships. These echoes from the natural environment are called clutter, which creates confusion on the radar display. When a clutter and the target echo appear in the same radar resolution cell, it is difficult to distinguish the aircraft, because the magnitude of the strength of the clutter echoes is many times larger than the aircraft echoes. The most efficient technique for detecting moving targets when there is large clutter is Doppler processing. When there is relative motion between radar and the moving target then there is a change in the frequency of the radar echo signal which is called as Doppler frequency shift. This Doppler frequency shift is used to remove clutter from the moving targets depending upon their velocity differences. This chapter deals with pulse characteristics, Doppler processing MTI, PDR and MTDs. The pulse radar that uses the Doppler shift is either Moving Target Indication (MTI) radar or Pulse Doppler Radar (PDR). MTI radar is used to detect the targets in presence of large clutter. Pulse Doppler Radar when used along with MTI can measure Doppler shift and radial velocity of the target where both the radars use a coherent signal. The working principle of both radars depends upon the predetermined time delay produced by the echoes from stationary targets return towards the receiver. In MTI radar a slow time signal is processed in time domain and produces limited information at a very low computational cost. But in Pulse Doppler Radar, signal is processed in frequency domain and produces more information and greater SNR improvement but it requires greater computation cost.

14.2 INTRODUCTION TO PULSE, MTI, AND PULSE DOPPLER RADARS

The differences between the MTI and Pulse Doppler Radar (PDR) techniques are discussed in this section for a better understanding of the MTI and Pulse Doppler Radar principles that are presented in the following sections. The operating principle of these radars is compared with the pulse radar also. The Doppler processing block where we apply the MTI and Doppler processing techniques is shown in the generic block diagram of radars (Figure. 14.1).

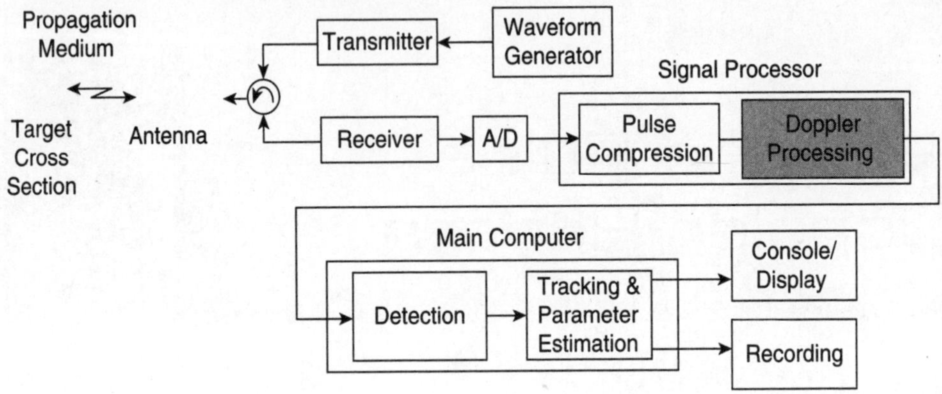

Figure 14.1 *MTI and Doppler Processing*

Pulse radars: Pulse radars transmit a high-frequency pulse (or signal) and receive the corresponding echoes before a new transmitted signal is sent out. They determine the azimuth, elevation, and range of the target from the measured antenna position and propagation time of the pulse (or signal). Direction, distance and the altitude of the target can also be found. The best examples of pulse radars are the weather radars.

MTI radars: The MTI radar's main objective is to reject the signals from stationary unwanted signals i.e., ground clutter, rain clutter, bird clutter, etc., that arrive as an echo along with the desired echo of the signal. In MTI, the processing system is used to eliminate unwanted clutter from the background and to detect moving targets even when the velocity of such targets is small relative to the radar platform. There are two basic types of MTI radars, namely,

- Coherent MTI and
- Non-coherent MTI

In coherent systems the receiver preserves the transmitted wave's phase in order to detect the Doppler shift in frequency. In coherent MTI systems, the Doppler shift in the echo signal from the moving target is used to differentiate it from stationary target. But in non-coherent MTI systems, the moving targets can be detected by observing the relative motion between the target and clutter background, and also by observing corresponding changes in amplitudes of pulses. Coherent detection requires dealing with the envelope of a signal, $g(t)$ and the phase of the sinusoidal carrier, $\phi(t)$. They need not be measured directly, but can be derived using inphase (I) and quadrature (Q) channels.

Pulse doppler radars: Compared with MTI, it is more advanced system. A pulse radar system makes use of Doppler Effect to obtain the target information (like velocity and amplitude) but not for clutter rejection purposes. There is a great dissimilarity between Pulse Doppler Radars and conventional pulse radars as this technique has been developed into many forms to get this distinction. The number of blind speeds can be reduced by using high PRF rates. Blind speed occurs if both Doppler frequency shift and PRF or PRF multiples are equal.

The differences between the MTI and Pulse Doppler Radar are given in Table 14.1:

Table 14.1 Differences between the MTI and Pulse Doppler Radars

MTI radar	Pulse doppler radar
1. Pulse radar that uses the Doppler frequency shift to differentiate moving target from fixed targets, operate at lower PRFs (< 4 KHz), and provide more accurate range resolution	1. Pulse radar that uses the Doppler frequency shift to differentiate moving target from fixed targets, operate at high PRFs (> 100 KHz), and provide better velocity discrimination and clutter rejection
2. MTI techniques only separate moving targets from clutter.	2. Besides eliminating clutter it also separates targets into different velocity regimes
3. Does not provide target velocity estimation	3. Provides good estimates of target velocity
4. Uses short waveforms (two or three pulses)	4. Uses long waveforms (many pulses, tens to thousands of pulses)
5. Has ambiguous Doppler measurement which is called *blind speeds* and unambiguous range measurement i.e., no second time-around echoes	5. PRF is high enough to operate with unambiguous Doppler (no blind speeds) but at the expense of range ambiguities
6. MTI radar uses magnetron oscillator as a source.	6. PDR uses klystron oscillator as a source.
7. Generally uses delay line cancellers	7. Generally uses range gated Doppler filters

14.2.1 Doppler Frequency

Terrestrial radars use MTI to distinguish moving targets of interest (e.g., Ship) from natural environment returns (e.g., from terrain and the ocean), called *radar clutter*. This is done using the Doppler-frequency shift of the received signals, where the Doppler-frequency shift is nothing but a shift added to the base signal when the target is moving towards or away from the radar. The clutter has only small radial velocity components due to the motion of vegetation or waves, while moving targets are likely to have larger radial velocities.

The relative motion between the wave source and observer causes difference in frequency at source and observer. This concept is most useful in astronomical measurements. When the radar pulse of frequency, f, impinges on a target moving with a certain radial velocity, the reradiated echo will be received at the radar with a frequency (f_r). The radial velocity v_r is nothing but the velocity of the target when radar and target are in line of sight (LOS).

The received frequency, $f_r = f + f_d$

where f_d is the Doppler frequency shift and is proportional to the radial velocity of the moving target. The relationship between the Doppler frequency and the target radial velocity (Doppler velocity), v_r is given by the equation

$$f_d = \frac{2v_r}{\lambda} \tag{14.1}$$

where λ is the wavelength of the radial carrier ($\lambda = c/f$) and c is the electromagnetic propagation velocity in a vacuum = 3×10^8 m/s.

v_r may be found from the successive range measurements, or from the Doppler-frequency shift of the return signal. From Eq. (14.1), we can say that the Doppler frequency is the frequency shift added to the base frequency of the radar. For example, if the radar base frequency is 2800 MHz, and the target is moving, the echo signal frequency is greater or less than 2800 MHz depending on whether the target is moving towards or away from the radar.

Proof for the relationship between Doppler frequency and radial velocity: The change in position of the reflecting object can be considered as angular velocity ($\omega = d\phi / dt$). Based on this, we can calculate the Doppler frequency. Let R be the distance of a target, v_r be the radial velocity and $2R$ be the roundtrip distance to the target which is equivalent to $2R/l$ wavelengths. In 2-way propagation, each wavelength corresponds to $2p$ radians of total phase change i.e., $(2R/l)\,2p = 4p\,R/l$ radians. If ϕ_0 is the phase of the transmitted signal, then the phase of the received signal is

$$\phi = \phi_0 + \frac{4\pi R}{\lambda}$$

If R changes, there will be a change in phase between pulses, which is $\omega = \frac{d\phi}{dt} = \frac{4\pi}{\lambda}\frac{dR}{dt} = \frac{4\pi}{\lambda} v_r$

By replacing the rate of change of phase $\omega = \frac{d\phi}{dt}$ with angular frequency, $\omega = 2\pi f_d$, we get

$$2\pi f_d = \frac{4\pi}{\lambda} v_r \Rightarrow f_d = \frac{2v_r}{\lambda}$$

14.2.2 Doppler Processing in CW, MTI, and PDRs

The Doppler processing shown in Figure 14.1 is mainly used to determine moving targets and stationary or moving clutter that are appearing in same time. The Doppler shift in CW radars can be determined by comparing the frequency of transmitted and echo signals. Using CW radar it is only possible to detect the direction and existence of a reflected object but not the range as there are no time marks to calculate the time interval. This restricts the CW radar to measure the speed of the moving objects based on Doppler Effect. In pulse radars, a single pulse is not sufficient to make a considerable frequency difference between the illuminated and echo signals using Doppler shift. By observing the variations in signal phase of successive echoes from the target, the Doppler shift is identified in MTI and Pulse Doppler Radars. The MTI filters out the received signal frequencies that have a low Doppler-frequency shift corresponding to the clutter. This is done by processing the phases of two or more successive pulses in a canceller.

14.2.3 MTI and Pulse Doppler Radar Transmitted Pulses and Pulse Processing

The rectangular blocks shown in Figure 14.2 are pulses. This figure illustrates the radar transmitted pulses, PRF, PRT, PRI, and CPI. The number of pulses being sent out is M, and the time between the pulses is T_r, called *pulse repetition interval* (PRI). The inverse of T_r is the pulse repetition frequency (PRF). PRF is generally of Hz or KHz. The time between the sending of all pulses (T_c) is referred to as *coherent processing interval* (CPI), and it is determined by multiplying the number of pulses with the pulse repetition interval (i.e. $T_c = M \times T_r$). M is generally 2, 3, or sometimes 4 for MTI.

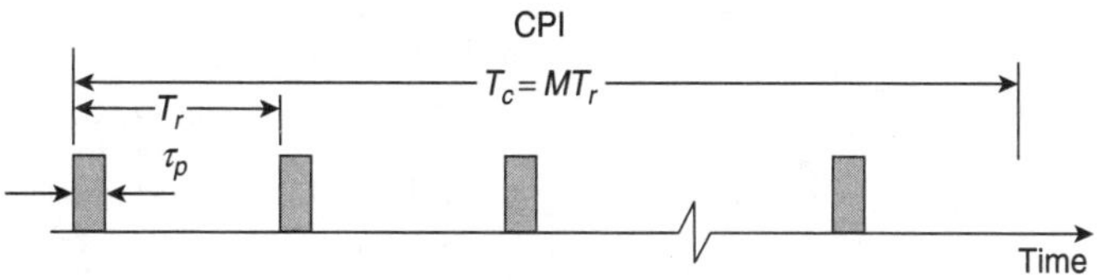

Figure 14.2 *A timing diagram of the radar transmitted pulses*

where τ_p = Pulse width

$B = 1/\tau_p$ =Bandwidth

T_r = Pulse repetition interval (PRI)

$f_r = 1/T_r$ =Pulse repetition frequency (PRF)

$\delta = \dfrac{\tau_p}{T_r}$ = Duty Factor of pulse waveform transmitted

$T_c = MT_r$= Coherent processing interval (CPI)

M = Number of pulses in the CPI, which may be equal to 2, 3, or 4 in case of MTI radar

Digital pulse processing: As shown in the receiver side of the radar block diagram in Figure 14.3 (a), the echoes that come out from the target are in analog form, and this will go through analog to digital (A/D) converter block. The A/D converter samples the analog-form signal, converts it into digital form, and stores it digitally. As mentioned earlier, in MTI and PDRs, the Doppler shift is found by observing the change in signal phase from consecutive echoes from the target. Therefore, we have to characterize the received echo by two numbers in the two-dimensional x axis and y axis. Since the received echo is a vector, it contains the amplitude and phase information. We can get information about both amplitude and phase by sampling the signal in-phase and quadrature channels.

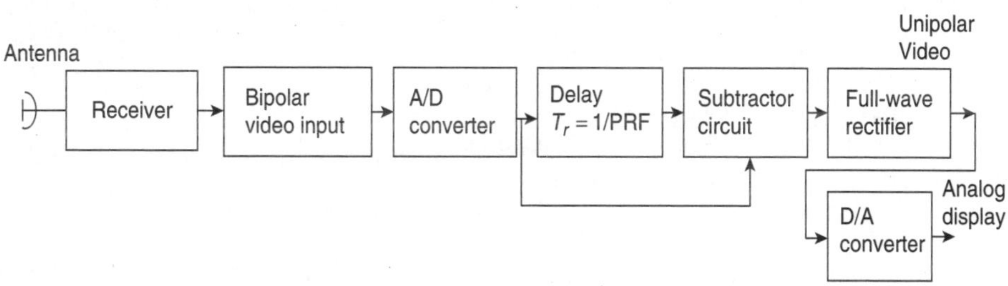

Figure 14.3 (a) *MTI receiver block diagram*

***I/Q* detection:** In *I/Q* detector, the received IF signal is divided into two channels, in-phase (*I*) channel and quadrature (*Q*) channel. In the *I* channel, the signal is in-phase and mixed with the transmitted signal, while in the *Q* channel, the signal phase is shifted by 90° and mixed with the transmitted signal. Irrespective of target movement with respect to transmitter, the two Doppler signals generated by the *I* and *Q* channels will have same frequency but have 180° phase difference between each other. This phase difference gives the direction of the target. The outputs from both the channels are sampled and complex fast Fourier Transform (FFT) is applied to determine the magnitude and phase of the combined Doppler spectrum. *I/Q* detection provide a gain of 3dB in SNR at the output of the Doppler spectral analysis function.

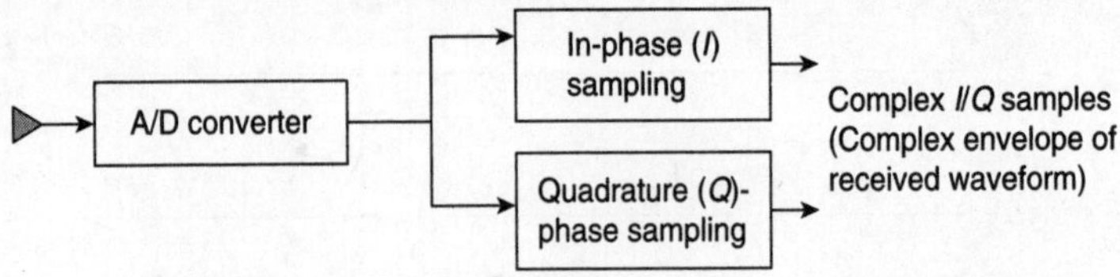

Figure 14.3 (b) *I/Q channel processing in MTI system*

Sampling of a received echo: Here, we are going to discuss how we are digitally storing information about the target, which is used for further processing either by using an MTI filter or by using Doppler filter banks. Let us suppose we are sending *M* pulses as shown in Figure 14.4 (a) one after another, which are separated by or delayed by one PRF. We send the pulse out and listen to the echo, which will be sampled at a radar receiver; each sample corresponds to some range. Each of the dots corresponds to a sample time (Figure 14.4 (a)). For example, if we consider the 12th sample, suppose it has a range of 8.3 km after the first pulse is sent, we will store this information in a matrix format as shown in Figure 14.4 (b) in which rows represent sample numbers and columns represent pulse numbers, which corresponds that pulse 1 at 12th sample we will store its range as 8.3km. Again, we will send a second pulse and we will get a range at each and every sample; this process will continue for *M* pulses; finally, after sending *M* pluses, we will get a whole bunch of range returns along the column, which is known as a *range bin*.

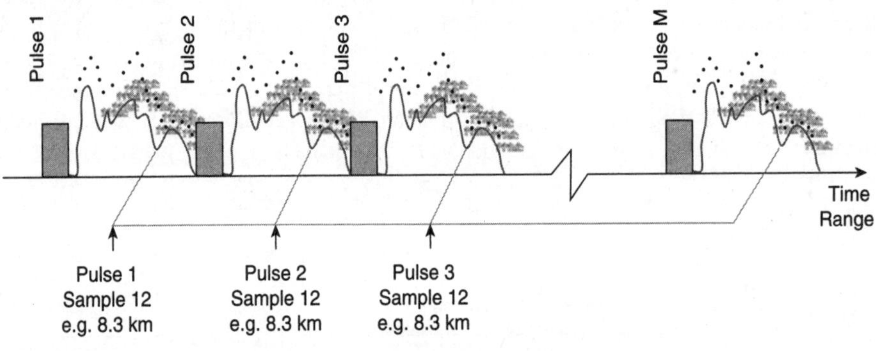

Figure 14.4 (a) *Sampling of the received echo*

Two-dimensional data matrix: Figure 14.4 (b) describes a two-dimensional matrix. The coherently demodulated baseband returns forms this two-dimensional matrix. We need to fill the matrix with real values of samples. For imaginary values another similar matrix is required. The range bins from the returns of the single pulse forms the samples in each column. The element in each column is complex number indicating the real and imaginary parts of the range bin. Thus, the measurements from the same range bin over successive pulses are represented in each row. The pulse number or range dimension is represented as vertical in Figure 14.4(b), and slow time or sample number dimension is represented as horizontal in Figure 14.4(b). The slow time is sampled at the pulse repetition interval of the radar.

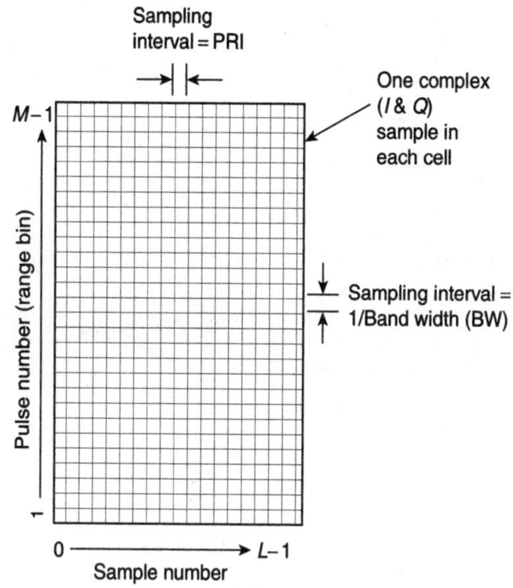

Figure 14.4 (b) *Two-dimensional matrix (each cell is one complex number)*

In MTI processing a slow time data sequence is passed through a linear filter to minimize the clutter component. It is shown in Figure 14.4(c). The output of this linear filter is a new slow time sequence containing some noise components gives one or more targets. This signal is applied to a detector which declares presence of target if the amplitude of filtered signal is greater than the detector threshold. Note that in MTI processing, detector output gives only information about presence or absence of target.

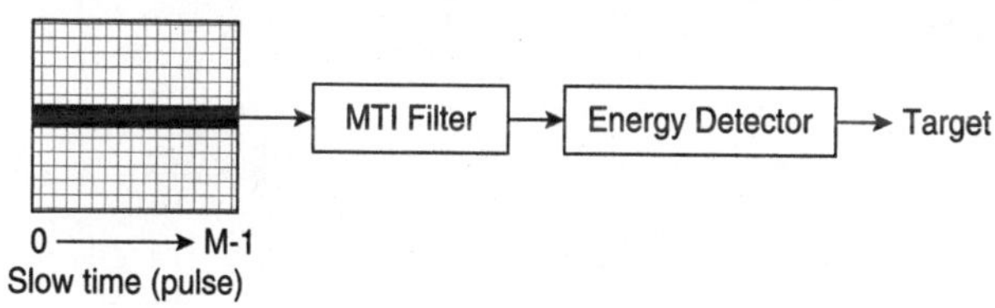

Figure 14.4 (c) *MTI filtering and detection process*

14.3 MTI RADARS

As shown in Figure 14.5, a delay line canceller and a phase comparator are used by the MTI radar to detect moving targets. The signals from the receiver and the transmitter are sent to the phase comparator where both the signals are compared. The output is then displayed in the display as shown in Figure 14.6. The figure shows that, for a fixed target, the amplitude is constant, whereas for moving targets, the amplitude changes continuously.

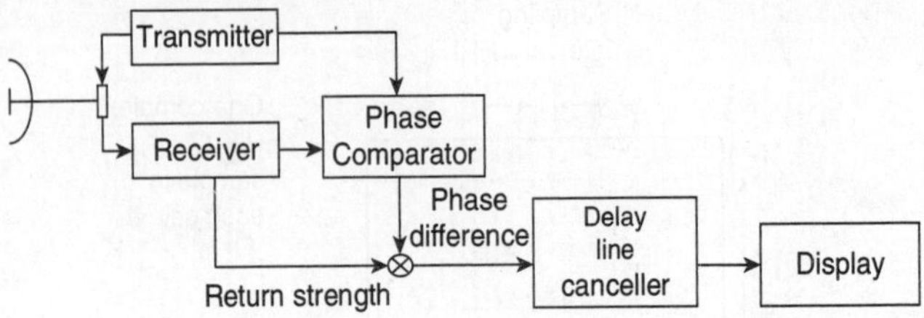

Figure 14.5 *MTI phase comparator radar system*

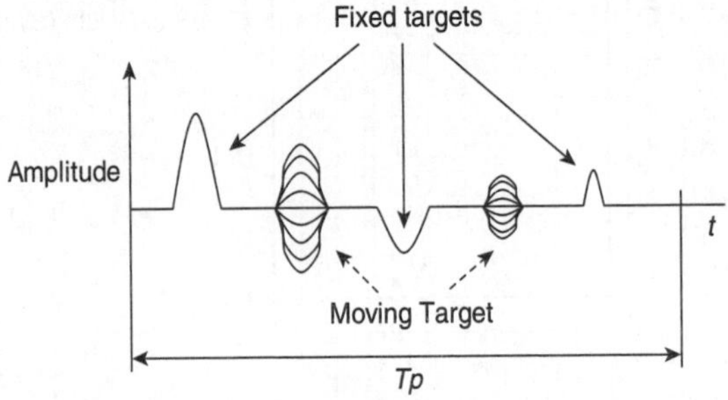

Figure 14.6 *Phase detector output for fixed and moving targets*

The output of phase comparison changes between its extreme values when the target range is varying. The phase shift of one full cycle is completed for one-half wavelength (of carrier frequency) change in range.

The output from the phase comparator is delayed for a period of 1 PRT. This time delayed output is then subtracted from the subsequent pulse, thus determining the target motion. If signals are from a stationary object, they will be similar for all pulses and are cancelled, as our objective is to detect moving object. The outputs from phase detector and delay line canceller (Figure 14.7) are compared to detect moving targets. Signals from moving targets are amplified as shown in Figure 14.8.

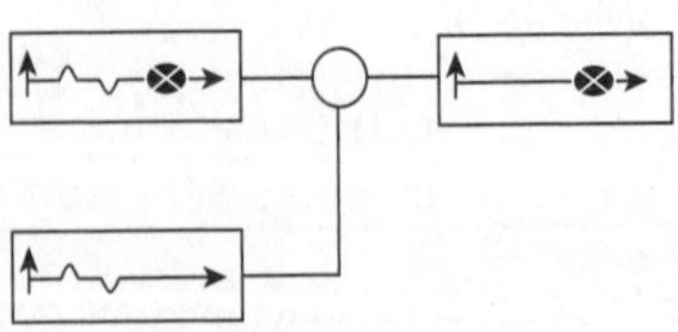

Figure 14.7 *Cancellation circuit of MTI processor*

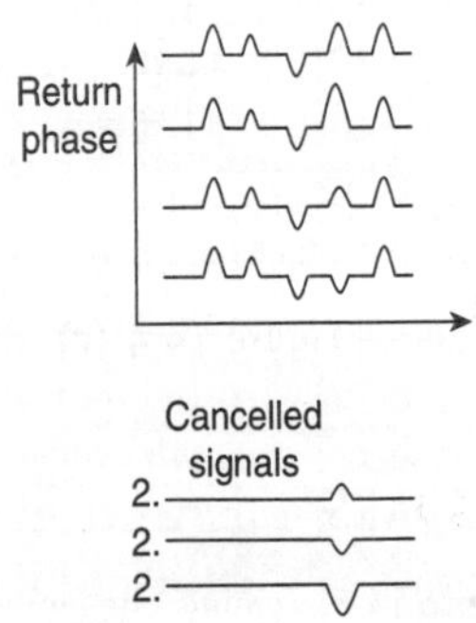

Figure 14.8 *Cancellation of echoes*

The MTI processor averages a sample obtained from the phase comparison output over a few cycles. The average value is zero for moving targets and for stationary targets, it is nonzero. Before showing the output, the average value is subtracted from output thus stationary targets are eliminated.

The range of stationary targets can not be changed where as for moving transmitters; returns from stationary objects on ground will have change in range. For moving transmittérs, the MTI systems should produce a modified input to the phase comparator. This modified input includes phase advance which gives information about the motion of transmitter. MTI radars are more suitable for search functions and also useful for clutter rejection. MTI radars typically classified as motion or movement detectors based on their limitations in unambiguous velocity measurements. For an aircraft flying at low level, MTI radar mode is used for searching aircrafts but it has some velocity ambiguities even though the clutter is reduced. So the radar uses higher PRF and pulse Doppler mode to reduce velocity ambiguities.

14.4 DELAY LINE CANCELLERS OR PULSE CANCELLERS

The delay line canceller is also known as a *transversal filter* or *tapped delay line filter* or *non-recursive filter,* or *moving average filter*, or *finite impulse response filter*. For moving targets, the amplitude continuously changes due to Doppler frequency shift. So for each pulse the amplitude varies. While filtering, one pulse is subtracted from the other pulse. Due to subtraction of pulses, fixed clutter echoes are cancelled and they are not detected; whereas in case of moving target echoes, they are not cancelled and are detected. To implement the above reasoning, a single delay line canceller, also referred to as the *two-pulse MTI canceller* or *first-order canceller*, is used to eliminate the clutter.

Figure 14.9 is a block diagram of a single delay line canceller, and Figure 14.10 is its spectral response. The input is a sequence of base-band complex (in-phase and quadrature phase, or I and Q) data samples from the same range bin over successive pulses, with a sampling interval T that is equal to the pulse repetition interval.

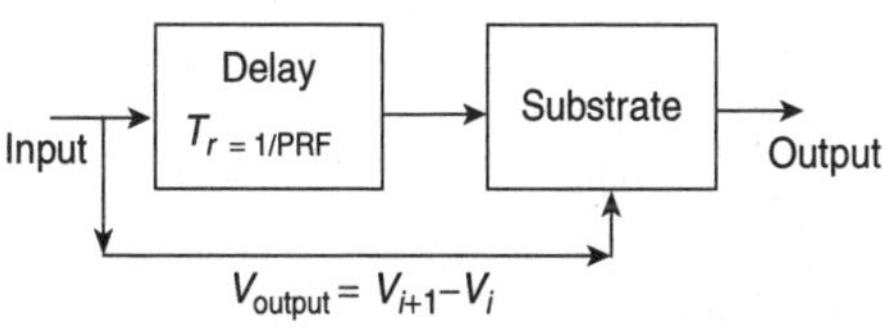

Figure 14.9 *Single delay line canceller*

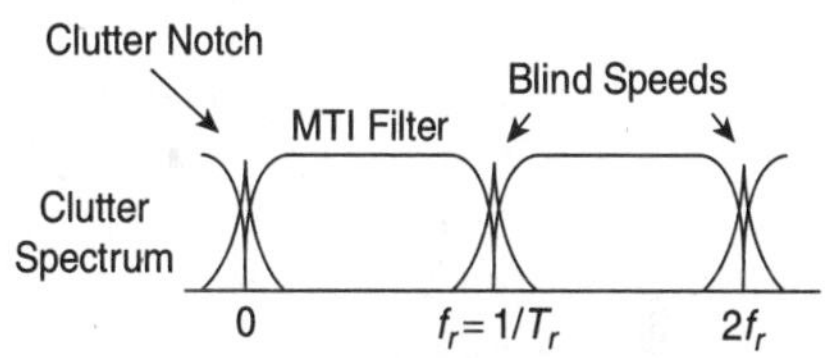

Figure 14.10 *Spectral response of single canceller for clutter*

The discrete time filter's frequency response is periodic and their periodicity (PRF) is 1/PRT Hz of the Doppler frequency shift. As shown in Figure 14.10, the MTI filters have nulls at zero frequency due to clutter notches and at Doppler frequencies which are multiples of the PRF due to blind speeds. So the MTI filter filters the echoes of the target moving with a radial velocity due to Doppler shift which is equal to the integer multiple of the PRF. The output of the single delay line canceller is zero, when the Doppler frequency of the target observed is

$$f_d = n/T_r = nf_r \tag{14.2}$$

where n is a integer (which is given as 0, ±1, ±2,...) and f_r is the pulse repetition frequency.

So from the Figure 14.10, we can observe that the delay line canceller cancels the d.c component caused by the clutter (n = 0) and it also rejects moving targets with doppler frequency (f_d) which is equal to PRF or multiples of PRF.

14.4.1 Blind Speeds

The spectral responses of the delay line canceller (or MTI filter) have nulls at Doppler frequencies that are multiples of the PRF. From this, it is concluded that Doppler frequencies which are multiples of PRF are attenuated. The echoes of the targets with corresponding velocities are subjected to attenuation and they are not detected by the radar. The *blind speeds* are defined as the target velocities which cause zero MTI response when the doppler frequencies are equal to PRF or multiples of PRF. These blind speeds occur at

$$v_{bn} = \frac{n\lambda f_r}{2} \tag{14.3}$$

where v_{bn} – Blind speed (m/s),

f_r – Pulse repetition frequency (PRF) (Hz) = n/T_r

λ – Transmitted signal wavelength (m),

$n = 0, \pm 1, \pm 2, \ldots$

The velocity of target is called the *blind speed* when the Doppler frequency (f_d) = $1/T_r = f_r$. The echoes of the moving target are continuously passed through the filter. The filter has iterative nature and has a succession of pass-bands which are similar to the spans of a bridge. The Doppler frequency measured during the generation of Doppler frequencies by the target velocities in the first pass-band represents the target's radial velocity. However, the Doppler frequency measured may not be due to radial velocity as it is measured by pulse radar. The Doppler frequency measured may be the difference of itself and multiple of PRF.

$$f_d + \text{nPRF} \tag{14.4}$$

Due to the presence of blind speeds within the Doppler frequency band, the ability of radars may be reduced. In order to get high first blind speed it is essential to operate with high PRF or with longer wavelength. But operating a radar with high PRF causes range ambiguities, which are not suitable for surveillance or long-range tracking radar applications. However, by operating radars at more than one PRF, we can achieve both range unambiguity and Doppler unambiguity.

14.5 STAGGERED PRFS TO INCREASE BLIND SPEED

The Pulse Doppler and MTI Radars cannot measure velocity at distinct values of PRF and measure ambiguous velocities because of their pulsed nature. This ambiguity takes place when PRF or multiples of PRF and Doppler frequency are equal. The velocity of first blind speed may exceed actual velocity for high PRF. So by using high PRF, the blind speeds can be avoided in pulsed Doppler radars.

However, the use of high PRF would cause a phenomenon called *range ambiguity*.

Range ambiguity: The *maximum unambiguous range*, R_{max} is defined as the highest range in which the echo of a transmitted pulse can reach the radar receiver before transmission of the next pulse. All targets at a range shorter than R_{max} = (C/2PRF = $CT_r/2 = C \times$ (maximum value of T_r)/2) are in a one-to-one correspondence with the range as measured by the radar.

However, targets at a range, R_n (= $C\,(T_r + \Delta T_r)/2$), that is, the targets are beyond the R_{max}, would appear to the radar to be at a range, R(= $C\,(\Delta T_r)$ / 2). Therefore, to avoid this range ambiguity, PRF should be low enough to ensure that all targets of interest are within R_{max}. The following example illustrates the range ambiguity with high PRF.

EXAMPLE PROBLEM 14.1

Show that if PRF is very high, which is the condition to avoid velocity ambiguity that causes ambiguous ranges?

Solution

Consider a radar operating at a frequency of 3 GHz, for a target moving with a velocity of 300m/sec; we can calculate the Doppler frequency as $f_d = \frac{2v_r}{\lambda}$

$$f_d = 2v_r/\lambda = 600/0.1 = 6 \text{ KHz.}$$

A radar designed for a 140 Km range, corresponding to an echo return time (where $t = 2R/c$, c = velocity of light) of 1 msec, must have a PRF (where PRF = $1/T_r$, $T_r = t$) less than 1 KHz in order to avoid range ambiguity. However, in such a case, in order to avoid velocity ambiguity, the maximum acceptable Doppler frequency should be equivalent to 1 KHz, that is only 1/6th of the required Doppler frequency. Alternatively, high PRF waveforms, which increase the blind speeds, are extremely ambiguous in range.

■

Range and Doppler ambiguities for different PRFs are summarized in Table 14.2. However, a radar system such as Pulse Doppler Radars (PDR) utilizes high PRFs and is discussed in the following sections:

Table 14.2 PRF ambiguities

PRF	Range ambiguous	Doppler ambiguous
Low PRF	No	Yes
Medium PRF	Yes	Yes
High PRF	Yes	No

Staggered PRF: Radars can detect target using constant and/or varying (agile) PRFs. Low PRF causes velocity ambiguities and high PRF causes range ambiguities. To avoid blind speeds, MTI radars use PRF agility which is known as *PRF staggering* or *staggered PRF*. The use of staggered PRFs raises the first blind speed (or null response) significantly without significantly degrading the ambiguous range. PRF agility is also used to avoid range and Doppler ambiguities.

PRF staggering can be achieved on either a pulse-to-pulse or CPI-to-CPI (CPI stands for Coherent Processing Interval) basis. The CPI-to-CPI case is used in Pulse Doppler Radars. In pulse to pulse, the pulse repetition interval varies from one pulse to another pulse within a single coherent processing interval (CPI). In general, pulse–pulse is used only in low PRF modes in which no range ambiguities are expected.

14.6 DOUBLE DELAY LINE CANCELLER

The single canceller frequency response is shown in Figure 14.11. The periodicity of the response is $1/T$. The rectified sine shape has a finite slope at $f = 0$, resulting in fast rise of the response. The resulting attenuation of the low frequencies may not be adequate for few applications in radars. This means that it is a poor approximation to an ideal high-pass filter for clutter suppression. The next step in MTI filtering is the double canceller (three-pulse canceller).

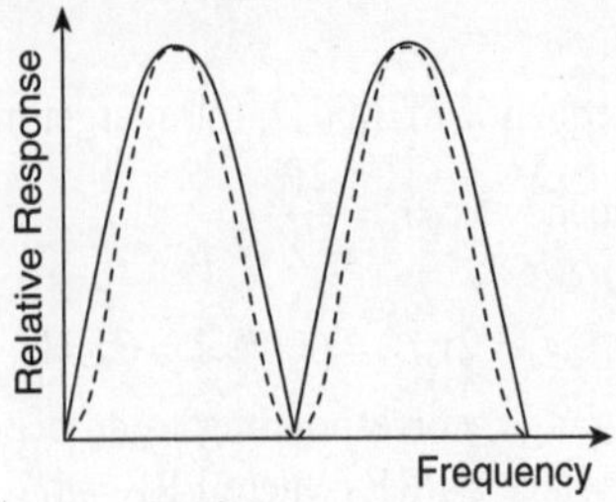

Figure 14.11 *Frequency response of a single canceller (solid line) and double canceller (dashed line)*

When the two single cancellers are cascaded, the frequency response characteristics of the single delay line canceller are improved or widened. The two cascaded single line cancellers are known as a double canceller (or a three-pulse canceller). The null depth and width in the vicinity of zero doppler can be improved by the Doppler three-pulse canceller and only two subtractions per output sample are required. Even though null depth of zero Doppler is improved, there is still a large variation in the filter gain or attenuation for moving targets at various Doppler shifts away from the zero Doppler (Figure 14.12 (a)). Clutter cancellation is done better by a double delay canceller with a broad spectrum than the single delay MTI.

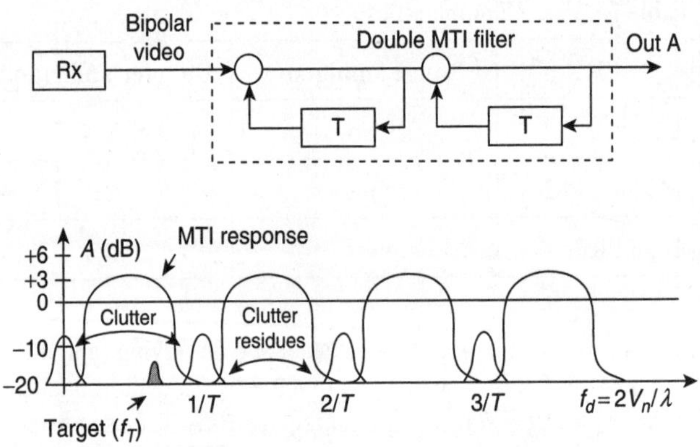

Figure 14.12 (a) *A double delay canceller*

$$\text{3-Pulse MTI } V_{output} = V_i - 2V_{i-1} + V_{i-2} \tag{14.5}$$

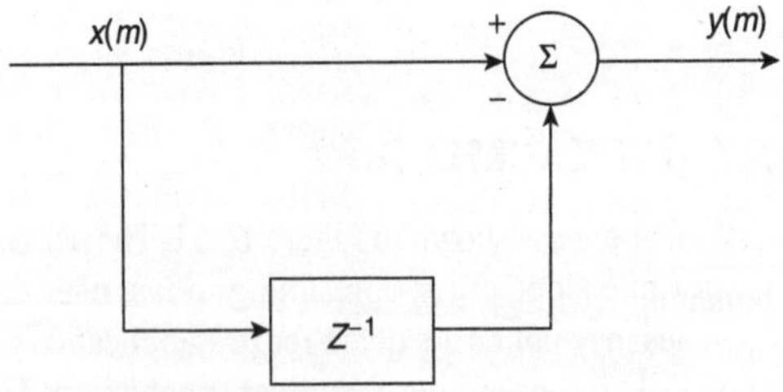

Figure 14.12 (b) *2-pulse canceller*

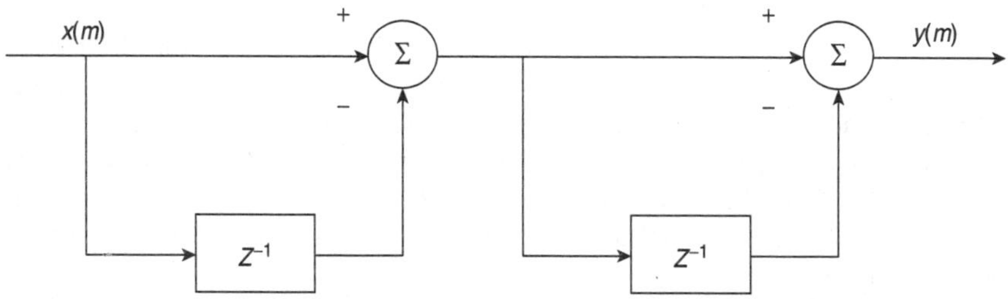

Figure 14.12 (c) *3-pulse canceller*

The transfer function of the 2-pulse canceller is $H(z) = 1 - z^{-1}$
The transfer function of the 3-pulse canceller is $H(z) = 1 - 2z^{-1} + z^{-2}$
Three-pulse canceller gives a wider clutter notch and higher clutter attenuation.

Such cascading can be done for N-pulse cancellation, by cascading $N - 1$ two-pulse canceller sections. Hence, the transfer function of the N-pulse canceller is,

$$H_N(z) = (1 - z^{-1})^{N-1} \tag{14.6}$$

EXAMPLE PROBLEM 14.2

With the specifications mentioned: wavelength = 0.1 m, PRF = 200 Hz. Calculate the blind speeds for a radar.

Solution

Data given: $\lambda = 0.1$ m, PRF $= f_r = 200$ Hz
Blind speeds occur at $v_{bn} = n\lambda f_r / 2$

$v_{bn} = n/2\ (0.1 \times 200) = 10n$ m/sec, where n can be any integer value

Therefore, the radar would be blind in terms of velocity to all targets having relative radial velocities of 10, 20, 30 m/sec, and so on. ■

14.7 MTI RADAR PERFORMANCE ANALYSIS

MTI filters are used to attenuate the clutter. But the MTI filters attenuate or amplify the echoes of the target along with clutter due to blind speeds. To evaluate the performance of various MTI filters used in MTI radars, four quantities are introduced. They are as follows: clutter attenuation, MTI improvement factor, sub-clutter visibility, and Canceller ratio.

1. **Clutter attenuation (CA)**

   It measures only the reduction in clutter power at the output of the MTI filter compared with the input that can be represented by the ratio of the clutter power at the filter input to the clutter power at the filter output. It is very easy to measure clutter attenuation.

$$CA = \frac{\text{clutter power into canceller or filter}}{\text{clutter power remaining after cancellation}} = \frac{\int_{-\infty}^{\infty} S_c(\omega)\, d\omega}{\int_{-\infty}^{\infty} S_c(\omega) \left|H_c(\omega)\right|^2 d\omega} \tag{14.7}$$

   where $S_c(\omega)$ is the clutter power spectral density, and $H_c(\omega)$ is the filter characteristic.

2. **Sub-clutter visibility (SCV)**
The ability of a radar to detect non-stationary targets implanted in a strong clutter background with a given signal-to-clutter ratio (SCR) is called as Sub-clutter visibility (SCV). It is a more complex measure that should be taken into account for the detection of false-alarm probabilities and detector characteristics. The MTI is intended to detect a target that produces a signal which is weaker than the clutter signal. The sub-clutter visibility is defined as the clutter-to-signal ratio present at the input of the MTI that permits target detection at the output for some probabilities of detection and false alarm.
3. **MTI Improvement Factor**
The efficiency of an MTI filter is usually measured by its improvement factor, *I*, which quantifies the increase in signal-to-clutter ratio due to MTI filtering. In mathematical form, the *improvement factor* is defined as

$$\text{MTI improvement factor} = I = \frac{(\text{signal / clutter})_{\text{out}}}{(\text{signal / clutter})_{\text{in}}} = \frac{S_0 / C_0}{S_i / C_i} = \frac{C_i}{S_i}\frac{S_0}{C_0} \tag{14.8}$$

As mentioned earlier, S_0/C_0 should be equal to the minimum signal-to-noise ratio (SNR), so that

$$I = \frac{C_i}{S_i}(SNR)_{\min} = (SCV)\,(SNR)_{\min} \tag{14.9}$$

Another way of expressing the MTI improvement factor, *I*, in terms of clutter attenuation is
I = MTI improvement factor = CA × G
where CA = clutter attenuation = C_i/C_0
$G = S_o/S_i$ = average power gain over all possible Doppler values
The improvement factor depends on internal elements (receive-transmit circuit's phase and stability in time) and external elements (stability of the clutter that should be cancelled).
4. *Canceller ratio* is defined as the ratio of the canceller voltage amplification to the gain of single unprocessed pulse.

14.8 TYPES OF MTI RADARS

There are two basic types of MTI radars, namely coherent and non-coherent MTI. The coherent MTI radar differentiates moving targets from stationary targets by using the Doppler shift passed on to the reflected signal by a moving target. The non-coherent radar detects the moving targets by the relative motion between the target and the clutter background and, thus by the corresponding amplitude changes from pulse to pulse or from one antenna scan to the next. In *Coherent systems*, the phase of the transmitted wave should be saved so that it is used by the receiver to detect Doppler shift in frequency, but it is not required in non-coherent systems.

The composite echo signal fluctuates in both phase and amplitude when it is reflected from a moving target and clutter. The Doppler component of moving target can be identified by the coherent Pulse Doppler radar and MTI radar using the fluctuations in phase of the signal. The MTI radars are categorized into two types.

- Non-coherent MTI radars
- Coherent MTI radars

14.8.1 Non-Coherent MTI Radars

A *non-coherent radar* is defined as an MTI radar which uses amplitude fluctuations instead of phase fluctuations. An internal coherent reference signal or a phase detector is not needed by non-coherent MTI radar. The IF amplifier used in this radar should be linear and should have large dynamic range. It can be logarithmic so that logarithmic-gain characteristic provides protection from saturations. Amplitude detector is used after the IF amplifier and this detector is a conventional detector. The local oscillator need not be very stable. Figure 14.13 shows the block diagram of noncoherent radars.

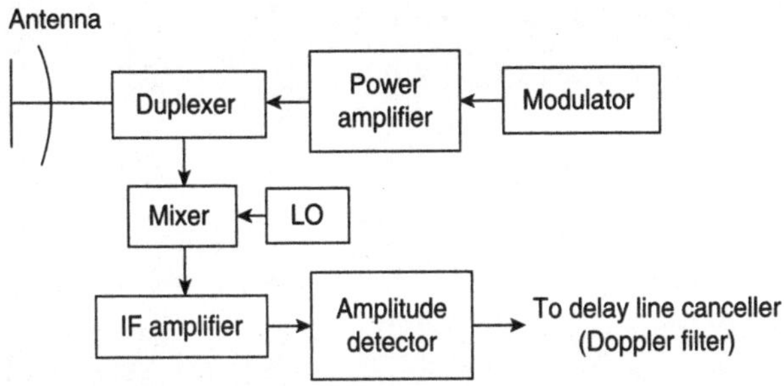

Figure 14.13 *Block diagram of non-coherent MTI radar*

The major advantage of a non-coherent MTI is that it is simple and it is mostly used in applications where space and weight are limited. The improvement factor of the non-coherent MTI, is not good. Clutter itself is the reference signal in the non-coherent radar. It will not detect moving targets, if clutter is not present.

14.8.2 Coherent MTI Radars

Coherent MTI radars can be of two types:

- MTI radar using a power amplifier as a transmitter
- MTI radar using a magnetron oscillator in place of an amplifier

MTI radars (Using power amplifier as transmitter)

A simple block diagram of an MTI radar is shown in Figure 14.14. The power amplifier is used as the transmitter in this type of MTI radar. Two local oscillators provide reference signals to the mixer. The coherent reference is provided by an oscillator called *Coho*, meaning coherent oscillator which is a stable oscillator. It has the same frequency of the IF used in the receiver. The output of the Coho f_c is also mixed with the stable local oscillator frequency, besides providing reference signals. The other local oscillator should be a stable oscillator and is called a *stable local oscillator* (STALO).

For MTI radar, the local oscillator's stability of superhetrodyne receiver must be more than the stability of local oscillator of radar which is not having Doppler. There may be a possibility of appearing uncancelled clutter residue at output of delay line canceller. This may lead to wrong detection of moving target although only clutter is present.

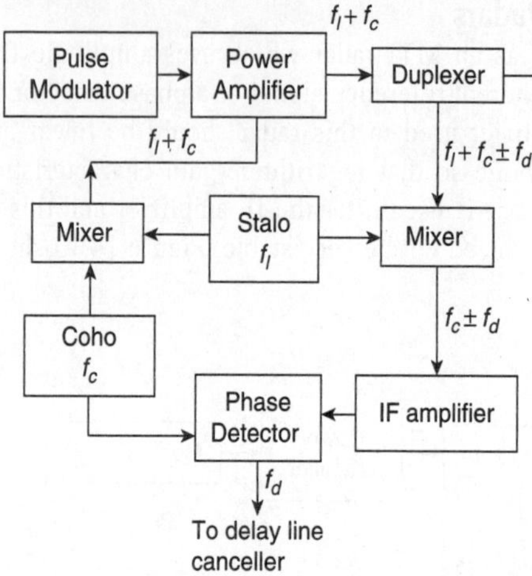

Figure 14.14 *Block diagram of MTI Radar (using power amplifier as transmitter)*

Instead of an amplitude detector as in the non-coherent radar, there exists a phase detector after the IF stage. This act is similar to a mixer in which the received signal coming from IF and the reference signal from the Coho is mixed to produce the difference between their frequencies. This difference is the Doppler frequency. The input signal applied to the power amplifier is nothing but the sum of Coho f_c and Stalo signals f_l. This is accomplished in the mixer as shown in Figure 14.14.

MTI radars using a magnetron oscillator in place of an amplifier

The block diagram of MTI radar is illustrated in Figure 14.15. In this magnetron oscillator acts as a transmitter. The phase of IF beat signal depends on the phase of Coho. This IF beat signal is produced by mixing transmitted signal and Stalo output. The phases of Coho and transmitted pulse are interrelated. The Coho signal can be used as the reference signal for echoes of a particular transmitted pulse. The COHO lock pulse is produced by the transmitted pulse. To again relock the phase of CW Coho another IF locking pulse is generated and continues upto next locking pulse. The block diagram of MTI radar with power oscillator transmitter is shown in Figure 14.15.

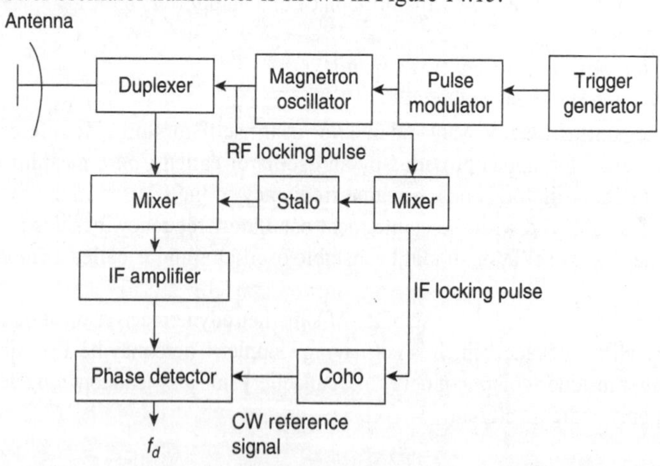

Figure 14.15 *Block diagram of MTI radar with power oscillator transmitter*

The two methods described earlier are not the only ones that are used for obtaining coherent reference signals in the MTI. The various arrangements may be classified accordingly: whether (1) the transmitter locus the oscillator or vice versa; (2) the locking takes place at RF or IF; and (3) the echo and the reference signals are accomplished at RF or IF. This results in eight possible combinations.

Advantages of MTI radars

- Elimination of clutter signals.
- Detection of smaller moving targets in the presence of clutter echoes.
- Minimizes the noise effect
- Useful range can be increased for a given power.

Disadvantages of MTI radars

- Low sub-clutter visibility
- High equipment instabilities cause lower improvement factor of MTI radars.
- The MTI radar can be limited by internal fluctuations of clutter (clutter from trees, rain, chaff, and sea).
- Due to limiter in MTI radars, performance may be degraded.

Applications of MTI radars

- Space-borne applications
- Unmanned aerial vehicles
- Some applications relative to ground MTI are locating, tracking, classifying, and identifying moving vehicles.
- Maritime moving target indicators (MMTIs)

14.8.3 Limitations to MTI Performance

The stability of the radar components and the velocity spread of the clutter are the cause for the limitations of MTI processing. Targets traveling tangentially to the radar will have little or no radial-velocity component, and will be cancelled along with the clutter. The minimum detectable target velocity (MDV) and the clutter cancellation ratio are the characteristics of the MTI processors.

In MTI processing, the repeated measurements of a stationary target give the same echo amplitude and phase. These successive echo samples, when subtracted from one another, any effect which is internal or external to the radar that causes the received echo from a stationary target to vary should be cancelled; which results in imperfect cancellation, limiting the improvement factor.

Another limiting factor is phase drift in either transmitter or receiver. This can occur due to instability in Coho used either as a part of the waveform generator, as we know that the phase is measured by considering the reference signal; if the reference phase changes between measurements, the apparent measured phase will change, resulting again in imperfect cancellation of two measurements when subtracted.

Due to radar system instabilities, there exist other limitations in MTI performance of radars. They are instability in transmitter or oscillator frequencies, transmitter phase drift, coherent oscillator locking errors, PRI jitter, pulse width jitter, and quantization noise. The external factor effecting the performance of MTI radar is the width of the clutter spectrum.

14.9 PULSE DOPPLER RADARS

In Pulse Doppler Radars, the pulses are transmitted in coherent bursts. Coherency involves extraction of the phase information of the transmitted signal and utilizes them in processing the received signals. A Fourier-transform type algorithm is used for processing the returned signals to separate the received signal into a series of spectral bands. The bands of the Doppler shift of clutter are rejected and the bands of Doppler shift of potential targets are used for detections. The radial velocity and Doppler-frequency shift are given by the spectral bands of Doppler shift of potential targets.

The Pulse Doppler Radar uses the Doppler shift to discriminate moving targets from stationary clutter just like MTI radars. We already know that the low PRF radar has a unambiguous range but results in blind speeds (due to velocity ambiguities), whereas a high PRF radar can prevent blind speeds but results in range ambiguities. The filter banks, which are implemented with a discrete Fourier transform rather than delay line cancellers to remove the clutter can be utilized by the Pulse Doppler Radar. The improvement factor is a function of the size of the Fourier transform and the window function used.

Principle of operation of Pulse Doppler Radar: PDRs use a higher PRF to avoid velocity ambiguities. Any unwanted non-moving echoes are selectively eliminated, by routing the Doppler shift through a frequency filtering device, called *BPF*. The function of BPF is it allows passage of only those signals that possess frequency characteristics within the narrow band for which the filter was designed. Filters are sensitive to velocity changes in the order of 3 meters/sec. hence, signals from stationary targets does not reach the indicator.

The targets of different velocity characteristics are distinguished by a series of BPFs, as shown in Figure 14.16.

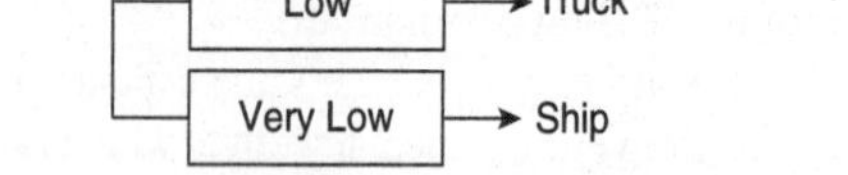

Figure 14.16 *Doppler frequency gates for separation of targets*

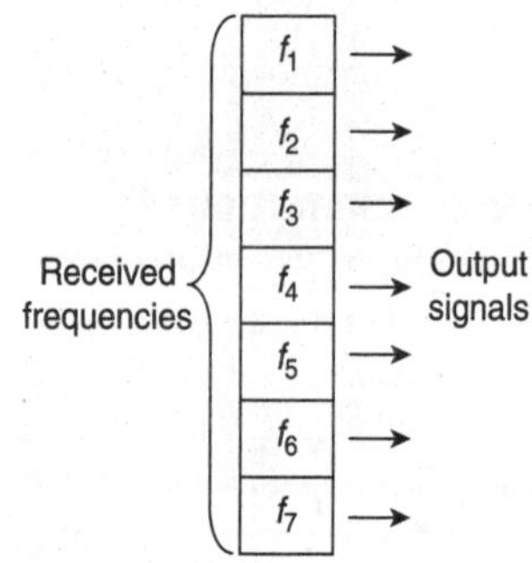

Figure 14.17 *Doppler filter bank*

In airborne applications, the motionless targets look like moving with a relative velocity of aircraft which is transmitting signals. So we selectively "run off" the appropriate BPF that corresponds to this velocity, in order to display only moving targets. PDRs are designed for deriving the target velocity information in this manner and these radars have advantages compared with CW systems. One advantage is that, they can measure range as well as velocity, as they are pulse modulated. Another is that only one antenna is required by a pulse Doppler system in place of two, and the transmission power is larger.

Doppler filter bank: A collection of filters that are used for detecting targets is called Doppler filter bank. Practically several banks of bandpass filters whose outputs are linked to an indicator are used by Pulse Doppler Radars. Radar receives the signals from many sources. These signals are sorted in the Doppler filters bank depending on their Doppler frequency.

The filters in the bank are designed in such a way that, it passes narrow band frequencies, f_i ideally (Figure 14.17). The output signals are obtained by ignoring the filter side lobes if the received signals are within specific range of frequencies called passband of the filter. The output obtained is more, if the received signal's frequency is nearer to the center frequency as shown in Figure 14.18. Inorder to reduce the overlapping of each passband, the filters in the bank are tuned to a center frequency.

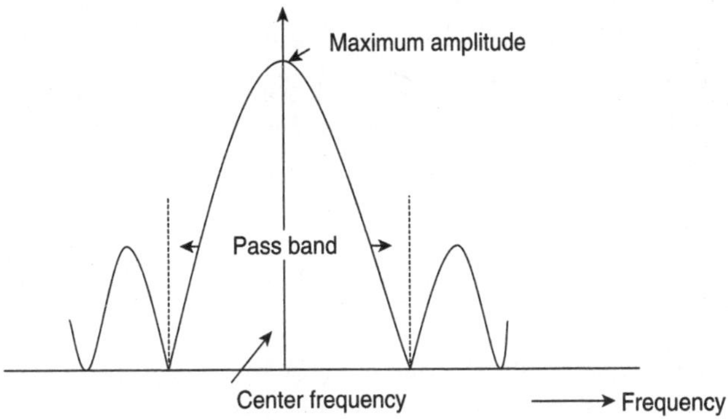

Figure 14.18 *Response of single doppler filter*

Filter bandwidth: The signals applied to the filter are integrated to obtain the selectivity. This is called the *filter bandwidth*, and it defines the minimal resolvable Doppler shift. The frequency band passed by the filter depends mainly on length of the integration time tint (on the order of millisec). This is shown in Figure 14.19. The optimum 3 dB bandwidth of pass band filter is approximately equal to $1/t_{int}$ or $BW_{3dB} = 1/t_{int}$.

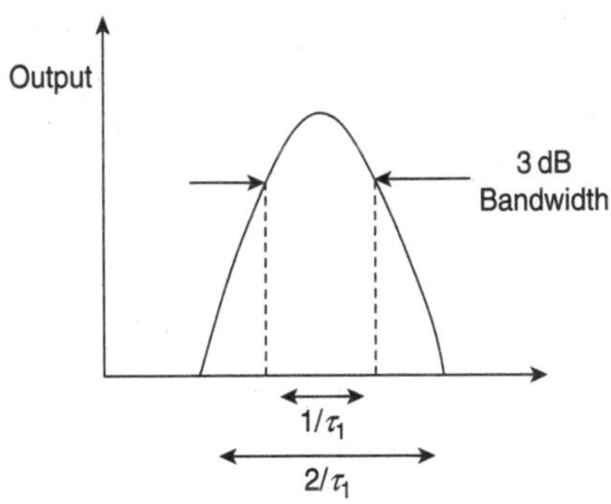

Figure 14.19 *Graphical reprsentation of 3 dB bandwidth*

In general, by sampling the return data the filtering is achieved. The sampling is done using discrete fourier transform, implemented in the form of FFT. Displaying the echo on the PPI display after colour coding is the important application. The Doppler shift is categorized into various factors like positive, zero, and negative. These are then associated with colors. In the same way the target relative velocity and bearing are displayed.

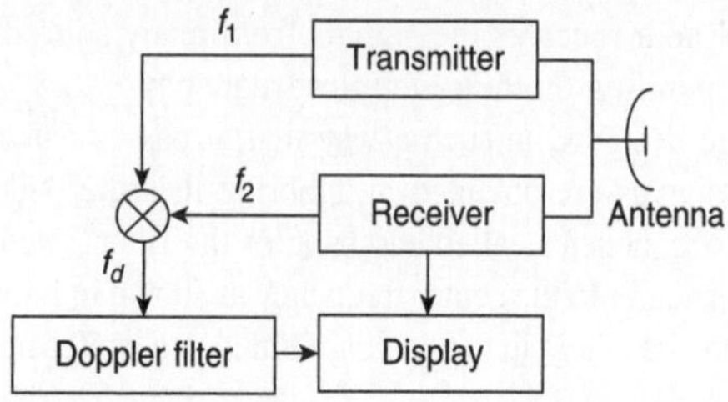

Figure 14.20 *Pulse Doppler Radar system*

Figure 14.20 shows a Pulse Doppler Radar systems which are employed in several military applications such as the standard weather radar in detecting the wind movement and tornadoes. For example, the relative motion of winds in the storm systems can be detected and information about it can be shown for that particular geographical location.

Advantages of Pulse Doppler Radars:

- High accuracies because of unambiguous Doppler (no blind speeds) and a relatively high pulse repetition frequency
- High signal-to-noise ratio compared with MTI radars
- Good resolution even in the presence of multiple targets
- Increases detection range by improving sub-clutter visibility

Disadvantages of Pulse Doppler Radars:

- It is costly and complex.
- Pulse Doppler Radar takes long time to cover the entire area above the horizon. So a fan beam antenna is used to cover the area since it takes less time.
- Antenna rotation must be slow enough so that all the returns are processed for atleast 3 different PRFs that corresponds to the maximum detection range.

Applications of Pulse Doppler Radars:

- It is used to detect the target and estimation of target motion.
- It can also be used to detect space targets such as satellites and astronomical bodies

14.10 MOVING TARGET DETECTOR (MTD)

The Doppler processing system used in modern radars *is referred to as moving target detector* (MTD). Example of such radar *is airport surveillance radar (ASR)*. This MTD improves the performance of radar in target detection. Figure 14.21 shows the block diagram of MTD. The input to the MTD is I/Q digitized radar echoes from each range cell. A standard 3-pulse canceller forms the starting of the upper channel. For pulse Doppler analysis output from the 3-pulse canceller is free of clutter and is given to 8-point FFT. Two PRFs are used in a block-to-block stagger to extend the unambiguous velocity region since two PRFs cause less velocity ambiguities. The frequency domain weighting is an implementation of time domain windowing of the data. The individual FFT samples are applied to a 16-range-bin CFAR (Constant False Alarm Rate) threshold detector. The different thresholds are chosen for each frequency bin. To detect the crossing targets, lower channel uses a site specific zero-velocity filter. The zero velocity filter isolates the echo from the clutter and low Doppler targets, and then the output is applied to clutter map followed by a threshold detector. The generic moving target detector is shown in Figure 14.22.

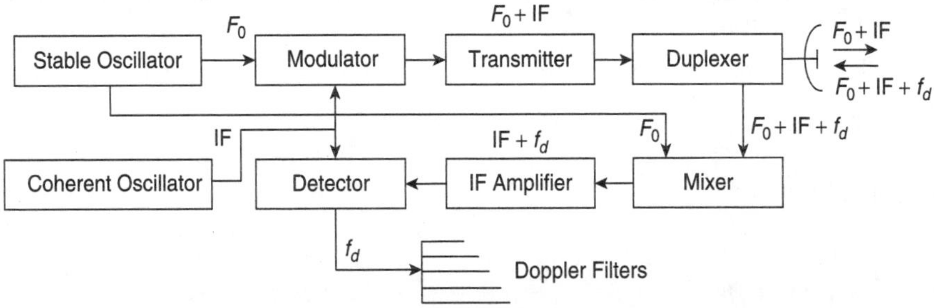

Figure 14.21 *Block diagram of Pulse Doppler Radar*

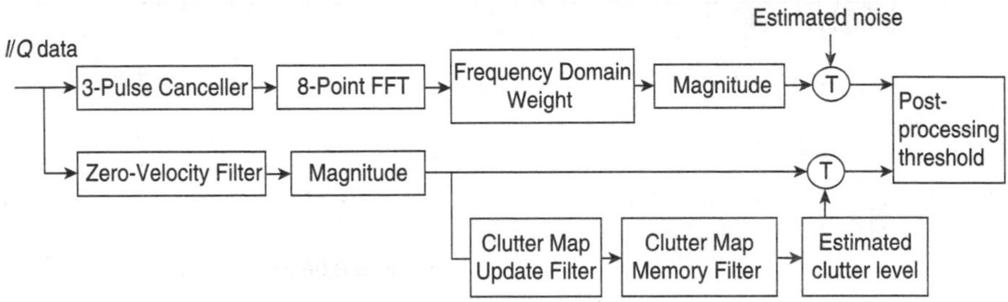

Figure 14.22 *Moving target detector*

14.10.1 Comparison of Moving Target Indicator and MTD

Two types of techniques are used for processing in radars. They are MTI processing which is simple but older processing technique and MTD processing which is complex technique. The features of MTI and MTD are shown in Table 14.3.

Table 14.3 Important features of MTI and MTD

MTI	MTD
Sensitivity is less	Sensitivity is higher
MTI has blind speeds effect and can be reduced by using a number of staggered PRFs	MTD has no blind speeds effects
Range ambiguities can be avoided using low PRFs	Doppler frequency ambiguities can be avoided using high PRFs
Moving targets are identified by using delay line cancellers	Moving targets are identified by using range gate Doppler filters
Inter-clutter visibility is possible only with detailed maps to switch in linear video	Inter-clutter visibility is possible
Tangential course scatterers are not seen in MTI region	Tangential course scatterers are seen in regions where there are no clutter
It has a single narrow notch which leads to poor suppression of rain and snow effects	The effects of rain and snow can be minimized by using cell averaging thresholds

EXAMPLE PROBLEM 14.3

Calculate the second blind speed of an MTI radar whose operating wavelength is 6 cm and PRF is 2000 MHz.

Solution

Given data $\lambda = 6$ cm, PRF = 2000 MHz

Blind speed $v_{bn} = \dfrac{n\lambda}{2} f_r$

2nd blind speed, that is, $n = 2$ is $v_{b2} = \dfrac{2\times 6\times 10^{-2}\times 2000}{2}$ =120 m/sec= 432 Kmph.

■

EXAMPLE PROBLEM 14.4

An MTI radar operates at 5 GHz with a PRF of 900 PPS. Calculate the lowest three blind speeds of this radar.

Solution

Given data f = 5 GHz, PRF = 900 PPS

$$\lambda = c/f = \frac{3\times 10^8}{5\times 10^9} = 6\times 10^{-2}\ \text{m} = 0.06\ \text{m}$$

Blind speed $v_{b1} = \dfrac{\lambda}{2} f_r = \dfrac{0.06\times 900}{2} = 27\ \text{m/sec} = 27\times 60\times 60\times 1/103 = 97.2\ \text{kmph.}$

$$v_{b2} = \frac{2\lambda}{2} f = 2\times 97.2 = 194.4\ \text{kmph.}$$

$$v_{b3} = \frac{3\lambda}{2} f = 3\times 97.2 = 291.6\ \text{kmph.}$$

■

EXAMPLE PROBLEM 14.5

An MTI radar operates at a PRF of 1.5 KHz. Its operating wavelength is 6 cm. Determine the lowest blind speed.

Solution

PRF = 1.5 KHz

$\lambda = 6$ cm

The blind speed is given by $v_b = n\dfrac{\lambda}{2}\times PRF$

$n = 1$ gives the lowest blind speed.

$$v_{b1} = 1\times \frac{0.06}{2}\times 1.5\times 10^3 = 45\ \text{m/s}$$

■

EXAMPLE PROBLEM 14.6

Two MTI radars have the same PRF but their operating frequencies are different. Determine the ratio of operating frequencies of these two MTI radars, if the first MTI radar's first blind speed and second MTI radar's third blind speed are the same.

Solution

The first blind speed of the first MTI radar is given by

$$v_{b1} = \frac{\lambda_1}{2} \times PRF_1$$

The third blind speed of the second MTI radar is given by

$$v_{b3} = \frac{\lambda_2}{2} \times PRF_2$$

Here, $PRF_1 = PRF_2 = PRF$
If $v_{b1} = v_{b3}$, we have

$$\frac{\lambda_1}{2} \times PRF_1 = \frac{\lambda_2}{2} \times PRF_2$$

$$\frac{\lambda_1}{2} \times PRF = \frac{3\lambda_2}{2} \times PRF$$

$$\frac{\lambda_1}{\lambda_2} = \frac{1}{3}$$

$$\frac{f_2}{f_1} = \frac{1}{3}.$$

■

EXAMPLE PROBLEM 14.7

An MTI radar system operating at 20 GHz and a repetition rate of 1500 Hz receives echoes from an aircraft that is approaching the radar with a radial velocity component of 1 km/sec. Determine the radial velocity component measured by the radar.

Solution

Given that,
For an MTI radar systems,
Operating frequency, $f_0 = 20$ GHz
Pulse Repetition rate, $f_p = 1500$ Hz
Radial velocity component of an approaching target, $V_r' = 1$ km/sec
Doppler frequency shift, $f_d = nf_p = n \times 1500 = 1500\,n = 1500, 3000, 4500$ Hz ...
Then, the radial component velocity of the aircraft with respect to the radar or the radial component velocity which is not detected by the radar system is

$$V_r = \frac{f_d c}{2f_0} = \frac{1500n \times 3 \times 10^8}{3 \times 2 \times 10^9} = 0.075n\,\text{km/sec}$$

For, $n = 1$, the radial component velocity which is not detected by the radar system is

$$V_r = 0.075\,n\ \text{km/sec}$$

Then, the radial velocity component which is measured by the radar is equal to the radial component velocity of the approaching aircraft minus radial component velocity which is not detected by the radar system i.e.,

$$V_r'' = V_r' - V_r = 3 - 0.075 = 2.925\ \text{km/sec}$$

But the generalized expression for the radial component velocity measured by the radar is,

$$V_r'' = 3 - 0.075\,n\ \text{km/sec}$$

SUMMARY

1. The pulse radar that uses the Doppler shift is either a Moving Target Indication (MTI) radar or a Pulse Doppler Radar (PDR).
2. The effect of blind speeds can be reduced by operating at more than one PRF.
3. The MTI radar's main objective is to reject the signals from stationary unwanted signals, i.e., ground clutter, rain clutter, bird clutter, etc.
4. An MTI based on a delay line canceller operates by taking the difference of the amplitudes of successive pulses.
5. MTI filters can be implemented using delay line cancellers.
6. Blind speed occurs if both Doppler frequency shift and PRF or PRF multiples are equal.
7. In the pulsed Doppler radar, Doppler data are obtained using range gates and Doppler filters.
8. In pulse radars, the Doppler shift is not enough to produce a measurable frequency difference between the transmitted and received signals for a single pulse.
9. Doppler signals' output from I and Q channels will have an identical frequency regardless of whether the target is approaching or receding; their phase relationship with each other will be reversed, and so, direction information can be obtained.
10. To avoid velocity ambiguities, PDRs use a higher PRF. By routing the Doppler shift through a frequency filtering device, called BPF, any unwanted non-moving returns are selectively eliminated.
11. The MTD also achieves a narrower notch at zero-velocity blind speeds.
12. The MTD processor that is based on digital technology uses a 3-pulse canceller which is followed by an 8-pulse FFT Doppler filter bank with weighting to reduce side lobes.
13. In MTD, true Doppler frequency is possible using coincidence or Chinese remainder theorem.

OBJECTIVE-TYPE QUESTIONS

1. The characteristic feature of coherent MTI radars is that the
   (a) transmitted signal should be out of phase with the reference signal in the receiver.
   (b) the transmitted signal should be equal in magnitude to the reference signal.
   (c) the transmitted signal should be coherent with the reference signal in the receiver.
   (d) the transmitted signal should not be equal to the reference signal in the receiver.
2. The Doppler frequency shift produced by a moving target may be used in pulse radar to
   (a) combine moving targets from desired stationary objects.
   (b) determine the relative velocity of a target.
   (c) separate desired moving targets from desired stationary objects.
   (d) determine the displacement of a target.

3. The limitation of the pulse MTI radar that does not occur with the CW radar
   (a) is blind speed (b) is delay lines
   (c) requires more operating powers (d) requires complex circuitry
4. The output of the MTI receiver phase detector should be quantized into a sequence of digital words by using
   (a) a digital quantizer (b) a digital phase detector
   (c) digital delay lines (d) a digital filter
5. A simple MTI delay line canceller is an example of
   (a) a frequency domain filter (b) a high-pass filter
   (c) an active filter (d) a time-domain filter
6. The effect of blind speed can be significantly reduced in
   (a) pulse MTI radars (b) delay line cancellers
   (c) staggered PRF–MTI (d) pulse cancellers
7. It provides stated probabilities of detection and false alarm:
   (a) clutter attenuations (b) clutter ratio
   (c) cancellation ratio (d) clutter visibility factor
8. An MTI radar that uses amplitude fluctuations is
   (a) coherent (b) a pulse Doppler
   (c) non coherent (d) a CW radar
9. The blind speeds of two independent radars operating at the same frequency will be different if their
   (a) amplitudes are different (b) blind speeds are different
   (c) pulse repetition frequencies are different (d) pulse intervals are different
10. The clutter-rejection notches may be widened by passing the output of the delay line canceller through a
   (a) Coho (b) Stalo
   (c) second delay line canceller (d) pulse canceller
11. Two- and three-pulse cancellers are examples of
   (a) CW radars (b) FM–CW radars
   (c) MTI filters (d) pulse radars
12. The following techniques are Doppler filtering techniques that reject stationary clutter and where radial velocity is not measured:
   (a) moving target indicator (MTI) (b) CW radar
   (c) MTD (d) pulse Doppler
13. The following are regions of Doppler space where targets with those Doppler velocities cannot be detected:
   a) blind speeds (b) delay line cancellers
   (c) staggered PRF (d) clutter visibility factor

14. Ambiguities in range and Doppler velocity can be resolved by transmitting multiple bursts of pulses
    (a) with constant PRF (b) with different PRFs
    (c) with two PRFs (d) none
15. Unambiguous range measurements and ambiguous velocity measurements are caused by
    (a) low PRF (b) high PRFs
    (c) medium PRFs (d) high PRF
16. Ambiguous range measurements and ambiguous velocity measurements are caused by
    (a) low PRF (b) high PRFs
    (c) medium PRFs (d) high PRF
17. Very ambiguous range measurements and unambiguous velocity measurements are caused by
    (a) low PRF (b) high PRFs
    (c) medium PRFs (d) high PRF
18. Staggering or changing the time between pulses will
    (a) raise the blind speed (b) lower the blind speed
    (c) maintain the same blind speed (d) none
19. The following radar does not provide the target velocity estimation:
    (a) Pulse Doppler Radar (b) MTI (c) both (d) none

ANSWERS TO OBJECTIVE-TYPE QUESTIONS

1. (c) 2. (b) 3. (a) 4. (c) 5. (d) 6. (c) 7. (d) 8. (c) 9. (c) 10. (c)
11. (c) 12. (a) 13. (a) 14. (b) 15. (a) 16. (c) 17. (d) 18. (a) 19. (b)

REVIEW QUESTIONS

1. Explain MTI radars with a block diagram.
2. Write short notes on delay line canceller and discuss the limitations of single delay line canceller.
3. Explain non-coherent MTI radars with a block diagram.
4. Explain the basic principle of MTI radars.
5. What is the difference between a pulse radar and a Pulse Doppler Radar?
6. What do you understand by blind speed? How can it be eliminated?
7. What are the limitations of MTI radars?
8. List the differences between MTI and MTD radars.
9. What are the methods used to eliminate blind speed in MTI radars?
10 What are COHO and STALO in MTI radars?

11. Explain blind speed and the methods for reducing the effects of blind speed.

12. Define MTI improvement factor.

13. An MTI radar operates at 5 GHz with a PRF of 800 PPS. Calculate the lowest three blind speeds of this radar.

14. An MTI radar operates at 4.8 GHz with a PRF of 600 Hz. Calculate the lowest blind speed of the radar.

15. The MTI radar is used by a traffic control police to measure the speed of vehicles. If the Doppler frequency shift measured from the moving vehicle is 2 kHz. Calculate the speed of vehicle, if radar is operating at 1 GHz with PRF of 2000 Hz.

16. An MTI radar system operating at 5 GHz and a repetition rate of 1000 Hz receives echoes from an aircraft that is approaching the radar with a radial velocity component of 1000 m/sec. Determine the radial velocity component as measured by the radar.

17. An MTI radar system operating at 10 GHz and a repetition rate of 2000 Hz receives echoes from an aircraft that is approaching the radar with a radial velocity component of 1km/sec. Determine the radial velocity component measured by the radar.

Tracking Radars

15

15.1 INTRODUCTION

The primary objective of all radars is to search a given volume of space and to detect targets without previous knowledge of the target's location or presence. The two basic functions that are common in almost all radars are search/detect and track. Tracking is an essential component of many radar systems. The exact position of the target is required for most of the applications of radar technology. A typical area where you would notice a radar system is at an airport. At smaller airports, the antennas are much more visible and can be seen rotating. Most of the antennas at larger airports are protected by domes. Targets such as an aircraft or a ship are usually struck by the radar signal many times. This is due to many scans by the radar antenna.

Tracking radars lock on to a target and track it for a certain distance or for a certain time period. Target tracking is important in military and Airport Surveillance Radars (ASRs). Military applications of radar systems include gun control, missile guidance, and airspace surveillance linking with spacecraft and/or satellites. ASRs play a major role in airports for aircraft traffic control, and they utilize tracking function as a means of controlling incoming and departing aircrafts.

Radars are used for many purposes such as tracking and surveillance. Characteristically, Tracking radars and surveillance radars are different. They are used for long range surveillance and the surveillance can be done within a small volume-scanning time. They are also used for tracking extremely active and small targets. They are advantageous, as one or more targets can be detected with much greater accuracy than with search radars. The main aim of the tracking radar is to continuously track the target and find out its location or the course in which it is moving. Once the radar detects the target in the given volume of space, it will track the target by estimating the target parameters accurately. To find the path of the target and its destination where it is going, the radars determine the coordinates of the moving target continuously. There are three tracking techniques: (1) range tracking, (2) velocity tracking, and (3) angle tracking. This chapter describes the search and tracking radar system, various scanning and tracking techniques, conical scan, sequential lobing, and monopulse (both phase comparison and amplitude comparison), tracking accuracy, frequency agility, track-while-scan (TWS) radars, phased array radars, and so on.

15.2 SEARCH AND TRACKING RADAR SYSTEM

By following the history of the target's position, the course and speed of the target can be calculated. In order to determine the position of target, radars search the airspace using Search radars. The Search radars search the airspace with their wide beam width. Once the position of the target is known, then the Tracking radars track the desired targets with its narrow beam width which is also known as pencil beam. From the movement of the targets they find the azimuth angle, elevation angle, direction and speed. When the course and speed information is known, it also predicts the location of the target in subsequent observations.

Any radar system is generally required to perform one of the following tasks: (i) either search for the target or (ii) track the targets once they have been pinpointed. In some applications, the two tasks can be executed by single radar in others they can be performed by two separate radars. As explained above, search radar uses wide beam antenna patterns and the tracking radar uses narrow beam antenna patterns. If a single radar is used then narrow beam antenna patterns are used for both searching and tracking purposes. The pencil beam antenna patterns take long time to search a target so we use a separate search radar and a separate tracking radar. Sometimes tracking radar must be used to search the airspace when the presence of a target is suspected by using special antenna patterns like helical, T.V. raster, cluster, and spiral patterns.

15.2.1 Search Radar System

Search radars are mainly designed to search the targets without any earlier knowledge about the target existence. They search the targets with wide beam antenna patterns within a given volume of a solid angle and within a given slant range in a specified amount of time. If the narrow beam antenna patterns are used rather than wide beam antenna patterns then the scanning time increases. So we prefer wide beam antenna patterns in search radars. To obtain wide beam antenna patterns we use two antennas, one move in azimuth angle and other moves in elevation angle. In this way they scan a large area in a limited amount of time. The applications of Search radar is that it provides detection and surveillance of submarines, surface vessels, and aircraft. The radar using this search radar principle is the air traffic control radar which is used at military airports.

15.2.2 Tracking Radar System

The tracking radar system continuously tracks the target and determines the position or direction of a target. The position or direction of target is measured in range, azimuth angle, elevation angle, and velocity. The future position and direction of target is also predicted by continuously tracking these measured parameters. To predict the future positions of the target continuously, an accurate estimation of all or part of the following parameters from the radar observations is essential. The target parameters to be estimated are as follows: (i) range R, (ii) angle (azimuth (θ) and elevation (ϕ)), (iii) radial velocity, and (iv) RCS.

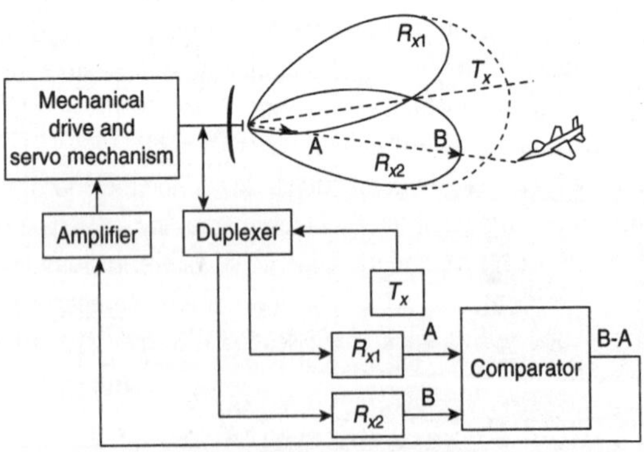

Figure 15.1 *Antenna tracking*

Many tracking radar systems track a single target by continually pointing the antenna beam at the target and controlling the antenna pointing angle and range measurement to coincide with the target position. Figure 15.1 shows tracking of target using only one antenna. The single antenna produces three beams. One beam is produced by the transmitter. The other two beams are produced by the receivers that are symmetrical and their gains are equal. The two receivers track the target simultaneously and the outputs are matched over a dynamic range. Depending on the signal from the receiver, the location of target is known. If the signals from the receivers are matched then target is in line with the antenna radiation. The two signals received from the receivers are sent to the comparator. The comparator compares the two signals and exhibits which signal is greater than the other signal. It gives zero if they are equal. If the signal from receiver 2 is greater than that from receiver 1 then the target is to the right of the axis of antenna radiation. Later it passes to the amplifier where the amplifier amplifies the difference level of signal. The amplified signal drives the antenna mechanically to the right or left to reduce the original difference level of signal. These types of tracking radars are mostly used in defense radar systems for tracking enemy targets and directing defensive missiles.

15.2.3 Differences Between Search and Tracking Radar

The differences between search radar and tracking radar are given in Table 15.1.

Table 15.1 Differences between search radar and tracking radar

Search radar	Tracking radar
1. Search radars search the targets within a given volume of a solid angle and within a given slant range in a specified amount of time 2. They search the targets with wide beam antenna patterns 3. Search radar is used in aircrafts and surveillance applications	1. The tracking radar continuously tracks the target and determines the position or direction of a target in range, azimuth angle, elevation angle, and velocity. 2. They track the targets with narrow beam antenna patterns which is known as pencil beam antenna patterns. 3. Tracking radar is used in defense for tracking enemy targets, directing defensive missiles and prediction of the target.

15.3 VARIOUS SCANNING AND TRACKING TECHNIQUES

The main objective of scanning and tracking radar are (i) detection of targets and the measurement of their position (range, angle, and velocity) and (ii) tracking of a detected target with continuous measurement of the target position and velocity. Radars can track the target in range, in angle, or in Doppler shift (or velocity) or any combination of the three relative to the radar location. However, in general, all tracking radars use angle measurement with overlapped measurements. The radar first searches a given volume of interest by pointing its antenna in a series of beam positions that collectively cover the volume of interest. The beam scanning can be done mechanically or electronically or both ways.

Antennas used for scanning can be of two types: (i) mechanically scanning antennas and (ii) electronically scanning antennas.

Mechanically scanning antennas: A mechanically scanned antenna moves through the volume continuously. One or more pulses are transmitted in one beam position, and the received data are examined to detect the presence of targets by using the threshold detection techniques. Once the radar detects the target in the given volume of space, radar will track the target by estimating the targets parameters accurately. Correlation of the detections (i.e., associate detections with specific targets) obtained from scan to scan to see that which one belongs to same physical object results in a track.

Mechanically scanning antennas are moved mechanically to cover a given surveillance area for radio ranging and detection applications. For example, the surveillance and tracking radar used in the air traffic control purpose is a monopulse, all-weather radar with mechanical beam steering (Figure15.2 (a)). The performance of mechanically scanned antennas is not meeting the requirements of some applications such as satellite-borne applications or for tracking moving targets such as missiles.

Source: upload.wikimedia.org

Figure 15.2 (a) *Mechanical beam steering ATC radar*

Electronically scanning antennas or Phased array antennas: With the advent of electronically controlled phase shifters and switches, antennas that can scan electronically without the need of mechanical motion have been developed. These electronically scan antennas are also known as *phased array antennas*.

Principle of operation: When the energy from each radiating element arrives at the point is in phase then the electromagnetic energy received at a point in space from two or more closely spaced radiating elements is maximum. A phased array antenna electronically joins element patterns to focus the radar beam in a particular direction as shown in Figure 15.2 (b). It is made of many radiating elements. All the radiating elements have a phase shifter each. The signal produced from each radiating element is phase shifted to form electronically scanned beams which creates constructive/destructive interference, in order that the beams are guided to preferred direction.

The *element channels* are defined as the individual paths which are formed by dividing radar waveform, produced by radio-frequency (RF) source and each channel contains a phase shifter and an amplifier. When all the phase shifters of the array are correctly arranged, the array produces a main beam focusing in the preferred direction.

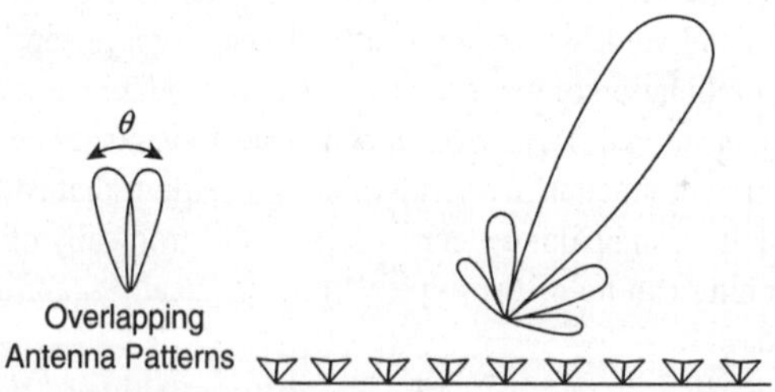

Figure 15.2 (b) *General concept of a phased array antenna*

15.4 RANGE TRACKING

In many tracking radars such as Approach Surveillance Radar (ASR), the target is continuously tracked in range as well as angle. *Range tracking* is defined as tracking the moving target continuously in range. The range of the target is measured by calculating the round-trip delay of the transmitted pulses.

The position of the moving target keeps on changing with time, so the range tracking radar should be continuously updated with the new position of the target so that target is not missed. Range tracking is implemented using a split-gate system, in which two range gates (early and late) are used (see Figure 15.3 (a)).

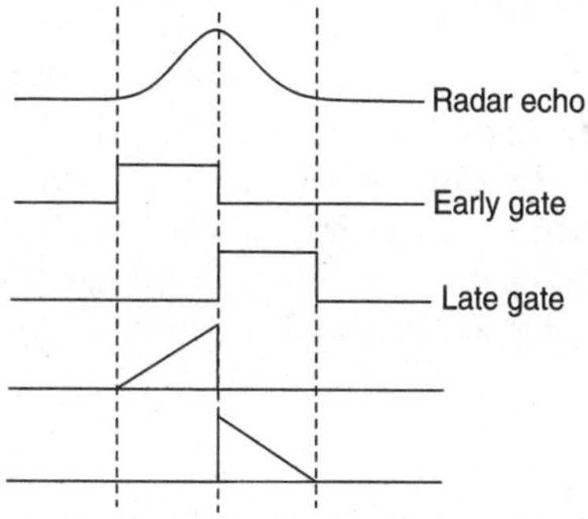

Figure 15.3 (a) *Split-gate system*

Range tracking is implemented similar to dual-beam angle tracking. After measuring the range of the target, the tracking radar predicts the range of the target on the next pulse. The range on first pulse is compared with the predicted range on next pulse using two range windows called the *early* and *late range gates*. The concept of split-gate tracking is illustrated in Figure 15.3 (b). The early gate pulse starts when the radar echo starts and closes at half of radar echo signal duration. The late gate pulse starts at the center of the radar echo signal and ends at the end of the echo signal. So the echo signal duration and center of the pulse signal should be sent to the range tracker to start the early and late gates at the start and center of the expected echo. This process is commonly known as the *designation process.*

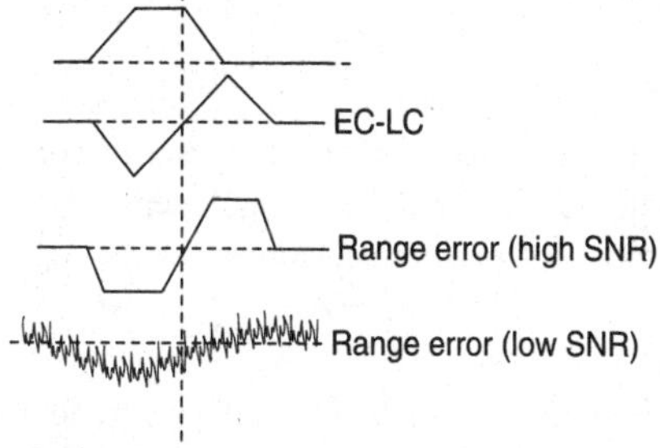

Figure 15.3 (b) *Range error signals at the output of a split gate*

During the positive voltage of the radar echo signal, the early gate pulse exists and during the negative voltage of the radar echo signal, the late gate pulse exists. The difference signal, obtained from comparing the early gate pulse and late gate pulse, is sent to an integrator to produce an error signal. The error signal will not be produced, if the pulse of early gate starts when the radar echo starts and the pulse of late gate starts at the center of the radar echo signal. Similarly if the gates are not started as discussed above, the integrator produces an error signal. Then pulses of the gates are moved to right or left based on the sign of the error signal.

Range gates or range bins

Since the received echo is a vector, the use of a complex detector allows measurement of the received signal amplitude and phase. For each pulse transmitted in radar, a series of complex (I and Q) samples of the echo corresponding to successive range intervals will be collected at the output of the receiver. By Nyquist criteria, these range samples are collected at a rate equal to the pulse bandwidth or greater; this dimension is referred to as *fast time*. Range samples are also referred to as *range gates* or *range bins*.

15.5 ANGLE TRACKING

Angle tracking is concerned with generating continuous measurements of the target's angular position in the azimuth and elevation angles. The target tracking is achieved by keeping the antenna beam's main axis (or angle indicator) on the target angle. Special antenna is used for tracking the target in angle tracking radar systems. The antenna should be in line of sight with the moving target all the time, to predict target position. For this purpose we use a simple lobe switching arrangement and tracking is performed in azimuth angle. The lobe switching arrangement shifts the position of the beam in the horizontal plane. So the antenna position is updated and moved in the direction of target depending on the error signal. The error signal is obtained from the difference in the strength of the returned signals from the positions of the two lobes.

The main aim of the angle tracking system is to maintain the bore-sight axis of the antenna beam aligned with the target. The ability of the radar to find the exact angle of the target depends on the size of the antenna beam employed. An error signal will be generated if the antenna beam's main axis is not exactly on the target. The deviation of the antenna beam's main axis from the target is corrected using the error signal. Depending on the sign of the error signal, the antenna beam's main axis is moved right or left to align with the moving target. There are three methods for generating the error signals for tracking:

- Sequential lobing
- Conical scan
- Monopulse tracking

Modern radar systems mostly use monopulse tracking techniques since they use only one pulse to determine target's position and obtain better angular measurements compared to sequential lobing and conical scanning tracking techniques which requires several radar pulses to determine target's position. They are also susceptible to errors due to RCS fluctuations.

15.5.1 Sequential Lobing

Sequential lobing is the foremost angular tracking techniques to determine the angular position of the target. It is also referred to as *lobe switching* or *sequential switching* since the angular position of target is determined by switching the lobe positions of the beam. Sequential lobing technique uses the symmetrical pencil beam i.e., azimuth and elevation beam widths are equal and the tracking is performed by continuously switching the pencil beam between two pre-determined symmetrical positions about the Antenna's Line of Sight axis (LOS). If the target is in line of sight with the antenna, the signal strength observed in each beam will be equal. If the target is on the beam axis of left or right of the beam position, then the signal strength observed in each beam will not be equal. So the two beam positions are moved towards the left or right until the signals in each of the beams are equal in strength.

The LOS is also called the *radar tracking axis*, as shown in Figure 15.4. The radar tracking axis is taken as reference to find the target position. Each position of the target as shown in Figure 15.4 on the beam corresponds to a voltage value. The bars A and B as shown in Figure15.4 (a) are the voltage values of the echoes from beam positions A and B, respectively. The difference of these two voltage amplitudes gives the angular measurement error. If the target is on the tracking axis (as shown in Figure 15.4 (a)) then the voltage values of the echoes of beam A and B are equal and the error is zero. If the target is on the beam axis of beam position A (as shown in Figure 15.4 (b)), then the voltage value of echo A is high compared to voltage value of echo B. So the beam is moved to the direction in which the amplitude of the voltage is larger (that is in direction of beam position A) to make the voltage difference zero.

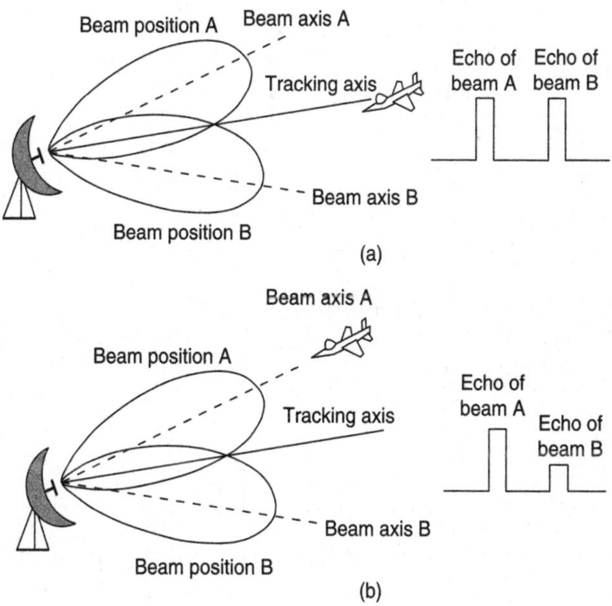

Figure 15.4 *Sequential lobing Tracking Technique (a) Target is located on the tracking axis; (b) Target is located off the tracking axis.*

15.5.2 Conical Scan

Conical scanning technique is another method to determine the angular position of the target. It does not use lobes as in sequential switching; it tracks the target by continuously rotating the antenna at an offset angle around an axis. The pencil beam forms a conical shape when it is rotated around the axis, so it is termed as conical scanning. The angle error is detected and it generates a correction voltage that is proportional to the tracking error with a sign showing its direction. The error measurement is sent to the servo mechanism which moves the antenna in the direction of the target. This tracking signal can be refined to predict the future target motion as well.

Figure 15.5 shows a typical conical scan beam. *Squint angle* is defined as the angle between the rotating axis and the beam axis. *Conical scan frequency* is defined as the frequency where the amplitude of the echo signal is modulated. The beam scan frequency is denoted as ω_S (radians per second).

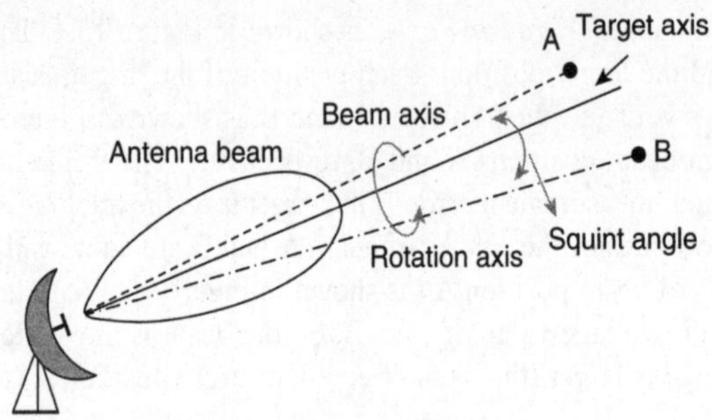

Figure 15.5 *Conical scan beam*

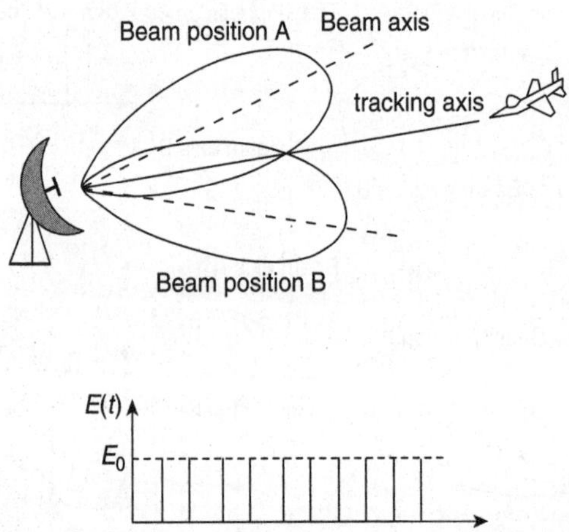

Figure 15.6 *Error signal produced when the target is on the tracking axis for a conical scan*

The beam rotation frequency is same as conical scan frequency. The Modulation occurs as the squinted beam rotates and the target offsets from the rotation axis. The location of the target is given by the phase of conical scan modulation. The elevation-angle error and the azimuth-angle error are combined by the error signal from the modulated signal. The antenna is positioned by providing these error signals into elevation and azimuth servo motors. If the antenna is on tracking axis, the amplitude of conical-scan modulation is zero.

In the above case 15.6, the target is considered to be on track. Let us now consider the case shown in Figure 15.7. The amplitude of the target echoes will change from the maximum values at position B to the minimum values at position A. In other words, the amplitude of echoes of the target will be at their maximum when the beam is at position B and at position A, the amplitude of the echoes of the target will be at their minimum.

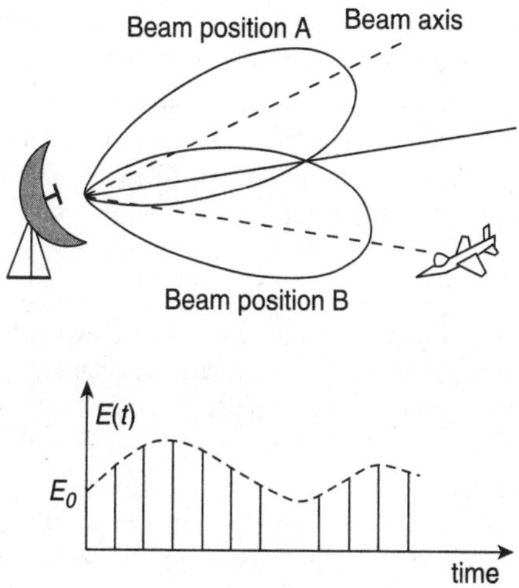

Figure 15.7 *Error signal produced when the target is on the beam axis of B for conical scan*

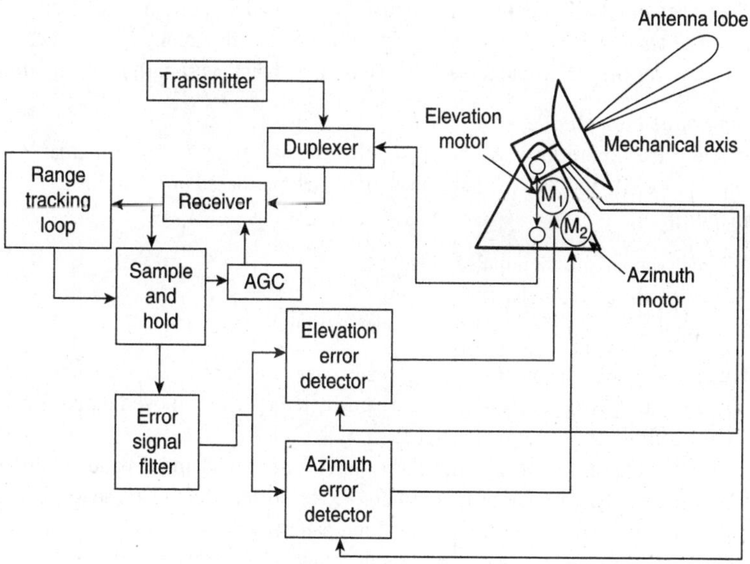

Figure 15.8 *Conical scan tracking*

Conical scan tracking technique is shown in Figure 15.8. The echo signal which is received by the receiver is sent to the range-tracking loop, which gives the time t of the maximum signal. The sample and hold circuit is enabled to sample and memorize the signal amplitude. Using received series of echo pulses from the sample and hold circuit, a sine wave can be reconstructed. There exists a proportionality between the amplitude of the sine wave and the angular deviation from bore-sight to the target, and the error direction is specified by its phase. The angle error detectors (azimuth and elevation) are phase-sensitive and indicate the position of the antenna beam. They are employed to compare the error signal with the elevation and azimuth reference signals. The two sinusoidal signals having same beam scan frequency and 90 degrees out of phase are known as reference signals, which are produced by the generator, the part of motors that rotates the antenna feed. There is a proportionality between the angle error detector's dc output magnitude (azimuth and elevation) and the angle error, and the direction of the error is indicated by its sign. The antenna elevation and azimuth motors are driven by these outputs.

Limitations of a conical scan:

- The angle errors due to RCS fluctuations are caused when there is a change in target RCS
- The comparisons of the amplitudes of the signals are not accurate
- Electronic counter measures affects this technique

15.5.3 Monopulse Tracking

In the tracking techniques explained above, scanning requires more than one pulse to generate the error signals. The variations in the radar echo pulses degrade the performance of the tracking accuracy. Signal fading occurs due to the pulse-to-pulse target fluctuations resulting in missed detections and wrong removal of tracks. So a single pulse is used for tracking the target to avoid such problems. Thus *Monopulse tracking technique* enters here. It uses a single pulse to track the target from the angular position of the target and it also derives error signal. So it is widely used in modern radar systems for angle tracking. The term *monopulse* refers to single pulse hence the tracking technique is referred as *Monopulse tracking technique*. Monopulse tracking is also known as *simultaneous lobbing*.

Concept of monopulse technique

A monopulse radar uses more than one beam simultaneously to measure the angular position of the target on a single pulse. Monopulse tracking can be of three types based on the utilization of the received signal information.

(i) Amplitude comparison monopulse radar
(ii) Phase comparison monopulse radar
(iii) Combination of amplitude and phase comparison monopulse radar

In this section, the concept of the monopulse is explained by considering two overlapping beams that are simultaneously generated and processed from the same antenna. The two overlapping beams generated are symmetrically offset from the normal antenna.

Figure 15.9 (a) and (b) shows the patterns of the sum of the two beams (Σ) and the difference of the two beams (Δ), respectively. The sum beam (or an on-axis beam) is formed, if the two beams are added together. If the two beams are subtracted, the resulting pattern will have a positive lobe on one side of the axis, a negative lobe on the other, and a null on the on-axis beam. The positive and negative lobes in the difference beam are 180° out of phase.

The radar should transmit a Σ pattern to put maximum power on the target and simultaneously receive both Σ and Δ patterns. The displacement of the target from the center of Σ beam is the monopulse error, is determined by comparing the outputs of Σ and Δ beams, and is given by

$$\text{monopulse error} = \frac{|\Delta(\theta)|}{|\Sigma(\theta)|}\cos\beta$$

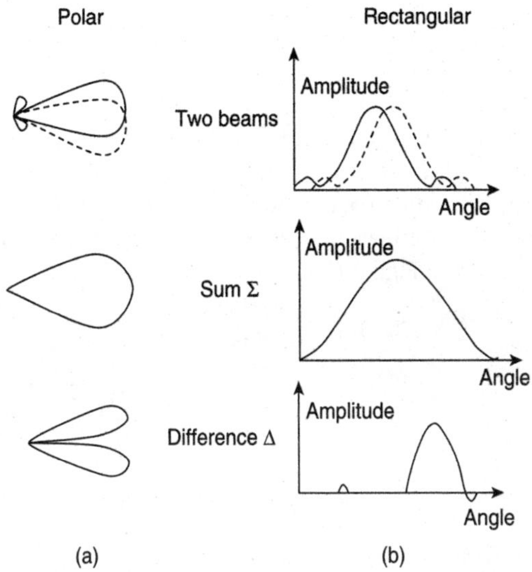

Figure 15.9 *Monopulse antenna beam patterns: (a) in polar form and; (b) rectangular coordinates*

Monopulse error signal is a ratio of the Δ beam over the Σ beam and is linear with the 3 dB beam width.

The error signal is used to re-steer the antenna bore-sight on to the target: The error signal is a function of the target's angle, θ with regard to the antenna bore-sight and the phase angle β between the Σ and Δ beam outputs, which is designed to be nearly 0° or 180°, depending on from which side of the center, the target signal is entering. In a practical radar, the two offset received beams are generated by using two feeds that are slightly displaced from the focus of a parabolic reflector (Figure 15.10).

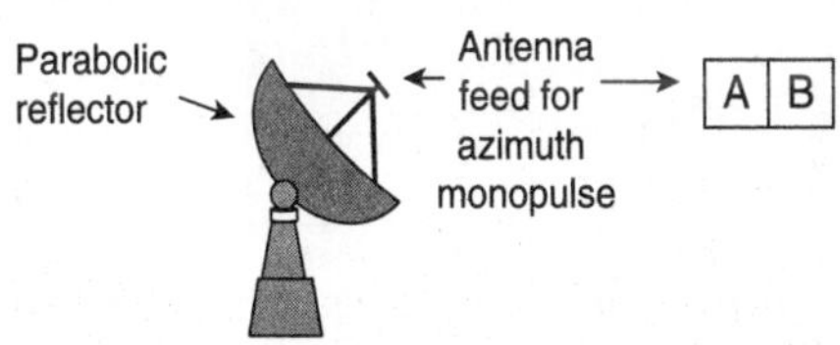

Figure 15.10 *Antenna horn feed – 2 feed horns (1D-azimuth)*

(i) Σ signal = $A + B$
(ii) $\Delta = A - B$

The sum and difference signals can be generated by feeding the signals (A and B of horn feed) to the Magic Tee waveguide junction or Hybrid Ring junction (or rat race) as shown below:

- When a signal is given as input at port A, it reaches the output port Δ by two separate paths. The two separate paths have the same path length ($3\lambda/4$) and they are reinforced at port Δ.
- When a signal is given as input at port B, it reaches the output port Δ by two separate paths. The two separate paths have different path lengths ($5\lambda/4$ and $\lambda/4$) and they are reinforced at port Δ.

- Paths from port A to port Δ and from port B to port Δ differ by 1/2 wavelength. If signals of same phase are entered at A and B, the outputs at port Σ and port Δ are the sum and difference of the signals.

In monopulse tracking technique, in two dimensions, two overlapping beams are generated by using two feeds to a parabolic dish. These are fed to a magic T or rat race junction that produces sum (S) and difference (D) patterns. The sum pattern is used to transmit the radar signals and sum and difference channels are used to processes the received echoes. The generation of sum and difference beams from Magic tee and rat race are shown in Figure 15.11.

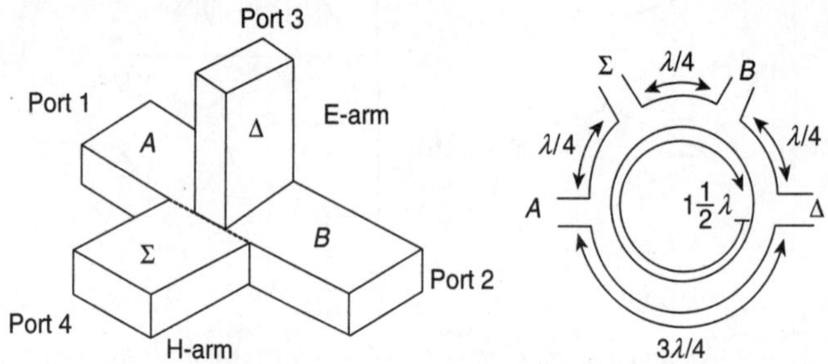

Figure 15.11 *Generation of sum and difference beams from Magic tee and rat race*

15.5.3.1 Amplitude comparison monopulse radar

The simplest and most reliable monopulse tracking system is the amplitude comparison monopulse tracking system. In amplitude monopulse comparison systems, the beams are formed at the same time. There are two overlapping squinted beams from separate feeds, pointing in somewhat different directions as seen in Figure 15.12 (a). Both the beams receive the echo of the target simultaneously.

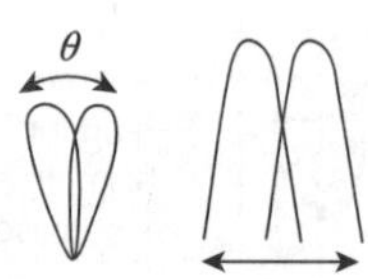

Figure 15.12 (a) *Two overlapping squinted antenna beams in polar and rectangular coordinates*

The outputs of the antenna are the difference and sum of these two squinted beams. The outputs are connected to a magic tee, which produces sum and difference signals. These antenna outputs help the system to track a target. The sum pattern is used to transmit the radar signals, while the sum and difference channels are used to processes the received echoes in the monopulse systems. Two antenna beams are set at an angle as shown in Figures 15.12 (b) and (c). The sum channel is used for detection since it has a higher signal-to-noise ratio and it is not used to measure the angle, since the sum beam is wider than the individual beams. The angular error is obtained from the difference of the amplitudes of these two beams (i.e. the difference beam). The direction of the angular error is obtained by comparing the phase of the sum pattern with the difference pattern. The sign of the error signal is determined by the phase-sensitive detector and is sent to the servo motors so that they could

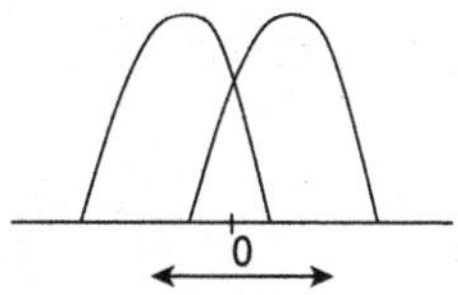

Figure 15.12 (b) *Overlapping antenna patterns*

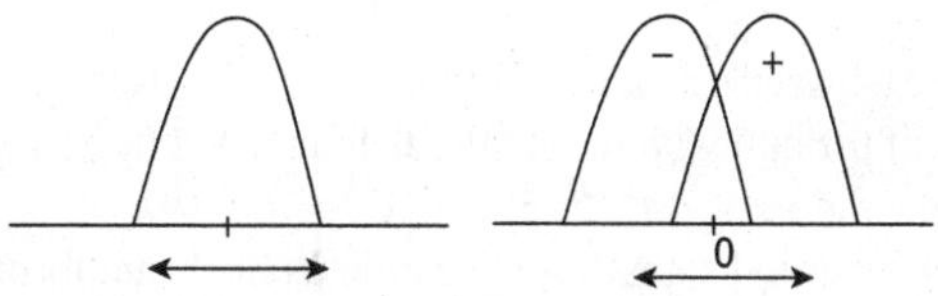

Sum (transmit and receive) Difference (receive only)

Figure 15.12 (c) *Sum and difference patterns*

drive the antenna in the particular direction to place the target in the line of sight with the antenna. The presence of the phase-sensitive detector does not signify that the system make use of the phase information contained within the echo to derive angular information. Considering the ratio Δ/Σ, an amplitude comparison monopulse can be improved. The ratio Δ/Σ normalizes the difference channel by the sum channel. Due to this a quotient is obtained, which is not dependent on the signal strength and linear against the angle error over a wide range of angles. In the past, by processing the signals with logarithmic amplifiers and then taking differences the division of signals is done but now it can be performed digitally. An amplitude comparison monopulse tracking technique is shown in Figure 15.13.

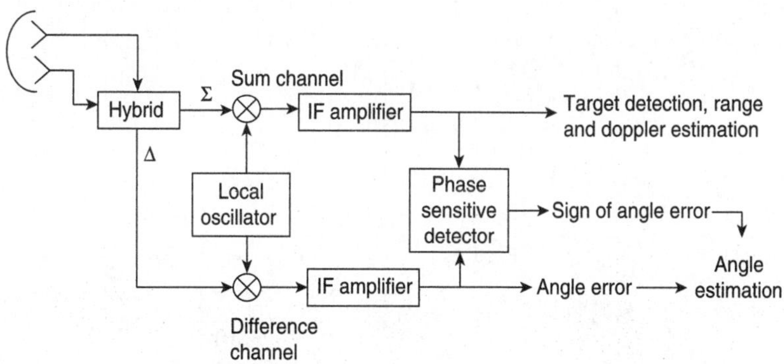

Figure 15.13 *Amplitude comparison monopulse tracking technique*

The angle-error signal is generated from the ratio of the difference pattern to sum pattern. The null in the center of the difference pattern can be placed on the target. The tracking process uses the voltage ratio:

$$\frac{\Delta}{\Sigma} = \frac{\text{difference voltage}}{\text{sum voltage}} \tag{15.1}$$

The complex voltage in Eq. (15.1) above can be separated to its real and imaginary components:

$$\mathrm{Re}\{\Delta/\Sigma\} = \frac{|\Delta|}{|\Sigma|}\cos\delta \tag{15.2}$$

$$\mathrm{Im}\{\Delta/\Sigma\} = \frac{|\Delta|}{|\Sigma|}\sin\delta \tag{15.3}$$

where δ is the relative phase between sum and difference channels. Usually, only Re $\{\Delta/\Sigma\}$ is processed, because it defines the side of the null with the sign of the ratio, and the target information can be found only from the real part of the ratio.

The sum channel has an enhanced signal-to-noise ratio as the beam formed is a combination of the signal power of the two individual beams. The detection of target and the measurement of range and Doppler information can be done using this combined beam. The gain of the sum channel beam offers the monopulse an SNR which is more advantageous than the earlier techniques where the target was looked off the bore-sight as the target is faced directly by it.

15.5.3.2 Two-dimensional monopulse tracking

In all practical applications, the target's position is measured in both azimuth and elevation angles. The monopulse tracking procedure described earlier is used to find the angular error in one dimension (i.e., in azimuth plane) only. To find the angular error in two coordinates (i.e., in azimuth and elevation angles), it is required to receive Σ and Δ beams in both azimuth and elevation planes.

Monopulse tracking radars use a single reflector and four feed horns. The monopulse antenna is divided into four quadrants. The different sections of the reflector are illuminated by the four feed horns thus forming two overlapping antenna beams for two orthogonal axes. The following signals are formed from the received signals of these four quadrants (Figure 15.14 (a)):

(i) Σ signal
(ii) Δ_{AZ}
(iii) Δ_{EL}

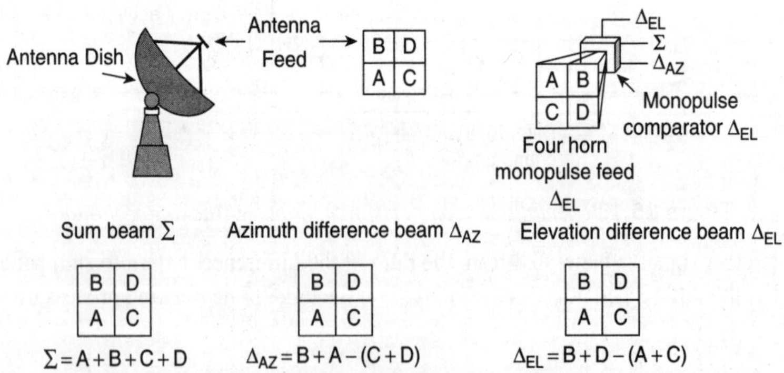

Figure 15.14 (a) *4 feed horns 2D azimuth and elevation – Feed arrangements of monopulse systems*

Figure 15.14 (b) shows a typical monopulse antenna pattern. The four conical scan beam positions are the four beams A, B, C, and D. The monopulse antenna pattern can be obtained using four feeds, especially horns. Four signals must have the same phase and different amplitudes for amplitude monopulse processing.

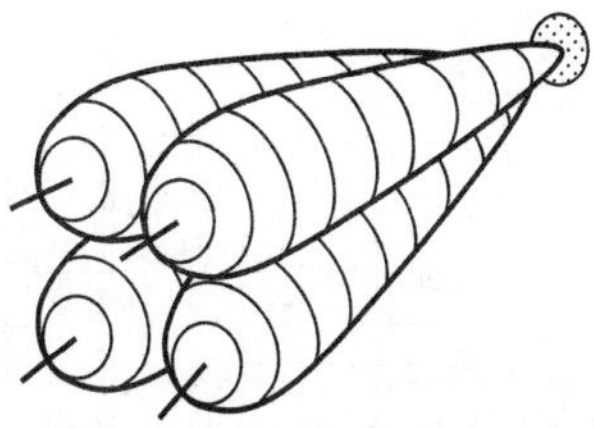

Figure 15.14 (b) *Monopulse antenna pattern*

In the 2D amplitude monopulse technique, circle that is located at the center of the antenna's tracking axis (Figure 15.14 (b)) represents the target echo signal, where the four beams are represented by the four quadrants (Figure 15.15). When the target is on the antenna's tracking axis, an equal amount of energy is received by the four horns. When the target is off the tracking axis (Figures 15.15 (b–d)), the energy in the different beams gets unbalanced. The servo-control system is operated by an error signal which is created due to this unbalanced energy. A sum Σ and two difference Δ (azimuth and elevation) antenna patterns are obtained as a result of Monopulse processing. The ratio of Δ/Σ can be obtained by determining the angle of the signal. The amount of target displacement off the tracking axis can be sensed by comparing the amplitudes and phases of all beam echoes of the radar continuously.

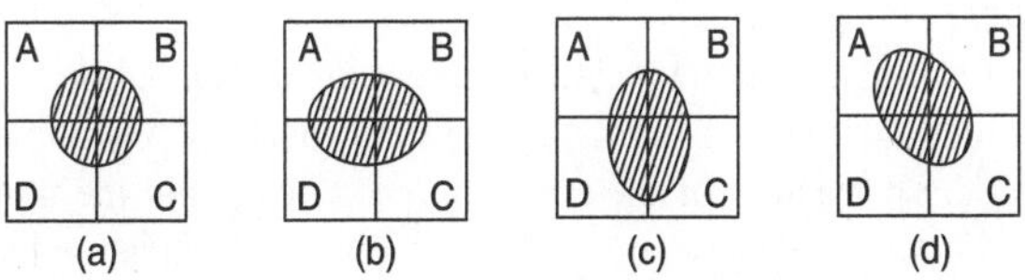

Figure 15.15 *Illustration of monopulse concept (a) Target on the tracking axis; (b)–(d) Target is off the tracking axis*

The parabolic reflector produces four partially overlapping beams with four feeds. A space-fed phased array antenna usually known as Cassegrain reflector is illuminated by the four feeds. The three received signals Σ (summation of four echo signals), Δ_{AZ} (difference pattern in one plane, where the sum of two adjacent feeds is subtracted from the sum of the other two adjacent feeds), and Δ_{EL} (difference pattern in orthogonal plane is obtained in a similar manner as), are shown in Figure 15.14 (a). The three signals Σ, Δ_{AZ}, and Δ_{EL} that are generated in the antenna are amplified by the three receivers matched in amplitude and phase, which means that the amplitude relationships and the relative phase shifts are maintained within very narrow tolerances. The received echo signals are entering into a sum channel through a circulator and difference channels. They are amplified by an intermediate frequency amplifier in three channels. The sum channel Σ has an improved signal-to-noise ratio since the beam it forms is a combination of the signal power of two individual beams by which the target is detected and the range is measured. For automatic detection circuits and for range tracking, AGC signals are required. So the AGC signals are also generated by the sum signal. The error voltage is produced by the difference channels Δ_{AZ} and Δ_{EL} and it is roughly proportional to the angular deviation of the target from the bore-sight.

The phase detector shown in Figure15.16 is used to determine the sign of the error voltage depending on the phase shift between the Σ and Δ_{AZ} (azimuth angle), Σ and Δ_{EL} (elevation angle).

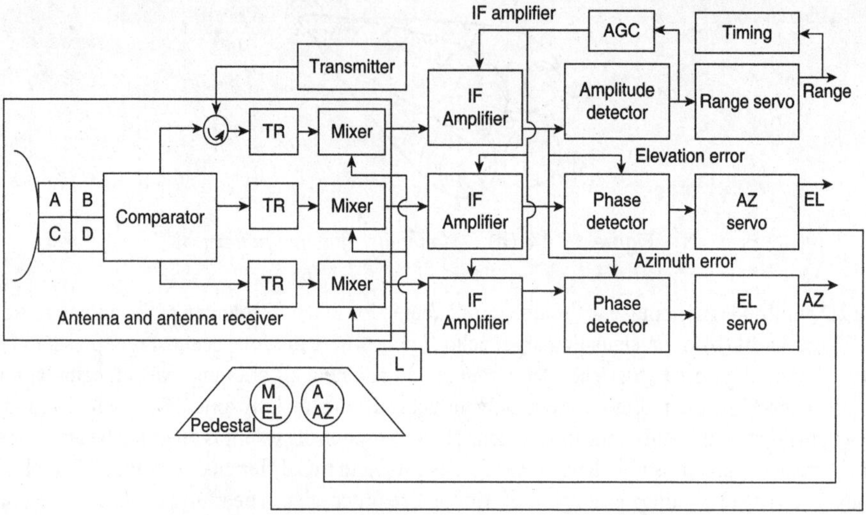

Figure 15.16 *Block diagram of monopulse radar*

15.5.3.3 *Phase comparison monopulse*

The second monopulse tracking technique is called *phase comparison*. This kind of a monopulse system is similar to the amplitude comparison monopulse system, but the two antenna beams point in the same direction. Hence, the projection of the target on each beam and the amplitudes of the returns from the target received by each of these antennas are the same. This is the main difference between phase comparison monopulse systems and amplitude comparison ones. By comparing the phase of the echoes from two separate beams (or two antennas) rather than the amplitudes of the echoes from the two squinted beams the angle of arrival from a target can be determined. The phase difference is zero as the echo reaches the bore-sight at the two antennas simultaneously. The arrival of the target at an angle q to bore-sight at one antenna will be later than the other due to the extra distance travelled by the echo from a target.

As in amplitude comparison monopulse, here also the target angular coordinates are derived from one sum (Σ) and two difference (Δ) channels. The major difference between phase comparison monopulse and amplitude comparison monopulse is that in the phase comparison monopulse, the signals have the same amplitude and different phases. Whereas, the four signals in the amplitude comparison monopulse are similar in phases and different in amplitudes. In this phase comparison techniques, two antennas will receive the signals from the target in the same direction as shown in Figure 15.17 (a). The signals produced by the target are with the same amplitude but with different phases (on axis phases equal). An error signal for each coordinate can be calculated by using the phase difference between the signals generated in the antenna elements. The signals received from the target at two antennas are at an angle θ. The geometry of the received signals is shown in Figure 15.17 (b).

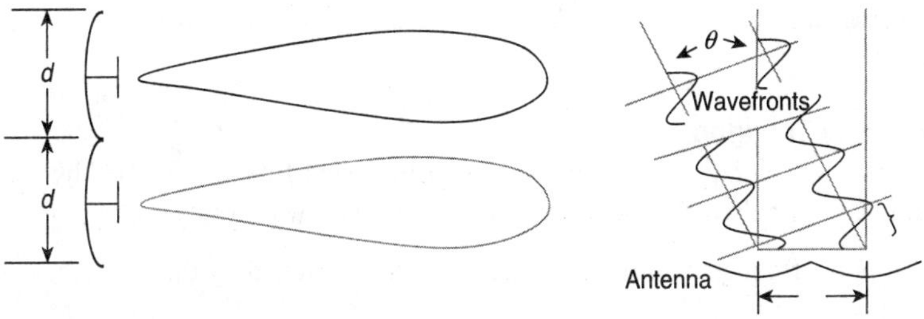

Figure 15.17 (a) *Two antennas radiating identical beams in the same direction*

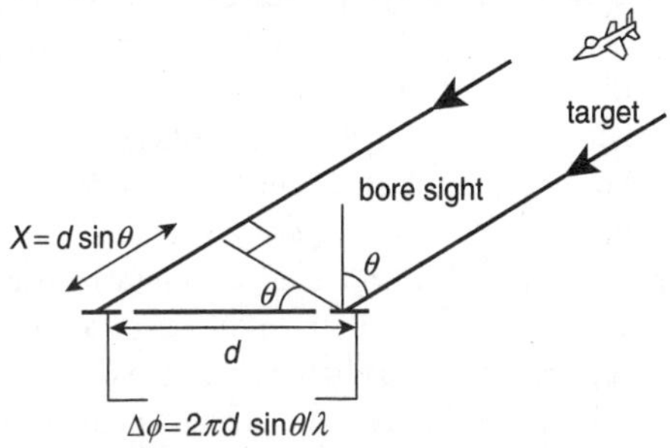

Figure 15.17 (b) *Geometry of the signals at the two antennas when received from a target at an angle θ*

The extra distance that one echo signal travels with regard to the other from the target is given by $X = d\sin\theta$. The received target echo varies in phase and is given by

$$\Delta\phi = 2\pi\left(\frac{d}{\lambda}\right)\sin\theta \tag{15.4}$$

where λ is the wavelength, and the phase difference $\Delta\phi$ is used to determine the angular target location.

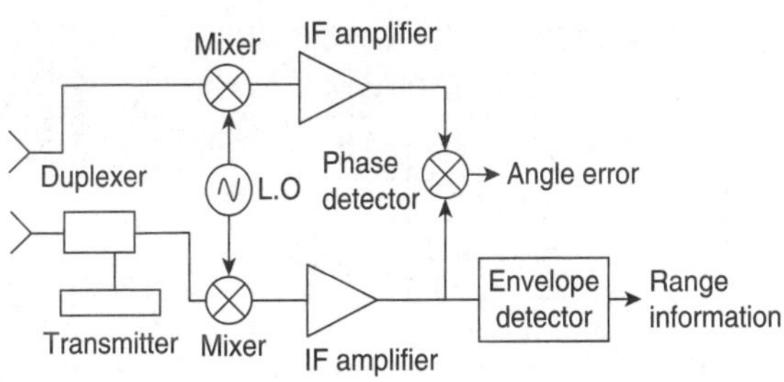

Figure 15.18 *Block diagram of phase comparison monopulse tracking radar*

The block diagram of phase comparison monopulse tracking radar is shown in Figure 15.18. The antenna position is shifted if there is a phase shift in the mixer and IF amplifier stages. When compared to amplitude comparison monopulse it is not easy to maintain highly constant bore-sight which may seriously affect the performance and also it is difficult to provide preferred antenna illumination for difference and sum signals. The phase comparison system becomes more capable to bore-sight change because of its mechanical loading or sag, differential heating, and so on, which is a result of longer paths between antenna outputs and comparator circuitry. This results in serious performance degradation.

15.5.4 Velocity Tracking

Doppler information, which can be achieved by a narrow band filter, is obtained by using Tracking radars similar to CW or pulsed Doppler radar systems. This has two advantages:

- If the Doppler frequency shift is higher than the clutter, then the signal-to-noise ratio is increased
- It is used to detect a target from a group where all the targets have same angle and range

The radar which estimates the target's velocity by using Doppler effect is called a Doppler radar. The estimation of target's velocity using Doppler effect is performed by sending a microwave signal to the target, capturing its reflection, and evaluating the change in the frequency of the returned signal due to the movement of the target. The direct and absolutely accurate measurements of the radial velocity component of a target relative to the radar can be obtained due to these variations.

When the target is approaching, the received frequency is higher than the emitted frequency; it is the same when the target is passing by; and when the target is moving back, it is lower. The change in frequency also depends on the direction of the movement of wave source with respect to the observer. If the source is moving towards or away from the observer then the change in frequency is maximum. It shrinks when the angle between direction of waves and direction motion increases. This continues until the source is moving at right angles. After it reaches right angles then there is no shift.

There will be a shift in the carrier frequency of the received signal, when the target is moving relative to the radar and this effect is called the *Doppler effect*. This shift in frequency is the Doppler shift, and it is a measure of the velocity of the target.

Some radars track the target in velocity (Doppler-frequency shift). There are two ways to measure Target radial velocity. Either from the Doppler-frequency shift of the received signal or from the multiple range measurements. Accuracy of the measurements that are made by the Doppler frequency shift is better when compared with non-coherent processing of range measurements. The movement of the reflecting or transmitting source results in Doppler effect which is responsible for frequency shifts of the received signal. The Doppler tracking machine is shown in Figure 15.19

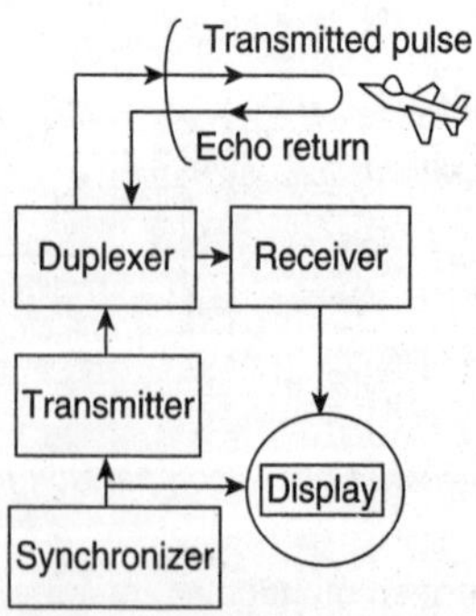

Figure 15.19 *Doppler tracking machine*

The Doppler error is produced at the intermediate frequency in a discriminator or a split-filter system that is very similar to the split-gate range error discriminator. The split-filter error generator is shown in Figure 15.20.

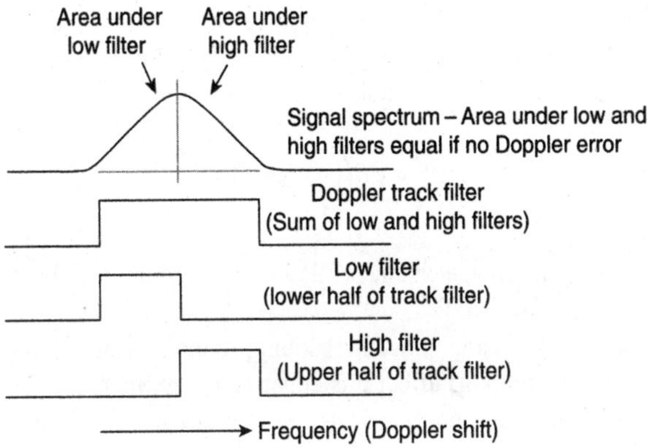

Figure 15.20 *Doppler split-filter error development*

The Doppler track error is represented by the difference between the target's intermediate frequency and the system's nominal intermediate frequency. After the error is filtered and amplified in a servo, it is used to change the receiver local oscillator frequency until the Doppler-shifted signal is at the nominal IF. The Doppler shift is the amount that the local oscillator has to be "pulled" from its nominal value. If the local oscillator is lower in frequency than the transmit frequency, the Doppler shift is

$$f_d = f_T - f_c - f_{LO} \tag{15.5}$$

where f_d = The Doppler shift (Hz)

f_T = The transmit frequency (Hz)

f_c = The COHO frequency, which is the nominal IF (Hz)

f_{LO} = The local oscillator frequency when the track error is zero (Hz)

The Doppler shift frequency $f_d = \dfrac{2v_r}{\lambda}$

where f_d is the Doppler-frequency shift, and λ is the radar signal wavelength.

The target radial velocity v_r can be measured by a coherent radar from the Doppler-frequency shift of the received signal. It can also be obtained by dividing the difference of the two range measurements by the time between the measurements:

$$v_r = \frac{R_1 - R_2}{t_1 - t_2}$$

where R_1 and R_2 are the ranges, and t_1 and t_2 are the respective measurement times.

15.6 TRACKING ACCURACY

The ability of the radar is determined by the accuracy with which the radar tracks the target. Tracking accuracy is a function of the radar, the target, and the propagation medium. The angular accuracy of the tracking radar is calculated by both electromechanical forces which were used to control the antenna for turning the gear and the SNR. Regardless of the tracking method used, the angular accuracy $\delta\theta$ should be related to the beam width and should be fundamentally limited by

$$\delta\theta \sim \nabla\theta / \left(\sqrt{2 \times \mathrm{SNR}}\right) \tag{15.6}$$

However, the use of sum and difference channels allows this to be improved such that $\delta\theta/K$ may be achievable, where K is the slope of the ∇/Σ curve near $\theta = 0$.

Eq. (15.6) is derived only by using sequential lobing, conical scanning, or monopulse tacking techniques. The amplitude fluctuations affect this equation approximation when the first two tracking techniques are used but this does not occur in case of monopulse tracking technique. The tracking noise is also increased due to RCS fluctuations, multi-paths, and changes in the atmospheric propagation.

At short ranges, the accuracy of the system is restricted by the finite size of the target. Tracking radars operate at comparatively short wavelengths since they use narrow pencil beams. At these short wavelengths, the target RCS operate in optical region and it behaves similar to many independent scatters, each of which contributes in a complex wave in both amplitude and phase to the overall RCS and causes an effect on optical glinting. At any given instant, the strongest scattering facet may lie elsewhere on the object and the radar may start to wander off the central point when we define a *center of gravity* of the target and attempt to track the independent scatters.

15.6.1 Limitations of Tracking Accuracy of Radars

The following are some of the limitations of tracking the accuracy of radars:

- Amplitude fluctuations
- Angle fluctuations
- Receiver and servo noise

Amplitude fluctuations: A complex moving target can be considered as a number of independent scattering elements. For example an aircraft or a ship. The echo signal is obtained by the vector addition of the contributions from the individual scatterers. Because of motion of the target or turbulence (as in the aircrafts), the relative phase and amplitude relationships of the target position changes with regard to the radar. As a result, the contributions from the individual scatterers change. In the design of the lobe switching radar and the conical-scan radar amplitude fluctuations of the echo signal are significant but in case of monopulse tracker, they are less consequent. A finite time is needed by conical-scan tracker and the lobe-switching tracker to get the angle error measurement.

The effect of amplitude noise on tracking can be minimized by selecting a conical-scan frequency at a low value of amplitude noise. The amplitude fluctuation noise can be eliminated with filters or AGC, if they occur in conical-scan or lobe-switching frequencies.

A usual scan frequency may be in the range of 30 Hz. With an increase in frequency, target amplitude noise decreases so higher frequencies are also be used as scan frequency.

The echo amplitude fluctuations does not affect the tracking accuracy of radars if they are operated with pulse repetition frequencies from 1000 Hz to 4000 Hz and a lobing or scan rate of one-quarter of the PRF.

Angle fluctuations: The apparent radar-reflecting center is the direction of the antenna when there is no error signal. This center drifts from one point to another point if there are changes in the target with respect to the radar. Noisy or jittered angle tracking happens due to the random drifting of the apparent radar-reflecting center. This is called *angle noise*, *angle scintillations*, *angle fluctuations,* or *target glint*. In general, the apparent center of reflection might not correlate to the target center.

All angle tracking techniques are affected by this angle fluctuation. At short range and for large targets like missiles, angular fluctuations greatly affect the tracking accuracy. Whereas, in most occasions, the angular fluctuations produced by small targets at a long range may be of little significance.

Receiver and servo noise: The receiver noise power is another important factor affecting the tracking accuracy. There exists an inverse proportionality between accuracy of the angle measurement and the square root of the signal-to-noise power ratio. The angular error is caused by the receiver noise. It is proportional to the square of the target's distance. This is due to the fact that the signal-to-noise ratio is proportional to $1/R^4$. Servo noise results because of the backlash and compliance in the gears, shafts, and structures of the mount in the tracking servo mechanism. The amount of this servo noise does not dependent upon the range because it is independent of the target echo.

15.7 FREQUENCY AGILITY

The process of changing the radar frequency from pulse to pulse is known as Frequency agility. On a pulse-to-pulse basis, a change in radar carrier frequency, within a band that is more than 10% of the carrier frequency exists due to this technique. Performance is improved with frequency agility than with fixed frequency operation, even though it entails severe technical difficulties.

These are as follows:

- An increase in range, leads other parameters being equal, of up to 35%
- Clutter can be reduced in radars without employing MTI technique
- In tracking radars, it reduces Glint and noddling
- It is also used in reduction or nullification of lobing
- It helps in reducing the jamming of the signals

To give uncorrelated measurements the frequency channels should be at a distance of target decorrelation bandwidth and the frequencies chosen should be at a distance of at least a radar bandwidth. This separation of frequency Δf should satisfy the following condition.

$$\Delta f \geq c/(2l) \text{ Hz} \tag{15.7}$$

where, l is the dimension of the target in the range.

Frequency agility is mostly used in defense ships for intercepting incoming missiles flying closer to the sea surface and with small cross section.

15.8 TRACK WHILE SCAN (TWS)

Track while scan system combines search and track functions and automatically tracks multiple targets at the same time, i.e., it performs multi-tasking and multi-target tracking. The target parameters between scans can be estimated by tracking radar systems. This is done by sampling each target once between two successive scans by specially designed smoothing and prediction filters. This phenomenon is known as Track while scan. The limited beam steering is one of the characteristic of TWS which is performed by either a mechanically steerable beam or a combined mechanically steerable antenna. The limitation of TWS radars is that the radar antenna is positioned mechanically even though they have the capacity for increased target-handling.

The range tracking system performance using gates is already explained above. To differentiate the targets from one another, a TWS system may use range, angle, Doppler or elevation gates. The multi-function operations, such as detection, tracking, and discrimination can be performed by Modern radar systems.

Operation

TWS radar starts a separate track file, whose main components are position, velocity, and acceleration, when a target is detected. To start a track file, at least one target should be confirmed. The target's future parameters are estimated by processing the subsequent detections from that target.

The radar gathers accurate tracks of valid targets from such successive scans. There are three main processes in this and those are as follows:

- **Preprocessing:** After successive scans, the targets with same range, range rate, and angle are combined and the detections are converted onto a reference co-ordinate system.
- **Correlation:** Based on earlier observations, the next target position is predicted using some prediction techniques such as kalman filter and these predictions are compared with measurements. When the prediction lies within specified limits, then the new observations made will be a part of the track file. Difficulties arise only when the closely spaced targets are detected and have to be resolved by employing high level statistical metrics.
- **Track creation or deletion:** A new track is started, when there is a new observation which is not a part of a track. If the correlation is zero then the second observation confirms the track and if the correlation is bad then the second observation deletes the track. The track is dropped, if there is no new detection in that track after few scans.

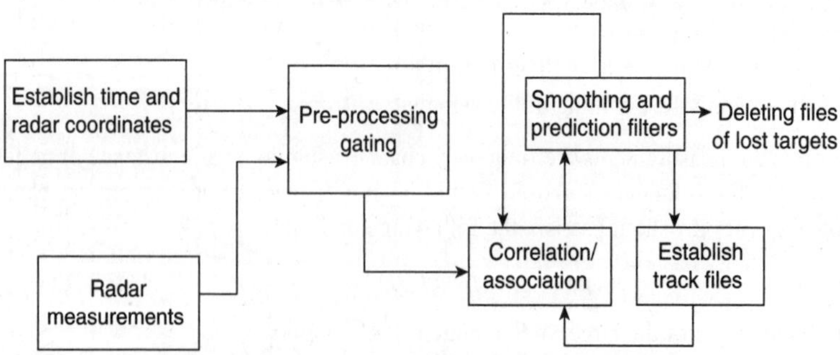

Figure 15.21 *Track-while-scan radar systems*

TWS radars should determine whether the detection made is of a new target or that of an older target from earlier scans in contrast to the single target tracking systems. This differentiation is done by correlation and association algorithms. Correlation process prevents redundant tracks, by correlating the new detection with all previous detections. If multiple correlation of a single detection, i.e., correlation of one detection with more than one track occurs, then a set of association rules will be implemented, so that it can be assigned to the proper track.

TWS data processing block diagram is shown in Figure 15.21. There are two methods for filtering and prediction purpose, one method is to use $\alpha\beta$ filters for the polar co-ordinates (R, θ) and the other method is to convert the measured positions from polar to Cartesian (x, y) before filtering.

15.9 PHASED ARRAY RADARS

Most modern radar systems employ phased array antenna technology, providing an ability to scan the antenna beam position electronically. Phased array radars use phased array antennas that are capable of steering the beam electronically in space when compared to other radars. Due to this flexibility, versatility, energy management increases. This also provides good tracking and searching functions which helps antennas in covering larger areas quickly. As a large number of targets cannot be tracked by a single antenna simultaneously and accurately, an array that consists of a large number of individual radiators placed in a particular way should be employed where a moving beam can be produced by a stationary antenna.

By varying phase difference in every antenna feeders beam steering can be obtained. Across the antenna aperture there should be good control over the phase of energy.

Mechanical scanning methods are naturally slow. This is because, of the antenna inertia and inflexibility in beam positioning in this method. Hence, a large amount of power is required by mechanical scanning methods to respond quickly while handling a large number of high-speed targets.

The problems such as inertia, time lags, and vibrations that occur in mechanical systems can be rectified in electronic beam scanning and radar beams can be positioned almost immediately. Electronic scanning phased array tracking radars can track multiple targets at the same time by measuring each target without echo or other signal sources.

The beam can be steered by phased array antenna by means of electronic control. The radiators can be placed on any different types of surfaces, such as

- **Linear array:** The arrays of elements are arranged in a straight line in one dimension.
- **Planar array:** The arrays of elements are arranged in a plane in two dimensions either in rectangular, square, or circular aperture.
- **Conformal array:** The arrays of elements are distributed on a non-planar surface.

The Figure 15.22 illustrates the block diagram of phased arrays. Passive phased arrays are shown in Figure 15.22 (b). The passive phased arrays mechanically scanned antennae are similar to each other except that passive phased arrays contain an electronically controlled phase shifter behind each radiating element. The receiver and transmitter are placed along a central feed network. Active phased array is shown in Figure 15.22 (a) where T/R (transmitter and receiver) module is placed in a similar manner as phase shifters. Both these arrays have many advantages when compared to mechanically steered arrays. They are as follows:

1. Reduce radar cross-sections
2. Excellent beam agility
3. Highly reliable, as there are no moving parts in them

Only the phase shifters are active elements in the passive phased array. Inspite of their failure, the antenna's performance degrades if 5% of the phase shifters fail and so time will be allotted to replace failed phase shifters.

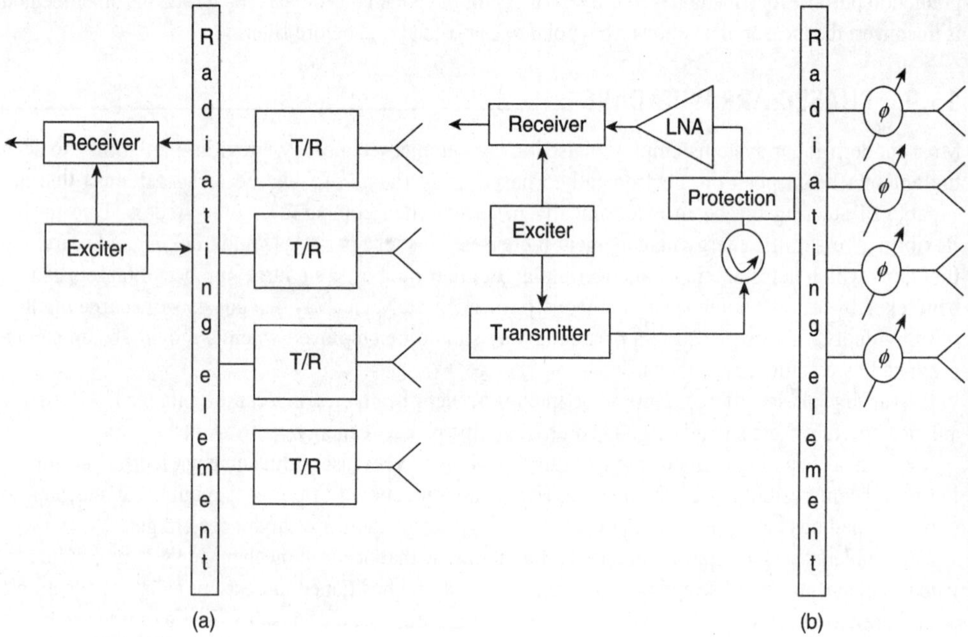

Figure 15.22 *Block diagram of phased arrays (a) Active phased arrays; (b) Passive phased arrays*

15.10 RADAR DISPLAYS

The device used to display radar information is known as a *radar indicator* or a *radar display*. Two types of radar videos are generally used. They are (i)raw video and (ii) synthetic video.

Raw video: To display raw videos, oscilloscopes are used which display the amplified detected target echo signals. These are controlled by operator to detect the clutter signals and target noise.

Synthetic video: Synthetic video generates its own symbol for each target which helps in cleaning the display by removing noise and clutter.

Display

Display helps in displaying Clear and crisp target information. The final output of the receiver part is displayed on the cathode ray tube or on any other display device. Most common radar displays are *PPI, A scope, B scope, C scope,* and *E scope* (Figure 15.23).

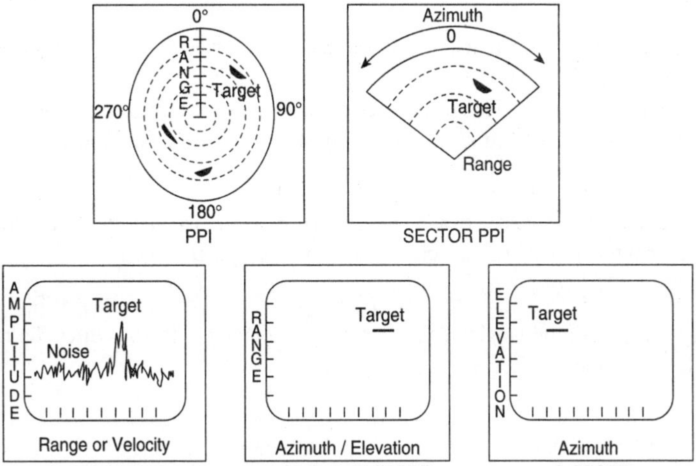

Figure 15.23 *Common radar displays*

PPI

The plan position indicator (PPI) is the most common type of CRT display. It is also known as p*olar plot indicator*, as it plots the location of the target in azimuth and range in polar coordinates.

As shown in Figure 15.24, the PPI is a circular display. At the center, a sweep rotates along with the transmitter antenna showing an image under the radars coverage area. The return for a particular angle is displayed along the sweep in the display.

Figure 15.24 shows simulated and real displays of PPI.

An intensity-modulated display is a kind of display device where the amplitude of the receiver output modulates the electron-beam intensity (*z* axis) as the electron beam is made to sweep outward from the center of the tube. Simulated and real displays of PPI are as shown in Figure 15.24.

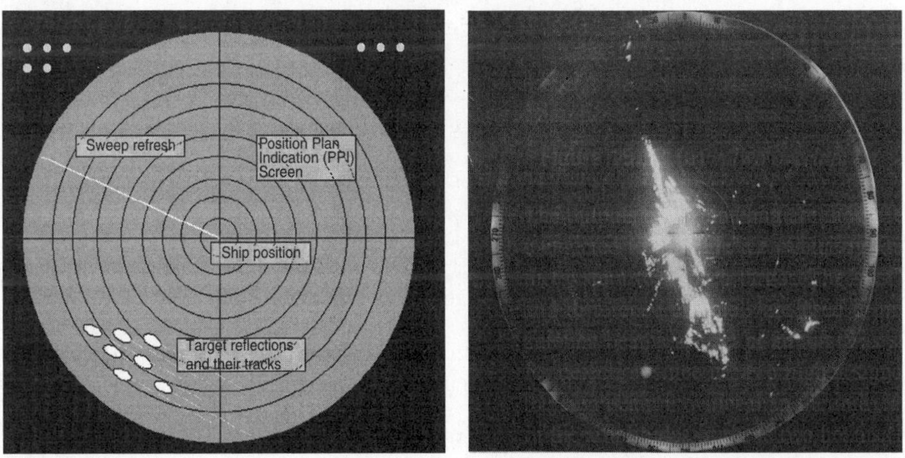

Source: www.aero.web.org

Figure 15.24 *Plan position indicators showing a simulated and a real display*

A-scope

The target signal amplitude versus range or velocity can be displayed in A-scope. A-scope also displays the targets detected along the pencil beam for selected range limits.

A-scope is similar to synthetic video displays and is also known as *range scope* (R-Scope).

A-scope is a deflection-modulated display and is mostly used for tracking-radars.

B-scope

B-scope displays target range versus target azimuth angle. B-scope is similar to PPI displays with limited dynamic range except that B-scope uses rectangular, rather than polar coordinates.

C-scope

C-scope displays target azimuth angle versus target elevation angle. It gives direction to the target up and to the right, but not the true range.

E-scope

It displays target elevation versus target range. Both B-scope and E-scope are similar. But only difference is in E-scope elevation is used instead of azimuth. The movement of the blip from the center of display indicates target's position with respect to antenna's beam axis.

EXAMPLE PROBLEM 15.1

For a 1 GHz base frequency and a target radial speed of 10 knots, determine the Doppler shift.

Solution

Given data, $v_r = 10$ knots (note that 1 knot = 1.852 km/hr)

$f = 10^9$ Hz

$c = 3 \times 10^8$ m/s

$$f_d = \left(\frac{2v_r}{\lambda}\right) = \left(\frac{2 \times 10 \times 1.852 \times 5/18}{0.3}\right) = 34.295 \text{ Hz}$$

■

SUMMARY

1. The tracking radar continuously tracks the target to determine the location or direction of target.
2. There are three types of tracking techniques. They are range tracking, velocity tracking, and angle tracking.
3. Search radars is used many applications, most of them are used to provide detection of ships and aircrafts and surveillance of submarines operating under snorkel conditions.
4. Tracking radar systems measure the target's relative position in certain parameters (they are range, azimuth angle, elevation angle, and velocity) using these parameters they can predict their future values.
5. Antennas used for scanning can be of two types: (i) mechanically scanning antennas and (ii) electronically scanning antennas.

6. Angle tracking systems track the target angular position in the azimuth and elevation angles.
7. Three methods are used for generating the error signals for tracking: (1) sequential lobing, (2) conical scan, (3) monopulse tracking.
8. Monopulse scanning is the most efficient and robust tracking technique. In all modern radars, angle tracking is usually done with monopulse tracking.
9. Track-while-scan (TWS) radars track many targets while scanning a limited airspace to find others.
10. Electronic scanning phased array tracking radars can track multiple targets simultaneously by measuring each target without echo or other signal sources.
11. The Plan Position Indicator (PPI) is the most common type of CRT display. It is also known as *polar plot indicator,* as it plots the location of the target in azimuth and range in polar coordinates.

OBJECTIVE-TYPE QUESTIONS

1. In the following scan radar, the direction of the antenna beam does not coincide with the bore-sight, but revolves around, it seeking the target direction.

   (a) lobe switching (or sequential lobing) (b) monopulse
   (c) conical scan (d) none of the above

2. Identify the radar where the antenna lobe can assume four different positions around the bore-sight.

   (a) lobe switching (or sequential lobing) (b) monopulse
   (c) conical scan (d) none of the above

3. The following technique keeps the beam pointed at the target to improve angle accuracy and it is based on the principle that the radar receiver will get maximum returned signal strength.

   (a) lobe switching (or sequential lobing) (b) monopulse
   (c) conical scan (d) none of the above

4. The tracking technique that derives angle error information on the basis of a single pulse is known as

   (a) lobe switching (or sequential lobing) (b) monopulse
   (c) conical scan (d) none of the above

5. The following radar requires a separate receiver for each channel, and it improves the performance of the conical scan and sequential lobing whose performance degrades with time-varying radar returns.

   (a) Pulse radar (b) monopulse
   (c) MTI (d) none of the above

6. Which radarscope plots target echo amplitude versus range on rectangular coordinates for some fixed direction? It is also used primarily for tracking radar applications than for surveillance radars.

   (a) PPI scope (b) B scope
   (c) A scope (d) none of the above

7. The most common type of CRT display that maps the location of the target in azimuth and range in polar coordinates is

(a) PPI scope (b) B scope
(c) A scope (d) none of the above

8. A surveillance radar that develops tracks on the targets it has detected is called a

(a) tracking radar (b) scanning radar
(c) track-while-scan radar (d) none of the above

9. The process of changing the radar frequency from pulse to pulse is known as

(a) frequency agility (b) blind speed
(c) target velocity (d) none of the above

10. Range estimation accuracy improves with

(a) decreasing bandwidth (b) increasing bandwidth
(c) constant bandwidth (d) none of the above

11. The radar that uses more than one beam simultaneously to measure the angular position of the target on a single pulse is

(a) lobe switching (b) sequential lobing
(c) conical scan (d) monopulse

12. The PPI scan-type indicator can indicate

(a) range of the target only (b) direction of the target
(c) Both a and b (d) Range, speed, and direction

13. Doppler effect is used in

(a) MTI (b) CW (c) FM (d) all of the above

ANSWERS TO OBJECTIVE-TYPE QUESTIONS

1. (b) 2. (a) 3. (b) 4. (b) 5. (b) 6. (c) 7. (a) 8. (c) 9. (a)
10. (b) 11. (d) 12. (c) 13. (d)

REVIEW QUESTIONS

1. Briefly explain the tracking radar and search radar systems.

2. Explain the following antenna tracking mechanisms:

(a) sequential lobing and (b) conical scanning

3. Explain with the help of a block diagram amplitude comparison monopulse radars for extracting error signals in both elevation and azimuth.

4. Explain phase comparison monopulse tracking radar technique.
5. Explain how tracking in range is achieved using a split-gate range.
6. Explain in detail the limitations of tracking accuracy.
7. Explain the working of a monopulse radar with the help of a block diagram.
8. Describe the various methods of monopulse tracking techniques.
9. Define tracking in range and explain the split-gate tracker method.
10. Explain the block diagram of an amplitude comparison in the monopulse in one angle coordinate.
11. Draw the block diagram of a conical-scan radar, one-coordinate monopulse tracking, and explain its operation.
12. With a neat diagram explain the operation of sequential lobing and a conical scan.
13. What are the various methods of acquisition before tracking a target with radars? Explain in detail.

Detection of Signals in Noise and Radar Receivers

16

16.1 INTRODUCTION

Radar receiver extracts the target information by processing the detected signal or echo (that is signal reflected from the target). The detection of a signal and extraction of information of the object from the detected echo are two separate operations. In general radar detects the signals in the presence of noise. The common problems associated during detecting radar signals in the presence of noise are detection of weak signals and extraction of target information from the weak signals. In all radar receivers, the matched filter, which is an optimum filter, is used for the detection of weak signals in the presence of noise. The matched filter is referred to as an optimum filter because the filter output (that is peak signal power divided by average noise power) is equal to twice the input (that is input signal power divided by the input noise power). The linear receiver characteristics of a matched filter that improves the signal-to-noise ratio (SNR), various methods for the detection of desired signals and the rejection of undesired noise, noise figure, phased array radars, and its characteristics are discussed in this chapter in detail.

16.2 MATCHED-FILTER RECEIVER

The target-detection capability of a radar can be improved by improving the power of a detector. The power can be increased by increasing the SNR of a radar receiver. The SNR can be increased by using matched filter, which is an optimum filter. The matched filter accomplishes this by matching the receiver transfer function to the input signal. Therefore, the matched filter is very important for the design of almost all radar receivers. In the radar receiver block diagram, the matched filter is included before the signal processor section. Matched-filter section is placed between the A/D converter and the detector sections (as shown in Figure 16.1(a)).

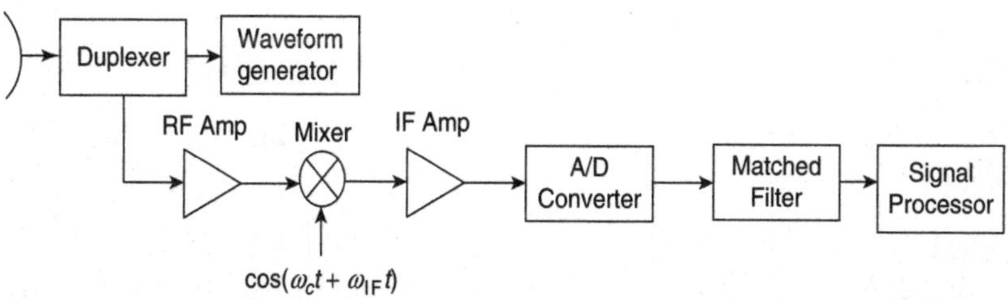

Figure 16.1 (a) *Block diagram of a radar receiver*

Matched Filter Frequency Response Function

The matched filter increases the SNR so as to improve the target-detection capability. The output peak SNR is given by $2E/N_0$, where E is the received signal energy and N_0 is the noise power. The matched filter used in the radar receiver is shown in Figure 16.1(b).

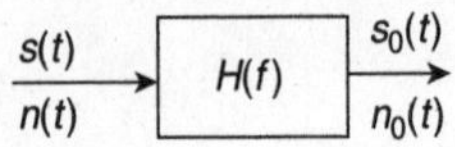

Fig. 16.1 (b) *Matched-filter block diagram*

where,

$h(t)$ = filter impulse response,

$s(t)$, $n(t)$ = input signal and noises to the matched filter, respectively

$s_0(t)$ = output of the matched filter obtained by convolution of input signal with the impulse response $h(t)$

$n_0(t)$ = noise at the output of the matched filter

$H(f)$ = matched-filter transfer function, which is the Fourier transform of impulse response, $h(t)$

So, we design optimum-matched filter to maximize SNR. In the design, the input signal $s(t)$ is a deterministic signal and noise signal $n(t)$ is a random process. The output of the matched filter is obtained by the convolution of input signal and matched-filter impulse response. The noise output is obtained by the convolution of input noise and matched-filter impulse response.

$$s_o(t) = s(t) * h(t) \quad n_0(t) = n(t) * h(t)$$

where $*$ denotes convolution.

The matched filter is designed in the frequency domain using the Fourier transforms to avoid the complex multiplications. When the input to the radar receiver is purely sinusoidal, then its frequency response is denoted as $H(f)$. $H(f)$ relates the amplitude and phase of the output of the system with the input of that system and is given below as $H(f) = \Im[h(t)]$. In frequency domain, the output signal and noise is obtained as:

$$S_o(f) = H(f)S(f) \quad N_o(f) = |H(f)|^2 N(f)$$

By applying inverse Fourier transform to the above output signal equation, we obtain the output signal in time domain as:

$$s_0(t) = \int_{-\infty}^{\infty} S(f)H(f)e^{j2\pi ft} df \tag{16.1}$$

The peak signal power P_S is given as $P_S = |s_0(t)|^2$ and noise is $P_N = \int_{-\infty}^{\infty} N_0(f)df$

So, the matched-filter impulse response is obtained from the above equations as:

$$h(t) = \max(\text{SNR}) = \max \frac{P_S}{P_N} = \max \frac{|s_0(t)|^2}{\int_{-\infty}^{\infty} N_0(f)df} = \max \frac{\left| \int_{-\infty}^{\infty} S(f)H(f)e^{j2\pi f t_0} df \right|^2}{\int_{-\infty}^{\infty} |H(f)|^2 N(f)df} \tag{16.2}$$

The noise spectral density is given as $N(f) = KTF_n G_a$

where F_n is noise factor

T is temperature coefficient

K is Boltzmann's constant

G_a is constant and it is the gain of all receiver components between the antenna and the input to the matched filter

By substituting noise spectral density in Eq. (16.2), the matched-filter impulse response is obtained as:

$$h(t) = \max \frac{\left| \int_{-\infty}^{\infty} S(f)H(f)e^{j2\pi f t_0}\,df \right|^2}{KTF_n G_a \int_{-\infty}^{\infty} |H(f)|^2 df} \tag{16.3}$$

The maximized value can be obtained by using Cauchy–Schwartz inequality which is given as:

$$\left| \int_a^b A(f)B(f)df \right|^2 \le \left(\int_a^b |A(f)|^2 df \right)\left(\int_a^b |B(f)|^2 df \right) \tag{16.4}$$

with equality when $A(f) = KB^*(f)$ (16.5)

By using Cauchy–Schwartz inequality considering $A(f) = H(f)$ and $B(f) = S(f)e^{j2\pi f t_0}$, the maximization can be solved as:

$$\frac{\left| \int_{-\infty}^{\infty} S(f)H(f)e^{j2\pi f t_0}\,df \right|^2}{KTF_n G_a \int_{-\infty}^{\infty} |H(f)|^2 df} \le \frac{\left(\int_{-\infty}^{\infty} |H(f)|^2 df \right)\left(\int_{-\infty}^{\infty} |S(f)|^2 df \right)}{KTF_n G_a \int_{-\infty}^{\infty} |H(f)|^2 df} \le \frac{\left(\int_{-\infty}^{\infty} |S(f)|^2 df \right)}{KTF_n G_a}$$

We have found the maximum value of P_S/P_n. That is, we have solved part of the maximization problem. To find the *h(t)* that yields the maximum P_S/P_n we invoke the second part of the Cauchy–Schwartz inequality given in Eq. (16.5) and it can be given'as

$$H(f) = G_a S^*(f)e^{-j2\pi f t_m} \tag{16.6}$$

where t_m is the time when the matched-filter output is maximum.

$S^*(f)$ is the complex conjugate of input signal's spectrum.

We may observe that from Eq. (16.6), $|H(f)| = |G_a S(f)|$. In other words, the matched- filter frequency response has the same shape as the frequency spectrum of the signal. They simply differ by a scaling factor $|G_a|$. This is the reason we term $h(t)$ as a *matched filter.*

Matched filter impulse response

In terms of impulse response *h(t),* the output of matched filter can be represented as

$$h(t) = \int_{-\infty}^{\infty} H(f)e^{j2\pi f t}\,df \tag{16.7}$$

As we know impulse response is nothing but output of a matched filter when the input is impulse. From Eq. (16.6), replacing H(f) in Eq. (16.7) we get

$$h(t) = G_a \int_{-\infty}^{\infty} S^*(f)\exp(-j2\pi f(t_m - t))df \tag{16.8}$$

since

$$S^*(f) = S(-f) \quad \text{we get} \quad h(t) = G_a \int_{-\infty}^{\infty} S(f)\exp(j2\pi f(t_m - t))df$$

$$h(t) = G_a \cdot s^*(t_m - t) \tag{16.9}$$

Thus, $h(t)$ is the conjugate of a scaled (by G_a), time reversed (because of the $-t$) and shifted (by t_m) version of the transmit signal, $s(t)$. The received waveform $s(t)$ is shown in Figure 16.2 (a) and the impulse response $h(t)$ (which is the image of $s(t)$) is shown in Figure 16.2(b).

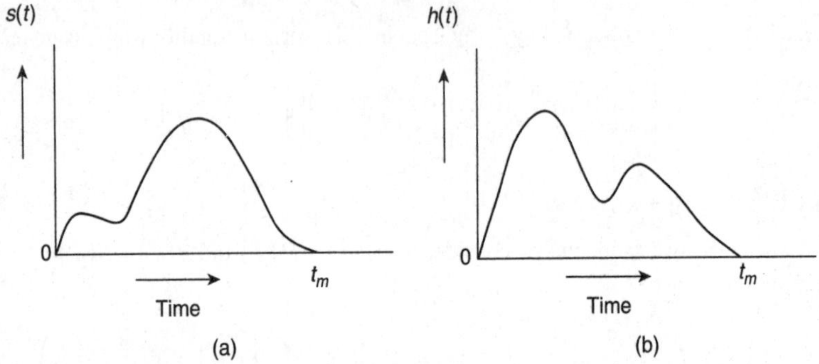

Figure 16.2 *(a) Received waveform of matched filter; (b) Impulse response of matched filter*

Receiver Bandwidth

The frequency response of matched filter is relative to the band pass characteristics of the receiver. The S/N ratio is minimum if bandwidth is wider, since more noise is introduced by wide bandwidth. If we go for narrow bandwidth, again S/N ratio is reduced because it reduces the signal along with noise.

So the bandwidth of a radar should be chosen in such a way that it should reduce the signal loss and S/N ratio should be optimum. To achieve this, bandwidth of the radar should be equal to reciprocal of the pulse width.

Efficiency of non-matched filters: In practice, the matched filter cannot be ideal. Therefore, it is essential to measure the efficiency of non-matched filter in comparison with an ideal filter. The efficiency of non-matched filter can be given as

$$\text{Efficiency} = \frac{\text{peak SNR from the non-matched filter}}{\text{peak SNR from the matched filter}} \tag{16.10}$$

The peak SNR from the matched filter is ($2E/N_0$).
Therefore,

$$\text{Efficiency} = \frac{N_0}{2E} \times (\text{peak singal-to-noise ratio from the non matched filter})$$

With the input as the rectangular pulse of width τ, the obtained efficiency plots for single-tuned (RLC) resonant filter and a rectangular-shaped filter of half-power bandwidth B_τ are shown in Figure 16.3.

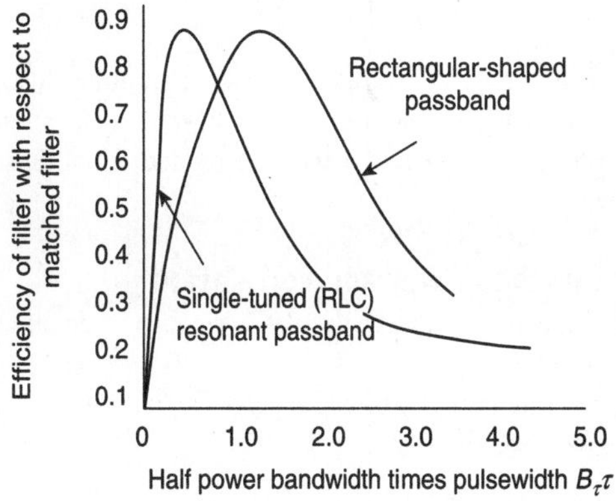

Figure 16.3 *Efficiency of a single-tuned resonant filter and a rectangular shaped filter*

For $B_\tau \approx 0.4$, the efficiency of the single-tuned filter is maximum and its corresponding loss in SNR is obtained as 0.88 dB in comparison with matched filter. The different combinations of filters and pulse shapes are considered and their values of B_τ which maximize the SNR are listed in Table 16.1.

Table 16.1 Efficiency of non-matched filters

Input signal	Filter	Optimum $B_\tau\tau$	Loss in SNR compared to matched filter (dB)
Rectangular pulse	Rectangular	1.37	0.85
Rectangular pulse	Gaussian	0.72	0.49
Gaussian pulse	Rectangular	0.72	0.49
Gaussian pulse	Gaussian	0.44	0 (Matched)
Rectangular pulse	Single-tuned circuit	0.4	0.88
Rectangular pulse	Two cascaded single-tuned circuits	0.613	0.56
Rectangular pulse	Five cascaded single-tuned circuits	0.672	0.5

16.3 CORRELATION DETECTION

As we know, in radar the range of the target can be found by calculating the time difference between transmitted signal and received echo signal, that is, we have to find out the time of arrival of echo signal which can be done by the process of correlation. This is nothing but a measure of similarity between two wave forms (in radar, the transmitted signal and the received echo). For two wave forms, s_1 and s_2, mathematically the correlation is given by

$$R(t) = \int_{-\infty}^{\infty} s_1(\tau)s_2(t+\tau)\mathrm{d}\tau \tag{16.11}$$

From the above relation, it is clear that correlation depends on time shift,'t' of the waveforms. This leads to the maximum correlation at some points and zero correlation at some other points. We can say that two wave forms are correlated or coherent if they are related, if not those two wave forms are said to be uncorrelated or incoherent.

Relationship between matched filter and correlation

Here we need to show that the output of the matched filter is proportional to the input signal and correlate the same signal with some delay.

Let us consider $s_1(\tau)$ and $s_2(\tau)$ are transmitted and received signals, respectively. The cross-correlation between the two signals is given by

$$R(t) = \int_{-\infty}^{\infty} s_1(\tau)s_2(\tau - t)\mathrm{d}\tau \tag{16.12}$$

Now at the matched filter side, the input to the matched filter $s_1(\tau)$ is given by

$$s_1(\tau) = s_2(\tau) + n(\tau) \tag{16.13}$$

where $n(\tau)$ = noise signal

$s_2(\tau)$ = received signal with SNR of (E/N_0)

Now, according to the principle of matched filter, the output of the filter with transfer function $h(t)$ is given as

$$s_0(t) = \int_{-\infty}^{\infty} s_1(\tau)h(\tau - t)\mathrm{d}\tau \tag{16.14}$$

The transfer function of matched filter can be given as

$$h(\tau) = s_2(t_1 + \tau) \tag{16.15}$$

By substituting Eq. (16.15) in Eq. (16.14), we get

$$s_0(t) = \int_{-\infty}^{\infty} s_1(\tau)s_2(t_1 - (t - \tau))d\tau \tag{16.16}$$

$$s_0(t) = \int_{-\infty}^{\infty} s_1(\tau)s_2(\tau - (t - t_1))d\tau \tag{16.17}$$

From the Eqs. (16.12) and (16.17), we have

$$s_0(t) = R(t - t_1) \tag{16.18}$$

From Eq. (16.18), we can say that the matched filter output is the cross-correlation between the transmitted signal and the received signal with noise.

There are two types of correlations: autocorrelation and cross-correlation

i. Autocorrelation

It is the measure of coherence of a signal with itself and it is maximum when time shift is zero or multiple of its time period. Normally in radar, autocorrelation is computed because transmitted signal and received signals are equal. Therefore at radar receiver, the two signals are continuously matched. The instant at which autocorrelation is maximum is the point at which the signal is arrived. For this computation, matched filter is used. From the above discussion, it is clear that correlation is the optimal technique where the required signal can be differentiated from noise. *Detecting known waveform form noise is known as matched filtering.*

ii. Cross-correlation

Cross-correlation measures the coherence between two pulses that can be used as a basis for radar receiver and is explained with the help of Figure 16.4 as shown below.

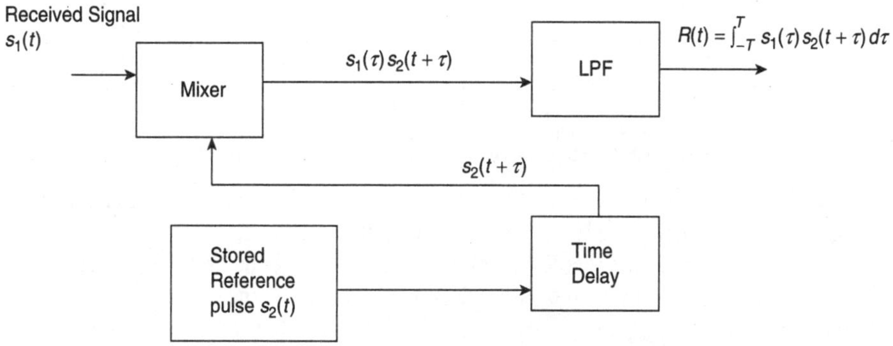

Figure 16.4 *Block diagram of cross-correlation detector*

Assume $s_1(t)$ is the received signal, which is a combination of pulses of sine wave and noise. The reference waveform $s_2(t)$ is stored in the receiver that is similar to the expected signal $s_1(t)$. Here, we are taking cross-correlation rather than autocorrelation because the received signal consists of required signal along with noise that is random in nature. So, we cannot perform autocorrelation. The receiver will compute cross-correlation for various values of time delay. If the computed value exceeds a particular threshold value, then we can say that target has been detected.

16.4 DETECTION CRITERIA

The detection is the process of finding the existence of target in antenna beam. Usually an operator can decide the presence of target by viewing display, where it is brighter when compared to the surrounding background. This target detection is performed automatically at receiver part of the modern radar systems, which can be accomplished by setting a threshold signal level by considering current interference (for example, external noise, clutter, and internal receiver thermal noise). This concept is shown in Figure 16.5. The detection in Sample 50 may in fact be from a target, or it may be a large noise spike, creating a false alarm.

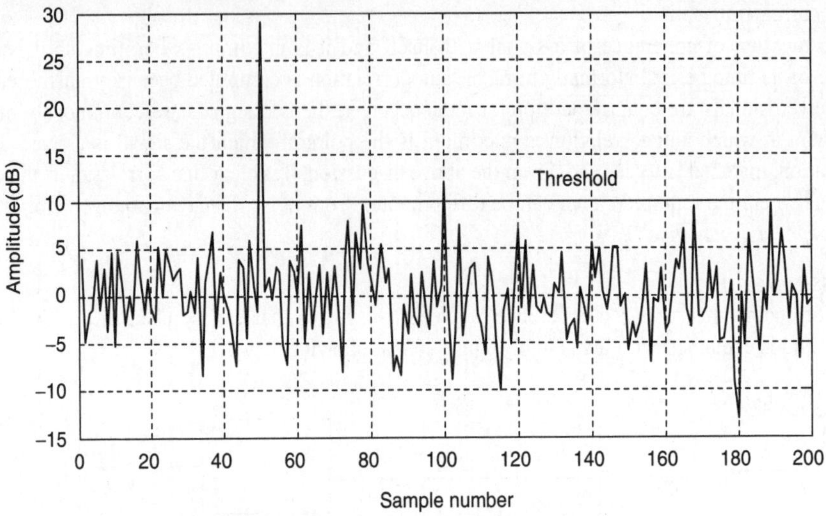

Figure 16.5 *Concept of threshold detection*

Normally after the processing of received signal, the detection process is performed. For detecting the target signal and to reject noise, the signal should be larger than the noise.

The threshold level can be set at a fixed voltage, which is above the noise level to maintain the probability of a false alarm (threshold crossing due to noise alone) at an acceptable level. In general, interference in radars is not only due to receiver noise but also due to clutter and noise jamming. The interference because of jamming and clutter will be quite variable as a function of range, angle, and Doppler cells in the vicinity of a target. Hence, the interference level can vary during operation and fixed threshold level is not possible. Particularly in modern radars, the threshold is often adaptive, automatically adjusted to the local interference level, which results in constant false alarm rate (CFAR). When the target is present, the received signal amplitude must exceed the threshold level.

Till now the target detection is explained based on amplitude. But there is a case where the clutter is a dominant source and especially if its amplitude exceeds that of a target signal. Here for detection, spectral signal processing is employed (moving target indication (MTI) or pulse-Doppler processing) to reduce the clutter level below that of the target signal. If the jamming is dominant interference and its level exceeds that of the target, angle-of-arrival processing (for example, sidelobe cancellation and adaptive beam forming) can be used. System suffering from significant clutter and jamming interference may use a combination of both, called space-time adaptive processing (STAP).

16.5 AUTOMATIC DETECTION

When an operator views a plan position indicator (PPI) display, then the operator "integrates" the combination of the echo pulses obtained from the target in his/her view. Though the operator's integration is as effective as automatic integrator, most of the times it is not possible to maintain the same efficiency because the performance is limited by fatigue, boredom, overload of the operator, and also the integrating characteristics of the phosphor of the CRT display. In the electronic automatic detection, the operations performed and detection decisions to be taken are made independent of the operator's intervention. For further processing, the data detected by automatic detector are either sent to an operator or to a omputer. Considering many aspects, automatic detection requires a much better receiver design than system. The manual detection limits the usage of some automatic devices nevertheless the operator ight have better discrimination capabilities compared to automatic methods for sorting clutter and nterference. But when we consider a case of large number of targets, the automatic, computer-based ecision devices operate with greater efficiency than an operator.

Automatic detection of radar signals involves the following steps:

- It scans the area to search and track the target and the coverage area is quantized into range, angle, and also resolution cells.
- It samples the output of the range and resolution cells such that there is at least one sample per cell. But in practical cases, there may be more than one sample per cell.
- It converts the samples of analog into digital.
- It uses signal processing in the receiver such that the signal processing helps in the removal of noise and clutter echoes.
- The samples at each resolution cell are integrated.
- When the receiver is unable to remove all the clutter and interference, then CFAR is used so that it automatically raises the threshold value to remove the interference and clutter.
- The clutter echoes can be removed by using clutter maps.
- The threshold is detected using automatic tracker so that the clutter echoes are removed and target echoes are further processed and
- The range and angle of target are measured when the target is detected.

6.6 MINIMAL DETECTABLE SIGNAL IN THE PRESENCE OF RECEIVER NOISE

eceiver noise is defined as the unwanted signal that reduces the efficiency of the radar receiver. Although e radar operates in noise-free environment, the unavoidable noise components exist due to thermal oise (which exists due to the conduction of electrons). For a receiver, the noise figure is given as:

$$F_n = \frac{\text{Noise due to actual device}}{\text{Noise due to an ideal device}} = \frac{N_0}{kT_0BG_a} \tag{16.19}$$

here N_0 is output noise from the receiver, G_a is gain of the device, B is bandwidth, T_0 is room temperare (290 K) and k = Boltzmann constant.

The noise figure can also be defined as a measure of the degradation of SNR, as the signal passes rough the receiver.

$$F_n = \frac{S_i/N_i}{S_0/N_0} \tag{16.20}$$

So from Eqs. (16.19) and (16.20), the input detectable signal is given as:

$$S_i = \frac{KT_0B_nF_nS_0}{N_0}$$

The minimum detectable signal $S_{\min}$ is detected when the ratio of output SNR is minimum which is iven as:

$$S_{\min} = KT_0B_nF_n\,(S_0/N_0)_{\min} \tag{16.21}$$

Hence the radar equation is obtained as below.

$$R_{\max} = \left[\frac{P_t G A_e \sigma}{(4\pi)^2 K T_0 B_n F_n \left(S_0 / N_0 \right)_{\min}} \right]^{1/4} \quad (16.22$$

16.7 CONSTANT FALSE ALARM RATE (CFAR) RECEIVER

Without any target signal, there is an opportunity for an interfering noise voltage to be interpreted a a target signal. False alarm is defined as the detection of a noise signal as target signal when the nois signal exceeds the threshold voltage. The fraction of the detection tests in which a false alarm occur is called the probability of false alarm, P_{FA} and it is defined by the Gaussian probability density func tion False alarm rate (FAR) is defined as the number of false alarms occurring within a given tim interval.

$$\mathrm{FAR} = \frac{P_{\mathrm{FA}} R}{T_R} = D P_{\mathrm{FA}} \quad (16.23$$

where R represents the number of resolution cells collected over a specific time interval T_R and $D = R/T$ is the number of detected decisions per unit time.

False alarms occur when the noise voltages exceed a certain threshold level due to the presence c clutter echoes and noise-jamming signals. Depending on the threshold values, false alarms occur whic causes the detection of target impossible. Let us consider different threshold levels (TH_1, TH_2, TH_3, an TH_4) as shown in Figure 16.6. As the threshold levels vary, the probability of detection of targets an probability of false alarms also vary.

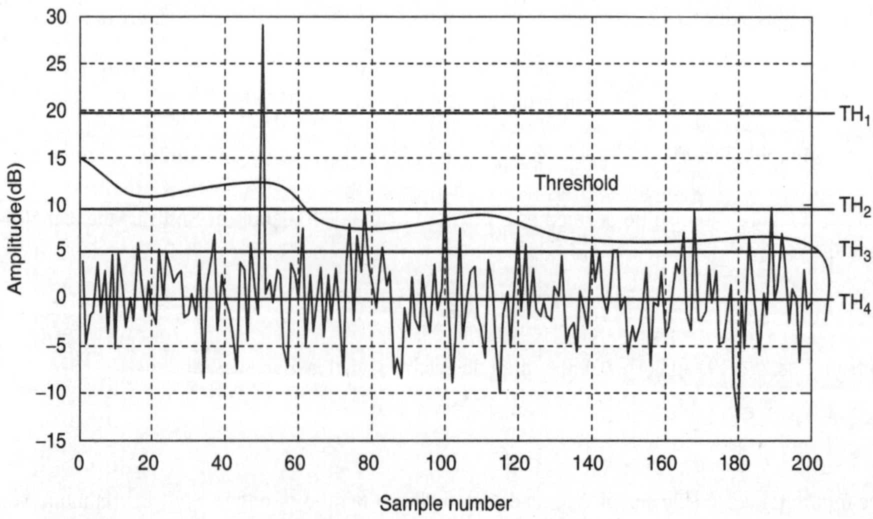

Figure 16.6 *Different threshold levels*

When the threshold level is at TH_1 (that is the threshold level is too high even the target signals are eglected), then the probability of detection will be too low (approximately 20%) and the probability f false alarm is zero. When the threshold level is at TH_2 (that is the threshold level is chosen to be ptimal), then the probability of detection is high (approximately 80%) and probability of false alarm s low (approximately 10%). When the threshold level is at TH_3 (that is the threshold level varies), then he probability of detection of targets is high (this is called as CFAR as the interference levels increase, he threshold level is automatically adjusted). If the threshold level is at TH_4 (that is the threshold level s too low), then the probability of detection of targets is low and the probability of false alarms is high.

Automatic detection and tracking (ADT) system can handle many targets and is efficient in determining the false alarms. However, this system is of limited capability and requires more computations to recgnize and discard the false alarms. So, for proper working of ADT and to avoid clutter and external noise eaching the automatic tracking computer, CFAR receiver is used. The CFAR receiver automatically aises the threshold level so that the external noise and clutter echo do not overload the automatic tracker. he CFAR can be accomplished at an expense of a lower probability of the detection for desired targets. n some cases when there is non-uniform clutter, then CFAR can also produce false echoes, suppress earby targets, and degrade the range resolution. The CFAR receiver's threshold (R_{TH}) is expressed as:

$$R_{TH} = \kappa\sigma \tag{16.24}$$

here κ is CFAR constant and σ is statistic data associated with the interference.

EXAMPLE PROBLEM 16.1

ow much the receiver threshold should be taken to avoid false alarms when the interference power at he output of the receiver is 3×10^{-3} and the CFAR constant is 0.005?

olution

nterference power at the output of the receiver = $\sigma = 3 \times 10^{-3}$

$$\text{CFAR constant} = \kappa = 0.005$$

eceiver threshold is the product of CFAR constant and interference power.

$$R_{TH} = \kappa\sigma = 3\times10^{-3}\times0.005$$
$$= 15\times10^{-6}\,\text{V}$$

■

EXAMPLE PROBLEM 16.2

etermine FAR in the region of 200 resolution cells over 12 ms interval, when the probability of the alse alarm rate is 0.09 ms.

olution

o. of resolution cells = $R = 200$

robability of false alarm = $P_{FA} = 0.09$ ms

ime interval = $T_R = 12$ ms

$$\textit{alse Alarm Rate}(FAR) = \frac{\textit{Probability of False Alarm} \times \textit{No. of Resolution cells}}{\textit{Time interval}}$$

$$\text{FAR} = \frac{P_{FA}R}{T_R} = \frac{200\times0.09\times10^{-3}}{12\times10^{-3}} = 1.5$$

■

16.7.1 Cell-Averaging CFAR (CA-CFAR)

A cell-averaging CFAR (CA-CFAR) is shown in Figure 16.7, which is widely used for the detection i radar receivers. In CA-CFAR, the threshold level is calculated by considering the noise around the cel under test and it can be done by calculating the average power level of block of cells around the tes cell. The CA-CFAR takes integrated outputs from several range cells as shown in Figure 16.7, which ar summed up and multiplied by the CA-CFAR constant to establish the detection threshold.

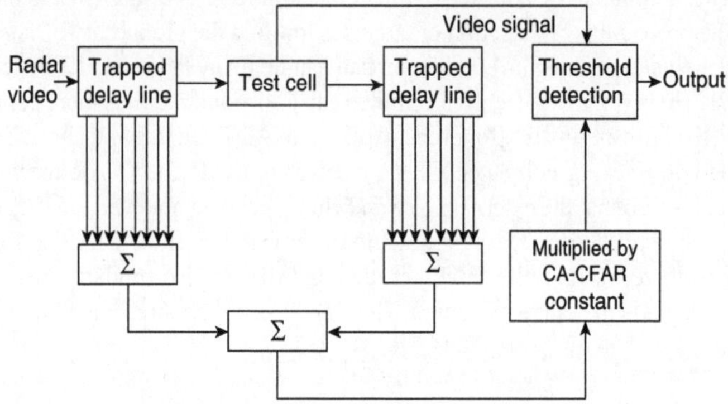

Figure 16.7 *A block diagram of the CA-CFAR*

To estimate the threshold accurately, the cells that are adjacent to the cell under test (CUT) are ignored These ignored cells are known as guard cells. Using this estimate, a target can be detected if the powe level of the CUT is greater than both the power levels of guard cells and local average power level.

16.8 DETECTORS

The detector extracts the modulating signal (that is, pulse modulated on to the carrier) from the modu lated carrier. Usually, the pulse radar rejects carrier and extracts the amplitude modulation using enve lope detector.

Normally detector characteristics will affect the received signal, which is a combination of require signal and noise. The detector characteristics like the probability of detection (based on the probabilit of false alarm), number of hits integrated, and SNR should be considered by a good radar detector.

There are several detectors used in a radar receiver. Some of them are discussed below.

Optimum Envelope Detector Law

The detector extracts the modulating signal (that is, pulse modulated on to the carrier) from the modu lated carrier. The envelope detector consists of three main sections. They are: IF amplifier, video ampli fier, and rectifying element. The IF amplifier exhibits bandpass filter (BPF) and video amplifier exhibit lowpass filter (LPF) characteristics. If the relation between the input and output is linear for positive volt age signals, and zero for negative voltage signals, then the detector is said to be linear. If the output of th detector is square of the input applied to the detector, then the detector is referred to as square-law detec tor. The square-law detector works on the principle of the rectifying element and the video integrator.

Logarithmic Detector

A detector is said to be logarithmic detector, or logarithmic receiver if the receiver's output and logarith mic value of the input envelope are proportional to each other. *Logarithmic detector is applicable* wher large variations of input signals are expected. In non-MTI receivers, these detectors are used to avoi eceiver saturation and to minimize unwanted clutter echoes. The MTI improvement factor is limited lue to non-linear characteristic, so that we do not use logarithmic characteristic in MTI receivers. The ise of logarithmic characteristics in receivers causes loss in detectability.

:oherent Detector

t is a single-channel detector same as I, Q detector's in-phase channel. However, the reference signal ias same phase and frequency as that of the input signal as shown in Figure 16.8. In general, the coher-:nt detector converts a high-frequency carrier to *dc* or low- frequency component which does not extract nformation from the carrier. The SNR from a coherent detector is 1 to 3 dB greater than the normal :nvelope detector. However, since the received radar signal phase information is rarely known, single-:hannel coherent detectors are not generally used in radar.

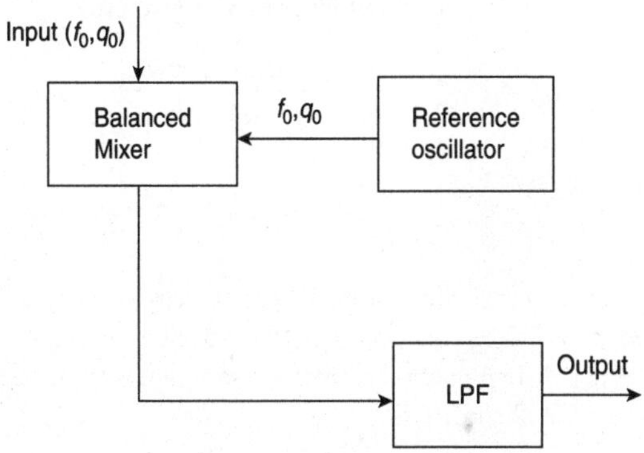

Figure 16.8 *A simple block diagram of coherent detector*

[, Q Detector

3y using I, Q detector, all components of the signal can be recovered from which sufficient information ibout the signal can be extracted. In the I and Q (in-phase and quadrature) channels, it is noticed that vhen a single-phase detector is fed by a coherent reference, it produces a significant loss in signal and his in turn depends on the relative time (or "phase") of the pulse train and the Doppler-shifted echo ignal. The blind phase is referred to as the loss in MTI radar. If we use a quadrature channel (Q chan-iel) which is 90° out of phase with the in-phase channel (I channel) in parallel, then it avoids the losses lue to the blind phase. Most of the modern signal processing analysis is adopting I and Q channels (as hown in Figure 16.9) as a receiver model when the Doppler frequency is needed to be extracted.

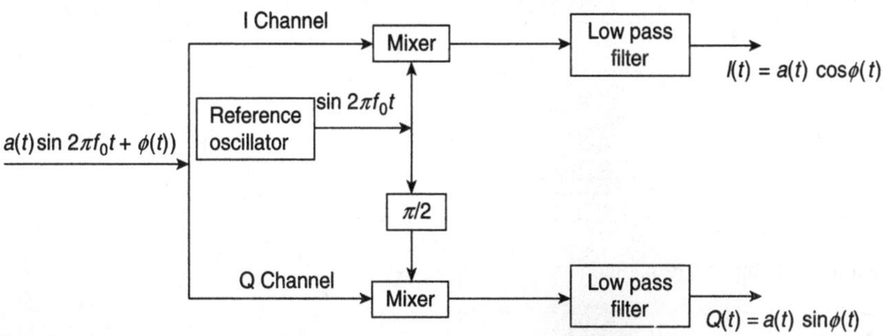

Figure 16.9 *I, Q detector*

16.9 VARIOUS NOISE COMPONENTS

Noise is an undesired signal that interferes with the ability of a receiver to detect the target in a system. The various noise components in a radar receiver are listed below.

- Noise factor (NF)
- Noise figure (F_n)
- Equivalent noise temperature (T_e) and
- System noise temperature (T_s)

16.9.1 Noise Factor (NF)

The NF of a device determines the amount of additional noise the device contributes to the noise that i from the source. The noise factor at a given input frequency, is given as

$$\text{NF} = \frac{\text{available output noise power}}{\text{available output noise due to source}} \quad (16.25)$$

The noise factor is a dimensionless ratio.

16.9.2 Noise Figure

The noise figure is used to characterize noise properties of devices by using effective noise temperature and gain and it is denoted as (F_n). The conversion of NF to decibel (dB) gives noise figure. The minimum noise figure of a device is 1. In modern receivers, noise figure is expressed technically in decibels which has typical values between 8 and 10 dB.

$$F_n = \frac{\text{noise power out of the actual device}}{\text{noise power out of an ideal device}} = \frac{P_n(\text{actual device})}{P_n(\text{ideal device})} \quad (16.26)$$

The noise power into the device is given by

$$P_n(\text{in}) = kT_0B \quad (16.27)$$

where B = bandwidth, T_0 = room temperature (290 K) and k = Boltzmann constant.

To compute the noise out of the ideal device, we assume that the device does not generate its own noise. So, the noise power out of the ideal device is given by

$$P_n(\text{ideal device}) = GP_n(\text{in}) = kT_0BG, \quad (16.28)$$

where G is gain of the device.

The noise power out of actual device is obtained as

$$P_n(\text{actual}) = GP_n(\text{in}) + kT_eBG = kT_0BG + kT_eBG \quad (16.29)$$

So, the noise figure is given as

$$F_n = \frac{P_n(\text{actual device})}{P_n(\text{ideal device})} = \frac{kT_0BG + kT_eBG}{kT_0BG} = 1 + \frac{T_e}{T_0} \quad (16.30)$$

Equivalent noise temperature is given as

$$T_e = T_0(F_n - 1) \quad (16.31)$$

Noise figure of overall cascaded components

The noise figure in radar systems is due to the existence of several devices that are accommodated at receiver section. In order to measure this noise figure of cascaded components, a method is required. Consider Figure 16.10; a noise source is cascaded to several devices. For the purpose of calculating noise figure, the source has an effective temperature T_0 and bandwidth B. The devices are represented by their gain G, noise figure F_n, and bandwidth B.

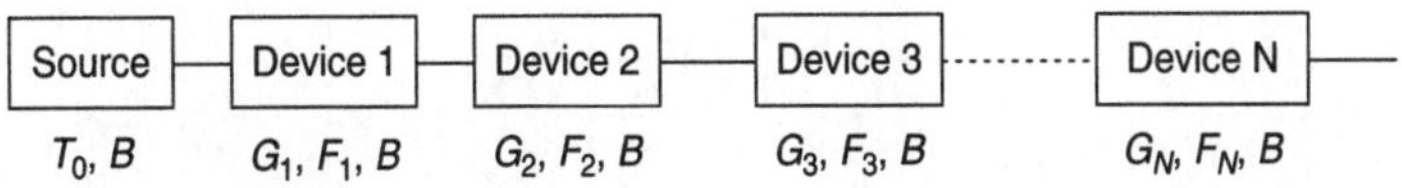

Figure 16.10 *Block diagram for computing system noise figure*

The overall noise figure of the N cascaded components is derived by considering first device, next first and second devices later first, second and third devices and so on. This helps us to develop a pattern that we can extend to N devices. After computing noise figure of the N devices, the effective noise temperature is also calculated for each device. Thus, the effective noise temperature of device p is

$$T_{\mathrm{ep}} = T_0(F_{\mathrm{np}} - 1) \tag{16.32}$$

For Device 1, the input noise power is

$$P_{n1}(\mathrm{in}) = kT_0B$$

For ideal Device 1, the noise power is

$$P_{n1}(\text{ideal device}) = G_1P_{n1}(in) = kT_0BG_1$$

For Device 1, the actual noise power is

$$P_{n1}(\text{actual}) = G_1P_{n1}(in) + kT_{e1}BG_1 = kT_0BG_1 + kT_{e1}BG_1 .$$

The system noise figure from the source through Device 1 is given as:

$$F_{n1} = \frac{P_{n1}(\text{actual device})}{P_{n1}(\text{ideal device})} = \frac{kT_0BG_1 + kT_{e1}BG_1}{kT_0BG_1} = 1 + \frac{T_{e1}}{T_0} = F_1 \tag{16.33}$$

For Device 2, the input noise power is

$$P_{n2}(\mathrm{in}) = kT_0BG_1 + kT_{e1}BG_1$$

For ideal cascade of Devices 1 and 2, the noise power is given as:

$$P_{n2}(\text{ideal device}) = G_1G_2P_{n1}(in) = kBG_1G_2T_0$$

The actual noise power out of Device 2 is

$$\begin{aligned} P_{n2}(\text{actual}) = G_2P_{n2}(in) + kT_{e2}BG_2 &= kBG_1G_2(T_0 + T_{e1}) + kT_{e2}BG_2 \\ &= kBG_1G_2(T_0 + T_{e1} + \frac{T_{e2}}{G_1}) \end{aligned}$$

The system noise figure from the source through Device 2 is

$$F_{n2} = \frac{P_{n2}(\text{actual device})}{P_{n2}(\text{ideal device})} = \frac{kBG_1G_2\left(T_0 + T_{e1} + \dfrac{T_{e2}}{G_1}\right)}{kBG_1G_2T_0} = 1 + \frac{T_{e1}}{T_0} + \frac{1}{G_1}\frac{T_{e2}}{T_0} \quad (16.34)$$

We know that $T_{e2} = T_0(F_2 - 1)$ from Eq. (16.32).

Hence
$$F_{n2} = F_1 + \frac{F_2 - 1}{G_1} \quad (16.35)$$

We can see that the second device noise figure can be minimized using the gain of the first device. Further, we will see how the total system noise figure can be computed from the source to the third device.

An *ideal* cascade of first, second, and third devices produces noise power as given below.

$$P_{n3}(\text{ideal device}) = G_1G_2G_3P_{n1}(in) = kBG_1G_2G_3T_0.$$

The actual noise power out of Device 3 is

$$P_{n3}(\text{actual}) = kBG_1G_2G_3\left(T_0 + T_{e1} + \frac{T_{e2}}{G_1}\right) + kT_{e3}BG_3$$

$$= kBG_1G_2G_3\left(T_0 + T_{e1} + \frac{T_{e2}}{G_1} + \frac{T_{e3}}{G_1G_2}\right)$$

The system noise figure from the source through Device 3 is

$$F_{n3} = \frac{P_{n3}(\text{actual device})}{P_{n3}(\text{ideal device})} = \frac{kBG_1G_2G_3\left(T_0 + T_{e1} + \dfrac{T_{e2}}{G_1} + \dfrac{T_{e3}}{G_1G_2}\right)}{kBG_1G_2G_3T_0}$$

$$= 1 + \frac{T_{e1}}{T_0} + \frac{1}{G_1}\frac{T_{e2}}{T_0} + \frac{1}{G_1G_2}\frac{T_{e3}}{T_0} \quad (16.36)$$

We know that
$$T_{ep} = T_0(F_{np} - 1)$$

Hence
$$F_{n3} = F_1 + \frac{F_2 - 1}{G_1} + \frac{F_3 - 1}{G_1G_2} \quad (16.37)$$

From Eq. (16.37), we can say that the noise figure of Device 3 is reduced by the factor G_1G_2 (gains of preceding two devices). The system noise figure from the source through device N is given as:

$$F_{nN} = F_1 + \frac{F_2 - 1}{G_1} + \frac{F_3 - 1}{G_1G_2} + \ldots\ldots\ldots\ldots + \frac{F_N - 1}{G_1G_2G_3\ldots\ldots G_{N-1}} \quad (16.38)$$

16.9.3 Noise Temperature

In the branch of electronics, the noise temperature is the way of expressing the available noise power in a particular system, which can be introduced by a component or source. Normally power spectral density of noise can be expressed in terms of temperature as given below.

$$P = BkT_e \tag{16.39}$$

where P is the power (in watts)

B is the total bandwidth (Hz) over which the noise power is measured

k is the Boltzmann constant (1.381×10^{-23} joules per degree Kelvin)

T_e is the noise temperature (in Kelvin)

The noise temperature can be expressed in noise figure as

$$T_e = T_0(F_n - 1)$$

If there are N devices cascaded, then the overall noise temperature is given as:

$$T_e = T_{e1} + \frac{T_{e2}}{G_1} + \frac{T_{e3}}{G_1G_2} + \ldots\ldots\ldots\ldots + \frac{T_{eN}}{G_1G_2G_3\ldots\ldots G_{N-1}} \tag{16.40}$$

16.9.4 System Noise Temperature

System noise temperature is defined as the summation of effective noise temperature (T_e) and antenna temperature (T_a), which is given as:

$$T_s = T_a + T_e \tag{16.41}$$

If there are N devices cascaded, then the overall system temperature is given as:

$$T_s = T_a + T_{e1} + \frac{T_{e2}}{G_1} + \frac{T_{e3}}{G_1G_2} + \ldots\ldots\ldots\ldots + \frac{T_{eN}}{G_1G_2G_3\ldots\ldots G_{N-1}} \tag{16.42}$$

EXAMPLE PROBLEM 16.3

If the noise figure of a radar receiver is 3.5 dB, determine the reduction (measured in dB) in the signal to noise ratio at the output compared to the signal to noise ratio at the input.

Solution

Given that for a radar receiver, noise figure = 3.5 dB.

Reduction in the SNR of the output compared to the signal to noise ratio of the input =?

Then, for any radar receiver, the noise figure is defined as (using equation 16.20),

$$F_n = \frac{(S/N)_o}{(S/N)_i}$$

$$\Rightarrow \frac{(S/N)_O}{(S/N)_i} = 3.5\,\text{dB} = 10^{(0.35)} = 2.2387$$

$$\Rightarrow \frac{(S/N)_O}{(S/N)_i} = 0.8059$$

$$\Rightarrow 1 - \frac{(S/N)_O}{(S/N)_i} = 1 - 0.8059 = 0.1941$$

$$\Rightarrow 1 - \frac{(S/N)_O}{(S/N)_i}(\text{dB}) = 10\log(0.1941) = -16.3938$$

Therefore, the reduction in the SNR at the output compared to the SNR at the input is – 16.3938 (negative sign indicates reduction). ■

EXAMPLE PROBLEM 16.4

A radar receiver is connected to a 40 Ω resistance antenna that has an equivalent noise resistance of 20 Ω. Calculate the noise figure of the receiver and the equivalent noise temperature of the receiver.

Solution

Given that,

For a radar receiver,

Antenna resistance, $R_a = 40\ \Omega$

Equivalent noise resistance, $R_{eq} = 20\ \Omega$

Noise figure of the receiver, $F = ?$

Equivalent noise temperature of the receiver, $T_{eq} = ?$

Then, the expression for noise figure of a receiver in terms of equivalent noise resistance R_{eq} and antenna resistance R_a is given by:

$$\begin{aligned} F_n &= 1 + \frac{R_{eq}}{R_a} \\ &= 1 + \frac{20}{40} \\ &= 1 + 0.5 \\ &= 1.5 \\ &= 4.05\,\text{dB} \\ \therefore F_n &= 4.05\,\text{dB} \end{aligned}$$

The expression for equivalent temperature of the receiver in terms of noise figure is given by:

$$T_e = T_0(F_n - 1)$$

Assume that the ambient temperature at receiver, $T_0 = 290\,\text{K}$

$$\begin{aligned} \Rightarrow T_e &= 290(1.5 - 1) \\ &= 290 \times 0.5 \\ &= 145\ \text{K} \\ \therefore T_e &= 145\,\text{K} \end{aligned}$$

■

EXAMPLE PROBLEM 16.5

Three network units, each of 7 dB noise figure and 9 dB, 6 dB, and 4 dB gains, respectively are cascaded. Determine the overall noise figure of the system.

Solution

Given that:

The network contains three cascaded units, I, II, and III with gains and noise figure as,

$$G_1 = 9\ \text{dB} = 7.9438$$

$$F_1 = F_2 = F_3 = 7 \text{ dB} = 5.0119$$

$$G_2 = 6 \text{ dB} = 3.98$$

$$G_3 = 4 \text{ dB} = 2.5119$$

As we know that the noise figure of "*N*" networks in cascade is given as,

$$F_0 = F_1 + \frac{F_2 - 1}{G_1} + \frac{F_3 - 1}{G_1 G_2} + \ldots\ldots\ldots\ldots + \frac{F_N - 1}{G_1 G_2 \ldots\ldots G_{N-1}}$$

$$F_0 = F_1 + \frac{F_2 - 1}{G_1} + \frac{F_3 - 1}{G_1 G_2}$$

$$F_0 = 5.0119 + \frac{5.0119 - 1}{7.9433} + \frac{5.0119 - 1}{7.9433 \times 3.98}$$

$$F_0 = 5.0119 + \frac{4.0119}{7.9433} + \frac{4.0119}{7.9433 \times 3.98}$$

$$F_0 = 5.6439 \text{ dB}$$

$$F_0 = 17.306$$

■

16.10 DUPLEXER

Duplexer is an electronic switch that facilitates a single antenna to be used for both transmission and reception. But the problem arises when an antenna is switched between the transmitting and receiving modes; that is the switching system has to ensure that the maximum use is made of the available energy. To overcome the problem, a switch is to be used to change the receiver connection to transmitter during the transmitted pulse and to the receiver during the echo pulse. In this scenario, a mechanical switch cannot be used practically, as the switching operation is to be done in few microseconds. So, electronic switches must be used.

16.10.1 Balanced Duplexer

The balanced duplexer can work at higher frequencies by having higher bandwidth. Balanced duplexer contains two sections of waveguides. They are joined along one of their narrow walls having a slot cut in the common wall. This provides coupling between the two waveguide sections. This is based on the short-slot hybrid junction. In each section of wave guide, one TR tube is used. In the transmit condition (as shown in Figure 16.11(a)), the power is evenly divided into each waveguide by first hybrid junction which is on the left. The incident power is reflected out of antenna when both TR tubes break down as shown in Figure 16.11(a). The short-slot hybrid junction possesses a property that its phase is increased by 90° every time the power passes through the slot in any of the directions; the power traveled is indicated by the solid lines. In case if any power leaks through the TR tubes (indicated by the dotted lines), then it is directed to the arm and matched with the dummy load but not to the receiver. Isolation of 20 to 30 dB is provided by the hybrid junctions along with the attenuation offered by the TR tubes.

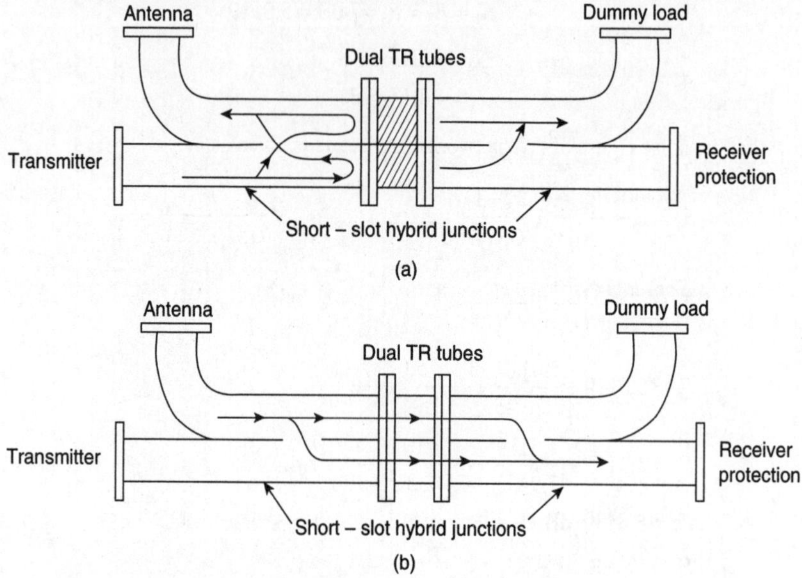

Figure 16.11 *Balanced duplexer using dual TR tubes and two short-slot hybrid junctions. (a) Transmit condition; (b) Receive condition*

During receive period, the signal enters the receiver through the duplexer. In this case, the TR tubes do not fire as shown in Figure 16.11(b). The received power through the antenna will be divided equally at the first junction and they recombine in the receiver arm as they are in the same phase. Due to 180° phase difference between the signals entering into the dummy load arm, no signal exists in that arm. The balanced duplexer gained popularity because of its good power-handling capability and wide bandwidth.

16.10.2 Branched Type Duplexers

Typical branched duplexer components are shown in Figure 16.12(a). These duplexers were used during World War II period and are still in use today. Branched type duplexer (as shown in Figure 16.12(a)) consists of ATR (anti-transmit receive) and TR (transmit receive) switches, which are nothing but gas discharge tubes. In general, TR provides receiver protection and switching and ATR directs the echo power in to receiver. During transmission, both the ATR and TR switches get activated and provide very low impedances at the waveguide walls. Therefore, the transmitted signal energy passes through the antenna with minimum attenuation. During transmission period, the TR unit protects the receiver by not allowing the transmitted signal leakage into the receiver channel.

During reception, the signal from the antenna passes through the TR switch. On reception, the TR switch is in inactive state (that is, it is matched to the transmission line impedance). Simultaneously, ATR is also in inactive state to maintain high impedance to receive the signal in the direction of the transmitter, which leads to minimization of the loss of return signal energy in that direction.

The advantages of the branched duplexer are: simple, compact, and low cost. The disadvantage is limited operating bandwidth (about 5%) because element spacings are significant.

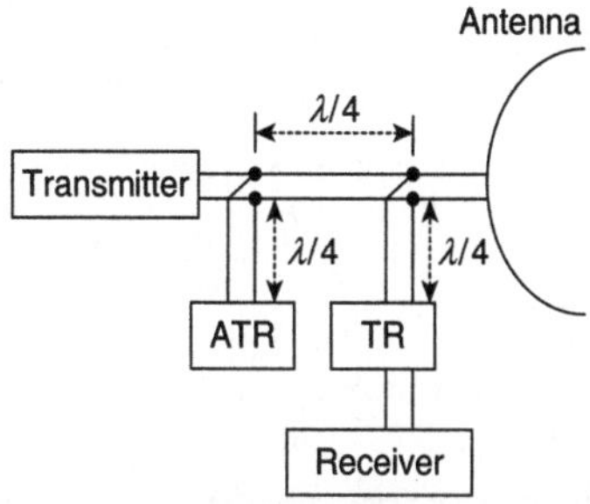

Figure 16.12 (a) *Branched duplexer*

Nowadays branched duplexers are obsolete in all types of radars. Mostly circulator type duplexers are used in all radars as shown in Figure 16.12(b). In the circulator type duplexer, the transmitter is connected at port 1, the antenna at port 2, and the receiver at port 3. When the transmitter sends a signal, the signal goes to the antenna with great ease and does not enter into the receiver because of the isolation of the circulator. When the signal comes back to the antenna, it goes directly to the receiver and not to the transmitter, because of the circulator operation.

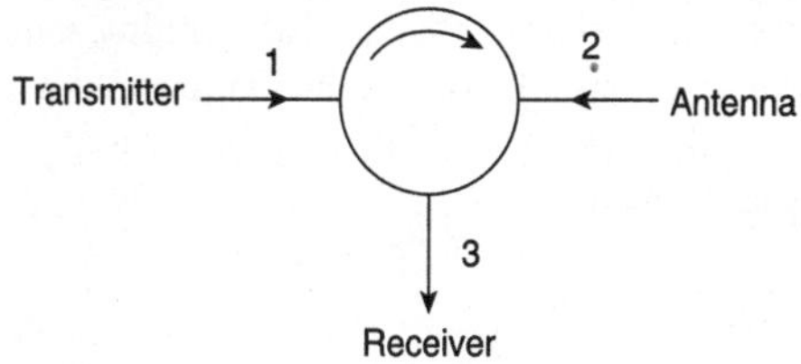

Figure 16.12 (b) *Circulator as a duplexer*

In radar, there should be zero isolation between the transmitter and receiver. It should be evident that to have the required isolation in the circulator, the transmitter, the receiver, and the antenna all must be well matched to the circulator. Still small part of the input signal emerges from port 3, and the ratio of this signal to the input signal is called isolation and it is expressed in decibels as:

$$I(\text{dB}) = 10\log_{10}\left(\frac{P_{\text{out}3}}{P_{\text{in}1}}\right) \tag{16.43}$$

So, the transmitter is turned off while receiving the signals from the antenna and the receiver is turned off while transmitting the signals.

16.11 INTRODUCTION TO PHASED ARRAY ANTENNAS

For achieving high directivity and large gain, a single antenna is not sufficient. So, groups of radiating elements are used and it is commonly known as an array. Number of elements present in an array may vary from a few units to thousands of units. The radiating elements are present in an array. The array can be dipoles, microstrip antennas, open-ended waveguides, slotted waveguides, and so on. The resultant radiation pattern shape and direction is obtained by relative phases and amplitudes of each radiating element.

Different Types of Phased Arrays

The groups of radiating elements can be arranged in different ways as explained below:

Linear array: The antenna elements are arranged in a straight line and spaced equally in one dimension.

Planar array: In planar array, elements are placed on a plane in two dimensions. A planar array can be defined as a linear array of linear arrays.

Conformal array: Here the elements are distributed on a non-planar surface.

16.12 PARALLEL AND SERIAL-FEED ARRAY

Parallel feed: If we apply variable phase shifts at each element of an array, then it is known as parallel-feed array. For different phase shifts to be applied at each element, variable phase shifters can be used to steer the beam in required directions.The phase difference between each element of antenna array is

$$\phi = 2\pi \frac{d}{\lambda} \sin\theta \tag{16.44}$$

It is sometimes called corporate feed because it resembles a tree-like structure if we use a series of power splitters like hybrid junctions as shown in Figure 16.13(a). In parallel- feed array, equal spacing between each antenna element and transmitter or receiver is required and the phase at each element should be same (except that introduced by phase shifters). If the loss at each element in a parallel feed is L_p, then the entire loss of parallel feed is also L_p.

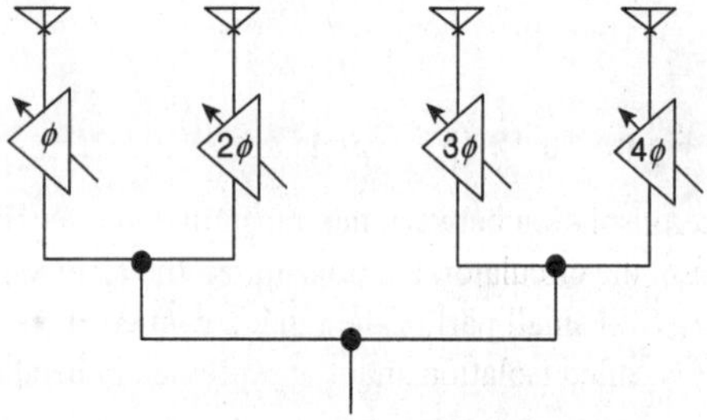

Figure 16.13 (a) *Parallel-feed*

Series Feed

In series feed, the radiating elements are connected in series as shown in Figure 16.13(b). Each phase shifter has the same phase, that is only one phase has to be applied when compared to $(N - 1)$ phase shifts applied in parallel-feed array, which reduces the overall system complexity. In general, series-feed arrays are frequency-sensitive and lead to bandwidth restrictions. If the frequency is changed, then the phase at each element changes proportionally to the length of the feed line, which leads antenna beam to be steered which is used in frequency scanning arrays.

The disadvantage of series-feed array is its high loss. If loss present at a single element in series-feed array is L_s, then total loss is $(N - 1) L_s$. Here the system complexity is less when compared to parallel-feed array.

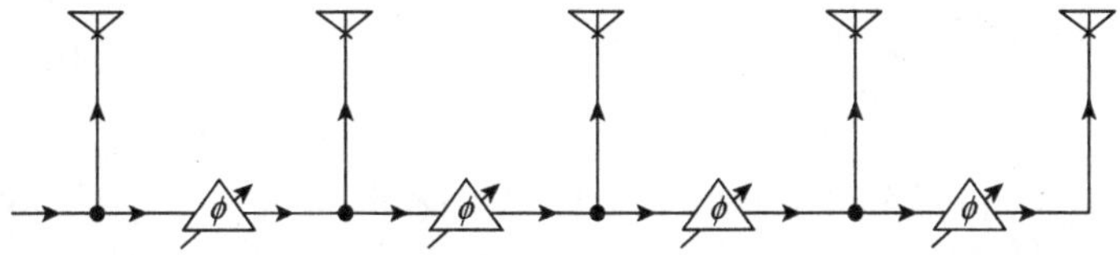

Figure 16.13 (b) *Series-feed*

16.13 RADIATION PATTERN OF PHASED ARRAY ANTENNAS

By considering the propagation of electric field from a set of radiating elements present in an array, antenna radiation pattern can be derived. Let us consider receiving linear array of N elements, which are equally spaced at a distance d, radiating an equal amplitude a_0 with a phase progression difference δ between adjacent elements as shown in Figure 16.14.

The phase difference in adjacent elements may be given as:

$$\Delta\phi = 2\pi\frac{d}{\lambda}\sin\theta \tag{16.45}$$

where, λ is the wavelength of the received signal.

θ is the angle between the boresight and incoming wave.

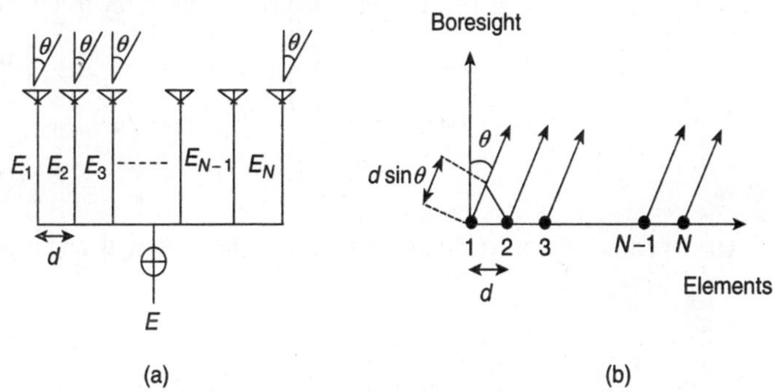

Figure 16.14 *Geometry of a linear array antenna: (a) a linear array configuration and (b) line representation of radiator*

A phase progression δ is introduced between the adjacent elements of phased array to provide progressive beam scanning. Therefore, total phase difference (ψ) of the radiating fields due to the adjacent elements can be rewritten as:

$$\psi = \Delta\phi + \delta = 2\pi\frac{d}{\lambda}\sin\theta + \delta \tag{16.46}$$

The sum of all voltages from the individual elements when the phase difference between adjacent elements is ψ can be expressed as a geometric series as shown below.

$$E = a_0\left(1 + e^{j\psi} + e^{2j\psi} + ... + e^{j(N-1)\psi}\right) \tag{16.47}$$

Assume, $a_0 = 1$ and multiplying Eq. (16.47) by $e^{j\psi}$, we get

$$Ee^{j\psi} = \left(e^{j\psi} + e^{2j\psi} + e^{3j\psi} + ... + e^{jN\psi}\right) \tag{16.48}$$

Assume $a_0 = 1$ in Eq. (16.47) also and subtracting Eq. (16.48) from Eq. (16.47) and dividing it by $(1 - e^{j\psi})$, we get

$$E = \frac{1-e^{jN\psi}}{1-e^{j\psi}} = \frac{e^{\frac{jN\psi}{2}}\left[e^{-\frac{jN\psi}{2}} - e^{\frac{jN\psi}{2}}\right]}{e^{\frac{j\psi}{2}}\left[e^{-\frac{j\psi}{2}} - e^{\frac{j\psi}{2}}\right]} \tag{16.49}$$

Equation (16.49) can be written as,

$$E = \left\{e^{\frac{j(N-1)\psi}{2}}\right\}\left[\frac{\sin\left(\frac{N\psi}{2}\right)}{\sin\left(\frac{\psi}{2}\right)}\right] \tag{16.50}$$

Here, the first term in curly brackets represents the phase of the field shifted $(N - 1)\psi/2$ and the second term in square brackets represents the amplitude factor or simply array factor, $f_a(\psi)$.

$$f_a(\psi) = \frac{\sin\left(\frac{N\psi}{2}\right)}{\sin\left(\frac{\psi}{2}\right)} \tag{16.51}$$

By considering the magnitude of Eq. (16.50), we get the array field strength and relating ψ with the physical dimension of the array antenna, we get

$$|E(\theta)| = \left|\frac{\sin\left(\frac{N\psi}{2}\right)}{\sin\left(\frac{\psi}{2}\right)}\right| = \left|\frac{\sin\left(N\left\{\frac{\pi d}{\lambda}\sin\theta + \frac{\delta}{2}\right\}\right)}{\sin\left(\frac{\pi d}{\lambda}\sin\theta + \frac{\delta}{2}\right)}\right| \tag{16.52}$$

The above expression represents voltage distribution, which can be converted to array radiation pattern, by considering the normalized square of the amplitude.

$$G(\theta) = \frac{|E(\theta)|^2}{N^2} = \left[\frac{\sin\left(N\left\{\frac{\pi d}{\lambda}\sin\theta + \frac{\delta}{2}\right\}\right)}{N\sin\left(\frac{\pi d}{\lambda}\sin\theta + \frac{\delta}{2}\right)}\right]^2 \tag{16.53}$$

Figure 16.15 shows the normalized radiation pattern of the array of six elements illuminated uniformly, with inter-element spacing (d) of half-wave and two-phase progression differences (δ) of 0° and 0.2°.

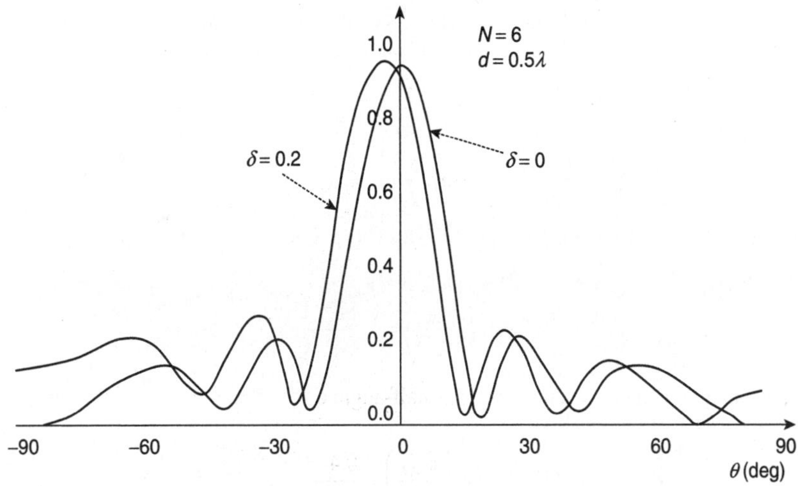

Figure 16.15 *Normalized electric field response of a linear array antenna with two phase progressions of 0° and 0.2°.*

When directive elements are used, the resulting radiation pattern can be given as:

$$G(\theta) = G_i(\theta)\left[\frac{\sin\left(N\left\{\frac{\pi d}{\lambda}\sin\theta + \frac{\delta}{2}\right\}\right)}{N\sin\left(\frac{\pi d}{\lambda}\sin\theta + \frac{\delta}{2}\right)}\right]^2 \tag{16.54}$$

where $G_i(\theta)$ is the individual element factor or radiation pattern of an individual element.

Two-dimensional Radiation Patterns

In a two-dimensional rectangular planar array, the resultant radiation pattern can be written as the product of the radiation pattern in the two planes (horizontal and vertical).

By considering Eq. (16.54) and neglecting phase progression effect, the product radiation pattern $G(\theta)\big|_{n,m}$ can be represented as,

$$G(\theta)\big|_{n,m} = \left|\frac{\sin\left(n\frac{\pi d}{\lambda}\sin\theta_n\right)}{n\sin\left(\frac{\pi d}{\lambda}\sin\theta_n\right)}\right|^2 \left|\frac{\sin\left(m\frac{\pi d}{\lambda}\sin\theta_m\right)}{m\sin\left(\frac{\pi d}{\lambda}\sin\theta_m\right)}\right|^2 \tag{16.55}$$

where

n = number of vertical elements which gives azimuth angle θ_n,
m = number of horizontal elements which gives elevation angle θ_m

16.14 BEAMWIDTH

The angle between the directions on either side of the main beam of antenna radiation pattern, at which the intensity drops one-half of the value is known as antenna beamwidth that is array beam width (θ_{BW}), can be calculated by finding out the half-power (– 3 dB) points of the main beam. This is accomplished by equating the amplitude factor of all the elements in (Eq. (16.51)) to $N/\sqrt{2}$

$$\frac{\sin\left(\frac{N\psi}{2}\right)}{\sin\left(\frac{\psi}{2}\right)} = \frac{N}{\sqrt{2}} \tag{16.56}$$

The solution to the above equation was given by Hansen as below if the number of elements N is greater than 7.

$$\sin\left(\frac{\theta_{BW}}{2}\right) = \frac{0.4429\lambda}{Nd} \tag{16.57}$$

The above equation can be split into terms of half-angle of two incoming signals θ_1 and θ_2 as shown below.

$$\sin\theta_2 = \sin\left(\frac{\theta_{BW}}{2}\right) = \frac{0.4429\lambda}{Nd} \tag{16.58}$$

$$\sin\theta_1 = \sin\left(\frac{-\theta_{BW}}{2}\right) = -\frac{0.4429\lambda}{Nd} \tag{16.59}$$

By considering small angle approximation, the beamwidth is

$$\theta_{BW} = \theta_2 - \theta_1 = \frac{0.8858\lambda}{Nd} = \frac{0.8858\lambda}{D} \tag{16.60}$$

where D is the array aperture

From Eq. (16.60), it is clear that increase in array physical length results in the decrease in antenna beamwidth for a given propagation wavelength.

16.15 BEAM STEERING

Beam steering is nothing but changing the direction of main beam of antenna radiation pattern, which can be done by changing the relative time delays or relative phase between the elements as shown in Figure 16.16.

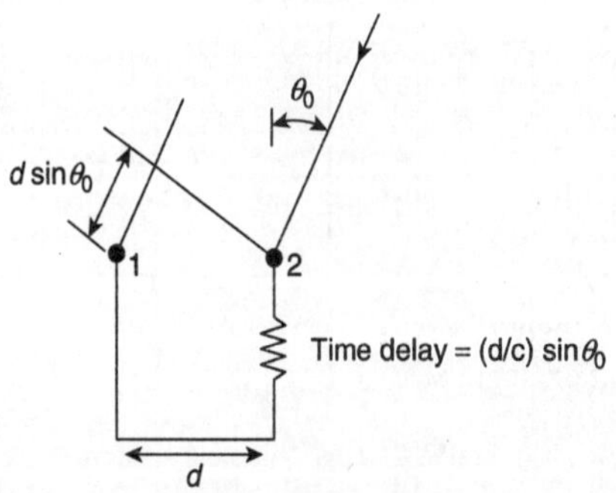

Figure 16.16 (a) *Beam steering based on true time delay*

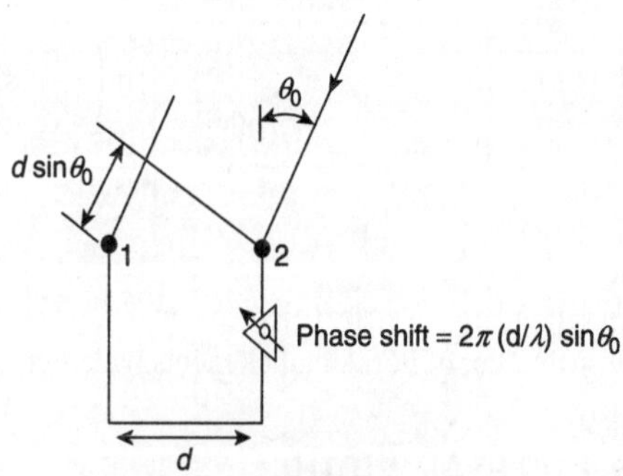

Figure 16.16 (b) *Beam steering using a phase shifter that is variable over the range from 0 to 2π*

Figure 16.16 consists of two antenna elements 1 and 2 that are spaced d apart. Let us consider a signal from a direction θ_0 relative to the normal of two elements will be arrived at element 2 first, rather than element 1. If the signal is delayed at element 2 for a time $\Delta T = (d/c)\sin\theta$, then there will be a coincidence with the signal that is present at element 1, which can add or subtract. If they add together, then the main beam of this antenna array will be placed in direction of θ_0. However, we can steer the beam by applying phase shift $\theta = 2\pi f_0 \Delta T$ at each element of the array.

If we want to have all signals from a linear array of same phase, then the phase shift to be applied at each individual element of linear array in $m\psi$, where m is an integer ranging from 0 to $(N-1)$. By substituting $\delta = -2\pi \frac{d}{\lambda}\sin\theta$ in (Eq. (16.54)), a beam steered at angle θ_0 will have a normalized radiation pattern given as below.

$$G(\theta) = \frac{\sin^2\left(N\frac{\pi d}{\lambda}[\sin\theta - \sin\theta_0]\right)}{N^2\sin^2\left(\frac{\pi d}{\lambda}[\sin\theta - \sin\theta_0]\right)} \tag{16.61}$$

At $\sin\theta = \sin\theta_0$, the maximum radiation pattern occurs. A plot of normalized radiation pattern of Eq. (16.61) is shown in Figure 16.17 for six antenna elements spaced half-wave (that is $d = 0.5\,\lambda$) and scanned off the boresight by 3°.

If we want to go for beam steering, the argument of the function (u) is the difference of the sines of the looking (θ) and steering angles (θ_0). Steering results either right (θ_0) or left ($-\theta_0$) shift off the boresight with no distortion in the electric field strength. This results in reduced sidelobe amplitudes.

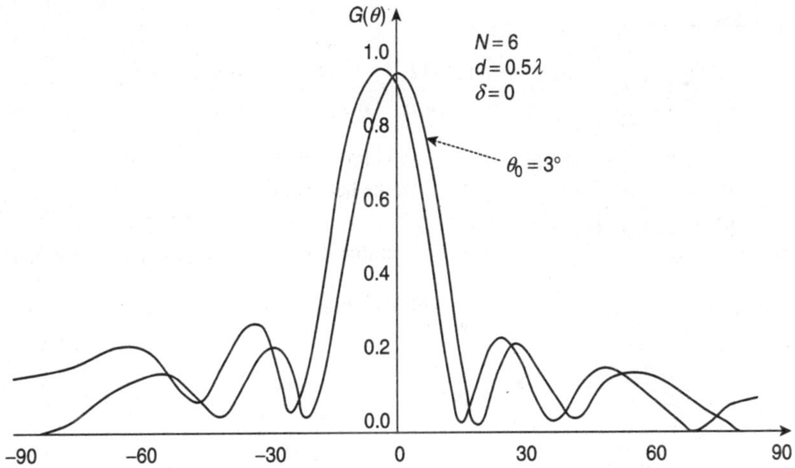

Figure 16.17 *Radiation pattern of a linear antenna steered off boresight*

Variation of Beamwidth with Respect to Beam Steering Angle

In general, if the beam of phased array scanned from broad-side direction then its half-power beamwidth will be increased, this is given by the following relation.

$$\text{beamwidth} \; \alpha \frac{1}{\cos\theta_0} \tag{16.62}$$

where θ_0 is the angle measured from the normal to the antenna.

The above-mentioned relation can be proved by considering Eq. (16.61) and replace the sine in the denominator with its argument as shown below.

$$G(\theta) = \frac{\sin^2 u}{u^2} \tag{16.63}$$

where $$u = N\frac{\pi d}{\lambda}[\sin\theta - \sin\theta_0]$$

The value of radiation pattern $G(\theta)$ will be half of its value when $u = \pm 0.443\pi$

Now the angles corresponding to half-power points are denoted as θ_+ and θ_- at $u = +0.443\pi$ and at $u = -0.443\pi$, respectively. In u, the term $\sin\theta - \sin\theta_0$ can be expressed as:

$$\sin\theta - \sin\theta_0 = \sin(\theta-\theta_0)\cos\theta_0 - [1-\cos(\theta-\theta_0)]\sin\theta_0 \tag{16.64}$$

The second term in the above expression can be neglected if θ_0 is very small which results as:

$$\sin\theta - \sin\theta_0 = \sin(\theta-\theta_0)\cos\theta_0$$

With this approximation, the two angles corresponding to the 3 dB points can be given as:

$$\theta_+ - \theta_0 = \sin^{-1}\frac{0.4429\lambda}{Nd\cos\theta_0} \approx \frac{0.4429\lambda}{Nd\cos\theta_0} \tag{16.65}$$

$$\theta_- - \theta_0 = \sin^{-1}\frac{-0.4429\lambda}{Nd\cos\theta_0} \approx -\frac{0.4429\lambda}{Nd\cos\theta_0} \tag{16.66}$$

Then the half-power beamwidth is given as:

$$\theta_B = \theta_+ - \theta_- = \frac{0.443\lambda}{Nd\cos\theta_0} + \frac{0.443\lambda}{Nd\cos\theta_0} = \frac{0.886\lambda}{Nd\cos\theta_0} \tag{16.67}$$

From Eq. (16.67), we can conclude that beamwidth is inversely proportional to $\cos\theta_0$.

Equation (16.67) is not valid at large angles when the beam is scanned from broad side because of the increase in mutual coupling effects.

If we want to apply Eq. (16.67) for uniform line-source distribution with a cosine on a pedestal aperture illumination of the form $a_0 + a_1\cos\left(\frac{2\pi n}{N}\right)$ of N elements, which are separated by distance d, then the beamwidth is given as:

$$\theta_B \approx \frac{0.886\lambda}{Nd\cos\theta_0}\left[1 + 0.636\left(\frac{2a_1}{a_0}\right)^2\right] \tag{16.68}$$

where a_0 and a_1 are constants

n = position of the element = 0, ±1, ±2, …± $(N-1)/2$.

EXAMPLE PROBLEM 16.6

Ten linear array antennas are placed at a distance of 0.1 mm. The angle between the boresight and incoming waves is 30°. Find the array factor when the wavelength of the received signal is 0.5 cm.

Solution

Given distance = $d = 0.1$ mm

Wavelength = $\lambda = 0.5$ cm

Angle = $\theta = 30°$

$$\text{Phase difference} = \psi = 2\pi \frac{d}{\lambda} \sin\theta = \frac{2\pi \times 0.1 \times 10^{-3} \sin 30°}{0.5 \times 10^{-2}} = 0.0628$$

$$\text{Array factor} = \frac{\sin\left(\frac{N\psi}{2}\right)}{\sin\left(\frac{\psi}{2}\right)} = \frac{\sin\left(\frac{10 \times 0.0628}{2}\right)}{\sin\left(\frac{0.0628}{2}\right)} = \frac{5.48 \times 10^{-3}}{5.48 \times 10^{-4}} = 10$$

■

EXAMPLE PROBLEM 16.7

Fifteen linear array elements are placed at a distance of 0.3 mm. The wavelength of the received signal is 0.6 cm. Find the array aperture and beamwidth of the phased array antenna.

Solution

Given distance = $d = 0.3$ mm

Wavelength = $\lambda = 0.6$ cm

No. of array elements = $N = 15$

$$\text{Array aperture} = D = Nd = 15 \times 0.3 \times 10^{-3} = 4.5 \times 10^{-3}$$

$$\text{Beamwidth} = \theta_{\text{BW}} = \frac{0.8858\lambda}{D} = \frac{0.8858 \times 0.6 \times 10^{-2}}{4.5 \times 10^{-3}} = 1.18°$$

■

16.16 APPLICATIONS OF PHASED ARRAY ANTENNAS

The main applications of phased array antennas are listed below.

Broadcasting

In AM broadcasting, phased arrays are used by broadcasting stations for enhancing the signal strength and also VHF phased arrays are used for FM broadcasting. The usage of phased array antennas increases the antenna gain thus strengthening the emitted RF energy towards the horizon, which leads to increase stations broadcast range.

Naval Usage

In navy, phased array radar systems are used by many warships for finding ships, finding aircrafts, missiles, and for missile uplink capabilities, because of the capability of steered beam of phased array antennas.

Space Probe Communication

In spacecrafts also, phased array antennas are used. One of the examples is the MESSENGER spacecraft, the planet Mercury in this spacecraft-radiating elements are linearly polarized.

Optics

Optical phased arrays can be constructed at visible or infrared spectrum of electromagnetic waves, which can be used in laser beam steering, telecommunications, and holography.

Radio-frequency Identification (RFID)

In recent years, RFID systems uses phased array antennas for increasing the reading capability of passive UHF tags.

16.17 ADVANTAGES AND DISADVANTAGES OF PHASED ARRAY ANTENNAS

Advantages

- By using phased array antenna we can get high gain with low side lobes.
- The resultant beam from an array can be moved continuously or in discrete steps without moving the entire antenna structure which can be done in a few seconds.
- Array antennas have the ability to provide an agile beam i.e. ability to generate many independent beams from same antenna aperture under the control of computer.
- By using array antennas we can perform both surveillance and tracking at the same time.
- Fault of any single component in array doesn't cause the failure of entire system.
- Based on requirement, a particular aperture distribution can be obtained by using array antennas which results in the feasibility of reduction in side lobes.
- Spill over loss can be avoided in array antennas which are present in lens or reflector antennas.

Disadvantages

- The main limitation of array antenna is its limited coverage due to single plane aperture.
- Cost and complexity are more due to the presence of more number of antenna elements.
- Care must be taken for maintaining negligible phase at transmission lines, amplifiers, mixers, and other components used in array.

SUMMARY

1. The detection of signal using radar in the presence of noise and extraction of information of the object (range, velocity and so on) are two separate operations involved in radar.
2. The detectability of a target using radar can be improved through maximization of output peak-signal-to-noise (power) ratio of a radar receiver, which can be done by using matched filter.
3. The bandwidth of radar should be equal to reciprocal of the pulse width (τ). $B = 1/\tau$
4. Autocorrelation is the measure of coherence of signal itself, which is maximum when time shift is zero or multiple of its time period.
5. By using correlation, detecting known waveform from noise is known as matched filtering.
6. The Noise Factor (NF) of a device determines the amount of additional noise the device contributes to the noise source.

7. Noise figure is used to measure reduction of the SNR due to components in the RF signal chain or it is a measure of amount of decrease of the SNR.

8. When a single antenna is used both as a transmitter and receiver, in a radar system, an electronic switch is required and these types of switching systems are called duplexers.

9. A group of radiating elements is commonly known as an array.

10. **Linear array:** Here antenna elements arranged in a straight line in one dimension, which are spaced equally.

11. **Planar array:** In planar array, elements are placed on a plane in two dimensions (rectangular, square, or circular aperture). A planar array can be defined as a linear array of linear arrays.

12. **Conformal array:** Here the elements are distributed on a non-planar surface.

13. **Parallel feed:** If we apply variable phase shifts at each element of an array, then it is known as parallel feed.

14. **Series feed**: The radiating elements are connected in series. Each phase shifter has the same phase.

15. The angle between the directions on either side of the main beam of antenna radiation pattern, at which the intensity drops to one-half the value is known as antenna beam width, that is, array beam width (θ_{BW}).

16. Beam steering is nothing but changing the direction of main beam of antenna radiation pattern, which can be done by changing the relative time delays or relative phase between the elements.

OBJECTIVE-TYPE QUESTIONS

1. The network whose frequency response function maximizes the SNR is known as:

   (a) correlation receiver (b) cross-correlation
   (c) matched filter (d) none of the above

2. Autocorrelation is a measure of_____ of the signal itself.

   (a) coherence (b) non-coherence
   (c) both a and b (d) none of these

3. In which of the following detectors, output follows the input linearly

   (a) square-law detector (b) linear detector
   (c) envelope detector (d) none of the above

4. _____________ is the optimal technique for detecting the known waveform in random noise.

   (a) correlation (b) convolution
   (c) both a and b (d) none of the above

5. The noise temperature of an ideal resistor is _____________

   (a) actual temperature of that resistor (b) increases with increase in temperature
   (c) decreases with increase in temperature (d) none of the above

6. Group of radiating elements is known as:

(a) cluster (b) array
(c) both a and b (d) none of these

7. Applying variable phase shifts at each element of an array is known as:

(a) parallel feed (b) serial feed
(c) both a and b (d) none of these

8. Beam steering can be done by:

(a) changing the relative time delays (b) changing the relative phase shifts
(c) both a and b (d) none of these

9. A linear array of linear arrays is known as

(a) linear array (b) planar array
(c) conformal array (d) both a and c

10. ______________ can be used as a duplexer.

(a) Isolator (b) Circulator
(c) Gyrator (d) Attenuator

ANSWERS TO OBJECTIVE TYPE QUESTIONS

1. (c)	2. (a)	3. (b)	4. (a)	5. (a)
6. (b)	7. (a)	8. (c)	9. (b)	10. (b)

REVIEW QUESTIONS

1. Write a short note on the detection of signals in noise.
2. Derive the expression for frequency response function of the matched filter.
3. What is meant by correlation? Explain cross-correlation with the help of neat block diagram.
4. Explain detection criteria in radar signals.
5. What is automatic detection? Explain the steps involved in this process.
6. Write about: a) branch type duplexer b) balanced duplexer.
7. Explain about different types of feeding arrangements used in phased array antennas.
8. Write about radiation pattern of phased array antennas with suitable equations.
9. Write a short note on applications of phased array antennas.
10. Write about: a) beam steering and b) beamwidth of phased array antennas.

11. If the noise figure of a receiver is 2.5 dB, determine the reduction (measured in dB) in the signal to noise ratio at the output compared to the signal to noise ratio at the input.

12. A radar receiver is connected to a 30 Ω resistance antenna that has an equivalent noise resistance of 25 Ω. Calculate the noise figure of the receiver and the equivalent noise temperature of the receiver.

13. Three network units, each of 5 dB noise figure and 15 dB, 20 dB, and 6 dB gains, respectively are cascaded. Determine the overall noise figure of the system.

14. Discuss in detail the quantitative analysis of receiver noise and hence derive the expression for minimum detectable signal.

15. Describe the different noise components present in radar systems.

16. Define noise figure. Derive an expression for the noise figure of two networks that are in cascade.

17. Define and distinguish the terms: noise figure, noise temperature, and system noise temperature, and system noise temperature of receivers.

Microwave Experiments

17

17.1 INTRODUCTION

A total of nine microwave laboratory experiments are described in this chapter. Each experiment begins with the aim of the experiment, equipment required, microwave bench setup, initial settings of the equipment, and experimental procedure and ends with observations and results. The microwave bench available in the laboratory has Reflex Klystron as a microwave source and is designed to accommodate an X-band signal (8–12 GHz). In case of the Gunn diode oscillator characteristics experiment, the Gunn diode oscillator is used in place of Reflex Klystron. The initial settings of the equipment such as Klystron power supply (or Gunn diode power supply in the case of the Gunn diode oscillator used as a microwave source) and the VSWR meter remain the same in all the experiments that use this equipment. The first five experiments describe the characterization of microwave sources (Reflex Klystron and Gunn diode) and microwave components (directional coupler, Magic Tee, and attenuator). The other three experiments use the same microwave bench setup with Klystron (or Gunn diode) for the measurement of various parameters (frequency, wavelength, VSWR, and impedance) at X-band frequencies. The ninth experiment describes the horn antenna radiation pattern. In all these experiments, a VSWR or a micro ammeter or a CRO can be used to read (or measure) the detected signal from the detector mount/crystal detector output, which is taken from the slotted-line waveguide section.

17.2 REFLEX KLYSTRON CHARACTERISTICS

Aim: To determine the frequency tuning range and to observe the mode characteristics of the Reflex Klystron tube using the microwave bench setup.

Equipment required: Klystron power supply, Klystron tube with mount, isolator, frequency meter, variable attenuator, detector mount, VSWR meter, oscilloscope, micro ammeter, BNC cable, waveguide stands, and accessories.

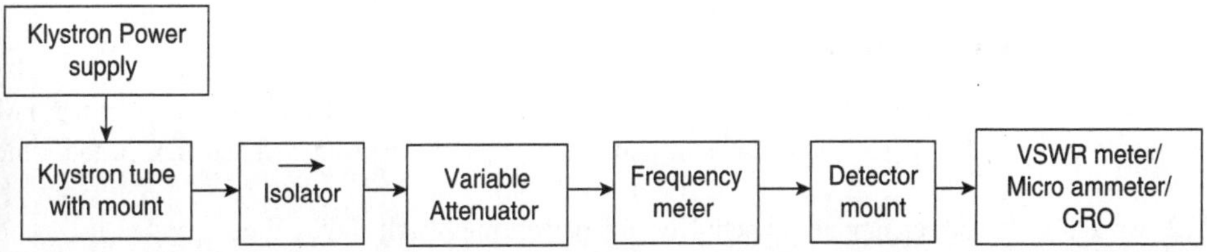

Figure 17.1 (a) *Block diagram of microwave bench setup for Reflex Klystron characteristics*

Source: 2.imimg.com

Figure 17.1 (b) *Schematic of microwave bench setup with Reflex Klystron as microwave source*

Initial settings of the experiment:

1. Klystron power supply:
   Beam voltage – **OFF,** Beam voltage knob – **fully anti clockwise**
   Repeller voltage – **fully clockwise**
   Mod-switch – **AM**
   Amplitude knob – **fully clockwise**
   Frequency knob –**mid position**
2. VSWR meter:
   Range dB – **40 dB / 50 dB,** Input switch – **low impedance**
   Meter switch – **normal,** Gain (coarse and fine) – **mid position approximately**

Note: In the following experimental procedure, steps 1 to 5 are common in all experiments that use the Reflex Klystron tube (the procedure differs in the Gunn diode characteristics experiment). In addition, we can use either VSWR or Micro Ammeter for finding the frequency of Reflex Klystron/Gunn diode oscillator.

Experimental procedure:

1. Connect the equipment as shown in Figure 17.1 (a).
2. The variable attenuator and the reflector voltage should be kept at their maximum positions and Klystron power supply to its minimum.
3. Rotate the knob of the frequency meter to one side fully.
4. Switch on the Klystron power supply and cooling fan.
5. Rotate the beam voltage knob slowly clockwise up to 300 V.
6. Vary the repeller voltage slowly and watch the micro ammeter. Set the repeller voltage for maximum deflection in the meter.
7. Tune the plunger of the Klystron mount to get the maximum output if necessary.
8. Slowly turn the knob of the frequency meter until a dip is observed in the micro ammeter/VSWR meter. The frequency reading between two horizontal lines and the vertical marker in a direct reading-type wave meter gives the corresponding frequency.
9. Change the reflector voltage and read the current and frequency for each reflector voltage.

Klystron mode observations on CRO:

1. Set the *mode* selector switch of Klystron power supply unit to ***FM-MOD*** position. Keep the FM amplitude, and FM frequency knobs at mid position. Let the remaining knobs are in the same position.
2. Switch on the power supplies to both Klystron power supply unit and oscilloscope.
3. Set the Klystron power supply unit meter selector switch to beam voltage position and set the anode beam voltage to approximately 300 V using the beam voltage control knob.

4. Keep the amplitude knob of the FM modulation to the maximum position in the modulation unit and rotate the reflector voltage knob in anti clockwise direction to get the modes on the oscilloscope as illustrated in Figure 17.2.
5. In the figure, the reflector voltage and output powers are represented in the horizontal and vertical axes respectively.
6. By changing the reflector voltage and amplitude of FM modulation, we can study all the Klystron modes on the oscilloscope.

Expected graph:

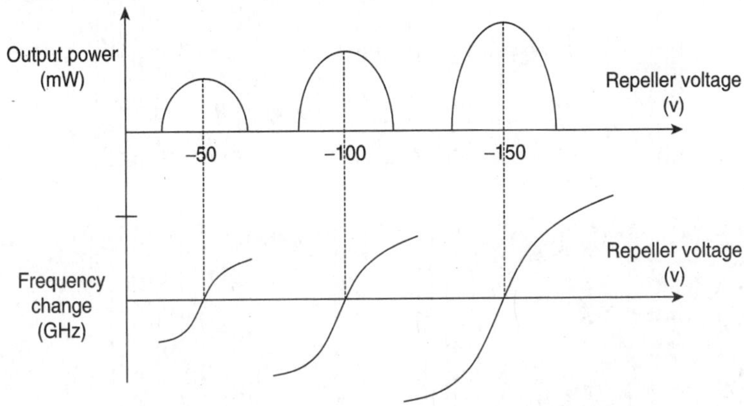

Figure 17.2 *Reflex Klystron characteristics*

Observation table: Beam voltage = 300 V, Beam current = 20 mA

Repeller voltage (V)	Current (μA)	Frequency (GHz)
40	0	9.93
50	240	9.95
60	0	9.98
90	0	9.935
100	295	9.95
110	0	9.985
140	0	9.94
150	450	9.95
160	0	9.99

Result: The mode characteristics of the Reflex Klystron tube are observed.

17.3 GUNN DIODE CHARACTERISTICS

Aim: To observe the V-I characteristics of the Gunn diode using a microwave bench setup.

Equipment required: Gunn power supply, Gunn oscillator, PIN modulator, isolator, frequency meter, detector mount, variable attenuator, VSWR meter, oscilloscope, BNC cable, waveguide stands, and accessories.

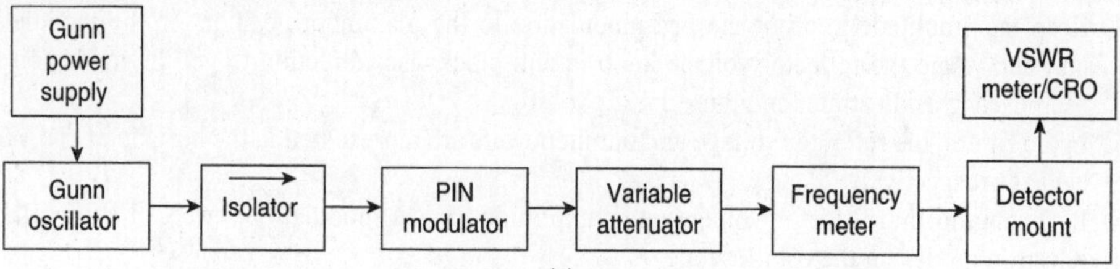

(a)

Source: scientechworld.com

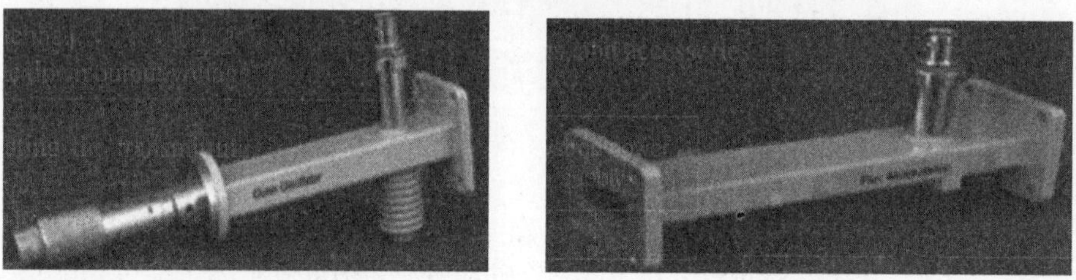

Source: www.hik-consulting.pl

(b)

Figure 17.3 *(a) Block diagram for measurement of Gunn diode characteristics; (b) Schematic of microwave bench setup with Gunn diode as microwave source*

Initial settings of the experiment:

1. Gunn power supply: Meter switch – **OFF**
   Gunn bias knob – **Fully anti clockwise**
   PIN bias knob – **Fully anti clockwise**
   PIN mode frequency – **Any position**

Experimental procedure:

1. Connect the equipment as shown in Figure 17.3 (a).
2. First set the variable attenuator for maximum attenuation.
3. Switch on the Gunn power supply.
4. Vary the Gunn bias voltage in steps and observe the respective Gunn diode current readings through the meter switch and digital panel meter.
5. Plot the voltage and current readings on the graph (Figure 17.4).
6. At maximum current, measure the threshold voltage.

Observation table:

Gunn bias voltage (V)	Gunn diode current (mA)
0.5	0.10
1.0	0.19
1.5	0.27
2	0.34
2.5	0.40
3	0.44
3.5	0.46
4	0.47
4.5	0.49
5	0.47
5.5	0.43
6	0.42
6.5	0.41

The threshold voltage of the Gunn diode oscillator = 4.5 V

Expected graph:

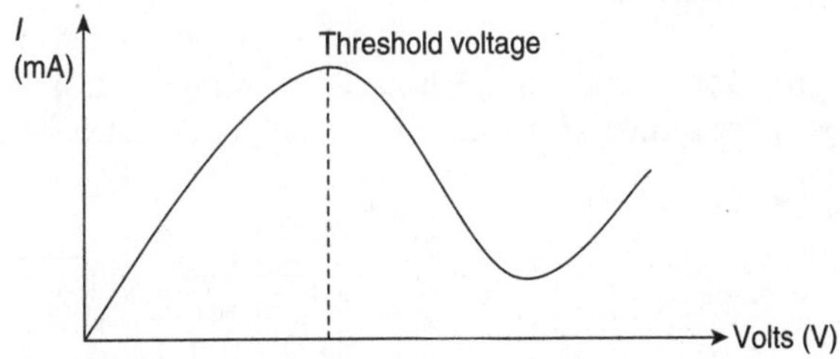

Figure 17.4 *V-I characteristics of Gunn oscillator*

Result: The V-I characteristics of the Gunn diode are observed.

17.4 MEASUREMENT OF ATTENUATION

Aim: To perform attenuation measurement of a network using a microwave bench setup.

Equipment required: Klystron power supply, Klystron tube with mount, isolator, frequency meter, variable attenuator, detector mount, power meter, BNC cable, waveguide stands, and accessories.

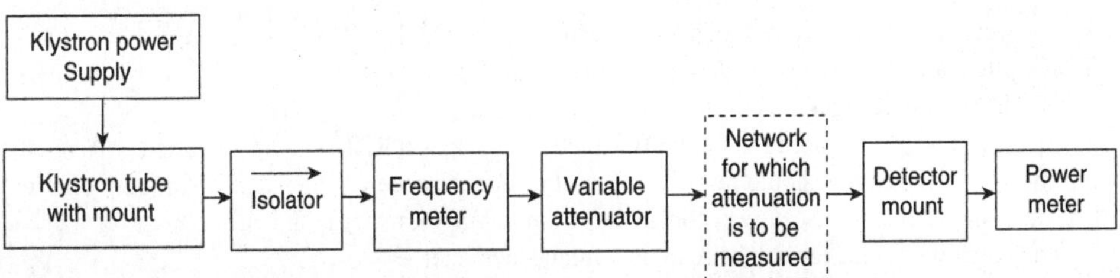

Figure 17.5 *Block diagram for measurement of attenuation*

Initial settings of the experiment: (same as in Section 17.2)

Experimental procedure:

1. The meter reading (P_1) is usually set at a maximum value with the network out of circuit by adjusting the variable attenuator in the line.
2. The network is placed in between the variable attenuator and the detector mount; power is measured on the meter (P_2).
3. Find out the attenuation value for different positions of the meter reading.

Observations: Beam Voltage = 300 V, Beam Current = 20 mA, $P_1 = 60$ W, and $P_2 = 30$ W

$$\text{Attenuation} = 10\log_{10}\left(\frac{P_1}{P_2}\right) = 10\log_{10}\left(\frac{60}{30}\right) = 3 \text{ dB}$$

Result: The attenuation of the test attenuator is determined.

17.5 MEASUREMENT OF FREQUENCY AND WAVELENGTH

Aim: To determine the frequency and wavelength of a rectangular waveguide operating in the TE_{10} mode using the microwave bench setup.

Equipment required: Klystron power supply, Klystron tube with mount, isolator, frequency meter, variable attenuator, slotted-line section, crystal detector, VSWR meter, matched load, BNC cable, waveguide stands, and accessories.

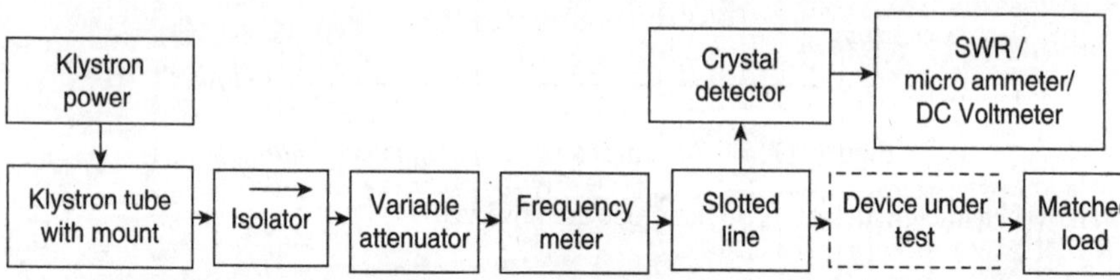

Figure 17.6 (a) *Block diagram for measurement of frequency and wavelength*

Initial settings of the experiment: (same as in Section 17.2)

Experimental procedure (Cavity wave-meter method):

1. The frequency meter cavity should be of resonance at an unknown frequency.
2. The attenuator should be adjusted to provide about 6–10 dB loss.
3. The 1KHz square-modulated wave should be applied.
4. Amp/Freq for peak reading in the VSWR meter (0° mark on RHS) should be varied.
5. The size of the cavity should be adjusted until resonance occurs. When a small amount of microwave energy enters the cavity, a considerable dip in VSWR meter is shown.
6. The unknown frequency can be read on a frequency meter.

Experimental procedure (Slotted-line method):

1. The slotted line is terminated by a short circuit.
2. Positions of two adjacent nulls are accurately located by moving the probe along the slotted line.
3. The position of nulls in the vernier scale of the slotted line (d_1 and d_2) are read.
4. Since two nulls are separated by $\lambda_g/2$, find the λ_g, using formula $\lambda_g = 2(d_1 - d_2)$.
5. Calculate the frequency by the following equation $f = \frac{c}{\lambda} = \sqrt{\left(\frac{1}{\lambda_g^2} + \frac{1}{\lambda_c^2}\right)}$

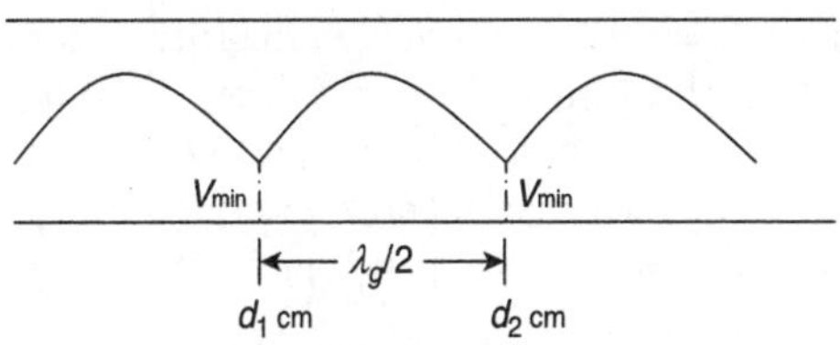

Figure 17.6 (b) *SWR pattern*

Observations: Beam voltage = 300 V, Beam current = 20 mA

$$d_1 = 10.5 \text{ cm, and } d_2 = 8.6 \text{ cm}$$
$$\lambda_g = 2(d_1 - d_2) = 2(10.5 - 8.6) = 3.8 \text{ cm}$$

For the TE_{10} mode, $a = 2.3$cm, and the cutoff wavelength $\lambda_c = 2a = 2 \times 2.3 = 4.6$ cm

$$f = \frac{c}{\lambda} = c\sqrt{\left(\frac{1}{\lambda_g^2} + \frac{1}{\lambda_c^2}\right)} = 3\times10^8\sqrt{\left(\frac{1}{(3.8)^2} + \frac{1}{(4.6)^2}\right)} = 10.2 \text{ GHz}$$

Result: The frequency and wavelength of a rectangular waveguide in the TE_{10} mode are determined.

17.6 DIRECTIONAL COUPLER CHARACTERISTICS

Aim: To measure the coupling factor, directivity, isolation, and insertion loss of a multi-hole directional coupler using the microwave bench setup.

Equipment required: Klystron power supply, Klystron tube with mount, isolator, frequency meter, variable attenuator, MHD coupler, crystal detector, micro ammeter, matched load, BNC cable, waveguide stands, and accessories.

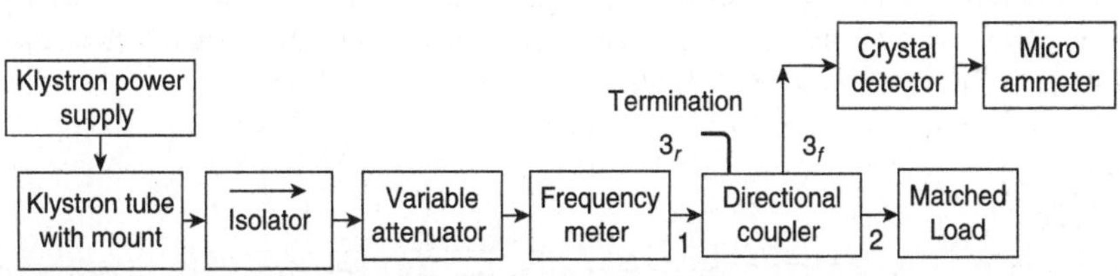

Figure 17.7 *Block diagram for measuring the characteristics of directional coupler*

Figure 17.8 *Directional coupler*

Initial settings of the experiment: (same as in Section 17.2)

Experimental procedure:

1. Measure I_1 by connecting the crystal detector/micro ammeter at port 1.
2. Insert the directional coupler between the frequency meter and crystal detector.
3. Connect matched load at port 2 and measure O/P at port 3 (I_{3f}).
4. Connect matched load at port 1 and input at port 2; measure O/P at port 3 (I_{3r}).
5. Connect matched load at port 3 and input at port 1; measure O/P at port 2 (I_2).

Observations: Beam Voltage = 300 V, Beam Current = 20 mA,

$$I_1 = 420\ \mu A,\ I_2 = 400\ \mu A,\ I_{3f} = 70\ \mu A,\ I_{3r} = 1\ \mu A.$$

$$\text{Coupling factor in dB} = 20\log\left(\frac{I_1}{I_{3f}}\right) = 20\log\left(\frac{420}{70}\right) = 15.56\ \text{dB}$$

$$\text{Isolation in dB} = 20\log\left(\frac{I_1}{I_{3r}}\right) = 20\log\left(\frac{420}{I}\right) = 52.46\ \text{dB and}$$

$$\text{Insertion Loss in dB} = 20\log\left(\frac{I_1}{I_2}\right) = 20\log\left(\frac{420}{400}\right) = 0.42\ \text{dB}$$

$$\text{Directivity in dB} = 20\log\left(\frac{I_{3f}}{I_{3r}}\right) = 20\log\left(\frac{70}{1}\right) = 36.9\ \text{dB}$$

Note: These characteristics are also measured in terms of voltage using the voltmeter instead of ammeter.

Result: The coupling factor, directivity, and insertion loss of the directional coupler are measured.

17.7 HORN ANTENNA RADIATION PATTERN

Aim: To plot the radiation pattern of a horn antenna using a microwave bench setup.

Equipment required: Klystron power supply, Klystron tube with mount, isolator, frequency meter, detector mount, variable attenuator, oscilloscope, horn antenna (transmitting and receiving), antenna rotating device, BNC cable, waveguide stands, and accessories.

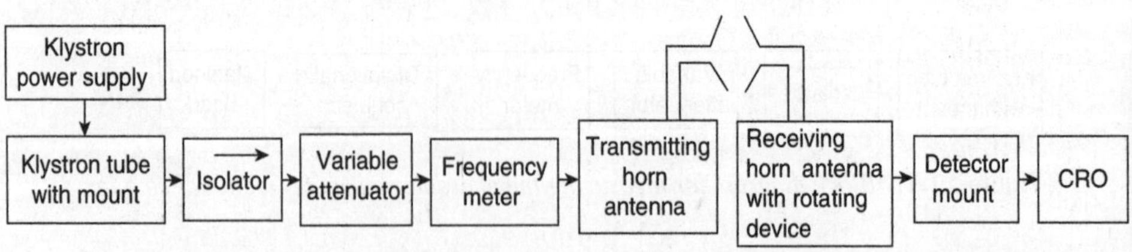

Figure 17.9 (a) *Block diagram of radiation pattern measurement of horn antenna*

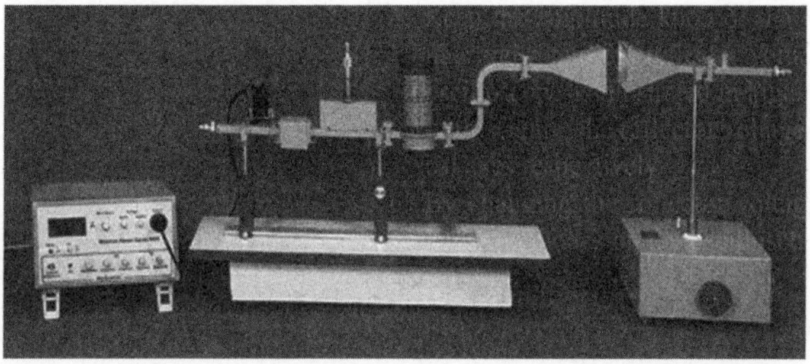

Source: 3.imimg.com

Figure 17.9 (b) *Schematic of microwave bench setup for radiation pattern measurement of horn antenna*

Initial settings of the experiment: (same as in Section 17.2)

Experimental procedure:

1. Adjust the orientation of the receiver antenna for polarization of the test antenna.
2. Turn the transmitting antenna to the left in 5° steps to full orientation and note down output voltage readings from CRO.
3. For obtaining the radiation pattern, repeat the above procedure after rotating the transmitting antenna and receiver horn about the axis.
4. Draw the radiation pattern, that is, output voltage versus angle, as shown in Figure 17.10.

Model graph:

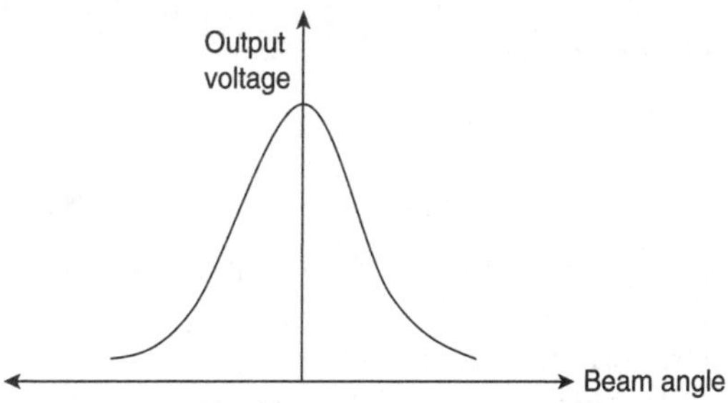

Figure 17.10 *Radiation pattern of horn antenna*

Observations: Beam voltage = 300 V, Beam current = 20 mA, Repeller Voltage = 117 V

Beam angle	Output voltage (mV)
–20°	0.4
–15°	12.5
–10°	32.5
–5°	80
0°	120

(*Continued*)

(*Continued*)

Beam angle	Output voltage (mV)
5°	84
10°	27.5
15°	15
20°	0.4

Result: The radiation pattern of a horn antenna is plotted.

17.8 MAGIC TEE CHARACTERISTICS

Aim: To observe the operation of Magic Tee and to calculate coupling co-efficient and isolation using microwave bench setup.

Equipment required: Klystron power supply, Klystron tube with mount, isolator, frequency meter, variable attenuator, Magic tee, crystal detector, VSWR meter, matched loads, BNC cable, waveguide stands, and accessories.

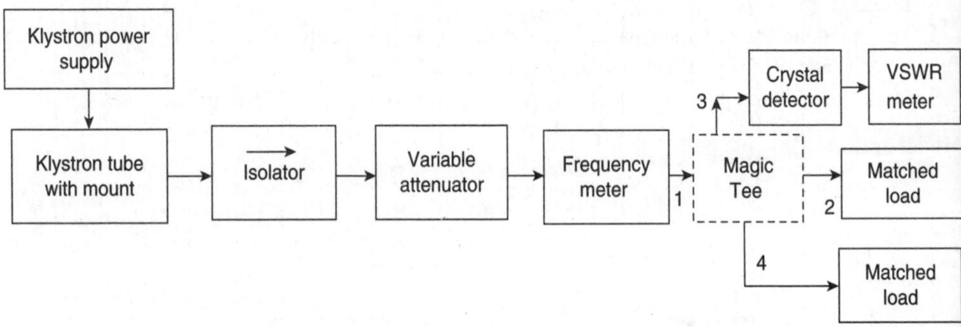

Figure 17.11 *Block diagram for measurement of scattering parameters of Magic Tee*

Initial settings of the experiment: (same as in Section 17.2)

Experimental procedure:

1. Connect the crystal detector to the VSWR meter and set any reference in the VSWR meter, for example, let it be P_3.
2. Insert the magic tee between the variable attenuator and crystal detector.
3. Connect matched loads at ports 1 and 2, input at port 3, and measure output at port 4.
4. Determine the isolation between port 3 and port 4 as P_3–P_4.
5. The same experiment may be repeated for other ports also.

Observations: Beam voltage = 300 V, Beam current = 20 mA, P_3 = 61dB, and P_4 = 32 dB

Isolation loss (α) = $P_3 - P_4$ = 29 dB
Coupling co-efficient, $C_{ij} = 10^{(-\alpha/10)}$. Therefore, $C_{34} = 10^{(-29/10)} = 0.0012$

Result: The coupling co-efficient and isolation loss of the magic tee are determined.

17.9 VSWR MEASUREMENT

Aim: (a) To determine the voltage standing-wave ratio for VSWR < 10.
(b) To determine the voltage standing-wave ratio for VSWR > 10.

Equipment required: Klystron power supply, Klystron tube with mount, isolator, frequency meter, variable attenuator, slotted-line section, crystal detector, VSWR meter, movable short/termination, S-S Tuner, BNC cable, waveguide stands, and accessories.

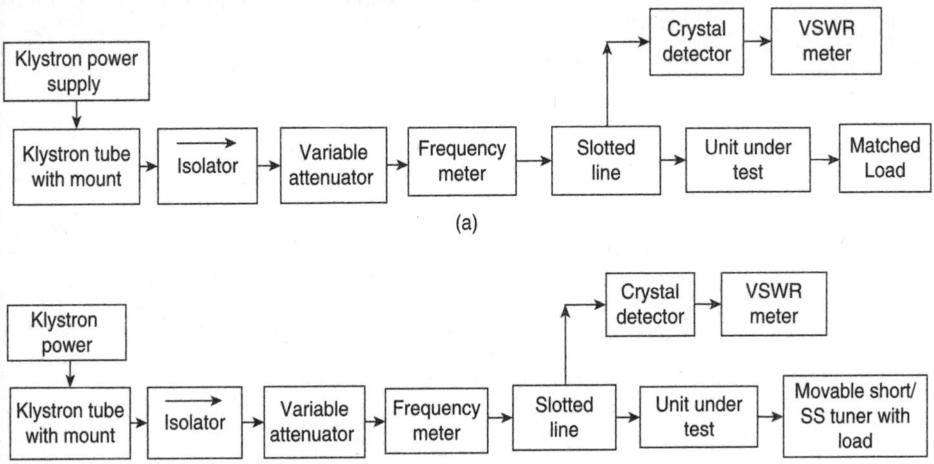

Figure 17.12 *(a) Block diagram for measurement of low VSWR (< 10); (b) Block diagram for measurement of high VSWR (> 10)*

Initial settings of the experiment: (same as in Section 17.2)

Experimental procedure for measurement of low VSWR < 10:

1. Adjust the attenuator to give an adequate reading on the maximum reading of the VSWR meter.
2. Adjust the probe carriage on the slotted line to get a maximum reading on the meter. This full-scale reading is noted down (V_{max}).
3. Adjust the probe carriage on the slotted line to get the minimum reading on the meter (V_{min}).
4. Note down the readings and compute VSWR from the following equation:

$$\text{VSWR} = \frac{V_{max}}{V_{min}}$$

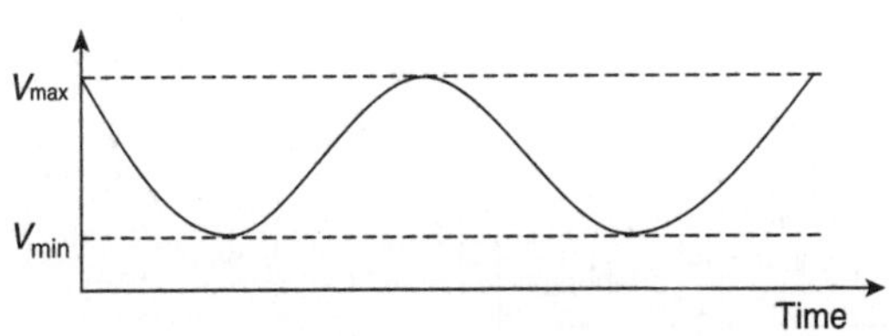

Figure 17.13 VSWR *pattern*

Experimental procedure for measurement of high VSWR > 10 (double minimum method)

1. Insert the probe to a depth, and then move it to a point where the minimum can be read.
2. Then, move the probe to a point where the power is twice the minimum. Let this position be denoted by d_1.
3. Next, move the probe to twice power point on the other side or the minimum. Let this position be denoted by d_2.
4. Replace the S-S tuner and termination by movable short.
5. Measure the distance between two successive minimum positions of the probe. The guide wavelength ($\lambda_g = 2(d_1 - d_2)$) is twice this distance.
6. Compute the VSWR from the equation below:

$$\text{VSWR} = \frac{\lambda_g}{\pi(d_1 - d_2)}$$

Observation table for low VSWR:

Repeller voltage(V)	V_{min}(V)	V_{max}(V)	VSWR
80	0.1	0.2	2
110	0.4	0.8	2
150	0.6	1.0	1.66
175	0.4	0.6	1.5
200	0.2	0.6	3

Observations for high VSWR: Beam Voltage = 300 V, Beam Current = 20 mA,

$d_1 = 10.8$ cm and $d_2 = 8.4$ cm, $\lambda_g = 2(d_1 - d_2) = 2(10.8 - 7.4) = 6.8$ cm

$d_1 = 8.5$ cm, and $d_2 = 7.9$ cm (Using S-S tuner) $\text{VSWR} = \frac{\lambda_g}{\pi(d_1 - d_2)} = \frac{6.8}{\pi(8.5 - 7.9)} = 3.6$

Result: Low VSWR and high VSWR of a microwave bench setup are measured.

17.10 IMPEDANCE MEASUREMENT USING REFLEX KLYSTRON

Aim: To measure an unknown impedance using a microwave bench setup.

Equipment required:

Klystron power supply, Klystron tube with mount, isolator, frequency meter, variable attenuator, slotted-line section, VSWR meter/micro ammeter, BNC cable, waveguide stands, and accessories

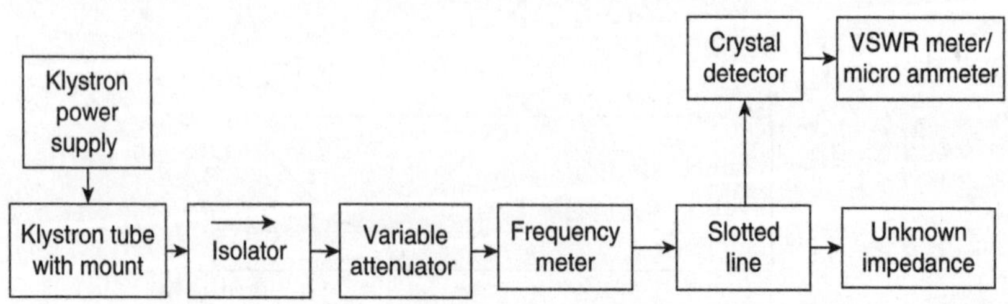

Figure 17.14 *Block diagram for measurement of impedance*

nitial settings of the experiment: (same as in Section 17.2)

xperimental procedure:

1. Terminate the load with an unknown impedance and measure VSWR and λ_g.
2. Note down the position of minimum with the help of a probe around the center of the slotted section = d_1.
3. Replace the load with short-circuited termination.
4. Move the probe carriage to a new standing wave minimum = d_2, shift in minimum = $d_2 - d_1$.
5. Draw a VSWR circle on the Smith chart.
6. Draw a line from center of circle to impedance value from which admittance and reactance ($Z = R \pm jx$) are calculated.

bservations: Beam Voltage = 300 V, Beam Current = 20 mA, $a = 2.282$ cm, $b = 1.016$ cm, VSWR $S_0) = 4$, $d_1 = 7.9$ cm, $d_2 = 8.7$ cm, and $d_3 = 11$ cm

Shift = $d_2 - d_1 = 0.8$. $\lambda_c = 2a = 2 \times 2.82 = 4.564$ cm, $\lambda_g = 2(d_2 - d_3) = 2.3$ cm

$$\lambda_0 = \frac{\lambda_g \lambda_c}{\sqrt{\left((\lambda_g)^2 + (\lambda_c)^2\right)}} = \frac{2.3 \times 4.564}{\sqrt{\left((2.3)^2 + (4.564)^2\right)}} = 2.05 \text{ cm}$$

$$Z_0 = 377 \times \left(\frac{b}{a}\right) \times \left(\frac{\lambda_g}{\lambda_0}\right) = 377 \times \left(\frac{1.016}{2.282}\right) \times \left(\frac{2.3}{2.05}\right) = 188.31$$

$$\text{Distance from Load} = \frac{d}{\lambda_g} = \frac{0.8}{2.3} = 0.3478$$

he measurement is performed in the following way: The difference between reference minima and inima position obtained from an unknown load is to be found and is denoted as "d." Assuming "1" as e center; draw a circle of radius equal to VSWR (S_0) on the Smith chart. Join the center with a point n the circumference of the Smith chart toward load side whose distance is equal to d/λ_g. Find the point here it cuts the drawn circle. The co-ordinates of this point will show the normalized impedance of the ad. From the Smith chart,

$$\text{VSWR } (S_0) = 4, Z_{\text{Norm}} = \left(\frac{Z_L}{Z_0}\right) = 0.8 + j\,1.05$$

$$Z_L = Z_{\text{Norm}} Z_0 = (0.8 + j1.05) \times 188.31 = (150.65 + j197.73) = 248.58\angle 52.7^\circ\ \Omega$$

esult: Unknown impedance is measured using a microwave bench setup.

UMMARY

1. Microwave laboratory experience is important because it helps in understanding the procedure to use microwave test equipment.

2. The parameters that can be conveniently measured at microwave frequencies are frequency, wavelength, power, attenuation, VSWR, impedance, and characteristics of microwave oscillators.

3. In a direct reading-type wave meter, frequency is directly read between two horizontal lines an vertical markers.

4. The waveguide slotted line section is used to know the behavior of standing waves in the waveguide and to measure VSWR, guided wavelength (λ_g), and impedance.

5. The VSWR meter is used to measure the SWR in conjunction with a slotted waveguide section.

6. Microwave frequency can be measured by two methods: slotted line and resonant cavity (wav meter).

7. Frequency of the source (Klystron or Gunn diode oscillator) is determined by measuring the wave length of a standing wave created in a waveguide due to mismatched load.

APPENDIX A

Glossary of Terms

Ampere's Law relates the total magnetomotive force along a closed loop and the net current enclosed by the loop.

Antenna is a method that is used for converting the guided waves present in a waveguide, feeder cable, or transmission line into radiating waves traveling in free space, or vice versa.

Attenuator is a passive device that is used to reduce the strength or amplitude of a signal.

Attenuation constant (α) represents attenuation of the wave on the transmission line. For a positive α, the amplitude exponentially decreases as a function of distance. The unit of attenuation constant is nepers/m.

Azimuth (or bearing) resolution is the ability of a radar system to separate objects at the same range but at slightly different bearings from a reference point. The degree of bearing resolution depends on the radar beam width and the angular distance between the targets.

Beam width is defined as the angle between the half-power (3 dB) points of the main lobe. The unit is degree.

Backward-wave Oscillator is one of the devices that efficiently converts the energy of an electron beam into electromagnetic radiation at microwave frequencies. Unlike the TWT, the BWO has no attenuator. As a result, the RF signals that travel backward, that is, toward the cathode, are not suppressed, and the helix is terminated with matching impedance. The output is taken from the end of the helix near the electron gun.

Bandwidth (BW) is the frequency difference between the upper and lower frequencies of the electromagnetic radiation pattern. It is expressed in units of Hertz (Hz).

BARITT diode is an improved version of IMPATT diodes. The ionization technique of IMPATT makes it too noisy; therefore, it is avoided in BARITT devices. It is replaced by barrier-injected minority carriers that are generated from forward-biased junctions instead of being generated from the impact ionization of the avalanche region as in the IMPATT diode.

Biot–Savart's Law relates the magnetic intensity at any point as being due to a steady current in an infinitely long, straight wire, and the distance from the point to the wire.

Blind speed The delay-line canceller of the MTI radar not only eliminates the dc component caused by clutter ($n = 0$), but also unfortunately rejects any moving target whose doppler frequency happens to be the same as the PRF or a multiple thereof. Those relative target velocities that result in a zero MT response are called *blind speeds*.

Boundary conditions are the set of conditions specified for the behavior of electric and magnetic field at the interface between two different media.

Bolometer It is a power sensor whose resistance changes with a change in temperature as it absorb the microwave power. It is a short, thin, metallic wire sensor with a positive temperature coefficient c resistance.

Calorimeter It is a convenient device setup that is used for measuring high power at microwave fre quencies and involves the conversion of microwave energy into heat, absorbing the heat in a fluid an determining the temperature.

Cavity resonators are formed by placing the perfectly conducting sheets of the rectangular or circula waveguide on the two end sections, and, hence, all the sides are surrounded by the conducting walls, thu forming a cavity. The electromagnetic energy is confined within this metallic enclosure, and this acts as resonant circuit. There are two types of cavity resonators. They are rectangular cavity resonator and circula cavity resonator. The quality factor of a cavity resonator may by high—several thousands or even more.

Characteristic impedance is the ratio of voltage applied to the current in an infinite line. When the lin is terminated with the characteristic impedance, the impedance measured at any point on the line is th same as the terminating impedance.

Charge is a quantity of electricity that is determined by the product of an electric current and the tim for which it flows, measured in coulombs.

Conservation of energy states that the total inflow of energy into a system should be equal to the tota outflow of energy from the system plus the change in the energy contained within the system. (Energ can be converted from one form to another, but it cannot be created or destroyed.)

Coulomb's Law gives the relationship between the force and two charges, their magnitude, and th distance between them.

Circulator is microwave device consisting of three or more ports that is used for coupling energy i only one direction around a closed loop. Microwave energy is applied to one port and passed to anothe with minimum or no attenuation; however, the signal will be greatly attenuated on its way to a third por The primary application of a circulator is a duplexer, which allows a single antenna to be shared by transmitter and a receiver.

Crystal diode Similar to a schottky diode, a crystal diode has a metal semiconductor junction. It is unipolar device and is, hence, free from the ill effects of minority carrier storage that are present in a con ventional PN diode. It depends on the pressure of contact between a semiconductor crystal and a whiske made of gold-plated tungsten.

Cross-field Amplifier (CFA) is a microwave power amplifier. It is a cross between the TWT and magne tron in its operation; that is, it has a magnetron structure which provides an interaction between crosse dc electric and magnetic fields on one hand and an RF field on the other hand. It also uses a slow-wav structure as in TWT to provide a continuous interaction between the electron beam and a moving R field.

Coherence A radar is said to be coherent if the phase of any two transmitted pulses is consistent; that is, there is a continuity in the signal phase from one pulse to the next. Coherency can be achieved by using a Stable Local Oscillator (STALO). A radar is said to be coherent-on receive or quasi-coherent if it stores in its memory a record of the phases of all transmitted pulses.

Coherent Oscillator (COHO) provides a reference signal that effectively remembers the phase of each of the transmitted and received pulses.

Current is the rate of transfer of charge with regard to time (across a reference point or surface).

Current density is a vector whose magnitude and direction are the current per unit area and the direction of current flow at a point in space, respectively.

Curl of a vector is a measure of the rotation of the field (or represents the circulation per unit area of the field).

Cutoff frequency is the frequency below which the wave propagation ceases.

Cutoff wavelength is the wavelength corresponding to the cutoff frequency. It is the wavelength below which there is wave propagation and above which there is no wave propagation.

Clutter Objects of interest such as aircraft are called *targets*, and objects that can potentially get confused with targets such as reflections from the ground, buildings, or the sea are called *clutters*.

Delay-line canceller The main function of the delay-line canceller is to act as a filter by rejecting the dc component of the clutter.

Diffraction is the bending of EM waves as they propagate through an aperture or around the edge of an object. The amount of diffraction present depends on the size of the aperture, relative to the wavelength of the wave.

Diffusion is a process by which a substrate is redistributed from an area of a relatively high concentration to an area of a relatively low concentration due to random thermal motion. The larger the concentration gradient, the faster diffusion occurs for a given temperature; conversely, the higher the temperature, the faster diffusion occurs for a given concentration gradient.

Dispersive material is the one in which the material constants are functions of frequency, $\varepsilon(\omega)$ or $\mu(\omega)$.

Displacement current is a quantity that is defined in terms of the rate of change of electric flux density.

Divergence computes how much a vector field converges to or diverges from a given point.

Divergence Theorem relates the surface integral to the volume integral.

Dominant mode in a particular rectangular waveguide is the mode having the lowest cutoff frequency (or longest cutoff wavelength). Dominant modes for TE and TM waves are TE_{10} and TM_{11}, respectively. TE_{10} ($m = 1$, $n = 0$) is the dominant mode of a rectangular waveguide, because the TE_{10} mode has the lowest attenuation of all the modes in a rectangular waveguide, and its electric field is definitely polarized in one direction everywhere. The dominant mode for a circular waveguide is defined as the lowest order mode having the lowest root value. The dominant modes for TE and TM waves in a circular waveguide are TE_{11} (lowest root value of 1.841) and TM01 (lowest root value of 2.405), respectively.

Doppler effect This measures the difference between the frequency at which sound or light waves leave a source and that at which they reach an observer. The change in the frequency is called *Doppler frequency shift* and is caused by relative motion of the observer and the wave source.

Degenerate modes Some of the higher-order modes having the same cutoff frequency are called *degenerate modes*. In a rectangular waveguide, TE_{mn} and TM_{mn} modes (both $m \neq 0$ and $n \neq 0$) are always degenerate.

Directional couplers allow us to sample or monitor the frequency level and/or power level of a given signal as it moves from one point to another.

Doppler filter bank This is a collection of filters that is used for detecting targets. In actual practice, Pulse-Doppler radars use several banks of bandpass filters whose outputs are linked to an indicator. The radar receives signals from many sources simultaneously, which are then sorted out on the basis of Doppler frequency in a bank of Doppler filters.

Duplexer is a switch (usually a circulator) that alternately connects the transmitter or the receiver to the antenna. It protects the receiver from the high power output of the transmitter. During the transmission of an outgoing pulse, the duplexer will be aligned to connect the transmitter to the antenna for the duration of the pulse.

Duty factor/Duty cycle Duty cycle is the amount of time that a radar transmits compared with its listening to receiving time. The ratio is sometimes expressed in percent. It can be determined by multiplying PRF and pulse width or by dividing the pulse width by PRT. It does not have any units.

Electric field is the region of space surrounding the electric charge.

Electric field intensity is the electric force exerted on a unit test charge in an electric field. It is also known as *electric field strength*.

Electric flux is an imaginary path or a line which is drawn in such a way that its direction at any point is the direction of the electric field at that point and is equal to the number of electric lines of force crossing the surface.

Electric flux density or electric field displacement is a measure of the number of electric flux lines passing through a given area.

Electric potential at a point is the work done in moving a unit positive charge from infinity to that point against the electric field.

Electromagnetic field theory is the study of the electric and magnetic phenomena caused by electrical charges at rest or in motion.

Electromagnetic (EM) waves are electric and magnetic fields oscillating at a particular frequency. Once an EM wave is launched, it becomes self propagating. EM waves propagate in free space as well as inside material media. The behavior of EM fields is described by Maxwell's equations.

Electrostatic field is a time-independent electric field that is produced by stationary charges.

Evanescent mode When the operating frequency is lower than the cutoff frequency, the propagation constant becomes real, that is, $\gamma = \alpha$. The wave cannot be propagated. This non-propagating mode is known as *evanescent mode*.

Epitaxial layer is a single crystal layer formed on top of a single crystal substrate. The doping level of the epitaxial layer is different from the substrate on which the epitaxial layer is formed.

Etching This is the process of removing a material by a chemical reaction.

Faraday's Law relates the net electromotive force (EMF) in a closed loop and the rate of change of magnetic flux (ϕ_B) enclosed by the loop.

Faraday's rotation law If a circularly polarized wave were made to pass through a ferrite rod that has been influenced by an axial magnetic field B, the axis of polarization gets tilted in the clockwise direction and the amount of tilt depends on the strength of the magnetic field and the geometry of the ferrite.

Ferrites are ceramic materials that possess a high resistivity and which behave nonreciprocally when they are embedded in a magnetic field. Ferrite devices such as isolators, circulators, attenuators, phase shifters, modulators, and switches are based on these properties.

Flanges are used to connect waveguide sections to one another or to terminate waveguides.

Frequency (f) refers to the number of completed wave cycles per second. Frequency is expressed in units of Hertz (Hz).

Frequency agility is the process of changing the radar frequency from pulse to pulse.

Gauss' Law relates the total outward electric displacement through any closed surface surrounding charges and the total charge enclosed in that surface.

Gradient of a scalar A is a measure of the rate of change of A with regard to distance in a particular direction. ($\hat{\imath}$) is the projection of grad A onto that direction.

Group velocity is the velocity with which the overall shape of the wave's amplitudes known as *envelope* of the wave propagates through space.

Gunn effect or Bulk effect This takes place when a DC bias voltage is applied to the contacts of n-type GaAs or InP. Gunn found the following: (i) Current first rises linearly from zero; (ii) Then, it begins to oscillate when a certain threshold is reached; and (iii) Time period of the oscillation is equal to the travel time of the electron from the cathode to the anode. This is known as *Gunn effect* or *bulk effect*.

Ridley–Watkins–Hilsum (RWH) theory Gunn diode, which is made of *n*-doped semiconductor material (e.g. GaAs or InP), is characterized by two valleys in their conduction bands with different mobility. The two-valley model is also called the *Ridley–Watkins–Hilsum (RWH) theory*.

Gunn diode The device that shows the Gunn effect is known as the *Gunn diode*. Gunn diodes are usually fabricated using N-type semiconductor materials (Eg. GaAs, InP); thus, they should be associated with electrons rather than with holes.

Guided wavelength is the wavelength of electromagnetic energy conducted in a waveguide.

Gyrator is a two-port device that provides a relative phase shift of 180^0 for transmission from port 1 to port 2 as compared with the phase for transmission from port 2 to port 1.

Hartree voltage is the critical voltage at which energy transfer to the wave ceases and the magnetron stops operating.

H Plane or the E Plane At microwave frequencies, waveguide tees have two possible forms—the H Plane or the E Plane. These two junctions split power equally, but due to the different field configurations at the junction, the electric fields at the output arms are in phase for the H-Plane tee and are anti-phase for the E-Plane tee.

Hull cutoff condition determines the anode voltage or magnetic field necessary to obtain nonzero anode current as a function of the magnetic field or anode voltage in the absence of an electromagnetic field.

Hybrid Ring (Rat-race Junction) This is a type of coupler that is used in RF and microwave systems. In its simplest form, it is a 3dB coupler and is, thus, an alternative to a magic tee. Unlike magic tees, a rat race needs no matching structure.

HEMT stands for *High Electron Mobility Transistor*. In these devices, a junction is formed between two different semiconductor materials having various band gaps. This results in the formation of low potential on one side of the junction. Electrons will concentrate in this low potential region, and they will travel through the un-doped material. This results in an increase of the mobility of carriers in the un-doped material.

Iris (or windows) Fixed or adjustable projections from the walls of waveguides are used for impedance matching purposes.

IMPATT diode is a solid-state microwave device that operates with a reverse bias which is sufficient to cause an avalanche breakdown. This is a high power diode and a very powerful microwave source that is used in high-frequency electronics and microwave devices. The IMPATT diode exhibits a dynamic negative resistance that is required for microwave oscillation and amplification applications.

Impedance matching If the load impedance is not equal to the source impedance, all the power that is transmitted from the source will not reach the load end, and, hence, some power is wasted. This is called *impedance mismatch condition*. So, for proper maximum power transfer, the impedances at the sending and receiving ends are matched. This is called *impedance matching*.

Isolator is a two-port non-reciprocal transmission device that is used to isolate one component from the reflections of the other components in the transmission line.

Intrinsic impedance is the ratio of the perpendicular components of the electric field and magnetic field phasors.

Klystron is a vacuum electron device that is used for transforming DC energy into RF energy, and it may be either an oscillator or an amplifier. Klystrons make use of the transit-time effect by varying the velocity of an electron beam.

Linear beam tubes In these tubes, the electrons emitted from the cathode are accelerated by an applied anode voltage. The resultant electron beam has a kinetic energy that is determined by the anode voltage. A portion of the kinetic energy contained in the electron beam is converted into microwave energy when an applied RF input interacts with the electron beam. The microwave energy is extracted at the RF output port. Examples are Klystron, TWT, BWO, and BWA.

Line integral of any function is an integral taken along a line, where the function has a continuously varying value along that line.

Lorentz force law relates the electromagnetic force F on a test charge at a given point and time, charge q, and velocity v.

Loop is used to introduce the magnetic field into a waveguide. It is generally mounted at a distance of $\lambda/2$.

Magic Tee The combination of the H-Plane and the E-Plane tees forms a hybrid tee. It allows for the realization of a four-port component, which could perform the vector sum (Σ) and the difference (Δ) of two coherent microwave signals. This device is known as the *magic tee*.

Magnetic field lines are always closed lines (as compared with electric field lines that start and end on charges.) that are caused by ferromagnetic material, a permanent magnet, or an electric current (moving electric charges).

Magnetic flux lines are the distribution of magnetic field lines.

Magnetic flux density is a measure of the number of magnetic flux lines passing through a given area.

Magnetic field strength is an auxiliary vector field that is used in describing magnetic phenomena, whose curl, in the case of static charges and currents, equals the free current density vector, independent of the magnetic permeability of the material. It is also known as *magnetic field*, *magnetic field intensity*, *magnetic force*, *magnetic intensity*, and *magnetizing force*.

Magnetron is a crossed-field device, because both magnetic and electric fields are employed in its operation, and they are produced in perpendicular directions so that they cross. Therefore, the flow of electrons is perpendicular to both the fields. In magnetrons, the anode and cathode are cylindrical and concentric. The magnetic field causes the electrons that are emitted from the cathode to move in curved paths. Magnetrons use various shapes of cavities to build oscillations and power.

Maxwell's equations in integral form describe the relations of the field vectors with regard to charge and current densities over an extended region of space. The integral form can be derived from the differential form through the use of Stokes' and divergence theorems. These equations are used to solve electromagnetic problems with complete symmetry.

MESFET (Metal Schottky FET) This is a field-effect transistor whose gate structure consists of a metallic Schottky barrier.

Microwave tubes are constructed so as to overcome the limitations of conventional VHF and UHF tubes. The basic operating principle of microwave tubes involves the transfer of power from the source of the dc voltage to the source of the ac voltage by means of a current density-modulated electron beam.

Modes The EM wave propagates along a waveguide in the form of some definite field patterns (or configurations) called *modes*.

Mode jumping The resonant modes of magnetrons are very close to each other. As a result, there is every possibility that one resonant frequency (or mode of operation) gets shifted easily to another; this is called *mode jumping*.

π-mode To avoid the mode jumping problem in magnetrons, resonant frequencies will be separated as widely as possible. The best desired mode is the π-mode, where adjacent blocks of the anode become positive and negative, respectively.

MMIC If a circuit is integrated directly on the surface of a semiconductor substrate, it is called a *monolithic integrated circuit*. Up to about 2 GHz, the monolithic circuits are made on Si; at higher frequencies, the substrate is usually GaAs and these circuits are called *monolithic microwave integrated circuits* (MMICs).

Monopulse The term *monopulse* is used, because the information obtained from a single pulse tells the radar about the angular position of the target from the radar and how far the beam is from the target axis that provides an error signal, which drives the radar beam toward the target. In all modern radars, *angle tracking* is usually done with monopulse tracking, which is also known as *simultaneous lobbing*.

Moving Target Indication (MTI) A processor that distinguishes moving targets from clutter by virtue of differences in their spectra is called an *MTI*. The MTI radar uses low pulse repetition frequency (PRF) to avoid range ambiguities, but these radars can have Doppler ambiguities. An MTI based on a delay-line canceller operates by taking the difference of the amplitudes of successive pulses into consideration.

Moving target detector (MTD) is a term that is applied to the Doppler processing system which is used in modern radars called *airport surveillance radars*. The MTD provides good target detection performance.

Nonreciprocal devices The devices possessing the feature that the forward characteristics are not equal to the reverse characteristics are called *non-reciprocal devices*.

Negative resistance is defined as that feature of a device which causes the current through it to be 180^0 out of phase with the voltage across it.

Network analyzer measures both amplitude and phase of a signal over a wide frequency range. It requires an accurate reference signal and a test signal.

Radiation pattern is a representation of the radiation characteristics of an antenna, which is a function of elevation angle, azimuth angle for a constant radial distance and frequency.

Oxidation This is the reaction of a material with oxygen, which results in the formation of a compound of the material and oxygen. Usually, oxygen (O_2) or water vapor (H_2O) is reacted with silicon (Si) at high temperatures to form SiO_2.

Omni directional antenna This is an antenna that radiates equally in all directions (nondirectional) or an antenna whose radiation pattern shows equal radiation in all horizontal directions.

Parametric amplifier is so called, because in this amplifier, a parameter is made to vary with time. It is also called a *reactance amplifier*, as the underlying principle of operation is based on reactance.

Phase This is the time difference between two different signals at the same frequency.

Phase shifter is a two-port component that provides a fixed or variable change in the phase of the traveling wave.

Pulsed radar transmits high-power, high-frequency pulses toward the target. Then, it waits for the echo of the transmitted signal for sometime before it transmits a new pulse. Choice of pulse repetition frequency decides the range and resolution of the radar. Target range and bearings can be determined from the measured antenna position and the time of arrival of the reflected signal.

Two broad categories of pulse radar employing Doppler shifts are MTI and PDR.

Pulse Doppler Radar Contrary to the MTI radar, the pulse Doppler radar uses high PRF to avoid Doppler ambiguities, but it can have numerous range ambiguities.

Propagation constant gives the manner in which the wave is propagated along a transmission line and specifies the variation of voltage and current in the transmission line as a function of distance. The propagation constant is a complex quantity and is expressed as $\gamma = \alpha + j\beta$. The real part is called the *attenuation constant*, α, whereas the imaginary part of the propagation constant is called the *phase constant*, β.

Permeability is the degree of magnetization of a material in response to a magnetic field and is given by the ratio of magnetic flux density and magnetic field intensity.

Permittivity of a material is the degree to which it can resist the formation of an electric field within it and is equal to the ratio between the electric flux density and the electric field strength generated by an electric charge in the material.

Phase constant (β) defines the phase variation of the wave due to spatial variation. Since for a wave the phase change over a wavelength (λ) is 2π, the phase change over a path length is $2\pi/\lambda$. The unit of the phase constant is radians per meter.

Phase velocity is the speed at which the phase of any one frequency component of the wave travels.

Phase center of an antenna is the reference point that minimizes the phase difference over the main beam. For example, the phase center of the feed antenna and the focal point of the reflector that is illuminated by the feed should coincide.

Phased Array Radar is capable of steering the beam electronically in space. This provides greater flexibility and makes the system increasingly versatile by being able to carry out better energy management in the volume of space and to optimize the search and track functions.

PIN diode It is a variation of the conventional PN junction diode with a small layer of an intrinsic semiconductor between P and N layers. This long intrinsic region makes the device capable of withstanding high breakdown voltages. The PIN diode can be used as a microwave device only above 200 MHz.

Plane wave is a wave in which at every moment, the displacements and velocities of the particles in a medium (for mechanical waves) or the strengths of the electric and magnetic fields (for electromagnetic waves) are the same at all points lying in any plane that is perpendicular to the direction of the wave's propagation.

Polarization is the direction of the electric field vector, *E*. Polarization may be linear, where *E* is always in the same direction, or it may be circular (or elliptical), where *E* and *H* rotate as the wave propagates. Polarization information is required to describe how an EM wave is transmitted and received.

- **A Circularly polarized** wave is one in which the resultant electric field vector remains constant in length but rotates around in a circular path as time progresses.
- **An Elliptically polarized** wave is one in which the resultant electric field vector does not remain constant in length but rotates around in an elliptical path as time progresses.
- **A Horizontally polarized** wave is one in which the electric field is aligned in parallel with the horizontal axis as time progresses.
- **A Vertically polarized** wave is one in which the electric field is aligned in parallel with the vertical axis as time progresses.

Poynting vector The magnitude and direction of the energy flux (i.e., energy per unit cross-section area, W/m^2) are given by the Pointing vector, P, which is the vector cross-product of E and H. For time-harmonic fields, the phasor Poynting vector is $E \times H^*$. The average power density is ½Re($E \times H^*$), where H^* is the complex conjugate of H.

PPI The most common form of CRT display used in radars is the plan position indicator (PPI), which maps in polar coordinates the location of the target in azimuth and range.

Pulse width Pulse width is the time interval between the leading edge and the trailing edge of a pulse at a point where the amplitude is 50% of the peak value. It is expressed in units of microseconds.

PRF Pulse repetition frequency is the number of peak power pulses transmitted per second. The PRF is primarily used for knowing the maximum range at which targets are expected.

PRT Pulse repetition time is the time interval between two peak pulses. During each PRT, the radar radiates energy only for seconds and listens for target returns for the rest of the PRT.

Probe is a $\lambda/4$ vertical antenna that is inserted in the waveguide at a distance of $\lambda/4$ from the closed end and the center of the broader dimension of the waveguide.

Radar Range The most important feature of a radar is its ability to determine the range of a target by measuring the time it takes for the transmitted RF signal to propagate at the speed of light to the target and back to the radar and then divides that time in two.

Radial velocity Radial velocity is nothing but the rate of change of a range over a period of time; it can be measured from the Doppler frequency shift.

Radome (or radar dome) Antennas of ground-based radars are often subject to severe weather conditions. So, some enclosure is needed for antennas to survive and wherein to perform under adverse weather conditions. These enclosures are called radome.

Radiation pattern is a representation of the radiation characteristics of an antenna, which is a function of elevation angle and azimuth angle for a constant radial distance and frequency.

Refraction is the bending of EM waves at the interference of two different dielectric materials,

Reflection The re-radiation of an EM wave from the surface of a matter of an object is called *scattering* or *reflection* of the incident wave.

Reflex Klystron is a single-cavity, variable-frequency microwave generator of low power and low efficiency.

Refractive index is a function of the material electric and magnetic energy storage properties characterized by its relative dielectric permittivity (ε_r) and permeability (μ_r), respectively.

Range Gate A movable gate is used to select radar echoes from a very short-range interval. A gate voltage is used to select radar echoes from a very short-range interval.

Radar Range Equation The radar range equation is a basic relationship that permits the calculation of the received echo signal strength, if certain parameters of the radar transmitter, antenna, propagation path, and target are known.

Range ambiguity (R_{max}) is the range beyond which targets appear as second-time-around echoes and is also known as *maximum unambiguous range*. All targets at a range shorter than R_{max} ($= c$ PRT/2) are in a one-to-one correspondence with the range as measured by the radar.

Range Resolution This is the ability of a radar to distinguish between targets that are close together.

Range Cell In a radar, a range cell is the smallest range increment that the radar is capable of detecting. If a radar has a range resolution of 50 yards and a total range of 30 nautical miles (60,000 yds), there are 60000/50 = 1,200 range cells.

Radar Cross-section is defined as the measure of reflective strength of the target. RCS is a function of the geometric cross-section, reflectivity, and directivity of a target.

Reciprocity A network is reciprocal if the power transfer and the phase do not change when the input and output are interchanged.

Reentrant cavities (or irregular-shaped resonators) are used in the place of tuned circuits at microwave frequencies. These are easily integrated into the structure of a microwave device.

Rotary joint allows for signal transmission between a fixed ground component and a rotating antenna for radar systems.

Resonator is a tuned circuit that resonates at a particular frequency at which the energy stored in the electric field is equal to the energy stored in the magnetic field. Resonators are built by using (i) lumped elements such as L and C; (ii) distributed elements such as sections of coaxial lines; and (iii) rectangular or circular waveguides.

Resonant circuits are circuits that offer a high impedance or low impedance (for parallel and series resonance, respectively) to the source at a particular frequency of operation.

Resonant frequency of microwave resonator is the frequency at which the energy in the resonator attains maximum value, that is, twice the electric energy or magnetic energy. The quality factor, Q of a resonator is a measure of frequency selectivity of the resonator. It is defined as

$Q = 2\pi \times$ Maximum energy stored/Energy dissipated per cycle.

Scalar is a quantity, such as voltage and length, that is completely specified by its magnitude and has no direction.

Scattering matrix or S matrix This matrix describes the relationship between the voltage waves incident on the ports and those reflected from the ports.

Scattering parameters are used to describe the behavior of a network at microwave frequencies.

Schottky Barrier Diode The Schottky barrier diode is a simple metal-semiconductor boundary with no P-N junction. A depletion region between the metal contact and the doped semiconductor region offers little capacitance at microwave frequencies. This diode finds use as detectors, mixers, and switches.

Skin effect is a phenomenon seen in conductors. When time-varying fields are present in a material that has high conductivity, the fields and currents tend to be confined to a region near the surface of the material.

Slotted section with line carriage is a microwave sectioned coaxial line connecting a coaxial E-field probe that penetrates inside a rectangular waveguide slotted section. The longitudinal slot is cut along the center of the waveguide's broad walls. The probe is made to move along the slotted wall that samples the electric field proportional to probe voltage.

Smith Chart is a special polar diagram that is useful in transmission line and waveguide design. It helps in calculating parameters such as impedances, admittances, and reflection coefficients.

Spectrum analyzer is a broadband super-heterodyne receiver that is also used to display a wave in the frequency domain; apart from power measurements, side bands can also be observed.

Source-free region is the one in which the charge density and current density are considered as the sources of electric and magnetic fields. Hence, the source-free region is where $\rho = J = 0$.

Solenoid is a coil of wire wound in the form of a cylinder.

STALO MTI system distinguishes moving targets by virtue of the target's Doppler frequency. Consequently, phase coherence within the radar system should be held in close tolerance. This coherence is provided by a Stable Local Oscillator (STALO). STALO translates the signal from Radio Frequency (RF) to an Intermediate Frequency (IF).

Stokes' Theorem relates the surface integral to the line integral.

Strapping To prevent mode jumping, strapping is used in magnetrons. It consists of two rings of heavy gauge wire connecting alternate anode poles. It provides a phase difference of 2π radians for the modes other than the π mode, thus preventing the occurrence of other modes, except the π mode.

Standing-wave ratio (SWR) The ratio of the maximum to minimum magnitudes of voltage or current on a line having standing waves is called standing-wave ratio (SWR). The range of values of the standing-wave ratio is theoretically 1 to infinity.

Step-recovery diode This is a PN diode whose construction is similar to that of a varactor diode. These are also known as *snap-off* diodes. These are usually made of silicon and operate under the forward bias condition.

Staggered PRF radars can make use of constant and/or varying (agile) PRFs. Since constant low PRF causes ambiguous target velocities and high PRF causes ambiguous ranges, MTI radars use PRF agility to avoid blind speeds. This kind of agility is known as *PRF staggering* or *staggered PRF*.

Subclutter Visibility describes the radar's ability to detect non-stationary targets that are embedded in a strong clutter background.

Surface integral is a definite integral taken over a surface (which may be a curve set in space); it can be considered the double integral analog of the line integral.

Time-harmonic Fields Representation is necessary to relate the instantaneous value of the vector fields to their complex spatial equivalents. This can be performed using the exponential function, $e^{j\omega t}$, as a basis function.

Tee junction In MW circuits, a waveguide or a coaxial junction with three independent ports is referred to as *tee junction* (e.g.: E-Plane Tee, H-Plane Tee).

TEM wave or mode The electric and magnetic fields of this wave have only transverse components; that is, in the direction of propagation, Hz = Ez = 0. TEM waves may propagate at all frequencies, so the TEM mode has no cutoff frequency.

TM wave or mode This wave has only a magnetic field in the transverse plane; that is, in the direction of propagation, Hz = 0, and Ez ≠ 0. These are sometimes referred to as *E waves*.

TE wave or mode This wave has an electric field only in the plane which is transverse to the direction of propagation; that is, the longitudinal components, Ez = 0 and Hz ≠ 0. These are sometimes referred to as *H waves*.

Hybrid wave or mode This wave is characterized as having both Ez ≠ 0 and Hz ≠ 0 and, therefore, is a combination of both TE and TM waves.

TRAPATT A slightly modified structure of the IMPATT diode that can be used at low frequencies is the TRAPATT diode, where TRAPATT stands for *trapped plasma avalanche triggered transit mode.*

Transit time is the time required for the electrons to travel from the cathode to the anode plate.

Track while scan (TWS) is a method of combining the search and track functions and of tracking multiple targets. It is capable of automatically tracking many targets simultaneously. TWS radar systems sample each target once per scan interval, and they use sophisticated smoothing and prediction filters to estimate the target parameters between scans.

Transferred Electron Devices (TEDs) are one of the important microwave devices. They are bulk devices having no junctions or gates as compared with microwave transistors that operate with either junctions or gates.

Tunable detector Tunable detectors are used to demodulate signals and to couple the required output to a high-frequency scope analyzer.

Traveling-wave tube (TWT) is a linear beam tube in which the interaction between the electron beam and the RF field is continuous over the full length of the tube. They are high-gain, low-noise, wide-bandwidth microwave amplifiers, capable of gains of 40 dB or more, with bandwidths more than an octave.

Tunnel diode impurity concentration in the tunnel diode is greatly increased as compared with the PN diode in which the width of the barrier is less than $3A^0$, because of which there is a large probability that particles will tunnel through the potential barrier. This is known as *tunneling*.

Unit vector along vector *A* is defined as a vector whose magnitude is unity and its direction is along vector *A*.

Varactor diode is a variable-capacitance junction diode. These two terminal devices are also called *varicaps*. This special type of the PN junction is designed to operate in the microwave range.

Vector is a quantity, such as electric field, that is completely specified by a magnitude and a direction.

Volume integral is an integral of a function of several variables with regard to volume measure taken over a three-dimensional subset of the domain of the function.

VSWR meter is a highly sensitive, high-gain, high-theta, low-noise voltage amplifier that is tuned normally at a fixed frequency of 1KHZ, of which microwave signals are modulated. This meter indicates calibrated VSWR reading for any loads.

Wave is a physical phenomenon that happens at one place at a given time and is later reproduced at other places and other times, with the time delay being proportional to space separation from the beginning.

Wavelength (λ) This is the distance from wavecrest to wavecrest (or trough to trough) along the direction of travel of electromagnetic waves and is called *wavelength*. The unit of wavelength is generally centimeters.

Waveguide is a hollow metallic tube. Since it is a completely enclosed system, no radiation loss occurs. In addition, the resistive loss of the inner conductor and the dielectric loss of the insulator supporting the inner conductor for the coaxial line are eliminated in waveguide. In hollow waveguides with only one conductor, the TEM mode with zero cutoff frequency does not exist and only TE and TM modes with nonzero cutoff frequency exist.

Waveguide bends or joints are used to change the shape or direction of a waveguide; so, we use bends, joints, or twists.

Wave number (β) specifies the direction of propagation whose magnitude is $2\pi/\lambda$ with units of rad/m, whereas βl specifies the phase change. Wave number is also known as a *phase shift constant.*

Wavemeter It is a device that is used for frequency measurement in microwaves. It has cylindrical cavity with a variable short-circuit termination. It changes the resonant frequency of the cavity by changing the cavity length.

Z parameters are also known as *impedance parameters*.

APPENDIX B

The Decibel (dB)

The unit of power is Watt (W). One watt is equal to one ampere (A) of current flowing through an electrical potential difference of one volt (V). However, microwave engineers have to deal with quantities that are very large or very small and the ratio of two quantities. In this context, the relative unit gives more information. Hence, power is expressed in decibel (**dB**), which is the ratio of power at a point of measurement to reference power (say input power). For example, a cellular base station might transmit approximately 80 W of power in a macro cell. The mobile phone receives power of about 0.000 000 002 W only, which is 0.0000000025% of the transmitted power. When expressed in dBs, which is logarithmic ratio, it gives more convenience for calculation. Since the logarithmic scale converts the multiplication/ division into addition/ subtraction, it will be very convenient to understand the relativeness on a log scale. This is thc reason that we go for computations in dBs. For example, the mobile base station in our example transmits at + 49 dBm, whereas the mobile phonc receives – 57 dBm, and produces a power difference of + 49 dBm – (–57 dBm) = 106 dBm.

The dB is used for stating the gain or loss of one device (*P*1) with regard to another device (*P*2). The formula for calculating gain or loss in dB is derived in the following manner:

1 bel is a ratio of 10:1 between two power levels. Hence, a power ratio of 200:20 is 1 bel (10:1), 200:40 is 5 bels (5:1), and 200:10 is 2 bels (20:1).

$$\text{bel} = \log (P1/P2)$$
$$\textbf{decibel (dB)} = 10 \times \log (P1/P2) = 10 \times \log (P_{out}/P_{in})$$

The dB by itself is not an absolute number, but a ratio. In general, a dimensionless quantity *P* is expressed in decibels (denoted by *P* dB) as

$$P\ \text{dB} = 10 \log_{10} P$$

In general, characters are added to the "dB" to denote the reference quantity.

For example, **dBW** is Decibels relative to 1 W and **dBm** is Decibels relative to 1 mW. Hence, if *P* is in measured watts:

$$P\ \text{dBW} = 10 \log (P/1) \text{ and } P\ \text{dBm} = 10 \log (P/0.001)$$

For amplifiers, a common reference unit is dBm, with 0 dBm being equal to 1 mill watt. As that of a amplifier with an output power of 30 dBm refers to 1 Watt.

In general, we use quantity in terms of dBW, dBm, or dBμ, which, in turn, mean

Power (dB or dBW) = $10 \times \log_{10}$ (Power in Watts)
Power (dBm) = $10 \times \log_{10}$ (Power in milli (m) Watts)
Power (dBμ) = $10 \times \log_{10}$ (Power in micro (μ) Watts)

Relationship Between Power and Voltage

Power = V^2 / R

Power in dB = $10 \times \log_{10}$ (Power in Watts) = $10 \times \log_{10} (V^2 / R) = 20 \times \log_{10}$ (Voltage / $\sqrt{R}$)

DB Representation for Antennas

dBd is used to represent GAIN (a ratio) relative to a dipole antenna:

Power (in maximum direction) of a dipole antenna = P_{dip}
Power (in maximum direction) of some other antenna = P_{ant}
Gain of that antenna relative to a dipole = P_{ant} / P_{dip}
Gain (dBd) = $10 * \log_{10} (P_{ant} / P_{dip})$

dBi is the unit that is used to measure the gain of an isotropic antenna. It states the gain of an antenna as referenced to an isotropic source. The greater the dBi value, the higher the gain and, as such, the more acute the angle of coverage. Gain of the antenna with regard to an isotropic antenna is given in dBi = 10 log (power max / power isotropic)

Linear Function	dB Function
1	0 dB
* 2	+ 3 dB
/ 2	− 3 dB
* 10	+ 10 dB
* 100	+ 20 dB
/ 10	− 10 dB
/ 100	− 20 dB

Common dB Factors

Example:
1 Watt = 0 dB
2 Watt = 0 + 3 dB = 3 dB
20 Watt = 3 dB + 10 = 13 dB

Comparison of MilliWatts and Decibel Change (relative to 1 mW)

converting dBm to milliWatts: dBm = $(\log_{10}$ (mW))*10
converting milliWatts to dBm: mW =10 (dBm/10)

The differences between values can become extremely large or small and more difficult to deal with. It is easier to say that a 100 mW signal decreased by 70 decibels than to say that it decreased to 0.00001 milliWatts.

$$10 \log(100/0.00001) = 70 \text{ dB}$$

The following Table B.1 gives the power in milliWatts and the corresponding decibel change:

Table B.1 Power in milliWatts and corresponding decibel change

dB change	–40	–30	–20	–10	0	+10	+20	+30	+40
mW	0.0001	.001	.01	.1	1	10	100	1000	10000

APPENDIX C

Doppler Frequency Shift

DOPPLER SHIFT

Suppose the target in the radar field of view is moving with a velocity component, v, toward the radar; from the theory of special relativity, the received frequency is given by

$$f_r = \left(\frac{1 + \frac{v}{c}}{1 - \frac{v}{c}} \right) f$$

where f is the transmitted frequency, and c is the speed of light.
For the receding target, v should be replaced by $-v$.

The equation can be simplified by

$$f_r = \left(1 + \frac{v}{c}\right)\left(1 - \frac{v}{c}\right)^{-1} f$$

$$= \left(1 + \frac{v}{c}\right)\left[1 + \frac{v}{c} + \left(\frac{v}{c}\right)^2 + \cdots\right] f$$

$$= \left[1 + 2\left(\frac{v}{c}\right) + 2\left(\frac{v}{c}\right)^2 + \cdots\right] f$$

Discarding all second-order and higher terms in $\left(\frac{v}{c}\right)$ leaves

$$f_r = \left[1 + 2\left(\frac{v}{c}\right)\right] f$$

The difference, f_d, between the transmitted and received frequencies is called the *Doppler frequency or Doppler shift.*

In the case of the approaching target, the Doppler shift is

$$f_d = \left(\frac{2v}{c} f\right) = \frac{2v}{\lambda}$$

where λ is the transmitted wavelength.

If the angle between the velocity vector of the target and the radar line of sight (LOS) is ψ, then

$$f_d = \frac{2v}{\lambda} \cos \psi.$$

Doppler Frequency Shift

For stationary and separate EM wave receivers and wave transmitters, such as sound or light, the frequency or wavelength generated by the wave source (f_s) will be the same at the receiver site (f_r); that is, $f_r = f_s$ or $\lambda_r = \lambda_s$, where the subscripts *r* and *s* refer to the wave receiver and wave source, respectively.

Doppler Frequency

When the wave source moves, either toward or away from the receiver, there will be a frequency difference between the wave source and the wave receiver. This frequency difference or shift is called the *Doppler frequency*.

Here, we will consider three cases, depending on whether the source and receiver are stationary or moving.

Case 1: Source is Approaching the Stationary Receiver

Let us consider the wave source is approaching the receiver at a speed of v_S.

For one source wave period T_S, which equals $1/f_S$, the source will move at a distance of $d = v_S T_S$.

The received wavelength λ_r is related to the source wavelength λ_s by

$$\begin{aligned}\lambda_r &= \lambda_s - d \\ &= \lambda_s - v_S T_S \\ &= \lambda_s - \frac{v_S \lambda_S}{c} \\ &= \frac{\lambda_s (c - v_s)}{c}\end{aligned}$$

where c is the speed of the wave, which is in this case the speed of light.

Since $\lambda = c/f$, the received frequency can be computed from the source frequency by

$$f_r = \frac{f_S c}{c - v_S}$$

The frequency difference between the receiver and the source, or Doppler frequency f_D, is

$$\begin{aligned}f_d &= f_r - f_s \\ &= \frac{f_S v_S}{c - v_S}\end{aligned}$$

The positive value of f_d means that the received frequency is higher than the frequency emitted by the approaching source. If the receiver is stationary, then, after the source passes the receiver, the speed of approach v_s becomes negative, and the frequency recorded by the receiver becomes lower than the frequency emitted by the now-receding source. The Doppler frequency, caused by relative movement between the source and the receiver, is also called the *Doppler frequency shift*.

Case 2: Source is Stationary, but the Receiver is Approaching the Source

The same principle applies when the source is stationary but the receiver is approaching it at the speed v_r. For one receiver wave period T_r, which equals $1/f_r$, the receiver will move at a distance of $d = v_r T_r$. This means that the receiver will receive the source wave by less distance d than when the receiver is stationary.

The new received wavelength is related to the source wavelength by

$$\begin{aligned}\lambda_r &= \lambda_s - d\\ &= \lambda_s - v_r T_r\\ &= \lambda_s - \frac{v_r \lambda_r}{c}\end{aligned}$$

$$\lambda_r = \frac{\lambda_S c}{c + v_r}$$

Since the moving receiver now determines the period of the wave, the received frequency is related to the source frequency by

$$f_r = \frac{f_S (c + v_r)}{c}$$

The frequency difference between the receiver and the source or Doppler frequency is

$$\begin{aligned}f_d &= f_r - f_s\\ &= \frac{f_S v_r}{c}\end{aligned}$$

Case 3: Both Source and Receiver are Moving

Now consider the case when both the wave source and the wave receiver are moving and v_S and v_r are the speeds with which they are approaching each other.

The source moves toward the receiver at the speed v_s for a time interval $t = T = 1/f_s$, with a distance $d1 = v_S T$.

The receiver moves toward the source at the speed v_r for the same time interval T, with a distance $d2 = v_r T$.

The received wavelength λ_r is related to the source wavelength λ_s by

$$\begin{aligned}\lambda_r &= \lambda_s - d_1 - d_2\\ &= \lambda_s - v_S T - v_r T\\ &= \lambda_s - \frac{v_S \lambda_S}{c} - \frac{v_r \lambda_S}{c}\\ &= \lambda_s \left(1 - \frac{v_S}{c} - \frac{v_r}{c}\right)\end{aligned}$$

The received frequency is related to the source frequency by

$$f_r = \frac{f_S}{1-(v_s+v_r)/c}$$
$$= f_S\left(1-\frac{v_{rs}}{c}\right)^{-1}$$

where $v_{rs} = v_r + v_s$. For $v_{rs} << c$.

$$f_r \approx f_S\left(1+\frac{v_{rs}}{c}\right)$$

The frequency difference between the receiver and the source, or Doppler frequency, is

$$f_d = f_r - f_s$$
$$= \frac{f_S v_{rs}}{c}$$
$$= \frac{v_{rs}}{\lambda_S}$$

Notice that the value of v_{rs}, and therefore f_d, is positive if the wave source and wave receiver are moving toward each other; otherwise, it is negative.

Physical Constants, Factors for Converting Measurements, and Measurement Unit Prefixes

APPENDIX D

Physical Constants are given in Table D.1. Factors for converting measurements into the metric system are given in Table D.2. Conversion in the opposite direction can be done by division by the given factor. Measurement unit prefixes and their values are given in Table D.3.

Table D.1 Physical constants

Constant	Symbol	Value
Velocity of light in vacuum	c	~3×10^8(m/s)
Electron charge magnitude	e	1.602×10^{-19}(C)
Mass of electron	m_e	9.11×10^{-31}(Kg)
Impedance of free space	Z_0	376.7 Ω
Boltzmann constant	k	1.38×10^{-23}(J/K)
Permittivity of free space	ε_0	$8.854 \times 10^{-12} \approx 10^{-9}/36\pi$ F/m
Relative permittivity of air	ε_r	1 F/m
Permeability of free space	μ_0	$4\pi \times 10^{-7}$ H/m $\approx 1.25 \times 10^{-6}$ H/m
Intrinsic impedance of free space	η_0	120π or 377 (Ω)

Table D.2 Factors for converting into the Metric System

To convert this unit	To this unit	Multiply by
Yards	Meters	0.9144
Foot	Meters	0.3048
Inches	Meters	0.02540
Miles	Kilometers	1.609
Nautical miles	Kilometers	1.852
Kilofeet	Kilometers	0.3048
Miles/hour	Meters/second	0.4470
Nautical miles/hour (or) knots	Meters/second	0.5144
Kilometers/hour	Meters/second	0.2777
Pounds	Kilograms	0.4536
Hours	Seconds	3,600
Degrees	Radians	$\pi/180$

Table D.3 Measurement unit prefixes and values

Prefix	Abbreviation	Meaning
Deci	*d*	10^{-1}
Centi	*c*	10^{-2}
Milli	*m*	10^{-3}
Micro	μ	10^{-6}
Nano	*n*	10^{-9}
Pico	*p*	10^{-12}
Tera	*T*	10^{12}
Giga	*G*	10^{9}
Mega	*M*	10^{6}
Kilo	*K*	10^{3}
Hecto	*h*	10^{2}
Deka	*da*	10^{1}

APPENDIX E

Manley-Rowe Relations

MR power relations are general power relations that are useful in predicting whether power gain is possible in any non-linear reactance. They represent conservation of energy. They considered the circuit as shown in Figure 1. It consists of resistive loads in a series with band pass filters connected in parallel with a lossless nonlinear capacitance. These filters reject power at all frequencies other than at their respective signal frequencies. A signal generator (v_s) and a pump generator (v_p) at their respective frequencies are connected as shown. The non-linear capacitance generates frequencies at harmonics of f_s and f_p ($mf_s \pm nf_p$), where m and n are integers.

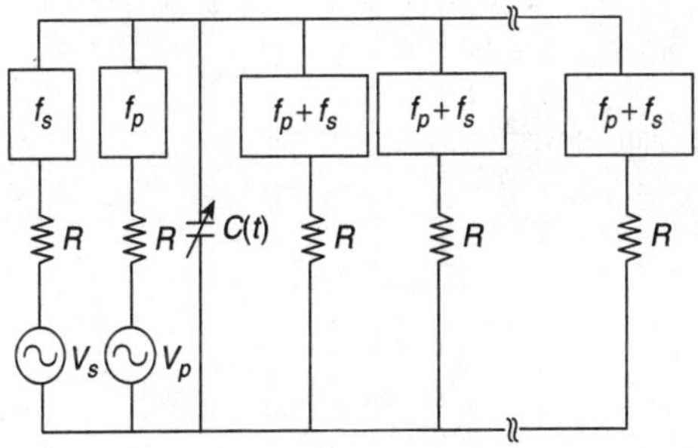

Figure 1 *Equivalent circuit of Manley–Rowe derivation*

Manley and Rowe related the input power at frequencies f_s and f_p to the output power at frequencies $mf_s \pm nf_p$.

The voltage across the capacitance is the sum of signal and pump voltages and is given by

$$v = v_p + v_s$$

$$(V_p \cos \omega_p t + V_s \cos \omega_s t) = V_p \frac{(e^{j\omega_p t} + e^{-j\omega_p t})}{2} + V_s \frac{(e^{j\omega_s t} + e^{-j\omega_s t})}{2}$$

The charge Q on the capacitor is a function of voltage and, hence, can be expanded in Taylor series in v to obtain

$$Q = Q(0) + \frac{\partial Q}{\partial v} v + \frac{1}{2} \frac{\partial^2 Q}{\partial v^2} v^2 + \cdots \tag{1}$$

Equation (1) contains all powers of v; therefore, Q will have frequencies at all harmonics of f_p and f_s. The currents that pass through C will contain all harmonics (As current is a function of Q). This means that voltage developed across the capacitor will also have all harmonics of f_s and f_p. Therefore, Q and V are given by

$$Q = \sum_{n=-\infty}^{\infty} \sum_{m=-\infty}^{\infty} Q_{nm} e^{j(n\omega_p + m\omega_s)t}$$

$$V = \sum_{n=-\infty}^{\infty} \sum_{m=-\infty}^{\infty} V_{nm} e^{j(n\omega_p + m\omega_s)t}$$

Since Q and V are real, $Q_{-n-m} = Q^*_{nm}$ and $V_{-n-m} = V^*_{nm}$

The current through C is the total change Q with time and is given by

$$I = \frac{dQ}{dt} = \sum_{n=-\infty}^{\infty} \sum_{m=-\infty}^{\infty} j(n\omega_p + m\omega_s) Q_{nm} e^{j(n\omega_p + m\omega_s)t}$$

$$= \sum_{n=-\infty}^{\infty} \sum_{m=-\infty}^{\infty} I_{nm} e^{j(n\omega_p + m\omega_s)t} \text{ where } I_{nm} = j(n\omega_p + m\omega_s) Q_{nm}$$

Since C is pure reactance, there will be no net power into and out of C; that is, the time average power due to interacting harmonics should be zero.

$$P_{m,n} = (V_{m,n} I^*_{m,n} + V^*_{m,n} I_{m,n})$$

$$= (V^*_{-m,-n} I_{-m,-n} + V_{-m,-n} I^*_{-m,-n})$$

$$= P_{-m,-n}$$

Then, conservation of power can be written as

$$\sum_{m=-\infty}^{\infty} \sum_{n=-\infty}^{\infty} P_{m,n} = 0 \tag{2}$$

Multiply Eq. (2) by a factor of

$$\frac{(m\omega_s + n\omega_p)}{(m\omega_s + n\omega_p)} \text{ gives}$$

$$\frac{(m\omega_s + n\omega_p)}{(m\omega_s + n\omega_p)} \sum_{m=-\infty}^{\infty} \sum_{n=-\infty}^{\infty} P_{m,n} = 0$$

$$\omega_s \sum_{m=-\infty}^{\infty} \sum_{n=-\infty}^{\infty} \frac{mP_{m,n}}{m\omega_s + n\omega_p} + \omega_p \sum_{m=-\infty}^{\infty} \sum_{n=-\infty}^{\infty} \frac{nP_{m,n}}{m\omega_s + n\omega_p} = 0 \tag{3}$$

Equation (3) will hold good for any arbitrary values of w_p and w_s only if

$$\sum_{m=-\infty}^{\infty} \sum_{n=-\infty}^{\infty} \frac{mP_{m,n}}{m\omega_s + n\omega_p} = 0$$

$$\sum_{m=-\infty}^{\infty} \sum_{n=-\infty}^{\infty} \frac{nP_{m,n}}{m\omega_s + n\omega_p} = 0 \tag{4}$$

Equation (4) can be written as

$$\sum_{m=0}^{\infty}\sum_{n=-\infty}^{\infty}\frac{mP_{m,n}}{m\omega_s+n\omega_p}+\sum_{m=0}^{\infty}\sum_{n=-\infty}^{\infty}\frac{-mP_{m,n}}{-m\omega_s-n\omega_{sp}}=0 \tag{5}$$

As $P_{m,n}=P_{-m,-n}$
Equation (5) can be written as

$$\sum_{m=0}^{\infty}\sum_{n=-\infty}^{\infty}\frac{2mP_{m,n}}{m\omega_s+n\omega_p}=0$$

Therefore,

$$\sum_{m=0}^{\infty}\sum_{n=-\infty}^{\infty}\frac{mP_{m,n}}{m\omega_s+n\omega_p}=0 \tag{6}$$

Similarly,

$$\sum_{m=-\infty}^{\infty}\sum_{n=0}^{\infty}\frac{nP_{m,n}}{m\omega_s+n\omega_p}=0 \tag{7}$$

These Eqs. (6) and (7) are standard forms of Manley–Rowe power relations.

Abbreviations

A

AC	Alternating Current
A/D	Analog to Digital
AGC	Automatic Gain Control
ARSR	Air-route Surveillance Radar
ASR	Approach Surveillance Radar
ATC	Air Traffic Control
ATD	Avalanche Transit-time Devices

B

BARITT	Barrier Injection Transit Time
BW	Bandwidth
BWA	Backward Wave Amplifier
BWO	Backward Wave Oscillator

C

CdTe	Cadmium Telluride
CF	Coupling Factor
CFA	Crossed-field Amplifier
CFAR	Constant False Alarm Rate
CFO	Crossed-field Oscillator
CRT	Cathode Ray Tube
CPI	Coherent Processing Interval
CVD	Chemical Vapor Deposition
CW	Continuous Wave

D

DC	Direct Current
DFT	Discrete Fourier Transform

E

EMF	Electromagnetic Field
EMFTL	Electromagnetic Field Theory and Transmission Lines

F

FET	Field Effect Transistor
FFT	Fast Fourier Transform
FMCW	Frequency-Modulated Continuous Wave
FWCFA	Forward-wave Crossed-field Amplifier

G

GaAs	Gallium Arsenide
GAGAN	GPS-aided Geo-augmented Navigation
GHz	Giga Hertz
GPS	Global Positioning System

H

HF	High Frequency
HFET	Heterostructure FET
HMIC	Hybrid Microwave Integrated Circuit

I

IEC	Inter-electrode Capacitance
IEEE	Institute of Electrical and Electronics Engineers
ILS	Instrument Landing System
IMPATT	Impact Avalanche Transit Time
IPP	Inter-pulse Period
InP	Indium Phosphide
IS	Isolation
ITU	International Telecommunication Union

K

KW	Kilo Watt

L

LAN	Local Area Network
LI	Lead Inductance
LOS	Line of Sight
LPE	Liquid-phase Epitaxy
LSA	Limited Space-charge Accumulation

M

MBE	Molecular Beam Epitaxy
MDS	Minimum Detectable Signal
MDTV	Minimum Detectable Target Velocity
MF	Medium Frequency
MIC	Microwave Integrated Circuit
MIM	Metal Insulator Metal
MMIC	Monolithic Microwave Integrated Circuit
MMTI	Maritime Moving Target Indicator
MESFET	Metal Semiconductor FET
MOSFET	Metal Oxide Semiconductor FET
MSSR	Monopulse Secondary Surveillance Radar
MTD	Moving Target Detector
MTI	Moving Target Indicator
MW	Mega Watt

O

O type	Linear Beam Tubes

P

PAR	Precision Approach Radar
PDR	Pulse Doppler Radar
PP	Principle Polarization
PPI	Plan Position Indicator

PRF	Pulse Repetition Frequency
PRI	Pulse Repetition Interval
PRR	Pulse Repetition Rate
PRT	Pulse Repetition Time
PVD	Physical Vapor Deposition

R

Radome	Radar Dome
RCS	Radar Cross-section of Target
RF	Radio Frequency
R Scope	Range Scope
RWH	Ridley–Watkins–Hilsum

S

SHF	Super High Frequency / Extremely High Frequency
SiO_2	Silicon Dioxide
S/N	Signal-to-Noise Ratio
STALO	Stable Local Oscillator
SWR	Standing Wave Ratio

T

TCR	Temperature Coefficient of Resistance
TED	Transferred Electron Device
TWT	Traveling Wave Tube
TE	Transverse Electric
TEM	Transverse Electro Magnetic
TM	Transverse Magnetic
T/R	Transmit/Receive
TRAPATT	Trapped Plasma Avalanche Triggered Transit
TWS	Track While Scan

U

UHF	Ultra High Frequency

V

VHF	Very High Frequency
VOR	VHF Omni-directional Range
VPE	Vapor-phase Epitaxy
VTMs	Voltage-tunable Magnetrons

Bibliography

Baden Fuller, A. J. (1969). *Microwaves*. Pergamon, Oxford, 35.

Baker, C. J. and Hume, A. L. Multistatic radar sensing. *IEEE Aerospace and Electronic Systems Magazine*, Vol. 18, Issue 2, pp 3–6, Feb. 2003.

Balanis, C. A. (1989). *Advanced Engineering Electromagnetics*. JohnWiley & Sons, Inc., New York.

Balanis, C. A. (1982). *Antenna Theory, Analysis and Design*. Harper & Row, NewYork.

Barett, R. M. (1984). Microwave printed circuits – an historical perspective. *IEEE Transactions on Microwave Theory and Techniques,* 32(9): 983–990.

Barkat, M. (1991). *Signal Detection and Estimation*. Artech House, Norwood, MA.

Barton, D. K. (1964). *Radar Systems Analysis*. Prentice-Hall, Englewood Cliffs, NJ.

Barton, D. K. (2005). *Radar System Analysis and Modeling*. Artech House, Norwood, MA.

Bowness, C. (1958). Microwave ferrites and their applications. *Microwave J.*, 1, 13–21, July–August.

Brown, W. C. (1984). The Microwave Magnetron and Its Derivatives. *IEEE Trans. Devices*, Ed-31, No. 11, November, 1595–1605.

Bryant, G. H. (1988). *Principles of Microwave Measurements*. Peter Peregrinus, London.

Carpentier, M. H. (1988). *Principles of Modern Radar Systems*. Artech House, Norwood, MA.

Carrara, N. (1932). The Detection of Microwaves. *Proceedings of the Institute of Radio Engineers (IRE)*, Vol. 20, No. 10, pp. 1615 –1625.

Chodorow, M. and Wessel-Berg, T. (1961). A High-Efficiency Klystron With Distributed Inter-Action. *Ire Trans. Electron Devices*, Ed-8, 44–55, January.

Cohn, S. (1954). Characteristic Impedance of the Shielded-Strip Transmission Line. *IRE Trans. of Microwave Theory and Techniques*, MTT-2, No. 7, 52, July.

Collin, R. E. (1960). *Field Theory of Guided Waves*. McGraw-Hill, New York.

Collin, R. E. (1966). *Foundations For Microwave Engineering*. Chapter 9. Mcgraw-Hill Book Company, New York.

Collin, R. E. (1992). *Foundation for Microwave Engineering*. 2nd ed., McGraw-Hill.

Collin, R. E. (2001). *Foundations for Microwave Engineering*. 2nd ed., New York, IEEE Press.

Cook, C. E. and Bernfield, M. (1967). *Radar Signals: An Introduction to Theory and Application.* Academic Press, New York.

Curry, R. G. (2005). *Radar System Performance Modeling*. 2nd ed., Artech House, Norwood, MA.

Edde, B. (1993). *Radar - Principles, Technology, Applications*. Prentice-Hall, Englewood Cliffs, NJ.

Ewell, G. (1981). *Radar Transmitters.* McGraw Hill, New York.

Farrell, J. and Taylor, R. (1964). Doppler Radar Clutter. *IEEE Trans.,* Vol. ANE-Il, pp. 162–172, September Reprinted in Barton, D. K. (1978). CW and Doppler Radars. Vol. 7, Artech House, Norwood, Mass, sec. VI-2, pp. 351–361.

Fremlin, J. H., Gent, A. W., Petrie, D. P. R., Wallis, P. J., and Tomlin, S. G. *Principles of Velocity Modulation. IEEE Journal* 93, Part III A, 1946, pp. 875–917.

Galati, G. (Ed.) (1993). *Advanced Radar Techniques and Systems*. The Institution of Electrical Engineers, London.

Gandhi, O. P. (1981). *Microwave Engineering and Applications*. Chapters 9, 10, and 11. Pergamon Press, New York.

Gewartowski, J. W. and Watson, H. A. (1965). *Principles of Electron Tubes.* D. Van Nostrand Company, Inc.

Griffiths, H. D. and Long, B. A. (1986). Television-Based Bistatic Radar. *IEE Proceedings*, Vol. 133, Part F, No.7, pp 649–657, Dec.

Harrington, R. F. (1961). *Time-Harmonic Electromagnetic Fields*. McGraw-Hill, New York.

Heaviside, O. (1950). Electromagnetic Theory. vol. l. Reprinted by Dover, New York.

Helszajn, J. (1994). *Microwave Planar Passive Circuits and Filters*. John-Wiley & Sons.

Hutter, R. G. E. (1960). *Beam and Wave Electrons in Microwave Tubes*. D. Van Nostrand Company, Princeton, N.J.

Jackson, J. D. (1962). *Classical Electrodynamics*. John Wiley & Sons, Inc., New York.

Keister, F. Z. (1968). An Evolution of Materials and Processes for Integrated Microwave Circuits. *IEEE transaction on "Microwave Theory and Techniques"*, MTT 16, July.

Kingsley, S. and Quegan, S. (1992). *Understanding Radar Systems*. McGraw- Hill, New York.

Knott, E. F., Shaeffer, J. F., and Tuley, M. T. (1993). *Radar Cross Section*. 2nd edition, Artech House, Norwood, MA.

Konishi, Y. (1998). *Microwave Electronic Circuit Technology*. Marcel Dekker Inc.

Kraus, J. D. (1984). *Electromagnetics*. McGraw-Hill, New York.

Laverghetta, T. S. (1989). *Modern Microwave Measurements and Techniques*. Artech House, Dedham, MA.

Lax, B. and Button, K. J. (1962). *Microwave Ferrites and Ferromagnetic*. McGraw-Hill, New York.

Larue, A. D. and Rubert. R. R. (1964). Multi-Megawatt Hybrid Twt's at S-Band and C-Band. *Presented to the IEEE Electron Devices Meeting*, Washington, D.C., October.

Lee, T. H. (2004). *Planar Microwave Engineering.* Cambridge University Press, Cambridge.

Lord Rayleigh (1897) (1903). On the Passage of Electric Waves through Tubes. *Philosophical Magazine Series* 5 43:125–132. Reprinted in Collected Papers Cambridge Univ. Press.

Ludwig, R. P. (2000). *Bretchko, RF Circuit Design - theory and applications*. Prentice-Hall Inc.

Mahafza, B. R. (1998). *Introduction to Radar Analysis*, CRC Press, Boca Raton, FL.

Marcuvitz, N. (1951). *Waveguide Handbook*, McGraw-Hill, New York.

Mark A. Richards, James A. Scheer, William A. Holm, "*Principles of modern Radar*", Scitech Publications, USA, 2010.

Meyer, D. P. and Mayer, H. A. (1973). *Radar Target Detection: Handbook of Theory and Practice.* Academic Press, New York.

Michael O. Kolawole, "Radar Systems, Peak Detection and Tracking", Newnes, 2002.

Microwave and Wireless simplified. Second edition by Thomas S.Laverghetta.

Microwave Components and Devices by Chapman and Hall

Microwave Devices Circuits and Subsystems for Communication Engineering by GLOVER

Montgomery, C. G., Dicke, R. H., and Purcell, E. M. (1987) (1948). *Principles of Microwave Circuits*, McGraw-Hill, New York, and Peter Peregrinus, Stevenage.

Morchin, W. (1993). *Radar Engineer's Sourcebook*. Artech House, Norwood, MA.

Oldfield, L. C. (1985). Microwave Measurement, *IEE Electrical Measurement Series* 3 (Peter Peregrinus, London).

Oldham, W. G. (1983). *The fabrication of microelectronic circuits*. In S.M. Sze (Ed.). *VLSI Technology*. John Wiley & Sons, New York.

Ott, H. (2009). *Electromagnetic Compatibility Engineering*. John Wiley & Sons Ltd, Chichester.

Packard, K. S. (1984). The Origin of Waveguides: A Case of Multiple Rediscovery. *IEEE Transactions on Microwave Theory and Techniques,* 32(9): 961–969.

Paul, C. R. (2006). *Introduction to Electromagnetic Compatibility*. John Wiley & Sons Ltd, Chichester.

Pierce, J. R. (1950). *Traveling-Wave Tubes*. P. 210. D. Van Nostrand Company, N. J. On Electron Princeton.

Pozar, D. M. (1998). *Microwave Engineering*. 2nd edition, John-Wiley & Sons (3rd edition, 2005 is also available from the John-Wiley & Sons).

Pozar, D.M. (1998). *Microwave Engineering*, 2nd ed., New York: John Wiley &Sons.

Pucel, Robert, A. (1981). Design Considerations for Monolithic Microwave Integrated Circuits. *IEEE Transaction on Microwave Theory and Techniques*, MTT 29, No 6, July.

Ramo, S., Whinnery, J. R., and Van Duzer, T. (1984). *Fields and Waves in Communication Electronics*. John Wiley & Sons, Inc., New York.

Ringel, M. B., Mooney, D. H., and Long W. H. (1983). F-16 Pulse Doppler Radar (AN/APG- 66) Performance, *IEEE Trans.,* vol. AES-19, pp. 147-158, January.

Rizzi, P. A. (1988). *Microwave Engineering: Passive Circuits*. Prentice Hall, Englewood Cliffs, NJ, 367–404.

Saad, T. and Hansen, R. C. (1971). *Microwave Engineer's Handbook*. Vol. 1. Dedham, MA: Artech House.

Schleher, D. C. (1991). *Moving Target Indication and Pulsed Doppler Radar*. Artech House, Boston, MA.

Sinclair, M. W. (1985). Microwave Measurement. *IEE Electrical Measurement Series* 3 (Peter Peregrinus, London).

Sobol, H. (1971). Applications of Integrated Circuit Technology to Micro Wave Frequencies. *Proc. IEEE*, 59, *No* 8, *August*.

Southworth, G. E. (1950). *Principles and Applications of Waveguide Transmission*. Princeton, NJ: D. Van Nostrand Company.

Stratton, J. A. (1941). *Electromagnetic Theory*. McGraw-Hill, New York.

Welch, J. D. and Pratt, H. J. (1966). *Losses in Microstrip Transmission Systems for Integrated Microwave Circuit.*, NEREM Rec.

Whitaker, J. C. (2001). *Tubes*. pp. 7-169–7-183. In *The RF and Microwave Handbook,* Golio, M. (ed.) CRC Press, Boca Raton, FL.

Young, L. (1974). *Advances in Microwaves*. Academic Press, New York.

Index

S

T

U

V

W

Y

Z